PRINCIPLES OF ELECTRONIC COMMUNICATION

ANALOG AND DIGITAL

2nd Edition

PRADIP KUMAR GHOSH
Professor and Head
Department of ECE
Mody University of Science and Technology
Lakshmangarh, Sikar
Rajasthan

Principles of Electronic Communication (2nd edition)

UNIVERSITIES PRESS (INDIA) PRIVATE LIMITED

Registered office
3-6-747/1/A & 3-6-754/1, Himayatnagar, Hyderabad 500 029, Telangana, India
info@universitiespress.com; www.universitiespress.com

Distributed by
Orient Blackswan Private Limited

Registered office
3-6-752 Himayatnagar, Hyderabad 500 029, Telangana, India

Other offices
Bengaluru, Bhopal, Chennai, Guwahati, Hyderabad, Jaipur, Kolkata,
Lucknow, Mumbai, New Delhi, Noida, Patna, Vijayawada

First edition published by Universities Press, 2008
Reprinted 2009, 2012
Second Edition 2017

ISBN: 978-93-86235-10-7

Cover and book design
© Universities Press (India) Private Ltd 2017

501229

11112016

Typeset in Palantino 10/12 by
MacroTex Solutions, Chennai 600 091, Tamil Nadu, India

Printed at BB Press, Noida

Published by
Universities Press (India) Private Limited
3-6-747/1/A & 3-6-754/1, Himayatnagar, Hyderabad 500 029, Telangana, India
email: info@universitiespress.com

Dedicated to

my father Late Balak Chandra Ghosh

my wife Dipanjana and my son Arkapravo

Preface to the Second Edition

The analog and digital communication theory deals with how effectively and efficiently analog signals can be encoded digitally, transmitted through channels with minimum possible contamination, and received at the target end with minimum error. This second edition of *Principles of Electronic Communication – Analog and Digital* builds on the chapters of the first edition, revising the topics effectively to reflect contemporary developments in the subject. It has been made more comprehensive and student-friendly, providing adequate coverage of the topics of digital communication appropriate to serve as a textbook for graduate-level students. The key features that have been updated and revised chapter-wise are as follows.

Chapter 1: Section 1.1 includes elaborate definitions of energy and power for discrete-time (DT) signals along with numerical examples of even and odd DT signals. DT unit sample function and pulse signal, unit impulse in DT and its approximations by gate/ triangular/ two-sided exponential/ Gaussian pulse/ sine over argument, and exponential and sinusoidal sequences have been incorporated in Section 1.2. Power spectral density in Section 1.3 has been supported by numerical examples. One more sub-section on time–frequency duality with examples has been added in Section 1.4. The concept of analytic signals which was missing in the first edition has now been explained in Section 1.4.6. The decomposition of vectors in orthogonal vector space is reinforced with an example. Correlation and spectral density is explained with an additional numerical example.

Chapter 2: Section 2.2.6 provides mathematical models for fading channels; this is a new inclusion. Section 2.3.3 has been augmented with real filters (Butterworth and Chebyshev) and numerical examples.

Chapter 3: Section 3.6.2 on TV signals has been enhanced and numerical examples included.

Chapter 4: Section 4.2 now includes an explanation of deviation sensitivity. The phase-locked loop is explained in more detail and numerical examples provided. For completeness and to maintain continuity, a new Section 4.13 on frequency-division multiplexing with illustrations and examples has been added. Similarly, a new Section 4.14 deals with non-linear distortion and interference in FM signals.

Chapter 5: The concept of total probability that was missing in the first edition is now provided in Section 5.2.2. The discussion on Gaussian distribution in Section 5.4.4 has been completed by providing the plot of Q-function and explaining the sum of two Gaussian random variables (RVs). The spectral representation of random processes has been added in Section 5.5.3.

Chapter 6: Major revision in this chapter includes addition of two sections on autocorrelation of sum of two random processes (Section 6.13) and filtering of stationary process in a linear time-invariant (LTI) system (Section 6.14).

Chapter 7: Elucidation of the noise concept has been expanded to include band-pass noise in sinusoidal signals. Effect of low-pass receiver filter on coherent detection is illustrated in Section 7.3. Both single-sideband (SSB) noise spectrum and band-pass noise spectrum for double-sideband (DSB) case are covered in this section. The FM system with de-emphasis is incorporated in Section 7.5.3.

Chapter 8: Under signal reconstruction in Section 8.2.1, an outline on ideal interpolation is added while an example on aliasing error is illustrated in Section 8.2.2. A new topic on band-pass sampling is provided in Section 8.2.3. Section 8.4.1 on time-division multiplexing (TDM) has been expanded by incorporating data frame of TDM for dissimilar channels. The concept of scrambling was not discussed in the first edition; it has now been included as Section 8.4.4. In Figure 8.28, a sketch on pulse-code modulation (PCM) receiver has been added with adequate explanation. Noise effect in delta modulation (DM) systems is included in Section 8.10. The digital multiplexing hierarchy in Section 8.12 is expanded by incorporating explanation of multiplexing Consultative Committee for International Telephony and Telegraphy (CCITT) standard with diagram.

Chapter 9: In Section 9.2, the relations between Nat, Bit and Hartley measures of information have been delineated. A new topic on computation of maximum entropy for a given continuous random variable with fixed average power is added in Section 9.2.2. A note on channel capacity is provided in Section 9.3. Discrete channel model of Section 9.4.1 is expanded by including mutual information of continuous random variables. Computation of continuous channel capacity and derivation of Shannon's law is added in Section 9.4.2. Section 9.5.2 has been expanded by providing notes on entropy encoding scheme. Linear block encoding and decoding, and error detection by syndrome testing is elaborated in Section 9.6.1.

Chapter 10: New topics on M-ary digital PAM (ASK) signal and M-ary PSK with signal constellation have been added in Sections 10.2.1 and 10.3.4 respectively.

I hope that this edition will be as useful and attractive to students as the previous edition. I have incorporated the latest trends in this edition with a current perspective. PowerPoint slides and a solutions manual for the chapter-end exercises can be accessed online at www.universitiespress.com/pkghosh/principlesofelectroniccommunication.

I consider myself blessed to be associated with Mody University, Lakshmangarh, and am thankful to my fellow colleagues for providing effective feedback and suggestions. Finally, I wish to thank Universities Press for publishing this new edition.

I request all readers to forward their valuable feedback, comments or suggestions to me at pkghosh_ece@yahoo.co.in; this may help me immensely to improve the book further.

Pradip Kumar Ghosh
Lakshmangarh, Sikar
Rajasthan

Preface to the First Edition

This book has been written to serve the needs of a course on electronic communications systems as is taught at the graduate level in engineering and technological institutions in India. It provides a comprehensive coverage of the important theories and principles of electronic communication. The language has been kept simple and the contents so structured as to enable students to understand and assimilate the basics of the subject with ease within the time generally allotted for such a course. Each of the 13 chapters of the book has a fair number of insightful examples that demonstrate the practical relevance of various engineering concepts. Review exercises and multiple-choice questions included at the end of every chapter will be a further help to students in their preparation for semester examinations.

Chapter 1 provides the mathematical background necessary for dealing with signals in communications systems. Chapter 2 deals with the modelling and analysis of the linear systems, with emphasis on filters. Chapters 3 and 4 contain the treatment of amplitude modulation and angle modulation respectively, and the various demodulation schemes applicable to them.

Chapter 5 is on probability of random processes and has discussions on narrowband random processes. In Chapter 6, this random analysis is applied to the theory of noise. The effect of noise in continuous-wave linear and exponential modulations is dealt with separately in Chapter 7.

Chapter 8 is devoted to the study of analog and digital pulse modulation. It includes discussions on analog-to-digital conversion, uniform and non-uniform quantisation and quantisation noise. Information theory and coding is introduced in Chapter 9. Channel encoding, entropy coding and forward error correction are discussed in this chapter. Chapter 10 presents the various digital modulation and demodulation schemes for transmission and reception of digital signals over bandlimited channels – amplitude-shift keying (ASK), phase-shift keying (PSK), frequency-shift keying (FSK) and quadrature amplitude modulation (QAM). The chapter also includes error analysis of each scheme.

Chapter 11 presents the methods of digital data encryption and decryption for ensuring a secure communication link. Another security measure in the transmission of signals is the spread-spectrum technique, which is discussed in Chapter 12. Chapter 13 portrays the use of equalisers to combat the distortion of signals introduced by the processes in the transmission and receiving sections of the communication system and the physical channel itself.

Students who wish to pursue further studies in the subject will find the short bibliography provided in the book useful. The contents of this book are based primarily on the lecture notes I developed for teaching a course on electronics and communication engineering at the Murshidabad College of Engineering and Technology (MCET), Berhampur, West Bengal. It gives me immense pleasure to acknowledge the contributions of many individuals to the writing of this book. Firstly, I have been fortunate to come across several batches of excellent students at the National Institute of Science and Technology (NIST), Berhampur, Orissa, St Xavier's College, Kolkata, and the Murshidabad College of Engineering and Technology, Berhampur, West Bengal, whose responses to my lectures helped me in developing the approach I have adopted for each topic included here. I would like to express my sincere appreciation to my PhD thesis supervisor, Professor D Chattopadhyay of the Institute of Radio Physics and Electronics, Calcutta University, with whom I had the preliminary discussion on the writing of this book. I would like to thank all faculty members and staff of the ECE department of MCET for the academic environment which helped transform my dream of this book into reality. I am thankful to Dr A K Panda and Sri R K Das of NIST, Berhampur, Orissa, for the friendly and useful discussions I have had with them. Thanks are due also to my colleagues at the computer science department, St Xavier's College, Kolkata.

I am indebted to my parents for their blessings. Their good wishes and unstinted support bolstered my confidence while embarking on this venture. This book could hardly have been possible but for the endless support of my wife, and I thank her for her patience, cooperation and encouragement all through. Thanks are also due to my three-and-a-half-year-old tiny tot, Arkapravo, who was considerate enough to allow me to carry on my arduous task with his unusual patience and non-interference.

My thanks are due to the members of Universities Press (India) Private Limited, Hyderabad, for their interest in the manuscript and their involvement in bringing the book to this final form. I hope it will serve the students' needs – only then will I consider my entire effort worthwhile. I look forward to valuable feedback from readers on errors that might have crept in inadvertently or on any obvious lapses so that these can be taken care of in the next edition of the book.

Pradip Kumar Ghosh
Berhampur, Murshidabad

Contents

1 Signal Analysis

Signals arise in a variety of contexts. A signal refers to a set of information or data. There is an analytical framework for describing signals. In this chapter, we will discuss certain basic concepts of signals by introducing a mathematical model and developing an analytical framework for the same.

1.1 Classification of Signals

A signal is traditionally expressed as the time variation in the physical state of an object. It may be electrical or non-electrical in nature. For example, a speech signal can be represented mathematically by acoustic pressure as a function of time variable. A mathematical model of a signal is the functional relation of a process with respect to time. A current, for example, flowing in any branch of a circuit may depend on time t and can be expressed as $i(t)$. When the instantaneous values of a signal can be computed accurately, the signal is called a deterministic signal; otherwise it is a random or stochastic signal. The model $v(t) = V_0 \sin \omega t$ defines a deterministic signal. Strictly speaking, deterministic signals are non-existent in nature. The interaction of the signal source with the surrounding deviates the signal from the real one. Several important classes of signals are discussed below.

1.1.1 Continuous-time and discrete-time signals

A signal is defined to be a continuous-time signal if the independent variable t varies continuously. On the other hand, signals are defined only at discrete times. A signal of the telephone output as a function of time is an example of continuous-time signal. The stock market daily average is an example of a discrete-time signal.

To distinguish between continuous-time and discrete-time signals, the independent variable in continuous-time case will be denoted by the symbol t enclosed in parentheses ( ). The discrete-time independent variable will be denoted by n enclosed in parentheses [ ]. Figure 1.1 illustrates the continuous-time and discrete-time signals. Note that $v(t)$ in Fig. 1.1(a) is defined for the continuous values of the independent variable t and $v[n]$ in Fig. 1.1(b) is defined only at integer values of the independent variable.

1.1.2 Analog and digital signals

We very often misinterpret the continuous-time signal as an analog signal. The amplitude of an signal can take on any value in a continuous range. The analog signal may be continuous or discrete in the time domain. If the signal values are not available at all times but only at fixed set of points along the time axis, then the signal is referred to as discrete or sampled signal. Discrete signals are essentially analog signals as its amplitude can take on an infinite number of values. A digital signal, on the other hand, is one whose amplitude can take on only a finite number of values. Thus, for a discrete signal to qualify as a digital signal, the amplitude must take on some finite allowed values. An analog signal can be converted into a digital signal through sampling (discrete in time) and quantisation (discrete in amplitude). Figure 1.2 shows the analog and digital signals both in continuous-time and discrete-time.

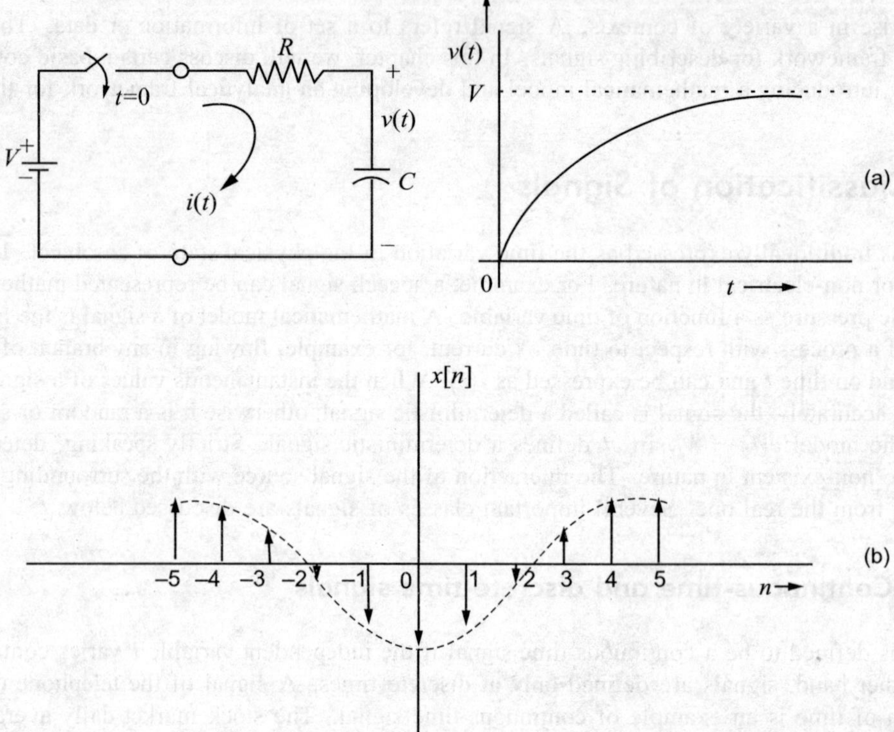

Fig. 1.1 *(a) Graphical representation of continuous-time signal. The signal here refers to the voltage that appears across the capacitor when charged to a dc voltage V through the resistance R. (b) Graphical representation of discrete-time signal. The dotted curve guides the eye to the continuous-time signal from which the samples have been taken.*

1.1.3 Periodic and aperiodic signals

An important class of signal that we very frequently encounter is the periodic signal. A signal $s(t)$ is said to be periodic if there is a positive value of T for which

$$s(t) = s(t+T) \quad \text{for all } t. \tag{1.1}$$

The smallest value of T for which the signal is unchanged by a time shift of T is the period of $s(t)$. One can readily observe that any periodic signal shifted by an integral multiple of T remains unaltered. That is, $s(t) = s(t + mT)$ for all t and for any integer m. The smallest positive value of T for which (1.1) is obeyed is the fundamental period. A signal that is not periodic is referred to as an aperiodic signal.

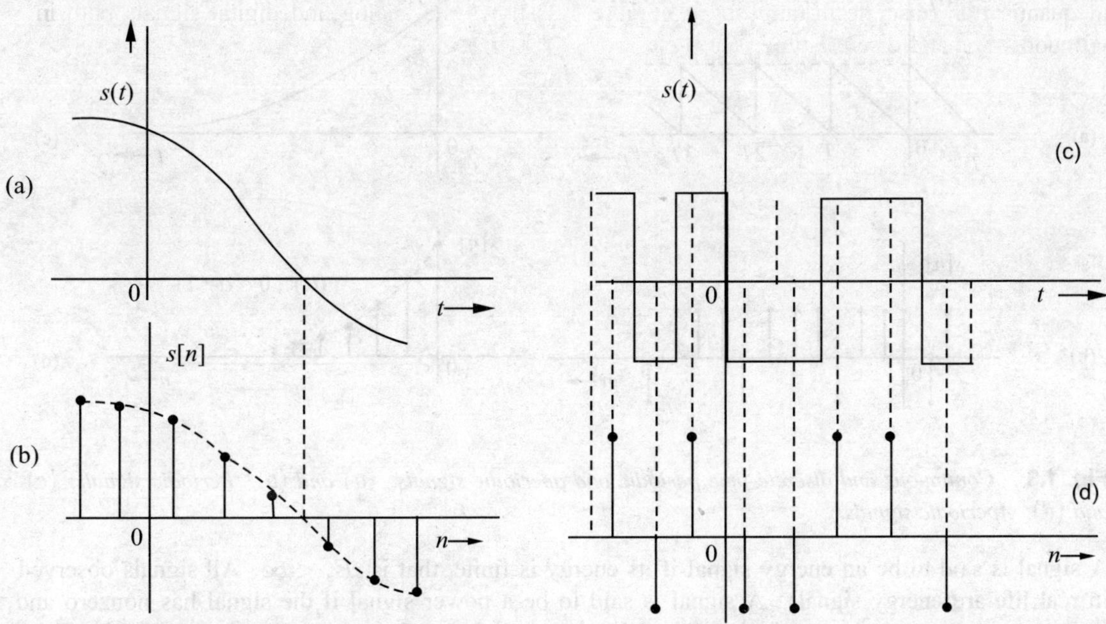

Fig. 1.2 *The sketches* (a) *and* (b) *refer to analog signal in continuous and discrete time, respectively.* (c) *digital continuous- and* (d) *digital discrete-time signals.*

The definition of periodic signals is also valid in discrete time. A discrete-time signal $s[n]$ is periodic with period N if the condition

$$s[n] = s[n + N] \tag{1.2}$$

is satisfied with smallest possible N. Examples of periodic and aperiodic signals are depicted in Fig. 1.3.

1.1.4 Energy and power signals

Consider a signal $s(t)$ defined over $-\infty < t < \infty$. Let $s^*(t)$ be the complex conjugate of $s(t)$. We define the 'norm' of $s(t)$ as

$$|s| = \sqrt{\int_{-\infty}^{\infty} s(t)s^*(t)dt}. \tag{1.3}$$

The square of the norm is termed the signal energy E_s.

Thus, $E_s = \int_{-\infty}^{\infty} |s^2(t)|dt.$ (1.4)

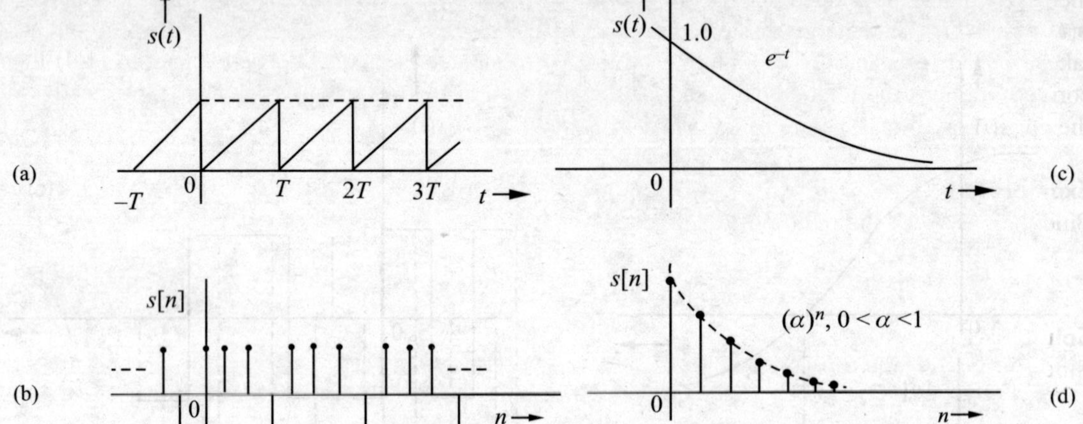

Fig. 1.3 *Continuous and discrete-time periodic and aperiodic signals. (a) and (b): Periodic signals; (c) and (d): Aperiodic signals.*

A signal is said to be an energy signal if its energy is finite, that is, $E_s < \infty$. All signals observed in real life are energy signals. A signal is said to be a power signal if the signal has nonzero and finite power. Since the power is defined as the time average of the energy, the total power P_s of the signal is given as

$$P_s = \lim_{T\to\infty} \frac{1}{T} \int_{-T/2}^{T/2} |s^2(t)|dt. \quad (1.5)$$

Alternately,

$$P_s = \lim_{T\to\infty} \frac{E_s}{T}. \quad (1.6)$$

Obviously, a finite energy signal has zero power since the averaging is over an infinitely large interval. In the case of discrete-time signals, the expressions for energy and power are given by

$$E_s = \sum_{n=-\infty}^{\infty} |x[n]|^2 \quad (1.7)$$

and

$$P_s = \lim_{N\to 0} \frac{1}{(2N+1)} \sum_{n=-N}^{N} |x[n]|^2. \quad (1.8)$$

Equation (1.4) does not give the actual signal energy as the signal energy depends on the load along with the signal strength. If $s(t)$ be a signal voltage across an R-ohm resistor, then current through the resistor is $s(t)/R$ and the instantaneous power is $p(t) = s^2(t)/R$. The energy dissipated is then $\int_{-\infty}^{\infty} \frac{s^2(t)}{R} dt = \frac{E_s}{R}$. Similarly, if $s(t)$ be the signal current flowing through the R-ohm resistor, then the voltage developed across the load is $s(t)R$ and so the instantaneous power dissipated is $p(t) = s^2(t)R$. The energy dissipated is then $\int_{-\infty}^{\infty} Rs^2(t)dt = RE_s$. To avoid confusion, one can take the load resistance $R = 1$ ohm. The energy dissipated is then E_s. The expression (1.4) then correctly defines the total signal energy dissipated in a normalised load (R=1 ohm), irrespective of the current or voltage nature of the signal $s(t)$.

Example 1.1

Find the energy of the radio pulse defined by

$$s(t) = V_0 \cos \omega_0 t, \quad 0 < t < \tau.$$

Solution

Since the function exists for the interval $(0, \tau)$, the energy of the pulse over this duration is

$$E_s = \int_0^{\tau} (V_0 \cos \omega_0 t)^2 dt = \frac{V_0^2}{2}\left[\tau + \frac{\sin 2\omega_0 \tau}{2\omega_0}\right] = \frac{V_0^2 \tau}{2}\left[1 + \frac{\sin 2\omega_0 \tau}{2\omega_0 \tau}\right].$$

When the duration $\tau \gg \frac{1}{\omega_0}$, then

$$E_s \approx \frac{V_0^2 \tau}{2}.$$

Example 1.2

Determine the energy of the signal sketched in the adjoining figure.

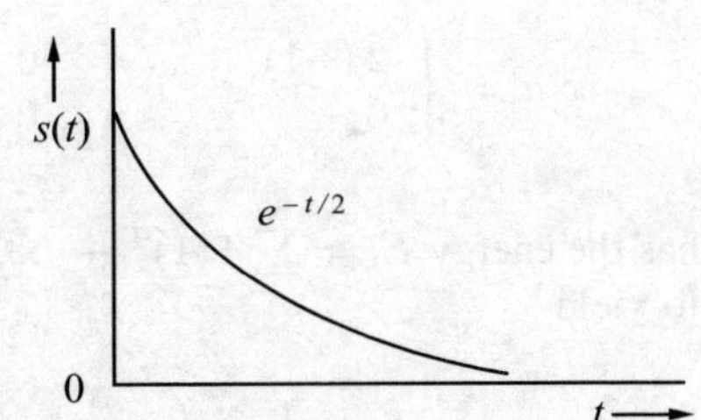

Solution

Here $s(t) = e^{-1t/2}$ so that

$$E_s = \int_{-\infty}^{\infty} s^2(t)dt = \int_0^{\infty} (e^{-t/2})^2 dt = \int_0^{\infty} e^{-t} dt = \left[\frac{e^{-t}}{-1}\right]_0^{\infty} = 1 - 0 = 1.$$

The of the discrete-time sequence as defined in Eq. (1.8) can also be related to its energy over the finite interval $-N \le n \le N$ as follows:

$$E_{s,N} = \sum_{n=-N}^{N} |s[n]|^2 \tag{1.8a}$$

What follows is that

$$P_{s,N} = \lim_{N\to\infty} \frac{1}{2N+1} E_{s,N} \tag{1.8b}$$

If the sequence be a periodic sequence with a period N, the can also be written as

$$P_s = \frac{1}{N}\sum_{n=0}^{N-1} |s[n]|^2 \tag{1.8c}$$

For a casual sequence $s[n]$, the expression for energy defined in (1.7) can be re-written as

$$E_s = \sum_{n=0}^{\infty} |s[n]|^2 \tag{1.8d}$$

We now give two illustrative examples of finding energy and power of discrete-time sequences.

For the sequence

$$x[n] = \begin{cases} \frac{1}{n}, \ n \geq 1 \\ 0, \ n \leq 0 \end{cases} \tag{1.8e}$$

the energy amounts to

$$E_x = \sum_{n=1}^{\infty} \left({}^{1}\!/_{n}\right)^2 = {}^{\pi^2}\!/_{6} \quad \text{(by applying Eq. (1.8d))}$$

However, the sequence

$$x[n] = \begin{cases} 2\,(-1)^n & , \quad n \geq 0 \\ 0 & , \quad n < 0 \end{cases} \tag{1.8f}$$

has the energy $E_x = \sum^{\infty} 4\,(1)^n \rightarrow \infty$. But the of this sequence (given by Eq. (1.8)) can be computed to yield

$$p_x = \lim_{N\to\infty} \frac{1}{2N+1}\left(4\sum_{n=0}^{N}(1)^n\right) = \lim_{N\to\infty} \frac{4\,(N+1)}{2N+1} = \frac{4}{2} = 2$$

1.1.5 Even and odd signals

A signal $s(t)$ is said to be an even signal if it is identical with its time-inverted counterpart. In other words, the mirror image of $s(t)$ about the vertical axis duplicates the same signal. In the continuous-time, a signal is even if

$$s(-t) = s(t). \tag{1.9}$$

A discrete-time signal $s[n]$ is even if

$$s[-n] = s[n]. \tag{1.10}$$

A signal is referred to as odd if

$$s(-t) = -s(t) \tag{1.11}$$

in continuous-time case, and

$$s[-n] = -s[n] \tag{1.12}$$

in discrete-time case. Observe that an odd signal must necessarily be zero at $t = 0$ or $n = 0$. Figure 1.4 gives the plots of odd and even signals, for the continuous-time and discrete-time signals.

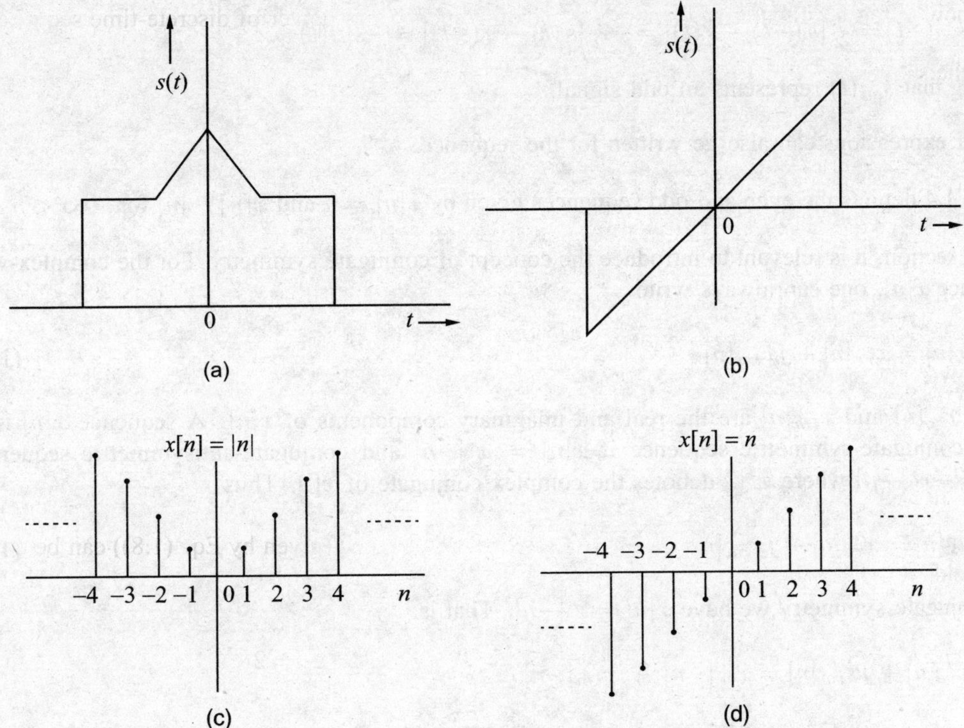

Fig. 1.4 *Even and odd signals; (a), (b) even and odd for continuous-time (c), (d) even and odd for discrete-time.*

We may write $s(t)$ as

$$\begin{aligned} s(t) &= s_{ev}(t) + s_{od}(t) \\ &= \frac{1}{2}\left[s\left(t\right) + s\left(-t\right)\right] + \frac{1}{2}\left[s\left(t\right) - s\left(-t\right)\right] \end{aligned} \tag{1.12a}$$

where the even and odd components are

$$s_{ev}(t) = \frac{1}{2}[s(t) + s(-t)] \tag{1.12b}$$

$$s_{od}(t) = \frac{1}{2}[s(t) - s(-t)] \tag{1.12c}$$

By (1.12b), one can write

$$s_{ev}(-t) = \tfrac{1}{2}[s(-t) + s(t)] = s_{ev}(t)$$

proving that $s_{ev}(t)$ is an even signal. Similarly, by (1.12c)

$$s_{od}(-t) = \tfrac{1}{2}[s(-t) - s(t)] = -\tfrac{1}{2}[s(t) - s(-t)] = -s_{od}(t)$$

proving that $s_{od}(t)$ represents an odd signal.

Similar expressions can also be written for the sequence.

Figure 1.4 depicts the even and odd sequences given by $x[n] = n$ and $x[n] = |n|$ for $-\infty < n < \infty$.

In this section, it is relevant to introduce the concept of conjugate symmetry. For the complex-valued sequence $x[n]$, one can always write

$$x[n] = x_{re}[n] + jx_{im}[n] \tag{1.12d}$$

where $x_{re}[n]$ and $x_{im}[n]$ are the real and imaginary components of $x[n]$. A sequence $x[n]$ is said to be conjugate-symmetric sequence if $x[n] = x^*[-n]$ and conjugate-antisymmetric sequence if $x[n] = -x^*[-n]$ where $x^*[.]$ denotes the complex conjugate of $x[.]$. Thus

$$x^*[n] = x_{re}[n] - jx_{im}[n] \tag{1.12e}$$

By conjugate symmetry we have $x[n] = x^*[-n]$. That is,

$$x_{re}[n] + jx_{im}[n] = x_{re}[-n] - jx_{im}[-n]$$

We thus have, $x_{re}[n] = x_{re}[-n]$ and $x_{im}[n] = -x_{im}[-n]$.

Hence, the real component is an even sequence and the odd component is an odd sequence.

Similarly, by conjugate anti-symmetry we have $x[n] = -x^*[n]$; that is,

$$x_{re}[n] + jx_{im}[n] = -\{x_{re}[-n] - jx_{im}[-n]\}$$

We thus have, $x_{re}[n] = -x_{re}[-n]$ and $x_{im}[n] = x_{im}[-n]$.

It can be shown that the signal $s(t) + s(-t)$ is always an even signal and the signal $s(t) - s(-t)$ is always an odd signal.

1.2 Some Ideal Signals

In this section, we introduce some basic signals that are of considerable importance in signal and system analysis. These include unit step, rectangular pulse, unit impulse, switching function, Gaussian pulse, exponential function, and the others. The unit impulse signal is the most important as this can be treated as the basic building block for the construction and representation of other signals.

1.2.1 Unit step function and signum function

One of the simplest functions is the unit step function $u(t)$, defined by

$$u(t) = \begin{cases} 1 & t \geq 0 \\ 0 & t < 0. \end{cases} \tag{1.13}$$

Figure 1.5 (a) is a sketch of the function $u(t)$. The unit step function shifted by t_0 (shown in Fig. 1.5(b)) may be expressed as

$$u(t - t_0) = \begin{cases} 1 & t \geq t_0 \\ 0 & t < t_0. \end{cases} \tag{1.13a}$$

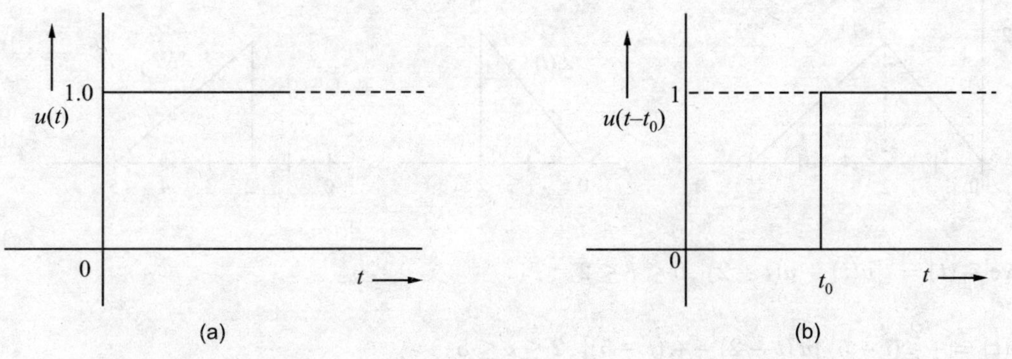

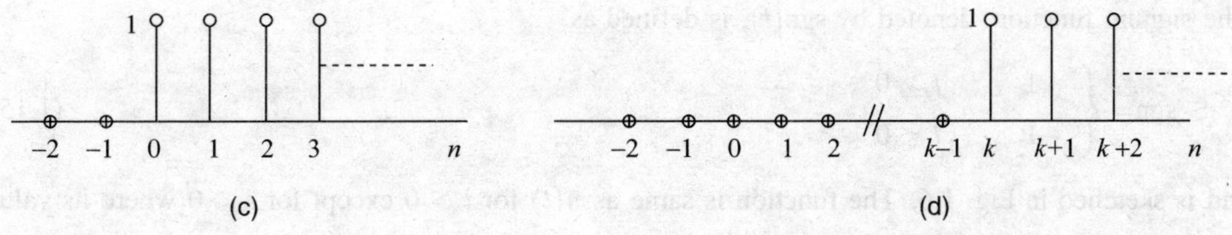

Fig. 1.5 *The unit-step and shifted unit-step function.* (a), (b); *continuous-time;* (c), (d) : *discrete-time sequence.*

If a signal is assumed to act on a network at $t = 0$ onward, we describe this effect by multiplying the signal by $u(t)$. For example, if we want a signal source $v(t)$ to start at $t = 0$, it can be described as $v(t)u(t)$.

The corresponding basic sequence in discrete-time case is the unit-step sequence, denoted by $u[n]$ and is defined by

$$u[n] = \begin{cases} 1, & n \geq 0 \\ 0, & n < 0 \end{cases} \tag{1.14}$$

The unit step shifted by k samples may be expressed as

$$u[n-k] = \begin{cases} 1, & n \geq k \\ 0, & n < k \end{cases} \tag{1.14a}$$

Figure 1.5 (c) and (d) show the functions defined above in (1.14) and (1.14a). For any discrete-time sequence $x[n]$ to be casual, we generally describe this by the product $x[n]u[n]$.

As an illustration, we can express the function $s(t)$ in terms of the unit-step function as follows : $s(t)$ can be decomposed into two components $s_1(t)$ and $s_2(t)$. Then we apply the property of $u(t)$, as shown in the following diagram.

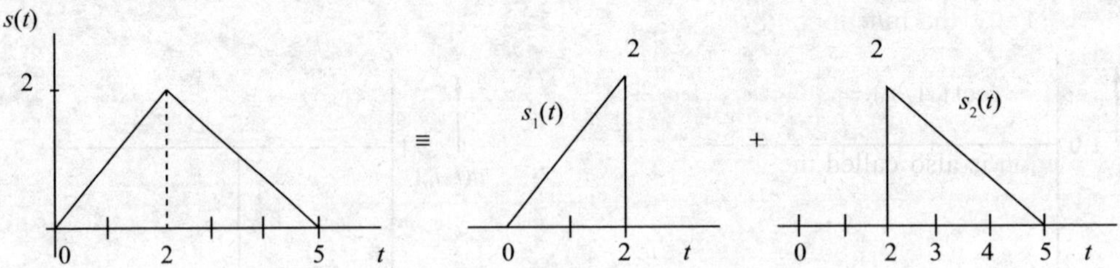

We have $s_1(t) = t[u(t) - u(t-2)], 0 \leq t \leq 2$

and $s_2(t) = -\dfrac{2}{3}(t-5)\left[u(t-2) - u(t-5)\right], 2 \leq t \leq 5$

The signum function, denoted by sgn(t), is defined as

$$\text{sgn} = \begin{cases} 1 & t \geq 0 \\ -1 & t < 0 \end{cases} \tag{1.15}$$

and is sketched in Fig. 1.6. The function is same as $u(t)$ for $t > 0$ except for $t < 0$ where its value is -1.

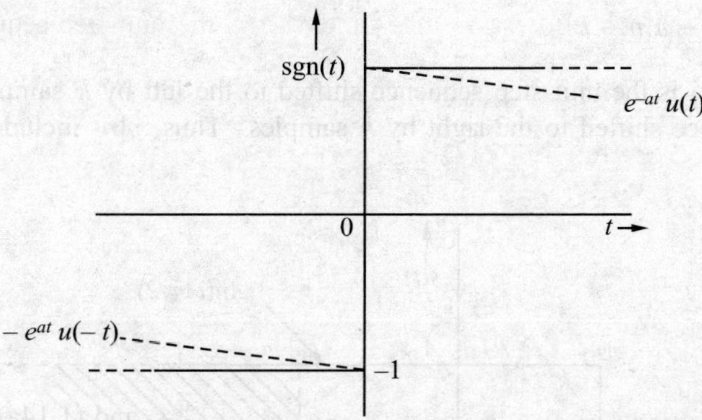

Fig. 1.6 *Signum function. This can be thought of as the sum of exponentials, namely,* $sgn(t) = \lim_{a \to 0} [e^{-at}u(t) - e^{at}u(-t)]$.

1.2.2 Rectangular pulse

A rectangular function of duration τ and amplitude A, symmetrical about $t = 0$ is shown in Fig. 1.7. Mathematically, the function is expressed as

$$s(t) = \text{rect}(A/\tau) = \begin{cases} A & |t| \geq \tau/2 \\ 0 & \text{otherwise.} \end{cases} \quad (1.16)$$

This function is also called the gating function.

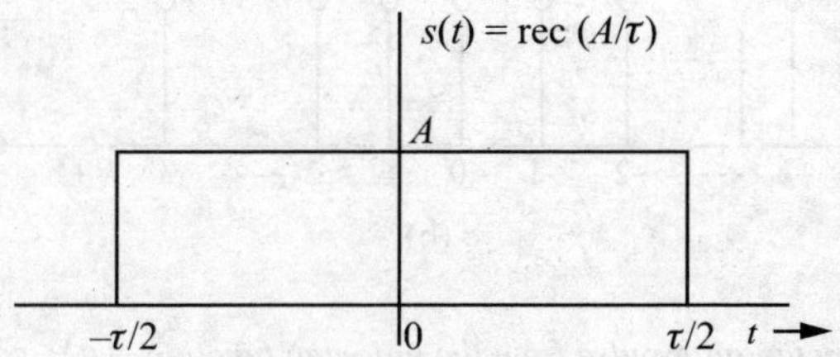

Fig. 1.7 *Rectangular pulse*

The rectangular pulse defined in (1.16) and represented in Fig. 1.7 can be constructed from the unit-step function. i.e.,

$$\pi(t/r) = \text{rect}(A/\tau) = A[u(t + \tau/2) - u(t - \tau/2)]. \quad (1.17)$$

The first term on the right-hand side in (1.17) gives the unit-step function starting at $t = -\tau/2$ while the second term represents the same starting at $t = \tau/2$. The difference between the two gives the rectangular pulse, as illustrated in Fig. 1.8.

Following equations (1.6) and (1.7), the rectangular pulse sequence $y[n]$ can be constructed as follows:

$$y[n] = u[n+k] - u[n-k] \tag{1.17a}$$

The sequence $u[n+k]$ is the unit-step sequence shifted to the left by k samples whereas $u[n-k]$ is the unit-step sequence shifted to the right by k samples. Thus, $y[n]$ includes $2k+1$ samples as shown in Fig. 1.8(b).

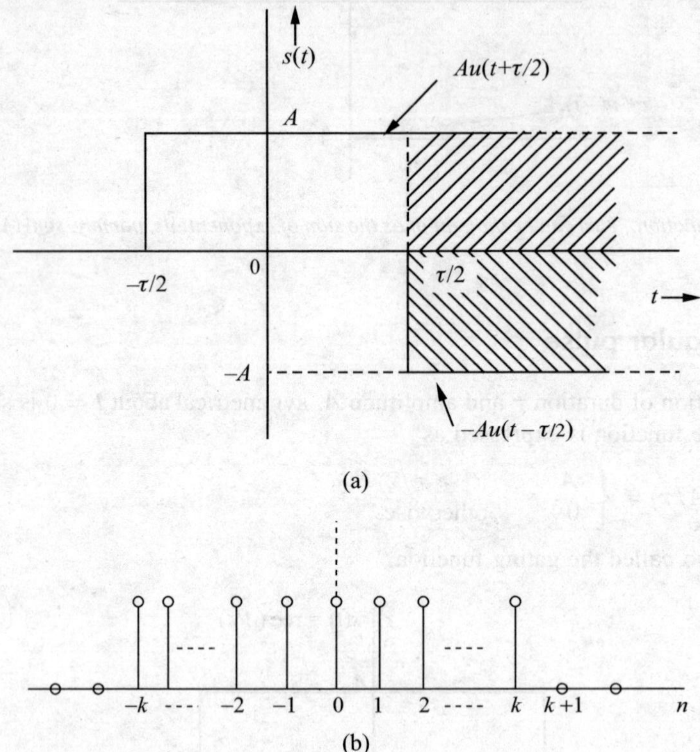

Fig. 1.8 *Construction of a rectangular pulse from the unit-step function.* (a): *continuous-time;* (b): *discrete-time.*

1.2.3 Unit impulse function or Dirac delta function

The unit impulse function $\delta(t)$ defined by Dirac another important function in the study of signals and system behaviour. The function is defined as

$$\begin{aligned} \delta(t) &= 0 \qquad t \neq 0 \\ &= 1 \qquad t = 0. \end{aligned} \tag{1.18}$$

Figure 1.9 shows the $\delta(t)$ function. The delta function can be visualised as a tall and narrow rectangular pulse of unit area (Fig. 1.9). If the duration of the rectangular pulse is made infinitesimally small, its height is increased without bound, keeping its area unchanged at unity. In the limit $\tau \to 0$, the narrow rectangular pulse approaches the delta function.

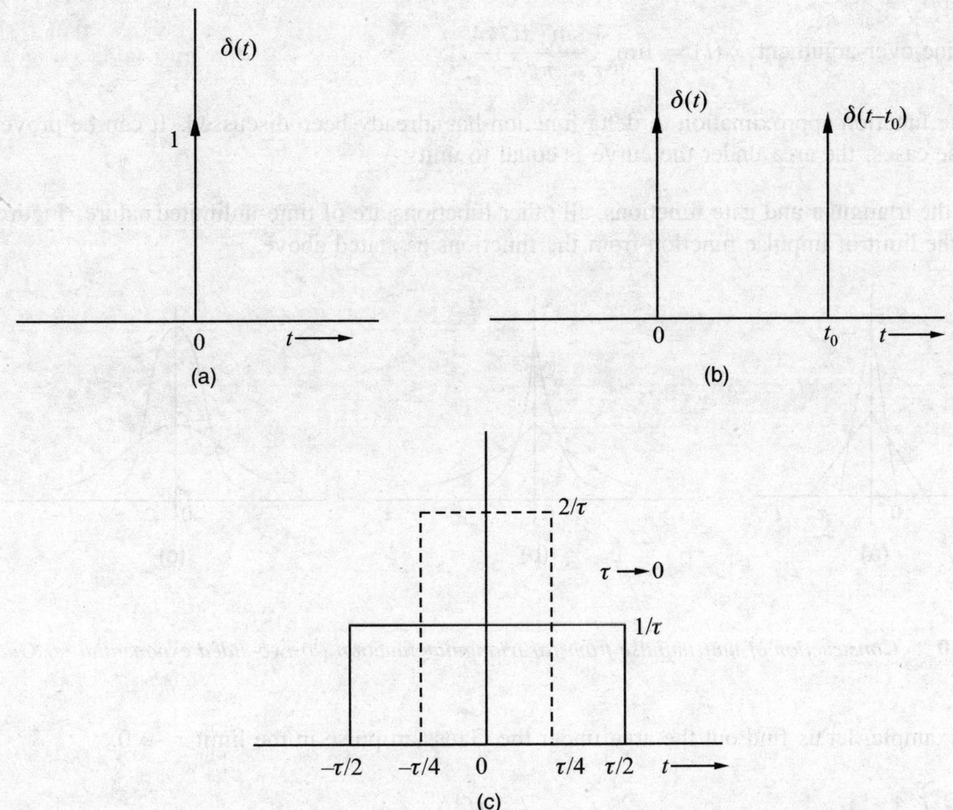

Fig. 1.9 *(a) Unit impulse, (b) Shifted unit impulse and (c) construction of unit impulse from* $\text{rect}(A/\tau)$ *as* $\tau \rightarrow 0$.

A unit impulse occurring at $t = t_0$ is expressed by $\delta(t - t_0)$ and is depicted in Fig. 1.9(b).

The gate function $\pi(t/\tau) = \text{rect}(t/\tau)$ is not the only one by which the delta function $\delta(t)$ can be approximated. We list here some more approximation to the delta function as the limit of some well-behaved functions:

Triangular function: $\delta(t) = \lim\limits_{\tau \to 0} \dfrac{1}{\tau} \Lambda\left(t/\tau\right)$

Two-sided exponential: $\delta\left(t\right) = \lim\limits_{\tau \to 0} \dfrac{1}{\tau} \exp\left(-2\left|t\right|/\tau\right)$

Gaussian function: $\delta\left(t\right) = \lim\limits_{\tau \to 0} \exp\left(-\pi\left(t/\tau\right)^2\right)$

Gate function: $\delta(t) = \lim_{\tau\to 0} \frac{1}{\tau}\mathrm{rect}\,(t/\tau)$

Sine-over-argument: $\delta(t) = \lim_{\tau\to 0} \frac{1}{\tau}\frac{\sin(\pi t/\tau)}{\pi t/\tau}.$

The gate function approximation to delta function has already been discussed. It can be proved that in all the cases, the area under the curve is equal to unity.

Expect the triangular and gate functions, all other functions are of time-unlimited nature. Figure 1.10 shows the limit of impulse function from the functions as stated above.

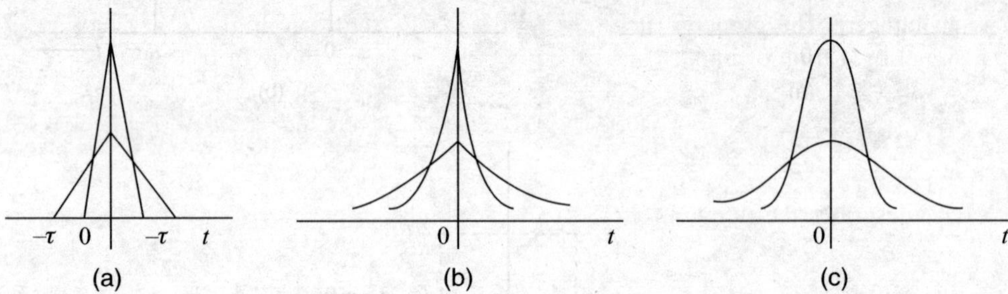

Fig. 1.10 *Construction of unit impulse from (a) triangular function (b) two-sided exponential (c) Gaussian function*

As an example, let us find out the area under the Gaussian pulse in the limit $\tau \to 0$.

$$\text{Area} = \int_{-\infty}^{\infty} \frac{1}{\tau}\exp\left(-\pi\,(t/\tau)^2\right) dt = \frac{2}{\tau}\int_{0}^{\infty} \exp\left(-\frac{\pi t^2}{\tau^2}\right) dt$$

$$\text{Let } \frac{\pi t^2}{\tau^2} = y, \text{ then } \frac{2\pi t}{\tau^2}dt = dy \text{ or, } dt = \frac{\tau^2}{2\pi}.\frac{dy}{\sqrt{\tau^2 y/\pi}} = \frac{\tau/2}{\sqrt{\pi}}y^{-1/2}dy$$

$$\text{Thus, area} = \frac{2}{\tau}\cdot\frac{\tau/2}{\sqrt{\pi}}\int_{0}^{\infty} y^{-1/2}e^{-y}dy = \frac{1}{\sqrt{\pi}}\Gamma(3/2) = \frac{1}{\pi}.\sqrt{\pi} = 1.$$

Readers are advised to verify that areas under each of the above functions are unity.

There is a relation between the unit-step function $u(t)$ and unit impulse function $\delta(t)$. In fact, the unit-step function is the running integral of the unit impulse. That is,

$$u(t) = \int_{-\infty}^{t} \delta(\tau)d\tau, \tag{1.19}$$

and hence the unit impulse is the first derivative of the unit-step function. That is,

$$\delta(t) = \frac{du}{dt}. \tag{1.20}$$

The unit impulse can be used to sample the value of a signal at an arbitrary point. For an impulse that exists at the point $t = t_0$,

$$s(t)\delta(t - t_0) = s(t_0)\delta(t - t_0). \tag{1.21}$$

Similarly, if $s(t)$ is multiplied by $\delta(t - nT)$, one can have

$$s(t)\delta(t - nT) = s(nT)\delta(t - nT), \tag{1.21a}$$

where n is an integer. The property of sampling a signal is shown in Fig. 1.11. This process of sampling a signal by a train of impulses is called impulse-train sampling. In fact,

$$\int_{-\infty}^{\infty} s(t)\delta(t)dt = s(0)\int_{-\infty}^{\infty} \delta(t)dt = s(0). \tag{1.22}$$

Equation (1.22) describes the sampling property of the impulse function.

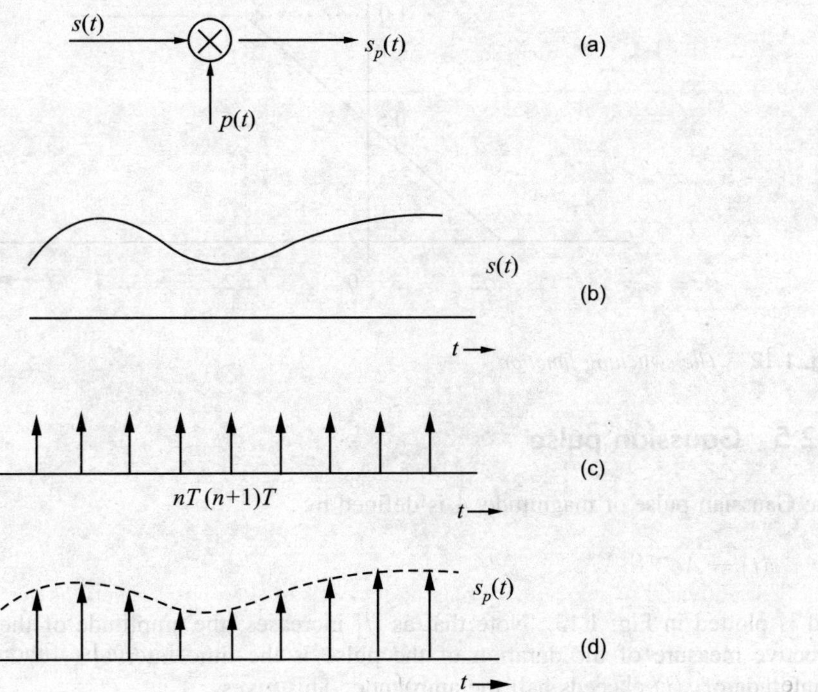

Fig. 1.11 *Sampling property of the unit impulse. (a) The sampler which multiplies a train of impulse by the continuous signal $s(t)$. The output gives the sampled signal $s_p(t)$. (b) Analog signal $s(t)$. (c) Impulse train $p(t)$. (d) Sampled signal $s_p(t)$.*

1.2.4 Switching function or Heaviside function

The switching or the Heaviside function describes a physical process that makes transition from one state to another. Consider a process that was initially at the zero-state described by 0 volt. Let the process make a transition to a new state described by +1 volt. The transition takes place in time τ. This switching process is depicted in Fig. 1.12. The process can be described by the equation

$$s(t) = \begin{cases} \frac{1}{2}\left(\frac{2t}{\tau}+1\right) & |t| < \tau/2 \\ 0 & \text{otherwise.} \end{cases} \tag{1.23}$$

When the transition takes place instantaneously (that is $\tau \rightarrow 0$), the function is the unit step function defined in (1.13).

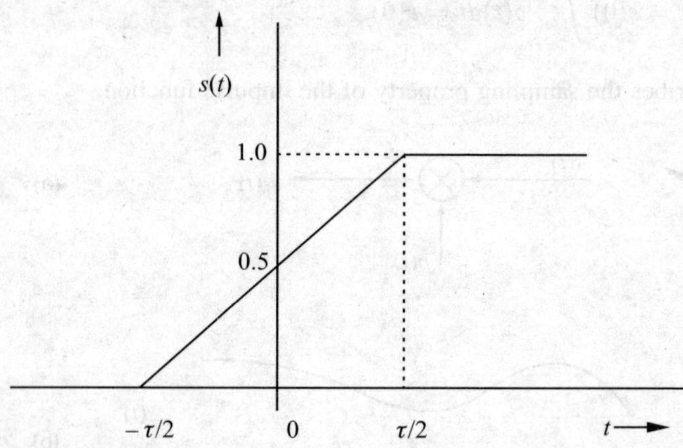

Fig. 1.12 *The switching function*

1.2.5 Gaussian pulse

The Gaussian pulse of magnitude A is defined by

$$s(t) = Ae^{-\pi t^2} \tag{1.24}$$

and is plotted in Fig. 1.13. Note that as $|t|$ increases, the amplitude of the signal decreases. An effective measure of the duration of the pulse is the time interval ($-\tau/2$ to $\tau/2$) in which the magnitude of $s(t)$ exceeds half the amplitude. This gives

$$\frac{1}{2}A = Ae^{-\pi(\tau/2)^2}$$

$$\text{or,}\quad \frac{\tau}{2} = \sqrt{\frac{\ln 2}{\pi}} = 0.469. \tag{1.25}$$

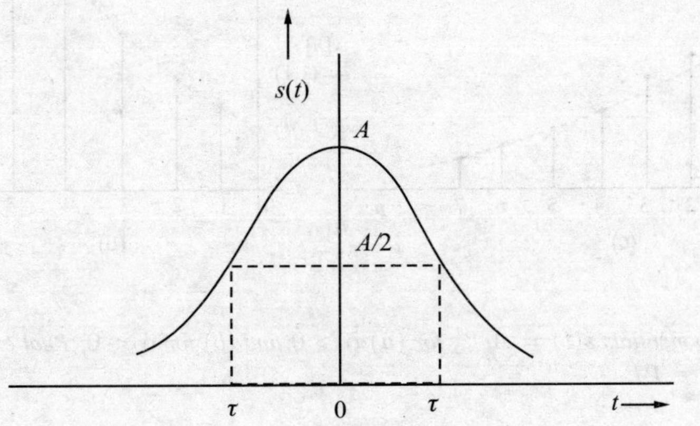

Fig. 1.13 *The Gaussian pulse*

1.2.6 Exponential function

The exponential signal is defined as

$$s(t) = Ae^{\alpha t}, \tag{1.26}$$

where A and α are in general complex numbers. For real exponential signals, A and α are real numbers. Figure 1.14 shows the representation of real exponential signals for α greater and less than zero.

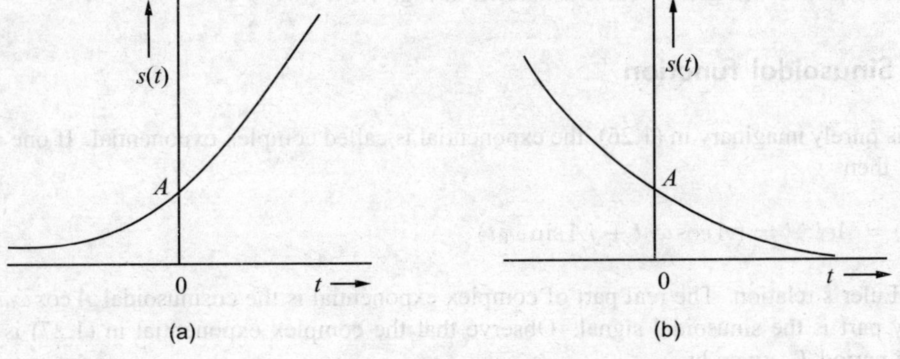

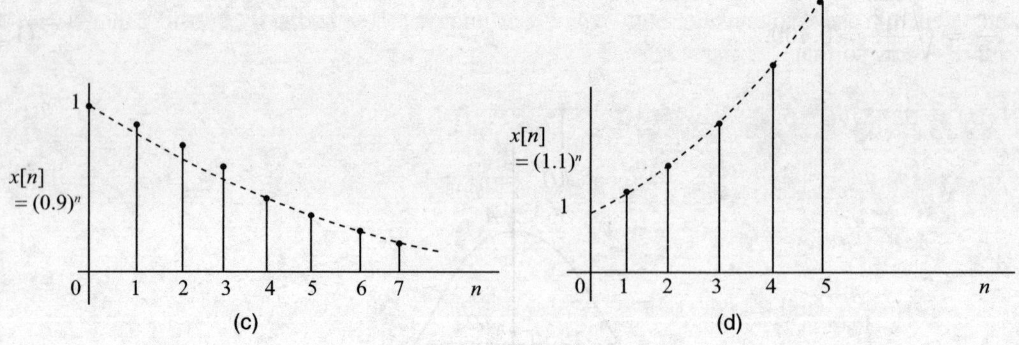

Fig. 1.14 *Real exponentials* $s(t) = Ae^{\alpha t}$ *for* (a) $\alpha > 0$ *and* (b) *for* $\alpha < 0$*; Real exponential* $x[n] = \alpha^n$ *for (c)* $\alpha = 0.8$ *(d)* $\alpha = 1.1$*.*

Exponential Sequences

Exponential sequences are important sequences in the discrete-time systems. The exponential function of the form $s(t) = e^{at}$ can always be expressed in the form $s(t) = b^t$ where a and b are related as $e^a = b$ or $a = lnb$.

In a similar fashion, as in Eq. (1.26), the exponential sequence can be formed by taking the n th sample value of some real or complex constant. That is,

$$x[n] = A\alpha^n, -\infty < n < \infty \tag{1.26a}$$

is the general form of the exponential sequence where A and α are real or complex constants. If we treat A and α as real numbers with $A = 1$ and $|\alpha| < 1$, then the sequence will decay with n. For $\alpha > 1$, the sequence will grow with n as shown in Fig. 1.14.

1.2.7 Sinusoidal function

When α is purely imaginary in (1.26), the exponential is called complex exponential. If one considers $\alpha = j\omega_0$, then

$$s(t) = Ae^{j\omega_0 t} = (A\cos\omega_0 t + jA\sin\omega_0 t) \tag{1.27}$$

by using Euler's relation. The real part of complex exponential is the cosinusoidal $A\cos\omega_0 t$ and the imaginary part is the sinusoidal signal. Observe that the complex exponential in (1.27) is periodic with time-period T_0 given by

$$T_0 = \frac{2\pi}{\omega_0}. \tag{1.28}$$

The most general case is to consider both A and α as imaginary. Specifically, consider that $A = a+jb$ and $\alpha = r + j\omega_0$ so that

$$
\begin{aligned}
s(t) &= (a + jb)\exp(r + j\omega_0)t \\
&= |A|e^{j\theta}e^{(r+j\omega_0)t} \\
&= |A|e^{rt}e^{j(\omega_0 t+\theta)}. \qquad (1.29)
\end{aligned}
$$

Here $|A| = \sqrt{a^2 + b^2}$ and $\theta = \arctan(b/a)$. Using Euler's relation,

$$s(t) = |A|e^{rt}\cos(\omega_0 t + \theta) + j|A|e^{rt}\sin(\omega_0 t + \theta). \qquad (1.30)$$

Observe that both the terms in (1.30) are sinusoidals multiplied by an exponential e^{rt}. The three cases corresponding to $r > 0$, $r < 0$ and $r = 0$ are shown in Fig. 1.15.

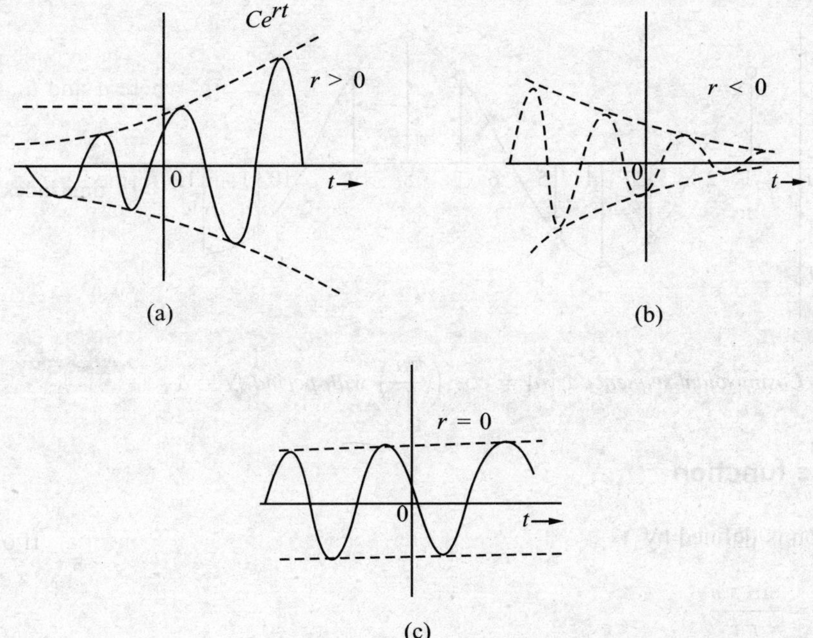

Fig. 1.15 *(a) Growing sinusoidal. The amplitude grows according to Ae^{rt}, $r > 0$. (b) Decaying sinusoidal where the amplitude decays according to Ae^{rt}, $r < 0$. (c) A sinusoidal whose amplitude is unaltered, $r = 0$.*

The sinusoidal sequence can be derived from the exponential sequence of Eq. (1.26a). By expressing $\alpha = e^{(\sigma_0+j\omega_0)}$ and $A = |A|e^{j\phi}$, we can write equation (1.26a) as

$$x[n] = A\alpha^n = |A|\, e^{j\varphi}.e^{(\sigma_0 + j\omega_0)n} = |A|\, e^{\sigma_0 n} e^{j(\omega_0 n + \varphi)} \tag{1.30a}$$

The above sequence is complex in general. The real and imaginary parts of $x[n]$ are given as

$$\left.\begin{aligned} x_{re}\,[n] &= |A|\, e^{\sigma_0 n} \cos\left(\omega_0 n + \varphi\right) \\ x_{im}\,[n] &= |A|\, e^{\sigma_0 n} \sin\left(\omega_0 n + \varphi\right) \end{aligned}\right\} \tag{1.30b}$$

The sinusoidal or cosinusoidal sequences defined above will remain constant ($\sigma_0 = 0$), decay ($\sigma_0 < 0$) or grow ($\sigma_0 > 0$). Without loss of generality, one may express the real sinusoidal sequence of constant amplitude by the equation

$$x[n] = A cos(\omega_0 n + \phi), -\infty < n < \infty \tag{1.30c}$$

Note here that the angular frequency ω_0 (radian/sample) has a different meaning than in Equations (1.29) or (1.30). If we assume $A = 1$, $\phi = 0$ and $\omega_0 = \pi/4$, the sequence becomes $x\,[n] = \cos\left(\frac{\pi}{4} n\right)$ and can be depicted in Fig. 1.15.1.

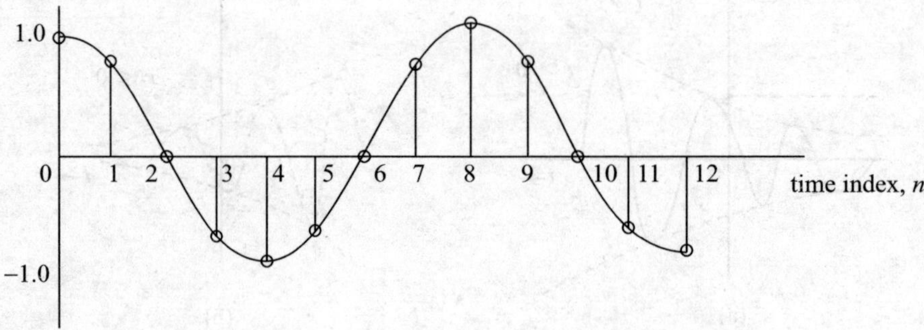

Fig. 1.15.1 *Cosinusoidal sequence* $x\,[n] = \cos\left(\dfrac{\pi n}{4}\right)$ *with period* $N = 8$

1.2.8 Sinc function

A sinc function is defined by

$$\text{sinc}(x) = \frac{\sin \pi x}{\pi x}, \tag{1.31}$$

which is an even function of x and having a maximum at $x = 0$ as shown in Fig. 1.16. The first zero-crossing occurs when $\pi x = \pi$, i.e. when $x = 1$. The function is oscillatory but its amplitude of oscillation decreases as $\frac{1}{\pi x}$. A closely related function is the sampling function defined as

$$S_a(x) = \frac{\sin x}{x}.$$

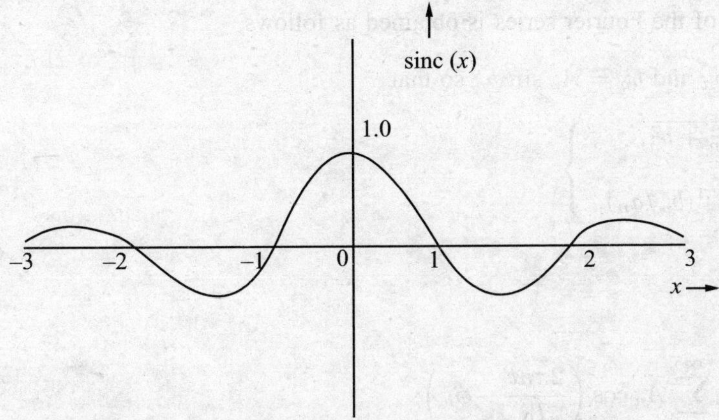

Fig. 1.16 *Sinc function*

1.3 Line Spectra and Fourier Series

In this section we will describe signals in the frequency domain. A suitable frequency-domain description is the line-spectrum which gives the plots of amplitude and phase against frequency. We begin with the Fourier series expansion of any periodic signal and then obtain the line spectrum.

1.3.1 Trigonometric Fourier series

A periodic function of time, $s(t)$, having a fundamental period T_0 can be represented as an infinite sum of sinusoidal waveforms. This summation, called a Fourier series, can be represented either in trigonometric form or in complex form (also called the exponential form). The trigonometric form of the Fourier series of $s(t)$ reads as

$$s(t) = a_0 + \sum_{n=1}^{\infty}\left(a_n \cos\frac{2\pi nt}{T_0} + b_n \sin\frac{2\pi nt}{T_0}\right), \tag{1.32}$$

where the constant a_0 is given as

$$a_0 = \frac{1}{T_0}\int_{-T_0/2}^{T_0/2} s(t)dt, \tag{1.33}$$

which gives the average value of $s(t)$. The series coefficients a_n and b_n are related to $s(t)$ by

$$a_n = \frac{2}{T_0}\int_{-T_0/2}^{T_0/2} s(t)\cos\frac{2\pi nt}{T_0}dt \tag{1.34}$$

and $$b_n = \frac{2}{T_0}\int_{-T_0/2}^{T_0/2} s(t)\sin\frac{2\pi nt}{T_0}dt. \tag{1.35}$$

An alternate form of the Fourier series is obtained as follows.

Let $a_n = A_n \cos \phi_n$ and $b_n = A_n \sin \phi_n$ so that

$$\left.\begin{aligned} A_n &= \sqrt{a_n^2 + b_n^2}, \\ \phi_n &= \tan^{-1}(b_n/a_n), \end{aligned}\right\} \tag{1.36}$$

and let $a_0 = A_0$.

Then (1.32) yields

$$s(t) = A_0 + \sum_{n=1}^{\infty} A_n \cos\left(\frac{2\pi nt}{T_0} - \phi_n\right). \tag{1.37}$$

It is observed from (1.32) and (1.37) that all the frequencies are integral multiples or harmonics of the fundamental frequency $f_0 = 1/T_0$. Thus the spectral lines of periodic signals have uniform spacing f_0. The concept of spectral lines is discussed in the following section.

Example 1.3

Find the Fourier series expansion of the periodic train of impulses.

Solution

A periodic impulse train can be written as

$$s(t) = A \sum_{k=-\infty}^{\infty} \delta(t - kT_0), \tag{1.38}$$

where A and T_0 denote respectively the strength and repetition period of the impulse. The coefficients a_0, a_n and b_n are obtained as

$$a_0 = \frac{1}{T_0}\int_{-T_0/2}^{T_0/2} A\delta(t)dt = \frac{A}{T_0},$$

$$a_n = \frac{2}{T_0}\int_{-T_0/2}^{T_0/2} A\delta(t)\cos\frac{2\pi nt}{T_0}dt = \frac{2A}{T_0},$$

$$b_n = \frac{2}{T_0}\int_{-T_0/2}^{T_0/2} A\delta(t)\sin\frac{2\pi nt}{T_0}dt = 0.$$

Thus, $$s(t) = \frac{A}{T_0} + \sum_{n=1}^{\infty}\frac{2A}{T_0}\cos\frac{2\pi nt}{T_0}. \tag{1.39}$$

Example 1.4

Find the Fourier series expansion of a rectangular pulse train as shown in Fig. 1.17.

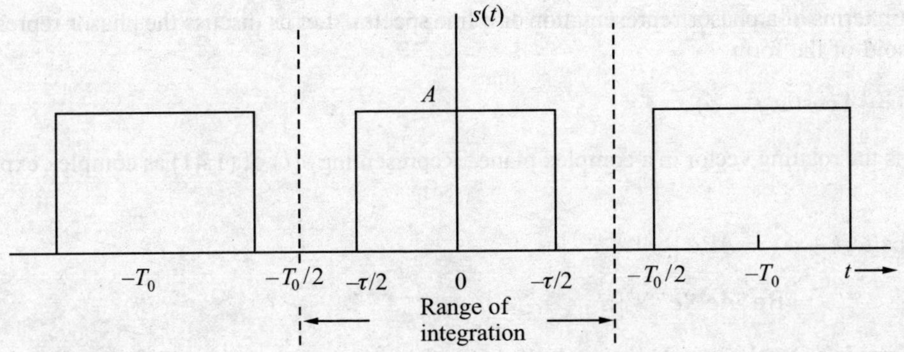

Fig. 1.17 *Rectangular pulse train of amplitude A, duration τ, and repeating every T_0 seconds*

Solution We select the range of integration over the central period $-T_0/2 \le t \le T_0/2$. In this range,

$$s(t) = \begin{cases} A, & |t| \le \tau/2 \\ 0, & |t| > \tau/2. \end{cases}$$

Thus,

$$a_0 = \frac{1}{T_0}\int_{-T_0/2}^{T_0/2} s(t)dt = \frac{1}{T_0}\int_{-\tau/2}^{\tau/2} Adt = \frac{A\tau}{T_0},$$

$$a_n = \frac{2}{T_0}\int_{-T_0/2}^{T_0/2} s(t)\cos\frac{2\pi nt}{T_0}dt = \frac{2}{T_0}\int_{-\tau/2}^{\tau/2} A\cos\frac{2\pi nt}{T_0}dt$$

$$= \frac{2A}{T_0}.\left[\frac{\sin\frac{2\pi nt}{T_0}}{\frac{2\pi n}{T_0}}\right]_{-\tau/2}^{\tau/2} = \frac{A}{\pi n}\times 2\sin\frac{\pi n\tau}{T_0}$$

$$= \frac{2A\tau}{T_0}.\left(\frac{\sin\frac{\pi n\tau}{T_0}}{\frac{\pi n\tau}{T_0}}\right),$$

$b_n = 0$ (since $s(t)$ is an even function).

$$\text{Thus, } s(t) = \frac{A\tau}{T_0} + \frac{2A\tau}{T_0}\sum_{n=1}^{\infty}\left(\frac{\sin\frac{\pi n\tau}{T_0}}{\frac{\pi n\tau}{T_0}}\cos\frac{2\pi nt}{T_0}\right). \tag{1.40}$$

1.3.2 Phasors and line spectra

The Fourier series in (1.37) indicates that a periodic signal $s(t)$ consists of an infinite summations of sinusoids of frequencies nf_0 whose amplitudes are A_n and phases are ϕ_n. We can express the sinusoids in terms of a phasor representation or a line spectra. Let us discuss the phasor representation of a sinusoid of the form

$$s(t) = A\cos(\omega_0 t + \phi). \tag{1.41}$$

A phasor is the rotating vector in a complex plane. Representing $s(t)$ of (1.41) as complex exponential, we get

$$\begin{aligned} A\cos(\omega_0 t + \phi) &= A\text{Re}\left[e^{j(\omega_0 t + \phi)}\right] \\ &= \text{Re}\left[Ae^{j\phi}e^{j\omega_0 t}\right]. \end{aligned} \tag{1.42}$$

Here 'Re' reads as 'real part of'. Figure 1.18 shows the phasor representation of $s(t)$ given by (1.41). A phasor diagram of a sinusoid provides the amplitude, phase angle, and the frequency of revolution.

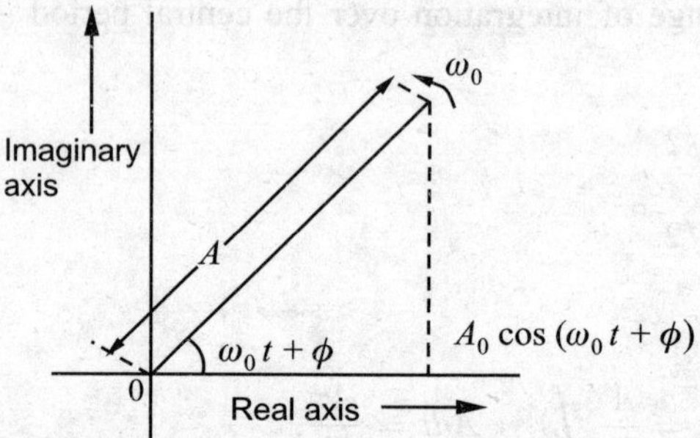

Fig. 1.18 *Phasor diagram for the sinusoid* $A\cos(\omega_0 t + \phi)$

An alternate representation of a sinusoid is the line spectrum which gives two plots: amplitude spectrum (amplitude versus frequency) and phase spectrum (phase angle versus frequency). For the sinusoid in (1.41), the amplitude and phase spectra are drawn in Fig. 1.19.

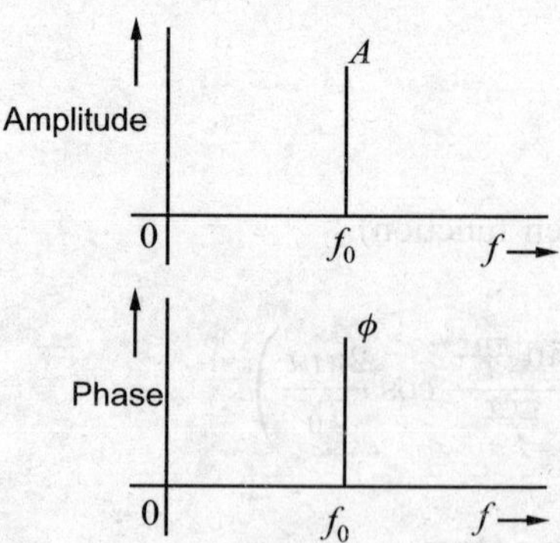

Fig. 1.19 *Line spectrum of* $A\cos(\omega_0 t + \phi)$

Example 1.5

Sketch the amplitude and phase spectra of the signal $s(t) = 4 - 10\cos(100\pi t - 45°) + 6\sin 200\pi t$.

Solution

Let us take the convention that the phase angle will be measured with respect to the cosine wave and the amplitude is considered to be always a positive quantity.

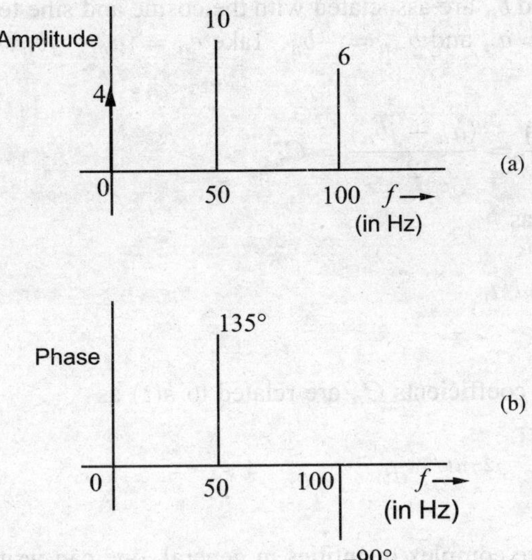

Fig. 1.20 *Line spectra of the signal: (a) amplitude spectrum (b) phase spectrum.*

The dc component has an amplitude of 4 unit. The term

$$-10\cos(100\pi t - 45°) = 10\cos(2\pi \times 50t - 45° + 180°)$$

$$= 10\cos(2\pi \times 50t + 135°),$$

whose spectrum consists of an amplitude of 10 units,located at the frequency of 50 Hz and a phase angle of 135°. Similarly, the term $6\sin 200\pi t = 6\cos(2\pi \times 100t - 90°)$ is of amplitude of 6 units and phase $-90°$ at a frequency of 100 Hz. The line spectra is drawn in Fig. 1.20. The spectra is one-sided as we have considered the positive-frequency only. The two-sided line spectrum involving both the positive and negative frequencies will be taken up later on.

1.3.3 Exponential form of Fourier series

The complex form of the Fourier series is readily obtained by applying Euler's formula in (1.32). Thus,

$$s(t) = a_0 + \sum_{n=1}^{\infty} \left[a_n \left(\frac{e^{j2\pi nt/T_0} + e^{-j2\pi nt/T_0}}{2} \right) + b_n \left(\frac{e^{j2\pi nt/T_0} - e^{-j2\pi nt/T_0}}{2j} \right) \right]$$

$$= a_0 + \sum_{n=1}^{\infty} \left[\left(\frac{a_n - jb_n}{2} \right) e^{j2\pi nt/T_0} + \left(\frac{a_n + jb_n}{2} \right) e^{-j2\pi nt/T_0} \right]. \tag{1.43}$$

Since the coefficients a_n and b_n are associated with the cosine and sine terms respectively, a_n is even and b_n is odd and so $a_{-n} = a_n$ and $b_{-n} = -b_n$. Take $c_n = (a_n - jb_n)/2$. So $C_n^* = (a_n + jb_n)/2$, the conjugate of C_n. Thus,

$$C_{-n} = \frac{(a_{-n} - jb_{-n})}{2} = \frac{(a_n + jb_n)}{2} = C_n^*.$$

Equation (1.43) then reads as

$$s(t) = \sum_{n=-\infty}^{\infty} C_n e^{j2\pi nt/T_0}, \tag{1.44}$$

where $C_0 = a_0$. The series coefficients C_n are related to $s(t)$ as

$$C_n = \frac{1}{T_0} \int_{-T_0/2}^{T_0/2} s(t) e^{-j2\pi nt/T_0} dt. \tag{1.45}$$

Since the coefficients C_n are complex quantities in general, one can write for C_n

$$C_n = |C_n| e^{j\phi_n}. \tag{1.46}$$

We observe that the series in (1.44) consists of phasors with amplitude and phase given by (1.46) at the frequencies $0, \pm f_0, \pm 2f_0, \cdots$. Thus the corresponding spectrum is a two-sided line spectrum.

Example 1.6
Find the exponential Fourier series of a rectangular pulse train of Example 1.4.

Solution
The trigonometric Fourier series has been derived (vide (1.40)). Rewriting the series,

$$s(t) = \frac{A\tau}{T_0} + \frac{2A\tau}{T_0} \sum_{n=1}^{\infty} \left(\frac{\sin \pi n\tau/T_0}{\pi n\tau/T_0} \right) \cos \frac{2\pi nt}{T_0}.$$

By direct calculation, the complex Fourier series coefficient C_n is obtained as

$$C_n = \frac{1}{T_0} \int_{-T_0/2}^{T_0/2} s(t) e^{-j2\pi nt/T_0} dt$$

$$= \frac{1}{T_0} \int_{-\tau/2}^{\tau/2} A e^{-j2\pi nt/T_0} dt = \frac{A}{-j2\pi n} \cdot \left[e^{-je\pi nt/T_0} \right]_{-\tau/2}^{\tau/2}$$

$$= \frac{A}{\pi n}\left[\frac{e^{j\pi n\tau/T_0} - e^{-j\pi n\tau/T_0}}{2j}\right] = \frac{A}{\pi n}\sin \pi n\tau/T_0$$

$$= \frac{A\tau}{T_0}\left(\frac{\sin \pi n\tau/T_0}{\pi n\tau/T_0}\right) = \frac{A\tau}{T_0}\sin c(n\tau/T_0). \tag{1.47}$$

Thus the series is

$$s(t) = \sum_{n=-\infty}^{\infty} \frac{A\tau}{T_0}\text{sinc}\frac{n\tau}{T_0}e^{j2\pi nt/T_0}. \tag{1.48}$$

Let us now draw the amplitude and phase spectra for a duty cycle of $\tau/T_0 = 1/4$. The first zero of the sinc function occurs when $n\tau/T_0 = 1$, that is, $n = T_0/\tau = 4$. This implies that the 4th harmonic is absent in the amplitude spectra. The line spectra according to (1.48) is plotted in Fig. 1.21. Note that the amplitude spectrum is an even function whereas the phase spectrum is an odd function of frequency.

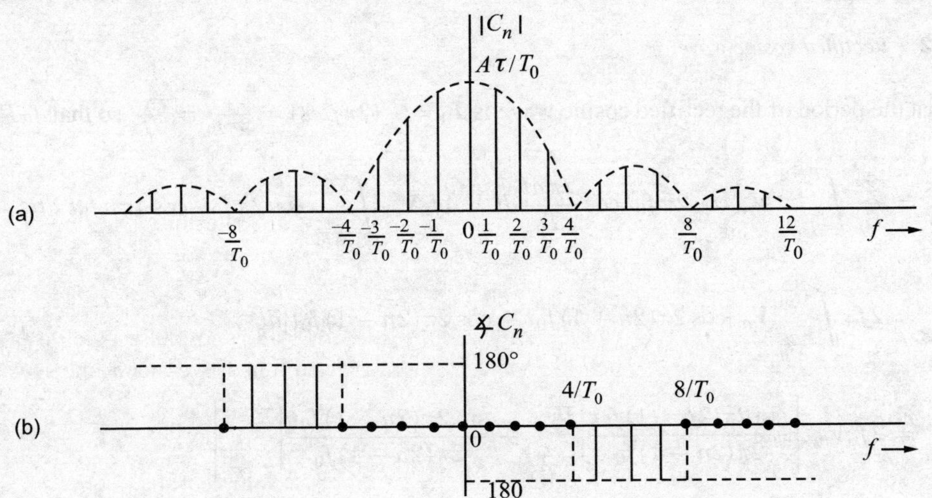

Fig. 1.21 *Line spectra of rectangular pulse with duty cycle = 1/4: (a) amplitude spectrum, (b) phase spectrum.*

Example 1.7
Find the Fourier series of a full-wave rectified cosine wave.

Consider the rectified cosine wave as shown in Fig. 1.22.

Consider the range of integration over the period $-T_0/2 \le t \le T_0/2$, where

$$s(t) = |V_m \cos 2\pi f_0 t|. \tag{1.49}$$

Here $a_0 = \frac{1}{T_0}\int_{-T_0/2}^{T_0/2} s(t)dt = \frac{V_m}{T_0}\int_{-T_0/2}^{T_0/2} \cos 2\pi f_0 t dt = \frac{V_m}{T_0}\left[\frac{\sin 2\pi f_0 t}{2\pi f_0}\right]_{-T_0/2}^{T_0/2}$

$$= \frac{V_m}{2\pi f_0 T_0}[\sin(\pi f_0 T_0) + \sin(\pi f_0 T_0)] = \frac{2 \times V_m}{2\pi \times \frac{1}{2}} = \frac{2V_m}{\pi}.$$

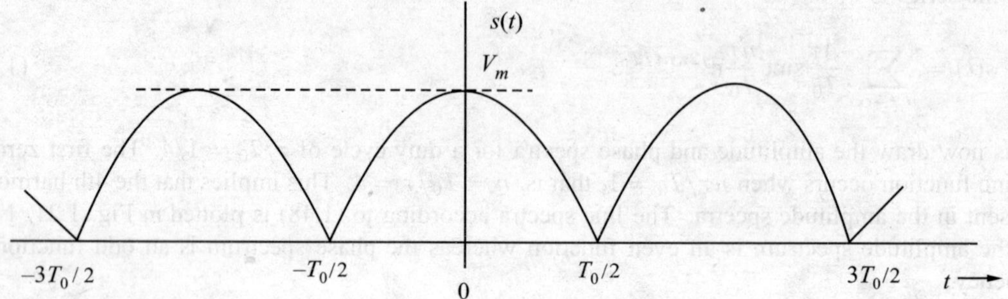

Fig. 1.22 *Rectified cosine wave*

(Note that the period of the rectified cosine wave is $T_0 = \frac{1}{2}(2\pi/\omega_0) = \frac{\pi}{2\pi f_0} = \frac{1}{2f_0}$ so that $f_0 T_0 = \frac{1}{2}$.)

$$a_n = \frac{2}{T_0}\int_{-T_0/2}^{T_0/2} V_m \cos 2\pi f_0 t \cos \frac{2\pi n t}{T_0} dt = 4 f_0 V_m \int_{-\frac{1}{4f_0}}^{\frac{1}{4f_0}} \cos 2\pi f_0 t \ \cos 4\pi n f_0 t \ dt$$

$$= 2 f_0 \int_{-\frac{1}{4f_0}}^{\frac{1}{4f_0}} V_m[\cos 2\pi(2n+1) f_0 t + \cos 2\pi(2n-1) f_0 t] dt$$

$$= 2 f_0 V_m \left[\frac{\sin 2\pi(2n+1) f_0 t}{2\pi(2n+1) f_0}\Bigg|_{-\frac{1}{4f_0}}^{\frac{1}{4f_0}} + \frac{\sin 2\pi(2n-1) f_0 t}{2\pi(2n-1) f_0}\Bigg|_{-\frac{1}{4f_0}}^{\frac{1}{4f_0}}\right]$$

$$= V_m \times 2 f_0 \left[\not{2} \times \frac{\sin 2\pi(2n+1) f_0 \times \frac{1}{4f_0}}{\not{2}\pi(2n+1) f_0} + \not{2} \times \frac{\sin 2\pi(2n-1) f_0 \times \frac{1}{4f_0}}{\not{2}\pi(2n-1) f_0}\right]$$

$$= \frac{2V_m}{\pi}\left[\frac{\sin(2n+1)\pi/2}{(2n+1)} + \frac{\sin(2n-1)\pi/2}{(2n-1)}\right]$$

$$= \frac{2V_m}{\pi}\left[\frac{(-1)^n}{2n+1} + \frac{(-1)^{n+1}}{2n-1}\right]$$

since the sine of any odd multiple of $\pi/2$ gives $+1$ or -1.

$b_n = 0$ as $s(t)$ is an even function.

The Fourier series is then

$$s(t) = \frac{2V_m}{\pi} + \sum_{n=1}^{\infty} \frac{2V_m}{\pi}\left[\frac{(-1)^{n+1}}{2n-1} + \frac{(-1)^n}{2n+1}\right]\cos 2n\omega_0 t. \quad (1.50)$$

From the series,

$$a_1 = \frac{2}{\pi}\left[\frac{1}{1} + \frac{-1}{3}\right] = \frac{4V_m}{3\pi}$$

$$a_2 = \frac{2}{\pi}\left[\frac{-1}{3} + \frac{1}{5}\right] = -\frac{4V_m}{15\pi}$$

$$a_3 = \frac{2}{\pi}\left[\frac{1}{5} + \frac{-1}{7}\right] = \frac{4V_m}{35\pi}, \text{ etc.}$$

The one-sided and two-sided amplitude spectra are shown in Fig. 1.23. The two-sided line spectrum include the negative frequencies and the amplitude is one-half of the original amplitudes at each frequency.

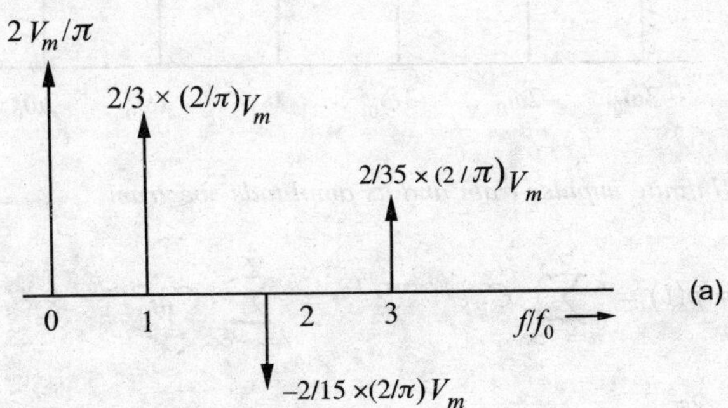

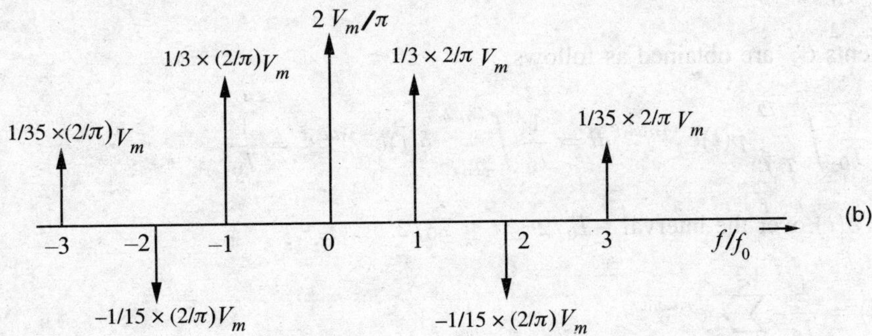

Fig. 1.23 *Amplitude spectrum of rectified sine wave: (a) one-sided, (b) two-sided.*

Example 1.8

Find the Fourier series and the spectra for the impulse train given by

$$p(t) = \sum_{n=-\infty}^{\infty} \delta(t - nT_0).$$

Solution

The exponential Fourier series is given by

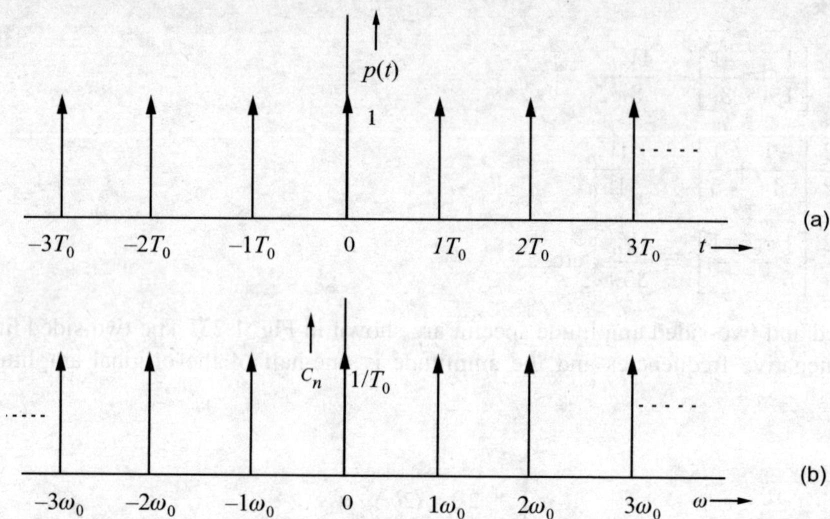

Fig. 1.24 *Infinite impulse train and its amplitude spectrum*

$$s(t) = p(t) = \sum_{n=-\infty}^{\infty} C_n e^{j2\pi nt/T_0} = \sum_{n=-\infty}^{\infty} C_n e^{jn\omega_0 t},$$

where $\omega_0 = \dfrac{2\pi}{T_0}$.

The coefficients C_n are obtained as follows.

$$C_n = \frac{1}{T_0}\int_{-T_0/2}^{T_0/2} p(t)e^{-jn\omega_0 t}dt = \frac{1}{t_0}\int_{-T_0/2}^{T_0/2} \delta(t)e^{-jn\omega_0 t} = \frac{1}{T_0},$$

since $p(t) = \delta(t)$ over the interval $-T_0/2 \leq t \leq T_0/2$.

Thus, $$p(t) = \frac{1}{T_0}\sum_{n=-\infty}^{\infty} e^{jn\omega_0 t} \tag{1.51}$$

which represents the Fourier series representation of the train of impulses. The spectrum is uniform for all the harmonics $n\omega_0$. The spectrum is real and so only the amplitude plot is shown in Fig. 1.24(b).

Example 1.9
Find the Fourier series expansion of a clipped sinusoid as shown in Fig. 1.25.

According to the figure,

$$s(t) = V_m \cos\omega_0 t - V_0, \quad t_1 < t < t_2 \tag{1.52}$$

and the wave is repeated every $T = 2\pi/\omega_0$ second. The coefficient a_0 is calculated as follows.

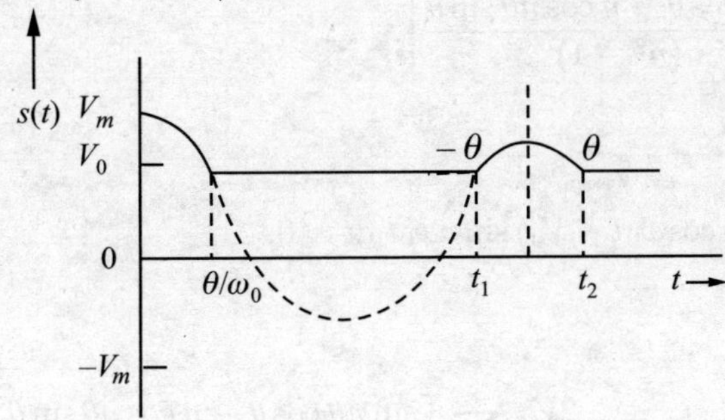

Fig. 1.25 *A clipped cosine wave*

$$a_0 = \frac{1}{T_0}\int_{t_1}^{t_2} (V_m \cos\omega_0 t - V_0)dt = \left[\frac{V_m \sin\omega_0 t}{\omega_0 T_0}\right]_{t_1}^{t_2} - \frac{V_0}{T_0}(t_2 - t_1)$$

$$= \frac{V_m}{2\pi}(\sin\omega_0 t_2 - \sin\omega_0 t_1) - \frac{V_0}{2\pi}(\omega_0 t_2 - \omega_0 t_1).$$

The cutoff angle θ is obtained from the figure as

$$V_m \cos\theta = V_0 \quad \text{or} \quad \theta = \cos^{-1}(V_0/V_m).$$

Then, $\quad \omega_0 t_2 = \theta, \quad \omega_0 t_1 = -\theta.$

So, $\quad a_0 = \dfrac{V_m}{2\pi}[\sin\theta - \sin(-\theta)] - \dfrac{V_0}{2\pi}[\theta - (-\theta)]$

$$= \frac{V_m}{\pi}\sin\theta - \frac{V_0}{\pi}\theta = \frac{V_m}{\pi}(\sin\theta - \frac{V_0}{V_m}\theta)$$

$$= \frac{V_m}{\lambda}(\sin\theta - \theta\cos\theta).$$

$$a_n = \frac{2}{T_0}\int_{t_1}^{t_2} (V_m \cos\omega_0 t - V_0)\cos n\omega_0 t \; dt$$

$$= \frac{V_m}{T_0}\int_{t_1}^{t_2} 2\cos\omega_0 t \; \cos n\omega_0 t \; dt - \frac{2V_0}{T_0}\int_{t_1}^{t_2} \cos n\omega_0 t \; dt$$

$$= \frac{V_m}{T_0}\int_{-\theta/\omega_0}^{\theta/\omega_0}[\cos(n+1)\omega_0 t + \cos(n-1)\omega_0 t]dt - \frac{2V_0}{T_0}\int_{-\theta_0/\omega_0}^{\theta/\omega_0}\cos n\omega_0 t\; dt$$

$$= \frac{2V_m}{\omega_0 T_0}\left[\frac{\sin(n+1)\theta}{n+1} + \frac{\sin(n-1)\theta}{n-1}\right] - \frac{2V_0}{T_0}\frac{2\sin n\theta}{n\omega_0}$$

$$= \frac{V_m}{\pi}\left[\frac{\sin(n+1)\theta}{n+1} + \frac{\sin(n-1)\theta}{n-1} - \frac{2\cos\theta\sin n\theta}{n}\right]$$

$$= \frac{2V_m}{\pi}\left[\frac{\sin n\theta\cos\theta - n\cos n\theta\sin\theta}{n(n^2-1)}\right].$$

Similarly,

$$b_n = \frac{2}{T_0}\int_{-\theta/\omega_0}^{\theta/\omega_0}(V_m\cos\omega_0 t - V_0)\sin n\omega_0 t\; dt = 0.$$

Thus,

$$s(t) = \frac{V_m}{\pi}(\sin\theta - \theta\cos\theta) + \frac{2V_m}{\pi}\sum_{n=2}^{\infty}\left[\frac{\sin n\theta\cos\theta - n\cos n\theta\sin\theta}{n(n^2-1)}\right]\cos n\omega_0 t. \qquad (1.53)$$

To verify our results, let us study the special case of half-wave rectifier where $V_0 = 0$, and $\theta = \pi/2$. So,

$$a_0 = \frac{V_m}{\pi}\left(\sin\frac{\pi}{2} - \frac{\pi}{2}\cos\frac{\pi}{2}\right) = \frac{V_m}{\pi},$$

$$a_n = \frac{2V_m}{\pi}\left(\frac{-n\cos n\pi/2}{n(n^2-1)}\right) = -\frac{2V_m}{\pi(n^2-1)}\cos n\pi/2.$$

Equation (1.53) does not give the value a_1; this is derived here as follows:

$$a_1 = \frac{2}{T_0}\int_{-\theta/\omega_0}^{\theta/\omega_0}(V_m\cos\omega_0 t - V_0)\cos\omega_0 t\; dt$$

$$= \frac{V_m}{T_0}\int_{-\theta/\omega_0}^{\theta/\omega_0}(1+\cos 2\omega_0 t)dt - \frac{2V_0}{T_0}\int_{-\theta/\omega_0}^{\theta/\omega_0}\cos\omega_0 t\; dt$$

$$= \frac{V_m}{T_0}\times\frac{2\theta}{\omega_0} + \frac{V_m}{2T_0\omega_0}\times 2\sin 2\theta - \frac{2V_0}{T_0\omega_0}\times 2\sin\theta$$

$$= \frac{V_m\theta}{\pi} + \frac{V_m}{2\pi}\sin 2\theta - \frac{2V_0}{\pi}\sin\theta$$

$$= \frac{V_m}{\pi}(\theta + \sin\theta\cos\theta - 2\sin\theta\cos\theta) = \frac{V_m}{\pi}(\theta - \sin\theta\cos\theta).$$

In this case,

$$a_1 = \frac{V_m}{\pi}\left(\frac{\pi}{2} - \sin\frac{\pi}{2}\cos\frac{\pi}{2}\right) = \frac{V_m}{2},$$

$$a_2 = \frac{-2V_m}{\pi(2^2-1)}\cos 2\times\frac{\pi}{2} = \frac{2V_m}{3\pi}$$

$$a_3 = \frac{-2V_m}{\pi(3^2-1)}\cos 3\times\frac{\pi}{2} = 0$$

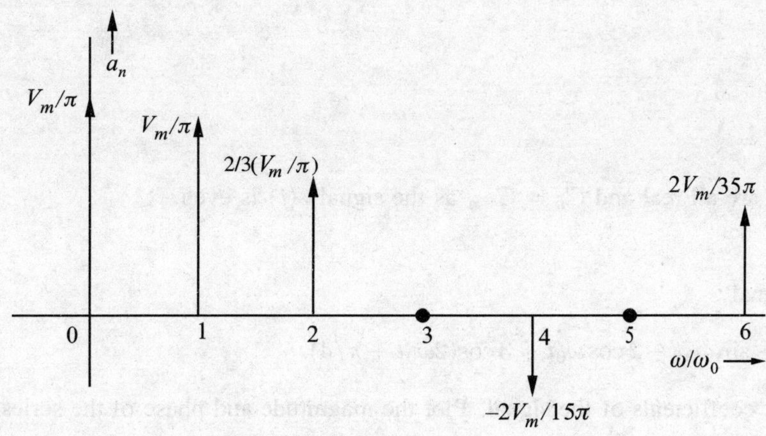

Fig. 1.26 *Amplitude spectrum of a half-wave rectified cosine wave*

$$a_4 = \frac{-2V_m}{\pi(4^2-1)}\cos 4\times\frac{\pi}{2} = -\frac{2V_m}{15\pi}$$

$$a_5 = 0,$$

$$a_6 = \frac{-2V_m}{\pi(6^2-1)}\cos 6\times\frac{\pi}{2} = \frac{2V_m}{35\pi}, \text{ etc.}$$

The corresponding amplitude spectrum is plotted in Fig. 1.26. Compare this spectrum with the spectrum of a full-wave cosine of Example 1.7.

Example 1.10
Find the exponential Fourier series coefficients for the signal $x(t)$ given by

$$x(t) = 1 + \frac{1}{2}\cos 2\pi t + \frac{1}{3}\cos 4\pi t + \frac{1}{4}\cos 6\pi t.$$

Solution
Using Euler's relation,

$$x(t) = 1 + \frac{1}{2}\left(\frac{e^{j2\pi t} + e^{-j2\pi t}}{2}\right) + \frac{1}{3}\left(\frac{e^{j4\pi t} + e^{-j4\pi t}}{2}\right) + \frac{1}{4}\left(\frac{e^{j6\pi t} + e^{-j6\pi t}}{2}\right)$$

$$= 1 + \frac{1}{4}\left(e^{j2\pi t} + e^{-j2\pi t}\right) + \frac{1}{6}\left(e^{j4\pi t} + e^{-j4\pi t}\right) + \frac{1}{8}\left(e^{j6\pi t} + e^{-j6\pi t}\right)$$

$$= \sum_{n=-3}^{3} C_n e^{j2\pi n f_0 t}, \text{ with } f_0 = 1.$$

Thus the coefficients are given as

$$C_0 = 1$$

$$C_1 = C_{-1} = \tfrac{1}{4}$$

$$C_2 = C_{-2} = \tfrac{1}{6}$$

$$C_3 = C_{-3} = \tfrac{1}{8}\,.$$

The coefficients are all real and $C_n = C_{-n}$ as the signal $s(t)$ is even.

Example 1.11
Consider the signal

$$x(t) = 1 + \sin\omega_0 t + 2\cos\omega_0 t + 3\cos(2\omega_0 t + \pi/4).$$

Find the Fourier coefficients of the signal. Plot the magnitude and phase of the series coefficients.

Solution
By Euler's relation,

$$x(t) = 1 + \frac{1}{2j}(e^{j\omega_0 t} + e^{-j\omega_0 t}) + \left(e^{j\omega_0 t} + e^{-j\omega_0 t}\right) + \frac{3}{2}\left(e^{j(2\omega_0 t + \pi/4)} + e^{-j(2\omega_0 t + \pi/4)}\right)$$

$$= 1 + \left(1 + \frac{1}{2j}\right)e^{j\omega_0 t} + \left(1 - \frac{1}{2j}\right)e^{-j\omega_0 t} + \left(\frac{3}{2}e^{j\pi/4}\right)e^{j2\omega_0 t} + \left(\frac{3}{2}e^{-j\pi/4}\right)e^{-j2\omega_0 t}.$$

The Fourier series coefficients are:

$$C_0 = 1$$

$$C_1 = 1 + \frac{1}{2j} = 1 - \frac{1}{2}j;\ \ |C_1| = \sqrt{\frac{5}{4}},\ \ \measuredangle C_1 = -\tan^{-1}\frac{1}{2}$$

$$C_{-1} = 1 - \frac{1}{2j} = 1 + \frac{1}{2}j;\ \ |C_1| = \sqrt{\frac{5}{4}},\ \ \measuredangle C_1 = +\tan^{-1}(\frac{1}{2})$$

$$C_2 = \frac{3}{2}e^{j\pi/4} = \frac{3}{2}(\cos\pi/4 + j\sin\pi/4) = \frac{3}{2\sqrt{2}}(1 + j);$$

$$|C_2| = \frac{3}{2\sqrt{2}} \times \sqrt{2} = \frac{3}{2};\ \ \measuredangle C_2 = \tan^{-1}(1) = \frac{\pi}{4}$$

$$C_{-2} = \frac{3}{2}e^{j\pi/4} = \frac{3}{2\sqrt{2}}(1-j); \quad |C_2| = \frac{3}{2}, \quad \measuredangle C_{-2} = \frac{-\pi}{4}$$

Observe that the magnitude plot is symmetric while the phase plot is antisymmetric (Fig. 1.27).

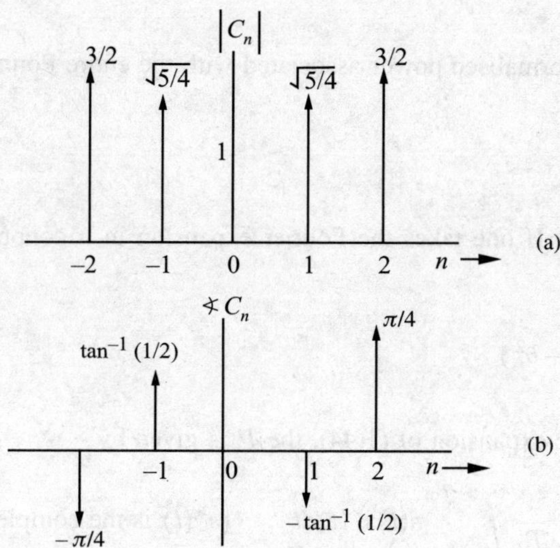

Fig. 1.27 *(a) Magnitude. (b) Phase angle.*

1.3.4 Normalised power: Parseval's power theorem

For a periodic signal $s(t)$, the is defined by (1.5). The power of $s(t)$ equals the power of its Fourier series. Each term in the Fourier series is a power signal. The total power in $s(t)$ is equal to the sum of the powers of its Fourier series components. To verify this result, consider first two terms of the Fourier expansion of (1.37), namely,

$$s_1(t) = A_1\cos(\omega_0 t - \phi_1) + A_2\cos(2\omega_0 t - \phi_2) \tag{1.54}$$

where $\omega_0 = 2\pi/T_0$. The normalised power of $s_1(t)$ is

$$\begin{aligned} P_1 &= \frac{1}{T_0}\int_{-T_0/2}^{T_0/2} s_1^2(t)\,dt \\ &= \frac{1}{T_0}\int_{-T_0/2}^{T_0/2} A_1\cos^2(\omega_0 t - \phi_1)dt + \frac{1}{T_0}\int_{-T_0/2}^{T_0/2} A_2^2\cos^2(2\omega_0 t - \phi_2)dt \\ &\quad + \frac{2}{T_0}\int_{-T_0/2}^{T_0/2} A_1A_2\cos(\omega_0 t - \phi_1)\cos(2\omega_0 t - \phi_2)\,dt. \end{aligned} \tag{1.55}$$

The first two terms yield $A_1^2/2$ and $A_2^2/2$ respectively, while the product term when integrated over a period, results in zero. This is because the terms are mutually orthogonal over the period of integration. Thus,

$$P_1 = \frac{A_1^2}{2} + \frac{A_2^2}{2}. \tag{1.56}$$

Extending this result, the normalised power associated with the entire Fourier expansion is

$$P = A_0^2 + \sum_{n=1}^{\infty} \frac{A_n^2}{2}, \tag{1.57}$$

where A_0^2 is the dc power. If one takes the Fourier expansion in trigonometric form in (1.32), the associated power is

$$P = a_0^2 + \sum_{n=1}^{\infty} \frac{1}{2}(a_n^2 + b_n^2). \tag{1.58}$$

For the exponential Fourier expansion of (1.44), the P is given by

$$P = \frac{1}{T_0}\int_t^{t+T_0} |s(t)|^2 dt = \frac{1}{T_0}\int_t^{t+T_0} s(t)s^*(t)dt \qquad [s^*(t) \text{ is the complex conjugate of } s(t)]$$

$$= \frac{1}{T_0}\int_t^{t+T_0} s(t)\left[\sum_{n=-\infty}^{\infty} C_n e^{j2\pi nt/T_0}\right]^* dt$$

$$= \frac{1}{T_0}\int_t^{t+T_0} s(t)\left[\sum_{n=-\infty}^{\infty} C_n^* e^{-j2\pi nt/T_0}\right] dt$$

$$= \sum_{n=-\infty}^{\infty}\left[\frac{1}{T_0}\int_t^{t+T_0} s(t)e^{-j2\pi nt/T_0}dt\right] c_n^* \qquad [\text{interchanging the order of integration and summation}]$$

$$= \sum_{n=-\infty}^{\infty} C_n C_n^*. \qquad [\text{vide } (1.45)]$$

$$\text{Thus, } P = \sum_{n=-\infty}^{\infty} |C_n|^2. \tag{1.59}$$

In the complex representation, we may say that the power associated with the spectral component at nf_0 is $C_n C_n^* = |C_n|^2$ and the power associated with the spectral component at $-nf_0$ is $C_n C_n^* = |C_n|^2$. Hence the total power is $2|C_n|^2$ associated with a particular frequency nf_0 (note that half of this power is associated with either of the spectral components at nf_0 or $-nf_0$). Similar to the amplitude spectrum we may construct one-sided and two-sided power spectra as shown in Fig. 1.28. The height of each vertical line is $|C_n|^2$ for two-sided spectrum and $2|C_n|^2$ for one-sided spectrum. The relation (1.59) is Parseval's power theorem.

1.3.5 Power spectral density

For a power signal, the power spectral density (PSD) represents the distribution of power in the frequency domain. According to (1.59), the power contributed by the nth spectral component is

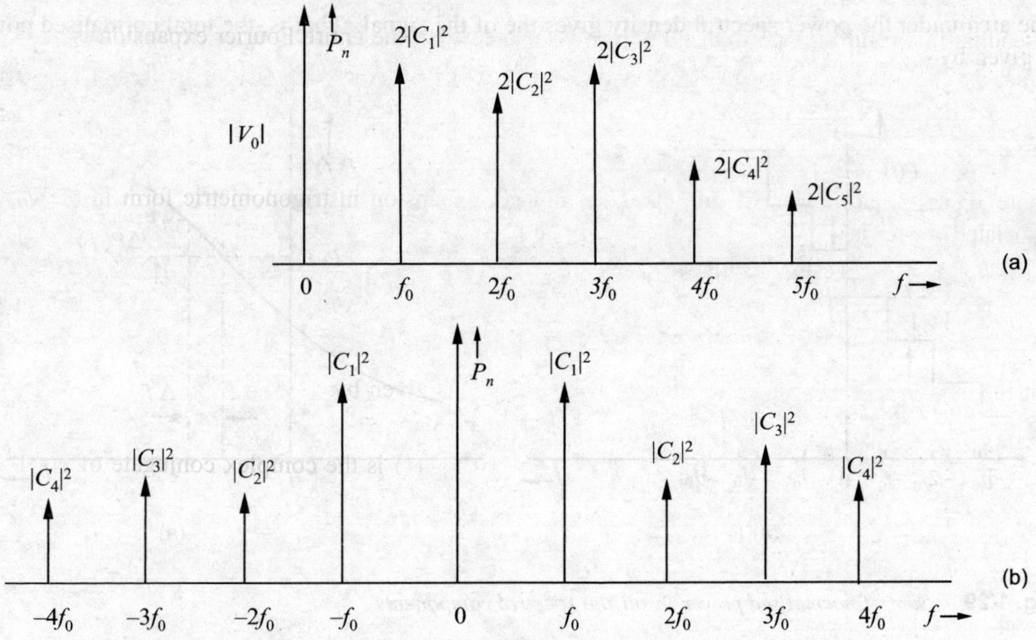

Fig. 1.28 *A power spectrum (P_n vs. f): (a) one-sided (b) two-sided.*

$P_n = |C_n|^2$. Suppose that we add the normalised powers contributed by each spectral line from the frequency $f = -\infty$ up to f. This summation $P(f)$ is a function of frequency and is plotted in Fig. 1.29. The shape of $P(f)$ is that of a staircase as $P(f)$ changes when a spectral line is added and remains constant in between any two spectral lines. Incremental change in normalised power at frequency f for a spectral bandwidth df is given by

$$dP(f) = \frac{dP(f)}{df} df. \tag{1.60}$$

We define the derivative $\dfrac{dP(f)}{df}$ as the power spectral density $G(f)$. Thus,

$$G(f) = \frac{dP(f)}{df}. \tag{1.61}$$

It is observed in Fig. 1.29 that $P(f)$ remains constant between the harmonic frequencies for which $\dfrac{dP(f)}{df} = 0$. At the harmonic frequency nf_0, the derivative in (1.61) gives an impulse. Thus the

PSD is given as

$$G(f) = \sum_{n=-\infty}^{\infty} |C_n|^2 \delta(f - nf_0). \tag{1.62}$$

The area under the power spectral density gives the of the signal. That is, the total normalised power is given by

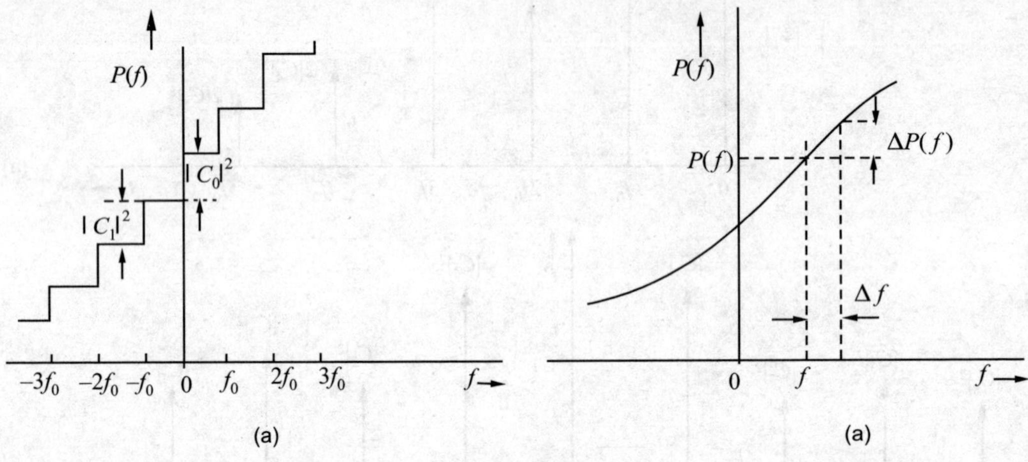

Fig. 1.29 *Sum of normalised power in all the spectral components*

$$P = \int_{-\infty}^{\infty} G(f)df = 2\int_{0}^{\infty} G(f)df, \tag{1.63}$$

the last step is for the real-valued signal.

The total power in the frequency range f_1 to f_2 is given by

$$P(f_1 \leq |f| \leq f_2) = \int_{-f_2}^{-f_1} G(f)df + \int_{f_1}^{f_2} G(f)df, \tag{1.64}$$

where the second term gives the power in the positive-frequency range f_1 to f_2 and the first term gives the power in the negative-frequency range $-f_2$ to $-f_1$.

Equation (1.59) and (1.63) give the average normalised power based on summation of spectral components and integration of power spectral density of the signal and these two will lead to the same .

Consider the cosine signal $s(t) = Acos2\pi f_0 t$. The Fourier series coefficients can be obtained as follows:

$$s\,(t) = \frac{A}{2}\left(e^{j2\pi f_0 t} + e^{-j2\pi f_0 t}\right)$$

Thus, by Eq. (1.44),

$$c_0 = 0,\ c_1 = \frac{A}{2} \text{ and } c_{-1} = \frac{A}{2} \text{ and all other coefficients are zero.}$$

The power spectral density by Eq. (1.62) is thus

$$G(f) = c_1^2 \delta(f - f_0) + c_{-1}^2 \delta(f + f_0)$$

$$= \left(\frac{A}{2}\right)^2 [\delta(f - f_0) + \delta(f + f_0)]$$

By Eq. (1.63),

$$\text{Normalised power } P = \int_{-\infty}^{\infty} G(f)\, df = \frac{A^2}{4} + \frac{A^2}{4} = \frac{A^2}{2}.$$

Similar results can also be obtained by time-averaging.

$$p = \frac{1}{T} \int_{-T/2}^{T/2} |s(t)|^2\, dt$$

$$= \frac{A^2}{2T} \cdot \int_{-T/2}^{T/2} (1 + \cos 4\pi f_0 t)\, dt$$

$$= \frac{A^2}{2T} \cdot T = \frac{A^2}{2},$$

since $T = 1/f_0$ (by Eq. (1.28)) and $\int_{-T/2}^{T/2} (1 + \cos 4\pi f_0 t)\, dt = 0$.

Example 1.12

Two-sided *PSD* of a signal waveform $s(t)$ is shown in the adjoining figure. Find the normalised average power in $s(t)$ over the frequency band from 5 kHz to 10 kHz.

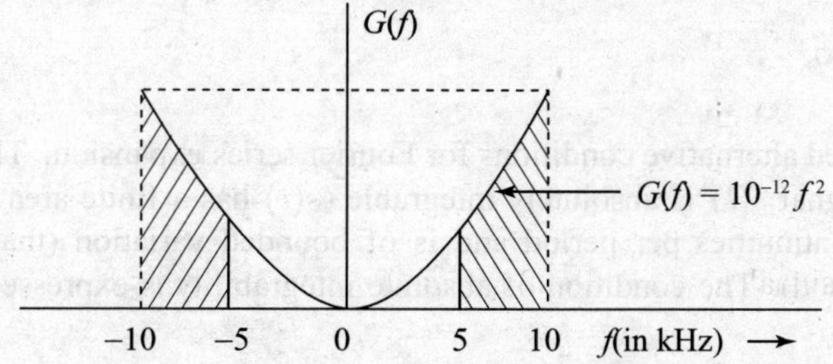

Here, by Eq. (1.64), the total power in the frequency range of 5 kHz to 10 kHz is given by

$$P = 2\int_{f_1}^{f_2} 10^{-12} f^2 df = 2 \times 10^{-12} \frac{f^3}{3}\Bigg]_{f_1}^{f_2} = \frac{2}{3} \times 10^{-12}\left[\left(10 \times 10^3\right)^3 - \left(5 \times 10^3\right)^3\right]$$

$$= \frac{2}{3} \times 10^{-12} \times 10^9 \,(1000 - 125)$$

$$= \frac{2}{3} \times 10^{-3} \times 875 \simeq 583\,\text{mW/Hz}.$$

1.3.6 Convergence of Fourier series: Dirichlet condition

Many large classes of periodic signals can be approximated by a Fourier series expansion. The question of validity of the Fourier series expansion now arises. Does the series expansion converge to the periodic signal? To understand the series convergence, let us approximate a periodic signal $s(t)$ by a finite number of terms. That is,

$$s_N(t) = \sum_{n=-N}^{N} C_n e^{j2\pi n f_0 t}. \tag{1.65}$$

The error arising out of this approximation is

$$e_N(t) = s(t) - s_N(t). \tag{1.66}$$

Under this truncation, the series $s_N(t)$ consisting of finite terms converges if the energy E_N associated with the error signal is as minimum as possible. Thus, for convergence

$$E_N = \lim_{N\to\infty} \int_{T_0} |e_N(t)|^2 dt$$

$$= \lim_{N\to\infty} \int_{T} |s(t) - s_N(t)|^2 dt = 0. \tag{1.67}$$

In other words, the periodic signal $s(t)$ must be square-integrable so that $s(t)$ possesses finite energy over a single period and so

$$\int_{T_0} |s(t)|^2 dt < \infty. \tag{1.68}$$

P. L. Dirichlet developed alternative conditions for Fourier series expansion. The Dirichlet conditions are as follows: The signal $s(t)$ is absolutely integrable ($s(t)$ has a finite area per period), has finite number of finite discontinuities per period and is of bounded variation (that is, finite number of extrema points per period). The condition of absolute integrability is expressed as

$$\int_{T_0} |s(t)| < \infty. \tag{1.69}$$

1.3.7 Properties of Fourier series

In this section we will discuss several important properties of Fourier series which may help to reduce the complexities in evaluating the Fourier series expansion.

Linearity

Let $x(t)$ and $y(t)$ denote two periodic signals with period T_0 and having Fourier series coefficients a_n and b_n, respectively. Any linear combination of these two signals of the form $Ax(t) + by(t)$ is also periodic with period T_0. The Fourier series coefficient of $Ax(t) + by(t)$ is given by

$$C_n = Aa_n + Bb_n. \tag{1.70}$$

The above property is extendable to the linear combination of any arbitrary number of signals with the same period.

Time shifting

If the periodic signal $x(t)$ is time shifted by t_0 to yield $y(t) = x(t - t_0)$, the Fourier series coefficient of $y(t)$ is $a_n e^{-j2\pi n f_0 t_0}$. The proof is as follows.
From (1.45),

$$b_n = \frac{1}{T_0}\int_{-T_0/2}^{T_0/2} x(t - t_0)e^{-j2\pi f_0 n t}dt.$$

Let $\tau = t - t_0$ so that $dt = d\tau$ and τ will range over an interval of duration T_0. Thus,

$$\begin{aligned} b_n &= \frac{1}{T_0}\int_{T_0} x(\tau)e^{-j2\pi n f_0(\tau+t_0)}d\tau = e^{-j2\pi n f_0 t_0}\frac{1}{T}\int_{T_0} x(\tau)e^{-j2\pi n f_0 \tau}d\tau \\ &= a_n e^{-j2\pi n f_0 t_0}. \end{aligned} \tag{1.71}$$

Note that the magnitude of the Fourier series coefficient has not changed as $|b_n| = |a_n|$.

Time scaling

Let $x(t)$ be periodic with period T_0. Then $x(\alpha t)$ is periodic with period T/α for a real and positive $\propto$. The Fourier series representation for $x(\alpha t)$ reads as

$$x(\alpha t) = \sum_{n=-\infty}^{\infty} C_n e^{j2\pi n f_0 \propto t}. \tag{1.72}$$

The coefficients C_n have not changed in time-scaling operation. The case $\alpha = -1$ is known as the time reversal operation. In this case,

$$x(\alpha t) = x(-t) = \sum_{n=-\infty}^{\infty} C_n e^{-j2\pi n f_0 t}. \tag{1.73}$$

Substituting $n = -m$, we get

$$y(t) = x(-t) = \sum_{m=-\infty}^{\infty} C_{-m} e^{j2\pi m f_0 t}. \tag{1.74}$$

Thus the Fourier series coefficients b_n are given as

$$b_n = C_{-n}. \tag{1.75}$$

An important case of time reversal operation is that if $x(t)$ is an even signal, that is, $x(t) = x(-t)$, then its Fourier coefficients are also even. That is, $C_{-n} = C_n$. For odd $x(t)$, $C_{-n} = -C_n$.

Multiplication

The product $x(t)y(t)$ of two periodic signals with period T is also periodic having the same period T. If a_n and b_n are the Fourier series coefficients of $x(t)$ and $y(t)$, then the coefficient C_n for the product signal $x(t)y(t)$ is given by

$$C_n = \sum_{l=-\infty}^{\infty} a_l b_{n-l}. \tag{1.76}$$

Expression (1.76) is derived as follows.

$$x(t)y(t) = \left(\sum_{p=-\infty}^{\infty} a_p e^{j2\pi p f_0 t}\right)\left(\sum_{q=-\infty}^{\infty} b_q e^{j2\pi q f_0 t}\right)$$

$$= (\ldots + a_{-1}e^{-j2\pi f_0 t} + a_0 + a_1 e^{j2\pi f_0 t} + \ldots) \times (\ldots + b_{-1}e^{-j2\pi f_0 t} + b_0 + b_1 e^{j2\pi f_0 t} + \ldots)$$

The nth component will have a coefficient given as

$$C_n = \ldots + a_{-1}b_{n+1} + a_0 b_n + a_1 b_{n-1} + \ldots + a_{n+1}b_{-1} + a_n b_0 + a_{n-1}b_1 + \ldots)$$

$$= \sum_{l=-\infty}^{\infty} a_l b_{n-l}.$$

Frequency shifting

For the signal $e^{jN(2\pi/T)t}x(t)$, the Fourier series coefficients are given by C_{n-N}, where C_n are the series coefficients of $x(t)$. To prove this, consider the signal

$$y(t) = e^{jN(2\pi/T)t}x(t)$$

$$= e^{jN(2\pi/T)t} \sum_{n=-\infty}^{\infty} C_n e^{j\frac{2\pi}{T}nt}$$

$$= \sum_{n=-\infty}^{\infty} C_n e^{j\frac{2\pi}{T}(n+N)t}.$$

Let $m = N + n$ so that

$$y(t) = \sum_{m=-\infty}^{\infty} C_{m-N} e^{j\frac{2\pi}{T}mt}$$

$$= \sum_{n=-\infty}^{\infty} C_{n-N} e^{j\frac{2\pi}{T}nt}. \quad (1.77)$$

Thus the Fourier series coefficients of $y(t)$ are given by C_{n-N}.

Differentiation and integration

The Fourier series coefficients C'_n for the signal $y(t) = \frac{dx(t)}{dt}$ is given by

$$C'_n = j\frac{2\pi}{T} nC_n, \quad (1.78)$$

which is obtained by differentiating $x(t)$ and collecting the coefficients of the harmonic exponents. That is,

$$y = \frac{dx(t)}{dt} = \frac{d}{dt} \sum_{n=-\infty}^{\infty} C_n e^{j\frac{2\pi}{T}nt} = j\frac{2\pi}{T} \sum_{n=-\infty}^{\infty} n.C_n.e^{j\frac{2\pi}{T}nt}.$$

Hence the result. Similarly, the signal $y(t) = \int_{-\infty}^{t} x(t)dt$ has the Fourier series coefficients $\frac{1}{jn\left(\frac{2\pi}{T}\right)} C_n$. The following example illustrates the time-shift and the differentiation property to evaluate the Fourier series coefficients of rectangular pulses.

Example 1.13

Find the Fourier series expansion for a rectangular pulse train and hence find the Fourier series expansion for the impulse train as shown in Fig. 1.30.

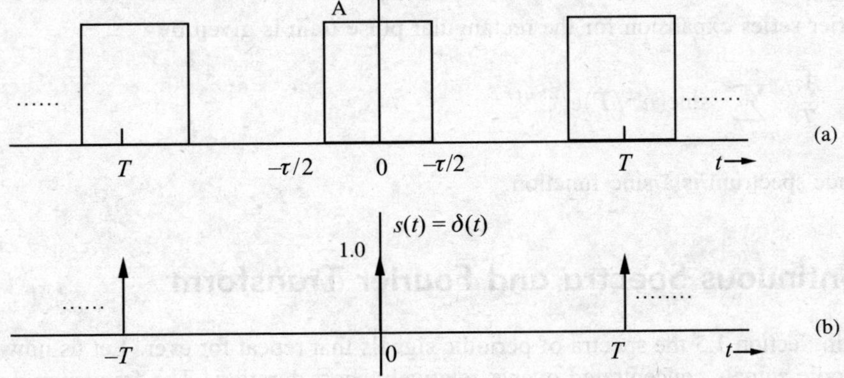

Fig. 1.30 *(a) Rectangular pulse train of amplitude A, duration τ and periodicity T, symmetrical about t = 0. (b) Unit impulse train.*

Solution

For a unit impulse train of period T, the Fourier series coefficients C_n are given by

$$C_n = \frac{1}{T}\int_T \delta(t)e^{-j2\pi nt/T}dt = \frac{1}{T}, \tag{1.79}$$

where the range of integration is $\frac{-T}{2} < t < \frac{T}{2}$.

For the rectangular pulse,

$$\begin{aligned} C_n &= \frac{1}{T}\int_{-T/2}^{T/2} s(t)e^{-j2\pi nt/T}dt = \frac{1}{T}\int_{-\tau/2}^{\tau/2} Ae^{-j2\pi nt/T}dt \\ &= \frac{1}{T}\frac{A}{-j2\pi n/T}[e^{-j2\pi nt/T}]_{-\tau/2}^{\tau/2} = \frac{A}{j2\pi n}(e^{j\pi n\tau/T} - e^{-j\pi n\tau/T}) \\ &= \frac{A}{n\pi}\sin(\pi n\tau/T) = \frac{A\tau}{T}\text{sinc}\left(\frac{n\tau}{T}\right) \end{aligned} \tag{1.80}$$

since the rectangular pulse approaches the unit impulse if $\tau \to 0$ with the condition that $A\tau \to 1$. Then,

$$\begin{aligned} C_n &= \lim_{\substack{A\tau\to 1 \\ \tau\to 0}} \frac{A\tau}{T}\text{sinc}(n\tau/T) = \lim_{\tau\to 0}\frac{1}{T}\text{sinc}(n\tau/T) \\ &= \frac{1}{T}\text{sinc}(0) = \frac{1}{T}. \end{aligned}$$

Thus, the Fourier series expansion for unit impulse train is

$$s(t) = \frac{1}{T}\sum_{n=-\infty}^{\infty} e^{j2\pi nt/T}. \tag{1.81}$$

The amplitude spectrum consists of impulses of strength $1/T$ located at the frequencies n/T where $n = 0, \pm 1, \pm 2, \cdots$.

The Fourier series expansion for the rectangular pulse train is given by

$$s(t) = \frac{A\tau}{T}\sum_{n=-\infty}^{\infty} \text{sinc}(n\tau/T)e^{j2\pi nt/T}. \tag{1.82}$$

The amplitude spectrum is a sinc function.

1.4 Continuous Spectra and Fourier Transform

We studied in Section 1.3 the spectra of periodic signals that repeat for ever. Let us now concentrate on non-periodic signals concentrated over a relatively short duration. The frequency-domain representation of these aperiodic signals will be a continuous spectrum in contrast to the discrete spectrum of the periodic case.

1.4.1 Fourier transform and Fourier integral

To introduce the Fourier transform, let us begin with the Fourier series representation of a periodic power signal $s(t)$ shown in Fig. 1.31. Here,

$$s(t) = \sum_{n=-\infty}^{\infty} C_n e^{j2\pi n f_0 t}$$

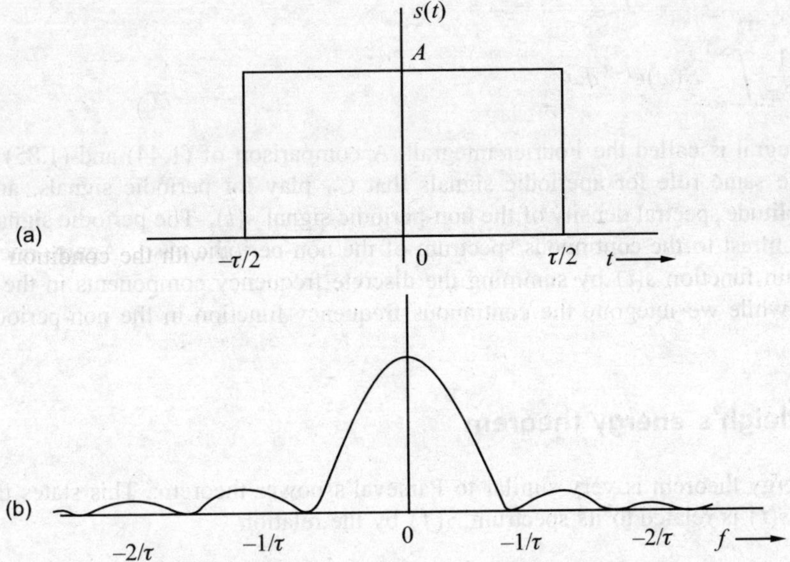

Fig. 1.31 *A train of pulse of finite duration which recurs with* T_0

$$s(t) = \sum_{n=-\infty}^{\infty} \left[\frac{1}{T_0} \int_{-T_0/2}^{T_0/2} s(t) e^{-j2\pi n f_0 t}\, dt \right] e^{j2\pi n f_0 t}. \tag{1.83}$$

An aperiodic signal can be thought of as a limiting case of a periodic signal when the period $T_0 \to \infty$, in which case the pulse centred at $t = 0$ remains in place and all others move outward. The spacing between adjacent spectral components $nf_0 (= n/T_0)$ and $(n+1)f_0$ becomes infinitesimal. We then replace the discrete variable nf_0 in (1.83) with a continuous variable f and the summation with integration. Thus, a Fourier series representation for the non-periodic case is

$$s(t) = \int_{\infty}^{\infty} \left[\int_{\infty}^{\infty} s(t) e^{-j2\pi f t} dt \right] e^{j2\pi f t} df = \int_{-\infty}^{\infty} S(f) e^{j2\pi f t} dt. \tag{1.84}$$

Note that for small frequency interval df, the spectral amplitudes C_n are replaced by $S(f)df$. Here $S(f)$ is called the Fourier transform of $s(t)$ and is denoted as

$$S(f) = \mathcal{F}[s(t)] = \int_{-\infty}^{\infty} s(t)e^{-j2\pi ft}dt, \tag{1.85}$$

which is a function of the continuous variable f. This transform is unique in the sense that no information is lost while taking the transform and therefore either $s(t)$ or $S(f)$ is good enough for describing the signal.

The time function $s(t)$ is recovered from $S(f)$ by the inverse Fourier transform

$$\begin{aligned} s(t) = \mathcal{F}^{-1}[S(f)] &= \int_{-\infty}^{\infty} S(f)e^{j2\pi ft}df \\ &= \frac{1}{2\pi}\int_{-\infty}^{\infty} S(\omega)e^{j\omega t}d\omega. \end{aligned} \tag{1.86}$$

The above integral is called the Fourier integral. A comparison of (1.44) and (1.85) indicates that $S(f)$ plays the same role for aperiodic signals that C_n play for periodic signals, and so $S(f)$ is called the amplitude spectral density of the non-periodic signal $s(t)$. The periodic signal forms a line spectrum in contrast to the continuous spectrum of the non-periodic case. Again, one can return to the time domain function $s(t)$ by summing the discrete frequency components in the periodic case (vide (1.84)), while we integrate the continuous frequency function in the non-periodic case (vide (1.86)).

1.4.2 Rayleigh's energy theorem

Rayleigh's energy theorem is very similar to Parseval's power theorem. This states that the energy E of a signal $s(t)$ is related to its spectrum $S(f)$ by the relation

$$E = \int_{\infty}^{\infty} S(f)S^*(f)df = \int_{-\infty}^{\infty} |S(f)|^2 df. \tag{1.87}$$

Note that the time domain integration of $|s(t)|^2$ also gives the total energy of the signal (vide (1.4)). The square of the amplitude spectrum $|S(f)|^2$ gives the distribution of energy in the frequency domain, and so is termed the energy spectral density (ESD). The energy in any differential frequency band df equals $|S(f)|^2 df$.

Example 1.14

Using Rayleigh's energy theorem, calculate the total energy contained in the rectangular pulse shown in Fig. 1.32 (a).

Here $s(t) = A\text{rect}(t/\tau)$. According to (1.4),

$$E = \int_{-\infty}^{\infty} |s(t)|^2 dt = \int_{-\tau/2}^{\tau/2} A^2 dt = A^2\tau,$$

which is the energy of the rectangular pulse. To calculate the energy in the frequency-domain analysis, let us find the energy spectral density $|S(f)|^2$. Now

$$S(f) = \mathcal{F}[s(t)] = \int_{\infty}^{\infty} s(t)e^{-j\omega t}dt = \int_{-\tau/2}^{\tau/2} Ae^{-j\omega t}dt = A\tau\text{sinc}(f\tau). \tag{1.88}$$

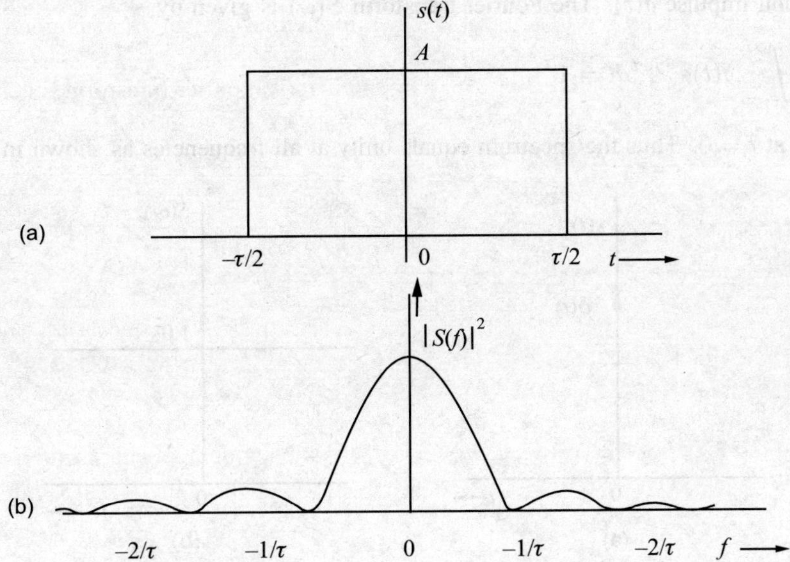

Fig. 1.32 *(a) Rectangular pulse of height A and duration τ. (b) Energy spectral density $S(f)|^2$ of the rectangular pulse.*

The spectrum is a real function of frequency. The energy spectrum $|S(f)|^2$ is plotted in Fig. 1.32 (b) which shows the zero-crossing taking place at $\pi f\tau = n\pi$, that is, at $f = \frac{1}{\tau}, \frac{2}{\tau}$, etc. By (1.87),

$$\begin{aligned} E &= \int_{-\infty}^{\infty} A^2\tau^2\text{sinc}^2(f\tau)df = A^2\tau^2 \int_{-\infty}^{\infty} \frac{\sin^2(\pi f\tau)}{(\pi f\tau)^2}df \\ &= \frac{A^2\tau^2}{\pi\tau}\int_{-\infty}^{\infty} \frac{\sin^2 x}{x^2}dx, \qquad \text{where } x = \pi f\tau \\ &= \frac{A^2\tau}{\pi} \times \pi = A^2\tau. \end{aligned}$$

Observe that in computing E the time-domain integration of $|s(t)|^2$ is easier. The energy in the first lobe (that is in the frequency band $|f| < 1/\tau$) is

$$E = \int_{-1/\tau}^{1/\tau} A^2\tau^2\text{sinc}^2(f\tau)df = 0.92A^2\tau,$$

which is more than 90% of the total energy. Thus a rough estimate of the bandwidth of a rectangular pulse of width τ is $1/\tau$ Hz.

1.4.3 Examples of Fourier transforms

Unit impulse

Let $s(t)$ be a unit impulse $\delta(t)$. The Fourier transform $S(\omega)$ is given by

$$S(\omega) = \int_{-\infty}^{\infty} \delta(t) e^{-j\omega t} dt = 1 \tag{1.89}$$

since $\delta(t) = 1$ at $t = 0$. Thus the spectrum equals unity at all frequencies as shown in Fig. 1.33.

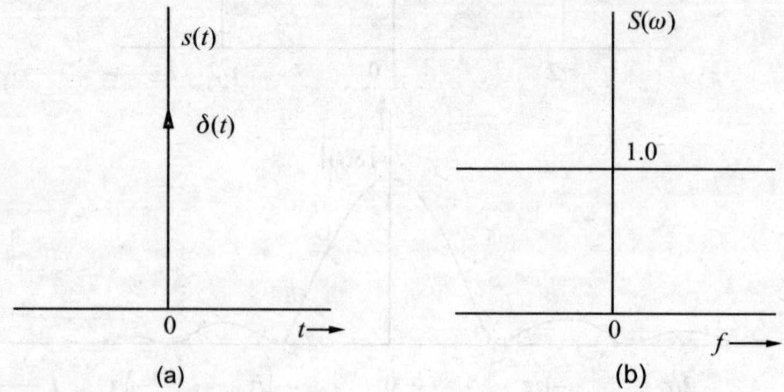

Fig. 1.33 *Unit impulse and its spectrum*

Constant (dc) signal

Let $s(t) = V$ be a constant dc signal. By (1.85), the Fourier spectrum is

$$S(f) = V \int_{-\infty}^{\infty} e^{-j\omega t} dt = V\delta(\omega). \tag{1.90}$$

The spectrum has a single frequency ($\omega = 0$) as shown in Fig. 1.34.

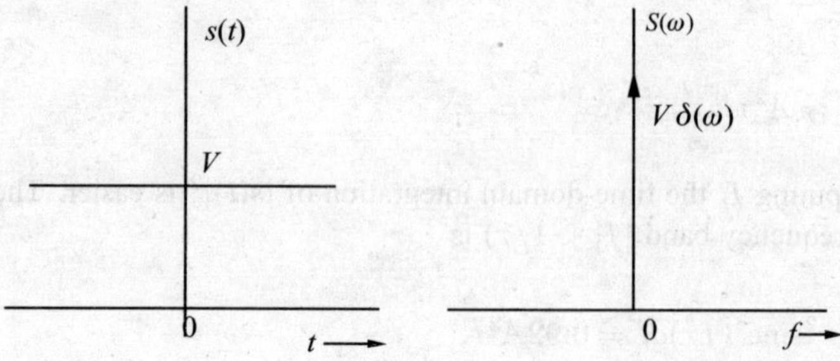

Fig. 1.34 *Constant signal and its Fourier transform*

Sinusoid signal
Let $s(t) = V \cos \omega_0 t$ be a sinusoid of frequency ω_0. Using Euler's relation,

$$s(t) = \frac{V}{2}\left(e^{j\omega_0 t} + e^{-j\omega_0 t}\right).$$

Thus,

$$\begin{aligned} S(f) &= \int_{-\infty}^{\infty} \frac{V}{2}\left(e^{j\omega_0 t} + e^{-j\omega_0 t}\right) e^{-j\omega t} dt \\ &= \frac{V}{2}\left[\int_{-\infty}^{\infty} e^{-j2\pi(f-f_0)t} dt + \int_{-\infty}^{\infty} e^{-j2\pi(f+f_0)t} dt\right] \\ &= \frac{V}{2}[\delta(f - f_0) + \delta(f + f_0)]. \end{aligned} \tag{1.91}$$

The spectrum consists of two impulses located at f_0 and $-f_0$ (see Fig. 1.35).

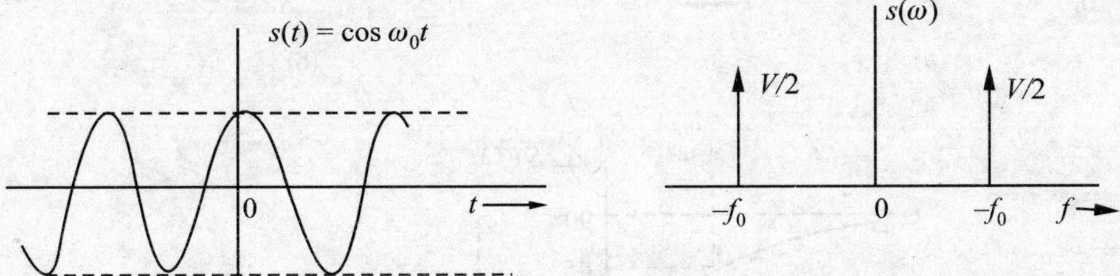

Fig. 1.35 *Cosine function* $s(t) = \cos \omega_0 t, -\infty \leq t \leq \infty$, *and its spectrum*

Exponential pulse
Let $s(t)$ be an exponential pulse shown in Fig. 1.36 (a) which decays exponentially with a time constant $1/\alpha$. Thus,

$$s(t) = \begin{cases} Ae^{-\alpha t} & t > 0 \\ 0 & t < 0. \end{cases} \tag{1.92}$$

Taking its Fourier transform,

$$\begin{aligned} S(f) &= \int_{-\infty}^{\infty} s(t) e^{-j\omega t} dt \\ &= \int_{0}^{\infty} Ae^{-\alpha t} e^{-j\omega t} dt \\ &= A\int_{0}^{\infty} e^{-(\alpha + j\omega)t} dt = \frac{A}{\alpha + j\omega}. \end{aligned} \tag{1.93}$$

The spectrum is a complex function. The magnitude and phase spectrum are obtained as

$$|S(f)| = \frac{A}{\sqrt{\alpha^2 + \omega^2}} = \frac{A}{\sqrt{\alpha^2 + 4\pi^2 f^2}} \tag{1.94}$$

$$\text{and } \measuredangle S(\omega) = -\text{arc}\tan\frac{\omega}{\alpha} = -\text{arc}\tan\frac{2\pi f}{\alpha}, \tag{1.95}$$

which are plotted in Fig. 1.36. The phase spectrum includes all angles from −90° to +90°.

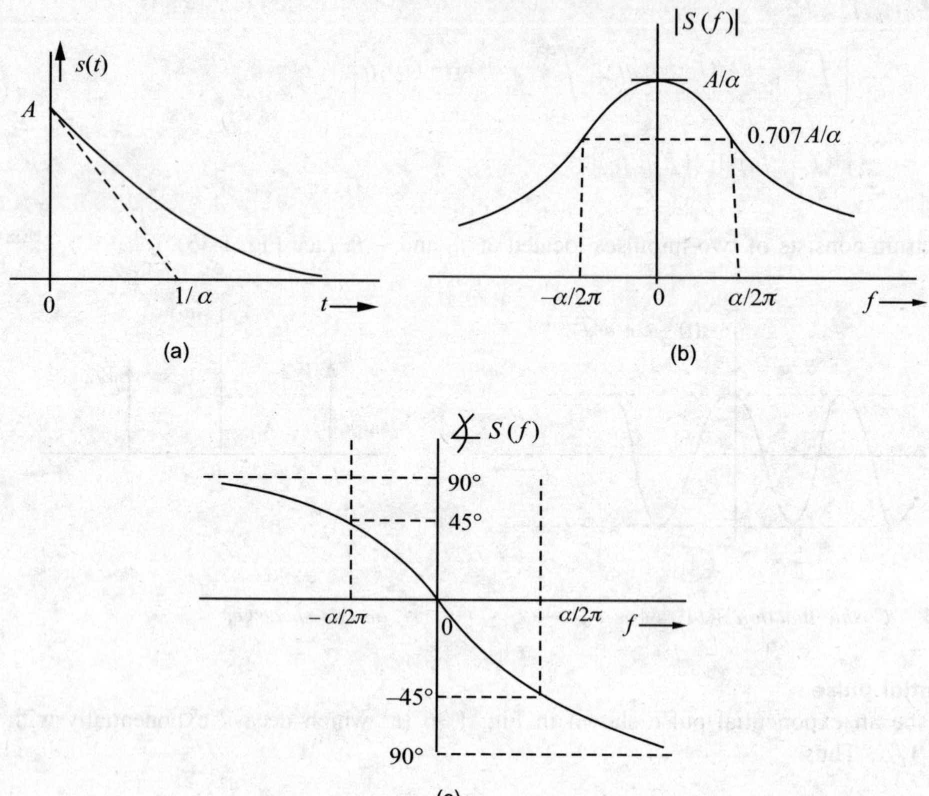

Fig. 1.36 *Exponential pulse: (a) waveform, (b) amplitude spectrum, (c) phase spectrum.*

The spectrum of a unit-step function $u(t)$ can be obtained from this result. In the limiting case of $\alpha \to 0$, the exponential function approaches a unit-step function. That is,

$$u(t) = \lim_{\alpha \to 0} e^{-\alpha t}, \quad t > 0. \tag{1.96}$$

A direct substitution of $\alpha = 0$ in (1.93) gives $S(f) = \frac{1}{j2\pi f}$, which is valid for all frequencies except

$f = 0$. The spectrum at $f = 0$ is obtained as follows.

$$\frac{1}{\alpha + j\omega} = \frac{\alpha}{\alpha^2 + \omega^2} - j\frac{\omega}{\alpha^2 + \omega^2}.$$

Now $\displaystyle\int_{-\infty}^{\infty} \frac{\alpha}{\alpha^2 + \omega^2} d\omega = \pi.$

Hence, $\displaystyle\lim_{\alpha \to 0} \frac{\alpha}{\alpha^2 + \omega^2} = \pi\delta(\omega)$. Thus the total spectrum of the unit-step function is

$$S(f) = \pi\delta(\omega) + \frac{1}{j\omega}. \tag{1.97}$$

Example 1.15 Spectrum of a modulated signal

Multiplying a signal $s(t)$ by a sinusoid $\cos \omega_c t$ translates its spectrum in frequency f_c in both positive and negative direction. The product signal $s(t) \cos \omega_c t$ is called the modulated signal. Let $s(t)$ have the spectrum of Fig. 1.37, bandlimited in f_m. The Fourier transform of $s(t)$ is $S(f)$. The spectrum of the product signal is obtained as follows:

$$s(t) \cos \omega_c t = \frac{1}{2}\left(e^{j\omega_c t} + e^{-j\omega_c t}\right) s(t).$$

Thus,

$$\begin{aligned}
\mathcal{F}[s(t) \cos \omega_c t] &= \frac{1}{2}\int_{-\infty}^{\infty} s(t)e^{j\omega_c t}e^{-j\omega t}dt + \frac{1}{2}\int_{-\infty}^{\infty} s(t)e^{-j\omega_c t}e^{-j\omega t}dt \\
&= \frac{1}{2}\int_{-\infty}^{\infty} s(t)e^{-j(\omega-\omega_c)t}dt + \frac{1}{2}\int_{-\infty}^{\infty} s(t)e^{-j(\omega+\omega_c)t}dt \\
&= \frac{1}{2}[S(f - f_c) + S(f + f_c)].
\end{aligned} \tag{1.98}$$

The spectrum of the product signal (Fig.1.37) consists of two patterns of the same form as $S(f)$ but shifted to the right and left by an amount f_c.

An inspection of the spectrum of Fig. 1.37 reveals that each of the two spectral patterns is one-half the amplitude of the spectral pattern $S(f)$; they also have a spectral width of $2f_m$. The doubled spectral width can be interpreted in terms of the appearance of the negative-frequency portion of $S(f)$ at positive frequencies.

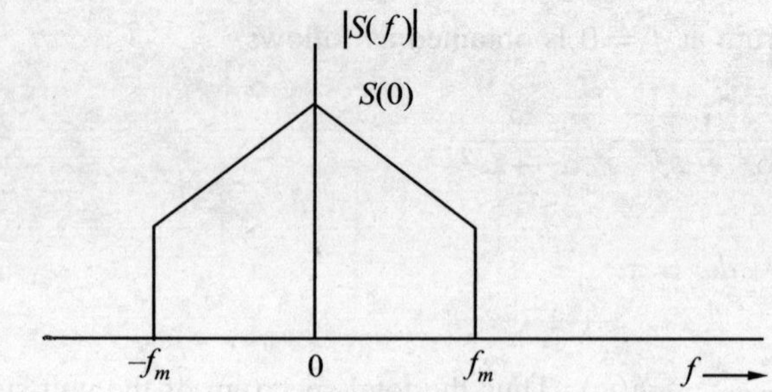

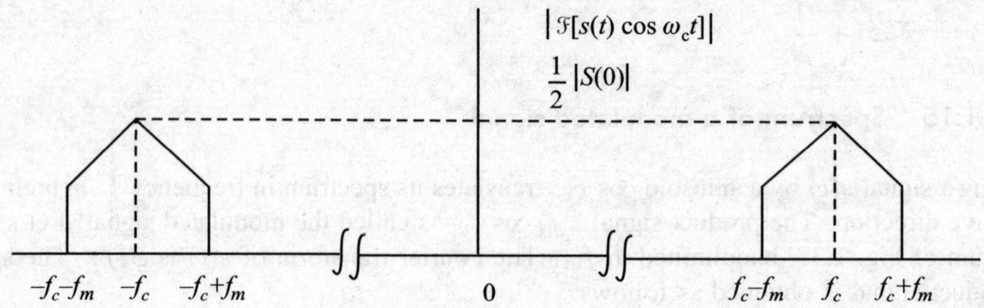

Fig. 1.37 *(a) Amplitude spectrum of $s(t)$ bandlimited to f_m. (b) Amplitude spectrum of $s(t)$ multiplied by $\cos \omega_c t$.*

Example 1.16

Find and sketch the amplitude spectrum of the product signal $s(t) \cos \omega_c t$ where $s(t)$ itself is a sinusoidal $s(t) = \cos \omega_m t$.

Let $v(t) = s(t) \cos \omega_c t$. Then $V(f)$ is given by

$$V(f) = \frac{1}{2} S(f - f_c) + \frac{1}{2} S(f + f_c).$$

Since $s(t) = \cos 2\pi f_m t$, its spectrum consists of impulses of strength $\frac{1}{2}$ located at $\pm f_m$. Thus,

$$S(f) = \frac{1}{2}[\delta(f + f_m) + \delta(f - f_m)]. \tag{1.99}$$

So,

$$V(f) = \frac{1}{4}\delta(f + f_c + f_m) + \frac{1}{4}\delta(f + f_c - f_m) + \frac{1}{4}\delta(f - f_c + f_m) + \frac{1}{4}\delta(f - f_c - f_m). \tag{1.100}$$

The spectrum shown in Fig. 1.38 consists of four spectral lines corresponding to the positive frequencies $f_c \pm f_m$ and negative frequencies $-f_c \pm f_m$.

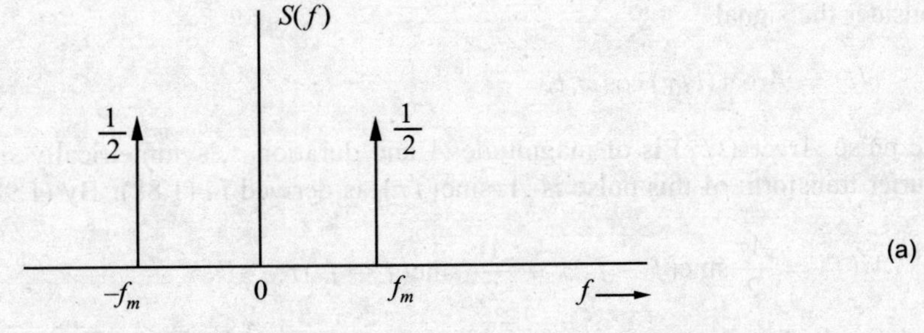

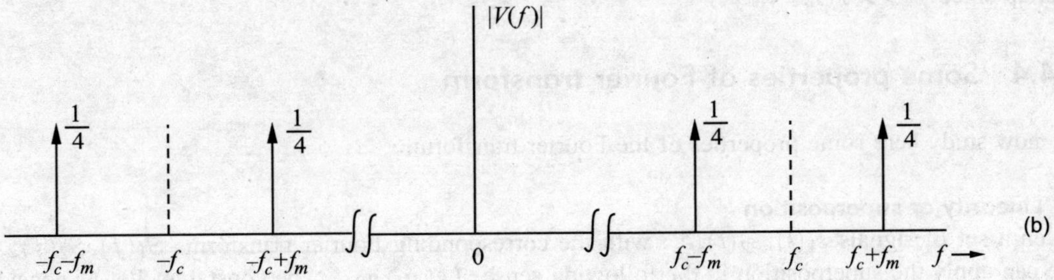

Fig. 1.38 *Amplitude spectrum:* (a) *two-sided spectrum of* $S(f) = \cos\omega_m t$, (b) *two-sided spectrum of the product* $s(t)\cos\omega_c t$.

Example 1.17

Find the spectrum of a finite duration sinusoid.

A sinusoid $\cos\omega_c t$ when multiplied by a rectangular pulse forms a truncated waveform shown in Fig. 1.39.

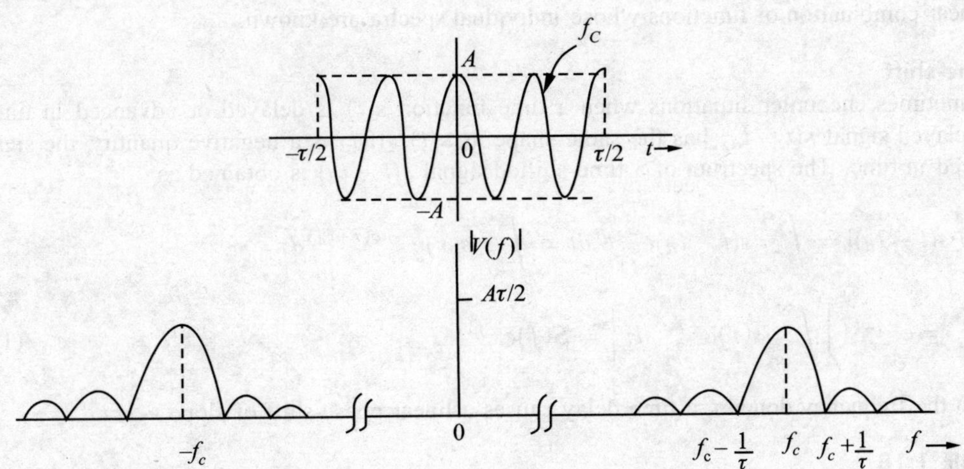

Fig. 1.39 (a) *Finite duration sinusoid* (b) *amplitude spectrum.*

Consider the signal

$$v(t) = A\text{rect}(t/\tau)\cos\omega_c t. \tag{1.101}$$

The pulse Arect (t/τ) is of magnitude A and duration τ, symmetrically situated about $t = 0$. The Fourier transform of this pulse is $A\tau\text{sinc}(f\tau)$ as derived in (1.88). By (1.98), we have immediately

$$V(f) = \frac{A\tau}{2}\text{sinc}(f - f_c)\tau + \frac{A\tau}{2}\text{sinc}(f + f_c)\tau. \tag{1.102}$$

The magnitude spectrum plotted in Fig. 1.39 is continuous. The two sinc functions have negligible overlap since $f_c >> 1/\tau$.

1.4.4 Some properties of Fourier transform

We now study here some properties of the Fourier transform.

(a) Linearity or superposition

Given a set of signals $s_1(t), s_2(t), \ldots$ with the corresponding Fourier transforms $S_1(f), S_2(f), \ldots$ one can apply the superposition in the following sense. Let $a_1, a_2, \ldots$ be constants. Let us consider the signal

$$s(t) = a_1 s_1(t) + a_2 s_2(t) + \ldots \tag{1.103}$$

The spectrum of the combined signal is then

$$S(f) = \mathcal{F}[s(t)] = a_1\mathcal{F}[s_1(t)] + a_2\mathcal{F}[s_2(t)] + \ldots \tag{1.104}$$

The implication of the superposition theorem greatly simplifies the spectral analysis of a signal which is a linear combination of functions whose individual spectra are known.

(b)Time-shift

We sometimes encounter situations when a time function $s(t)$ is delayed or advanced in time. A time-delayed signal $s(t - t_d)$ has the same shape as $s(t)$. If t_d is a negative quantity, the signal is advanced in time. The spectrum of a time-shifted signal $s(t - t_d)$ is obtained as

$$\begin{aligned}\mathcal{F}\, s(t - t_d)] &= \int_{-\infty}^{\infty} s(t - t_d)e^{-j\omega t}dt = \int_{-\infty}^{\infty} s(\tau)e^{-j\omega(\tau + t_d)}d\tau \\ &= e^{-j\omega t_d}\left[\int_{-\infty}^{\infty} s(\tau)e^{-j\omega\tau}d\tau\right] = S(f)e^{-j\omega t_d}.\end{aligned} \tag{1.105}$$

Thus in the frequency domain, a time delay causes a linear phase-shift of slope $-2\pi t_d$.

Example 1.18

Find the Fourier transform of $s(t)$ which is the linear combination of two signals $s_1(t)$ and $s_2(t)$ as sketched in Fig. 1.40.

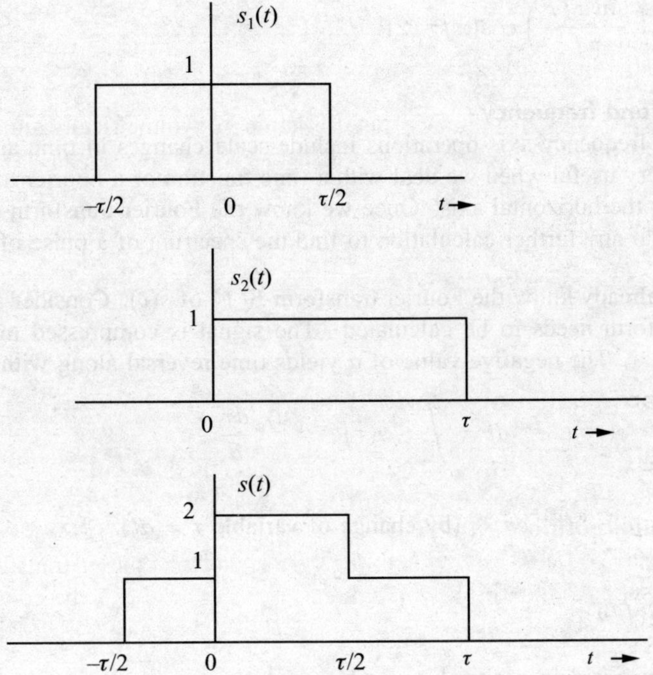

Fig. 1.40 $s_1(t), s_2(t)$ *and* $s(t)$ *of Example 1.18*

The function $s(t)$ is expressed as

$$s(t) = \begin{cases} 1 & -\tau/2 < t < 0 \\ 2 & 0 < t < \tau/2 \\ 1 & \tau/2 < t < \tau \\ 0 & \text{otherwise.} \end{cases}$$

Observe that $s_2(t) = s_1(t - \tau/2)$. Applying superposition and time-shift property, the spectrum of $s(t)$ is obtained as

$$\begin{aligned} S(f) &= \mathcal{F}[s_1(t)] + \mathcal{F}[s_2(t)] \\ &= S_1(f) + e^{-j\omega\tau/2}\, S_1(f) \\ &= (\tau \operatorname{sinc} f\tau)(1 + e^{-j\pi f\tau}), \quad \text{since } S_1(f) = \tau \operatorname{sinc}(f\tau). \\ &= (\tau \operatorname{sinc} f\tau)\, e^{-j\pi f\tau/2} \left(e^{j\pi f\tau/2} + e^{-j\pi f\tau/2} \right) \\ &= 2\tau \operatorname{sinc} f\tau \cos(\pi f\tau/2) e^{-j\pi f\tau/2} \end{aligned}$$

$$\therefore \quad S(f) = 2\left(\frac{\sin \pi f\tau}{\pi f}\right)\cos(\pi f\tau/2)e^{-j\pi f\tau/2}.$$

(c) Scaling in time and frequency

Other time-axis and frequency-axis operations include scale changes in time and frequency. These two properties are very useful when we deal with a time function or a Fourier transform and expand or compress it along the horizontal axis. Once we know the Fourier transform of a pulse of certain width, we need not do any further calculation to find the spectrum of a pulse of any width.

Suppose that we already know the Fourier transform $S(f)$ of $s(t)$. Consider a scaled signal $s(at)$ whose Fourier transform needs to be calculated. The signal is compressed in time if $a > 1$ and expanded if $0 < a < 1$. The negative value of a yields time reversal along with time scaling.

$$\mathcal{F}[s(at)] = \int_{\infty}^{\infty} s(at)e^{-j\omega t}dt = \int_{-\infty}^{\infty} s(\tau)e^{-j\omega\tau/a}\frac{d\tau}{a}$$

$$\frac{1}{a}\int_{-\infty}^{\infty} s(\tau)e^{-j(\omega/a)\tau}d\tau \quad \text{(by change of variable } \tau = at).$$

So, $\mathcal{F}[s(at)] = \dfrac{1}{a}S(f/a).$ (1.106)

which shows that compressing a signal expands its spectrum with the corresponding reduction in amplitudes. As for example, if $a = 3$, the time axis is compressed by a factor of 3, the frequency axis is expanded by the same factor and the amplitude of the spectrum is also reduced by 3. The case $a = -1$ implements time-reversal operation and we get from (1.106)

$$\mathcal{F}[s(-t)] = -S(-f). \tag{1.107}$$

This shows that the magnitude of the spectrum is unchanged but the spectra are reversed, indicating that the phases of each harmonic components are shifted by 180°.

The frequency scaling results from the time-scaling property. If one replace a by $1/a$ in (1.106), one readily proves the frequency scaling.

That is,

$$\mathcal{F}[s(t/a)] = aS(af)$$

or, $\mathcal{F}^{-1}[S(af)] = \dfrac{1}{a}s(t/a).$ (1.108)

(d) Differentiation and integration

We often encounter certain processes that involve differentiation and integration in the time domain of a signal. The Fourier transform of a derivative of a signal is derived as follows.

$$s(t) = \int_{-\infty}^{\infty} S(f)e^{j\omega t}dt.$$

So, $$\frac{ds(t)}{dt} = \frac{d}{dt}\left[\int_{-\infty}^{\infty} S(f)e^{j\omega t}\,dt\right] = \int_{-\infty}^{\infty} S(f)\frac{d}{dt}(e^{j\omega t}).dt$$

$$= \int_{-\infty}^{\infty} [j\omega S(f)]e^{j\omega t}\,dt,$$ which is the inverse Fourier transform of $j\omega S(f)$.

Thus,

$$\mathcal{F}\left[\frac{ds}{dt}\right] = j\omega S(f). \tag{1.109}$$

Extending the result to n differentiation,

$$\mathcal{F}\left[\frac{d^n s(t)}{dt^n}\right] = (j\omega)^n S(f). \tag{1.110}$$

Differentiation enhances the high frequency components of a signal.

The Fourier transform of the integration of a signal is derived as follows.

$$\mathcal{F}\left[\int_{-\infty}^{t} s(t)dt\right] = \int_{-\infty}^{\infty} e^{-j\omega t}\left[\int_{-\infty}^{t} s(t)dt\right] dt = S_1(\omega), \text{ say.}$$

Then, $\mathcal{F}\left[\frac{d}{dt}\int_{\infty}^{t} s(t)dt\right] = j\omega S_1(\omega)$ (by (1.109)].

Therefore, $j\omega S_1(\omega) = \mathcal{F}\,[s(t)] = S(\omega)$.

That is, $$S_1(\omega) = \frac{1}{j\omega}S(\omega). \tag{1.111}$$

These two theorems are applicable only when the differentiated or integrated signals are Fourier transformable.

Example 1.19

Apply differentiation theorem on Fourier transform to find the Fourier transform of the triangular pulse shown in Fig. 1.41. The triangular pulse shape is expressed by

$$s(t) = \begin{cases} A(1 - \frac{|t|}{\tau}) & |t| < \tau \\ 0 & \text{otherwise.} \end{cases} \tag{1.112}$$

Successive differentiations of $s(t)$ are also shown. The second derivative $\frac{d^2 s}{dt^2}$ is a sequence of impulses (vide Fig. 1.41(c)). Thus,

$$\frac{d^2 s(t)}{dt^2} = \frac{A}{\tau}\delta(t+\tau) - \frac{2A}{\tau}\delta(t) + \frac{A}{\tau}\delta(t-\tau)$$

$$= \frac{A}{\tau}[\delta(t+\tau) - 2\delta(t) + \delta(t-\tau)]. \tag{1.113}$$

By (1.110),

$$\mathcal{F}\left[\frac{d^2 s(t)}{dt^2}\right] = (j\omega)^2 S(\omega) = -\omega^2 S(\omega).$$

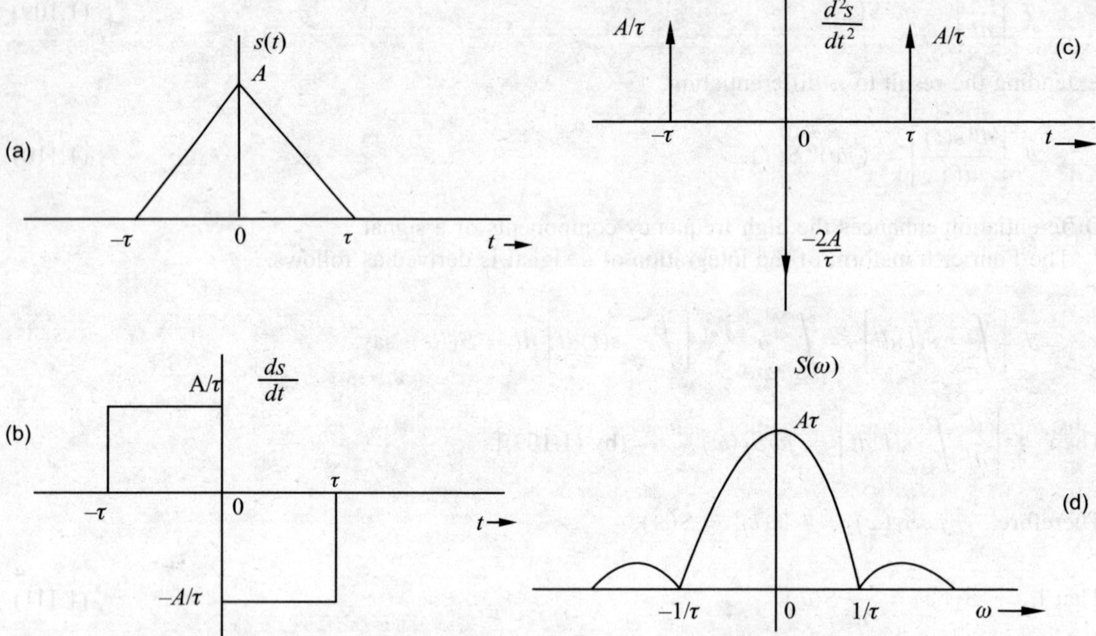

Fig. 1.41 *Differentiation theorem to find the Fourier transform of* (a) *triangular wave,* $(b)\&(c)$ *plot the first and second derivatives,* (d) *spectrum of the wave of figure (a).*

Hence, $\quad -\omega^2 S(\omega) = \dfrac{A}{\tau}(e^{j\omega\tau} - 2 + e^{-j\omega\tau}) = \dfrac{A}{\tau}\left(2\cos\omega\tau - 2) = (-\dfrac{2A}{\tau}\sin^2(\omega\tau/2\right).$

That is,

$$S(\omega) = \frac{2A}{\omega^2\tau}\sin^2(\omega\tau/2) = \frac{A\tau}{2}\frac{\sin^2(\omega\tau/2)}{(\omega\tau/2)^2}$$

$$= \frac{A\tau}{2}\text{sinc}^2(f\tau). \tag{1.114}$$

A direct calculation also gives the same result.

(e) Time–Frequency Duality

An interesting comparison can be made in the pair of Fourier integrals defined by (1.84) and (1.85). The two equations differ only by the variable of integration and sign in the exponent. This parity leads to the time–frequency duality theorem. The theorem states that if $s(t)$ and $S(f)$ denote the Fourier transform pair $s(t) \rightleftharpoons S(f)$, and if there is a time function $x(t)$ related to the frequency function $S(f)$ by the relation

$$x(t) = S(t)$$

then,

$$F[x(t)] = X(f) = s(-f)$$

The theorem can be proved as follows:

By Eq. (1.85),

$$s(t) = \int_{-\infty}^{\infty} S(f)e^{j2\pi ft}df = \int_{-\infty}^{\infty} S(\lambda)e^{j2\pi\lambda t}d\lambda \tag{1.114a}$$

by introducing dummy variable λ replacing f. Also,

$$\begin{aligned} F\left[x\left(t\right)\right] &= \int_{-\infty}^{\infty} x\left(\lambda\right)e^{-j2\pi f\lambda}d\lambda \qquad \text{(replacing t by the dummy } \lambda) \\ &= \int_{-\infty}^{\infty} S\left(\lambda\right)e^{j2\pi\lambda(-f)}d\lambda \end{aligned} \tag{1.114b}$$

Comparing Eqs. 1.114(a) and 1.14(b),

$$F\left[x\left(t\right)\right] = s\left(-f\right)$$

As an example, if we take $s(t)$ to be a rectangular pulse $s(t) = A\pi(t/\tau)$, its spectrum would be a sinc-pulse $S(f) = A\tau\ \text{sinc}(f\tau)$. If we assume another time-domain function $x(t) = B\ \text{sinc}(f\tau)$, then by the duality theorem, its spectrum can be obtained as $x(f) = \frac{B}{2W}\pi\left(t/2w\right)$. Readers are advised to prove this.

(f) Frequency translation

We consider a signal $s(t)$ with Fourier transform $S(f)$. If the signal is multiplied by $e^{j2\pi f_c t}$, this is equivalent to translating its spectrum by an amount f_c on the frequency scale.

The spectrum of $s\left(t\right)e^{j2\pi f_c t}$ is obtained as

$$F\left[s(t)\, e^{j2\pi f_c t}\right] = \int\limits_{-\infty}^{\infty} s(t)\, e^{j2\pi f_c t}.e^{-j2\pi f t} dt = \int\limits_{-\infty}^{\infty} s(t)\, e^{-j2\pi (f-f_c)t} dt$$

$$= S(f - f_c) \tag{1.114c}$$

Similarly, the spectrum of $s(t)\, e^{-j2\pi f_c t}$ is equivalent to $S(f + f_c)$.

That is,

$$F\left[s(t)\, e^{-j2\pi f_c t}\right] = S(f + f_c) \tag{1.114d}$$

Eq. (1.114c) represents spectrum shift towards right and Eq. (1.114d) represents spectrum shift towards left.

1.4.5 Real and imaginary parts of the spectrum

For a signal $s(t)$, the Fourier transform $S(f)$ is in general, a complex quantity. Thus,

$$S(f) = S_e(f) + jS_O(f), \tag{1.115}$$

where $S_e(f)$ and $S_O(f)$ are the real and the imaginary parts of $s(t)$ and are given as

$$S_e(f) = \int_{\infty}^{\infty} s(t) \cos \omega t dt,$$

$$S_O(f) = -\int_{-\infty}^{\infty} s(t) \sin \omega t dt. \tag{1.116}$$

The time-domain signal $s(t)$ is obtained by the inverse Fourier transform.

$$s(t) = \int_{-\infty}^{\infty} \left[S_e(f) + jS_O(f)\right] e^{je\pi f} df$$

$$= \int_{-\infty}^{\infty} \left[S_e(f) + jS_O(f)\right] (\cos 2\pi f t + j \sin 2\pi f t) df$$

$$= \int_{-\infty}^{\infty} \left[S_e(f) \cos 2\pi f t - S_O(f) \sin 2\pi f t\right] df$$

$$+ j \int_{-\infty}^{\infty} \left[S_e(f) \sin 2\pi f t + S_O(f) \cos 2\pi f t\right] df$$

For a real $s(t)$,

$$\int_{-\infty}^{\infty} S_e(f)\sin 2\pi ftdt + \int_{-\infty}^{\infty} S_O(f)\cos 2\pi ftdt = 0.$$

Since $\sin 2\pi ft$ is an odd function and $\cos 2\pi ft$ an even function, the real part of the spectrum $S_e(f)$ must be an even function of frequency while $S_O(f)$ must be an odd function. That is,

$$\text{Re}\,[S(f)] = S_e(f) \text{ and } \text{Im}\,[S(f)] = S_O(f).$$

If $s(t)$ is an even function such that $s(t) = s(-t)$, then $s(t)\cos\omega t$ is even and $s(t)\sin\omega t$ is odd. Hence by (1.116), $S_O(f) = 0$ so that

$$S(f) = S_e(f) = \int_{-\infty}^{\infty} s(t)\cos\omega tdt$$

$$= 2\int_{0}^{\infty} s(t)\cos\omega tdt. \tag{1.117}$$

For $s(t)$ being an odd function,

$$S(f) = S_O(f) = -j2\int_{0}^{\infty} s(t)\sin\omega tdt. \tag{1.118}$$

Example 1.20

Consider a signal $v(t)$ defined by

$$v(t) = as(t) + bs(-t),$$

where $s(t)$ is a real signal. Find the Fourier transform of $v(t)$ in terms of even and odd parts of $S(f)$.

Solution

$\mathcal{F}[s(-t)] = -S(-f)$ and since $s(t)$ is real, $S^*(f) = S(-f)$.

So, $V(f) = aS(f) - bS(-f)$

$$= aS(f) - bS^*(f)$$

$$= a[S_e(f) + jS_O(f)] - b[S_e(f) - jS_O(f)]$$

$$= (a-b)S_e(f) + j(a+b)S_O(f).$$

1.4.6 Analytic signal

The spectrum of the baseband signal is concentrated around zero frequency. Carrier modulation is adopted for sending the signal. The spectrum of such modulated signals is limited in an interval of bandwidth centred about the carrier frequency. Such signal is referred to as a bandpass signal.

Double side band (DSB) amplitude modulation is an example of bandpass signal (to be discussed in Chapter 3 in detail). It is sometimes convenient to represent such bandpass signal as an equivalent low-pass signal and apply simple mathematics to analyse the system. Analytic signal comes in this treatment.

We consider a real-valued bandpass signal $s(t)$ with the frequency band around a carrier frequency f_c, as shown in Fig.1.42. The bandwidth of the bandpass signal is taken as B Hz.

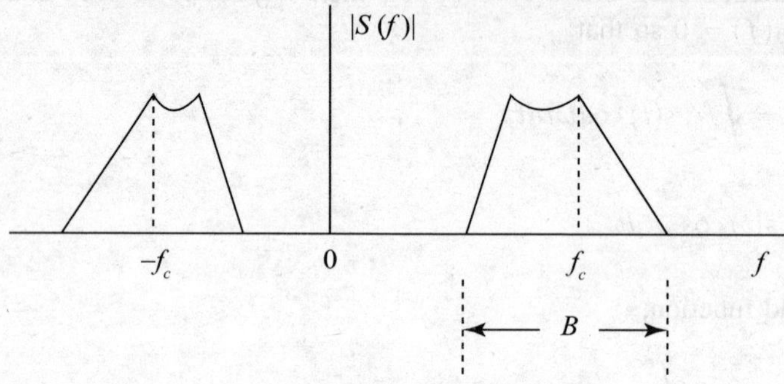

Fig. 1.42 *Spectrum of bandpass signal*

When the bandpass is much smaller than the carrier frequency, that is, when $B/f_c << 1$, the signal is referred to as narrowband signal. The spectrum contains both negative and positive frequencies in $s(t)$. If one consider a signal $s_+(t)$ that contains only the positive frequency in $s(t)$, then $s_+(t)$ is called the analytic signal of $s(t)$. To build up a mathematical framework of such signal, we proceed in the frequency-domain. If $S_+(f)$ denotes the spectrum referred to all positive frequency, then

$$S_+(f) = 2S(f)u(f) \tag{1.118a}$$

where $u(f)$ is the unit-step function in f and the factor 2 accounts for the consideration of negative frequency spectrum in the positive frequency region. By taking the inverse transformation,

$$\begin{aligned} s_+(t) &= F^{-1}[S_+(f)] \\ &= \int_{-\infty}^{\infty} S_+(f) e^{j2\pi ft} df \\ &= \int_{-\infty}^{\infty} 2u(f) S(f) e^{j2\pi ft} df \\ &= F^{-1}[2u(f)] * F^{-1}[S(f)] \end{aligned} \tag{1.118b}$$

(Since the spectrum of the product of two time-domain signals is the product of the spectra i.e. $F[X(t) * Y(t)] = X(f)Y(f)$), the operation * denotes convolution operator.

Since $f\left[2u\left(t\right)\right]=\delta\left(t\right)+\dfrac{j}{\pi t}$ (vide Eq 1.97), one can write

$$s_{+}\left(t\right)=\left[\delta\left(t\right)+\frac{j}{\pi t}\right]*s\left(t\right)$$

$$=\delta\left(t\right)*s\left(t\right)+j\frac{1}{\pi t}*s\left(t\right)$$

$$=s\left(t\right)+j\frac{1}{\pi t}*s\left(t\right)$$

$$=s\left(t\right)+j\widehat{s}\left(t\right) \tag{1.118c}$$

where

$$\widehat{s}\left(t\right)=\frac{1}{\pi t}*s\left(t\right)=\frac{1}{\pi}\int_{-\infty}^{\infty}\frac{s\left(\tau\right)}{t-\tau}d\tau \tag{1.118d}$$

denotes the Hilbert transform of $s(t)$ (to be detailed in Section 2.3.4). The analytic signal is the bandpass signal. The equivalent low-pass signal $s_l(t)$ of $s_+(t)$ can be obtained by frequency translation (vide Eq. 1.114(d)). Thus

$$s_l\left(t\right)=s_{+}\left(t\right)e^{-j2\pi f_c t}=\left[s\left(t\right)+j\widehat{s}\left(t\right)\right]e^{-j2\pi f_c t} \tag{1.118e}$$

In frequency domain, Eq. (1.118e) reads as

$$S_l\left(f\right)=S_f\left(f+f_c\right) \tag{1.118f}$$

In one step ahead, Eq. (1.118e) can be written as

$$[s(t)+j\hat{s}(t)]=s_l(t)e^{j2\pi f_c t} \tag{1.118g}$$

Since $s(t)$ and $\hat{s}(t)$ are both bandpass signals, $s(t)+j\hat{s}(t)$ is a complex-valued bandpass signal. The equivalent low-pass signal $s_l(t)$, when multiplied by $e^{j2\pi f_c t}$ results in the bandpass signal with the spectrum centred around f_c. Since any complex-valued signal can be expressed in terms of real and imaginary components, one can write the equivalent low-pass analytic signal as

$$s_l(t)=x(t)+jy(t) \tag{1.118h}$$

where $x(t)$ and $y(t)$ are real-valued function. Then by (1.18g)

$$s(t)+j\hat{s}(t)=[x(t)+jy(t)]\left(\cos 2\pi f_c t+j\sin 2\pi f_c t\right)$$

Thus, $s(t)=x(t)\cos 2\pi f_c t-y(t)\sin 2\pi f_c t$ (1.118i) and

$$\hat{s}(t) = x(t) \sin 2\pi f_c t + y(t) \cos 2\pi f_c t \tag{1.118j}$$

The interpretation of (1.118i) is straight forward. We have started with the real-valued bandpass signal $s(t)$. The right hand side of Eq. (1.118i) must be bandpass signal and so $x(t)$ and $y(t)$ must be low-pass signals. The low frequency signal components modulate the amplitudes of the carriers $\cos 2\pi f_c t$ and $\sin 2\pi f_c t$. As the carriers are in phase quadrature, the low-frequency components are named as quadrature components. That is, $x(t)$ is the in phase and $y(t)$ is the quadrature component of $s(t)$. Thus, Eq. (1.118i) gives the quadrature representation of the bandpass signal $s(t)$.

Another possible representation of the bandpass signal is as follows:

By Eq. (1.11h), $s_i(t) = c(t)e^{j\theta(t)}$ (1.118k)

where $c(t) = \sqrt{x^2(t) + y^2(t)}$ and $\theta(t) = \tan^{-1} \dfrac{y(t)}{x(t)}$

Thus, by (1.118c),

$$s(t) = R_e[s_t(t)]$$

$$s_l(t) = R_e[s_l(t)e^{j2\pi f_c t}] \quad \text{(by (1.118e))}$$

$$= R_e[c(t)e^{j\theta(t)}e^{j2\pi f_c t}] \quad \text{(by (1.118k)}$$

$$= c(t)\cos(2\pi f_c t + \theta(t)) \tag{1.118l}$$

The signal $c(t)$ (low frequency signal) is called the envelope of $s(t)$ and $\theta(t)$ is the phase of $s(t)$. So Eq. (1.118l) represents the envelope-phase representation of the bandpass signal.

1.5 Signal Space Analysis

The geometrical representation of signals provides a useful tool for the study of digital data transmission. Our study in this section includes the correspondence between signals and vectors, orthogonality, and Gram–Schmidt procedure for orthogonalisation.

1.5.1 Signals as vectors

A signal space is a multidimensional Euclidean space consisting of a set of real-valued energy signals $s_1(t), s_2(t), \ldots, s_M(t)$ of duration T. Each of these signals can be represented by a set of N basis vectors $\phi_1(t), \phi_2(t), \ldots, \phi_N$. Thus,

$$s_1(t) = s_{11}\phi_1(t) + s_{12}\phi_2(t) + \ldots + s_{1N}\phi_N(t)$$
$$s_2(t) = s_{21}\phi_1(t) + s_{22}\phi_2(t) + \ldots + s_{2N}\phi_N(t)$$
$$\vdots \qquad \vdots \qquad \vdots \qquad \vdots$$
$$s_M(t) = s_{M1}\phi_1(t) + s_{M2}\phi_2(t) + \ldots + s_{MN}\phi_N(t),$$

or in closed form:

$$s_i(t) = \sum_{j=1}^{N} s_{ij}\phi_j(t), \begin{cases} i = 1, 2, \ldots, M \\ 0 \le t \le T. \end{cases} \tag{1.119}$$

The basis functions $\phi_j(t)$ are orthonormal with the properties

$$\int_0^T \phi_i(t)\phi_j(t)dt = \begin{cases} 1, & \text{if } i = j \\ 0, & \text{if } i \neq j. \end{cases} \tag{1.120}$$

The coefficients of expansion of (1.119) are given as

$$s_{ij} = \int_0^T s_i(t)\phi_j(t)dt \quad \begin{cases} i = 1, 2, \ldots, M \\ j = 1, 2, \ldots, N. \end{cases} \tag{1.121}$$

Obviously, any signal in the set $\{s_i(t)\}$ can be completely determined by the coefficient vector s_i defined by

$$s_i = \begin{bmatrix} s_{i1} \\ s_{i2} \\ \vdots \\ s_{iN} \end{bmatrix}, i = 1, 2, \ldots M. \tag{1.122}$$

This vector is the signal vector and the set of signal vectors in (1.122) specify a set of M points in the space called the signal space. The signal space of all real energy signals is closed in the sense that any linear combination of signal vectors is another vector in the space.

To complete the geometric structure of the N-dimensional space, we require the measures of vector lengths and angles between the vectors. The length or the norm of a signal vector s_i is defined to be the square root of the signal energy and is denoted by the symbol $||s_i||$. Thus,

$$||s_i|| \equiv \sqrt{E_i} = \left[\int_0^T s_i^2(t)dt\right]^{1/2}. \tag{1.123}$$

With (1.119),

$$||s_i|| = \left\{ \int_0^T \left[\sum_{j=1}^{N} s_{ij}\phi_j(t) \right] \left[\sum_{k=1}^{N} s_{ik}\phi_k(t) \right] dt \right\}^{1/2},$$

or, rearranging,

$$||s_i|| = \left\{ \sum_{j=1}^{N} \sum_{k=1}^{N} s_{ij} s_{ik} \int_0^T \phi_j(t)\phi_k(t) dt \right\}^{1/2},$$

$$= \left\{ \sum_{j=1}^{N} s_{ij}^2 \right\}^{1/2} \qquad i = 1, 2, \ldots, M. \tag{1.124}$$

Since $\phi_j(t)$ and $\phi_k(t)$ are orthonormal, the nonzero value of the integration on the right-hand side of (1.124) is obtained for $j = k$.

To measure the angle θ_{ik} between two signal vectors s_i and s_k, we define the inner product or scalar product of them over (0, *T*) as

$$(s_i, s_k) \equiv \int_0^T s_i(t) s_k(t) dt. \tag{1.125}$$

The cosine of the angle θ_{ik} is defined as

$$\cos\theta_{ik} = \frac{(s_i, s_k)}{||s_i|| \; ||s_k||}. \tag{1.126}$$

Equation (1.26) can be used to measure the angle between the vectors.

Example 1.21

Consider the signals $s_1(t), s_2(t)$ and $s_3(t) = s_1(t) + s_2(t)$. Construct a vector diagram. Find the energy when the vectors $s_1(t)$ and $s_2(t)$ are orthogonal.

Solution

Figure 1.43(a) shows the vector diagram.

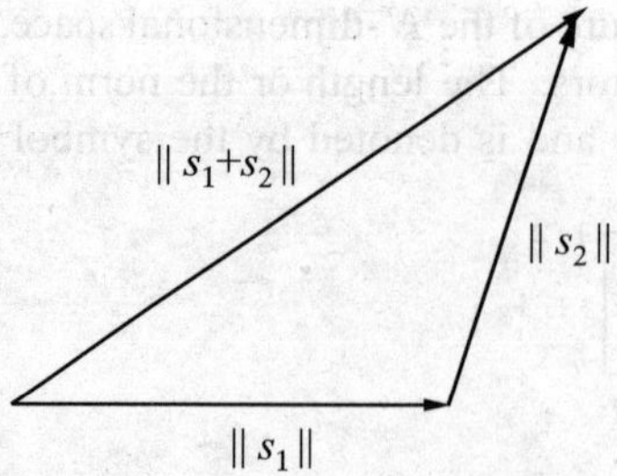

Fig. 1.43 (a) *Vector representation of signals of Example 1.21*

The energy of the resultant signal of $s_1(t)$ and $s_2(t)$ is given by (vide (1.123))

$$\begin{aligned}||s_1 + s_2||^2 &= (s_1 + s_2,\ s_1 + s_2)\\ &= (s_1, s_1) + (s_1, s_2) + (s_2, s_1) + (s_2, s_2)\\ &= ||s_1||^2 + 2(s_1, s_2) + ||s_2||^2. \end{aligned} \tag{1.127}$$

As $s_1(t)$ and $s_2(t)$ are orthogonal, $\theta_{12} = 90°$. Therefore, $(s_1, s_2) = 0$ and

$$||s_1 + s_2||^2 = ||s_1||^2 + ||s_2||^2 = \text{signal energy.}$$

Note that orthogonality in signal space corroborates the superposition of energy.

Example 1.22

Figure 1.43 illustrates the representation of two signals ($M = 2$) in a two-dimensional space ($N = 2$). Find the coefficients of expansion and the distance between the points A and B.

Solution

From the figure,

$$\begin{aligned}s_1(t) &= s_{11}\phi_1(t) + s_{12}\phi_2(t)\\ &= 1 \times \phi_1(t) + 3 \times \phi_2(t).\end{aligned}$$

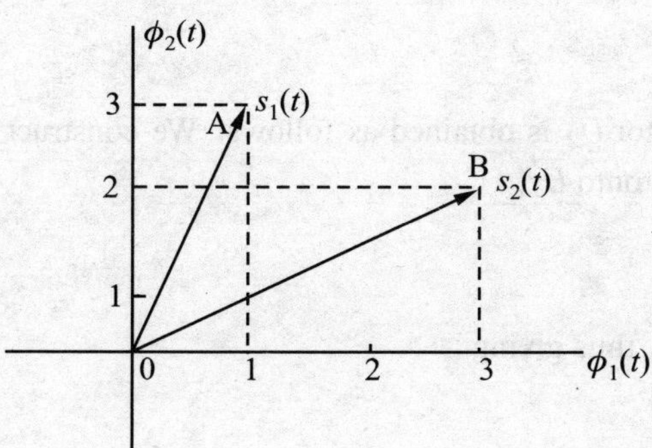

Fig. 1.43 (b) *Vector representation of signals of Example* 1.22

$$\begin{aligned}s_2(t) &= s_{21}\phi_1(t) + s_{22}\phi_2(t)\\ &= 3 \times \phi_1(t) + 2 \times \phi_2(t).\end{aligned}$$

Thus, $s_{11} = 1, s_{12} = 3, s_{21} = 3$, and $s_{22} = 2$.

The distance d_{ij} between the points representing the signal vectors s_i and s_j is given by

$$d_{ij}^2 = ||s_i - s_j||^2 = \sum_{k=1}^{N}(s_{ik} - s_{jk})^2. \tag{1.128}$$

Here $i = 1, j = 2, N = 2$ and so

$$d_{12}^2 = \sum_{k=1}^{2}(s_{1k} - s_{2k})^2 = (s_{11} - s_{21})^2 + (s_{12} - s_{22})^2 = (1-3)^2 + (3-2)^2 = 5.$$

So, $d_{12} = \sqrt{5}$ units.

1.5.2 Gram–Schmidt procedure

We have explained the correspondence between signals and vectors. To represent the signal in the vector space one needs the orthonormal basis functions. The Gram–Schmidt procedure helps in constructing the orthonormal basis, which is outlined below.

Given a set of n-dimensional (or n-component) vectors $\overrightarrow{V_i}, 1 \le i \le n$ one can always construct a set of orthonormal vectors following Gram–Schmidt procedure.

The first orthonormal vector would be

$$\overrightarrow{U_1} = \frac{\overrightarrow{V}_1}{||\overrightarrow{V}_1||}$$

The second orthonormal vector $\overrightarrow{U_2}$ is obtained as follows: We construct a vector $\overrightarrow{U_2'}$ by subtracting from $\overrightarrow{V_2}$ the projection of $\overrightarrow{V_2}$ onto $\overrightarrow{U_1}$:

$$\overrightarrow{U_2'} = \overrightarrow{V_2} - (\overrightarrow{V_2} \cdot \overrightarrow{U_1})\,\overrightarrow{U_1}$$

Normalise $\overrightarrow{U_2'}$, with its norm, thus giving

$$\overrightarrow{U_2} = \frac{\overrightarrow{U'}_2}{||\overrightarrow{U'}_2||}$$

Similarly by subtracting from $\overrightarrow{V_3}$, the projection of $\overrightarrow{V_3}$ onto $\overrightarrow{U_1}$ and $\overrightarrow{U_2}$ we get

$$\overrightarrow{U_3'} = \overrightarrow{V_3} - (\overrightarrow{V_3} \cdot \overrightarrow{U_1})\,\overrightarrow{U_1} - (\overrightarrow{V_3} \cdot \overrightarrow{U_2})\,\overrightarrow{U_2}$$

Normalising $\overrightarrow{U_3'}$,

$$\overrightarrow{U_3} = \frac{\overrightarrow{U'}_3}{||\overrightarrow{U'}_3||}$$

The process continues for all remaining vectors.

We illustrate the process with an example. Let us define three 3-dimensional vectors $\overrightarrow{V_1} = [0\,0\,1]$, $\overrightarrow{V_2} = [0\,1\,1]$ and $\overrightarrow{V_3} = [1\,0\,1]$.

Thus, $\overrightarrow{V_1} = 0.\overrightarrow{e_1} + 0.\overrightarrow{e_2} + 1.\overrightarrow{e_3} = \overrightarrow{e_3}$

$$\overrightarrow{V_2} = \overrightarrow{e_2} + \overrightarrow{e_3},$$

$$\overrightarrow{V_3} = \overrightarrow{e_1} + \overrightarrow{e_3}.$$

We first introduce the procedure of constructing orthonormal vectors in the vector space and then extend the idea to the signal space. An n-dimensional vector $\overrightarrow{v} = [v_1, v_2 \ldots v_n]$ can be represented as a linear combination of unit vector or basis vector $\overrightarrow{e_i}$

$$\overrightarrow{V} = \sum_{i=1}^{n} v_i e_i, \quad 1 \le i \le n$$

where v_i is the projected length of $\overrightarrow{v}$ onto $\overrightarrow{e_i}$. The inner product of such vectors $\overrightarrow{V_1} = [v_{11}, v_{12} \ldots v_{1n}]$ and $\overrightarrow{V_2} = [v_{21}, v_{22} \ldots v_{2n}]$ can be defined as

$$\overrightarrow{V_1} \cdot \overrightarrow{V_2} = (v_{11}, v_{21}, +v_{12}, v_{22} + \cdots + v_{1n} v_{2n}) = \sum_{i=1}^{n} v_{1i}, v_{2i}$$

When $\overrightarrow{V_1}$ and $\overrightarrow{V_2}$ are orthogonal then $\overrightarrow{V_1} \cdot \overrightarrow{V_2} = 0$

The norm of a vector $\overrightarrow{V}$ is denoted by $\left\|\overrightarrow{V}\right\|$ and is defined as

$$\left\|\overrightarrow{V}\right\| = (\overrightarrow{V} \cdot \overrightarrow{V})^{1/2} = \sqrt{\sum_{i=1}^{n} v_1^2} = \text{length of} \overrightarrow{V}.$$

Orthonormal: A set of m vectors is said to be orthonormal if the vectors are orthogonal and each vector has unit length (norm).

Two n-dimensional vectors $\overrightarrow{V_1}$ and $\overrightarrow{V_2}$ must satisfy the triangle inequality:

$$\left\|\overrightarrow{V_1} + \overrightarrow{V_2}\right\| \le \left\|\overrightarrow{V_1}\right\| + \left\|\overrightarrow{V_2}\right\|,$$

the equality holds if $\overrightarrow{V_1}$ and $\overrightarrow{V_2}$ are colinear, that is $\overrightarrow{V_1} = a\overrightarrow{V_2}$.

Also, Cauchy–Schwarz inequality states that

$$\left\|\overrightarrow{V_1} \cdot \overrightarrow{V_2}\right\| \le \left\|\overrightarrow{V_1}\right\| \cdot \left\|\overrightarrow{V_2}\right\|,$$

the equality holds for $\overrightarrow{V_1} = a\overrightarrow{V_2}$.

Now, $\overrightarrow{U_1} = \dfrac{\overrightarrow{V_1}}{\left\|\overrightarrow{V_1}\right\|} = \dfrac{\overrightarrow{e_3}}{\|\overrightarrow{e_3}\|} = \overrightarrow{e_3}$

$$\overrightarrow{U_2'} = \overrightarrow{V_2} - (\overrightarrow{V_1} \cdot \overrightarrow{U_1})\, \overrightarrow{U_1}$$

$$= (\overrightarrow{e_2} + \overrightarrow{e_3}) - [(\overrightarrow{e_2} + \overrightarrow{e_3}) \cdot \overrightarrow{e_3}]\, \overrightarrow{e_3}$$

$$= (\overrightarrow{e_2} + \overrightarrow{e_3}) - \overrightarrow{e_3} = \overrightarrow{e_2}$$

$$\overrightarrow{U_2} = \frac{\overrightarrow{U_2'}}{\|U_2'\|} = \frac{\overrightarrow{e_2}}{1} = \overrightarrow{e_2}$$

Similarly,

$$\overrightarrow{U_3'} = \overrightarrow{V_3} - (\overrightarrow{V_3} \cdot \overrightarrow{U_1})\, \overrightarrow{U_1} - (\overrightarrow{V_3} \cdot \overrightarrow{U_2})\, \overrightarrow{U_2}$$

$$= (\overrightarrow{e_1} + \overrightarrow{e_3}) - [(\overrightarrow{e_1} + \overrightarrow{e_3}) \cdot \overrightarrow{e_3}]\, \overrightarrow{e_3} - [(\overrightarrow{e_1} + \overrightarrow{e_3}) \cdot \overrightarrow{e_2}]\, \overrightarrow{e_2}$$

$$= (\overrightarrow{e_1} + \overrightarrow{e_3}) - \overrightarrow{e_3} - 0.\overrightarrow{e_2} = \overrightarrow{e_1}$$

So, the orthonormal vectors are

$$\overrightarrow{U_1} = \overrightarrow{e_3} = [0\ 0\ 1]$$

$$\overrightarrow{U_2} = \overrightarrow{e_2} = [0\ 1\ 0]$$

and $\overrightarrow{U_3} = \overrightarrow{e_1} = [1\ 0\ 0]$.

The above concept is now extended to the signal as follows.

In the first step, select from the set any signal $s_1(t)$ having finite energy E_1. We then have

$$s_1(t) = s_{11}\phi_1(t). \tag{1.129}$$

$\phi_1(t)$ is a normalised function so that $||\phi_1|| = 1$ and thus

$$\phi_1 = \frac{s_1(t)}{s_{11}} = \frac{s_1(t)}{\sqrt{E_1}}. \tag{1.130}$$

In the second step, select another signal $s_2(t)$ for which we set to zero all the expansion coefficients except s_{21} and s_{22}. Thus,

$$s_2(t) = s_{21}\phi_1(t) + s_{22}\phi_2(t). \tag{1.131}$$

We define the coefficient s_{21} as

$$s_{21} = \textstyle\int_0^T s_2(t)\phi_1(t)dt. \tag{1.132}$$

(The integral equals s_{21} as $\phi_1(t)$ and $\phi_2(t)$ are orthonormal.)
Since s_{21} is known, one can evaluate s_{22} as follows:

$$s_{22}\phi_2(t) = s_2(t) - s_{21}\phi_1(t). \tag{1.133}$$

Squaring and integrating over the interval $(0, T)$,

$$\int_0^T s_{22}^2\phi_2^2(t)dt = \int_0^T \left[s_2(t) - s_{21}\phi_1(t)\right]^2 dt$$

$$\text{or,}\quad s_{22} = \left\{\int_0^T [s_2(t) - s_{21}\phi_1(t)]^2 dt\right\}^{1/2}. \tag{1.134}$$

At this stage both s_{21} and s_{22} are known and we can find $\phi_2(t)$ from (1.131) as

$$\phi_2(t) = \frac{1}{s_{22}}\left[s_2(t) - s_{21}\phi_1(t)\right]$$

$$= \frac{1}{s_{22}}\left[s_2(t) - \frac{s_{21}}{s_{11}}s_1(t)\right]. \tag{1.135}$$

In the third step, we write $s_3(t)$ as

$$s_3(t) = s_{31}\phi_1(t) + s_{32}\phi_2(t) + s_{33}\phi_3(t) \tag{1.136}$$

and set $s_{3N}, N > 3$ equals zero. The coefficients s_{31} and s_{32} are given by

$$s_{31} = \int_0^T s_3(t)\phi_1(t)dt \tag{1.137}$$

$$\text{and}\quad s_{32} = \int_0^T s_3(t)\phi_2(t)dt. \tag{1.138}$$

We now compute s_{33} as

$$s_{33} = \left\{\int_0^T [s_3(t) - s_{31}\phi_1(t) - s_{32}\phi_2(t)]dt\right\}^{1/2}. \tag{1.139}$$

The basis function $\phi_3(t)$ is given by

$$\phi_3(t) = \frac{s_3(t) - s_{31}\phi_1(t) - s_{32}\phi_{2)(t)}}{s_{33}}, \tag{1.140}$$

and has the property that $||\phi_3|| = 1$ and is orthonormal to both $\phi_1(t)$ and $\phi_2(t)$. We continue this procedure until all orthonormal functions are evaluated.

Example 1.23

Apply Gram–Schmidt procedure to obtain the orthonormal basis functions for the two signals shown in Fig. 1.44.

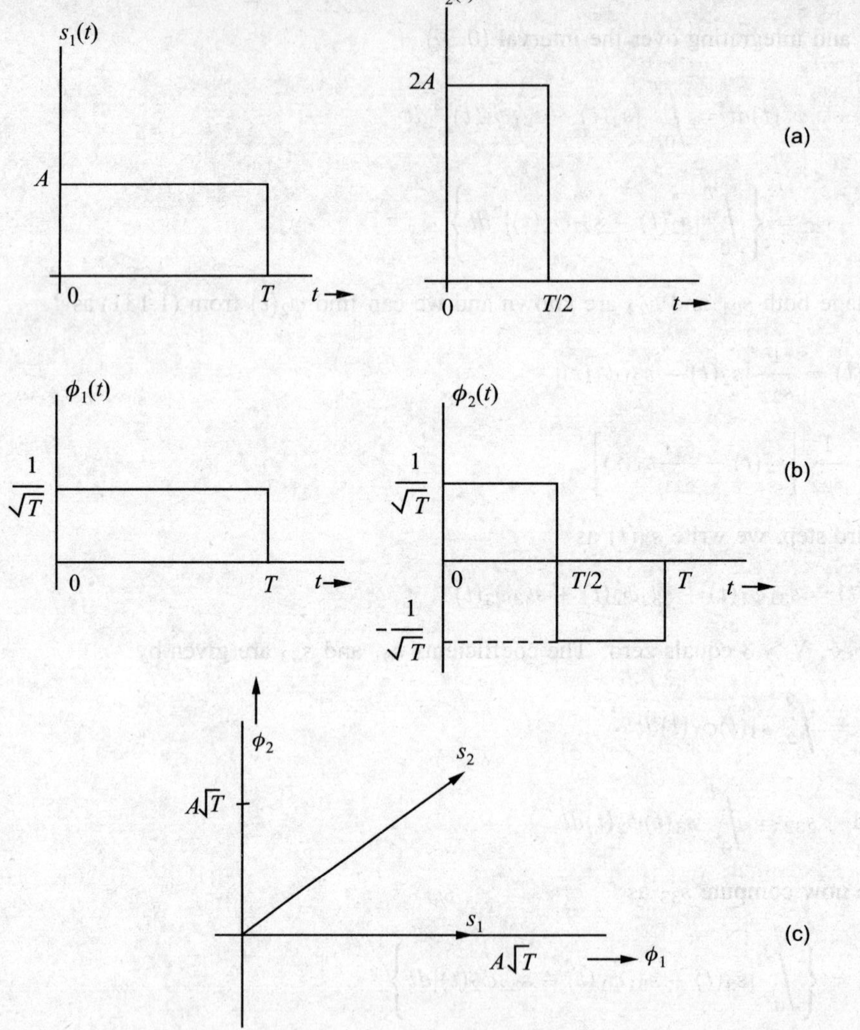

Fig. 1.44 *(a) Signal set $s_1(t)$ and $s_2(t)$; (b) basis functions $\phi_1(t)$ and $\phi_2(t)$ and (c) vector diagram.*

Solution

By (1.130),

$$s_{11} = \left[\int_0^T s_1^2(t)dt\right]^{1/2} = \left[\int_0^T A^2 dt\right]^{1/2} = A\sqrt{T}.$$

So, $\phi_1(t) = \dfrac{s_1(t)}{s_{11}} = \dfrac{s_1(t)}{A\sqrt{T}} = \dfrac{A}{A\sqrt{T}} = \dfrac{1}{\sqrt{T}}.$

From (1.132),

$$s_{21} = \int_0^T s_2(t)\phi_1(t)dt = \int_0^{T/2} 2A \times \frac{1}{\sqrt{T}}dt = \frac{2A}{\sqrt{T}}\frac{T}{2} = A\sqrt{T}.$$

By (1.134),

$$s_{22} = \left\{\int_0^T [s_2(t) - s_{21}\phi_1(t)]^2 dt\right\}^{1/2}.$$

The quantity $s_{21}\phi_1(t) = A\sqrt{T}.\dfrac{1}{\sqrt{T}} = A$. Hence $s_2(t) - s_{21}\phi_1(t)$ equals A for $0 < t < T/2$ and $-A$ for $T/2 < t < T$. Thus,

$$s_{22} = \left\{\int_0^{T/2} A^2 dt + \int_{T/2}^T (-A)^2 dt\right\}^{1/2} = \left[2A^2\frac{T}{2}\right]^{1/2} = A\sqrt{T}$$

From (1.135),

$$\phi_2(t) = \frac{1}{A\sqrt{T}}\left[s_2(t) - \frac{A\sqrt{T}s_1(t)}{A\sqrt{T}}\right] = \frac{1}{A\sqrt{T}}[s_2(t) - s_1(t)].$$

Hence, $s_1(t) = s_{11}\phi_1(t) = A\sqrt{T}\phi_1(t).$

Hence, $s_1(t)s_2(t) = s_{21}\phi_1(t) + s_{22}\phi_2(t).$

Hence, $s_1(t) = A\sqrt{T}\phi_1(t) + A\sqrt{T}\phi_2(t).$

Example 1.24
Consider two signals $s_1(t)$ and $s_2(t)$ given by

$$s_1(t) = \sqrt{2P_s}\cos 2\pi f_c t,$$

$$s_2(t) = \sqrt{2P_s}\cos(2\pi f_c t + \pi) = -\sqrt{2P_s}\cos 2\pi f c t,$$

where $0 \le t \le T$ and P_s is the signal power. Find the orthonormal coordinates and draw the signal-space diagram.

Solution
We choose the basis function as

$$\phi_1(t) = \sqrt{\frac{2}{T}}\cos 2\pi f_c t, \quad 0 \le t < T$$

which has unit energy, that is, $||\phi_1|| = 1$ since

$$||\phi_1|| = \left[\int_0^T \phi_1^2(t)dt\right]^{1/2} = \left[\int_0^T \frac{2}{T}\cos^2 2\pi f_c t dt\right]^{1/2} = \left[\frac{2}{T}\times\frac{T}{2}\right]^{1/2} = 1.$$

Then, $s_1(t) = \sqrt{\frac{2E_b}{T}}\cos 2\pi f_c t = \left(\sqrt{\frac{2}{T}}\cos 2\pi f_c t\right)\sqrt{E_b} = +\sqrt{E_b}\phi_1(t)$

and $s_2(t) = -\sqrt{E_b}\phi_1(t)$, where $E_b = \frac{P_s}{T}$ is the signal energy per bit. Since $s_1(t) = s_{11}\phi_1(t)$ and $s_2(t) = s_{21}\phi_1(t)$, the expansion coefficients s_{11} and s_{21} are given by

$$s_{11} = \int_0^T s_1(t)\phi_1(t)dt = \int_0^T \sqrt{E_b}\phi_1^2(t)dt = +\sqrt{E_b}$$

and $s_{21} = \int_0^T s_2(t)\phi_1(t)dt = -\sqrt{E_b}.$

The signal-space diagram is sketched in Fig. 1.45.

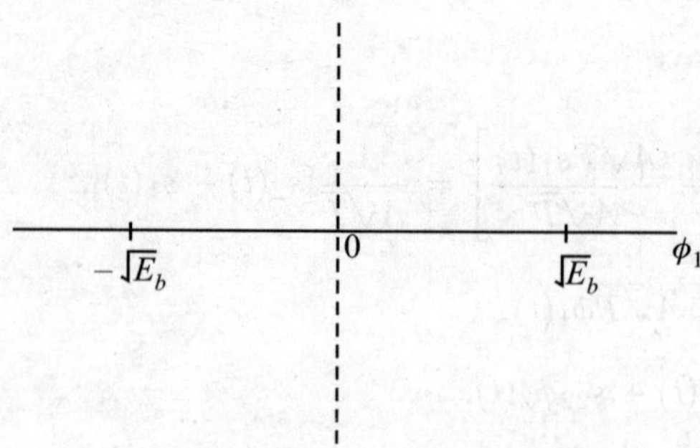

Fig. 1.45 *Signal-space diagram*

Observe that the message points are $2\sqrt{E_b}$ apart from each other. The two signals are known as binary phase-shift keying (BPSK) signals, to be discussed in a later chapter.

Example 1.25

Consider a quaternary phase-shift keying (QPSK) signal described by

$$s_i(t) = \begin{cases} \sqrt{2P_s}\cos[2\pi f_c t + (2i-1)\frac{\pi}{4}], & 0 \le t \le T, \quad i = 1, 2, 3, 4. \\ 0, & \text{elsewhere.} \end{cases}$$

Give a signal-space diagram for the signals.

Solution

Let us first find out the orthonormal basis functions.

$$s_i(t) = \sqrt{2P_s}\cos 2\pi f_c t \cos(2i-1)\frac{\pi}{4} - \sqrt{2P_s}\sin 2\pi f_c t \sin(2i-1)\frac{\pi}{4}.$$

We choose the basis functions as in Example 1.24.

$$\phi_1(t) = \sqrt{\frac{2}{T}}\cos 2\pi f_c t, \quad 0 \le t \le T$$

$$\phi_2(t) = \sqrt{\frac{2}{T}}\sin 2\pi f_c t, \quad 0 \le t \le T$$

Thus, $s_i(t) = \sqrt{E_b}\cos(2i-1)\frac{\pi}{4}.\phi_1(t) - \sqrt{E_b}\sin(2i-1)\frac{\pi}{4}\phi_2(t)$.

The signal vectors are

$$s_i = \begin{bmatrix} \sqrt{E_b}\cos(2i-1)\frac{\pi}{4} \\ -\sqrt{E_b}\sin(2i-1)\frac{\pi}{4} \end{bmatrix}, i = 1, 2, 3, 4.$$

The four message points have the coordinates $(s_{11}, s_{12}), (s_{21}, s_{22}), (s_{31}, s_{32})$ and (s_{41}, s_{42}). For $i = 1$ the coordinates are $s_{11} = \sqrt{E_b}\cos\frac{\pi}{4} = \sqrt{E_b/2}$ and $s_{12} = -\sqrt{E_b/2}$. For other i-values s_{i1} and s_{12} are summarised in the table. The signal-space diagram is shown in Fig. 1.46.

i	Phase	s_{11}	s_{12}
1	$\pi/4$	$+\sqrt{E_b/2}$	$-\sqrt{E_b/2}$
2	$3\pi/4$	$-\sqrt{E_b/2}$	$-\sqrt{E_b/2}$
3	$5\pi/4$	$-\sqrt{E_b/2}$	$+\sqrt{E_b/2}$
4	$7\pi/4$	$+\sqrt{E_b/2}$	$+\sqrt{E_b/2}$

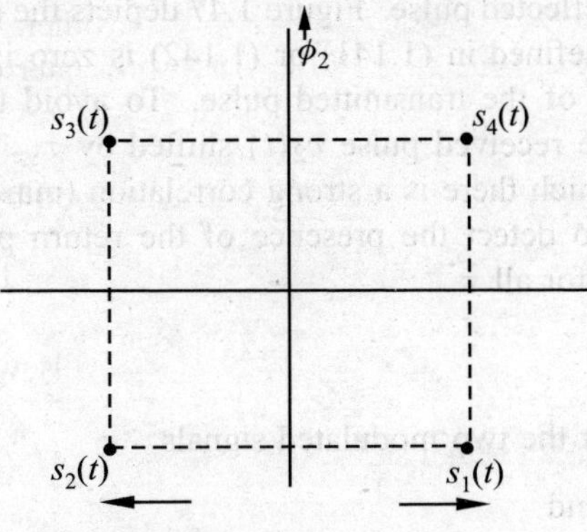

Fig. 1.46 *Signal-space diagram of the QPSK signal*

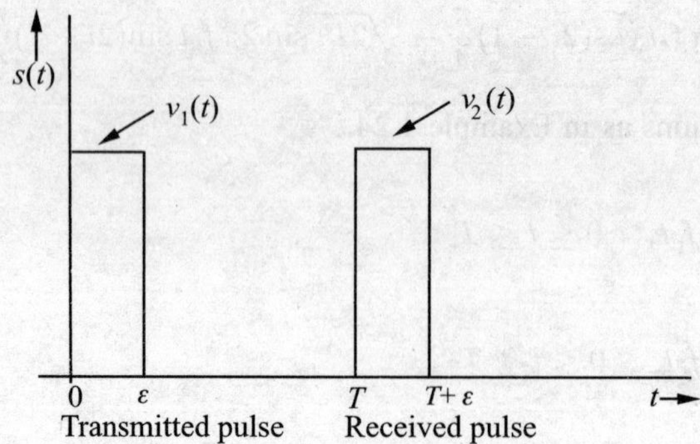

Fig. 1.47 *Explanation of correlation function*

1.6 Correlation Functions

The correlation between two waveforms measures their relatedness or similarity. The correlation $R_{12}(\tau)$ between two waveforms $v_1(t)$ and $v_2(t)$ is defined by

$$R_{12}(\tau) = \lim_{T\to\infty} \frac{1}{T}\int_{-T/2}^{T/2} v_1(t)v_2(t+\tau)dt. \tag{1.141}$$

The two signals need not be periodic. If they are periodic with the same period T_0, then

$$R_{12}(\tau) = \frac{1}{T_0}\int_{-T_0/2}^{T_0/2} v_1(t)v_2(t+\tau)dt. \tag{1.142}$$

The need for the parameter τ while comparing signals may be made clear with an example to follow. Consider the application of correlation to the case of signal detection in radar. A pulse is transmitted towards the target to be detected and gets reflected from it. The time delay between the transmitted and the reflected pulses gives the measure of the distance of the target. Had there been no target at all, there would have been no reflected pulse. Figure 1.47 depicts the positions of the transmitted and received pulses. The integral defined in (1.141) or (1.142) is zero if the received echo is detected at times beyond the duration ϵ of the transmitted pulse. To avoid this difficulty, we compare the transmitted pulse $v_1(t)$ with the received pulse $v_2(t)$ shifted by τ. Thus τ behaves as a searching parameter for some value of which there is a strong correlation (maximum overlap between the two waveforms) and we are able to detect the presence of the return pulse $v_2(t)$. The functions are uncorrelated when $R_{12}(\tau) = 0$ for all τ.

Example 1.26

Find the correlation function for the two modulated signals

$$v_1(t) = x(t)\cos 2\pi f_c t \quad \text{and}$$

$$v_2(t) = x(t)\sin 2\pi f_c t,$$

where $x(t)$ is a stationary process.

Solution
By (1.142),

$$R_{12}(\tau) = \frac{1}{T}\int_{-T/2}^{T/2} x(t)\cos 2\pi f_c t\; x(t+\tau)\sin 2\pi f_c(t+\tau)dt.$$

Here $T = \dfrac{2\pi}{\omega} = \dfrac{2\pi}{2\pi f_c} = \dfrac{1}{f_c}$.

$$\therefore \quad R_{12}(\tau) = \frac{1}{2T}\int_{-T/2}^{T/2} x(t)x(t+\tau)\left\{\sin(4\pi f_c t + 2\pi f_c \tau) + \sin 2\pi f_c \tau\right\} dt$$

$$= \frac{1}{2T}\int_{-T/2}^{T/2} x(t)x(t+\tau).$$

1.6.1 Cross-correlation of power signals

For a power signal $v(t)$, the P_v is defined as

$$P_v = \lim_{T\to\infty}\frac{1}{T}\int_{-T/2}^{T/2} v(t)v^*(t)dt = \lim_{T\to\infty}\frac{1}{T}\int_{-T/2}^{T/2} |v(t)|^2 dt. \tag{1.143}$$

Consider the composite signal

$$v(t) = v_1(t) + v_2(t) \tag{1.144}$$

where the individual signal $v_1(t)$ and $v_2(t)$ are power signals. The average power of $v(t)$ is given by

$$P_{12} = \lim_{T\to\infty}\frac{1}{T}\int_{-T/2}^{T/2} v(t)v^*(t)dt$$

$$= \lim_{T\to\infty}\frac{1}{T}\int_{-T/2}^{T/2} [v_1(t)+v_2(t)][v_1^* + v_2^*(t)]dt$$

$$= \lim_{T\to\infty}\frac{1}{T}\int_{-T/2}^{T/2} v_1(t)v_1^*(t)dt + \lim_{T\to\infty}\frac{1}{T}\int_{-T/2}^{T/2} v_2(t)v_2^*(t)dt$$

$$+ \lim_{T\to\infty}\frac{1}{T}\int_{-T/2}^{T/2} v_1(t)v_2^*(t)dt + \lim_{T\to\infty}\frac{1}{T}\int_{-T/2}^{T/2} v_1^*(t)v_2(t)dt$$

$$= P_1 + P_2 + 2\mathrm{Re}\left\{\lim_{T\to\infty}\frac{1}{T}\int_{-T/2}^{T/2} v_1(t)v_2^*(t)dt\right\}, \tag{1.145}$$

where P_1 and P_2 denote the average powers of the individual signals. In the last term we have used the following relation:

$$\left[\lim_{T\to\infty}\frac{1}{T}\int_{-T/2}^{T/2} s(t)dt\right]^* = \lim_{T\to\infty}\frac{1}{T}\int_{-T/2}^{T/2} s^*(t)dt.$$

Example 1.27

Evaluate the normalised power of $[v_1(t) + v_2(t+\tau)]$, where $v_1(t)$ and $v_2(t)$ are both power signals and τ is the independent variable representing time delay.

Solution

Normalised power

$$P_{12} = \lim_{T\to\infty}\frac{1}{T}\int_{-T/2}^{T/2} [v_1(t) + v_2(t+\tau)]\,[v_1^*(t) + v_2^*(t+\tau)]dt$$

$$= \lim_{T\to\infty}\frac{1}{T}\int_{-T/2}^{T/2} v_1(t)v_1^*(t)dt + \lim_{T\to\infty}\frac{1}{T}\int_{-T/2}^{T/2} v_2(t+\tau)v_2^*(t+\tau)dt$$

$$+ \lim_{T\to\infty}\frac{1}{T}\int_{-T/2}^{T/2} v_1(t)v_2^*(t+\tau)dt + \lim_{T\to\infty}\frac{1}{T}\int_{-T/2}^{T/2} v_1^*(t)v_2(t+\tau)dt$$

$$= P_1 + P_2 + 2\mathrm{Re}\left[\lim_{T\to\infty}\frac{1}{T}\int_{-T/2}^{T/2} v_1(t)v_2^*(t+\tau)dt\right]$$

$$= P_1 + P_2 + 2R_{12}(\tau).$$

If $v_1(t)$ and $v_2(t)$ are uncorrelated for all τ, then $R_{12}(\tau) = 0$ and $P_{12} = P_1 + P_2$, implying that superposition of powers holds for uncorrelated signals.

1.6.2 Autocorrelation of periodic signals

Suppose we correlate a signal with itself, that is, $v_1(t) = v_2(t)$. This generates the autocorrelation function $R_v(\tau)$ given by

$$R_v(\tau) = \lim_{T\to\infty}\frac{1}{T}\int_{-T/2}^{T/2} v_1(t)v_1(t+\tau)dt. \tag{1.146}$$

We list below a few properties of the autocorrelation function.

(a) When the independent variable τ is zero, then

$$R(0) = \lim_{T\to\infty}\frac{1}{T}\int_{-T/2}^{T/2} [v(t)]^2 dt = P_v, \text{ the signal power.}$$

That is, the similarity between $v(t)$ and $v(t+\tau)$ is maximum for $\tau = 0$ and the maximum value at the origin equals the signal power.

(b) Since $R(0)$ is maximum, for any τ,

$|R(\tau)| \leq R(0)$. For a proof, see Example 1.28.

(c) Autocorrelation is an even function of τ. That is,

$$R(\tau) = R(-\tau).$$

(d) If $v(t)$ is periodic, then $R(\tau)$ will have the same periodicity.

Example 1.28
Prove that $R(0 \geq R(\tau)$.

Solution
To prove this, consider the non-negative quantity

$$Q = \lim_{T\to\infty} \frac{1}{T} \int_{-T/2}^{T/2} [v(t) - v(t+\tau)]^2 \, dt \geq 0.$$

Expanding the terms and integrating,

$$Q = \lim_{T\to\infty} \frac{1}{T} \int_{-T/2}^{T/2} v^2(t)dt + \lim_{T\to\infty} \frac{1}{T} \int_{-T/2}^{T/2} v^2(t+\tau)dt - 2 \lim_{T\to\infty} \frac{1}{T} \int_{-T/2}^{T/2} v(t)v(t+\tau)dt$$

$$= P_v + P_v - 2R(\tau) \geq 0, \text{ where } P_v = R(0).$$

The result comes directly due to the fact that as integration is over the entire time axis, a time shift in $v(t)$ does not affect the value of the integral. That is, $v^2(t)$ and $v^2(t+\tau)$ will yield the same value after integration. Thus,

$$P_v - R(\tau) \geq 0 \quad \text{or} \quad R(0) \geq R(\tau).$$

Example 1.29
Find the autocorrelation function of the sinusoidal signal

$$s(t) = A\cos 2\pi f_c t.$$

Solution
The autocorrelation function $R(\tau)$ of $s(t)$ is

$$R(\tau) = \lim_{T\to\infty} \frac{1}{T} \int_{-T/2}^{T/2} A\cos(2\pi f_c t) A\cos\{2\pi f_c(t+\tau)\} \, dt$$

$$= \frac{A^2}{2} \lim_{T\to\infty} \frac{1}{T} \int_{-T/2}^{T/2} [2\cos(2\pi f_c t)\cos(2\pi f_c t + 2\pi f_c \tau)]\, dt$$

$$\frac{A^2}{2} \lim_{T\to\infty} \frac{1}{T} \int_{-T/2}^{T/2} [\cos(4\pi f_c t + 2\pi f_c \tau) + \cos(2\pi f_c \tau)]\, dt$$

$$= \frac{A^2}{2} \cos 2\pi f_c \tau \tag{1.147}$$

since the first term integrates to zero. Note that $R(\tau)$ has the periodicity of $1/f_c$, same as the original signal $s(t)$. The function $s(t)$ and its autocorrelation $R(\tau)$ are plotted in Fig. 1.48.

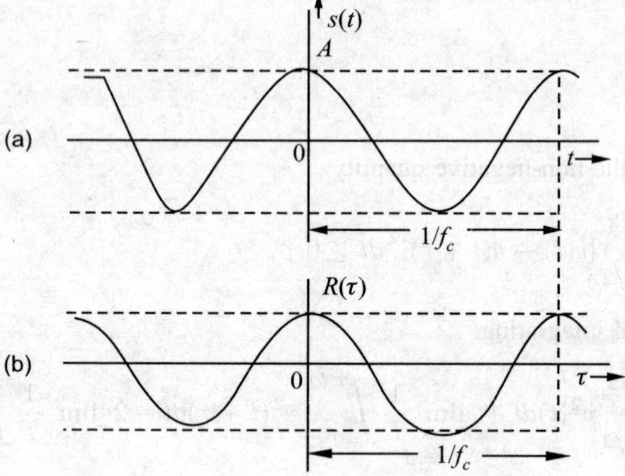

Fig.1.48 *(a) The signal $s(t) = A\cos\omega_c t$; (b) its autocorrelation $R(\tau)$.*

Readers can easily verify that if the signal $s(t)$ contains any random phase, the result will not change.

1.6.3 Correlation and spectral density

We will now establish the relation between the power spectral density of a signal and its correlation function. First consider the case of a periodic signal. By expressing the signal $s(t)$ in exponential Fourier series expansion,

$$s(t) = \sum_{n=-\infty}^{\infty} C_n e^{j2\pi nt/T},$$

the correlation function $R(\tau)$ reads as

$$R(\tau) = \frac{1}{T} \int_{-T/2}^{T/2} \left(\sum_{k=-\infty}^{\infty} C_k e^{j2\pi kt/T} \right) \left(\sum_{l=-\infty}^{\infty} C_l e^{j2\pi l(t+\tau)/T)} \right) dt$$

$$= \frac{1}{T} e^{j2\pi l\tau/T} \int_{-T/2}^{T/2} \sum_{k=-\infty}^{\infty} C_k e^{j2\pi kt/T} \sum_{l=-\infty}^{\infty} C_l e^{j2\pi lt/T} dt$$

$$= \frac{1}{T} e^{j2\pi l\tau/T} \sum_{k=-\infty}^{\infty} \sum_{l=-\infty}^{\infty} \int_{-T/2}^{T/2} C_k C_l e^{j2\pi(k+l)t/T} dt$$

$$= \frac{1}{T} e^{je\pi l\tau/T} \sum_{k,l=-\infty}^{\infty} C_k C_l \frac{e^{j\pi(k+l)} - e^{-j\pi(k+l)}}{j2\pi(k+l)/T}.$$

$$= e^{j2\pi l\tau/T} \sum_{k,l=-\infty}^{\infty} C_k C_l \frac{\sin \pi(k+l)}{\pi(k+l)}. \tag{1.148}$$

As k and l are integers, $\sin \pi(k+l)/\pi(k+l)$ is zero except when $k+l=0$. Thus,

$$R(\tau) = \sum_{l=-\infty}^{\infty} C_l C_{-l} e^{j2\pi l\tau/T} = \sum_{l=-\infty}^{\infty} |C_l|^2 e^{j2\pi l\tau/T}$$

$$= |C_0|^2 + 2\sum_{l=1}^{\infty} |C_n|^2 \cos 2\pi l\tau/T. \tag{1.149}$$

Now, taking the Fourier transform of $R(\tau)$, we get

$$\mathfrak{F}\,[R(\tau)] = \int_{l=-\infty}^{\infty} \left(\sum_{l=-\infty}^{\infty} |C_l|^2 e^{j2\pi l\tau/T} \right) e^{-j2\pi f\tau} d\tau.$$

Interchanging the order of summation and integration,

$$\mathfrak{F}\,[R(\tau)] = \sum_{l=-\infty}^{\infty} |C_l|^2 \int_{-\infty}^{\infty} e^{-j2\pi(f-l/T)\tau} d\tau$$

$$= \sum_{l=-\infty}^{\infty} |C_l|^2 \delta(f - l/T). \tag{1.150}$$

Observe that the right-hand side of (1.150) is the power spectral density $G(f)$ of the periodic waveform $s(t)$. Thus,

$$G(f) = \mathfrak{F}\,[R(\tau)]. \tag{1.151}$$

Equation (1.151) is known as Wiener–Kinchine theorem. For a real signal $s(t)$, its autocorrelation function $R_s(\tau)$ is real and even and so the power spectrum $G_s(f)$ is also even-symmetric, that is, $G_s(f) = G_s(-f)$.

The power spectrum may be continuous or impulsive or both based on the signal under consideration.

In Example (1.29), we have worked out the autocorrelation function $R_s(\tau)$ of a sinusoidal signal $s(t) = A\cos 2\pi f_c t$ as

$$R(\tau) = \frac{A^2}{2}\cos 2\pi f_c \tau.$$

The spectrum $G(f)$ of this signal now comes out as

$$G(f) \;= \mathfrak{F}[R(\tau)] = \frac{A^2}{4}\delta(f - f_c) + \frac{A^2}{4}\delta(f + f_c) \tag{1.151a}$$

which is imputative as plotted in Fig. 1.49.

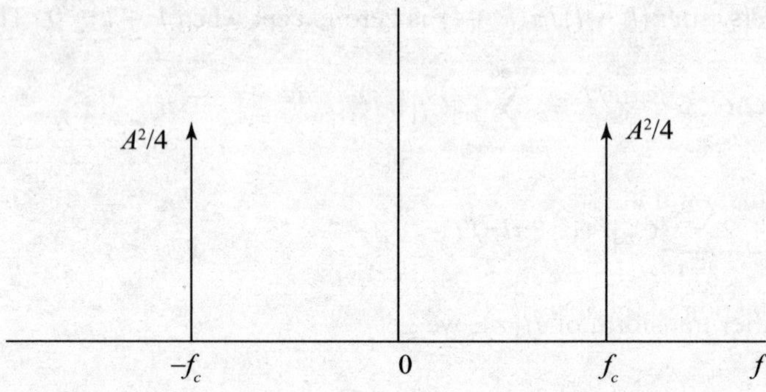

Fig.1.49 *Power spectra of a cosinusoid*

The total power will be thus

$$P = \int_{-\infty}^{\infty} G(f)\,df = \frac{A^2}{4} + \frac{A^2}{4} = \frac{A^2}{2}.$$

Another interesting result of correlation theorem is in the modulation operation. Let the signal $s(t)$ modulate the carrier amplitude $\cos(\omega_c t)$ giving the modulated signal

$$v(t) = s(t)\cos(2\pi f_c t) \tag{1.151b}$$

The autocorrelation $R_v(\tau)$ of $v(t)$ is

$$R_v(\tau) = \lim_{T\to\infty}\frac{1}{T}\int_{-T/2}^{T/2} s(t)s(t+\tau)\cos(2\pi f_c t)\cdot\cos[2\pi f_c t(t+\tau)]\,dt$$

$$= \lim_{T\to\infty} \frac{1}{T} \int_{-T/2}^{T/2} s(t)s(t+\tau)\,\cos(2\pi f_c t)\cdot[\cos 2\pi f_c t \cdot \cos 2\pi f_c \tau - \sin 2\pi f_c t \cdot \sin 2\pi f_c \tau]\,dt$$

$$= \lim_{T\to\infty} \frac{1}{T} \int_{-T/2}^{T/2} s(t)s(t+\tau)\,\left\{\frac{1}{2}\cos 2\pi f_c \tau \cdot [1+\cos 4\pi f_c t] - \frac{1}{2}\sin 2\pi f_c \tau \cdot \sin 4\pi f_c t\right\} dt$$

Thus,

$$R_v(\tau) = \frac{1}{2}\cos 2\pi f_c \tau \cdot R_s(\tau) \tag{1.151c}$$

The Fourier transformation yields

$$G_v(f) = \frac{1}{4}[G_s(f-f_c) + G_s(f+f_c)] \tag{1.151d}$$

This implies that modulation translates the power spectrum by $\pm f_c$ on the frequency axis.

Example 1.30

The autocorrelation function of an aperiodic power signal is given by

$$R(\tau) = e^{-\tau^2/2\sigma^2}.$$

Find the power spectral density and the normalised average power of the signal.

Solution

The correlation function is plotted in Fig. 1.50.

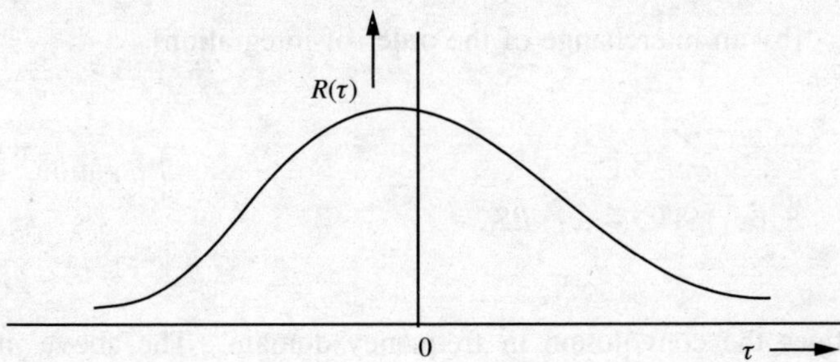

Fig. 1.50 *Autocorrelation function*

By (1.151), the autocorrelation and the power spectral density form a Fourier transform pair, that is,

$$G(f) = \int_{-\infty}^{\infty} R(\tau)e^{-j2\pi f\tau}\,d\tau$$

and $\quad R(\tau) = \int_{-\infty}^{\infty} G(f) e^{j2\pi f\tau} df.$

Hence, $\quad G(f) = \int_{-\infty}^{\infty} e^{-\tau^2/2\sigma^2} e^{-j2\pi f\tau} d\tau$

$$= \sqrt{2\pi\sigma^2} e^{-(2\pi f\sigma)^2/2}.$$

Normalised average power is given by (refer Section 1.6.2)

$$P_v = R(0) = e^0 = 1.$$

1.7 Convolution

Given two time functions $s_1(t)$ and $s_2(t)$ with their Fourier transforms $S_1(\omega)$ and $S_2(\omega)$ respectively, the spectrum of the product signal $s(t) = s_1(t)s_2(t)$ can be obtained by the convolution theorem. The Fourier transform $S(\omega)$ of $s(t)$ is

$$S(\omega) = \int_{-\infty}^{\infty} s_1(t) s_2(t) e^{-j\omega t} dt$$

$$= \frac{1}{2\pi} \int_{-\infty}^{\infty} s_1(t) \left[\int_{-\infty}^{\infty} S_2(\omega_1) e^{j\omega_1 t} d\omega_1 \right] e^{-j\omega t} dt,$$

[$s_2(t)$ is expressed by its spectrum $S_2(\omega)$; ω_1 is the dummy variable]

$$= \frac{1}{2\pi} \int_{-\infty}^{\infty} S_2(\omega_1) \left[\int_{-\infty}^{\infty} s_1(t) e^{-j(\omega-\omega_1)t} dt \right] d\omega_1$$

[by an interchange of the order of integration].

So,

$$S(\omega) = \frac{1}{2\pi} \int_{-\infty}^{\infty} S_2(\omega_1) S_1(\omega - \omega_1) d\omega_1. \tag{1.152}$$

Equation (1.152) defines the convolution in frequency domain. The above integral is called the convolution integral. Equivalently, one can write

$$S(\omega) = \frac{1}{2\pi} \int_{-\infty}^{\infty} S_1(\omega_1) S_2(\omega - \omega_1) d\omega_1 = S_1(\omega) * S_2(\omega). \tag{1.153}$$

The asterisk notation is read as '$S_1(\omega)$ is convolved with $S_2(\omega)$'. To derive the convolution in time domain, we begin by writing

$$
\begin{aligned}
s(t) &= \mathcal{F}^{-1}\left[S_1(\omega)S_2(\omega)\right] \\
&= \frac{1}{2\pi}\int_{-\infty}^{\infty} S_1(\omega)S_2(\omega)e^{j\omega t}d\omega \\
&= \frac{1}{2\pi}\int_{-\infty}^{\infty}\left[\int_{-\infty}^{\infty} s_1(\tau)e^{-j\omega\tau}d\tau\right] S_2(\omega)e^{j\omega t}d\omega \\
&= \int_{-\infty}^{\infty} s_1(\tau)\left[\frac{1}{2\pi}\int_{-\infty}^{\infty} S_2(\omega)e^{j\omega(t-\tau)}d\omega\right] d\tau.
\end{aligned}
$$

That is, $s(t) = \int_{-\infty}^{\infty} s_1(\tau)s_2(t-\tau)d\tau = s_1(t) * s_2(t)$ (1.154)

or equivalently, $s(t) = \int_{-\infty}^{\infty} s_2(\tau)s_1(t-\tau)d\tau.$ (1.155)

The above two equations define the convolution in time domain of two functions $s_1(t)$ and $s_2(t)$. The convolution itself is a function of time t. In many cases, the convolution integral is difficult to evaluate in compact form. Sometimes graphical method is applied to evaluate convolution simply by inspection.

1.7.1 Convolution theorem

The theorem states that the Fourier transform of a time function which is again the convolution of two functions, equals the product of their Fourier transforms. That is,

$$\mathcal{F}\left[s_1(t)^* s_2(t)\right] = S_1(\omega)S_2(\omega). \tag{1.156}$$

To prove this, we start with

$$
\begin{aligned}
\mathcal{F}\left[s_1(t)^* s_2(t)\right] &= \int_{-\infty}^{\infty} e^{-j2\pi t f t}\left[\int_{-\infty}^{\infty} s_1(\tau)s_2(t-\tau)d\tau\right] d\tau \\
&= \int_{-\infty}^{\infty} s_1(\tau)\left[\int_{-\infty}^{\infty} e^{-j2\pi f t}s_2(t-\tau)dt\right] d\tau \\
&= \int_{-\infty}^{\infty} s_1(\tau)e^{-j2\pi f\tau}\left[\int_{-\infty}^{\infty} s_2(t')e^{-j2\pi f t'}dt'\right] d\tau
\end{aligned}
$$

[A change of variable $t' = t - \tau$]

$$= \int_{-\infty}^{\infty} s_1(\tau)e^{-j2\pi f\tau}d\tau \int_{-\infty}^{\infty} s_2(t')e^{-j2\pi ft'}dt' = S_1(f)S_2(f).$$

That is,

$$\mathcal{F}[s_1(t) * s_2(t)] = S_1(f)S_2(f). \tag{1.157}$$

In the similar way, one can prove that

$$\mathcal{F}[s_1(t)s_2(t)] = S_1(f)^* S_2(f). \tag{1.158}$$

In other words, multiplication of two functions in the time domain appears as convolution in the frequency domain and vice versa.

Example 1.31

Find the convolution of the signal $s(t)$ with itself where $s(t)$ is a square pulse of unit amplitude and time-limited to the range between $t = -t_0$ and $t - t_0$, as shown in the following figure.

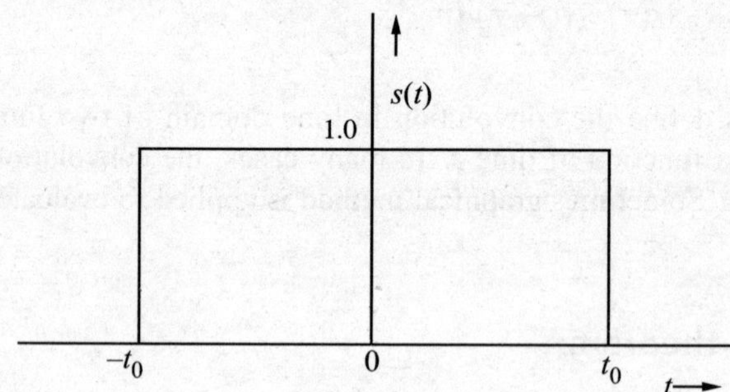

Solution

Expressing $s(t)$ in terms of unit-step function $u(t)$,

$$s(t) = u(t + t_0) - u(t - t_0).$$

By definition,

$$s(t) * s(t) = \int_{-\infty}^{\infty} s(\tau)s(t - \tau)d\tau.$$

Here, $\quad s(t - \tau) = u(t - \tau + t_0) - u(t - \tau - t_0).$

So, $\quad s(\tau)s(t - \tau) = [u(\tau + t_0) - u(\tau - t_0)][u(t - \tau + t_0) - u(t - \tau - t_0)]$

$$= u(\tau + t_0)u(t - \tau + t_0) - u(\tau + t_0)u(t - \tau - t_0)$$

$$- u(\tau - t_0)u(t - \tau + t_0) + u(\tau - t_0)u(t - \tau - t_0).$$

$$s(t) * s(t) = \int_{-\infty}^{\infty} u(\tau + t_0)u(t - \tau + t_0)d\tau - \int_{-\infty}^{\infty} u(\tau + t_0)u(t - \tau - t_0)d\tau$$

$$- \int_{\infty}^{\infty} u(\tau - t_0)u(t - \tau + t_0)d\tau + \int_{-\infty}^{\infty} u(\tau - t_0)u(t - \tau - t_0)d\tau.$$

Remembering that $u(\tau + t_0) = 0$ for $\tau < -t_0$ and $u(\tau - t_0) = 0$ for $\tau > t_0$, the convolution becomes

$$s(t) * s(t) = \int_{-t_0}^{\infty} u(t - \tau + t_0)d\tau - \int_{-t_0}^{\infty} u(t - \tau - t_0)d\tau$$

$$- \int_{t_0}^{\infty} u(t - \tau + t_0)d\tau + \int_{t_0}^{\infty} u(t - \tau - t_0)d\tau$$

since $u(t - \tau + t_0) = 0$ for $\tau > t + t_0$

and $u(t - \tau - t_0) = 0$ for $\tau > t - t_0$.

Thus,

$$\int_{-t_0}^{\infty} u(t - \tau + t_0)d\tau = \int_{-t_0}^{t+t_0} d\tau = t + 2t_0 \text{ provided that } t + t_0 > -t_0,$$

that is, $t > -2t_0$; otherwise the integral becomes zero. Similarly,

$$\int_{-t_0}^{\infty} u(t - \tau - t_0)d\tau = \int_{-t_0}^{t-t_0} d\tau = t \text{ provided that } t - t_0 > -t_0, \text{ that is, } t > 0.$$

$$\int_{t_0}^{\infty} u(t - \tau + t_0)d\tau = \int_{t_0}^{t+t_0} d\tau = t, \text{ provided that } t + t_0 > t_0, \text{ that is, } t > 0,$$

and

$$\int_{t_0}^{\infty} u(t - \tau - t_0)d\tau = \int_{t_0}^{t-t_0} d\tau = t - 2t_0, \text{ provided that } t - t_0 > t_0, \text{ that is, } t > 2t_0.$$

Hence,

$$s(t) * s(t) = (t + 2t_0)u(t + 2t_o) - 2tu(t) + (t - 2t_0)u(t - 2t_0).$$

The results of the convolution are sketched in Fig. 1.51.

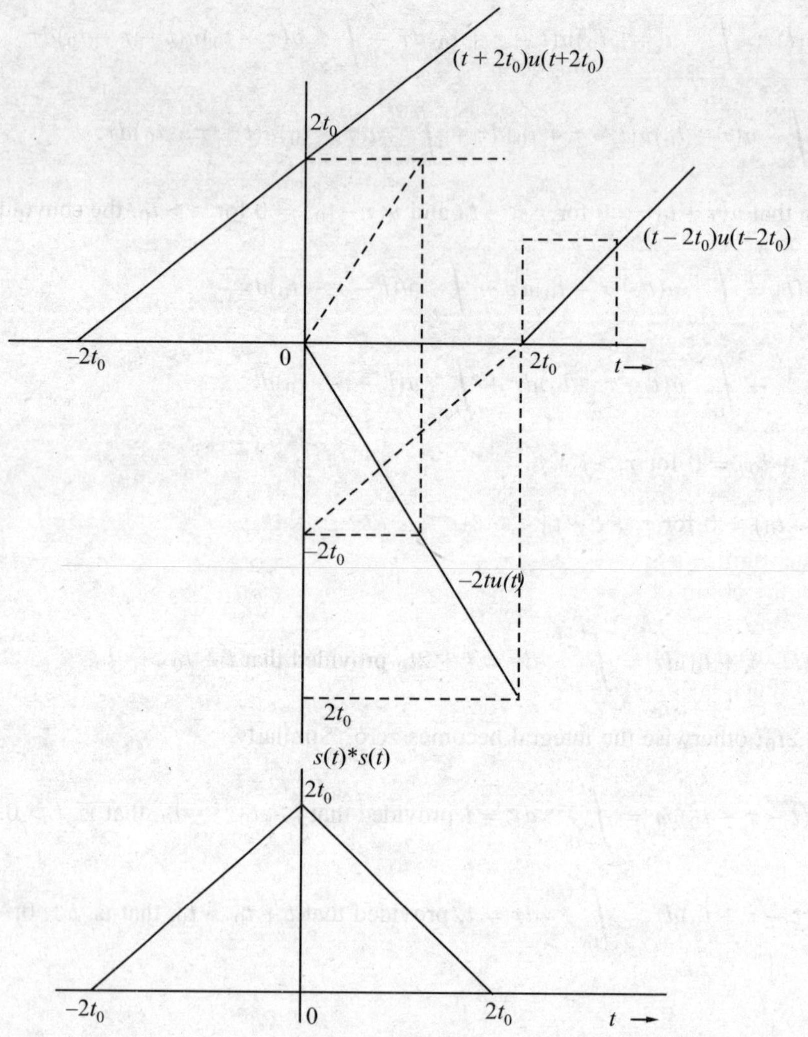

Fig. 1.51 *Graphical method of convolution. A square wave convolved with itself results in a triangular wave.*

Example 1.32

Find and sketch the convolution of a rectangular pulse with a triangular pulse shown in Fig. 1.52.

Here the functions $s_1(t)$ and $s_2(t)$ are given as

$$s_1(t) = A, \quad 0 < t < T$$

$$= 0, \quad \text{elsewhere}$$

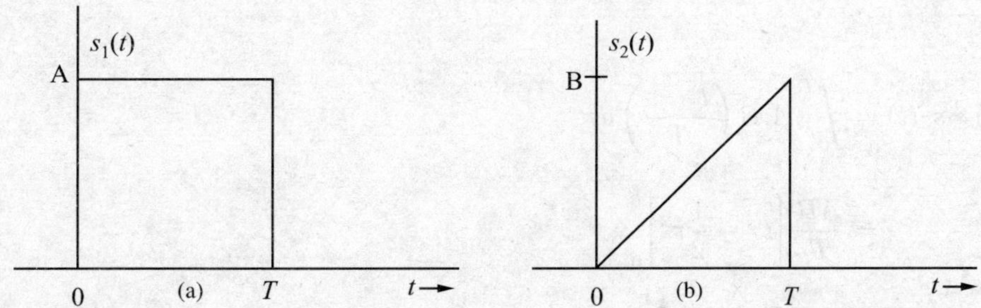

Fig. 1.52 (a) *A rectangular pulse;* (b) *triangular pulse.*

and

$$s_2(t) = \frac{Bt}{T}, \quad 0 < t < T$$
$$= 0, \quad \text{elsewhere.}$$

To obtain $s_2(t-\tau)$ as a function of τ, we reverse $s_2(t)$ in time to get $s_2(-t)$ and replace t with τ to get $s_2(-\tau)$. We then shift $s_2(-\tau)$ by t units along the τ-axis to yield $s_2(t-\tau)$. The quantity t is always measured from the origin of $s_1(\tau)$. A negative t indicates a left shift and a positive t a right shift.

The convolution $s_1(t) * s_2(t)$ is evaluated for $-\infty < t < \infty$. As illustrated in Fig. 1.53, the functions do not overlap for $t < 0$ and $t > 2T$ and hence $s_1(t) * s_2(t) = 0$ in those regions. When $0 < t < T$, the functions overlap.

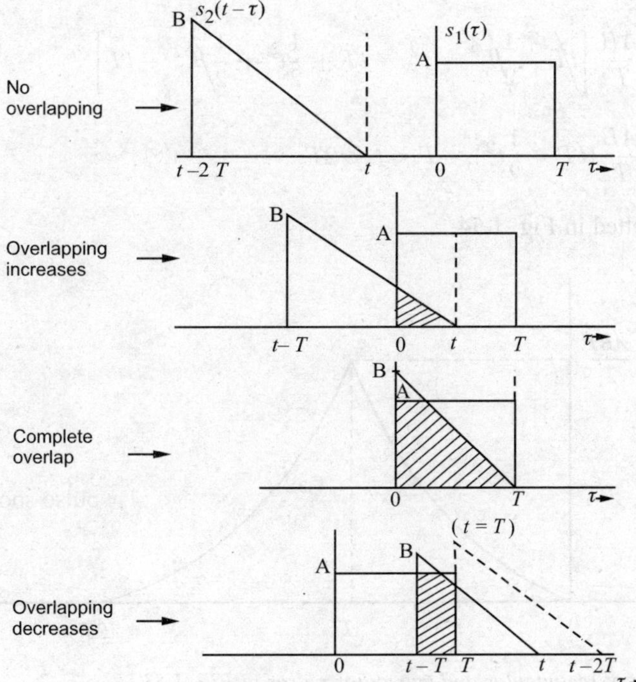

Fig. 1.53 *This shows how the overlapping zone changes as one signal is moved along the time axis.*

So,

$$s_1(t) * s_2(t) = \int_0^T A.B\left(\frac{t-\tau}{T}\right) d\tau$$

$$= \frac{AB}{T}\left[t\tau - \frac{1}{2}\tau^2\right]_0^t$$

$$= \frac{AB}{T}\left[t^2 - \frac{1}{2}t^2\right] = \frac{AB}{2T}t^2, \quad 0 < t < T.$$

When $t > T$, the functions overlap in the region $t - T < \tau < T$. So,

$$s_1(t) * s_2(t) = \int_{t-T}^T A.B\left(\frac{t-\tau}{T}\right) d\tau$$

$$= \frac{AB}{T}\left[t\tau - \frac{1}{2}\tau^2\right]_{t-T}^T$$

$$= \frac{AB}{T}\left[\left\{tT - \frac{1}{2}T^2\right\} - \left\{t(t-T) - \frac{1}{2}(t-T)^2\right\}\right]$$

$$= \frac{AB}{T}\left[tT - \frac{1}{2}T^2 - t^2 + tT + \frac{1}{2}t^2 + \frac{1}{2}T^2 - tT\right]$$

$$= \frac{AB}{T}(tT - \frac{1}{2}t^2), \quad T < t < 2T.$$

The convolution is plotted in Fig. 1.54.

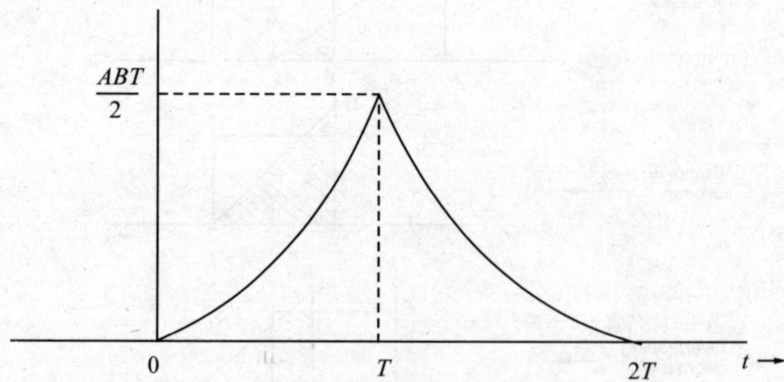

Fig. 1.54 *Convolution of rectangular and triangular pulses of Fig. 1.53*

Solved Numericals

1. A signal $s(t)$ is given by

$$s(t) = \cos^2(\omega_o t) + 2\cos(3\omega_o t) + 3\cos(4\omega_o t).$$

Find the Fourier Series coefficients of $s(t)$.

Solution

By Eq. (1.44), we have

$$s(t) = \sum_{n=-\infty}^{0} c_n e^{i\omega_o nt},$$

where $\omega_0 = \dfrac{2\pi}{T_0}$, to bring the fundamental period.

We may write by using Euler's relation

$$s(t) = \frac{1}{2}\{1 + \cos(2\omega_o t)\} + 2cos(3\omega_o t) + 3\cos(4\omega_o t)$$

$$= \frac{1}{2} + \frac{1}{2}\left(\frac{e^{i2\omega_0 t} + e^{-i2\omega_0 t}}{2}\right) + 2\left(\frac{e^{i3\omega_0 t} + e^{-i3\omega_0 t}}{2}\right) + 3\left(\frac{e^{i4\omega_0 t} + e^{-i4\omega_0 t}}{2}\right)$$

$$= \frac{1}{2} + \frac{1}{4}\left(e^{i2\omega_0 t} + e^{-i2\omega_0 t}\right) + \left(e^{i3\omega_0 t} + e^{-i3\omega_0 t}\right) + \frac{3}{2}\left(e^{i4\omega_0 t} + e^{-i4\omega_0 t}\right)$$

Comparing with exponential form of Fourier series,

$c_0 = \dfrac{1}{2}$, $c_2 = c_{-2} = \dfrac{1}{4}$, $c_3 = c_{-3} = 1$, $c_4 = c_{-4} = \dfrac{3}{2}$ and all other coefficients are zeros.

2. Find the Fourier transform of the signum function $(sgn(t))$ defined as

$$s(t) = sgn(t) = \begin{cases} 1, & t > 0 \\ 0, & t = 0 \\ -1, & t < 0 \end{cases}$$

Solution

The signum function violates the Dirichlet condition (Eq. (1.69)) and hence we cannot use integration (Eq. (1.85)) to directly find the Fourier transform. An alternative approach is to approximate the signum function as a sum of two limiting exponentials:

$$sgn(t) = \underset{a\to 0}{\text{Lt}}\left[e^{-at}u(t) - e^{at}u(-t)\right] \quad \text{with}\, a > 0$$

and to plot it as shown in the adjoining figure.

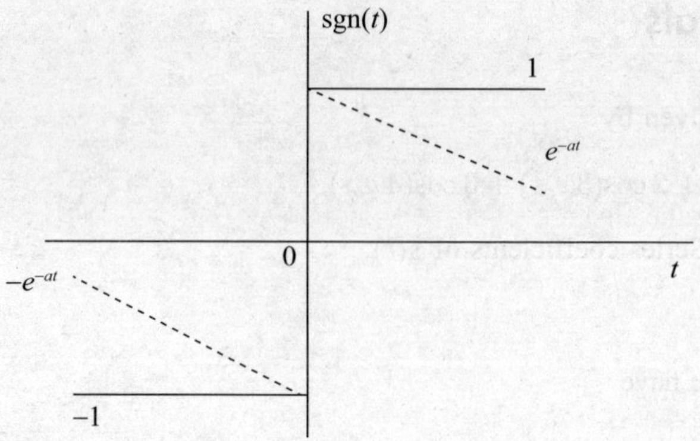

The first term shows exponential decay for $t > 0$ and the second term, for $t < 0$.

Now, taking the Fourier transform (Eq. 1.85).

$$F\left[sgn(t)\right] = \underset{a\to 0}{\text{Lt}}\left[\int_0^\infty e^{-at}e^{-j\omega t}dt - \int_{-\infty}^{0} e^{at}e^{-j\omega t}dt\right]$$

$$= \underset{a\to 0}{\text{Lt}}\left[\frac{1}{a+j\omega} - \frac{1}{a-j\omega}\right] \quad \text{(by Eq.1.93)}$$

$$= \underset{a\to 0}{\text{Lt}}\left(\frac{a-j\omega-a-j\omega}{a^2+\omega^2}\right)$$

$$= \underset{a\to 0}{\text{Lt}}\left(\frac{-2j\omega}{a^2+\omega^2}\right) = \frac{-2j\omega}{\omega^2}$$

$$= \frac{-2j}{w} = \frac{-j2}{2\pi f} = \frac{1}{j\pi f}.$$

(The limits of integration of the second item above is from $-\infty$ to 0 as $u(-t)$ is involved).

3. If a signal $s(t)$ has the Fourier transform denoted by $S(f)$, that is, $s(t) \leftrightarrow S(f)$ is a Fourier transform pair, then show that the signal $S(t)$ will have the Fourier transform given as $s(-f)$ (This is the duality property of Fourier transform).

Solution

By Eq. (1.85), we have

$$S(f) = \int_{-\infty}^{\infty} s(t)e^{-i2\pi f^t}dt$$

and by Eq. (1.86), the inverse Fourier transform formula

$$s(t) = \int_{-\infty}^{\infty} S(f)e^{j2\pi ft}df$$

By time reversing the signal above (replacing t by $-t$)

$$s(-t) = \int_{-\infty}^{\infty} S(f)e^{-j2\pi ft}df$$

In the above, interchange t and f to get

$$s(-f) = \int_{-\infty}^{\infty} S(t)e^{-j2\pi tf}dt$$

which shows that $S(t)$ has the Fourier transform $s(-f)$. Hence the proof.

4. Three vectors are given as

$$\bar{s}_1 = (3, -4, 0),\ \bar{s}_2 = (1, 0, 2) \text{ and } \bar{s}_3 = (4, 0, -3).$$

Find

(a) The distances between $(\bar{s}_1, \bar{s}_2)$, $(\bar{s}_2, \bar{s}_3)$, and $(\bar{s}_3, \bar{s}_1)$

(b) The norm of the vector $\bar{s}$ given by $\bar{s} = \bar{s}_1 + 2\bar{s}_2 + 3\bar{s}_3$

(c) Also find if any pair of these vectors is orthogonal

Solution

(a) By Eq. (1.128), the distance between $\bar{s}_i$ and $\bar{s}_j$ is given by $d_{ij} = \sqrt{\|\bar{s}_i - \bar{s}_i\|^2}$

So, $d_{12}^2 = \|\bar{s}_1 - \bar{s}_2\|^2 = \|3-1, -4-0, 0-2\|^2 = \|2, -4, -2\|^2 = 2^2 + (-4)^2 + (-2)^2 = 24$

$$\therefore d_{12} = \sqrt{24} = 2\sqrt{6}$$

$$d_{23}^2 = \|\bar{s}_2 - \bar{s}_3\|^2,$$

$$\bar{s}_2 - \bar{s}_3 = (1-4, 0-0, 2+3) = (-3, 0, 5)$$

$$\|\bar{s}_2 - \bar{s}_3\|^2 = (-3)^2 + 0^2 + 5^2 = 34$$

$$\therefore d_{23} = \sqrt{34}$$

Similarly, $d_{31} = \sqrt{(4-3)^2 + (0+4)^2 + (-3-0)^2} = \sqrt{1+16+9} = \sqrt{26}$

(b) The vector

$$\bar{s} = (3, -4, 0) + 2(1, 0, 2) + 3(4, 0, -3)$$

$$= (3, -4, 0) + (2, 0, 4) + (12, 0, -9) = (17, -4, -5)$$

So, $\|\bar{s}\| = \sqrt{(17)^2 + (-4)^2 + (-5)^2} = \sqrt{289 + 16 + 25} = 330$

(c) Now, we find the inner products of the pairs.

$$\langle \bar{s}_1, \bar{s}_2 \rangle = 3 \times 1 + (-4) \times 0 + 0 \times 2 = 3$$

$$\langle \bar{s}_2, \bar{s}_3 \rangle = 1 \times 4 + 0 \times 0 + 2(-3) = -2$$

$$\langle \bar{s}_3, \bar{s}_1 \rangle = 4 \times 3 + 0 \times (-4) + (-3) \times 0 = 12$$

So, the vectors are not orthogonal.

5. Consider three exponential signals

$$s_1(t) = e^{-at}u(t), s_2(t) = e^{-2at}u(t), s_3(t) = e^{-3at}u(t).$$

Apply Gram–Schmidt procedure to find 3-dimensional orthonormal basics.

Solution

Obviously the exponentials are not orthogonal. We may apply the procedure as outlined in Section 1.5.2 to find the basis $\phi_1(t), \phi_2(t)$ and $\phi_3(t)$.

By Eq. (1.129), $s_1(t) = s_{11}\phi_1(t)$

As $\phi_1(t)$ is a normalised function with $\|\phi_1\| = 1$, $s_{11} = \sqrt{\mathrm{E}_1}$, E_1 is the energy of $s_1(t)$ and is given by

$$E_1 = \int_0^\infty s_1^2(t)dt = \int_0^\infty e^{-2at}dt = \left.\frac{e^{-2at}}{-2a}\right|_0^\infty$$

$$= \frac{-1}{2a}(0 - 1) = \frac{1}{2a}$$

$$\phi_1(t) = \frac{s_1(t)}{s_{11}}$$

$$= \frac{s_1(t)}{\sqrt{E_1}} = \frac{e^{-at}u(t)}{\sqrt{1/2a}}$$

$$= \sqrt{2a} \cdot e^{-at}u(t)$$

Now, we express the second vector $s_2(t)$ as

$$s_2(t) = s_{21}\phi_1(t) + s_{22}\phi_2(t) \text{ (vide Eq. 1.131)}$$

The coefficient s_{21} is given by

$$s_{21} = \int_0^\infty s_2(t)\phi_1(t)dt = \int_0^\infty e^{-2at} \cdot \sqrt{2a}\, e^{-at}dt = 2a\int_0^\infty e^{-3at}dt = \frac{\sqrt{2a}}{3a} = \sqrt{\frac{2}{9a}}$$

$$\left(\text{since } \int_0^\infty \phi_1(t)\phi_2(t)dt = 0\right)$$

Since s_{21} is known, we can find s_{22} as

$$s_{22}\phi_2(t) = s_2(t) - s_{21}\phi_1(t)$$

Squaring and integrating,

$$s_{22}^2\int_0^\infty \phi_2^2(t)dt = \int_0^\infty s_2^2(t)dt + s_{21}^2\int_0^\infty \phi_1^2(t)dt - 2s_{21}\int_0^\infty s_2(t)\phi_1(t)dt$$

or,

$$s_{22}^2 = E_2 + s_{21}^2 - 2s_{21} \cdot s_{21} = E_2 - s_{21}^2,$$

where E_2 is the energy of $s_2(t)$ and $E_2 = \frac{1}{4a}$.

$$\therefore\ s_{22}^2 = \frac{1}{4a} - \frac{2a}{9a^2} = \frac{1}{4a} - \frac{2}{9a} = \frac{1}{36a}$$

$$\therefore\ s_{22} = \frac{1}{6\sqrt{a}}.$$

So,

$$\begin{aligned}\phi_2(t) &= \frac{1}{s_{22}}\left(s_2(t) - s_{21}\phi_1(t)\right)\\ &= 6\sqrt{a}\left(e^{-2at} - \sqrt{\frac{2}{9a}} \cdot \sqrt{2a}e^{-at}\right)u(t)\\ &= 6\sqrt{a}\, e^{-2at}\, u(t) - 4\sqrt{a}e^{-at}u(t)\end{aligned}$$

Now we express the third vector $s_3(t)$ as

$$s_3(t) = s_{31}\phi_1(t) + s_{32}\phi_2(t) + s_{33}\phi_3(t)$$

$\therefore$ The coefficient s_{31} is given by

$$s_{31} = \int_0^\infty s_3(t)\phi_1(t)dt = \int_0^\infty e^{-3at}\sqrt{2a}\,e^{-at}dt = \frac{\sqrt{2a}}{4a} = \sqrt{\frac{2a}{16a^2}} = \frac{1}{\sqrt{8a}}$$

$$\left(\text{since} \int_0^\infty \phi_1(t)\phi_2(t)dt = \int_0^\infty \phi_1(t)\phi_3(t)dt = 0\right)$$

and s_{32} is given by

$$s_{32} = \int_0^\infty s_3(t)\phi_2(t)dt$$

$$= \int_0^\infty e^{-3at}\left[6\sqrt{a}\,e^{-2at} - 4\sqrt{a}\,e^{-at}\right]dt$$

$$= \frac{6\sqrt{a}}{5a} - \frac{4\sqrt{a}}{4a} = \frac{6}{5\sqrt{a}} - \frac{1}{\sqrt{a}} = \frac{1}{5\sqrt{a}}$$

Since s_{31} and s_{32} are evaluated, one can find s_{33} as

$$s_{33}\phi_3(t) = s_3(t) - s_{31}\phi_1(t) - s_{32}\phi_2(t)$$

Squaring and integrating,

$$s_{33}^2 = E_3 + s_{31}^2 + s_{32}^2 - 2s_{31}^2 - 2s_{32}^2$$

$$= E_3 - s_{31}^2 - s_{32}^2 \text{ where } E_3 \text{ is the energy of } s_3(t) \text{ and } E_3 = \int_0^\infty \left(e^{-3at}\right)^2 dt = \frac{1}{6a}$$

$$= \frac{1}{6a} - \frac{1}{8a} - \frac{1}{25a} = \frac{1}{600a}$$

Thus, $s_{33} = \dfrac{1}{10\sqrt{6a}}$

$$\phi_3(t) = \frac{1}{s_{33}}\left(s_3(t) - s_{31}\phi_1(t) - s_{32}\phi_2(t)\right)$$

$$\phi_3(t) = 10\sqrt{6a}\left(e^{-3at} - \frac{1}{\sqrt{8a}}\sqrt{2a}\,e^{-at} - \frac{1}{5\sqrt{a}}\left(6\sqrt{a}\,e^{-2at} - 4\sqrt{a}e^{-at}\right)\right)u(t)$$

$$= \left[10\sqrt{6a}\,e^{-3at} - 5\sqrt{6a}\,e^{-at} - 12\sqrt{6a}\,e^{-2at} + 8\sqrt{6a}\,e^{-at}\right]u(t)$$

$$= \left(3\sqrt{6a}\,e^{-at} - 12\sqrt{6a}\,e^{-2at} + 10\sqrt{6a}\,e^{-3at}\right)u(t)$$

Thus,

$$\phi_1(t) = \sqrt{2a}\, e^{-at}\ u(t)$$
$$\phi_2(t) = \sqrt{a}(6e^{-2at} - 4e^{-at})u(t)$$
$$\phi_3(t) = \sqrt{6a}\left(3e^{-at} - 12e^{-2at} + 10e^{-3at}\right)u(t)$$

Verify: $\|\phi_1\| = \int\limits_0^\infty \phi_1^2(t)dt = 2a\int\limits_0^\infty e^{-2at}dt = \frac{2a}{2a} = 1$

$$\|\phi_2\| = a\int\limits_0^\infty \left[36e^{-4at} + 16e^{-2at} - 48e^{-3at}\right]dt$$
$$= a\cdot\left[\frac{36}{4a} + \frac{16}{2a} - \frac{48}{3a}\right] = 9 + 8 - 16 = 1$$
$$\|\phi_3\| = 6a\int\limits_0^\infty \left[9e^{-2at} + 144e^{-4at} + 100e^{-6at} - 72e^{-3at} - 240e^{-5at} + 60e^{-4at}\right]dt$$
$$= 6\left[\frac{9}{2} + \frac{144}{4} + \frac{100}{6} - \frac{72}{3} - \frac{240}{5} + \frac{60}{4}\right] = 6\left(\frac{9}{2} + 36 + \frac{50}{3} - 24 - 48 + 15\right)$$
$$= 127 - 126 = 1$$

Students can verify that the set $\{\phi_1(t), \phi_2(t), \phi_3(t)\}$ is orthogonal.

Exercises

Activity

1.1 A discrete-time signal $s[n]$ is given by $s[n] = a^n u[n]$, where n is an integer. Sketch the signal for $0 < a < 1$ and $a > 1$.

1.2 A continuous-time signal $s(t)$ is given by $s(t) = t^2$ in the interval $0 < t < 10$. Sketch the waveform after multiplying $s(t)$ by unit impulses $p(t) = \sum_{n=0}^{10} \delta(t - nT)$.

1.3 For an arbitrary signal $s(t)$, let us define $s_e(t)$ and $s_0(t)$ as

$$s_e(t) = \frac{s(t) + s(-t)}{2} \quad \text{and} \quad s_0(t) = \frac{s(t) - s(-t)}{2}.$$

(a) Show that $s_e(t)$ is an even function and $s_0(t)$ is an odd function.

(b) Find $s_e(t)$ and $s_0(t)$ for $s(t)$ a unit-step function $u(t)$.

1.4 Find the average normalised power in the waveform $x(t) = A\cos 2\pi f_0 t$. Use time averaging and summation of spectral components.

1.5 Evaluate the trigonometric Fourier series expansion of the function $s(t)$ shown in Fig. P1.1. The function repeats after every T_0 seconds.

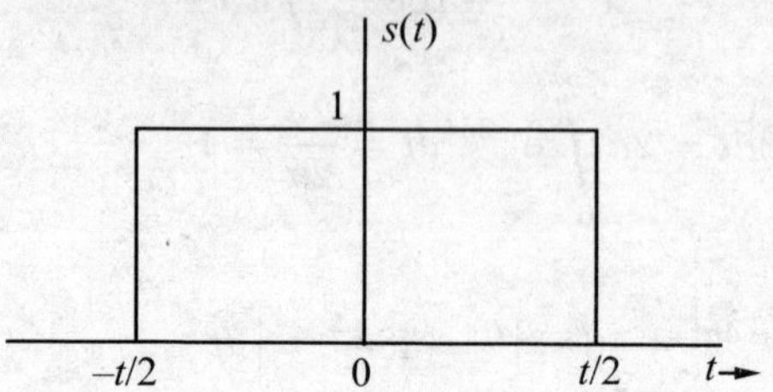

Fig. P1.1

1.6 Find the Fourier transform of $\cos 2\pi f_0 t$. Apply time-shift property to this result and find the Fourier transform of $\sin 2\pi f_0 t$.

1.7 Find the Fourier transform of a Gaussian pulse given as $s(t) = Ae^{-t^2/2a^2}, -\infty < t < \infty$.

1.8 Given that the Fourier transform of $s(t)$ is $S(f)$, find the Fourier transform of $\dfrac{s(t+T) - s(t)}{T}$.
From this result find the Fourier transform of $\dfrac{ds(t)}{dt}$.

1.9 Convolve the time function $s_1(t)$ with another time function $s_2(t)$ as shown in Fig. P1.2. Here $|\tau_2| > |\tau_1|$.

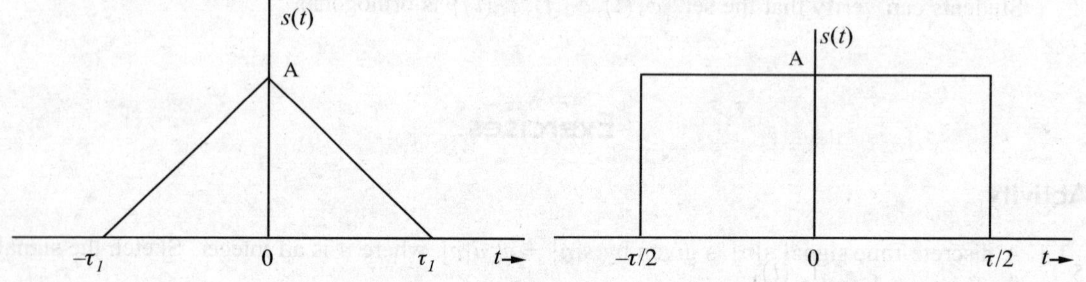

Fig. P1.2

1.10 A signal $s(t)$ has the spectrum bandlimited to f_M. By the convolution theorems, prove that the spectrum of $s^2(t)$ is bandlimited to $2f_M$.

1.11 Use convolution theorem to evaluate the integral

$$\int_{-\infty}^{\infty} \frac{\sin 4\pi\tau}{4\pi\tau} \frac{\sin \pi(t-\tau)}{\pi(t-\tau)} d\tau.$$

[Hint: Each time function is of the sinc type. So their spectrum is rectangular in shape. Convolve them in the frequency domain.]

1.12 Evaluate the following integrals.

(a) $\int_{-\infty}^{\infty} \delta(t-2)(t^3+4)dt$

(b) $\int_{--\infty}^{\infty} \delta(t-1)e^{-t^2}\, dt$

1.13 Find and sketch $S(f) = S_1(f) * S_2(f)$ where the spectra $S_1(f)$ and $S_2(f)$ are as shown in Fig. P.1.3

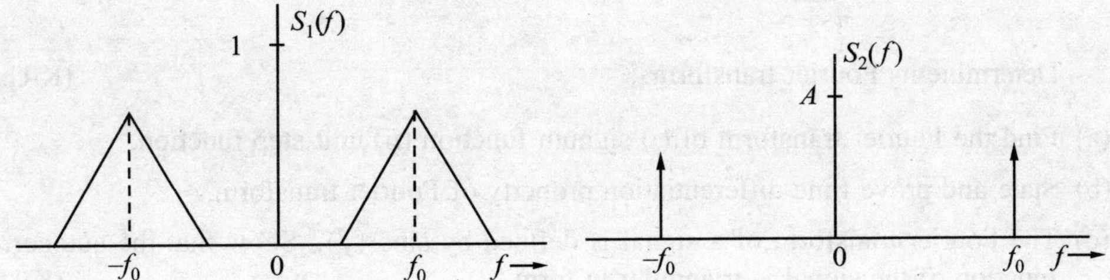

Fig. P1.3

1.14 Consider that the signal s(t) is represented in terms of orthonormal N-basis functions $\phi_i(t),\ i = 1, 2, \ldots, N$ as $s(t) = \sum_{i=1}^{N} s_i\phi_i(t)$, where

$$< \phi_i(t), \phi_j(t) >= \int_{-\infty}^{\infty} \phi_i(t), \phi_j(t)dt = \begin{cases} 1, & i = j \\ 0, & i \neq j \end{cases}$$

If $\hat{s}(t)$ does not form a complete set, then there will be an error whose energy is given by

$$E_e = \int_{-\infty}^{\infty} |s(t) - \hat{s}(t)|^2 dt$$

Find the coefficients $\{s_i\}$ of the expansion that minimises E_e.

1.15 For the periodic signal $s(t)$ with the period T_0, prove the following:

$c_n = \frac{2}{T_0}\int_0^{T_0/2} s(t)\cos(2\pi nt/T_0)dt$, if v(t) has even-symmetry, $c_n = -i\frac{2}{T_0}\int_0^{T_0/2} s(t)Sin(2\pi nt/T_0)dt$,

if v(t) has odd-symmetry where c_n is the coefficient of the Fourier series expansion of $s(t)$.

Review questions

1. (a) Find the Fourier transform of a gate function $\pi(t)$ defined by

$$\pi(t) = \begin{cases} 1 & |t| < 1/2 \\ 0 & |t| > 1/2. \end{cases}$$

(b) If a function $g(t)$ has its Fourier transform $G(\omega)$, then prove that $g(t-t_0)$ has its Fourier transform given by $G(\omega)e^{-j\omega t_0}$.

(c) Define convolution of two functions $f_1(t)$ and $f_2(t)$ and prove that $f_1(t) * f_2(t) = F_1(\omega)F_2(\omega)$, where $F_1(\omega)$ and $F_2(\omega)$ are the Fourier transforms of $f_1(t)$ and $f_2(t)$ respectively.

(d) A unit train of impulses $f(t)$ is defined as

$$f(t) = \sum_{n=-\infty}^{\infty} \delta(t - nT_0).$$

Determine its Fourier transform. (K.U. 2001)

2. (a) Find the Fourier transform of (i) signum function (ii) unit-step function.

(b) State and prove time differentiation property of Fourier transform.

(c) The Fourier transform of a signal is defined by sinc (f). Show that the autocorrelation function of the signal is triangular in form. (K.U.2001)

3. (a) Define power spectral density function (PSDF) and autocorrelation function (ACF) of a power signal. Find ACF and PSDF of the sinusoid

$$x(t) = A\cos(\omega_0 t + \theta).$$ (WBUT 2004)

4. (a) Define Fourier transform pair and obtain a relation between autocorrelation function and power spectral density of a periodic signal.

(b) Find the highest value of autocorrelation of a function $f(t)$ where

$$f(t) = 2\cos t + 3\cos 3t + 4\sin 4t.$$

(c) Explain the correlation between random variables. [Ber.U.1998)

5. (a) The signal $s(t) = 5\,\text{Cos}\,(2\pi \times 10^7 t)$ Volts is received with a delay of 1×10^{-9} second. What is the phase difference between the transmitted and received signal?

6. For the two signals $v(t)$ and $w(t)$ having odd symmetry, show that the convolution $v(t) * w(t)$ will also have odd-symmetry.

7. The convolution sum $y[n]$ of two sequences $x[n]$ and $h[n]$ is given by

$$y[n] = \sum_{k=0}^{\infty} x[k]h[n-k]$$

Find the deconvolution rule to find x[n], supposing that y[n] and h[n] are given.

8. Let $h(t)$ be a signal with Fourier transform $H(f)$. If $h(t)$ is time translated and scaled as

$$h_{a,\tau}(t) = h\left(\frac{t-\tau}{a}\right),$$

find $H_{a,\tau}(f)$ in terms of $H(f)$.

9. The signal $s(t)$ has the energy E_s. Find the energy of each of the following signals.

$$s(-t), s(t-t_d), -s(at), s\left(\frac{t-\tau}{a}\right), as(t)$$

Also, comment on signal energy for the time-scaled signal.

10. (a) If two signals are multiplied in time domain, what will be the effect on the output bandwidth?

 (b) Two identical signals are multiplied in the frequency domain. Discuss the effect on the duration of the output signal. Explain (a) and (b) with reasoning and analysis.

Multiple-choice questions

1. The energy of the signal $x(t) = 2e^{-t} - 6e^{-2t}, t > 0$ is

 (a) 3 joules (b) 19 joules (c) 16 joules (d) 6 joules

2. For the signal $x(t) = \sin \pi t$, the signal power is

 (a) 0.5 W (b) 1 W (c) 0.25 W (d) none

3. Consider the signal $\text{sinc}(t) = \dfrac{\sin \pi t}{\pi t}$.

 (i) Signal has even symmetry (iii) Signal has equi-spaced zero-crossings
 (ii) Signal has unit area (iv) It is an energy signal

 (a) Only (i) and (ii) correct (b) Only (ii) and (iii) correct

 (c) Only (iii) and (iv) correct (d) All the four statements are correct.

4. The Fourier transform of a sinc pulse is

 (a) uniform (b) rectangular (c) Gaussian (d) Rayleigh

5. An odd function $f(t)$ satisfies the Dirichlet's conditions. Then $f(t)$ can be expanded in a Fourier series having

 (a) only sine-terms (b) only cosine-terms (c) both sine and a constant terms

 (d) both cosine and a constant term.

6. The Fourier transform $F(j\omega)$ of an arbitrary signal has the property

 (a) $F(j\omega) = F(-j\omega)$ (b) $F(j\omega) = -F(-j\omega)$

 (c) $F(j\omega) = F^*(-j\omega)$ (d) $F(j\omega) = -F^*(-j\omega)$

7. Let $F(j\omega) = \frac{1}{j\omega} + \pi\delta(\omega)$. The function $f(t)$ equals

 (a) $\sin \omega t$ (b) $\cos \omega t$ (c) $\text{sgn}(t)$ (d) $u(t)$.

8. If $F(\omega)$ is the Fourier transform of a function $f(t)$, then $F(0)$ is

 (a) $\int_{-\infty}^{\infty} f(t)dt$ (b) $\int_{-\infty}^{\infty} |f(t)|^2 dt$ (c) $\int_{-\infty}^{\infty} |tf(t)|^2 dt$ (d) $\int_{-\infty}^{\infty} tf(t)dt$

9. The autocorrelation function $R_x(\tau)$ satisfies the property

   (a) $R_x(\tau) = -R_x(-\tau)$ (b) $R_x(\tau) = R_x = R_x(-\tau)$

   (c) $R_x(\tau) \geq R_x(0)$ (d) $R_x(\tau) \geq 1$

10. A rectangular pulse $f(t)$ exists for $0 \leq t \leq T$. Then $f(t) * f(t)$ is
    (a) a rectangular shape of width $2T$ centred at $t = 0$
    (b) a rectangular shape of width $2T$ centred at $t = T$
    (c) a triangle of width $2T$ peaked at $t = 0$
    (d) a triangle of width $2T$ peaked at $t = T$.

11. The convolution $y(t) = x(t) * x(-t)$
    (i) is nothing but the autocorrelation of $x(t)$ (ii) is always even symmetric
    (iii) has maximum at the origin.

    (a) Only (i) & (ii) correct (b) Only (ii) & (iii) correct
    (c) Only (i) & (iii) correct (d) All the three are correct

2 Signal Transmission through Linear Systems and Filters

Many physical systems are involved in transmitting a signal from the source to the destination. The signal that we receive may not be of the same form as the original signal; some kind of distortion takes place while transmitting, propagating, or receiving the signal. Filtering operations on signals employed in transmission networks also purposely distort a waveform. We will consider here the characterisation of systems and filters and their effects on signals and noise.

2.1 Linear Time-Invariant System

Consider a system as shown in Fig. 2.1 with an external input (or excitation) $r(t)$ and an output (or response) $s(t)$. The signal applied to the input of the system can be described either as a time-domain signal $r(t)$ or in terms of its spectrum $R(f)$. The system is said to be linear if the principle of superposition applies, that is, a linear combination of inputs, $a_1 r_1(t) + a_2 r_2(t) + \cdots +$, results in the output $a_1 s_1(t) + a_2 s_2(t) + \cdots$. Let us denote by $\widehat{T}$ the system function or operator which gives the functional relationship between the input and output such that

$$\widehat{T} r(t) = s(t). \tag{2.1}$$

The principle of superposition states that

$$\widehat{T}[a_1 r_1(t) + a_2 r_2(t) + \cdots] = a_1 s_1(t) + a_2 s_2(t) + \cdots \tag{2.2}$$

A system is said to be time-invariant or stationary if the characteristics of the system do not change with time. In other words, for a time-shifted input $r(t - t_0)$, the time-invariant system produces a time-shifted output $s(t - t_0)$, the shape of the waveform being unchanged. Thus,

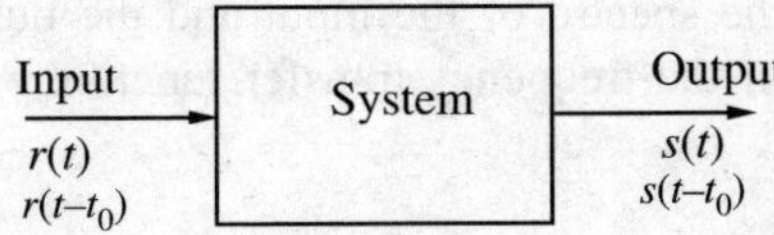

Fig. 2.1 *Linear system*

$$\widehat{T}r(t - t_0) = s(t - t_0). \tag{2.3}$$

The system that obeys both linearity and stationarity is termed a linear time-invariant system or LTI system for short. It is assumed that there is no internal stored energy at the time the input is applied so that the output is due entirely to the input.

2.1.1 Impulse response and frequency transfer function

In time domain, an LTI system is characterised by its impulse response $h(t)$ defined as the response for an input of unit impulse $\delta(t)$. Thus by (2.1),

$$h(t) = \widehat{T}\delta(t) \tag{2.4}$$

and by (2.3),

$$h(t - t_0) = \widehat{T}\delta(t - t_0). \tag{2.5}$$

Since the filtering property of the delta function for a continuous input $r(t)$ reads as $r(t) = r(t) * \delta(t)$, so the output is

$$\begin{aligned} s(t) = \widehat{T}r(t) &= \widehat{T}\int_{-\infty}^{\infty} r(\tau)\delta(t-\tau)d\tau \\ &= \int_{-\infty}^{\infty} r(\tau)\widehat{T}\delta(t-\tau)d\tau \\ &= \int_{-\infty}^{\infty} r(\tau)h(t-\tau)d\tau \qquad \text{(by (2.5))} \\ &= r(t) * h(t). \end{aligned} \tag{2.6}$$

Thus the response $s(t)$ of the LTI system to an arbitrary input $r(t)$ is obtained by convolving $r(t)$ with its impulse response $h(t)$. However, for higher order systems the time domain convolution becomes complicated. Since the convolution in time domain transforms into multiplication in the frequency domain, a clear picture of system response is obtained in the frequency domain analysis.

By taking the Fourier transform of (2.6) and applying the convolution property, we get

$$S(\omega) = R(\omega)H(\omega). \tag{2.7}$$

Here $R(\omega)$ and $S(\omega)$ represent the spectra of the input and the output, respectively. The quantity $H(\omega)$ is the frequency response or the frequency transfer function of the system and is given as

$$H(\omega) = \mathfrak{F}[h(t)] = \frac{S(\omega)}{R(\omega)}, \tag{2.8}$$

provided that $R(\omega) \neq 0$ for all ω. In general, $H(\omega)$ is complex and may be written as

$$H(\omega) = |H(\omega)|e^{j\theta(\omega)}, \tag{2.9}$$

where $|H(\omega)|$ is the amplitude response and $\theta(\omega)$ is the phase response of the system. Expressing (2.7) in polar form,

$$\begin{aligned} |S(\omega)|e^{j\theta_2(\omega)} &= |R(\omega)|e^{j\theta_1(\omega)}|H(\omega)|e^{j\theta_s(\omega)} \\ &= |R(\omega)||H(\omega)|e^{j[\theta_1(\omega)+\theta_s(\omega)]}, \end{aligned} \tag{2.10}$$

where θ_1, θ_2 and θ_s are the phase angles associated with $R(\omega)$, $S(\omega)$ and $H(\omega)$, respectively.

That is, the amplitude spectrum of the output is the product of the amplitude spectra of the input and the system function. Similarly, the phase spectrum of the output is the sum of the phase spectra of the input and the system function.

Example 2.1

An electrical circuit consists of a resistor R in parallel with an inductor L, excited by a current source $i(t)$ as shown in Fig. 2.2. Considering $i(t)$ as the input and $v(t)$ as the response, find the system function.

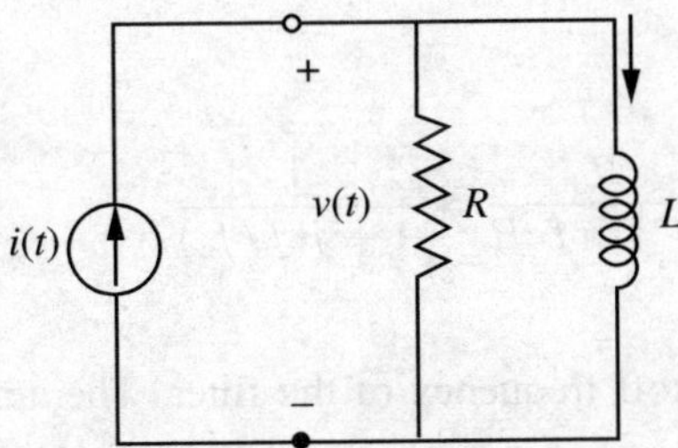

Fig. 2.2 *Electrical circuit to explain the system function*

Solution

The problem simply boils down to finding the input impedance of the circuit which is the system function. Thus,

$$H(\omega) = R \parallel (j\omega l) = \frac{j\omega LR}{R + j\omega L},$$

where ω is the frequency of the input current in rad/second. Clearly, the amplitude spectrum

$$|H(\omega)| = \frac{\omega LR}{\sqrt{R^2 + (\omega L)^2}} \tag{2.11}$$

and the phase spectrum

$$\theta(\omega) = \tfrac{\pi}{2} - \tan^{-1}(\omega L/R). \tag{2.12}$$

2.1.2 Response of first-order systems

A mathematical model is used to relate the output $s(t)$ to the input $r(t)$ of an LTI system. The input–output relationship of an LTI system can be described by a linear differential equation of the form

$$a_n \frac{d^n s(t)}{dt^n} + a_{n-1} \frac{d^{n-1} s(t)}{dt^{n-1}} + \cdots + a_1 \frac{ds(t)}{dt} + a_0 s(t)$$

$$= b_m \frac{d^m r(t)}{dt^m} + b_{m-1} \frac{d^{m-1} r(t)}{dt^{m-1}} + \cdots + b_1 \frac{dr(t)}{dt} + b_0 r(t). \tag{2.13}$$

Here the a's and b's are constant coefficients which depend on the system elements. The value n gives the order of the system. For first order system, $n = 1$ and the differential equation (2.13) reduces to

$$a_1 \frac{ds(t)}{dt} + a_0 s(t) = b_0 r(t). \tag{2.14}$$

An RC lowpass filter is an example of a first order system. Figure 2.3 shows the RC circuit with the input $r(t)$ and the output $s(t)$ taken across the capacitor. The frequency transfer function $H(f)$ is given by

$$H(f) = \frac{\frac{1}{j2\pi fc}}{R + \frac{1}{j2\pi fc}} = \frac{1}{1 + j2\pi fcR} = \frac{1}{1 + j(f/f_c)}, \tag{2.15}$$

where $f_c = \frac{1}{2\pi RC}$ is the 3 dB cutoff frequency of the filter. The amplitude $|H(f)|$ and phase $\phi(f)$ are given by

$$\left.\begin{aligned} |H(f)| &= \frac{1}{\sqrt{1+(f/f_c)^2}} \\ \phi(f) &= -\tan^{-1}(f/f_c), \end{aligned}\right\} \tag{2.16}$$

and are plotted in Fig. 2.4. To illustrate the frequency-selective property of this filter, let us consider an input $r(t)$ having frequency range $f << f_C$. Since $|H(f)| \approx 1$ and $\phi(f) \approx 0°$, the output signal

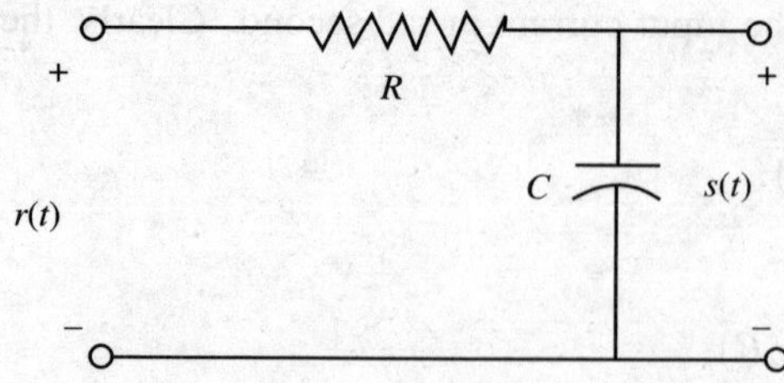

Fig. 2.3 *RC lowpass filter*

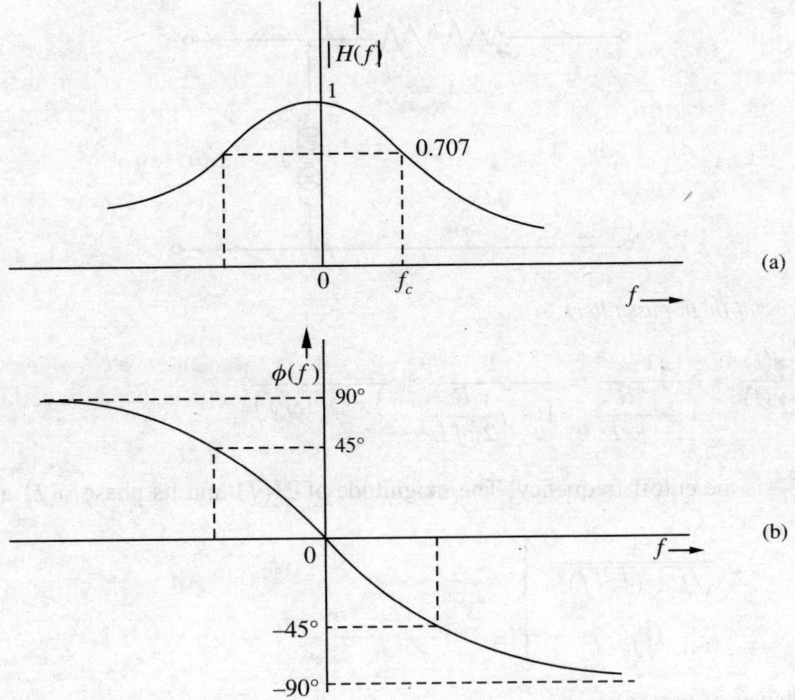

Fig. 2.4 *Transfer function of the RC filter : (a) magnitude, (b) phase.*

maintains its shape. As the frequency of the input signal increases, the transmitted signal through the filter becomes weak in amplitude in addition to nonzero phase change. When the highest frequency component in the spectrum of the input signal is much higher than f_c, the capacitor offers low impedance and the transfer function reduces to a low value. Note that at $f = f_c$, the magnitude of the transfer function drops to $1/\sqrt{2} = 0.707$ of its low-frequency value.

Example 2.2

For the circuit of Fig. 2.5,

(a) find the transfer function $H(f)$,
(b) sketch $|H(f)|$ versus f,
(c) what function is this circuit performing?

Solution

Here $\dfrac{S(\omega)}{R(\omega)} = \dfrac{j\omega L}{R + j\omega L}$.

For an input of the form $r(t) = e^{j\omega t}$,

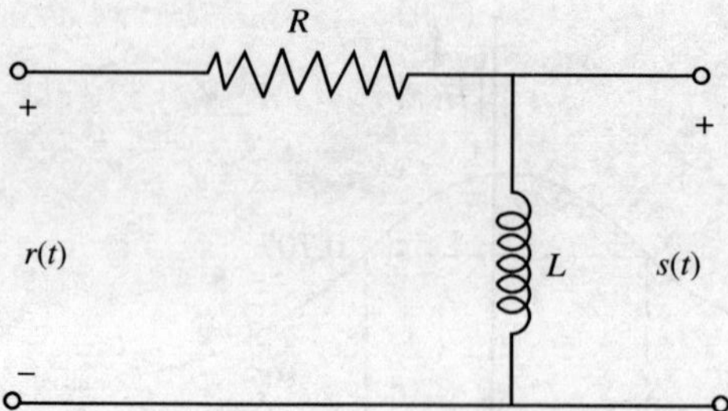

Fig. 2.5 *RL circuit (high-pass filter)*

$$H(f) = \frac{s(t)}{r(t)} = \frac{1}{1 + \dfrac{R}{j\omega L}} = \frac{1}{1 - j\dfrac{R}{2\pi f L}} = \frac{1}{1 - j(f_c/f)}, \tag{2.17}$$

where $f_c = \frac{R}{2\pi L}$ is the cutoff frequency. The magnitude of $H(f)$ and its phase $\phi(f)$ are given by

$$\left.\begin{aligned} |H(f)| &= \frac{1}{\sqrt{1 + (fc/f)^2}} \\ \phi(f) &= \tan^{-1}(f_c/f) \end{aligned}\right\}, \tag{2.18}$$

which is the answer of part (a).

The plot of $|H(f)|$ as a function of frequency is shown in Fig. 2.6.

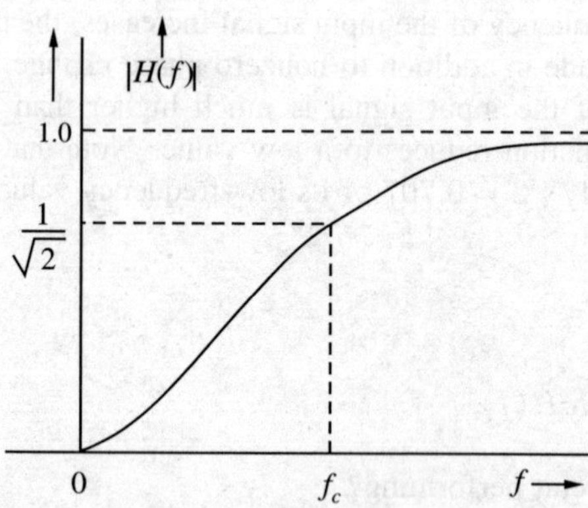

Fig. 2.6 *Plot of* $|H(f)|$ *vs.* *frequency of Fig. 2.5*

Example 2.3

The input–output relation in a transmission channel is described by

$$s(t) = ar(t) + r(t - T) + ar(t - 2T).$$

Show that the transfer function of the channel is given by

$$H(f) = (1 + 2a/\cos\omega T)e^{-j\omega T}.$$

Solution

Taking the Fourier transform on both sides of the input–output relation, we get

$$S(f) = aR(f) + R(f)e^{-j\omega T} + aR(f)e^{-j2\omega T}$$

$$\text{or,}\quad H(f) = \frac{S(f)}{R(f)} = a + e^{-j\omega T} + ae^{-j2\omega T}$$

$$= e^{-j\omega T}(ae^{j\omega T} + 1 + ae^{-j\omega T})$$

$$= (1 + 2a\cos\omega T)e^{-j\omega T}.$$

Students are advised to plot the magnitude and the phase spectrum and comment on the result.

Power and energy transfer through linear systems

Consider an LT1 system shown in Fig. 2.7 where $r(t)$ be any signal with spectrum $R(f)$. The output $s(t)$ is obtained by convolving $r(t)$ with the impulse response $h(t)$ of the system. That is, $s(t) = r(t) * h(t)$.

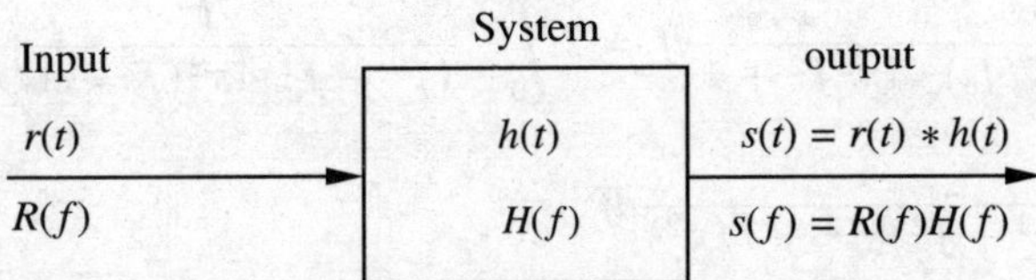

Fig. 2.7 *Input–output relation in both time and frequency domain of an LT1 system*

Taking the Fourier transform, the output spectrum $S(f)$ is obtained as the input spectrum $R(f)$ multiplied by the system transfer function $H(f)$. Thus,

$$S(f) = H(f)R(f). \tag{2.19}$$

Now, consider that $r(t)$ is an energy signal having energy spectral density

$$G_i(f) = |R(f)|^2. \tag{2.20}$$

Then output is also an energy signal whose spectral density is determined in terms of the input density function and the system function. That is,

$$G_0(f) = |S(f)|^2 = |R(f)|^2|H(f)|^2. \tag{2.21}$$

The total output energy E_s is given by

$$E_s = \int_{-\infty}^{\infty} |H(f)|^2 |R(f)|^2 df. \tag{2.22}$$

Example 2.4

The input to an RC lowpass filter is $r(t) = e^{-2t}u(t)$. Find the ratio of the output energy to the input energy in terms of f/f_c, where f_c is the 3 dB cutoff frequency of the filter.

Solution

The total energy of the input signal is

$$E_r = \int_{-\infty}^{\infty} |r(t)|^2 dt = \int_0^{\infty} (e^{-2t})^2 dt = \int_0^{\infty} e^{-4t} dt = \frac{1}{4}.$$

The energy spectral density of the input signal is

$$G_i(f) = |R(f)|^2 = \left|\frac{1}{j2\pi f + 2}\right|^2 = \frac{1}{4\pi^2 f^2 + 4},$$

and the magnitude squared of the system function

$$|H(f)|^2 = \frac{1}{1 + (f/f_c)^2}.$$

Thus total output energy

$$E_s = \int_{-\infty}^{\infty} \frac{1}{1 + (f/f_c)^2} \frac{1}{4\pi^2 f^2 + 4} df = 2\int_0^{\infty} \frac{f_c^2}{(f^2 + f_c^2)4\pi^2(f^2 + 1/\pi^2)} df$$

$$= \frac{f_c^2}{2\pi^2} \int_0^{\infty} \frac{1}{(f^2 + f_c^2)(f^2 + 1/\pi^2)} df$$

$$= \frac{(1/\pi^2 - f_c^2) fc^2}{2\pi^2} \int_0^{\infty} \left(\frac{1}{f^2 + f_c^2} - \frac{1}{f^2 + 1/\pi^2}\right) df$$

$$= \frac{(1/\pi^2 - f_c^2) f_c^2}{2\pi^2} \left[\frac{1}{f_c} \tan^{-1} \frac{f}{f_c} - \pi \tan^{-1} \pi f\right]_0^{\infty}$$

$$= \frac{(1/\pi^2 - f_c^2)^2 f_c^2}{2\pi^2} \left[\frac{1}{f_c} . \frac{\pi}{2} - \pi \frac{\pi}{2}\right] = \frac{f_c^2(\frac{1}{\pi^2} - f_c^2)}{4\pi} \frac{(1 - \pi f_c)}{f_c}$$

$$= \frac{f_c(1 - \pi f_c)^2 (1 + \pi f_c)}{4\pi^3}.$$

So, the ratio

$$\frac{E_s}{E_i} = \frac{f_c}{\pi^3}(1 - \pi f_c)^2 (1 + \pi f_c).$$

2.2 Signal Distortion

In all communication systems, the channel constitutes the link between the information source and the destination. The channel may consists of a pair of wires, coaxial cable, fibre optic cable, air, etc. All such channels have distributed resistance, capacitance, and inductance that reduce the strength of the output signal along with the possibility of altering its shape. In other words, the signal gets distorted while passing through the channel. In communication systems, a distortionless channel is often desired.

A signal is undistorted if its 'shape' is preserved. We know that the shape remains unchanged if the signal is delayed by finite time, or its strength is altered, or both. Thus the signal $s(t)$ is said to be a replica of another signal $r(t)$ provided that

$$s(t) = Ar(t - t_d), \tag{2.23}$$

is satisfied for constants A and t_d, where A is the gain and t_d is the delay. So, a distortionless channel is characterised by a system transfer function of the form

$$H(f) = Ae^{-j\omega t_d}, \tag{2.24}$$

meaning that $H(f)$ must have a constant amplitude response $|H(f)| = A$ (independent of frequency) and a negative linear phase-shift $\angle H(f) = -2\pi f t_d$. However, these two conditions can only be satisfied approximately in practice, and so signal transmission through any medium is always associated with some amount of signal distortion. The distortion can be categorised as (i) linear distortion and (ii) nonlinear distortion. Amplitude and phase distortion fall in the first category where the channels are assumed linear. Nonlinear distortion is significant at higher transmission frequencies; this occurs when the system includes nonlinear elements.

2.2.1 Amplitude distortion

Amplitude distortion results when $|H(f)| \neq A$, which means that the output frequency components are not in proper proportion. Excess attenuation or amplification results in certain range of frequencies in the signal spectrum. The amplitude distortion is described in the frequency domain by assigning amplitude variation $A(f)$ with frequency in the expression for $H(f)$.

Suppose that a transmission system has the transfer function shown in Fig. 2.8. Observe that $|H(f)| = 1$ for $f_1 \leq f \leq f_2$. Over this range amplitude distortion is absent. There is amplitude distortion for $f < f_1$ and $f > f_2$ as the low and high frequency components are attenuated.

2.2.2 Phase distortion

For a distortionless transmission channel, the phase characteristic is linear as shown in Fig. 2.9.

Since the phase variation is a result of time delay, the higher frequency components must undergo proportionately higher phase-shifts to obtain the same time delay. From (2.24), the phase of the transfer function is obtained as

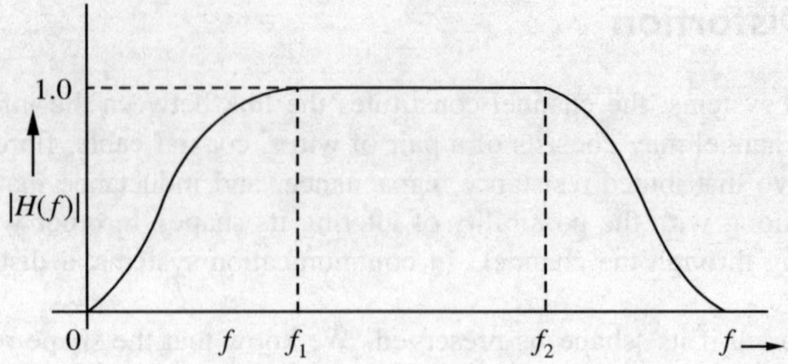

Fig. 2.8 *Transfer function of a model system*

$$\phi(f) = -\omega t_d = -2\pi f t_d, \tag{2.25}$$

which indicates that the time delay t_d is the negative of the slope of the phase response. That is,

$$t_d(f) = -\frac{d\phi}{d\omega} = -\frac{1}{2\pi f}\angle H(f). \tag{2.26}$$

When the slope is not constant, the time delay varies with frequency (the portion of the phase response beyond f_1 of Fig. 2.9) and distortion results.

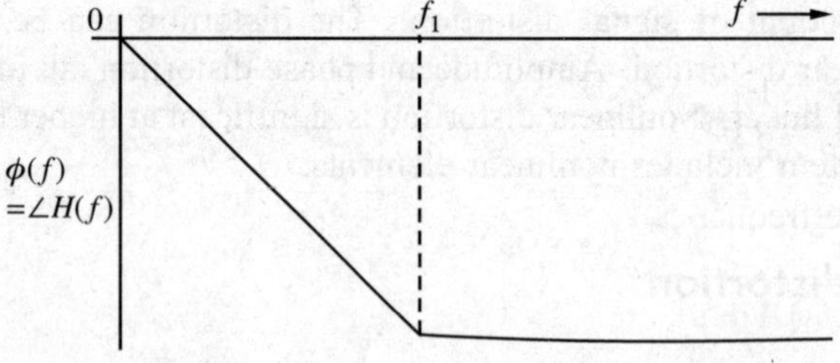

Fig. 2.9 *Phase characteristics of a model system. Up to frequency f_1, the phase characteristic is linear.*

The departure from linear phase can be characterised by variation of the slope of the phase characteristic and that of a line from origin to the point. Let us look at the phase characteristic of Fig. 2.10. We define the phase delay t_p and group delay t_g as

$$t_p(f) = -\frac{\phi(f)}{2\pi f}; \quad t_g(f) = -\frac{1}{2\pi}\frac{d\phi(f)}{df}. \tag{2.27}$$

Both the delays, t_p and t_g, equal the time delay t_d of the input signal for a distortionless channel. The nonlinear phase characteristic of Fig. 2.10 can be approximated as piecewise linear. Taking such an approximation over a relatively narrow band around f_0, the phase $\phi(f)$ can be written as

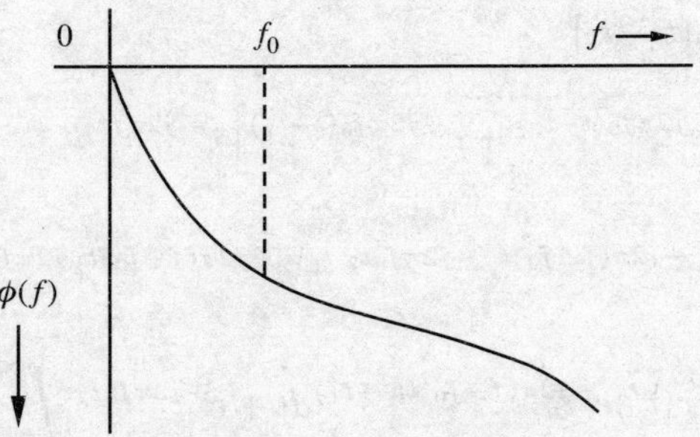

Fig. 2.10 *Nonlinear phase characteristics*

$$\phi(f) = \phi(f_0) + (f - f_0)\phi'(f_0)$$

$$= -2\pi[f_0 t_p(f_0) + (f - f_0)t_g(f_0)]. \tag{2.28}$$

The system function of the arbitrary channel can then be put as

$$H(f) = A\exp[-j2\pi\{f_0 t_p(f_0) + (f - f_0)t_g(f_0)\}]. \tag{2.29}$$

Now suppose that the input to this channel is a modulated wave of the form $r(t)\cos 2\pi f_0 t$ whose spectrum is

$$\mathfrak{F}[r(t)\cos 2\pi f_0 t] = \frac{1}{2}[R(f - f_0) + R(f + f_0)] \tag{2.30}$$

Considering the positive frequency component, the spectrum of the output is

$$S(f) = \frac{1}{2}R(f - f_0)A\exp[-j2\pi\{f_0 t_p(f_0) + (f - f_0)t_g(f_0)\}]. \tag{2.31}$$

Now,

$$\mathfrak{F}[r(t - t_0)\cos 2\pi f_0 t] = \frac{1}{2}\int_{-\infty}^{\infty} r(t - t_0)\left[e^{j2\pi f_0 t} + e^{-j2\pi f_0 t}\right]e^{-j2\pi f t}dt$$

$$= \frac{1}{2}\int_{-\infty}^{\infty} r(t - t_0)[e^{-j2\pi(f - f_0)t} + e^{-j2\pi(f + f_0)t}dt$$

$$= \frac{1}{2}\int_{-\infty}^{\infty} r(t')[e^{-j2\pi(f - f_0)(t' + t_0)} + e^{-j2\pi(f + f_0)(t' + t_0)})]dt'$$

$$= \frac{1}{2}\left[e^{-j2\pi(f - f_0)t_0}\int_{-\infty}^{\infty} r(t')e^{-j2\pi(f - f_0)t'}dt' + e^{-j2\pi(f + f_0)t_0}\int_{-\infty}^{\infty} r(t')e^{-j2\pi(f + f_0)t'}dt'\right]$$

$$= \frac{1}{2}\left[R(f - f_0)e^{-j2\pi(f - f_0)t_0} + R(f + f_0)e^{-j2\pi(f + f_0)t_0}\right].$$

Now, $\mathfrak{F}[r(t-t_1)\cos 2\pi f_0(t-t_2]$

$$= \frac{1}{2}\int_{-\infty}^{\infty} r(t-t_1)\left[e^{j2\pi f_0(t-t_2)} + e^{-j2\pi f_0(t-t_2)}\right] e^{-j2\pi f t} dt$$

$$= \frac{1}{2}\int_{-\infty}^{\infty} r(t-t_1)\left[e^{-j2\pi(f-f_0)t}e^{-j2\pi f_0 t_2} + e^{-j2\pi(f+f_0)t}e^{j2\pi f_0 t_2}\right] dt$$

$$= \frac{1}{2}\left[e^{-j2\pi f_0 t_2}\int_{-\infty}^{\infty} r(t')e^{-j2\pi(f-f_0)(t'+t_1)}dt + e^{+j2\pi f_0 t_2}\int_{-\infty}^{\infty} r(t')e^{-j2\pi(f+f_0)(t'+t_1)}dt\right]$$

$$= \frac{1}{2}e^{-j2\pi f_0 t_2}e^{-j2\pi(f-f_0)t_1}R(f-f_0) + \frac{1}{2}e^{j2\pi f_0 t_2}e^{-j2\pi(f+f_0)t_1}R(f+f_0). \tag{2.32}$$

The first term of (2.32) is very similar to $S(f)$ of (2.31). Thus the output of the channel can be written as

$$s(t) = Ar[t - t_g(f_0)]\cos 2\pi f_0[t - t_p(f_0)]. \tag{2.33}$$

Equation (2.33) indicates that the carrier has been delayed by $t_p(f_0)$. This time delay is the phase delay. The time delay $t_g(f_0)$ between the envelopes of the input and the output is called the envelope or group delay of the channel. Both the group and the phase delays are evaluated at the carrier frequency f_0.

The time-domain description of the linear distortion discussed above is due to the spreading of pulse. This causes overlapping with a neighbouring pulse thereby distorting their spectra; this is very much undesirable in time-division-multiplexed (TDM) systems. Example 2.5 explains this adverse effect.

Example 2.5

A communication channel behaves as cosine filter having transfer function

$$H(f) = (1 + a\cos\omega T)e^{-j\omega t_0}$$

in the passband. A pulse $p(t)$ is applied at its input. Find the channel output.

Solution

$$H(f) = (1 + a\cos\omega T)e^{-j\omega t_0}$$

$$= \left[1 + \frac{a}{2}(e^{j\omega T} + e^{-j\omega T})\right]e^{-j\omega t_0}$$

$$= e^{-j\omega t_0} + \frac{a}{2}e^{-j\omega(t_0-T)} + \frac{a}{2}e^{-j\omega(t_0+T)}.$$

If the input pulse $p(t)$ is bandlimited, the output is given by

$$S(f) = e^{-j\omega t_0}P(f) + \frac{a}{2}e^{-j\omega(t_0-T)}P(f) + \frac{a}{2}e^{-j\omega(t_0+T)}P(f)$$

so that

$$s(t) = p(t - t_0) + \frac{a}{2}p(t - t_0 + T) + \frac{a}{2}p(t - t_0 - T).$$

Here $P(f)$ is the Fourier transform of $p(f)$. The output then consists of the delayed input $p(t)$ and its two time shifted echoes. Figure 2.11 displays the effects of these echoes in the time domain.

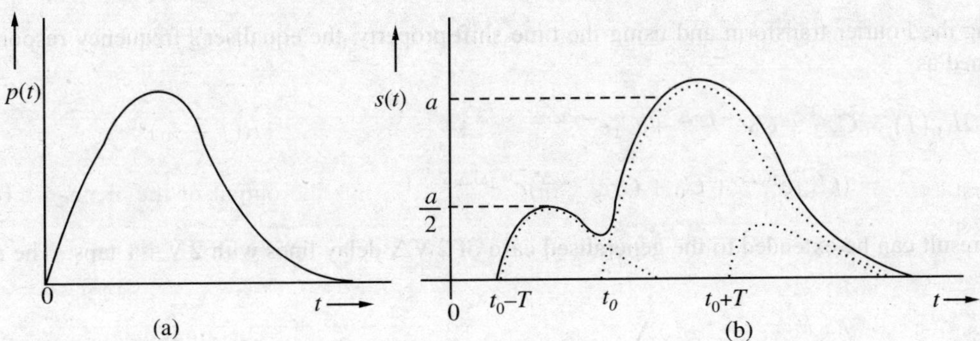

Fig. 2.11 *(a) The pulse $p(t)$. (b) The pulse gets distorted and spreaded as it passes through the channel.*

2.2.3 Equalisation to remove distortion

We have observed that a linear distortion by the transmission medium attenuates and disperses the pulse train. The distortion is primarily due to the attenuation of high frequency components, since $|H(f)|$ falls off at higher frequencies. An equaliser is a network cascaded with a practical channel that restores the high frequency components, thereby compensating for the distorted pulses. This can be achieved if the overall transfer function of the transmission channel and the equaliser satisfies (2.24). Thus,

$$H_{ch}(f)H_{eq}(f) = Ae^{-j\omega t_d}, \tag{2.34}$$

where $H_{ch}(f)$ and $H_{eq}(f)$ are the transfer functions of the channel and the equaliser, respectively. The equaliser transfer function compensates the channel distortion and is thus the inverse of the channel transfer function. That is,

$$H_{eq}(f) = \frac{Ae^{-j\omega t_d}}{H_{ch}(f)}. \tag{2.35}$$

Such an exact equaliser design is hardly possible but can be approximated so as to keep the distortion at an acceptable level.

Recent developments in the design of equalisers have been mainly with filters that contain only feed-forward elements (transversal equaliser) or both feed-forward and feedback elements (decision feedback equaliser). They can be preset or adaptive in nature.

To begin with, consider the transversal filter with a delay line of total delay 2Δ equipped with taps at each end and the middle, as shown in Fig. 2.12. The tap outputs are passed through adjustable tap gains C_{-1}, C_0 and C_1. The outputs are then summed to yield the final output.

If one sets the tap gains $C_0 = 1, C_{-1} = C_1 = 0$, then the filter output will be the input signal delayed by Δ, that is, $r(t - \Delta)$. The output $s(t)$ of Fig. 2.12 is

$$s(t) = C_{-1}r(t) + C_0 r(t - \Delta) + C_1 r(t - 2\Delta). \tag{2.36}$$

Taking the Fourier transform and using the time-shift property, the equaliser's frequency response is obtained as

$$\begin{aligned} H_{\text{eq}}(f) &= C_{-1} + C_0 e^{-j\omega\Delta} + C_1 e^{-j2\omega\Delta} \\ &= (C_{-1}e^{j\omega\Delta} + C_0 + C_1 e^{-j\omega\Delta})e^{-j\omega\Delta}. \end{aligned} \tag{2.37}$$

(The result can be extended to the generalised case of $2N\Delta$ delay lines with $2N + 1$ taps. The result is

$$H_{\text{eq}}(f) = \left(\sum_{n=-N}^{N} C_n e^{-j\omega n\Delta} \right) e^{-j\omega\Delta}. \tag{2.38}$$

The coefficients C_n's are calculated by expanding (2.35) in a Fourier series and comparing with (2.38). A detailed discussions on equalisers can be found in chapter 13.

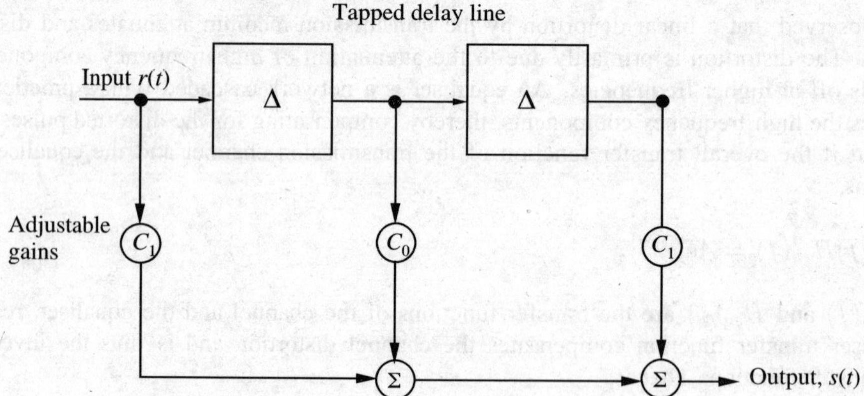

Fig. 2.12 *Tapped delay line equaliser*

2.2.4 Multipath effect

Multipath distortion takes place when there are two or more propagation paths of different delays between the transmitter and the receiver. For example, in radio systems, signals may be received by direct path and by reflections from other objects such as hills. Similarly, distortion is observed in a cable link where the signal is received in the form of direct wave in addition to reflections due to impedance mismatching along the path. In such cases, the transmission channel can be modelled as a number of channels in parallel with different time delays and attenuations. Let us consider the multipath distortion in a two-path system as depicted in Fig. 2.13. The upper path is the direct path which introduce a delay $t_1 = t_d$ and gain A_1, and the lower path corresponds to that of reflection. The gain and the time delay in this path are assumed to be A_2 and $(t_d + t_0)$, respectively. The channel output is then

$$s(t) = A_1 r(t - t_d) + A_2 r(t - t_2). \tag{2.39}$$

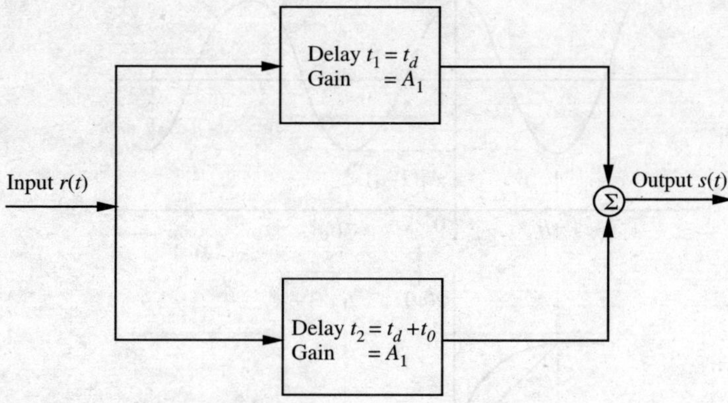

Fig. 2.13 *Two-path transmission*

The transfer function of such a channel is obtained as

$$\begin{aligned} H(\omega) &= A_1 e^{-j\omega t_1} + A_2 e^{-j\omega t_2} \\ &= A_1 e^{-j\omega t_1}\left(1 + \frac{A_2}{A_1} e^{-j\omega t_0}\right) \\ &= A_1 e^{-j\omega t_1}\left(1 + a e^{-j\omega t_0}\right). \end{aligned} \tag{2.40}$$

The magnitude of $H(\omega)$ is

$$\begin{aligned} |H(\omega)| &= A_1 \left[(1 + a\cos\omega t_0)^2 + a\sin^2\omega t_0\right]^{1/2} \\ &= A_1\sqrt{1 + a^2 + 2a\cos\omega t_0} \end{aligned}$$

and the phase is

$$\phi(\omega) = -\left[\omega t_d + \tan^{-1}\left(\frac{a\sin\omega t_0}{1 + a\cos\omega t_0}\right)\right].$$

The magnitude and phase spectra are plotted in Fig. 2.14. Observe that when the gains of both the paths are very close, i.e., $A_1 = A_2$ and $a = 1$, the signals cancel out each other at frequencies $\pi/t_0, 3\pi/t_0, \cdots$, called the multipath null frequencies. But at frequencies $\omega = n\pi/t_0$ (n even), the two signals are enhanced. Fading of transmitted signals by multipath effect can be partly addressed to by using a tapped delay-line equaliser discussed earlier.

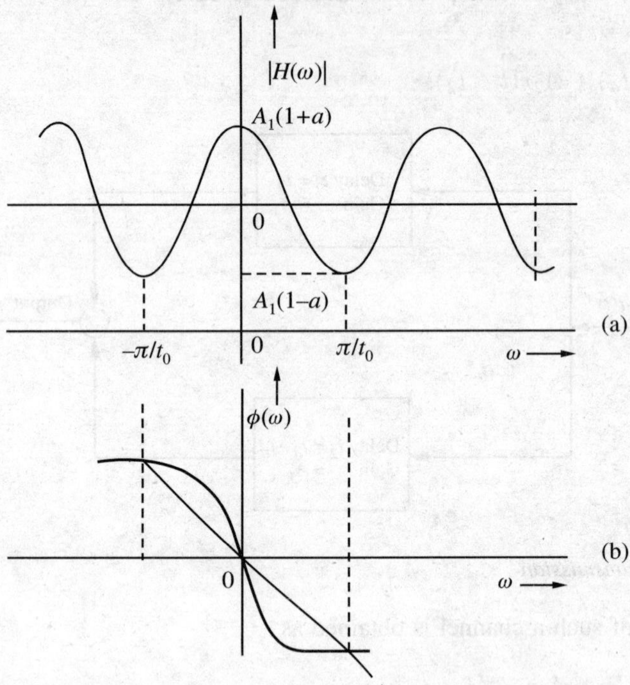

Fig. 2.14 *Spectrum of $H(\omega)$: (a) amplitude $|H(\omega)|$ vs. ω, (b) phase $\phi(\omega)$ vs. ω.*

Example 2.6

A transmission system suffering from multipath distortion has the transfer function $H_{ch}(\omega) = Ae^{-j\omega t_d}\left(1 + ae^{-j\omega t_0}\right)$, where t_d is the time delay in the direct path and t_0 is the difference between the time delays of the two paths. Design a tapped delay-line equaliser that overcomes the multipath distortion.

Solution

For distortionless channel we require that

$$H_{ch}(f)H_{eq}(f) = Ae^{-j\omega t_d}.$$

Thus equaliser transfer function becomes

$$H_{eq}(f) = \frac{1}{1 + ae^{-j\omega t_0}},$$

which, upon binomial expansion reduces to

$$H_{\text{eq}}(f) = 1 - ae^{j\omega t_0} + a^2 e^{-j2\omega_0 t} - a^4 e^{-j4\omega_0 t} + \cdots.$$

The expansion is possible as $|ae^{-j\omega t_0}| << 1$, and for the same reason we keep only the first three terms. Thus,

$$\begin{aligned} H_{\text{eq}}(f) &= 1 - ae^{-j\omega t_0} + a^2 e^{-j2\omega_0 t} \\ &= (e^{j\omega t_0} - a + a^2 e^{-j\omega t_0})e^{-j\omega t_0}. \end{aligned}$$

Comparing this with (2.37), we see that $C_{-1} = 1$, $C_0 = -a$, and $C_1 = a^2$, and $\Delta = t_0$. The required equaliser is shown in Fig. 2.15.

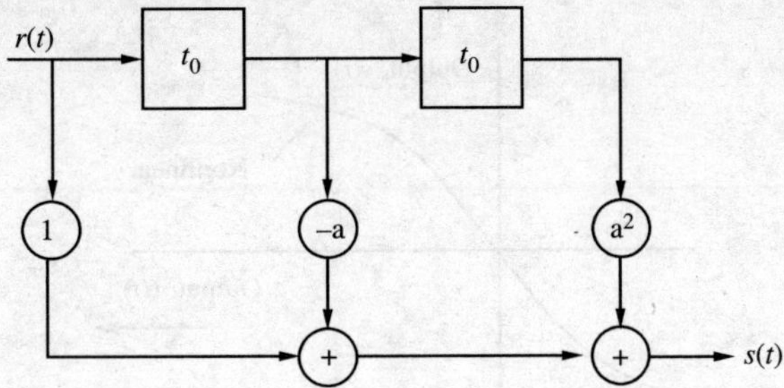

Fig. 2.15 *Tapped-delay line equaliser of Example 2.6*

Example 2.7

The transfer function of a certain channel is given by

$$H(\omega) = e^{-j(\omega t_0 + k\sin\omega T)}, \quad k << 1.$$

For an input pulse $r(t)$ bandlimited to f_m, what is the channel output $s(t)$?

Solution

Since $e^{-jx} \approx 1 - jx$ for small x,

$$\begin{aligned} H(\omega) &= e^{-j\omega t_0} e^{-jk\sin\omega T} \equiv e^{-j\omega t_0}\left[1 - jk\sin\omega T\right] \\ &= e^{-j\omega t_0} - jk\sin\omega T e^{-j\omega t_0}. \end{aligned}$$

Hence,

$$\begin{aligned} S(\omega) = H(\omega)R(\omega) &= \left[e^{j\omega t_0} - jk\sin\omega T e^{-j\omega t_0}\right] R(\omega) \\ &= R(\omega)e^{-j\omega t_0} - \frac{k}{2}\left[e^{j\omega T} - e^{-j\omega T}\right] R(\omega)\dot{e}^{-j\omega t_0} \\ &= R(\omega)e^{-j\omega t_0} - \frac{k}{2}R(\omega)e^{-j\omega(t_0 - T)} + \frac{k}{2}R(\omega)e^{-j\omega(t_0+T)}. \end{aligned}$$

Taking the inverse Laplace transform, the output is

$$s(t) = r(t - t_0) - \frac{k}{2}r(t - t_0 + T) + \frac{k}{2}r(t - t_0 - T).$$

2.2.5 Nonlinear distortion

So far we considered the channel to be linear, where the output is related to the input in a linear fashion. This is valid only for small amplitude signals. For large amplitudes the channel nonlinearities cannot be neglected. The transfer characteristic in such a case is nonlinear, of the form shown in Fig. 2.16. For small-amplitude signals, it is possible to consider the piece-wise linear approximation.

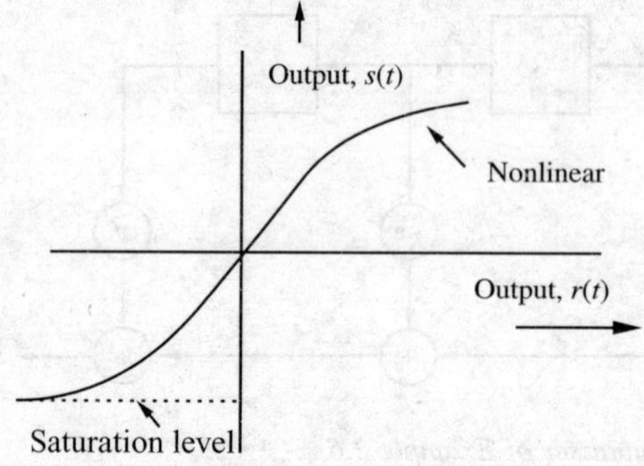

Fig. 2.16 *Nonlinear transfer characteristics*

In the following analysis, we will assume no memory.
In the general approach, we consider a power-series expansion of the curve:

$$s(t) = a_0 + a_1 r(t) + a_2 r^2(t) + a_3 r^3(t) + \cdots = \sum_{n=0}^{\infty} a_n r^n(t), \tag{2.41}$$

where

$$a_n = \frac{1}{n!}\frac{d^n s(t)}{dr^n(t)}\bigg|_{r(t)=0}$$

For $a_1 \neq 0$ and $a_k = 0$ for $k > 1$, the system is linear. The coefficients a_k for $k > 1$ give rise to nonlinear distortion. a_0 is the output dc offset level, $a_1(r)$ is the first order term (linear) and so on. The output spectrum $S(f)$ is obtained by the convolution theorem as

$$S(f) = a_1 R(f) + a_2 R(f) * R(f) + a_3 R(f) * R(f) * R(f) + \cdots, \tag{2.42}$$

where we have omitted the dc offset level.

Now, if the bandwidth of $r(t)$ is B Hz, then the bandwidth of $r^k(t)$ of $r^k(t)$ is kB kHz. So the output spectrum spreads beyond the input spectrum, that is, new frequency components are produced outside the input frequency band. At the same time, due to $a_k r^k(t)$, new frequency components are also added within the input frequency band which overlap with $R(f)$. Suppose that a cosine wave $r(t) = \cos\omega_0 t$ is applied at the input of the nonlinear system described by (2.41). Then the output is

$$s(t) = a_0 + a_1\cos\omega_0 t + a_2\cos^2\omega_0 t + a_3\cos^3\omega_0 t + a_4\cos^4\omega_0 t + \cdots$$

$$= a_0 + a_1\cos\omega_0 t + \frac{a_2}{2}(1+\cos 2\omega_0 t) + \frac{a_3}{4}(\cos 3\theta + 3\cos\theta) + \cdots$$

$$+ a_4\left(\frac{3}{8} + \frac{1}{2}\cos 2\theta + \frac{1}{8}\cos 4\theta\right) + \cdots$$

$$= \left(a_0 + \frac{a_2}{2} + \frac{3a_4}{8} + \cdots\right) + \left(a_1 + \frac{3a_3}{4} + \cdots\right)\cos\omega_0 t$$

$$+ \left(\frac{a_2}{2} + \frac{a_4}{4} + \cdots\right)\cos 2\omega_0 t + \cdots. \tag{2.43}$$

Observe that the nonlinear distortion creates a dc level $(a_0 + a_2/2 + 3a_4/8 + \cdots)$.

Also, channel nonlinearity results in frequency components which are integral multiples of the fundamental. A measure of the distortion caused by different harmonics is the ratio of the amplitude of the harmonic to the fundamental. So,

$$\text{second-harmonic distortion} = \frac{a_2/2 + a_4/4 + \cdots}{a_1 + 3a_3/4 + \cdots} \times 100\%.$$

The intermodulation distortion is obtained by using a two-tone signal as explained in Example 2.8.

Example 2.8

A signal $s(t) = \cos\omega_1 t + \cos\omega_2 t$ is applied to a nonlinear system described by (2.41). Find the output. Discuss the results.

Solution

$$= a_1[\cos\omega_1 t + \cos\omega_2 t] + a_2[\cos\omega_1 t + \cos\omega_2 t]^2 + a_3[\cos\omega_1 t + \cos\omega_2 t]^3$$

$$+ a_4[\cos\omega_1 t + \cos\omega_2 t]^4 + \cdots$$

$$= [a_1\cos\omega_1 t + a_1\cos\omega_2 t] + a_2[\cos^2\omega_1 t + \cos^2\omega_2 t + 2\cos\omega_1 t.\cos\omega_2 t]$$

$$+ a_3[\cos^3\omega_1 t + \cos^3\omega_2 t + 3\cos^2\omega_1 t\cos\omega_2 t + 3\cos\omega_1 t\cos^2\omega_2 t] + \cdots$$

$$= (a_1\cos\omega_1 t + a_1\cos\omega_2 t) + a_2\left[\frac{1}{2} + \frac{1}{2}\cos 2\omega_1 t + \frac{1}{2} + \frac{1}{2}\cos 2\omega_2 t\right.$$

$$+\cos(\omega_1+\omega_2)t+\cos(\omega_1-\omega_2)t]+a_3\left[\frac{3}{4}\cos\omega_1 t+\frac{1}{4}\cos 3\omega_1 t+\frac{3}{4}\cos\omega_2 t+\right.$$

$$\left.\frac{1}{4}\cos 3\omega_2 t+\frac{3}{2}\cos\omega_2 t+\frac{3}{4}\{\cos(2\omega_1+\omega_2)t+\cos(2\omega_1-\omega_2)t\}+\cdots\right]+\cdots.$$

Thus the output contains all the harmonics of f_1 and f_2 in addition to $f_1 \pm f_2$, $2f_1 \pm f_2$, $f_1 \pm 2f_2$, etc. The terms of the form $mf_1 \pm nf_2$ are due to intermodulation distortion. The term intermodulation suggests that the distortion term is present only when both inputs are present.

Example 2.9
A nonlinear transmission system is described by

$$s(t) = r(t) - 2r^3(t).$$

Evaluate the second and third harmonic distortion for the input $r(t) = A\cos\omega_0 t$.

Solution

$$\begin{aligned} s(t) &= A\cos\omega_0 t - 2A^3\cos^3\omega_0 t \\ &= A\cos\omega_0 t + \frac{2A^3}{4}(\cos 3\omega_0 t + 3\cos\omega_0 t). \\ &= \left(A - \frac{3A^3}{2}\right)\cos\omega_0 t - \frac{A^3}{2}\cos 3\omega_0 t. \end{aligned}$$

Since the second harmonic term $\cos 2\omega_0 t$ does not appear, the distortion is zero.

$$\text{Third harmonic distortion} = \left|\frac{A^3/2}{A - 3A^3/2}\right| \times 100\% = \left|\frac{A^2}{2 - 3A^2}\right| \times 100\%.$$

2.2.6 Mathematical models for fading channels

We have discussed so far, the signal degradation effects while transmitting information through the channel. The impulse response $h(t)$ gives an insight into such signal degradation. Besides distortion, noise from the transmission media may add to the signal and may cause annoying disturbances to the reception. In Section (2.2.4), we have discussed multipath effects through which copies of signal may be received via multiple paths. This section is devoted to the discussion on fading effects of the channel. Once the fading characteristic of the channel is provided, the quality of the received signal can be assessed and consequently one can design the demodulator/decoder accordingly. Also given the channel property, the type of the best-suited transmitting signal can be judged and accordingly, the design of transmitters/encoder becomes transparent.
We can classify the physical channel into three categories, namely.

- Additive noise channel
- Linear filter channel

- Linear Time-variant filter channel

Additive noise channel: This is the simplest model of transmission channel and assumes simple mathematical methods to analyse signal transmission through such channels. This is the native idea that signal amplitude gets attenuated while transmitting through the channel. Also, the signal gets corrupted by random noise process in the receiver, transmitter or the channel. In the simplest form, if a signal $x(t)$ undergoes attenuation in addition to the noise effect, the output $y(t)$ of the channel may be written as

$$y(t) = \alpha x(t) + n(t) \tag{2.43a}$$

where α is the attenuation factor. We have assumed here that the noise component $n(t)$ is additive in nature. In some situations, the noise impairment may be multiplicative. We note that $n(t)$ is used to model noise. In practice, noise is the outcome of random processes which cannot be modeled by deterministic signal $n(t)$. The analysis of noise is assisted by statistical analysis. However, the equation (2.43a) is the widely accepted. The model for a channel is shown in the figure.

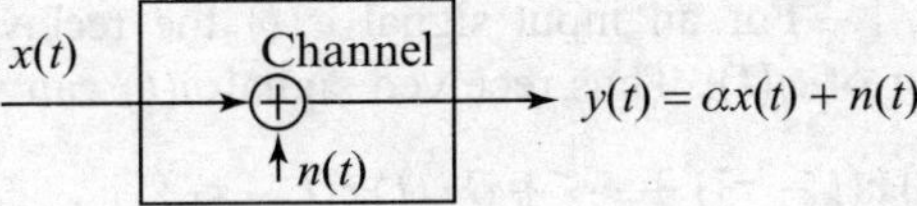

Linear filter channel: As the name implies, the channel behaves as a linear filter, allowing some specified frequency components to pass through. If we assume the impulse response of the filter as $h(t)$, then the output $y(t)$ can be obtained by the convolution integral of $h(t)$ with $x(t)$ as

$$\begin{aligned} y(t) &= x(t) * h(t), +n(t) \\ &= \int_{-\infty}^{\infty} h(\tau)\, x(t-\tau)\, d\tau + n(t) \end{aligned} \tag{2.43b}$$

where again the noise is assumed to be additive. The channel model is sketched in the figure.

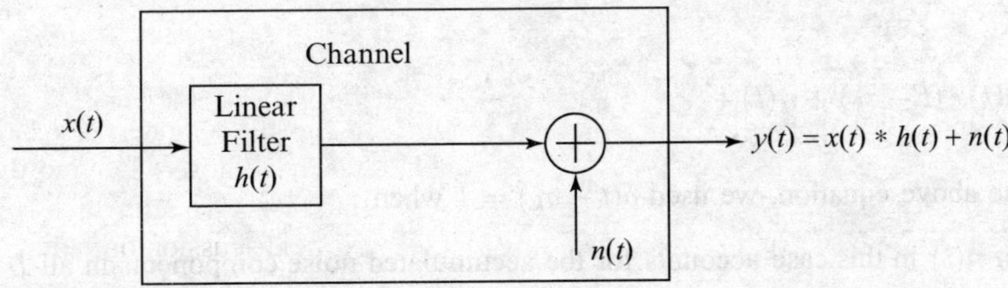

The model is appropriate when the input signal is constrained not to exceed the bandwidth limitations. One example is the two-wire telephone line where the input noise is limited within 20 kHz. We note in Eq. (2.43b) that the impulse response $h(t)$ is linear and time-invariant and the signal path is only one. The attenuation factor is implicit in $h(t)$.

Linear Time-variant Filter Channel: In contrast to the wireline channel as discussed above, there are some physical channels such as ionospheric channel that reflects the signals in different multiple paths. Thus, the receiver gets multiple copies of the same signal with different attenuation factor, delay and corrupted by noise of different strengths. Such a channel is characterised by time-varying impulse response $h(\tau;t)$ where $h(\tau;t)$ is the response of the channel at time t for an impulse inputted τ sec earlier. The channel output is thus modeled by the input–output relation.

$$y(t) = x(t) * h(\tau; t), +n(t)$$

$$\int_{-\infty}^{\infty} h(\tau; t)\, x(t-\tau)\, d\tau + n(t) \tag{2.43c}$$

The channel is assumed still linear as the correlation is a linear operation. Since $h(\tau;t)$ includes the effect of attenuation and delay of all possible multipaths, one can model $h(\tau;t)$ with simple reasoning as follows:

Let there be L independent multipath channels with the attenuation factors $\{\alpha_1(t), \alpha_2(t), \ldots, \alpha_L(t)\}$ and time delay $\{\tau_1, \tau_2, \ldots, \tau_L\}$. For an input signal $x(t)$ the receiver will get L-copies of the attenuated and delayed receiver of $x(t)$. Thus received signal $y(t)$ can written as

$$y(t) = \alpha_1(t)x(t-\tau_1) + \alpha_2(t)x(t-\tau_2) + \cdots + \alpha_L(t)x(t-\tau_L) \tag{2.43d}$$

So, the impulse response is time-varying and given by

$$h(\tau; t) = \alpha_1(t)\delta(t-\tau_1) + \alpha_2(t)\delta(t-\tau_2) + \cdots + \alpha_L(t)\delta(t-\tau_L)$$

$$= \sum_{k=0}^{L} \alpha_k(t)\delta(t-\tau_k) \tag{2.43e}$$

The received signal has the from (substituting 2.43e into 2.43c)

$$y(t) = \int_{-\infty}^{\infty} \sum_{k=1}^{L} \alpha_k(t)\delta(t-\tau_k)\; x(t-\tau)\, d\tau + n(t)$$

$$= \sum_{k=1}^{L} \alpha_k(t)x(t-\tau_k) + n(t) \tag{2.43f}$$

(in deriving the above equation, we used $\delta(t-\tau_k) = 1$ when $\tau = \tau_k$)

The noise term $n(t)$ in this case accounts for the accumulated noise components in all L multipaths.

To illustrate the filtering effect of the channel, we take an example of a radio channel with very few multipath components, that is, L of Eq. (2.43f) is a small number. Let the radio link be from an airplane to the base station. The link consists of a strong line-of-sight (LOS) link and a single multipath link with a delay τ_0 with respect to the LOS link. So, the received signal $y(t)$ can be written as

$$y(t) = \alpha_0 x(t) + \alpha(t)x(t-\tau_0(t)) \tag{2.43g}$$

where α_0 denotes the attenuation factor in the LOS-link, assumed constant and $\alpha(t)$ refers to the same in the time-varying multipath link. The impulse response of the link is thus

$$h(\tau;t) = \alpha_0\delta(t) + \alpha(t)\delta(t - \tau_0(t)) \tag{2.43h}$$

The time-varying transfer function of this channel is thus obtained by Fourier transform of Eq. (2.43h) and is written as

$$H(f;t) = \int_{-\infty}^{\infty} h(\tau;t)\, e^{-j2\pi ft} dt$$

$$= \int_{-\infty}^{\infty} \alpha_0\delta(t)\, e^{-j2\pi ft} dt + \int_{-\infty}^{\infty} \alpha_1(t)\, \delta(t - \tau_0(t))\, e^{-j2\pi ft} dt$$

$$H(f;t) = \alpha_0 + \alpha_1(\tau_0)\; e^{-j2\pi f\tau_0(t)} \tag{2.43i}$$

The time-varying nature of $H(f;t)$ is due to the delay $\tau_0(t)$ being time-varying component.

2.3 Filters

Filters are used traditionally to eliminate undesired frequency components from an information-bearing signal. Many communications systems include ideal distortionless filters. In this section, we will discuss ideal filters, explain the meaning of real filters, and study the effect of filtering on the signals.

2.3.1 Ideal filters

By definition, an ideal filter permits distortionless transmission over some specified frequency band (called the passband) and blocks transmission at all other frequencies (called the stopband). These two ranges of frequency are separated by the cutoff frequency of the filter. The filters may be classified as lowpass filter (LPF), highpass filter (HPF), bandpass filter (BPF), and band-reject or notch filter.

Ideal lowpass filter

An ideal lowpass filter has the characteristic of distortionless transmission of the input signal up to the cutoff frequency f_c, completely blocking the frequency components greater than f_c. The transfer function $H(f)$ of the ideal LPF is then given by

$$H(f) = \begin{cases} Ae^{-j\omega t_0} & |f| < f_c \\ 0 & \text{otherwise.} \end{cases} \tag{2.44}$$

Figure 2.17 shows the plot of the transfer function of (2.44). The impulse response $h(t)$ is obtained by the inverse Fourier transform of $H(f)$.

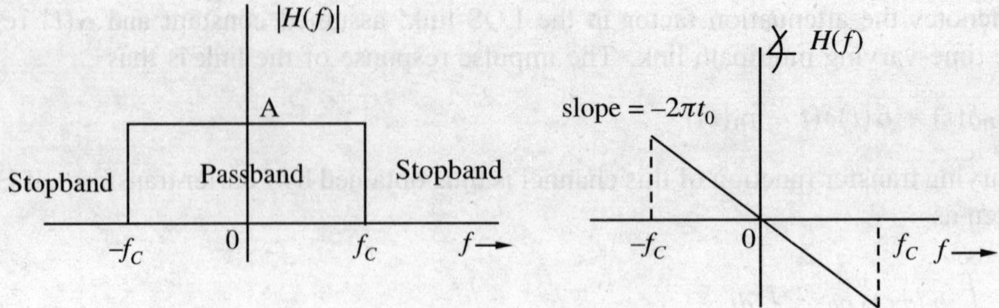

Fig. 2.17 *An ideal LPF transfer function: magnitude and phase angle.*

That is,

$$h(t) = \frac{1}{2\pi}\int_{-\infty}^{\infty} H(\omega)e^{j\omega t}d\omega = \frac{1}{2\pi}\int_{-f_c}^{f_c} Ae^{-j\omega t_0}e^{j\omega t}dt$$

$$= \frac{A}{2\pi}\int_{-f_c}^{f_c} e^{j\omega(t-t_0)}dt$$

or, $$h(t) = 2f_c A\frac{\sin 2\pi f_c(t-t_0)}{2\pi f_c(t-t_0)}$$

$$= 2f_c A\text{sinc}[2f_c(t-t_0)]. \tag{2.45}$$

Figure 2.18 shows the plot of the impulse response $h(t)$. The delay t_0 is proportional to the slope of the phase plot. The peak of $h(t)$ is proportional to the cutoff frequency f_c and the spacing between zero-crossings of $h(t)$ is inversely proportional to f_c. Note that $h(t)$ has nonzero values for $t < 0$. The ideal characteristic of the filter must be unrealisable since the filter cannot anticipate the response.

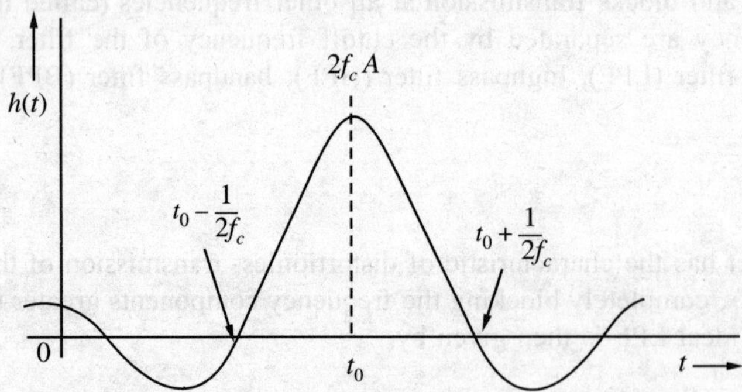

Fig. 2.18 *Impulse response of ideal LPF*

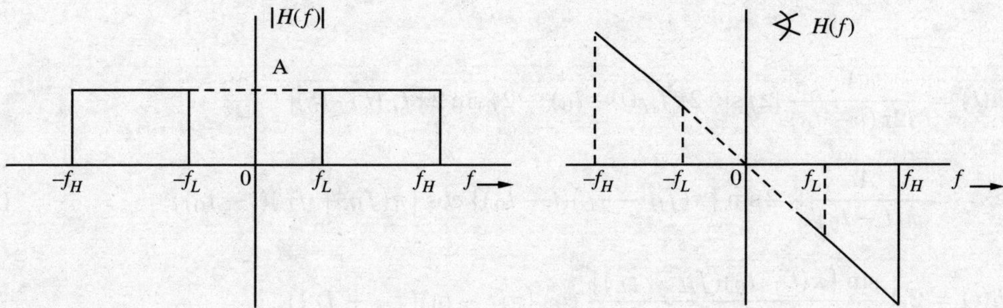

Fig. 2.19 *Ideal BPF transfer function*

Ideal bandpass filter

An ideal bandpass filter (BPF) has the characteristics of signal transmission over a specified range of frequencies between f_L and f_H and zero response outside the range. The transfer function of this filter is given by

$$H(f) = \begin{cases} Ae^{-j\omega t_0} & f_L < |f| < f_H \\ 0 & \text{otherwise.} \end{cases} \tag{2.46}$$

Figure 2.19 illustrates the amplitude and phase responses of this filter. Here, f_L and f_H denote, respectively, the lower and upper cutoff frequencies. The frequency band $B = f_H - f_L$ is the bandwidth or passband of the filter. The impulse response of the BPF is found by evaluating the inverse Fourier transform of $H(f)$ of (2.46), as follows:

$$h(t) = \mathfrak{F}^{-1}\left[H(f)\right] = \int_{-\infty}^{\infty} H(f)e^{j2\pi ft}df$$

$$= \int_{-f_H}^{-f_L} Ae^{j2\pi ft_0}\, e^{j2\pi ft}df + \int_{f_L}^{f_H} Ae^{-j2\pi ft_0}\, e^{j2\pi ft}df = I_1 + I_2,$$

where,

$$I_1 = \int_{-f_H}^{-f_L} Ae^{j2\pi f(t-t_0)}df = \frac{A}{j2\pi(t-t_0)}\left[e^{j2\pi f(t-t_0)}\right]_{-f_H}^{-f_L}$$

$$= \frac{A}{j2\pi(t-t_0)}\left[e^{-j2\pi f_L(t-t_0)} - e^{-j2\pi f_H(t-t_0)}\right]$$

and

$$I_2 = \int_{f_L}^{f_H} Ae^{j2\pi f(t-t_0)}df = \frac{A}{j2\pi(t-t_0)}\left[e^{j2\pi f(t-t_0)}\right]_{f_L}^{f_H}$$

$$= \frac{A}{j2\pi(t-t_0)}\left[e^{j2\pi f_H(t-t_0)} - e^{j2\pi f_L(t-t_0)}\right]$$

Thus,

$$h(t) = \frac{A}{j2\pi(t - t_0)} [2j \sin 2\pi f_H(t - t_0) - 2j \sin 2\pi f_L(t - t_0)]$$

$$= \frac{A}{\pi(t - t_0)} \times 2 \sin\{\pi(f_H - f_L)(t - t_0)\} \cos\{\pi(f_H + f_L)(t - t_0)\} \tag{2.47}$$

or, $$h(t) = 2A \frac{\sin\{\pi(t - t_0)(f_H - f_L)\}}{\pi(t - t_0)} \cos\{\pi(t - t_0)(f_H + f_L)\}.$$

In (2.47) the frequency of the cosine term is $(f_H + f_L)$, which is larger than that of the sine term of difference frequency $(f_H - f_L)$. The response $h(t)$ appears as the amplitude modulation of a high frequency carrier $(f_H + f_L)$ with a low frequency signal $\frac{\sin \pi(f_H - f_L)(t - t_0)}{\pi(t - t_0)}$, and is sketched in Fig. 2.20. Note that since $h(t)$ is nonzero for $t < 0$, the ideal BPF is also physically unrealisable.

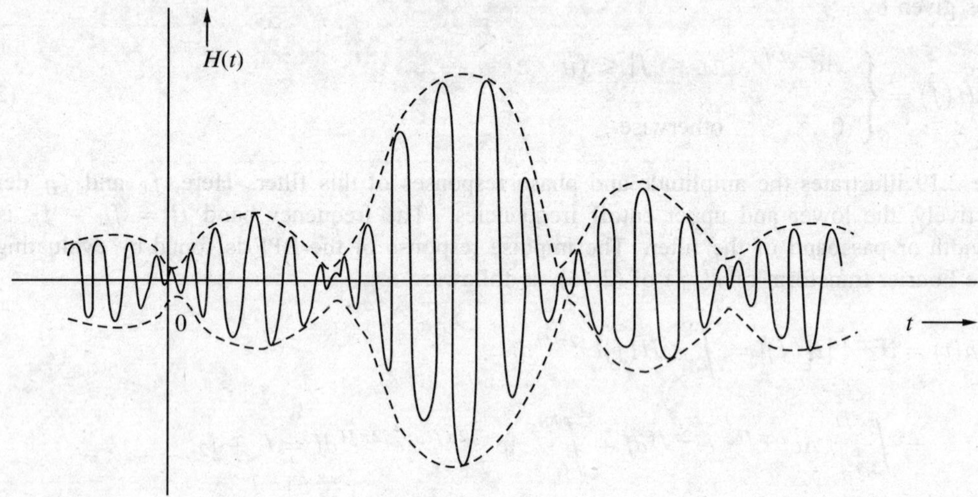

Fig. 2.20 *Impulse response of an ideal bandpass filter. The envelope corresponds to the sine term.*

Example 2.10

Find and sketch the impulse response of an ideal HPF defined by

$$H(f) = \begin{cases} Ae^{-j2\pi f t_0} & f_L \leq |f| < \infty \\ 0 & \text{otherwise.} \end{cases} \tag{2.48}$$

Solution

Figure 2.21 sketches the BPF frequency response. Now,

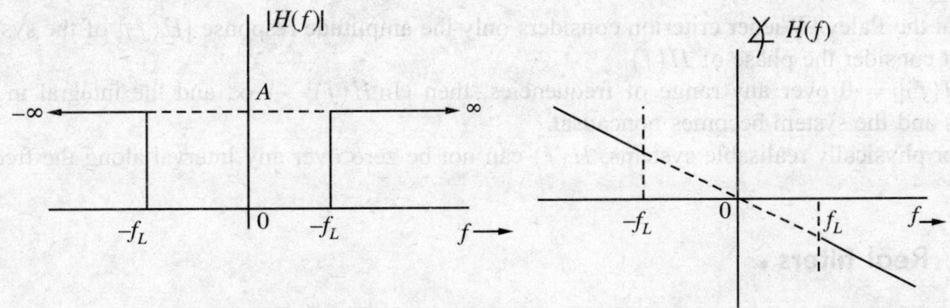

Fig. 2.21 *An ideal bandpass filter frequency response*

$$h(t) = \int_{-\infty}^{\infty} H(f)e^{j2\pi ft}df$$

$$= \int_{-\infty}^{-f_L} Ae^{-j2\pi ft_0}e^{j2\pi ft}df + \int_{f_L}^{\infty} Ae^{-j2\pi ft_0}e^{j2\pi ft}df$$

$$= \int_{-\infty}^{-f_L} Ae^{j2\pi f(t-t_0)}df + \int_{f_L}^{\infty} Ae^{j2\pi f(t-t_0)}d_f$$

$$= \frac{A}{j2\pi(t-t_0)}\left[e^{j2\pi f(t-t_0)}\right]_{-\infty}^{-f_L} + \frac{A}{j2\pi(t-t_0)}\left[e^{j2\pi f(t-t_0)}\right]_{f_L}^{\infty}$$

$$= \frac{A}{j2\pi(t-t_0)}\left[e^{-j2\pi f_L(t-t_0)} - e^{j2\pi f_L(t-t_0)}\right]$$

$$= \frac{-A}{\pi(t-t_0)}\sin 2\pi f_L(t-t_0).$$

2.3.2 Causality criterion

A system is said to be causal if its output at any instant depends only on the input at that time or prior to that time. That is, the output of a causal system will no longer depend on any future values of the input. Causal systems are physically realisable. In the time domain, the causality of a system is stated in terms of its impulse response $h(t)$ as

$$h(t) = 0 \quad \text{for} \quad t < 0.$$

In the frequency domain, the Paley–Wiener criterion states the condition of physical realisability of a system. Accordingly, the system function $|H(f)|$ is realisable if

$$\left.\begin{aligned} &\int_{-\infty}^{\infty} \frac{|\ln H(f)|}{1+(2\pi f)^2}\,df < \infty \\ \textit{and} \qquad &\textstyle\int_{-\infty}^{\infty} |H(f)|^2 df < \infty. \end{aligned}\right\} \tag{2.49}$$

Note that the Paley–Wiener criterion considers only the amplitude response $|H(f)|$ of the system, it does not consider the phase of $H(f)$.

If $|H(f)| = 0$ over any range of frequencies, then $|\ln H(f)| \to \infty$ and the integral in (2.49) diverges and the system becomes noncausal.

So for physically realisable systems, $H(f)$ can not be zero over any interval along the frequency axis.

2.3.3 Real filters

In practice, we cannot realise the filter characteristics that approach the ideal one. The amplitude response of real filters are gradual in contrast to the sharp changes at the cutoff frequency of ideal filters. Although $|H(f)|$ is quite small in real filters at the stopband, it is not zero as in the ideal situation. Consequently, real filters allow frequency components of small amplitudes to pass.

Lowpass filter

To begin our discussion, let us recall the analysis of the RC circuit in Section 2.1.2. The transfer function of this filter is given by the (2.15). Let us find its impulse response.

$$h_{\mathrm{RC}}(t) = \mathfrak{F}^{-1}[H(f)] = \int_{-\infty}^{\infty} \frac{1}{1 + j2\pi fCR} e^{j2\pi ft} df$$

$$= \frac{1}{CR} e^{-t/RC}. \tag{2.50}$$

But for an ideal lowpass filter,

$$h_{\mathrm{ideal}}(t) = 2f_c \frac{\sin 2\pi f_c(t - t_0)}{2\pi f_c(t - t_0)} \qquad \text{(vide (2.45)).}$$

The impulse responses of the two filters are sketched in Fig. 2.22 for comparison. We have taken the delay as t_0.

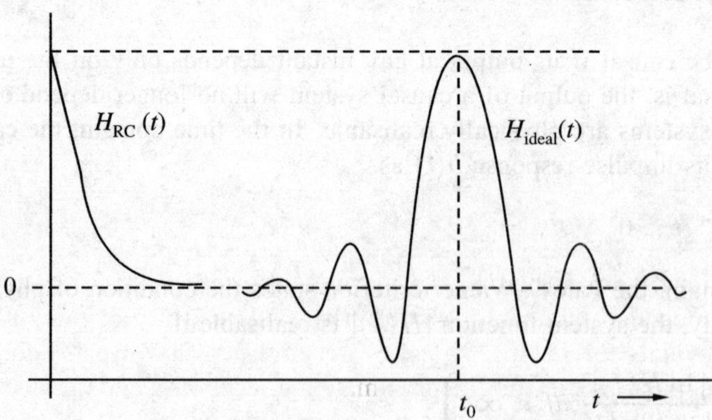

Fig. 2.22 *Impulse responses of ideal and real LPF*

At this point we can say that the RC network is not a good design for a lowpass filter. There are several approximations to the ideal LPF design. The simplest one is the Butterworth LPF design with reactive elements. The transfer function of such filter can be written in the form

$$H(f) = \frac{1}{P_n(jf/f_c)}, \tag{2.51}$$

where P_n is a complex polynomial called Butterworth polynomial and is defined as

$$|P_n(jf/f_c)|^2 = 1 + (f/f_c)^{2n}. \tag{2.52}$$

The number n refers to the order of the filter. Thus the amplitude response of the nth-order Butterworth LPF is approximated as

$$|H_n(f)| = \frac{1}{\sqrt{1+(f/f_c)^{2n}}}, \quad n \geq 1. \tag{2.53}$$

The Butterworth response for different values of n is sketched in Fig. 2.23. Note that the larger the n, the more closely does the curve approach the ideal lowpass characteristics (shown by the dotted curve marked $n = \infty$). The amplitude $|H_n(f)|$ equals $1/\sqrt{2}$ at $f = f_c$ for all values of n. This frequency is referred to as half-power, or the 3-dB bandwidth of the filter.

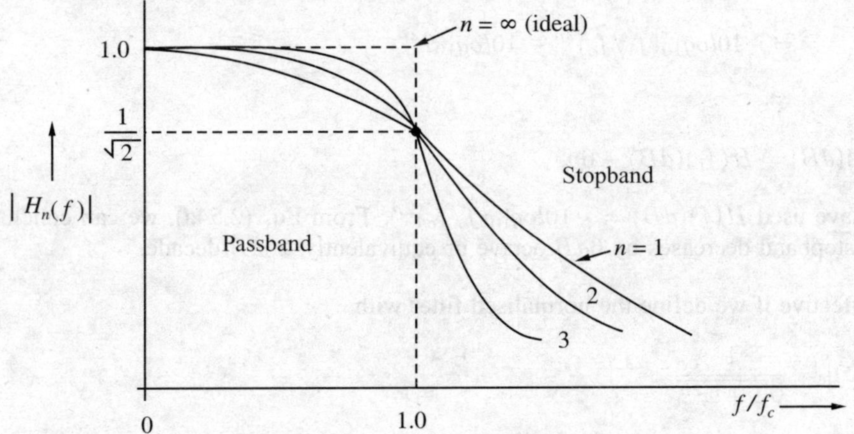

Fig. 2.23 *Butterworth response for different values of n*

Since the amplitude $|H_n(f)|$ and phase $\measuredangle H_n(f)$ are interrelated, we cannot make $|H_n(f)|$ to behave approximately ideally by increasing n. Because, then the slope of the phase will increase with n and may even become intolerable, particularly in the vicinity of the cutoff frequency f_c. The Butterworth polynomials for $n = 1, 2$, and 3 are tabulated in Table 2.1 for ready reference. Butterworth filters are popular as they best approximate the ideal ones and provide maximal flatness in its passband.

Table 2.1 *Butterworth polynomials*

n	$P_n(s),\ s = jf/f_c$
1	$s+1$
2	$s^2 + \sqrt{2}s + 1$
3	$(s+1)(s^2+s+1)$

For the case $f >> f_c$, (2.53) approximates as

$$|H_n(f)| = \frac{1}{(f/f_c)^n}. \tag{2.54}$$

Equation (2.54) tells us that for large n, the response decreases very quickly, closely resembling that of the ideal LPF. Thus for $f \to 0$ and $f >> f_c$, the filter shows very good response. However, in the vicinity of the cutoff frequency the response is poor due to phase distortion.

The decrease in the gain in stop band can be calculated as follows. We define $f_2 = 2f_1$ and $f_1 >> f_c$ in the stop band. The gain in dB at f_2 can be written as (from Eq. (2.54))

$$H(f_2)(dB) = -10log_{10}(f_2/f_c)^{2n} = -10\log_{10}\left(\frac{2f_1}{f_c}\right)^{2n}$$

$$= -10log_{10}(f_1/f_c)^{2n} - 10log_{10}2^{2n}$$

Thus,

$$H(f_2)(dB) = H(f_1)(dB) - 6n \tag{2.54a}$$

where we have used $H(f)(dB) = -10log_{10}(f/f_c)^{2n}$. From Eq. (2.54a), we can conclude that the gain in the stopband decreases by $6dB$/octave or equivalently, $20dB$/decade.

It is very effective if we define the normalised filter with

$$|H(j\omega)| = \frac{1}{\sqrt{1+\omega^{2n}}} \tag{2.54b}$$

where the frequency is normalised with respect to the half-power frequency ω_c being $\omega_c = 1$ radian/sec. This is a very effective approach in filter design because any desired transfer function for any value of ω_c can be obtained by simply frequency scaling. Here, we replace s by s/ω_c in the normalised transfer function.

The response of the filter shows maximally flat at $\omega = 0$ since first $(2n-1)$ derivatives of the amplitude response defined in Eq. (2.53) is zero at $\omega = 0$. Omitting suffix n of $H_n(f)$ of Eq.(2.53), we can write

$$|H(j\omega)|^2 = \frac{1}{1+\omega^{2n}} = H(j\omega)\cdot H(-j\omega)$$

In the s-domain $(s = j\omega)$.

$$H(s)\,H(-s) = \frac{1}{1+(s/j)^{2n}}$$

The poles of $H(s)$. $H(-s)$ are given by $1+(s/j)^{2n}=0$ or $s^{2n}+(j)^{2n}=0$ or, $s^{2n}=-(j)^{2n}$

Since $-1=e^{-j\pi}=e^{j\pi(2k-1)}$, k is any integer and $j=e^{j\pi/2}$, we have

$$s^{2n}=e^{j\pi(2k-1)}\cdot e^{j\pi/2})^{2n}=e^{j\pi(2k-1)}\cdot e^{j\pi 4}=e^{j\pi(2k-1+n)}.$$

Thus, the poles are located at

$$s_k=e^{j\frac{\pi}{2n}(2k+n-1)},\quad k=1,2,\ldots.,2n \tag{2.54d}$$

Note that there are total $2n$-poles of $H(s)H(-s)$; poles are of unit magnitude (unit circle on the s-plane) and separated by angle π/n. [The angular separation between the kth and $(k+1)$ at poles is

$$\Theta_k\sim\Theta_{k+1}=\frac{\pi}{2n}\,(2k+n-1)\sim\frac{\pi}{2n}\,(2\,(k+1)+n-1)=\pi\,/n]$$

Since the poles of $H(s)$ and $H(-s)$ are symmetrical on the unit circle, we can assume the poles of $H(s)$ as

$$s_k=e^{j\frac{\pi}{2n}(2k+n-1)},\quad k=1,2,\ldots,n \tag{2.54e}$$

The transfer function $H(s)$ thus

$$H(s)=\frac{1}{(s-s_1)(s-s_2)\cdots(s-s_n)} \tag{2.54f}$$

As an example, let us consider a Butterworth filter of order $n=1$. Thus, by (2.54e), $k=1$ and the poles is

$$s_1=e^{j\frac{\pi}{2}(2+1-1)}=e^{j\pi}=-1$$

So, $H\,(s)=\dfrac{1}{(s-s_1)}=\dfrac{1}{s+1}$ and $P_n(s)=(s+1)$

Similarly, for $n=2$, $k=1,2$, the poles are

$$s_k=e^{j\frac{\pi}{2.2}(2k+1)}=e^{j\frac{\pi}{4}(2k+1)}$$

So, $\quad s_1=e^{j\frac{\pi}{4}3}=\cos\dfrac{3\pi}{4}+j\sin\dfrac{3\pi}{4}=-\dfrac{1}{\sqrt{2}}+j\dfrac{1}{\sqrt{2}}$

$$s_2=e^{j\frac{\pi}{4}5}=\cos\frac{5\pi}{4}+j\sin\frac{5\pi}{4}=-\frac{1}{\sqrt{2}}-j\frac{1}{\sqrt{2}}$$

The Butterworth polynomial is then

$$P_n(s)=(s-s_1)\,(s-s_2)=\left(s+\frac{1}{\sqrt{2}}-j\frac{1}{\sqrt{2}}\right)\left(s+\frac{1}{\sqrt{2}}+j\frac{1}{\sqrt{2}}\right)=\left(s+\frac{1}{\sqrt{2}}\right)^2+\left(\frac{1}{\sqrt{2}}\right)^2$$

$$= s^2 + \sqrt{2}s + 1$$

In this way, students can verify the entries in Table 2.1. The poles of normalised low-pass Butterworth filter of various orders are plotted in the adjoining figure. Note that for

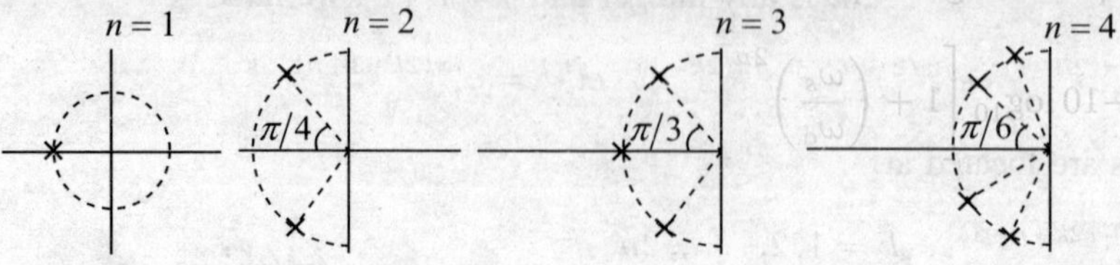

Figure plot of poles for filter order $n = 1$, 2, 3 and 4.

odd order ($n = 1$, 3, 5, ... etc), there must be a pole at $s = -1$.

The determination of order of the filter can be obtained from the specifications. Let G_p and G_s be the gains of the filter at the passband edge frequency ω_p and stop-band edge frequency ω_s, as shown in Fig. 2.24.

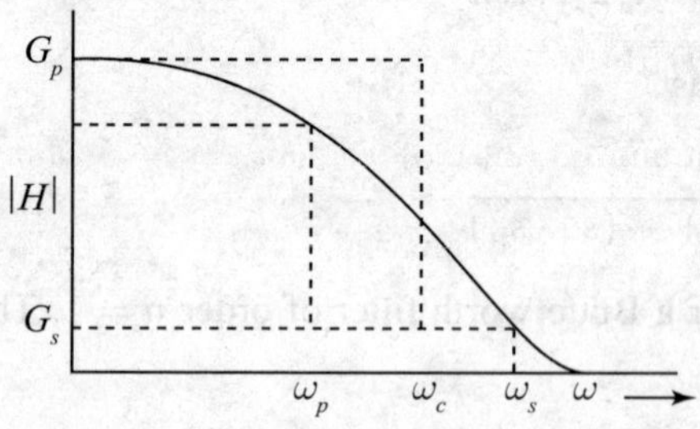

Fig. 2.24 *Filter response and specifications.*

At any frequency, $\omega = \omega_x$, the gain in dB is

$$G_x = 20\log_{10}|H(j\omega)| = 20\log_{10}\frac{1}{1 + (\omega_x/\omega_c)^{2n}}$$

$$= -10\log_{10}\left[1 + \left(\frac{\omega_x}{\omega_c}\right)^{2n}\right] \quad (2.54g)$$

Thus at the pass-band edge frequency ω_p

$$G_p = -10\log_{10}\left[1 + \left(\frac{\omega_p}{\omega_c}\right)^{2n}\right]$$

or,

$$(\omega_p/\omega_c)^{2n} = \left(10^{-G_p/10} - 1\right) \tag{2.54h}$$

and at the stopband edge frequency ω_s

$$G_s = -10\log_{10}\left[1 + \left(\frac{\omega_s}{\omega_c}\right)^{2n}\right]$$

or,

$$(\omega_s/\omega_c)^{2n} = \left(10^{-G_s/10} - 1\right) \tag{2.54i}$$

Dividing (2.54 h) by (2.54 i) and manipulating,

$$n = \frac{10\log_{10}\left[(10^{-G_p/10} - 1)/(10^{-G_s/10} - 1)\right]}{2\log_{10}(\omega_s/\omega_p)} \tag{2.54j}$$

The Chebyshev filters on the other hand, can attenuate unwanted frequencies more effectively than Butterworth filters. The transfer function of the Chebyshev LPF has the form

$$|H_n(f)| = \frac{1}{\sqrt{1+\epsilon^2 C_n^2(f/f_c)}}, \tag{2.55}$$

where ϵ^2 is a design constant and $C_n(f)$ is the Chebyshev polynomial of the first kind of degree n. The magnitude response of the normalised Chebyshev lowpass filter of order n is given by

$$|H_n(j\omega)| = \frac{1}{\sqrt{1+\epsilon^2 C_n^2(\omega)}} \tag{2.55a}$$

where ϵ^2 controls the height of the ripple. The Chebyshev polynomial $C_n(\omega)$ is given by

$$C_n(\omega) = \begin{cases} \cos(n\cos^{-1}\omega), & |\omega| < 1 \\ \cosh(n\cosh^{-1}\omega), & |\omega| > 1 \end{cases} \tag{2.55b}$$

The recursion relation of $C_n(x)$ is given by

$$C_n(x) = 2xC_{n-1}(x) - C_{n-2}(x) \tag{2.55c}$$

with $C_0(x) = 1$ and $C_1(x) = x$

There are two types of Chebyshev transfer functions.

Type 1: Magnitude characteristic is equiripple in the passband and monotonic in the stopband.

Type 2: Magnitude response is monotonic in the passband and equiripple in the stopband.

Using the recursion relation (2.55 c), the Chebyshev polynomial for different n can be calculated.

Table 2.2 shows few polynomials for up to $n = 7$.

Table 2.2 Chebyshev polynomial of different of different order

We observe that

$$|H_n(j0)| = \begin{cases} 1, & n = odd \\ \frac{1}{\sqrt{1+\varepsilon^2}}, & n = even \end{cases}$$

n	$C_n(\omega)$
0	1
1	ω
2	$2\omega^2 - 1$
3	$4\omega^3 - 3\omega$
4	$8\omega^4 - 4\omega^2 + 1$
5	$16\omega^5 - 20\omega^3 + 5\omega$
6	$32\omega^6 - 48\omega^4 + 18\omega^2 - 1$
7	$64\omega^7 - 112\omega^5 + 56\omega^3 - 7\omega$

Figure 2.25 shows the magnitude plots of low-pass Chebyshev filter of type 1 for $n = 6$ and 7. In both the cases, there are equiripples in the passband and monotonic decreasing in the stop band. For even n ($n = 6$, say), the zero-frequency response starts from $\frac{1}{\sqrt{1+\varepsilon^2}}$ while for odd n ($n = 7$, say), the zero-frequency response starts from unity.

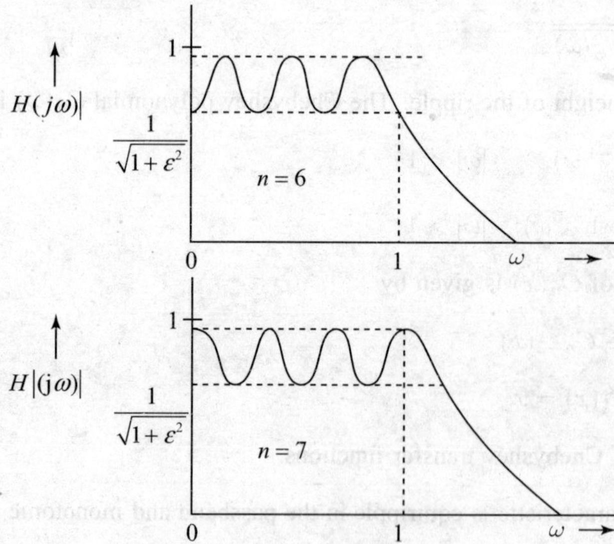

Fig. 2.25 *Chebyshev low pass filter(Type1) for filter order $n = 6$ and 7.*

The order of the filter can be obtained from the attenuation at a particular frequency in the stop band. For example, if we consider the stopband frequency $\omega = \omega_s$ at which the magnitude is G_s, one can write

$$|H\,(j\,\omega)|^2_{\omega=\omega_s} = \frac{1}{1+\epsilon^2 C_n^2(\omega_s/\omega_c)} = G_s^2$$

or, $$\frac{1}{1+\epsilon^2\left[\cosh\left(n\,\cosh^{-1}(\omega_s/\omega_c)\right)\right]^2} = G_s^2$$

Solving by making use of the approximation $\cosh^{-1}(x) = \ln\left(x+\sqrt{x^2-1}\right)$

$$n = \frac{\cosh^{-1}\left(\sqrt{1/G_s^2 - 1}\,/\,\epsilon\right)}{\cosh^{-1}(\omega_s/\omega_c)} \tag{2.55d}$$

Chebyshev polynomial of the first kind of order n is given by

$$C_n(f) = \cos(n\cos^{-1} 2\pi f). \tag{2.56}$$

The recursion relation of $C_n(x)$ is

$$C_n(x) = 2xC_{n-1}(x) - C_{n-2}(x)$$

where $C_0(x) = 1$, $C_1(x) = x$

This type of filter is also called equiripple filter as it has small amplitude ripples in the passband. Outside the passband, the filter provides sharpest transition for a given n. The Chebyshev response for different n is shown in Fig. 2.26 for a given k.

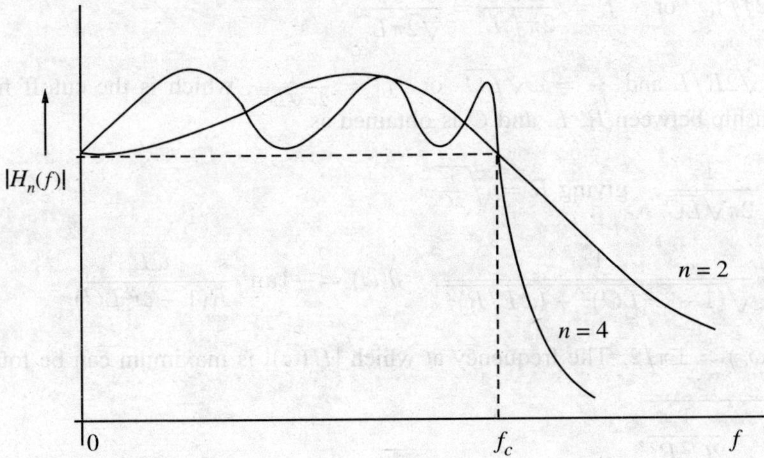

Fig. 2.26 *Chebyshev response for $n = 2$ and 4*

Example 2.11
For the circuit shown in Fig. 2.27, find and sketch the transfer function $H(f)$. What function is this circuit performing? Also find the impulse response.

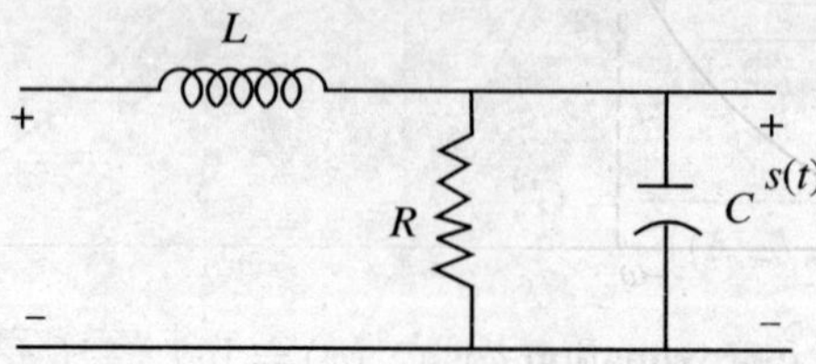

Fig. 2.27 *The circuit of Example 2.11*

Solution
The transfer function

$$H(f) = \frac{[(1/j\omega c)R]/[1/j\omega c + R]}{j\omega L + [(1/j\omega c)R]/[1/j\omega c + R]}$$

$$= \frac{R/(1 + j\omega CR)}{j\omega L + R/(1 + j\omega CR)} = \frac{R}{R - \omega^2 LCR + j\omega L} = [1 + j\omega L/R - (\omega\sqrt{LC})^2]^{-1}$$

which is of the form

$$H(f) = \left[1 + \sqrt{2} jf/f_c + (jf/f_c)^2\right]^{-1}. \tag{2.57}$$

Thus the circuit is a design for a Butterworth LPF of order $n = 2$. Comparing term by term,

$$\frac{\omega L}{R} = \sqrt{2} f/f_c \quad \text{or} \quad f_c = \frac{\sqrt{2} fR}{2\pi f.L} = \frac{R}{\sqrt{2}\pi L}.$$

That is, $\omega_C = \sqrt{2}R/L$ and $\frac{f}{f_c} = \omega\sqrt{LC}$ or $f_c = \frac{1}{2\pi\sqrt{LC}}$, which is the cutoff frequency. The required relationship between R, L, and C is obtained as

$$\frac{R}{\sqrt{2}\pi L} = \frac{1}{2\pi\sqrt{LC}}, \quad \text{giving } R = \sqrt{\frac{L}{2C}}.$$

Now, $|H(\omega)| = \dfrac{1}{\sqrt{(1 - \omega^2 LC)^2 + (\omega L/R)^2}}, \quad \theta(\omega) = -\tan^1 \dfrac{\omega L}{R(1 - \omega^2 LC)}.$ (2.58)

At $\omega = \omega_c$, $\theta(\omega_c) = \pm\pi/2$. The frequency at which $|H(\omega)|$ is maximum can be found as

$$\omega = \sqrt{\omega_c^2 - \frac{1}{2C^2R^2}},$$

which is less than $|\omega_c|$. Figure 2.28 gives the plots of $|H(\omega)|$ and $\sphericalangle H(\omega)$.

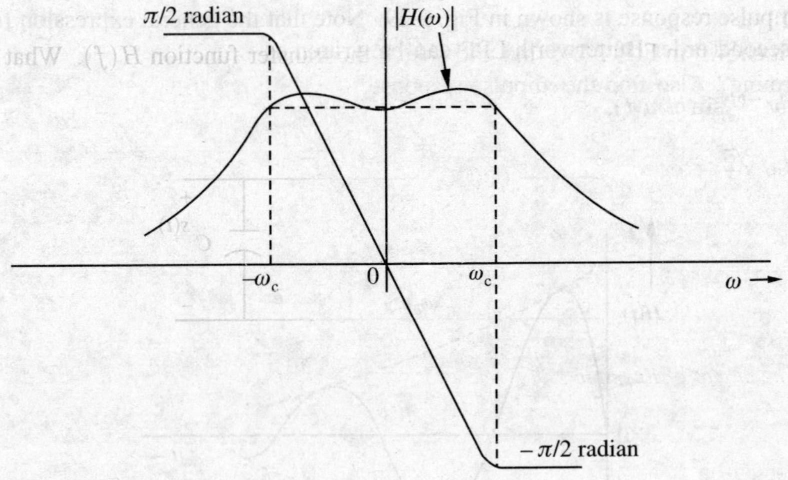

Fig. 2.28 *Magnitude and phase response of the transfer function*

Obviously, the network behaves as a lowpass filter. The impulse response of the system is obtained by taking the inverse Fourier transform of $H(\omega)$.

Since $LC = \frac{1}{\omega_c^2}$ and $\frac{L}{C} = 2R^2$, one can have

$L = \dfrac{\sqrt{2}R}{\omega_c}$ and $C = \dfrac{1}{\sqrt{2}\omega_c R}$. Thus,

$$H(\omega) = \frac{1}{(1-\omega^2/\omega_c^2) + j\sqrt{2}\omega/\omega_c} = \frac{\omega_c^2}{(\omega_c^2-\omega^2) + j\sqrt{2}\omega\omega_c}$$

$$= \frac{\omega_c^2}{(j\omega)^2 + 2j\omega\frac{\omega_c}{\sqrt{2}} + \left(\frac{\omega_c}{\sqrt{2}}\right)^2 + \frac{\omega_c^2}{2}}$$

$$= \frac{\omega_c^2}{\left(\frac{\omega_c}{\sqrt{2}} + j\omega\right)^2 + \left(\frac{\omega_c}{\sqrt{2}}\right)^2}$$

$$= \frac{\frac{\omega_c}{\sqrt{2}} \times \sqrt{2}\omega_c}{\left(\frac{\omega_c}{\sqrt{2}} + j\omega\right)^2 + \left(\frac{\omega_c}{\sqrt{2}}\right)^2}.$$

Using the property $\mathfrak{F}\left[e^{-at}\sin bt\right] = \dfrac{b}{(\omega+a)^2+b^2}$, we find that

$$h(t) = \sqrt{2}\omega_c e^{-\frac{\omega_c}{\sqrt{2}}t}\sin\left(\frac{\omega_c}{\sqrt{2}}t\right)u(t). \tag{2.59}$$

A plot of the impulse response is shown in Fig. 2.29. Note that the general expression for the impulse response of a second order Butterworth LPF can be written as

$$h(t) = 2be^{-bt} \sin bt \; u(t), \tag{2.60}$$

where $b = 2\pi f_c/\sqrt{2}$.

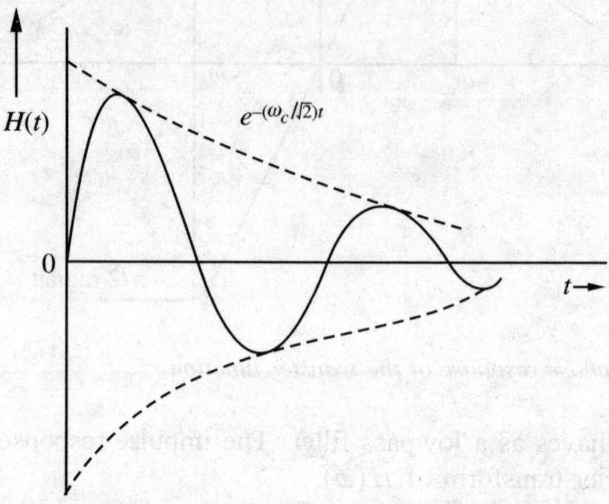

Fig. 2.29 *Impulse response of a practical LPF*

Example 2.12

Derive the expression for the amplitude and the phase response of an RLC bandpass filter shown in Fig. 2.30.

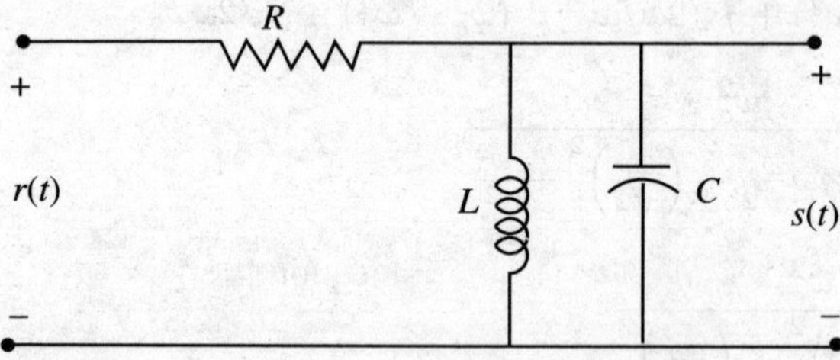

Fig. 2.30 *The circuit that approximates a bandpass filter*

Solution

The behaviour of the circuit can be physically explained as follows.

At very low frequencies the inductor behaves as short circuit and at very high frequencies the capacitor behaves as short circuit. The circuit response (across the parallel combination of L and C) thus reduces to zero at both extremes.

At the intermediate frequencies, the response is nonzero, showing maximum at some frequency. The transfer function is given by

$$H(f) = \frac{Z_{LC}}{R + Z_{LC}} = \frac{j\omega L/(1-\omega^2 LC)}{R + j\omega L/(1-\omega^2 LC)}$$

$$= \frac{j\omega L}{R(1-\omega^2 LC) + j\omega L}, \tag{2.61}$$

where $Z_{LC} = (j\omega L) \,||\, \left(\frac{1}{j\omega C}\right)$.

The magnitude and phase responses are given by

$$|H(f)| = \frac{1}{\sqrt{1 + R^2\left[\frac{1}{2\pi f L} - 2\pi f C\right]^2}} \tag{2.62}$$

and $\phi(f) = -\tan^{-1} R\left(\frac{1}{2\pi f L} - 2\pi f C\right)$.

The quantity $|H(f)|$ is maximum when $\frac{1}{2\pi f L} = 2\pi f C$, that is, when $f = \frac{1}{2\pi\sqrt{LC}} = f_0$ (say).

2.3.4 Hilbert transform and quadrature filters

The Hilbert transform of a time function is obtained by merely shifting the phase of positive frequency components by $-90°$ and that of negative frequency components by $+90°$. The network that implements this operation is called a quadrature filter. The transfer function $H_q(f)$ of the quadrature filter is given by

$$H_q(f) = -j \operatorname{sgn} f = \begin{cases} -j & f > 0 \\ +j & f < 0 \end{cases} \tag{2.63}$$

since a phase-shift of $\pm 90°$ is equivalent to multiplication by $e^{\pm j90} = \pm j$. Figure 2.31 shows the plot of $H_q(f)$. Note that the quadrature filter is an allpass filter, which simply shifts the phases. The impulse response of this system is given by

$$h_q(t) = \frac{1}{\pi t}, \tag{2.64}$$

which has been derived in Chapter 1. Since the output of any system is related to its input by the convolution with its impulse response, the output $\widehat{s}(t)$ of a quadrature filter is given by

$$\widehat{s}(t) = s(t) * h_q(t)$$

$$= s(t) * \frac{1}{\pi t} = \frac{1}{\pi}\int_{-\infty}^{\infty} \frac{s(\tau)}{t-\tau}\, d\tau \equiv H[s(t)]. \tag{2.65}$$

Note that the transform $\widehat{s}(t)$ is again in the time domain. The Hilbert transformation is an effective tool for the cases where the signals are required to be separated on the basis of their phases.

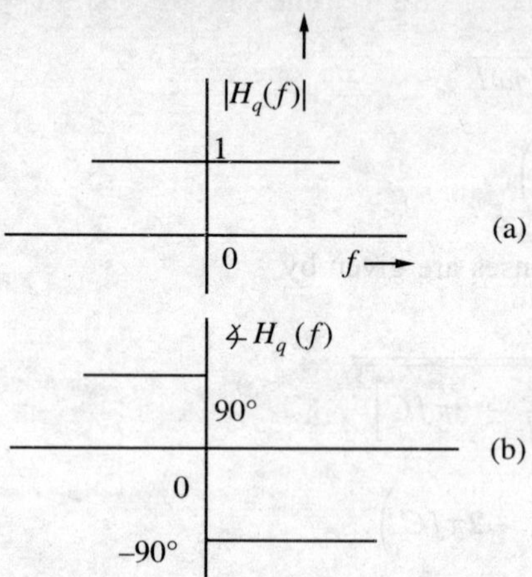

Fig. 2.31 *(a) Magnitude plot of $Hq(f)$. (b) Phase plot of $Hq(f)$.*

Example 2.13
Find the spectrum of the Hilbert transform of $s(t)$.

Solution
Since $\widehat{s}(t) = s(t) * h_q(t)$, so

$$\widehat{S}(f) = \mathfrak{F}[\widehat{s}(t)] = S(f)H_q(f), \quad \text{(by the convolution property)},$$
$$= (-j \operatorname{sgn} f)S(f).$$

Please note that the amplitude spectra of $s(t)$ and $\widehat{s}(t)$ are the same as $|-j \operatorname{sgn} f| = 1$ for all f.

Example 2.14
Find the Hilbert transform of the cosine signal $s(t) = \cos(\omega_0 t + \theta)$.

Solution
Since $s(t) = \cos(\omega_0 t + \theta) = \cos\omega_0(t + \theta/\omega_0)$ and $\mathfrak{F}[\cos\omega_0 t] = \frac{1}{2}[\delta(f - f_0) + \delta(f + f_0)]$, by the time shift property of Fourier transform,

$$S(f) = \frac{1}{2}[\delta(f - f_0) + \delta(f + f_0)]e^{j\theta\omega/\omega_0}.$$

Thus the Fourier transform of $\widehat{s}(t)$ is

$$\widehat{S}(f) = (-j\,\mathrm{sgn}\, f)\frac{1}{2}\left[\delta(f - f_0) + \delta(f + f_0)\right] e^{j\theta\omega/\omega_0}$$

$$= \frac{1}{2j}\left[\delta(f - f_0) + \delta(f + f_0)\right] e^{j\theta\omega/\omega_0}\,\mathrm{sgn}\, f.$$

The quantity on the right side is the spectrum of the sine wave $\sin(\omega_0 t + \theta)$.

Thus the Hilbert transform of a cosine wave is a sine wave.

Example 2.15
Find the Hilbert transform of $\sin\omega t$ and $\cos\omega t$.

Solution
Let an arbitrary signal be $s(t)$ with the Fourier transform $S(\omega)$. Thus,

$$s(t) = \frac{1}{2\pi}\int_{-\infty}^{\infty} S(\omega)e^{j\omega t}\,d\omega = \frac{1}{2\pi}\int_{-\infty}^{\infty} S(\omega)(\cos\omega t + j\sin\omega t)\,d\omega.$$

On the basis of Example 2.14, the conjugate signal $\widehat{s}(t)$ has the Fourier transform

$$\widehat{S}(\omega) = -j\,\mathrm{sgn}(\omega)S(\omega) = \begin{cases} -jS(\omega) & \omega > 0 \\ jS(\omega) & \omega < 0. \end{cases}$$

So,

$$\widehat{s}(t) = \frac{1}{2\pi}\int_{-\infty}^{\infty} -j\,\mathrm{sgn}\,(\omega)S(\omega)[\cos\omega t + j\sin\omega t]\,d\omega$$

$$= \frac{1}{2\pi}\int_{-\infty}^{\infty} \mathrm{sgn}\,(\omega)S(\omega)(\sin\omega t - j\cos\omega t)\,d\omega.$$

Comparing the right-hand sides of $s(t)$ and $\widehat{s}(t)$ in the above equation,

$$H[\cos\omega t] = \sin\omega t\,\mathrm{sgn}\,(\omega)$$

$$H[\sin\omega t] = -\cos\omega t\,\mathrm{sgn}\,(\omega).$$

2.4 Duration–Bandwidth Product

It sometimes happens that the mathematical model of a signal $s(t)$ in the time domain is complicated and the frequency domain description $S(f)$ proves very simple. The signal $s(t)$ and its spectrum $S(f)$ are related by a Fourier transform pair. The bandwidth of a signal refers to the range of frequency within which the spectrum amplitude is not smaller than some specified level. Signal bandwidth is strongly related to the duration of the signal pulse.

2.4.1 Bandlimiting and time-limiting

A signal $s(t)$ is bandlimited to W if the highest frequency component present in the signal does not exceed W. Thus,

$$S(f) = 0 \qquad |f| > W. \tag{2.66}$$

Similarly, the signal is time-limited provided

$$\begin{aligned} s(t) &\neq 0 \qquad t_1 < t < t_2 \\ &= 0 \qquad t < t_1 \text{ and } t > t_2 \end{aligned} \tag{2.67}$$

where t_1 and t_2 mark the times when $s(t)$ starts and ends, respectively.

An ideal lowpass filtering produces strictly bandlimited signals which are not realisable because they have infinite duration. Figure 2.32 shows a strictly bandlimited signal in both time and frequency domains. It can be concluded that a strictly bandlimited signal cannot be time-limited. Similarly, a strictly time-limited signal is unrealisable since the signal energy is distributed at arbitrarily high frequencies. Thus a strict time-limited signal cannot be bandlimited, as depicted in Fig. 2.33.

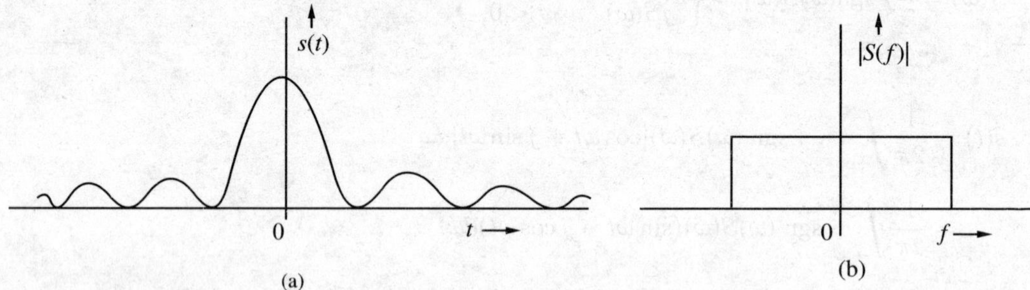

Fig. 2.32 *Bandlimited signal in (a) time domain, (b) frequency domain.*

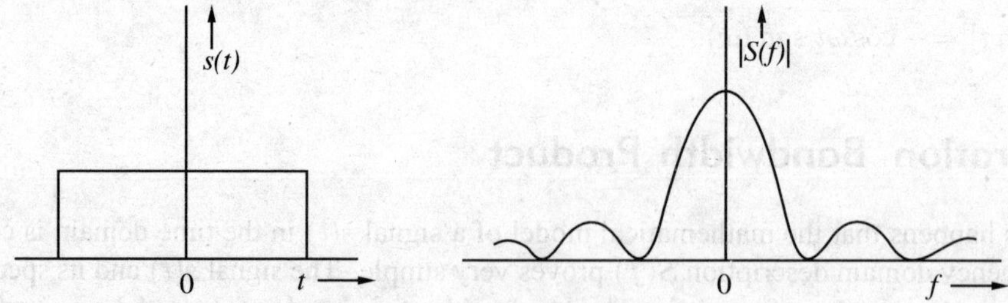

Fig. 2.33 *Time-limited signal in (a) time domain, (b) frequency domain.*

2.4.2 Unit-step response and rise time

Since the output signal spectrum of a filter is the product of the input signal spectrum and the system transfer function, the bandwidth of the output equals approximately the smaller of these two bandwidths. In particular, for a wideband input signal the output bandwidth is restricted by the channel bandwidth. A fast-varying signal such as a rectangular pulse is enriched with high-frequency components. If the transmission channel behaves as a lowpass filter, a significant portion of the high frequency components will be eliminated. This smoothing effect by the filter on a fast-varying input can also be viewed in the time domain.

Suppose a unit-step signal $s(t) = u(t)$ is applied to a lowpass filter having an impulse response $h(t)$. The step response $g(t)$ will be

$$g(t) = \int_{-\infty}^{\infty} h(\tau)u(t-\tau)d\tau = \int_{-\infty}^{t} h(\tau)d\tau \tag{2.68}$$

since $u(t-\tau) = 0$ for $\tau > t$. The result is obvious since a step function is the time integral of an impulse function and so the step response is the time integral of the impulse response. For an RC lowpass filter of Fig. 2.3, the step response $g(t)$ is given by

$$g(t) = (1 - e^{-t/RC})u(t) = (1 - e^{-2\pi f_c t})u(t), \tag{2.69}$$

where $f_c = \frac{1}{2\pi RC}$ is the 3 dB bandwidth. The impulse response of an ideal LPF has been derived in (2.45). Considering unit gain and zero time delay, (2.45) reads as

$$h(t) = 2f_c \text{sinc} 2f_c t. \tag{2.70}$$

Then,

$$g(t) = \int_{-\infty}^{t} 2f_c \,\text{sinc}\,(2f_c\tau)d\tau = \int_{-\infty}^{0} \text{sinc}\,\beta d\beta + \int_{0}^{2f_c t} \text{sinc}\beta d\beta, \tag{2.71}$$

where $\beta = 2f_c\tau$. The first integral is equal to $1/2$ and the second integral can be expressed in terms of the sine integral $Si(\theta)$ defined as

$$Si(\theta) = \int_{0}^{\theta} \frac{\sin x}{x} dx = \pi \int_{0}^{\theta/\pi} \text{sinc}\beta d\beta. \tag{2.72}$$

So, $$g(t) = \frac{1}{2} + \frac{1}{\pi} Si(2f_c t). \tag{2.73}$$

The step response $g(t)$ is shown in Fig. 2.34, showing overshoot, undershoot, precursor, and oscillations. If we include the filter delay t_0, the response, altogether will shift towards the right by t_0. Let us now define the rise time t_r of the step response, which is a measure of the fastness of the step response. In that sense, one can define the rise time as the time required for a signal to attain the final value from the initial state along a straight line, with a slope equalling the maximum slope of

$g(t)$. Since the slope $\frac{dg}{dt}(t)$ is nothing but $h(t)$, hence the maximum slope is the maximum value of $h(t)$, that is, $2f_c$. Thus the time rise is

$$t_r \approx \frac{1}{2f_c}. \tag{2.74}$$

The other definition of rise time is the usual $10 - 90\%$ rise time defined as the time interval between attaining 10% and 90% of $g(t)$. For an ideal LPF, the $10–90\%$ rise time is calculated as $t_r \approx 0.44/f_c$, which closely satisfies (2.74). Equation (2.74) illustrates an important relation between the rise time and the bandwidth. Although (2.74) has been derived for ideal LPF, the inverse relationship between the bandwidth and the rise time is applicable in the general case.

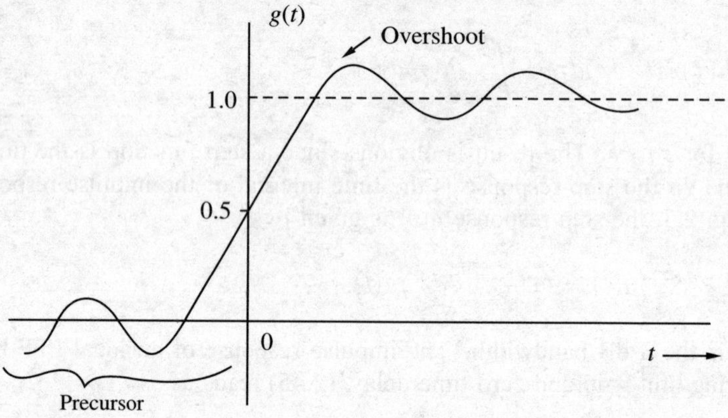

2.34 *Ideal LPF step response*

2.4.3 Bandwidth criteria for transmission channel

The term bandwidth applies to the description of a pulse in the frequency domain; the equivalent time domain concept is the pulse width. For a pulse $s(t)$ shown in Fig. 2.35, the pulse width τ is defined as the width of a rectangle (shown dotted) whose height equals $s(0)$ such that the area of the rectangle is the same as the area under $s(t)$.

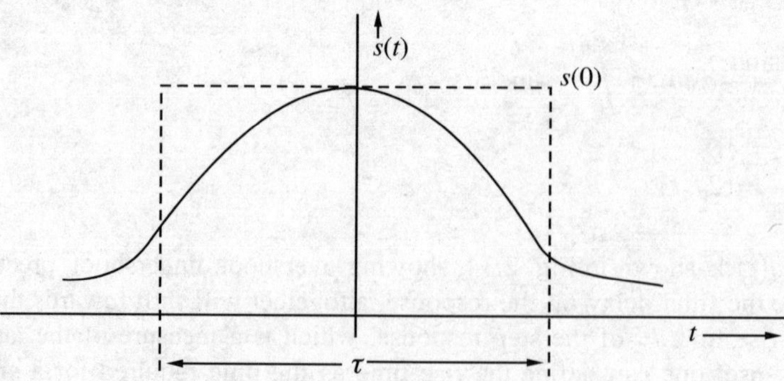

Fig. 2.35 *Defining pulse width*

Mathematically,

$$\tau s(0) = \int_{-\infty}^{\infty} s(t)dt$$

or $$\tau = \frac{\int_{-\infty}^{\infty} s(t)dt}{s(0)}. \tag{2.75}$$

The definition is meaningful only when $s(0)$ gives the maximum value of the pulse.

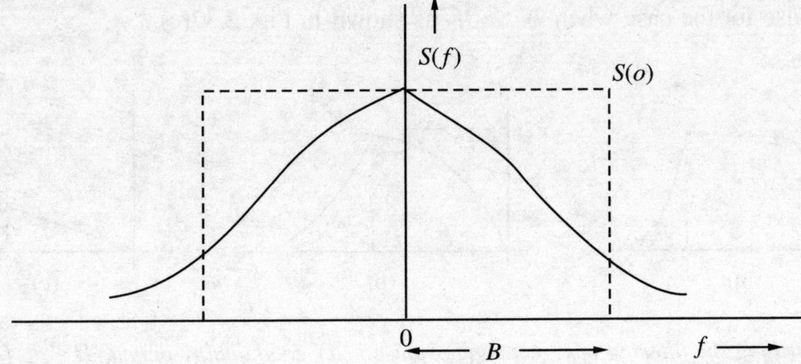

Fig. 2.36 *Spectrum of a time function*

There are a number of criteria for defining the bandwidth of a pulse. The most common definition of bandwidth is made clear by referring to Fig. 2.36. Suppose a time function has its Fourier transform as shown in the figure. The figure has no resemblance to the actual shape of the spectrum of a signal; it is only to help grasp the concept of bandwidth. We define the effective bandwidth B as

$$B = \frac{\int_{-\infty}^{\infty} S(f)df}{2S(o)}, \tag{2.76}$$

although the absolute bandwidth is theoretically infinite. The product of the pulse width and bandwidth is

$$\text{Pulse width} \times \text{Bandwidth} = \frac{\int_{-\infty}^{\infty} s(t)dt \times \int_{-\infty}^{\infty} S(f)df}{2s(o)S(f)}.$$

Since $s(o) = \int_{-\infty}^{\infty} S(f)e^{j2\pi ft}df\Big|_{t=0} = \int_{-\infty}^{\infty} S(f)df$

and $S(o) = \int_{-\infty}^{\infty} s(t)e^{-j2\pi ft}dt\Big|_{f=0} = \int_{-\infty}^{\infty} s(t)dt,$

$$\text{Pulse width} \times \text{Bandwidth} = \frac{1}{2}.$$

The above relation is of prime importance in communications engineering. As the pulse width decreases, its bandwidth broadens and the product has a value of the order of unity. The implication of the above result is that a fast time-varying signal requires more space on the frequency axis. We can see that the signal transmission rate through a channel cannot be increased to a large value as this rate is restricted by the bandwidth of the channel. The following examples clarify the filtering effect on a pulsed signal. Suppose we want to transmit a pulse of bandwidth B through an RC lowpass filter of bandwidth f_c. For the case where $B << f_c$, the output response is a fairly good approximation of the input pulse, as shown in Fig. 2.37(a). For the case where $B \approx f_c$, the output can still be recognised as that of a rectangular pulse (vide Fig. 2.37(b)). But it is hardly possible to recognise the pulse for the case when $B >> f_c$ as shown in Fig. 2.37(c).

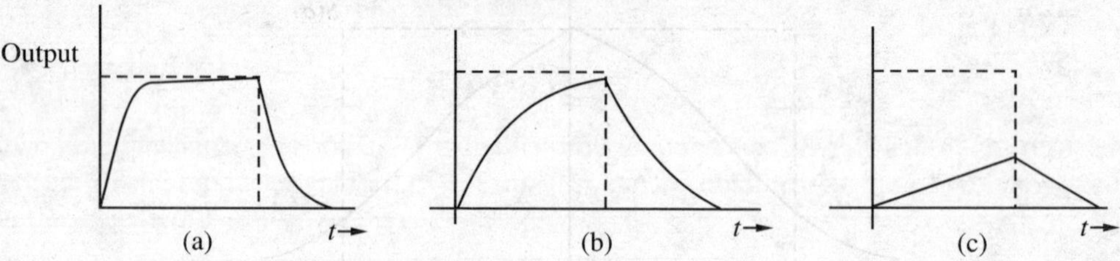

Fig. 2.37 *Illustrating the filtering of a rectangular pulse: (a) good quality output, $B << f_c$, (b) medium quality output, $B \approx f_c$, (c) worst-case output, $B >> f_c$.*

Solved Numericals

1. Find the energy of the Gaussian pulse.

$$s(t) = \frac{1}{\sqrt{2\pi\sigma^2}} e^{-t^2/2\sigma^2}$$

Solution

By direct integration,

$$\text{Energy in } s(t), E_s = \int_{-\infty}^{\infty} |s^2(t)| dt = \frac{1}{2\pi\sigma^2} \int_{-\infty}^{\infty} e^{-t^2/\sigma^2} dt$$

Let $x = t/\sigma$ so that $dt = \sigma dx$.

$$\text{Thus } E_s = \frac{1}{2\pi\sigma^2} \int_{-\infty}^{\infty} e^{-x^2} \sigma dx = \frac{1}{2\pi\sigma} \int_{-\infty}^{\infty} e^{-x^2} dx = \frac{\sqrt{\pi}}{2\pi\sigma} = \frac{1}{2\sigma\sqrt{\pi}}$$

$$\left(\text{since} \int_{-\infty}^{\infty} e^{-x^2} dx = \sqrt{\pi} \right)$$

2. Consider the signal $s(t) = e^{-at}u(t)$. Find the original bandwidth B in Hz that accommodates 99 % of the signal energy.

Solution

We have $s(t) = \dfrac{1}{i2\pi f + a}$

So, the energy spectral density (Eq. 2.20) is given by

$$G_s(f) = |s(f)|^2 = \frac{1}{(2\pi f)^2 + a^2}$$

The total energy E_s is given by

$$E_s = \int_{-\infty}^{\infty} s^2(t)dt = \int_{-\infty}^{\infty} e^{-2at}\, dt = \frac{1}{2a}$$

(E_s can be calculated as $E_s = \int_{-\infty}^{\infty} |s(f)|^2 df$)

Let B Hz be the required bandwidth that includes 99% of the total energy. Then the area under the energy spectral density from $-B$ Hz to B Hz would be

$$0.99 \times \frac{1}{2a} = \int_{-B}^{B} \frac{df}{(2\pi f)^2 + a^2}$$

Let $2\pi f = x$

$$\text{Then, } \frac{1}{2\pi} \cdot \int_{-2\pi B}^{+2\pi B} \frac{dx}{x^2 + a^2} = \frac{1}{2\pi a} \cdot \tan^{-1}[x/a]\,\Big|_{-2\pi B}^{+2\pi B}$$

$$= \frac{1}{2\pi a} 2\tan^{-1}\left(\frac{2\pi B}{a}\right) = \frac{1}{\pi a}\tan^{-1}\left(\frac{2\pi B}{a}\right)$$

$$\text{So, } \tan^{-1}\left(\frac{2\pi B}{a}\right) = \frac{0.99\pi}{2}$$

$$\text{or, } B = \frac{a}{2\pi}\tan\left(\frac{0.99\pi}{2}\right) = 10.13a\,\text{Hz}$$

In a similar way, the 90% bandwidth can be obtained as

$$B = \frac{a}{2\pi}\tan\left(\frac{0.9\pi}{2}\right) \approx a\text{Hz}$$

3. With reference to the Hilbert transform (Section 2.3.4), prove the following.
   a) When the signal $s(t)$ is a constant, its Hilbert transform is zero.
   b) Hilbert transform of an even signal is an odd signal and vice versa.

c) The original signal $s(t)$ and its Hilbert transform $\hat{s}(t)$ are orthogonal.

Solution

(a) when $s(t) = A$, a constant, then

$$S(f) = A\delta(f)$$

So, output of the Hilbert filter

$$\hat{S}(f) = S(f)H_q(f) = A\delta(f)H_q(f)$$

Since

$$H_q(f) = -j\text{sgn}f = \begin{cases} -j & f > 0 \\ 0 & f = 0 \\ -j & f < 0 \end{cases} \quad \text{(vide Eq. 2.63)}$$

we have $\hat{S}(f) = 0$, as $H_q(f) = 0$ for $f = 0$.

b) For an even signal, $s(t) = s(-t)$ and by Eq. (2.65), we may write the output signal

$$\hat{s}(t) = s(t) * h_q(t) = \int_{-\infty}^{\infty} s(\tau)h_q(t-\tau)d\tau$$

We may then write

$$\begin{aligned} s(-t) * h_q(-t) &= \int_{-\infty}^{\infty} s(\tau)h_q(-t-\tau)d\tau \\ &= \int_{-\infty}^{\infty} s(\tau)h_q[-(t+\tau)]\,d\tau \\ &= \int_{-\infty}^{\infty} s(\tau)h_q[(-t)-\tau]\,d\tau = \hat{s}(-t) \end{aligned}$$

Thus, $\hat{s}(t) = s(t) * h_q(t)$ and $\hat{s}(-t) = s(-t) * h_q(-t)$

Now, $h_q(t) = \dfrac{1}{\pi t}$, (vide Eq (2.64))

So, $h_q(-t) = -\dfrac{1}{\pi t} = -h_q(t)$.

Thus, for even $s(t)$,

$$\hat{s}(-t) = s(t) * [-h_q(t)] = -\hat{s}(t)$$

So, Hilbert transform $\hat{s}(t)$ is an odd function. When the signal $s(t)$ is odd, that is $s(-t) = -s(t)$, we have

$$\hat{s}(-t) = -s(t) * [-h_q(t)] = s(t) * h_q(t) = \hat{s}(t),$$

So in this case, Hilbert transform is an even function.

c) In part (b), we have proved that if s(t) is even, $\hat{s}(t)$ is odd and vice versa. That is, the product $s(t)\hat{s}(t)$ is always an odd function. So over infinite duration, the integration

$$\int_{-\infty}^{\infty} s(t)\hat{s}(t)dt = \int_{-\infty}^{0} s(t)\hat{s}(t)dt + \int_{0}^{\infty} s(t)\hat{s}(t)dt = 0$$

and hence, the signal and its Hilbert transform are orthogonal over infinite time interval.

4. A channel producing nonlinear distortion is given by the input-output relation

$$s(t) = r(t) + 10^{-4}r^2(t) \text{ (vide Eq. 2.41)}$$

Show that the bandwidth of the output signal s(t) is twice that of the input signal r(t). Also justify that simply low-pass filtering will not recover the original signal (because of this distortion).

Solution
Let $r(t)$ be band limited to B Hz, that is

$$R(f) = 0 \text{ for } |f| > B$$

Since $r^2(t) = r(t)r(t)$ and product in time domain manifests in convolution in the frequency domain, we have Fourier transform of $r^2(t) = R(f) * R(f)$ and thus $R(f) * R(f) = 0$ for $|f| > 2B$. In fact,

$$S(f) = R(f) + 10^{-4}R(f) * R(f).$$

Thus, the spectra will be the desired signal plus the extended spectra of the distorting signal. So simple low-pass filtering will not remove the interference due to overlapping spectrum.

As an illustration, if the input signal be a sync pulse, that is

$$r(t) = 2B \sin c(2\pi Bt),$$

then, its spectrum is

$$R(f) = \Pi\left(\frac{f}{2B}\right),$$

a rectangular shaped low-pass spectrum of height unity and frequency ranging from $-B$ to B Hz. The signal $r^2(t) = 4B^2 \sin c^2(2\pi Bt)$ will have its spectrum given by $R(f) * R(f) = 2B \cdot \wedge\left(\frac{f}{4B}\right)$, a triangularly shaped spectrum of peak 2B at $f = 0$ and ranging from $-2B$ to $+2B$ Hz.

$\left(\text{Since } A\text{sinc}^2(\pi A t) \leftrightarrow \wedge\left({}^{f}/_{2A}\right)\right)$

The resulting spectra are shown in the adjoining figures.

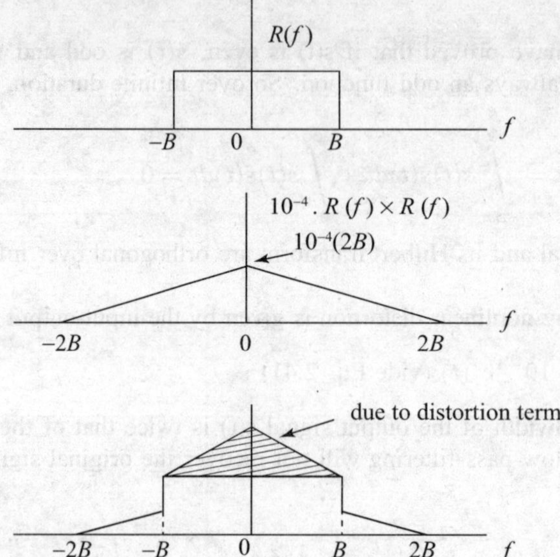

It is obvious that because of overlapping with the distortion term within the baseband range, the low-pass filtering will not reproduce r(t) exactly.

5. The frequency response of a distorted channel is given by

$$H(f) = \cos\pi f, \quad \frac{-1}{2} < f < \frac{1}{2}.$$

Find the channel impulse response $h(t)$.

Solution

Applying inverse Fourier transformation,

$$h(t) = \int_{-\infty}^{\infty} H(f)e^{i2\pi ft}\,df = \int_{-1/2}^{1/2} \cos(\pi f)e^{i2\pi ft}\,df$$

Now, $\cos\pi f = \frac{1}{2}\left(e^{i\pi f} + e^{-i\pi f}\right)$

So,

$$\cos\pi f \cdot e^{i2\pi ft} = \frac{1}{2}\left(e^{i\pi f} \cdot e^{i2\pi ft} + e^{-i\pi f} \cdot e^{i2\pi ft}\right)$$

$$= \frac{1}{2}\left(e^{i\pi f(1+2t)} + e^{i\pi f(1-2t)}\right)$$

Thus,

$$h(t) = \frac{1}{2}\int_{-1/2}^{1/2} e^{i\pi f(1+2t)}\, df + \frac{1}{2}\int_{-1/2}^{1/2} e^{i\pi f(1-2t)}\, df$$

$$= \frac{1}{2}.\frac{1}{j\pi(1+2t)}.e^{i\pi f(1+2t)}\Big|_{-1/2}^{1/2} + \frac{1}{2}.\frac{1}{-j\pi(1-2t)}.e^{-i\pi f(1-2t)}\Big|_{-1/2}^{1/2}$$

$$= \frac{1}{2i\pi(1+2t)}\cdot\left[e^{i\pi(1+2t)/2} - e^{-i\pi(1+2t)/2}\right] - \frac{1}{2i\pi(1-2t)}\cdot\left[e^{-i\pi(1-2t)/2} - e^{-i\pi(1-2t)/2}\right]$$

$$= \frac{1}{\pi(1+2t)}\sin\frac{\pi}{2}(1+2t) + \frac{1}{\pi(1-2t)}\sin\frac{\pi}{2}(1-2t)$$

$$= \frac{1}{\pi(1+2t)}\cos\pi t + \frac{1}{\pi(1-2t)}\cos\pi t$$

$$= \frac{1}{\pi}(\cos\pi t)\cdot\left[\frac{1-2t+1+2t}{1-4t^2}\right]$$

$$h(t) = \frac{2}{\pi}\frac{\cos\pi t}{1\ \ 4t^2}$$

Exercises

Activity

2.1 A system is described by the differential equation $\frac{dy(t)}{dt} + y(t) = x(t)$, where $x(t)$ and $y(t)$ are the input and output of the system. Check whether the system is linear on not.

2.2 Consider the network as shown in Fig. P2.1. Find and sketch $|H(f)|$ and $\measuredangle H(f)$ for the system.

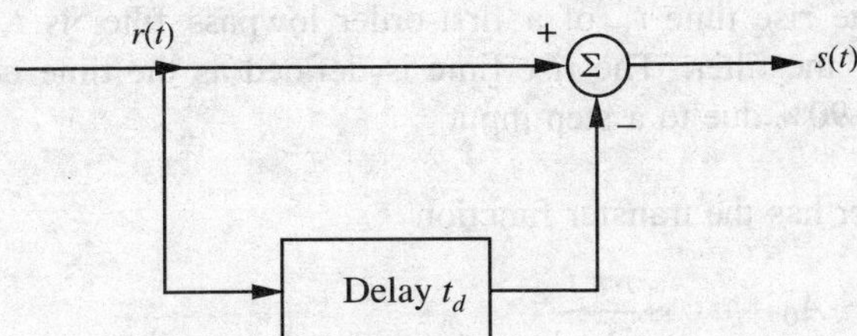

Fig. P2.1

2.3 Find the system function $H(f)$ of a system whose input $r(t)$ and output $s(t)$ are related as

$$\frac{ds}{dt}(t) + 2\pi s(t) = \frac{dr}{dt}(t) + 8\pi r(t).$$

2.4 Find the overall system function $H(f)$ of the negative feedback system shown in Fig. P2.2

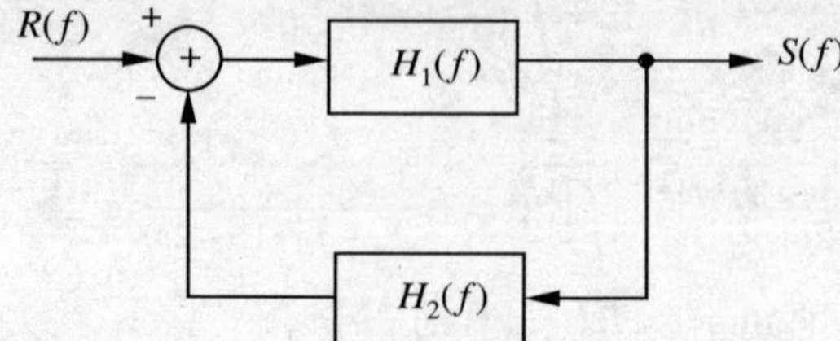

Fig. P2.2

2.5 The input to the RC network of Fig. P2.3 is a finite power signal with power spectral density $G_i(f) = \delta(f+1) + \delta(f-1)$. Find the average power at the output.

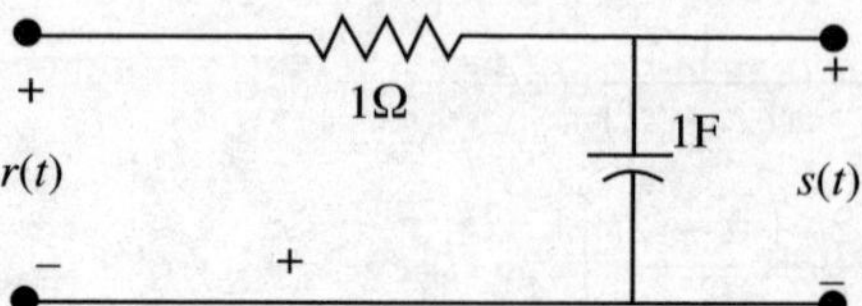

Fig. P2.3

2.6 Verify that a first order lowpass filter offers almost distortionless transmission if the input $r(t)$ is bandlimited to $WH_z << f_c$, where f_c is the filter's cutoff frequency.

2.7 A bandpass channel has the transfer function $H(f)$ given by

$$H(f) = \begin{cases} 1 & f_L < |f| < f_H \\ 0 & \text{otherwise.} \end{cases}$$

Prove that the output $s(t)$ of the filter is given by

$$s(t) \approx (f_H - f_L)|R(f_0)| \cos[2\pi f_0 t + \theta],$$

where the input spectrum $R(f_0)$ is evaluated at the centre frequency $f_0 = \frac{1}{2}(f_L + f_H)$ and θ is the phase of $R(f_0)$. Assume very narrow bandwidth, i.e., $(f_H - f_L) << f_0$.

2.8 Prove that the rise time t_r of a first-order lowpass filter is $t_r \approx 0.35/B$, where B is the bandwidth of the filter. The rise time is defined as the time needed to change the response from 10% to 90% due to a step input.

2.9 A cosine filter has the transfer function

$$H(f) = A_0 + a \cos \frac{n\pi f}{f_m}.$$

Find the impulse response $h(t)$.

2.10 Suppose $r(t) = 2\cos 2\pi f_0 t$ is applied to a nonlinear system described by $s(t) = 3r(t) - 2r^2(t) + r^3(t)$. Compute the second-and third harmonic distortion.

Review questions

1. (a) What is meant by a linear time-invariant (LTI) system? Prove that the commutative property of convolution holds in an LTI system.
   (b) Define impulse response of a system. Find the impulse response of the LTI system described by $y(t) = x(t - t_0)$.
   (c) Compute the impulse response of the inverse system of an accumulator. (K.U. 2002)

2. (a) Verify that the system described by the input–output relation $\dfrac{dy(t)}{dt} + ty(t) = x(t)$ is linear. Here $x(t)$is the input and $y(t)$ is the response.
   (b) Justify that series connections of linear time-invariant system produce another LTI system.

3. (a) Show that for a distortionless filter, the gain and group delay are constants over the nonzero range of the input signal spectrum.
   (b) Write down the transfer function of the nth-order Butterworth lowpass filter. Sketch the normalised amplitude response.
   (c) Show that the Butterworth polynomial is $1 + \sqrt{2}s + s^2$ for $n = 2$.

4. Obtain the conditions for distortionless transmission through a linear system. Find an expression for the transfer function of a practical lowpass filter as shown in Fig. P2.4.

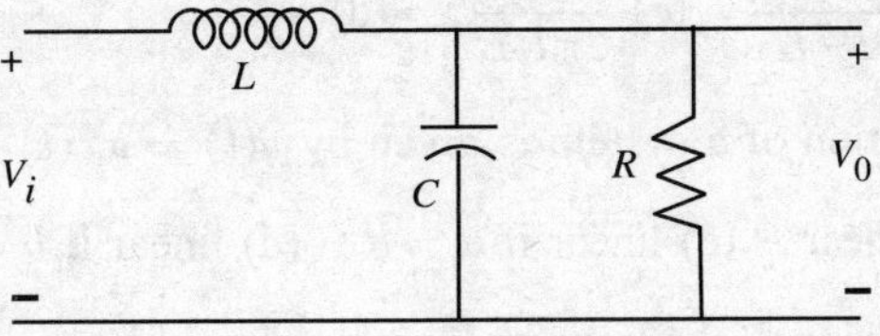

Fig. P2.4

Give an impulse response of an ideal lowpass filter. Is an deal LPF physically realisable? Answer, giving reason. [RPE 2004)

5. (a) If $H(j\omega)$ is the transfer function of a linear system, prove that $|H(j\omega)|$ is an even function and $|\theta(j\omega)|$ is an odd function.
   (b) A certain channel has ideal amplitude response but non-ideal phase response given by $|H(j\omega)| = 1$ and $\theta(\omega) = -\omega t_0 - k\sin\omega T$, where constant $k << 1$. Show that $\mathfrak{F}(t)$, the channel response to an input signal $g(t)$ bandlimited to β Hz, is given by $\mathfrak{F}(t) = g(t - t_0) + \frac{1}{2}[g(t - t_0 - T) - g(t - t_0 + T)]$. (WBUT 2005)

6. An input signal $s(t) = 5\cos\frac{\pi}{3}t + 10\cos\frac{\pi}{2}t$ is applied to a system with frequency response $H(f) = 2e^{-j2w}$. Find the steady-state output of the system. (Hint: $y(t) = \alpha x(t - t_d)$ implies $H(f) = \alpha \cdot e^{-jwt_d}$).

7. A signal is given as $s(t) = A\pi\left[\frac{t-\tau}{T}\right]$. Sketch the signal. Find the power spectral density and autocorrelation function of the signal.

8. Derive the relation between autocorrelation function and power spectrum density of an aperiodic signal. How are they related if the signal be a sequence (in discrete time)?

9. The autocorrelation function $R_{xx}[n]$ of a discrete{time sequence $x[n]$ is given by $R_{xx}[n] = (0.8)^{|n|}$. Find its power spectral density.

10. A system is defined by its input{output relation $y = f(x) = 5x + 10$. Comment on system linearity. Repeat this problem if $f(x) = x^3$.

Multiple-choice questions

1. The group delay $\tau(\omega)$ is related with the phase $\phi(\omega)$ as

   (a) $\tau(\omega) = \frac{d\phi}{d\omega}$ (b) $\tau(\omega) = -\frac{d\phi}{d\omega}$ (c) $\tau(\omega) = \frac{d^2\phi}{d\omega^2}$ (d) $\tau(\omega) = -\frac{d^2\phi}{d\omega^2}$

2. The impulse response of a filter, matched to rectangular pulse is

   (a) an impulse (b) a unit-step (c) a rectangular pulse (d) a triangular pulse

3. The $R - L$ lowpass filter has the cutoff frequency f_c

   (a) $\frac{1}{2\pi L/R}$ (b) $\frac{1}{2\pi R/L}$ (c) $\frac{1}{2\pi RL}$ (d) $\frac{RL}{2\pi}$

4. The input–output relation of a system is given by $y(t) = ax(t) + b$. The system is

   (a) linear (b) nonlinear (c) linear if $a = b$ (d) linear if $b = 0$

5. For a distortionless channel, the transfer function should have

   (a) Constant amplitude and nonlinear phase-shift
   (b) constant amplitude and linear phase-shift
   (c) frequency-dependent amplitude and linear phase-shift
   (d) frequency dependent amplitude and nonlinear phase-shift

6. Intermodulation distortion is

   (a) linear distortion (b) nonlinear distortion
   (c) a class of harmonic distortion (d) caused only when two or more signals are present

7. The width of a pulse is a doubled. The spectrum width will be

   (a) doubled (b) halved (c) unchanged (d) nothing can be said

8. For a transmitted signal of $s(t) = A\cos(2\pi \cdot 10^7 t)$, the received signal is $r(t) = A\cos\left(2\pi \cdot 10^7 t - \frac{\pi}{4}\right)$. The minimum distance between source and destination is

   (a) $\lambda/4$ (b) 4λ (c) $\lambda/8$ (d) None

9. A signal $s(t)$ is band-limited to B Hz. The signal $s^n(t)$ will have its band limited to

   (a) infinite bandwidth (b) B Hz (c) nB Hz (d) B/n Hz

10. An ideal low-pass filter should have its phase spectrum in the pass band

    (a) linear and positive slope (b) linear and negative slope (c) zero phase (d) None

3 Continuous-Wave (CW) Modulation

The purpose of a communications system is to transmit message signals over some distance through a communication channel. The message signal, also termed the information-bearing signal, is generated by the source of information. This original signal is also referred to as the baseband signal. The term baseband designates the band of frequencies occupied by the source signal in its original form. In the case of a speech (or audio), the sound generated by the vibration of vocal chords represents the baseband signal, bandlimited to around 15 kHz. These low frequency signals may be sent directly through wires over short distances as in the local loop of a telephone. But effective long-haul transmission usually requires frequency translation or modulation. Suppose we attempt to transmit an audio signal of frequency 10 kHz through air. This audio tone corresponds to a wavelength of 30 km in air. Since an antenna operates efficiently when its physical size is comparable to the wavelength of the signal being transmitted, a quarter-wavelength antenna would then have to be 7.5 km in size. This would be a bit of impractical. By a shift of the audio spectrum to some high frequency, the size of the antenna can be reduced proportionately. The other benefit of frequency translation is that it allows multiplexing of a number of distinct messages over a common channel. The distinct message signals are translated to different locations on the frequency axis, which the receiver can easily separate. The added advantage in multiplexing is that the modified signal is less susceptible to noise than the original signal.

Frequency translation also serves the purpose of converting a wideband signal into a narrowband. The terms wideband or narrowband do not mean the absolute bandwidth of the message signal; it simply refers to the fractional bandwidth B/f_c, where B is the bandwidth of the message signal and f_c is some high frequency at which the message is translated. For speech or music transmission, the baseband signal is usually confined within 100 Hz to 10 kHz, with a highest to lowest frequency ratio of 100. So the antenna size also differs by the same ratio at the two edges. This irregularity is removed, however, if the baseband spectrum is taken to some high frequency, say 100 kHz. The spectrum of the modified signal now occupies the bandwidth in the range of 100.1 kHz to 110 kHz, with a highest to lowest frequency ratio of 1.09. This ensures a uniform size for the antenna within the band.

We thus see that for effective communication by frequency translation, the transmission system requires a bandpass transfer function. Let us begin with the analysis of the bandpass systems and signals.

3.1 Bandpass Signal Transmission

The frequency response of a bandpass filter has been discussed in Chapter 2. A bandpass filter was defined in the frequency domain as

$$H_{\text{bp}}(f) = 0 \quad |f| < f_c - W \text{ and } |f| > F_c + W, \tag{3.1}$$

where $2W$ and f_c denote the bandwidth and centre frequency of the filter, respectively. The bandpass signal in the time domain can be written as

$$v_{\text{bp}}(t) = A(t)\cos[\omega_c t + \phi(t)], \tag{3.2}$$

where the time functions $A(t)$ and $\phi(t)$ denote respectively the amplitude and the phase. The quantity $A(t)$ is the slowly varying envelope of the bandpass signal; the frequency of $A(t)$ is W of (3.1). The phasor diagram of (3.2) is shown in Fig. (3.1), where the angular rotation at the frequency f_c is omitted. Now (3.1) can be put as

$$v_{\text{bp}}(t) = A(t)\cos\phi(t)\cos\omega_c t - A(t)\sin\phi(t)\sin\omega_c t$$

$$= x(t)\cos\omega_c t - y(t)\sin\omega_c t, \tag{3.3}$$

where $x(t) = A(t)\cos\phi(t), \quad y(t) = A(t)\sin\phi(t).$ (3.4)

Equation (3.3) is the quadrature-carrier representation of a bandpass signal. The functions $x(t)$ and $y(t)$ are called the in-phase and quadrature components, respectively. By applying modulation property, the Fourier transform of (3.3) gives

$$V_{\text{bp}}(f) = \frac{1}{2}[X(f - f_c) + X(f + f_c)] + \frac{j}{2}[Y(f - f_c) - Y(f + f_c)]. \tag{3.5}$$

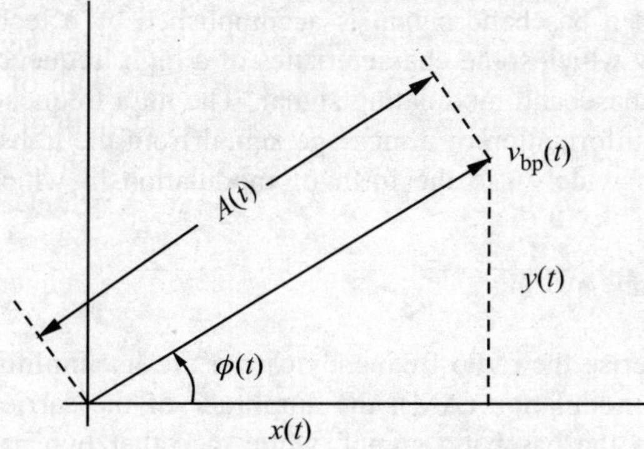

Fig. 3.1 *Phasor diagram of* $v_{\text{bp}}(t)$

Here $X(f)$ and $Y(f)$ are the spectra of $x(t)$ and $y(t)$, respectively. By comparing (3.5) with (3.1), one can argue that the in-phase and quadrature components must be lowpass signals:

$$X(f) = Y(f) = 0, \quad |f| > W. \tag{3.6}$$

The lowpass equivalent spectrum is then

$$v_{\text{lp}}(t) = \tfrac{1}{2}[X(f) + jY(f)]. \tag{3.7}$$

Only the positive frequency of $v_{\text{bp}}(f)$ has been considered in (3.7). The corresponding time function is

$$\begin{aligned} v_{\text{lp}}(t) &= \tfrac{1}{2}[x(t) + jy(t)] \\ &= \frac{1}{2}[A(t)\cos\phi(t) + jA(t)\sin\phi(t)] = \frac{1}{2}A(t)e^{j\phi(t)}. \end{aligned} \tag{3.8}$$

Again, from (3.2),

$$v_{\text{bp}}(t) = \text{Re}\left[A(t)e^{j[\omega_c t + \phi(t)]}\right] = 2\text{Re}[v_{\text{lp}}(t)e^{j\omega_c t}]. \tag{3.9}$$

Thus, a lowpass signal multiplied by a complex exponential results in a bandpass signal in the time domain. The frequency domain expression is

$$V_{\text{bp}}(f) = V_{\text{lp}}(f - f_c) + V_{\text{lp}}^*(-f - f_c), \tag{3.10}$$

where the term V_{lp}^* represents the negative frequency portion. Since the bandpass signals that we mostly encounter are real,

$$V_{\text{bp}}(f) = V_{\text{lp}}(f - f_c), \quad f > 0. \tag{3.11}$$

3.2 Amplitude Modulation

The frequency translation of a baseband signal is accomplished by a technique called modulation. Modulation is the process by which some characteristics of a high frequency signal called the carrier is changed according to the baseband modulating signal. The high frequency signal is named carrier since it is used to carry the information of a message signal from the transmitter to the receiver.

Communications systems widely use the form of modulation in which the carrier is simply a harmonic wave of the form

$$s_c(t) = A_c \cos[\omega_c t + \phi_c(t)]. \tag{3.12}$$

The parameters that characterise the radio frequency carrier are its amplitude A_c, frequency ω_c, and phase $\phi_c(t)$. In amplitude modulation (AM), the amplitude of the carrier wave is varied about a mean value in proportion to the baseband signal, while the other two parameters, ω_c and ϕ_c, are kept constant. The variation of ω_c or $\phi_c(t)$ of the carrier signal in accordance with the baseband modulating signal leads to frequency modulation or phase modulation, respectively.

Let the modulating signal be of the form

$$s_m(t) = A_m \cos(\omega_m t + \phi_m). \tag{3.13}$$

Without loss of generality, the phase angles ϕ_c and ϕ_m in (3.12) and (3.13) can be omitted. Since in an AM signal the amplitude of the carrier is made to vary proportionately to the baseband signal, the modulated carrier amplitude becomes

$$A(t) = A_c + ks_m(t) = A_c + kA_m \cos \omega_m t,$$

where k is the constant of proportionality that can be set equal to unity without loosing any information. The complete AM signal is then

$$\begin{aligned} s_{\text{AM}}(t) &= A(t) \cos \omega_C t = [A_c + ks_m(t)] \cos \omega_c t \\ &= (A_c + kA_m \cos \omega_m t) \cos \omega_c t = A_c \cos \omega_c t + ks_m(t) \cos \omega_c t \\ &= A_c(1 + m \cos \omega_m t) \cos \omega_c t. \end{aligned} \tag{3.14}$$

The quantity $m = \frac{A_m}{A_c}$ is a positive constant and is called the modulation index or depth of modulation. Figure 3.2 shows a typical message signal $s_m(t)$ and the resulting AM signal $s_{\text{AM}}(t)$ for two values of m, namely, $m < 1$ and $m > 1$. For mathematical convenience, we normalise the message signal such that $A_m \leq 1$.

We observe that the envelope of $s_{\text{AM}}(t)$ has the same shape as $s_m(t)$ provided that $f_c >> W$ and $m \leq 1$. Here W represents the message bandwidth. The condition $f_c >> W$ ensures that the carrier oscillates much faster than the modulating signal and only then can the envelope be visualised satisfactorily. The second condition, $m \leq 1$, ensures that $A_c(1 + m \cos \omega_m t)$ does not go negative. With $0 < m < 1$, the envelope varies between $A_{\text{max}} = A_c(1 + m)$ and $A_{\text{min}} = A_c(1 - m)$. With 100% modulation ($m = 1$), $A_{\text{max}} = 2A_c$ and $A_{\text{min}} = 0$. Note that overmodulation ($m > 1$) results in carrier phase reversal that yields distorted envelope.

For a modulating signal of $f_m = 1$ kHz with carrier frequency of $f_c = 100$ kHz, the ratio $f_c/f_m = 100$. So the carrier may not be resolvable on a CRO when observing few periods of the modulating signal.

3.2.1 AM spectra

The Fourier transform of (3.14) gives

$$S_{\text{AM}}(f) = \frac{A_c}{2}[\delta(f - f_c) + \delta(f + f_c)] + \frac{k}{2}[S_m(f - f_c) + S_m(f + f_c)]. \tag{3.15}$$

The positive-frequency half of $S_{\text{AM}}(f)$ is given by

$$S'_{\text{AM}}(f) = \frac{A_c}{2}\delta(f - f_c) + \frac{k}{2}S_m(f - f_c), \quad f > 0. \tag{3.16}$$

Figure 3.3 shows the spectra. The modulated waveform $s_{\text{AM}}(t)$ contains frequencies between $f_c - f_m$ and $f_c + f_m$. The baseband signal $s_m(t)$ contains frequencies between 0 and f_m. The modulated spectra are symmetrical about the carrier-frequency impulses at $\pm f_c$. The frequency band $|f_c|$ to $|f_c + f_m|$ is referred to as the upper sideband (USB) and the band $|f_c - f_B|$ to $|f_c|$ is referred to as the lower sideband (LSB). The hatched portions of Fig. 3.3(b) are the USB. Because of these two sidebands, the modulation is also called double-sideband (DSB) amplitude modulation.

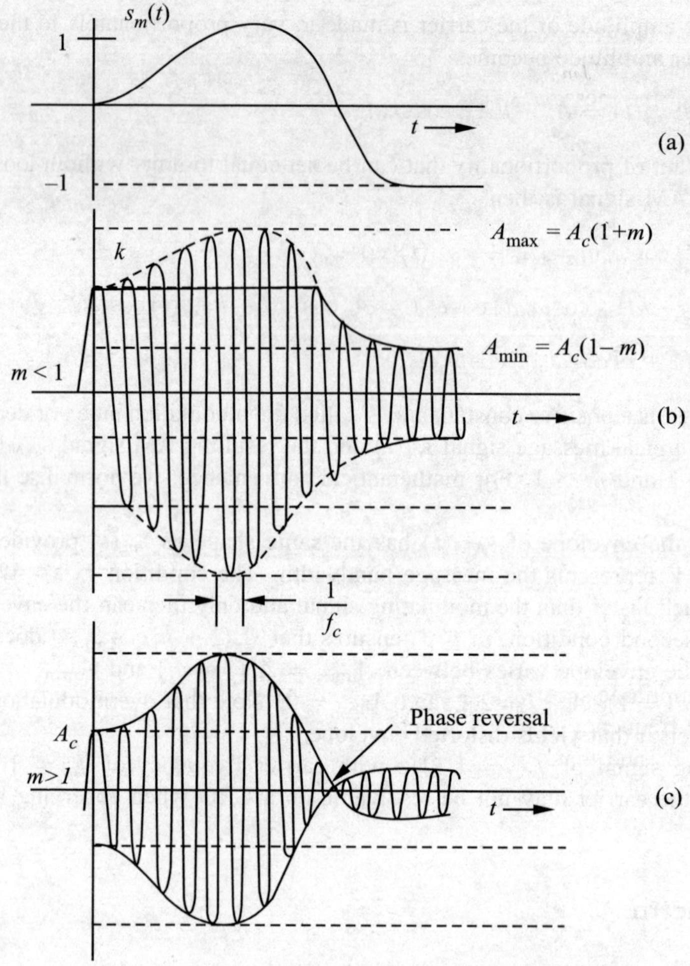

Fig. 3.2 *(a) Baseband; (b) AM signal with $m < 1$; (c) AM signal with $m > 1$.*

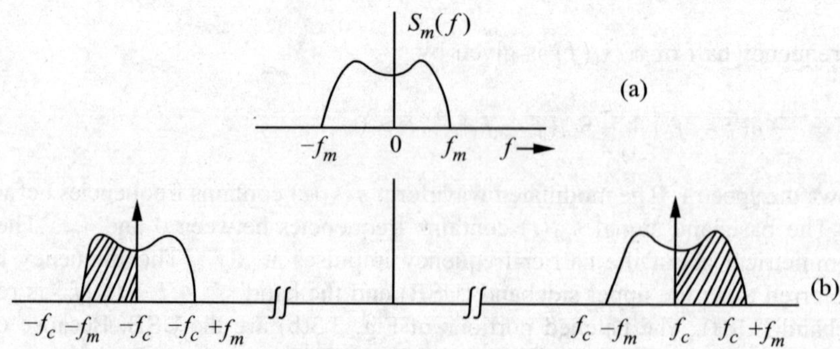

Fig. 3.3 *Fourier transform of: (a) baseband signal; (b) modulated wave.*

For positive frequencies, the highest frequency component in AM wave is $f_c + f_m$ and the lowest frequency component is $f_c - f_m$. That is why the transmission bandwidth B_T for the AM wave is twice the bandwidth of the message. That is,

$$B_T = 2f_m. \tag{3.17}$$

Example 3.1
Find the spectra of a tone-modulated DSB signal.

Solution
For a single tone at the frequency f_1, the unmodulated baseband $s(t)$ is given as

$$s(t) = A_m \cos 2\pi f_1 t$$

and hence the DSB with carrier waveform takes the form

$$\begin{aligned} s_{\text{AM}}(t) &= A_c[1 + m\cos 2\pi f_1 t]\cos 2\pi f_c t \\ &= A_c \cos 2\pi f_c t + \frac{mA_c}{2}[\cos 2\pi(f_c - f_1)t + \cos 2\pi(f_c + f_1)t]. \end{aligned} \tag{3.18}$$

The corresponding amplitude spectra for the tone-modulated AM wave are plotted in Fig. 3.4. The positive and negative frequency spectra are symmetrical. Observe that the amplitude of the carrier-frequency impulse is A_c and that of the side frequencies is $mA_c/2$.

AM power distribution

The average transmitted power, P_t, of the AM signal is given by

$$P_t = \langle s^2_{\text{AM}}(t)\rangle = \frac{1}{T}\int_0^T s^2_{\text{AM}}(t)dt.$$

Since, $s_{\text{AM}}(t) = A_c \cos\omega_c t + \dfrac{mA_c}{2}\left[\cos(\omega_c + \omega_m)t + \cos(\omega_c - \omega_m)t\right]$,

hence the instantaneous power $p(t)$ is

$$\begin{aligned} p(t) &= s^2_{\text{AM}}(t)/R \\ &= \frac{A_c^2}{R}\cos^2\omega_c t + \frac{1}{R}\left(\frac{mA_c}{2}\right)^2 [\cos^2(\omega_c + \omega_m)t + \cos^2(\omega_c - \omega_m)t \\ &\quad + 2\cos(\omega_c + \omega_m)t\cos(\omega_c + \omega_m)t] \\ &\quad + 2\frac{A_c}{R}\cos\omega_c t\left(\frac{mA_c}{2}\right)[\cos(\omega_c + \omega_m)t + \cos(\omega_c - \omega_m)t]. \end{aligned} \tag{3.19}$$

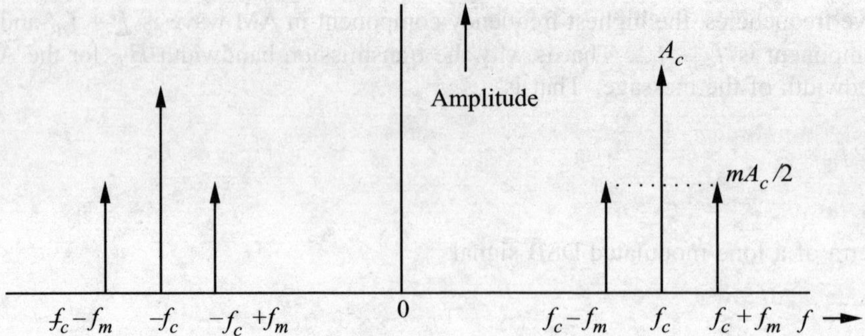

Fig. 3.4 *Line spectra for single-tone modulated AM wave*

To find the normalised average power over a period of the high-frequency carrier, we make use of the formula

$$P_{\text{av}} = \lim_{T\to\infty} \frac{1}{T}\int_0^T p(t)dt. \tag{3.20}$$

In doing so,

$$P_t = \frac{A_c^2}{R}\langle\cos^2\omega_c t\rangle + \frac{1}{R}\left(\frac{mA_c}{2}\right)^2 [(\cos^2(\omega_c+\omega_m)t) + (\cos^2(\omega_c-\omega_m)t)]$$

$$= \frac{1}{2R}A_c^2 + \frac{m^2A_c^2}{4R}\times 1 = \frac{A_c^2}{2R} + \frac{m^2A_c^2}{4R}$$

$$= \frac{A_c^2}{2R}\left(1+\frac{m^2}{2}\right).$$

That is, $P_t = P_c(1+\frac{m^2}{2})$, (3.21)

where $P_c = A_c^2/2R$ is the unmodulated carrier power. Thus the transmitted power is the sum of the power in the carrier and the two sidebands. Power per sideband is given by

$$P_{\text{usb}} = P_{\text{lsb}} = \frac{1}{2}P_c\frac{m^2}{2}. \tag{3.22}$$

Example 3.2

A load resistance of 50 Ω carries an AM voltage given by

$$v(t) = 2(1+0.8\cos(4\times10^3t))\cos(5\times10^6t) \text{ volts.}$$

Find (a) the carrier power, (b) the sideband power, and (c) the fraction of the total power in the carrier.

Solution

By comparing with (3.14),

$$\text{the carrier amplitude, } A_c = 2 \text{ volts,} \quad \text{modulation index, } m = 0.8.$$

$$\text{Carrier power, } P_c = \frac{A_c^2}{2R} = \frac{(2)^2}{2 \times 50} = 40 \text{ mW.}$$

$$\text{Total sideband power, } P_{\text{SB}} = P_c \frac{m^2}{2} = \frac{1}{2} \times (0.8)^2 \times 40 \text{ mW} = 12.8 \text{ mW.}$$

$$\text{Required fraction } \frac{P_c}{P_t} = \frac{1}{1 + m^2/2} = \frac{1}{1 + (0.8)^2/2} = 75.75\%.$$

Equation (3.21) tells that the power in the sidebands does not exceed 50% of the total power even for a 100% modulation. So, amplitude modulation is inefficient as far as power transmission is concerned since at least 50% of the total power resides in the carrier which carries no message information.

In spite of that, DSB transmission with full carrier is still in use because of the simple and inexpensive detector circuit at the receiver.

Sometimes, it is desirable to express power in terms of current. If I_t and I_c denote the measured total current and the carrier current respectively, then,

$$\frac{P_t}{P_c} = \frac{I_t^2 R}{I_c^2 R} = \left(\frac{I_t}{I_c}\right)^2 = (1 + m^2/2)$$

$$\text{so that} \quad I_t = I_c \sqrt{1 + m^2/2}. \tag{3.23}$$

When a number of sine waves modulate the carrier simultaneously to different depths of modulation, say $m_1, m_2, \cdots$, the total depth of modulation m_t is given by

$$m_t^2 = m_1^2 + m_2^2 + \cdots. \tag{3.24}$$

Correspondingly, the total transmitted power P_t becomes

$$P_t = P_c(1 + m_t^2/2) \tag{3.25}$$

provided that the combined voltage does not overmodulate the carrier.

In this context, we may define the modulation efficiency η as the ratio of the signal power to the total power. Thus,

$$\eta = \frac{P_c \times \frac{m^2}{2}}{P_c(1 + \frac{m^2}{2})} = \frac{m^2}{m^2 + 2}. \tag{3.26}$$

Thus, for a 100% modulation, η is 33%.

Example 3.3

A 400 watt carrier is modulated by a 1 kHz tone to a depth of 75%. Find the total sideband power and the modulation efficiency.

Solution

Here $P_c = 400$ W, $m = 3/4$.

Hence sideband power $= P_c \times \dfrac{m^2}{2} = 400 \times \left(\dfrac{1}{2}\right)\left(\dfrac{3}{4}\right)^2$

$= 112.5$ watts.

Modulation efficiency $\eta = \dfrac{m^2}{2 + m^2} = \dfrac{(3/4)^2}{2 + (3/4)^2}$

$= 0.22$ or 22%.

3.2.2 Double-sideband-suppressed carrier modulation

We have studied that the DSB signals consists of the spectra of carrier impulse and the two symmetric sidebands. The transmitted signal power can be saved by suppressing the carrier frequency component. This form of modulation is called the double-sideband-suppressed carrier (DSB-SC) modulation. The modulated wave for DSB-SC case is

$$s_c(t) = A_c s(t) \cos 2\pi f_c t. \tag{3.27}$$

The Fourier transform of (3.27) is

$$S_c(f) = \tfrac{1}{2} A_c \left[S(f - f_c) + S(f + f_c)\right]. \tag{3.28}$$

The spectrum of DSB-SC wave looks like an AM spectra without the carrier impulse. For the case of the message signal $s(t)$ being bandlimited to the interval $-f_m < f < f_m$, the DSB-SC spectra occupy the frequency range $f_c - f_m$ to $f_c + f_m$, thus giving a transmission bandwidth of $2f_m$.

Because of the carrier being suppressed, the total transmitted power is contained in the sidebands.

$$\text{Thus } P_t = P_{\text{USB}} + P_{\text{LSB}} = P_c \times \frac{m^2}{2}. \tag{3.29}$$

3.3 AM Modulators

This section is devoted to modulators and transmitters that generate AM or DSB-SC signals. Since the modulation process involves frequency translation, the modulators are necessarily nonlinear systems. We will discuss the product, gated, square-law and balanced modulators in this section. Product modulation performs the mathematical multiplication of two input signals. This is also known as an ideal mixer.

3.3.1 Product modulator

Figure 3.5 illustrates the schematic diagram of an amplitude modulator which is based on (3.14). A DSB suppressed carrier shown in Fig. 3.5(a) needs only the multiplier to produce the product $s_c(t) = kA_c s(t) \cos 2\pi f_c t$. The DSB with carrier needs an extra block that adds the carrier with the product as shown in Fig. 3.5(b).

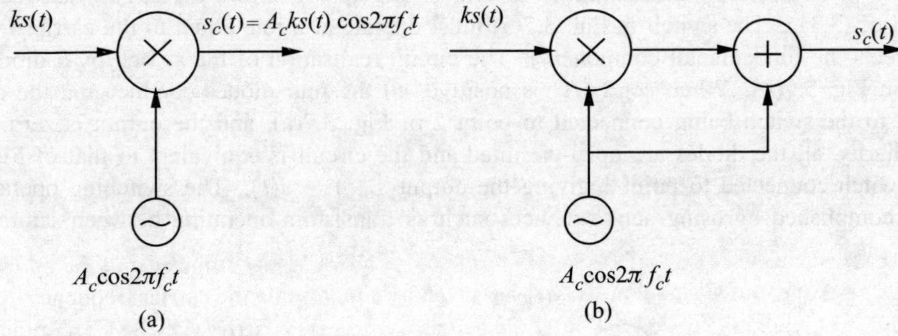

Fig. 3.5 *AM modulator: (a) DSB-SC (b) DSB with carrier.*

3.3.2 Gated modulator

The gated modulator is based on the fact that multiplication of the baseband signal $s(t)$ by any periodic signal yields a series of AM waves located at the frequencies that are multiples of the fundamental frequency of the periodic function. For a periodic signal $p(t)$ having fundamental period $1/f_c$, the Fourier series representation is

$$p(t) = a_0/2 + \sum_{n=1}^{\infty} a_n \cos(2\pi n f_c t).$$

The product signal is

$$s(t)p(t) = s(t)\left[\frac{a_0}{2} + \sum_{n=1}^{\infty} a_n \cos(2\pi n f_c t)\right]. \tag{3.30}$$

For a periodic pulse train of unit amplitude, we have

$$a_0 = 1 \text{ and } a_n = \frac{2\sin(n\pi/2)}{n\pi}$$

The first term gives the spectrum of $s(t)$ at the baseband. The second term gives the spectrum of $s(t)$ located at f_0 and so on. Each term is important for the message signal. But the AM waves at lower carrier frequencies are more significant since the Fourier series coefficients a_n decrease in magnitude with increasing n.

In particular, if one selects the signal $p(t)$ a periodic pulse train which is either 0 or 1, the multiplication can be viewed as a gating or switching operation. Figure 3.6 shows a multiplier followed by a bandpass filter. The spectra of $s(t)$ and the product $s(t)p(t)$ are also shown. If the BPF is tuned to the fundamental frequency f_e, the modulator output is then

$$s_c(t) = \frac{2}{\pi} s(t) \cos 2\pi f_e t. \tag{3.31}$$

The factor $\frac{2}{\pi}$ is obtained by evaluating a_1 of (3.30). Figure 3.7 shows the circuit that realises the AM signal of (3.31). The switch in Fig. 3.7(a) must operate at a rate equal to the carrier frequency (if one selects the fundamental component). The circuit realisation of the switch by a diode bridge is shown in Fig. 3.7(b). When $\cos 2\pi f_e t$ is positive, all the four diodes conduct and the circuit is equivalent to the switch being connected to point 2 of Fig. 3.7(a), and the output is zero. On the reverse polarity, all the diodes are open-circuited and the circuit is equivalent to that of Fig. 3.7(a) with the switch connected to point 1 giving the output $s_c(t) = s(t)$. The switching operation can also be accomplished by using active devices such as transistors operating between saturation and cutoff.

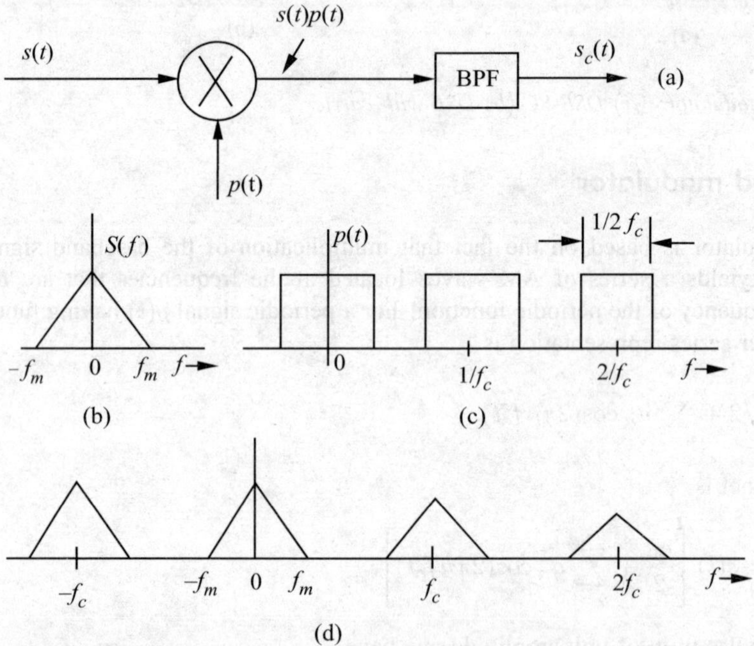

Fig. 3.6 *(a) A product modulator; (b) spectrum of the message signal; (c) periodic pulse train assumed to have even symmetry; and (d) the spectrum of the product $s(t)p(t)$.*

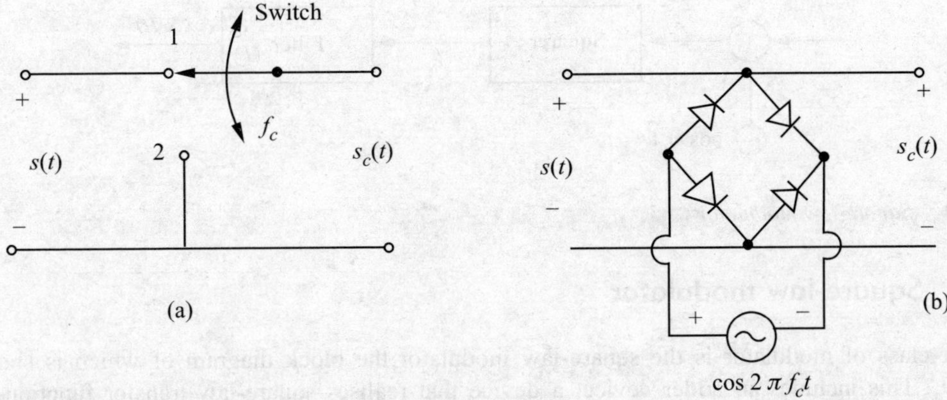

Fig. 3.7 *Gated modulator: (a) functional diagram; (b) realisation by a diode bridge.*

A ring modulator shown in Fig. 3.8 is a variation of the gated modulator for generating DSB-SC signals. A square-wave carrier $c(t)$ at the frequency t_c is fed into the centre taps of the input and output transformers. When $c(t)$ is positive, the top and bottom diodes D_1 and D_2 are ON and the inner diodes D_3 and D_4 are OFF. This results in $s_{\text{out}}(t) = s(t)$. On the other hand, when $c(t)$ is negative, the inner diodes are ON and the top and bottom diodes are OFF. In this case, $s_{\text{out}}(t) = -s(t)$. Observe that the output $s_{\text{out}}(t)$ is the gated version of the input. The DSB-SC signal can be obtained by passing $s_{\text{out}}(t)$ through a BPF centred at f_c.

As we have seen, the multiplication operation is obtained from the $n = 1$ term given by Eq. 3.31. This term generates up and down conversion signals at $f_c + f_m$ and $f_c - f_m$ if $s(t)$ is a baseband signal of spectrum extending from dc to f_m. In addition to this, the other frequency bands are also present in the output signal $s(t)p(t)$. These are at frequencies $|f = nf_c \pm f_m|$, $n = 3, 5, 7, \ldots$. The feed through term $\frac{1}{2}s(t)$ also appers at the output.

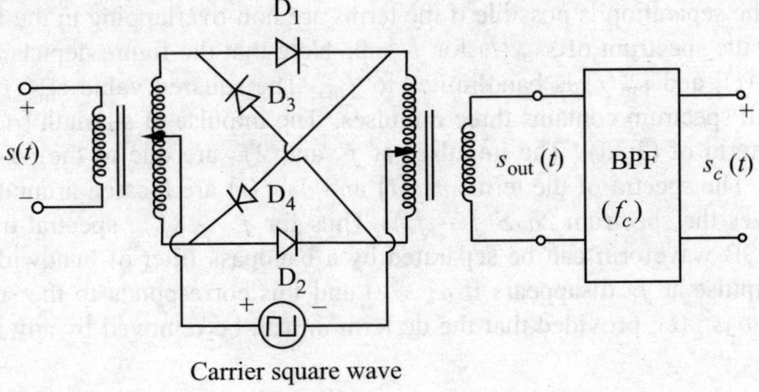

Fig. 3.8 *Ring modulator*

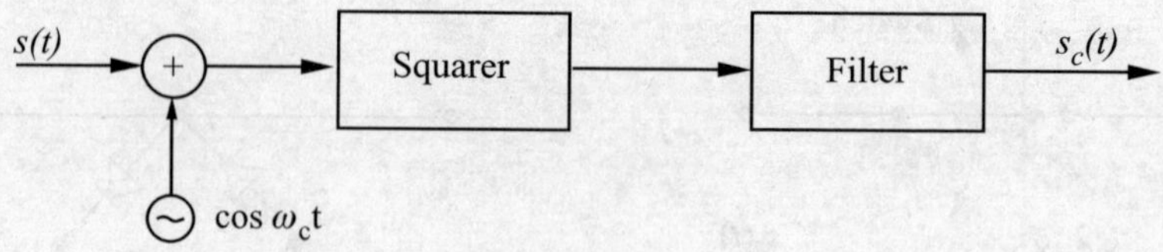

Fig. 3.9 *Square-law modulator*

3.3.3 Square-law modulator

Another class of modulator is the square-law modulator the block diagram of which is shown in Fig. 3.9. This includes an adder device, a device that realises square-law transfer function and a bandpass filter. The adder device can be active or passive; a summing operational amplifier can be used. The square-law transfer function should follow the following relation:

$$s_{\text{out}}(t) = a_0 + a_1 s_{\text{in}}(t) + a_2 s_{\text{in}}^2(t). \tag{3.32}$$

For an input $s_{in}(t) = s(t) + \cos\omega_c t$,

$$s_{\text{out}}(t) = a_0 + a_1\left[s(t) + \cos\omega_c t\right] + a_2\left[s(t) + \cos\omega_c t\right]^2$$

or, $$s_{\text{out}}(t) = a_0 + a_1 s(t) + a_2 s^2(t) + a_1\cos\omega_c t + a_2\cos^2\omega_c t + 2a_2 s(t)\cos\omega_c t$$

$$= a_0 + a_1 s(t) + a_2 s^2(t) + a_2\cos^2\omega_c t + [a_1 + 2a_2 s(t)]\cos\omega_c t.$$

$$s_{\text{out}} = \left(a_0 + \frac{a_2}{2}\right) + a_1 s(t) + a_2 s^2(t) + \frac{a_2}{2}\cos 2\omega_c t + a_1\cos\omega_c t + 2a_2 s(t)\cos\omega_c t. \tag{3.33}$$

The term $2a_2 s(t)\cos\omega_c t$ is the desired DSB-SC wave. We must find a way of separating this term from the others. The separation is possible if the terms are non-overlapping in the frequency domain. Figure 3.10 shows the spectrum of $s_{\text{out}}(t)$ for $f > 0$. Note that the figure depicts an arbitrary shape of $S_{\text{in}}(f) = \mathcal{F}[s_{\text{in}}(t)]$ and $s_{\text{in}}(t)$ is bandlimited to f_m. The squared value $s_{\text{in}}^2(t)$ is bandlimited to to $2f_m$. The overall spectrum contains three impulses. The impulse of strength $(a_0 + \frac{1}{2}a_2)$ at $f = 0$ is due to the first term of (3.33). The impulses at f_c and $2f_c$ are due to the $\cos\omega_c t$ and $\cos 2\omega_c t$ terms respectively. The spectra of the terms $a_1 s(t)$ and $a_2 s^2(t)$ are located around $f = 0$. The term $2a_2 s(t)\cos\omega_c t$ gives the spectrum $2a_2 S(f - f_c)$. Thus for $f_c > 3f_m$, spectral overlapping can be avoided and the DSB waveform can be separated by a bandpass filter of bandwidth $2f_m$ centred at f_c. The carrier impulse at f_c disappears if $a_1 = 0$ and this corresponds to the square-law transfer function $s_{\text{out}}(t) = a_2 s_{\text{in}}^2(t)$, provided that the dc term a_0 can be removed by any means.

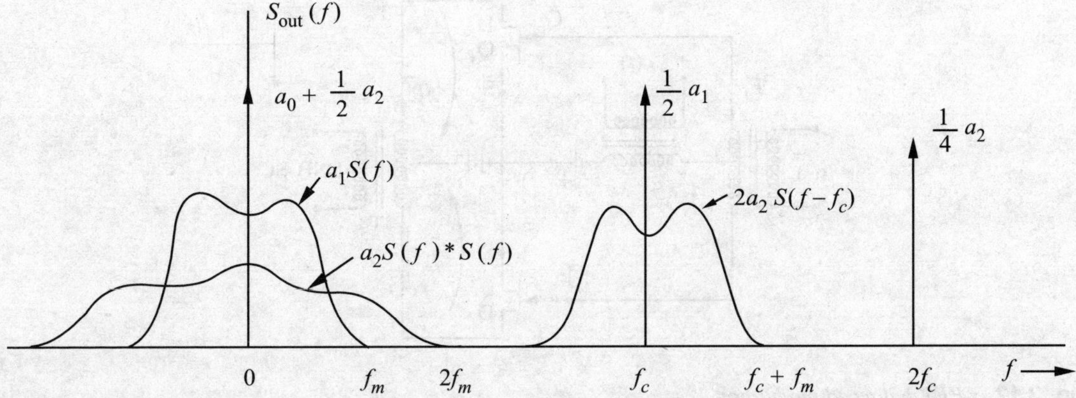

Fig. 3.10 *Spectra of (3.33)*

3.3.4 Balanced modulator

Square-law modulators are easy to build in the real world and very often they appear unintentionally in electronic circuits when the devices are driven into nonlinear region. In fact, the nonlinear devices follows the relationship

$$s_{\text{out}}(t) = a_0 + a_1 s_{\text{in}}(t) + a_2 s_{\text{in}}^2(t) + a_3 s_{\text{in}}^3(t) + \cdots \tag{3.34}$$

For a squarer, the series coefficients must satisfy the condition that $a_n << a_2$ for $n > 2$ or else $s_{\text{in}}(t)$ should be small enough such that $s_{\text{in}}^n(t) << s(t)$ for higher n. These constraints on the nonlinear device may be ignored if one constructs a balanced modulator configuration as shown in Fig. 3.11. Each branch performs the operation of amplitude modulation. Since the final output is obtained by taking the difference, there occurs a cancellation of the odd-power-terms of (3.34) while the other terms are doubled. Hence higher odd powers do not create undesired terms at the output. Since the first-order term is eliminated, the output is obviously a suppressed carrier DSB. Please observe that the dc term in the message signal $s(t)$ if any, is not cancelled out.

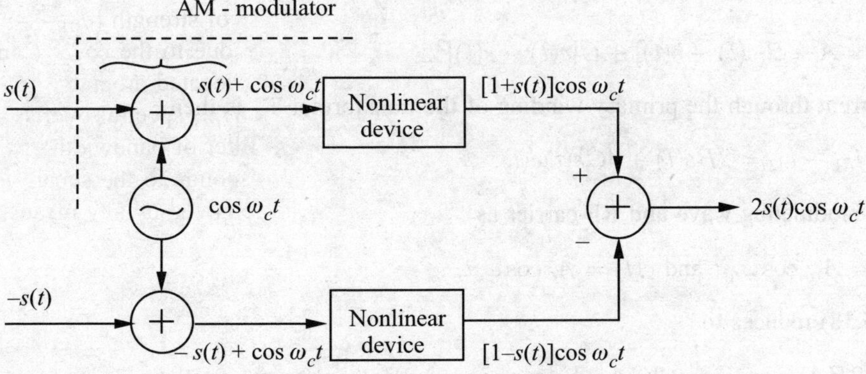

Fig. 3.11 *Balanced modulator*

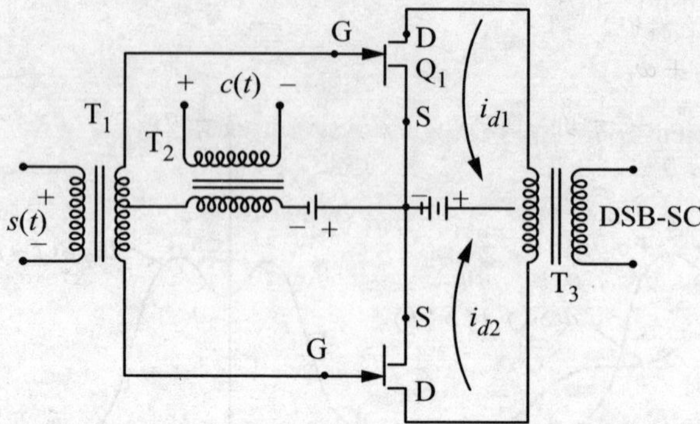

Fig. 3.12 *FET balanced modulator*

Figure 3.12 shows a balanced modulator using field-effect transistor (FET) as a nonlinear device. The drain-current–gate voltage $(v_g - i_d)$ characteristics for a FET may be written as

$$i_d = A + Bv_g + Cv_g^2, \tag{3.35}$$

where A is the quiescent dc drain current determined by bias voltage,
B is the transconductance,
C is a constant, determining second-order nonlinearity.

The baseband signal $s(t)$ is applied (at the primary of the transformer T_1) in a push-pull configuration so that at the gates of the transistors Q_1 and Q_2, $s(t)$ appears with 180° phase reversal. The RF carrier $c(t)$ is fed (through the transformer T_2) into the circuit in parallel so that at the gates it appears in phase. The modulated currents i_{d1} and i_{d2} are subtracted at the primary of the transformer T_3. The voltages at the gates of the FETs are

$$v_{g1} = c(t) + s(t) \quad \text{and} \quad v_{g2} = c(t) - s(t). \tag{3.36}$$

The corresponding drain currents are written as

$$\begin{aligned} i_{d1} &= A + B[c(t) + s(t)] + C[c(t) + s(t)]^2, \\ i_{d2} &= A + B[c(t) - s(t)] + C[c(t) - s(t)]^2. \end{aligned} \tag{3.37}$$

The net current through the primary winding of the transformer T_3 is then

$$i_d = i_{d1} - i_{d2} = 2Bs(t) + 4Cs(t)c(t). \tag{3.38}$$

Taking the modulating wave and RF carrier as

$$s(t) = A_m \cos \omega_m t \text{ and } c(t) = A_c \cos \omega_c t,$$

equation (3.38) reduces to

$$i_d = 2BA_m \cos \omega_m t + 2CA_m A_c \left[\cos(\omega_c + \omega_m)t + \cos(\omega_c - \omega_m)t\right]. \tag{3.39}$$

The input transformer T_1 is an audio transformer while T_2 and T_3 are RF transformers. So the audio frequency component will not appear at the secondary of T_3. Thus the output voltage

$$s_{out}(t) = k[\cos(\omega_c + \omega_m)t + \cos(\omega_c - \omega_m)t], \tag{3.40}$$

where k includes the current-to-voltage conversion factor. The output is DSB-SC.

Example 3.4

Referring to the schematic diagram in Fig. 3.11, find the output signal if the two nonlinear elements are unbalanced. Assume the transfer curves for them as $s_1 = a_1 s_i + a_2 s_i^2 + a_3 s_i^3$ and $s_2 = b_1 s_i + b_2 s_i^2 + b_3 s_i^3$.

Solution

The signals after AM modulators:

$$s_1(t) = a_1(s + \cos\omega_c t) + a_2(s + \cos\omega_c t)^2 + a_3(s + \cos\omega_c t)^3$$

$$= a_1(s + \cos\omega_c t) + a_2(s^2 + \cos^2\omega_c t + 2s\cos\omega_c t)$$

$$+ a_3(s^3 + \cos^3\omega_c t + 3s^2\cos\omega_c t + 3s\cos^2\omega_c t);$$

$$s_2 = b_1(s + \cos\omega_c t) + b_2(s^2 + \cos^2\omega_c t + 2s\cos_c t) + b_3(s^3 + \cos^3\omega_c t + 3s^2\cos\omega_c t + 3s\cos^2\omega_c t).$$

So, the output signal

$$s_0(t) = s_1(t) - s_2(t)$$

$$= (a_1 - b_1)s(t) + (a_1 - b_1)\cos\omega_c t + (a_2 - b_2)s^2(t)$$

$$+ (a_2 - b_2)\cos^2\omega_c t + 2(a_2 - b_2)s(t)\cos\omega_c t$$

$$+ (a_3 - b_3)s^3(t) + (a_3 - b_3)\cos^3\omega_c t + 3(a_3 - b_3)s^2(t)\cos\omega_c t$$

$$+ 3(a_3 - b_3)s(t)\cos^2\omega_c t$$

$$= \frac{1}{2}(a_2^2 - b_2^2) + (a_1 - b_1)s(t) + (a_2 - b_2)s^2(t)$$

$$+ (a_3 - b_3)s^3(t) + \left[(a_1 - b_1) + \frac{3}{4}(a_3 - b_3)\right]\cos\omega_c t$$

$$+ \left[2(a_2 - b_2) + \frac{3}{2}(a_3 - b_3)\right] s(t)\cos\omega_c t + 3(a_3 - b_3)s^2(t)\cos\omega_c t$$

$$+ \frac{1}{2}(a_2 - b_2)\cos 2\omega_c t + \frac{3}{2}(a_3 - b_3)s(t)\cos 2\omega_c t$$

$$+ \frac{1}{4}(a_3 - b_3)\cos 3\omega_c t.$$

3.4 Single-Sideband Amplitude Modulation

In conventional AM systems, the frequency range occupied by the transmitted signal is from $f_c - f_m$ to $f_c + f_m$ with a total transmission bandwidth of $2f_m$. Since the transmitted power and the channel bandwidth are two important resources, they should be utilised efficiently. The RF carrier does not bear any message information and so it may be suppressed as in DSB-SC, thereby saving at least 2/3 of the transmitted power. Also, the upper and lower sidebands are positioned symmetrically with respect to the carrier frequency. Either of the two contains all the essential information. This is the essence of single-sideband (SSB) transmission. The advantages of SSB transmission are the followings:

- *Conservation of bandwidth*: SSB transmission requires half the bandwidth of the conventional DSB transmission. The bandwidth consideration is stringent as the bandwidth in AM broadcasting is exhausted.
- *Conservation of power*: Only one of the sidebands is transmitted and the carrier is either suppressed or its power reduced. This requires much less transmission power.
- *Effect of sideband fading:* The sidebands in AM transmission may propagate through different paths. As a result, they are attenuated by different amounts. Any phase change of the carrier or the sideband components also distort the signal. Such happenings do not occur in SSB transmission. The sideband fading changes only the amplitude. Fortunately, such changes produce no significant distortion to cause loss of intelligence in the recovered signal.
- *Noise reduction*: Since the bandwidth is reduced, thermal noise power is also reduced to that extent.

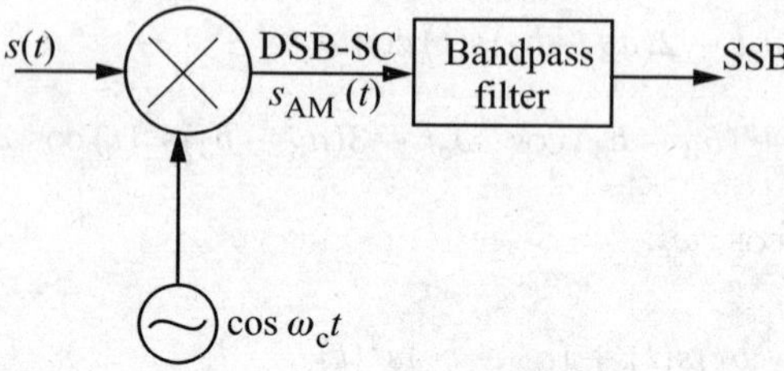

Fig. 3.13 *Single-sideband modulation*

Single-sideband is a technique that allows transmission of either upper or lower sidebands. Since the two sidebands are separated in frequency, we may generate the SSB signals by using the frequency-discrimination method. Figure 3.13 represents a single-sideband modulator which consists of two stages – one the product modulator that generates DSB-SC signal and the other the bandpass filter designed to suppress one sideband and pass the other. The bandpass filter must have a sharp selectivity since there is no guard band separating the two sidebands. If one selects the upper sideband, the desired passband of the bandpass filter should be f_c to $f_c + f_m$ and must block frequencies just below f_c. But a sharp cutoff at $f = f_c$ cannot be synthesised, rather there must be a filter's transition band. This constraint on the filter can be relaxed since many modulating signals of practical interest have negligible energy close to origin, as illustrated in Fig. 3.14(a). A value of $f_L = 300$ Hz is typical for audio signals. The spectrum of DSB signal after product modulator shows that there is a gap of $2f_L$ (from $f_c - f_L$ to $f_c + f_L$) about the carrier frequency (Fig. 3.14(b)). Since the message information in this range is negligible, the filter may be designed to have the transition region covering this void frequency range.

3.5 SSB Generation

The commonly used methods of generating SSB signals are the filter method and phase-shift method. An alternate method of SSB generation is due to Weaver; we will discuss all these methods here.

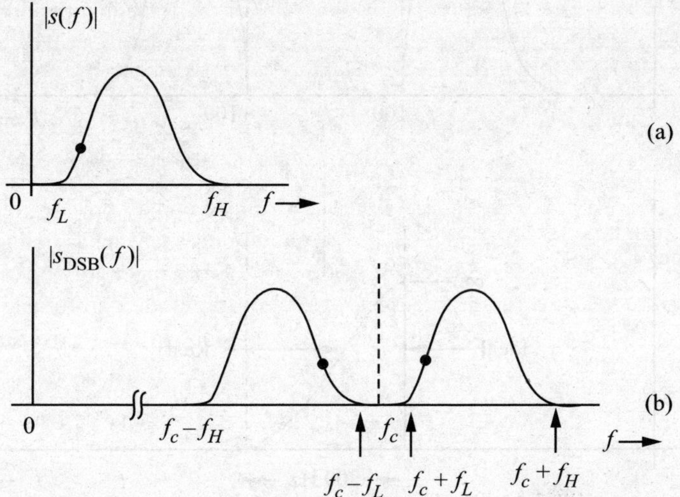

Fig. 3.14 (a) *Spectrum of the baseband signal for* $f > 0$, (b) *spectrum of DSB-SC for* $f > 0$.

3.5.1 Filter method

Figure 3.15 shows the block diagram representation of the SSB generation system. The modulation process involves frequency translation in two steps rather than one.

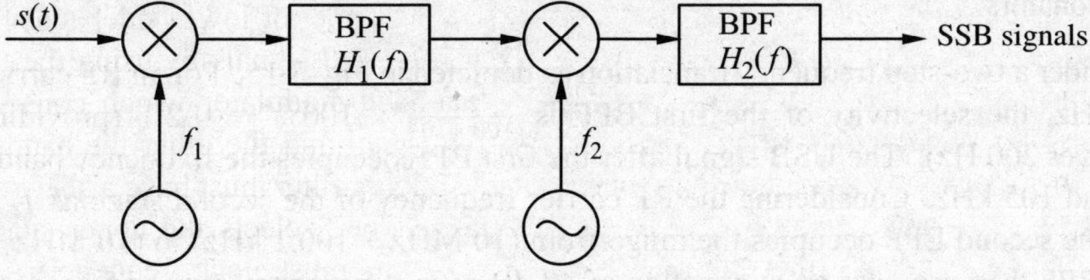

Fig. 3.15 *Filter method of SSB generation*

This is due to the fact that perfect selectivity can never be achieved in real filters.

Let us take the example of an audio signal having significant spectral range from 100 Hz to 5 kHz. With a carrier of frequency f_c, the USB occupies $f_c + 100\text{Hz}$ to $f_c + 5000$ Hz and the LSB occupies $f_c - 5000$ Hz to $f_c - 100$ Hz (vide Fig. 3.16).

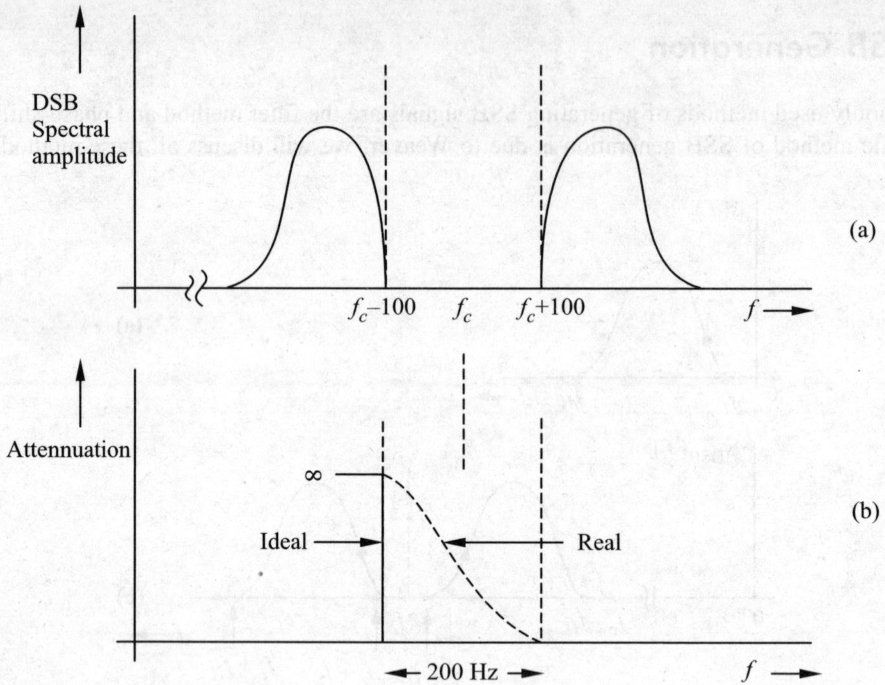

Fig. 3.16 *(a) DSB spectrum for a message signal having insignificant energy below 100 Hz; (b) bandpass filter attenuation characteristics.*

If the USB is to be retained, the filter must pass frequencies beyond $f_c + 100$ Hz and stop the frequencies below $f_c - 100$ Hz, as shown in Fig. 3.16(b). Thus the filter should have a transition region of 200 Hz centred at f_c. For $f_c = 10$ MHz, the filter selectivity is $\frac{200\text{Hz}}{10\times10^6\text{Hz}} \times 100\% = 0.002\%$. Such a sharp cutoff cannot be achieved in practice even for an attenuation of 40 dB. Conventional LC circuits do not offer such selectivity; highly selective filters can be realised in practice by means of crystal resonators.

Now consider a two-step frequency translation as depicted in Fig. 3.15. For an RF carrier frequency f_1 of 100 kHz, the selectivity of the first BPF is $\frac{200}{100\times10^3} \times 100\% = 0.2\%$ (providing a 40 dB attenuation over 200 Hz). The USB signal after the first BPF occupies the frequency band in between 100.1 kHz and 105 kHz. Considering the RF carrier frequency of the second stage as $f_2 = 10$ MHz, the USB of the second BPF occupies the range from (10 MHz + 100.1 kHz) to (10 MHz + 105 kHz). The second BPF then provides an attenuation of 40 dB over a frequency range of 2×100.1 kHz, i.e, 200.2 kHz. The filter selectivity is then $\frac{200.2\text{ kHz}}{10\text{ MHz}} \times 100\% = 2\%$, which can be realised.

As a rule of thumb, the filter transition region cannot be smaller than 1 percent of the centre frequency. That is, $\frac{2f_2}{f_L} \times 100 > 1$.

3.5.2 Phase-shift method

Another method of SSB generation is the phase-shift method. We begin with filter of Fig. 3.17 that generates the lower sideband signal. The passband of the LPF covers only the LSB of the DSB spectrum. We write the DSB spectrum as

$$S_{\text{DSB}}(f) = \frac{1}{2}\left[S(f - f_c) + S(f + f_c)\right], \tag{3.41}$$

where S is the spectrum of the message signal. The transfer function of the lowpass filter can be written as

$$H(f) = u(f + f_c) - u(f - f_c). \tag{3.42}$$

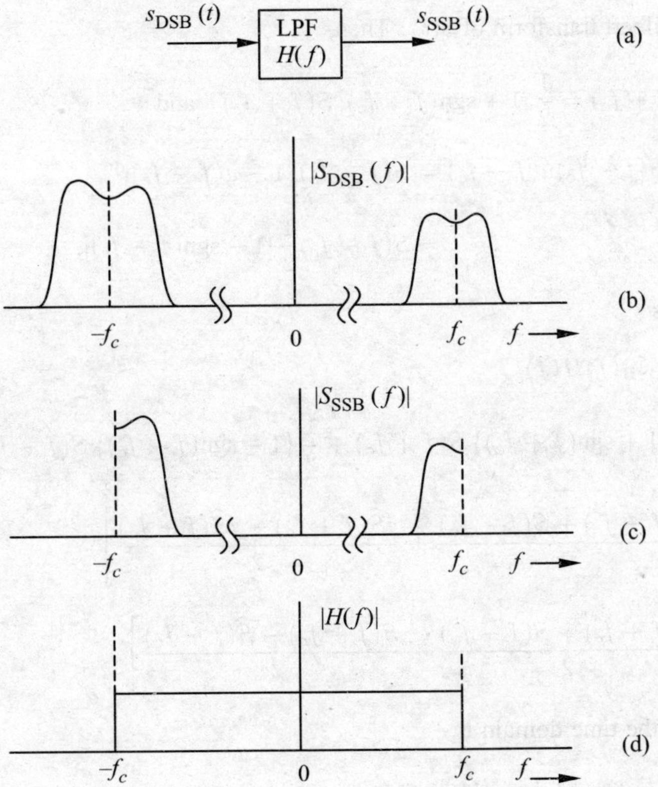

Fig. 3.17 (a) *SSB generation by lowpass filtering;* (b) *DSB spectrum;* (c) *SSB spectrum (lower sideband);* (d) *filter transfer function.*

The spectrum of the signal at the filter output is

$$
\begin{aligned}
S_{\mathrm{DSB}}(f)H(f) &= \frac{1}{2}\left[S(f-f_c)+S(f+f_c)\right]\left[u(f+f_c)-u(f-f_c)\right] \\
&= \frac{1}{2}\left[S(f-f_c)u(f+f_c)+S(f+f_c)u(f+f_c)\right. \\
&\quad \left. -S(f-f_c)u(f-f_c)-S(f+f_c)u(f-f_c)\right]. \\
&= \frac{1}{2}\left[S(f-f_c)+S(f+f_c)u(f+f_c)-S(f-f_c)u(f-f_c)\right]. \qquad (3.43)
\end{aligned}
$$

We recall that $u(f) = \frac{1}{2}[1+\mathrm{sgn}\,(f)]$ (3.44)

and $\mathcal{F}[\hat{s}(t)] = \dfrac{S(f)\,\mathrm{sgn}\,(f)}{j} = \hat{S}(f)$ (3.45)

where $\widehat{s}(t)$ is the Hilbert transform of $s(t)$. Then,

$$
S(f+f_c)u(f+f_c) = \frac{1}{2}\left[1+\mathrm{sgn}(f+f_c)\right]S(f+f_c) \quad \text{and}
$$

$$
\begin{aligned}
S(f-f_c)-S(f-f_c)u(f-f_c) &= S(f-f_c)\left[1-u(f-f_c)\right] \\
&= S(f-f_c)\frac{1}{2}\left[1-\mathrm{sgn}(f-f_c)\right].
\end{aligned}
$$

Thus (3.43) reads as

$$
\begin{aligned}
S_{\mathrm{LSB}}(f) &= S_{\mathrm{DSB}}(f)H(f) \\
&= \frac{1}{2}\left[\frac{1}{2}\left\{1+\mathrm{sgn}(f+f_c)\right\}S(f+f_c)+\frac{1}{2}\left\{1-\mathrm{sgn}(f-f_c)\right\}S(f-f_c)\right] \\
&= \frac{1}{2}\left[\frac{S(f+f_c)+S(f-f_c)}{2}+\frac{j\widehat{S}(f+f_c)-j\widehat{S}(f-f_c)}{2}\right] \\
&= \frac{1}{2}\left[\frac{S(f+f_c)+S(f-f_c)}{2}+\frac{\widehat{S}(f+f_c)-\widehat{S}(f-f_c)}{2j}\right]. \qquad (3.46)
\end{aligned}
$$

Hence the signal in the time domain is

$$
s_{\mathrm{LSB}}(t) = \tfrac{1}{2}s(t)\cos\omega_c t + \tfrac{1}{2}\hat{s}(t)\sin\omega_c t. \qquad (3.47)
$$

Again, if we subtract the LSB from the DSB signal, it yields the USB signal. Thus,

$$
s_{\mathrm{USB}}(t) = s(t)\cos\omega_c(t) - s_{\mathrm{LSB}}(t)
$$

$$= \frac{1}{2} s(t) \cos \omega_c t - \frac{1}{2} \hat{s}(t) \sin \omega_c t. \tag{3.48}$$

So, the general expression for the SSB signal is given by

$$s_{\text{SSB}}(t) = \frac{1}{2} s(t) \cos \omega_c t \pm \frac{1}{2} \hat{s}(t) \sin \omega_c t, \tag{3.49}$$

where the positive sign refers to the USB and the negative sign to the LSB signals. Therefore, the SSB signal consists of two DSB signals with quadrature carriers ($\cos \omega_c t$ in one case and $\sin \omega_c t$ in the other case), the modulating signals being $s(t)$ and $\widehat{s}(t)$. The SSB generator based on (3.49) is sketched in Fig. 3.18. Although simple in principle, the phasing method is less popular than the filter method. The quadrature phase-shifter $H_q(f)$ requires careful design. The phase-shift network will change the phase by exactly 90° but not the amplitude. But for the baseband signal frequencies covering more than five octaves, it is difficult to design a phase-shift network. An imperfect design may cause the unwanted sideband components and RF carrier to appear at the output.

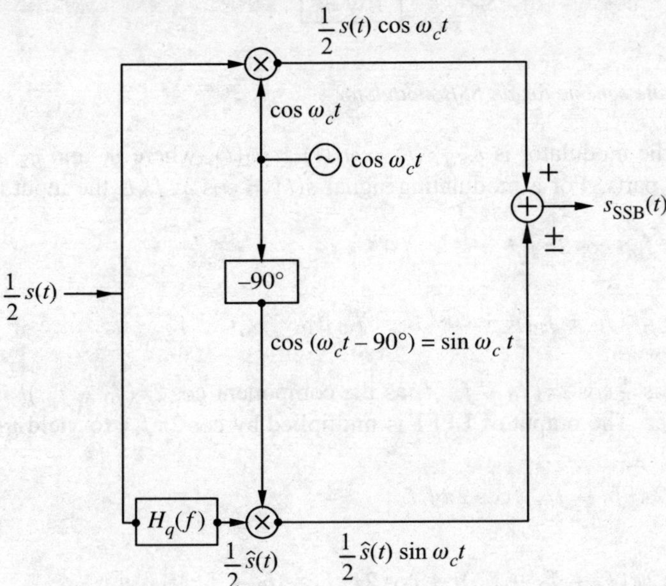

Fig. 3.18 *Phase-shift method for SSB generation. The block H_q performs the Hilbert transform.*

3.5.3 SSB generation by Weaver's method

Figure 3.19 shows the block diagram of the SSB modulator in an alternate method that avoids the quadrature phase-shifter. The scheme is very much similar to that of the phase-shift method. The exception in this scheme is the use of two carrier waves – one the audio subcarrier at the frequency ω_0 and the other the RF carrier at the frequency ω_c.

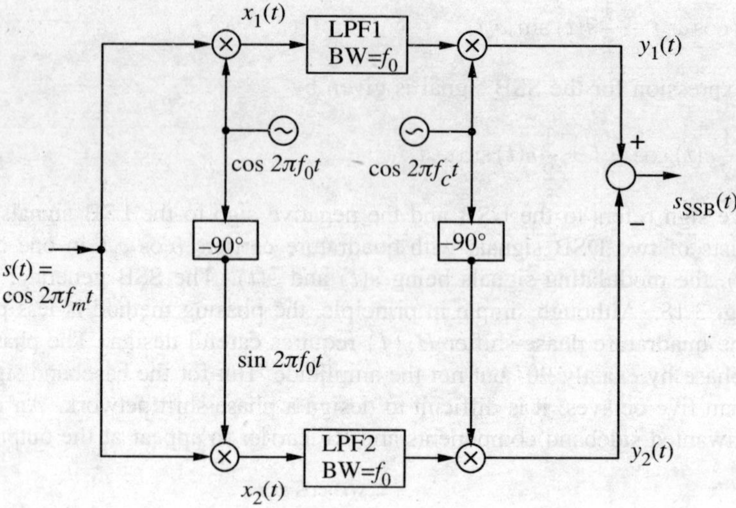

Fig. 3.19 *An alternate scheme for an SSB modulator*

The final output of the modulator is $s_{SSB}(t) = y_1(t) \pm y_2(t)$, where y_1 and y_2 are the signals from the upper and lower parts. For a modulating signal $s(t) = \cos 2\pi f_m t$, the input to the LPF1 is

$$x_1(t) = \cos 2\pi f_m t \cos 2\pi f_0 t$$

$$= \frac{1}{2}[\cos 2\pi(f_0 + f_m)t + \cos(f_0 - f_m)t].$$

The output of LPF1 is $\frac{1}{2}\cos 2\pi(f_0 - f_m)t$, as the component $\cos 2\pi(f_0 + f_m)t$ is beyond the cutoff frequency of the filter. The output of LPF1 is multiplied by $\cos 2\pi f_c t$ to yield $y_1(t)$. Thus,

$$y_1(t) = \frac{1}{2}\cos 2\pi(f_0 - f_m)t \cos 2\pi f_c t$$

$$= \frac{1}{4}[\cos 2\pi(f_c + f_0 - f_m)t + \cos 2\pi(f_c - f_0 + f_m)t]. \tag{3.50}$$

Similarly, the input to the LPF2 is

$$x_2(t) = \cos 2\pi f_m t \sin 2\pi f_0 t \text{ or, } x_2(t)$$

$$= \frac{1}{2}[(\sin 2\pi f_0 + f_m)t + \sin 2\pi(f_0 - f_m)t].$$

The output of the LPF2 is $\frac{1}{2}\sin 2\pi(f_0 - f_m)t$. This is multiplied by $\sin \omega_c t$ to yield $y_2(t)$. Thus,

$$y_2(t) = \frac{1}{2}\sin 2\pi(f_0 - f_m)t \sin 2\pi f_c t$$

$$= \frac{1}{4}\left[\cos 2\pi(f_c - f_0 + f_m)t - \cos 2\pi(f_c + f_0 - f_m)t\right]. \tag{3.51}$$

Thus the output signal is

$$s(t) = y_1(t) - y_2(t) = \frac{1}{2}\cos 2\pi(f_c + f_0 - f_m)t, \tag{3.52}$$

which corresponds to the lower sideband of the RF carrier of frequency $(f_c + f_0)$. Similarly, by adding $y_1(t)$ and $y_2(t)$, one can obtain the upper sideband of the RF carrier of frequency $(f_c - f_0)$.

3.6 Vestigial Sideband Modulation

Bandwidth conservation is a merit of SSB over DSB in addition to power saving. The disadvantage of SSB lies in the design of the SSB transmitter and receiver circuits. A sharp filter is needed to remove one of the sidebands. As we proceed to achieve a sharp cutoff, the phase characteristic develops ripples. That is, the SSB systems have poor low-frequency response. There are certain applications such as commercial television where the message signal has a large spectral width with significant low-frequency content. In such a case SSB will not work, particularly for low message frequencies. In contrast, DSB can be used efficiently but at the expense of double the transmission bandwidth. We therefore compromise between SSB and DSB with a new method namely vestigial sideband (VSB) modulation, where, instead of suppressing one of the sideband completely (as in SSB), we include a vestige or trace of that sideband. The VSB signals have approximately the same bandwidth as that of the SSB. At the same time, it is easier to construct a modulator for it since a sharp filter is not necessary as in SSB.

A popular method of VSB generation is to generate a DSB-SC and then pass it through a bandpass filter, as shown in Fig. 3.20. The BPF requires a special design. Assume that a vestige of the lower sideband is transmitted along with the full upper sideband. The transfer function $H(f)$ of the filter takes the shape as shown in Fig. 3.21. The shape of the characteristic is not important; what matters is that the response should have odd symmetry about the carrier frequency f_c and $|H(f_c)| = 1/2$. This implies that the sum of the responses at any two frequencies in the transition region, $f_c - f_L < f_c < f_c + f_L$, spaced equally apart from f_c, is unity. Outside the USB (that is, $|f| > f_c + W$, where W is the message bandwidth), the response may have any arbitrary shape. The information content in the frequency range up to f_L of the message signal is retained as vestige of the LSB. The sideband filter with the above specification is also called a VSB filter.

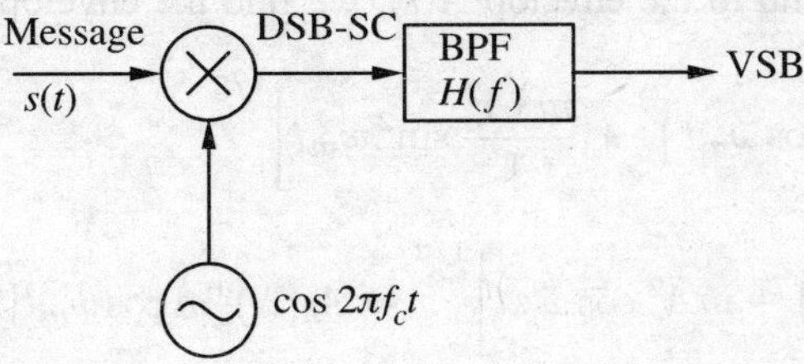

Fig. 3.20 *Generation of VSB signals*

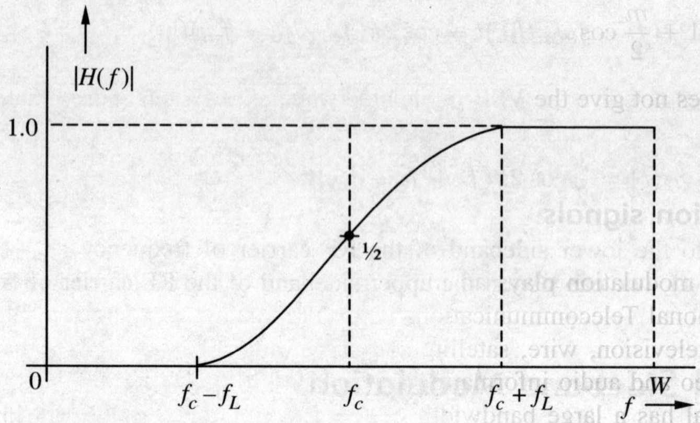

Fig. 3.21 *Normalised magnitude response of VSB filter for* $f > 0$

The transmission bandwidth of the VSB modulation is

$$B = W + f_L \approx W \tag{3.53}$$

when $f_L << W$, which is usually true. The VSB waveform in the time domain can be described as follows. Starting with the AM wave,

$$s_{\text{AM}}(t) = A_c(1 + m\cos\omega_m t)\cos\omega_c t$$

$$= A_c\cos\omega_c t + \frac{mA_c}{2}\cos(\omega_c + \omega_m)t + \frac{mA_c}{2}\cos(\omega_c - \omega_m)t,$$

and suppressing the LSB completely, the VSB-modulated signal approximates to

$$s_{\text{VSB}}(t) = A_c\cos\omega_c t + \frac{mA_c}{2}\cos(\omega_c + \omega_m)t$$

$$= A_c\cos\omega_c t + \frac{mA_c}{2}\cos\omega_c t\cos\omega_m t - \frac{mA_c}{2}\sin\omega_c t\sin\omega_m t$$

$$= A_c\left(1 + \frac{m}{2}\cos\omega_m t\right)\cos\omega_c t - \frac{mA_c}{2}\sin\omega_m t.\sin\omega_c t. \tag{3.54}$$

Since the detector will respond to the envelop $A(t)$, we find the envelope as

$$A(t) = \left[A_c^2\left(1 + \frac{m}{2}\cos\omega_m t\right)^2 + \frac{m^2A_c^2}{4}\sin^2\omega_m t\right]^{1/2}$$

$$= \left[A_c^2\left(1 + \frac{m^2}{4}\right) + mA_c^2\cos\omega_m t\right]^{1/2} \approx A_c[1 + m\cos\omega_m t]^{1/2}.$$

For not too large m,

$$A(t) \approx A_c\left(1 + \frac{m}{2}\cos\omega_m t\right). \tag{3.55}$$

Equation (3.55) does not give the VSB modulated wave exactly, but approximates it for $m << 1$ and $f_L << W$.

3.6.1 Television signals

Vestigial sideband modulation plays an important role in commercial television (TV) broadcasting. The ITU (International Telecommunications Union) regulates interstate and international communications by radio, television, wire, satellite and cable, and allocates a fixed bandwidth to each TV channel. Both video and audio information are included in this specified band of frequencies.

The video signal has a large bandwidth of about 4 MHz with significant low-frequency content that suggests VSB modulation. DSB-AM is not considered since this would require approximately 8 MHz of bandwidth. Likewise, SSB modulation is also not considered since a sharp filter is really difficult to achieve. It is difficult to control the phase and the amplitude characteristics of a signal simultaneously. For the audio signals, phase distortions do not produce serious problems because the human ear is not sensitive to such distortion. However, phase distortion in video applications results in shifting of the image on the screen. This phenomenon is commonly referred to as a ghost image, and is very much undesirable and annoying. The sound portion of the TV signal is broadcast using frequency modulation (FM). Figure 3.22 shows the magnitude spectrum of the TV signal. In PAL system, the sound carrier frequency f_s is located 5.5 MHz above the picture carrier frequency f_c. A filter called the sound trap separates the sound signal from the video signal. A frequency band of ± 50 kHz is allocated on each side of f_s for sound sidebands.

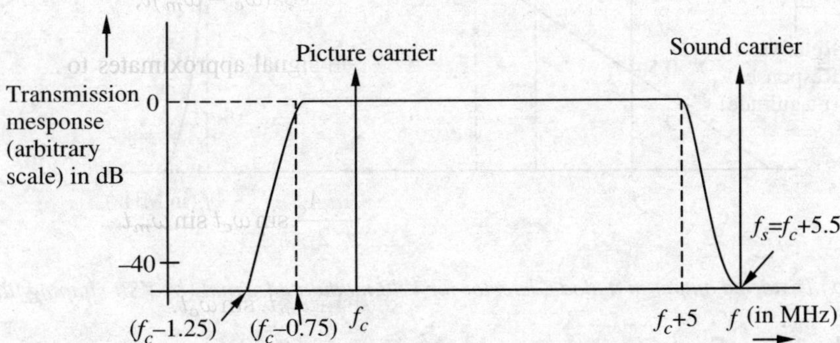

Fig. 3.22 *Frequency details of a TV signal*

The entire upper sideband of the picture is transmitted without attenuation up to 5 MHz and then gradually attenuated to zero to avoid interference with the sound signal. The lower sideband of the picture is sent without attenuation up to 0.75 MHz (in the baseband spectrum); the VSB filter then gradually attenuates the portion above 0.75 MHz with full attenuation at 1.25 MHz. The spectrum of the TV signal is thus confined to a bandwidth requirement of 5.5+1.25 = 6.75 MHz in contrast to a total bandwidth requirement of $2 \times 5.5 = 11.0$ MHz for DSB modulation. A saving of roughly 30% in bandwidth is thus achieved.

The carrier of the video signal is made large enough to use envelope detection for demodulation. The detected output produces waveform distortion due to the presence of the quadrature component (up to 1.25 MHz) of the VSB wave. A typical diode-demodulator output is depicted in Fig. 3.23(a). Observe that the relative amplitude is 2 over the frequency range f_c to $f_c + 0.75$ MHz since both the sidebands are transmitted. This non-uniformity is removed in a VSB-shaping filter having magnitude response as shown in Fig. 3.23(b).

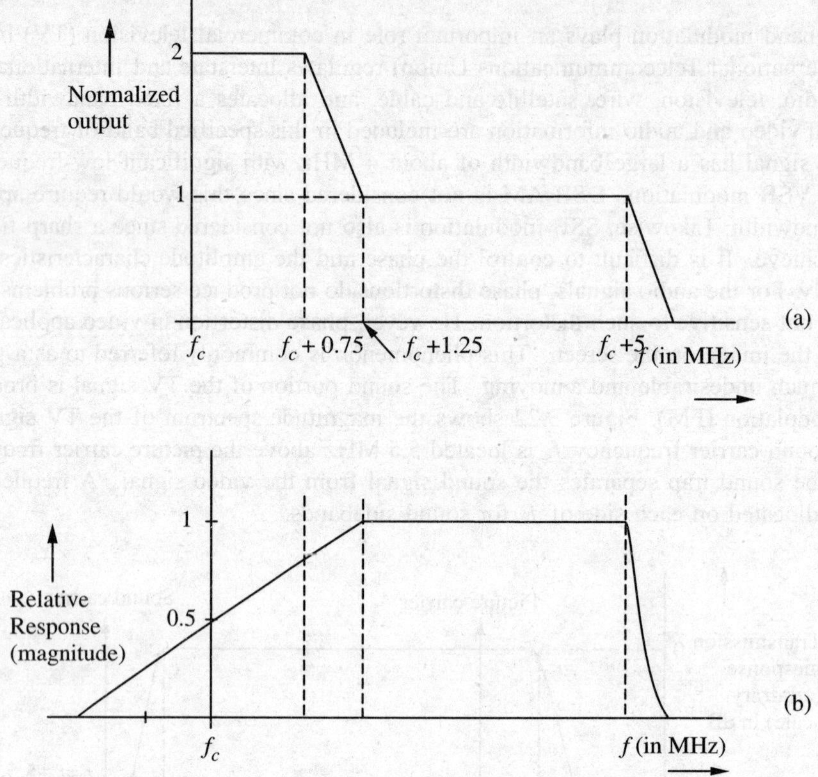

Fig. 3.23 *(a) Distorted output in a diode detector for VSB modulated signal (b) VSB-shaping filter response.*

3.6.2 Television Systems

There are several television standards used in different countries. These are National Television System Committee (NTSC), Sequential Couleur Avec Mémoire or Sequential Colour with Memory (SECAM) and Phase Alternating Line (PAL). The message transmitted by the television is a two-dimensional image (called frame) cast sequentially at a rate of 25 to 30 per second, to produce continuity due to persistence of vision. One image frame is a function of two spatial coordinates and the motion picture adds another dimension called the time. Thus, a TV signal is essentially a 3-dimensional signal. The TV signal consists of both video as well as audio signal (the combined signal is also referred to as composite video signal). The two-dimensional intensity pattern denoted by $I(x, y)$ is converted to proportional electrical signal $s(t)$ by scanning. Here, x and y are the horizontal and vertical coordinates of the still picture of a frame. To this electrical signal $s(t)$, blanking and synchronising pulses are added. This is what is called the video signal.

The blocks of monochrome TV transmitter are sketched in Fig. 3.24. The audio signal is generated separately by using FM modulation with a sound carrier f_s. The video modulator is a high-level AM modulator with the picture carrier f_c. The outputs of audio and video transmitters are combined together in a combining network and are radiated out.

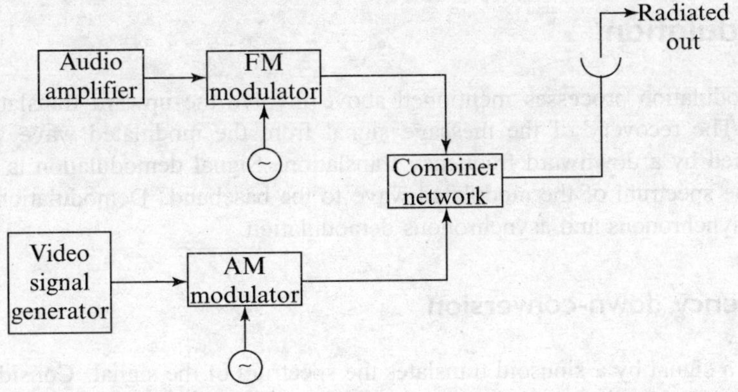

Fig. 3.24 *Monochrome TV transmitter*

The monochrome TV receiver is basically a super heterodyne receiver. The audio and video components are required to be separated and processed separately. A block diagram representation of the TV receiver is shown in Fig. 3.25. Since the audio part is frequency-modulated whereas the video part is USB-modulated, the corresponding detector circuits should be FM detector and AM detector, respectively. The output of the envelope detector $y(t)$ consists of both the audio component (FM audio) and the video component (VSB / AM video) and is given by

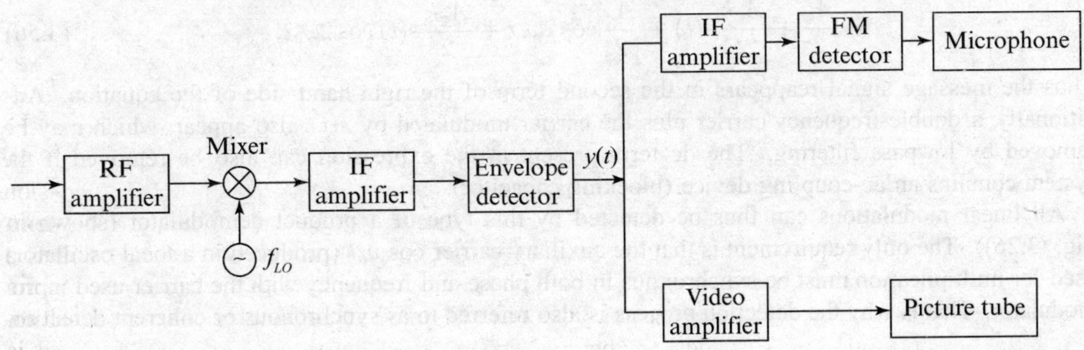

Fig. 3.25 *TV receiver*

$$y(t) = A_1\,(1 + m\,s(t)) + A_2\,\cos\,(\omega_a t + \phi(t)) \tag{3.55a}$$

Here A_1 and A_2 are the amplitudes of the video carrier and audio carrier respectively with $A_1 >> A_2$, since the transmitted audio power is only 10–20% of the video power. $s(t)$ is the video signal which is amplitude modulated to a depth of m. In the audio signal component, ω_a is the audio carrier frequency. The frequency-modulated audio component is $\phi(t)$. The video amplifier removes the audio component from $y(t)$ and applies to the picture tube. On the other branch of the receiver, the modulated audio component is detected in the FM detector.

3.7 Demodulation

All the linear modulation processes mentioned above involve the upward translation of the message spectrum. The recovery of the message signal from the modulated wave will therefore be naturally performed by a downward frequency translation. Signal demodulation is the process that down-converts the spectrum of the modulated wave to the baseband. Demodulation can be broadly categorised into synchronous and asynchronous demodulation.

3.7.1 Frequency down-conversion

Multiplication of a signal by a sinusoid translates the spectrum of the signal. Consider, for example, the AM wave given in (3.14).

$$s_{\mathrm{AM}}(t) = A_c(1 + ks(t))\cos\omega_c t.$$

Multiplying both sides by $\cos\omega_c t$, we get

$$\begin{aligned} s_{\mathrm{AM}}(t)\cos\omega_c t &= A_c[1 + ks(t)]\cos^2\omega_c t \\ &= \frac{A_c}{2}[1 + ks(t)](1 + \cos 2\omega_c t) \\ &= \frac{A_c}{2} + \frac{A_c k}{2}s(t) + \frac{A_c}{2}\cos 2\omega_c t + \frac{A_c k}{2}s(t)\cos 2\omega_c t. \end{aligned} \tag{3.56}$$

Thus the message signal reappears in the second term of the right-hand side of the equation. Additionally, a double-frequency carrier plus the carrier modulated by $s(t)$ also appear, which may be removed by lowpass filtering. The dc term present in the expression can also be removed if the system contains an ac-coupling device (blocking capacitor).

All linear modulations can thus be detected by this type of a product demodulator (shown in Fig. (3.26)). The only requirement is that the auxiliary carrier $\cos\omega_c t$ (produced in a local oscillator) used for multiplication must be synchronous in both phase and frequency with the carrier used in the modulator. That is why the detection process is also referred to as synchronous or coherent detection.

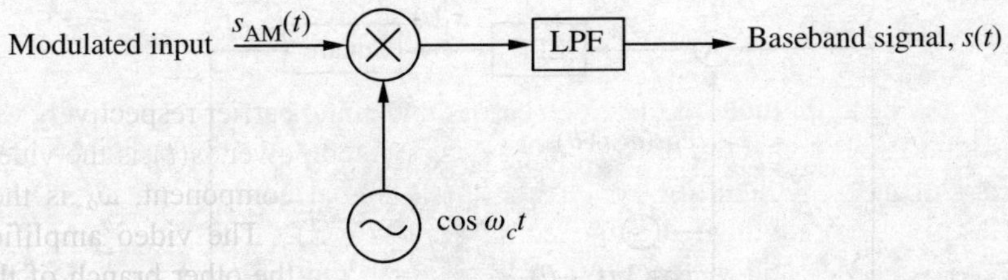

Fig. 3.26 *Product demodulator scheme*

Example 3.5

Suppose the locally generated carrier differs in phase by an angle θ. Discuss the adverse effect in demodulation. Repeat the problem for a frequency drift by an amount Δf of the local oscillator.

Solution

Multiplication of the AM wave by the auxiliary carrier $\cos(\omega_c t + \theta)$ results in the product signal

$$\begin{aligned} s_{\text{AM}}(t)\cos(\omega_c t + \theta) &= A_c[1 + ks(t)]\cos\omega_c t\cos(\omega_c t + \theta) \\ &= A_c[1 + ks(t)](\cos^2\omega_c t\cos\theta - \frac{1}{2}\sin 2\omega_c t\sin\theta) \\ &= A_c[1 + ks(t)]\left[\frac{1 + \cos 2\omega_c t}{2}\cos\theta - \frac{1}{2}\sin 2\omega_c t\sin\theta\right]. \end{aligned} \tag{3.57}$$

A close look into (3.57) reveals that the term that gives the message signal is $\frac{kA_c}{2}s(t)\cos\theta$. If θ varies with time, the signal strength will also vary with time and eventually it may disappear completely for $\theta = \pi/2$. For slowly-varying θ, we observe a fading signal.

It can also be proved in a similar way that for a frequency drift by the amount Δf, the detected signal will be proportional to $s(t)\cos(2\pi\Delta f t)$. The signal may be unacceptable if the frequency drift Δf is comparable to the frequencies present in the message signal. Multiplication by $\cos(2\pi\Delta f t)$ reflects a pair of tones $f_m + \Delta f$ and $f_m - \Delta f$ for a given tone f_m, which is highly undesirable. Even though the LPF is designed to pass frequencies only up to W (message bandwidth), it is reasonable to assume that the entire term passes through the filter since $\Delta f << W$.

One method to make the receiver insensitive to the change of phase is to employ quadrature receiver as outlined in Fig. 3.27. The outputs of the two lowpass filters are obtained as $x_1(t) = \frac{k}{2}s(t)\cos\theta$ and $x_2(t) = -\frac{k}{2}s(t)\sin\theta$. Squaring each term, adding, and finally taking the square-root,

$$\begin{aligned} s_0(t) &= \left[\left(\frac{k}{2}s(t)\cos\theta\right)^2 + \left(-\frac{k}{2}s(t)\sin\theta\right)^2\right]^{1/2} \\ &= \frac{k}{2}\sqrt{s^2(t)} = \frac{k}{2}s(t). \end{aligned} \tag{3.58}$$

We find here that demodulation is effective for $s(t) > 0$. In other words, only AM signals can be demodulated using this scheme.

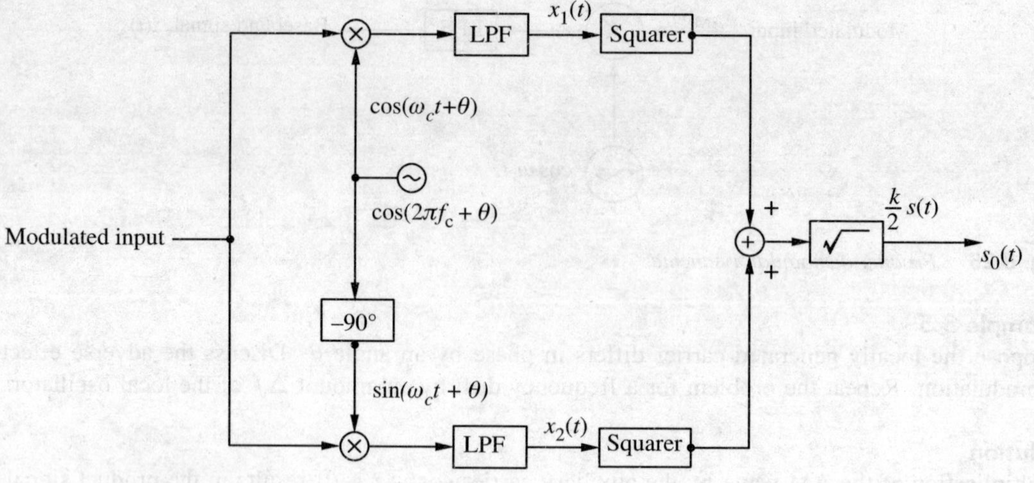

Fig. 3.27 *A quadrature receiver*

3.7.2 Coherent or synchronous detection

The stringent requirement for the product demodulator of Fig. 3.24 is that the auxiliary carrier should be synchronous (in phase and frequency) with the carrier at the transmitter. Frequency matching is possible if the transmitted AM signal contains a component at the carrier frequency and generates the auxiliary carrier from that. The scheme for carrier recovery from a DSB-SC wave is outlined in Fig. 3.28. Consider the DSB-SC signal input to the system:

$$s_c(t) = ks(t)\cos\omega_c t. \tag{3.59}$$

The signal is squared and passed through a narrow bandpass (high Q) filter tuned to $2\omega_c$. The squarer output is

$$\begin{aligned} x_1(t) &= s_c^2(t) = k^2 s^2(t)\cos^2\omega_c t \\ &= \frac{k^2}{2}s^2(t) + \frac{k^2}{2}s^2(t)\cos 2\omega_c t. \end{aligned} \tag{3.60}$$

For the message signal $s(t) = \cos\omega_m t$,

$$\begin{aligned} x_1(t) &= \frac{k^2}{4}(1+\cos 2\omega_m t) + \frac{k^2}{4}(1+\cos 2\omega_m t)\cos 2\omega_c t \\ &= \frac{k^2}{4}\left[(1+\cos 2\omega_m t + \cos 2\omega_c t + \frac{1}{2}\cos 2(\omega_c+\omega_m)t + \frac{1}{2}\cos 2(\omega_c-\omega_m)t\right]. \end{aligned} \tag{3.61}$$

The filter's output is $x_2(t) = \frac{k^2}{4}\cos 2\omega_c t$, which is then passed through a 2:1 frequency divider. The final output is the desired carrier of frequency ω_c. In this way the carrier of correct frequency is generated.

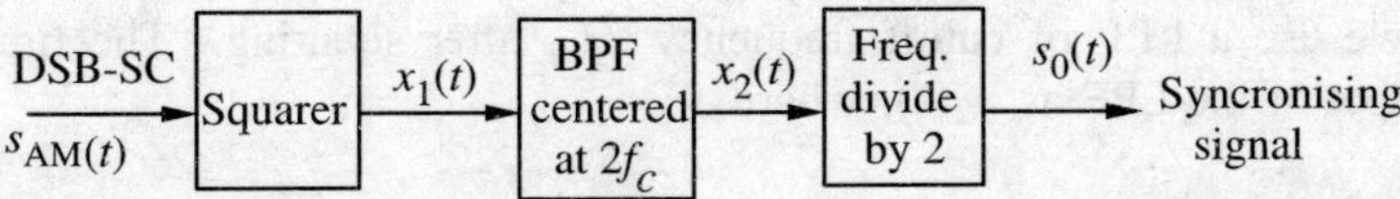

Fig. 3.28 *Carrier recovery in DSB-SC signal*

To ensure identical carrier frequencies both at the transmitter and the receiver, one can use highly stable crystal oscillators. But at very high carrier frequencies, the oscillator performance may not be satisfactory due to the extremely reduced size of the crystal in the oscillator. In such cases the suppressed-carrier systems may transmit carrier at a reduced level (usually about 20 dB). This pilot carrier is separated at the receiver by a narrow bandpass filter tuned to the carrier frequency, which when amplified serves the purpose of the LO. The scheme is known as homodyne detection, shown in Fig. 3.29.

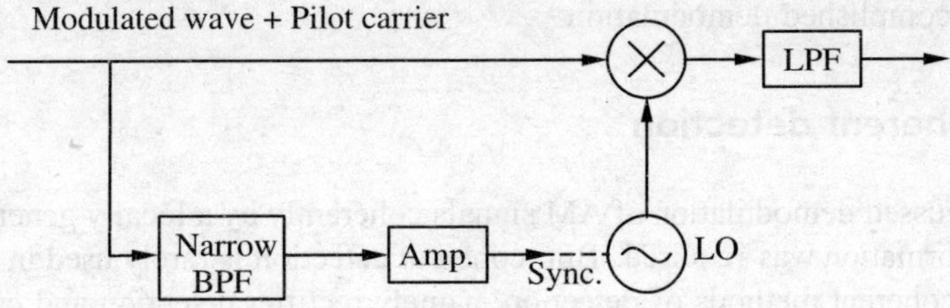

Fig. 3.29 *Homodyne detection*

3.7.3 Square-law detector

In the early days, AM signals were demodulated by a crystal detector followed by a lowpass filter, as shown in Fig. 3.30. The crystal acts as a squaring device. For the received AM signal, the squared signal $x_1(t)$ is given by

Fig. 3.30 *Square-law demodulator for AM signal*

$$x_1(t) = A_c^2(1 + ks(t))^2 \cos^2 \omega_c t = \frac{A_c^2}{2}[1 + ks(t)]^2(1 + \cos 2\omega_c t)$$

$$= \frac{A_c^2}{2}[1 + ks(t)]^2 + \frac{Ac^2}{2}[1 + ks(t)]^2 \cos 2\omega_c t. \tag{3.62}$$

The second term represents the wave around a carrier of frequency $2\omega_c$, which can easily be removed by a lowpass filter. The first term is $A_c^2/2[1 + k^2s^2(t) + 2ks(t)]$, which contains $s^2(t)$ and $s(t)$. Since the spectrum of $s^2(t)$ is bandlimited to $2f_m$ (as $s(t)$ is bandlimited to f_m), it overlaps that of $s(t)$. Naturally these two terms cannot be separated in the frequency domain. One can proceed therefore as follows.

Suppose that we use a LPF of cutoff frequency $2f_m$ after squaring. The first term of (3.62) appears at the output of the LPF as $x_2(t)$, where

$$x_2(t) = \frac{A_c^2}{2}[1 + ks(t)]^2. \tag{3.63}$$

We then take the square root of $x_2(t)$ to get $x_3(t)$ where

$$x_3(t) = 0.707A_c|1 + ks(t)|. \tag{3.64}$$

We have thus recovered the message signal.

The signal is not free from distortion. If the signal be a pure sinusoid, its magnitude would be a rectified sine wave of fundamental frequency equal to twice the original. All the other harmonics will also be present. The problem can be readily circumvented if the quantity kA_m does not exceed unity. In that case $[1 + ks(t)]$ never goes negative and the magnitude $|1 + ks(t)|$ equals $[1 + ks(t)]$. We have indeed accomplished demodulation.

3.7.4 Non-coherent detection

So far we have discussed demodulation of AM signals coherently by a locally generated carrier, where accurate timing information was required. But, coherent detection is rarely used in practice. We shall consider here non-coherent methods of detection, namely rectifier detection and envelope detection.

Rectifier detection

The rectifier can be of two types – half-wave and full-wave. Let us examine the half-wave rectifier detection shown in Fig. 3.31. Half-wave rectification is equivalent to the mathematical operation of multiplying an AM wave by a square wave (see Fig. 3.32). The rectified output is

$$x_1(t) = \{A[1 + ks(t)]\cos\omega_c t\}r(t), \tag{3.65}$$

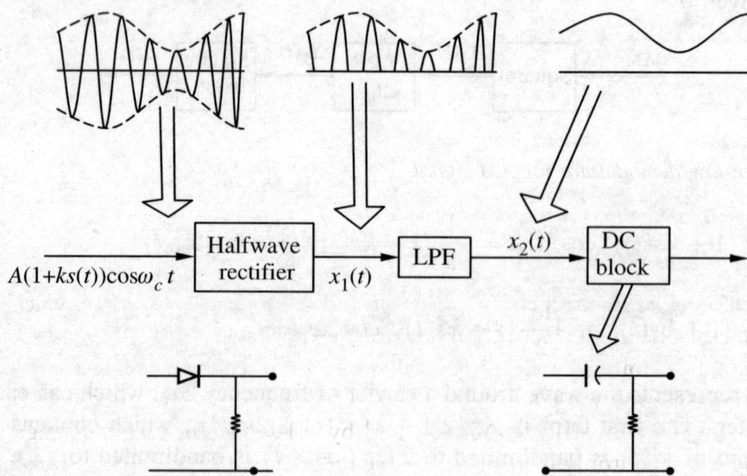

Fig. 3.31 *Half-wave rectifier detector for AM signal*

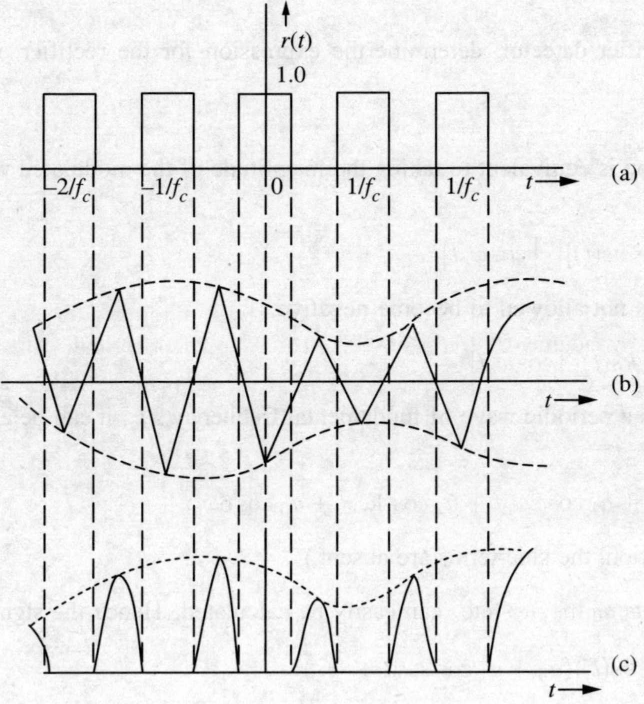

Fig. 3.32 *Square wave for half-wave rectification: (a) square wave; (b) AM signal; (c) rectified AM.*

where $r(t)$ is the square wave of the fundamental frequency ω_c whose Fourier series expansion is given by

$$r(t) = \frac{1}{2} + \frac{2}{\pi}(\cos\omega_c t - \frac{1}{3}\cos 3\omega_c t + \frac{1}{5}\cos\omega_c t - \cdots). \tag{3.66}$$

Thus,

$$\begin{aligned} x_1(t) &= \{A(1+ks(t))\cos\omega_c t\}\left[\frac{1}{2} + \frac{2}{\pi}(\cos\omega_c t - \frac{1}{3}\cos 3\omega_c t + \cdots\right] \\ &= \frac{1}{\pi}A[1+ks(t)] + \text{ other terms which will be blocked by the LPF.} \end{aligned} \tag{3.67}$$

Since the filter's cutoff frequency is just enough to pass the baseband spectrum, the output of the LPF is then

$$x_2(t) = \tfrac{1}{\pi}A[1+ks(t)]. \tag{3.68}$$

The dc term A/π is blocked by a blocking capacitor and demodulation is thereby accomplished.

The rectifier detector of Fig. 3.31 also demodulates the AM wave when the rectifier is full-wave; in this case the output is doubled. Note that full-wave rectification is mathematically equivalent to multiplication of the signal by a square wave of Fig. 3.33(a). The operation is equivalent to multiplication by $+1$ when the carrier is positive and by -1 when the carrier is negative. Thus the square wave should be synchronised in phase and frequency with the carrier. In other words, the rectifier detection is effectively a synchronous procedure since it implies multiplication by a square wave of matching carrier frequency.

Example 3.6

For a full-wave rectifier detector, determine the expression for the rectifier output and hence the output of the LPF.

Solution

Full-wave rectification is equivalent to taking the magnitude of the modulated wave, that is after the rectification:

$$x_1(t) = \big|A[1 + ks(t)]\big| \ \big|\cos\omega_c t\big|.$$

Since $A[1 + ks(t)]$ is not allowed to become negative,

$$x_1(t) = A[1 + ks(t)] \, |\cos\omega_c t|.$$

The term $|\cos\omega_c t|$ is a periodic wave of fundamental frequency $2f_c$; it can be expanded in a Fourier series to yield

$$|\cos\omega_c t| = a_0 + a_1\cos 2\omega_c t + a_2\cos 4\omega_c t + a_3\cos 6\omega_c t + \cdots.$$

(Being an even function, the sine-terms are absent.)

The series coefficients $a_0, a_1, \cdots$ etc. can easily be calculated. Hence the signal

$$x_1(t) = A[1 + ks(t)](a_0 + a_1\cos 2\omega_c t + \cdots)$$

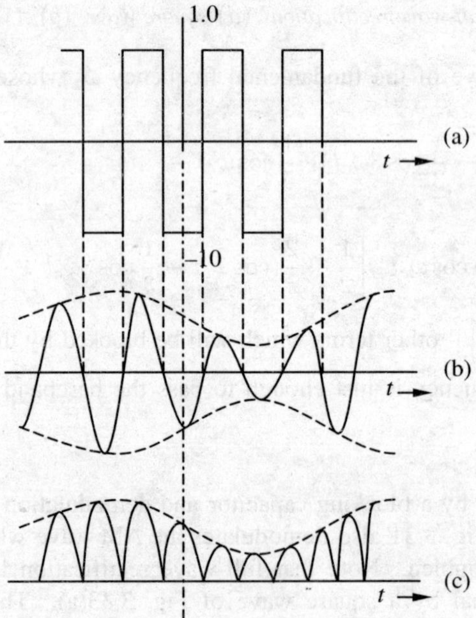

Fig. 3.33 *Square wave multiplication for full-wave rectifier: (a) square wave; (b) AM signal; (c) rectified AM.*

and the output of the LPF is

$$x_2(t) = Aa_0[1 + ks(t)].$$

Envelope detector

In an envelope detector the output follows the envelope of the modulated wave. The circuit shown in Fig. 3.34 performs envelope detection. With a proper choice of parameter values like the resistance, capacitance, and diode characteristics, the circuit will serve as an AM demodulator. To begin with, let us remove the resistor. The circuit then acts as a peak detector. The output in this case never decreases with time as there is no discharge path for the capacitor (see Fig. 3.35).

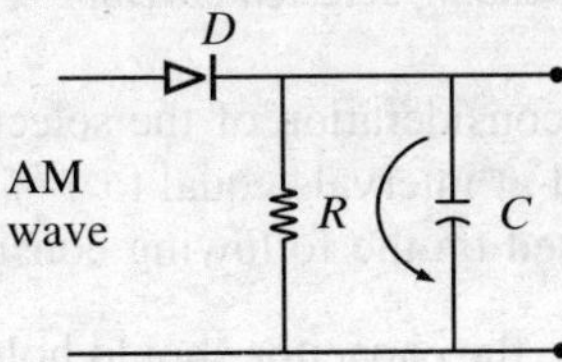

Fig. 3.34 *Envelope detector for AM wave*

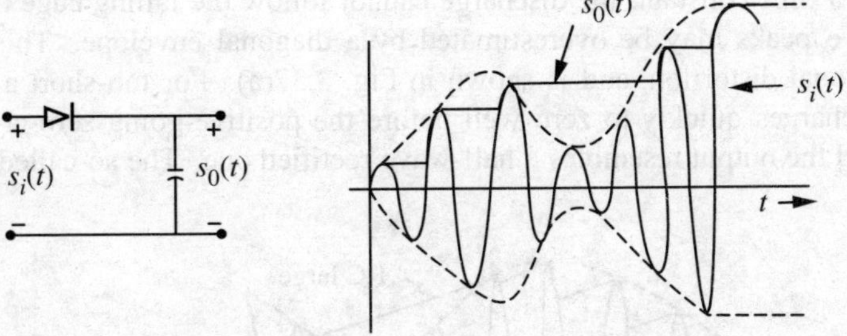

Fig. 3.35 *Peak detector and its output*

With the resistor in place as in Fig. 3.34, the capacitor finds a discharge path and the output decreases exponentially between the peaks of the AM wave. During the positive cycle of the input signal, the diode is forward-biased (zero resistance, ideally) and conducts. The capacitor charges up to the peak voltage (marked 1 in Fig. 3.36) of the input signal. The input signal then starts decreasing and the diode is reverse-biased (infinite resistance, ideally) and open. The capacitor discharges through the resistor at a slow rate until the input signal of the next cycle attains the capacitor voltage. At this point (marked 2) the diode again conducts and the capacitor charges to the next peak. The process is repeated again and again. We observe that the detected wave closely follows the envelope of the AM wave; it contains ripples at the carrier frequency owing to the discharging of the capacitor. The ripple can be removed with an audio filter.

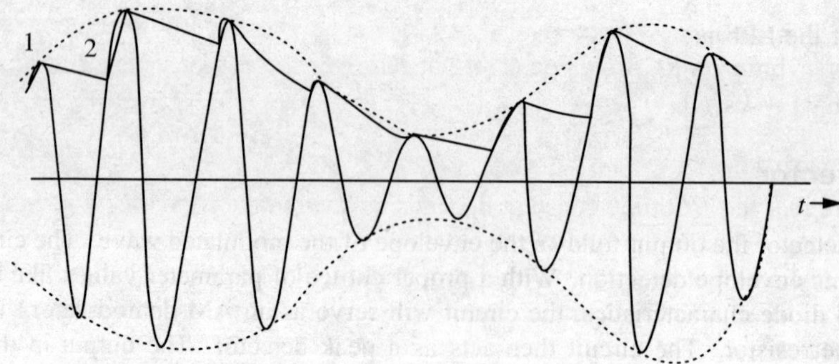

Fig. 3.36 *The AM wave and the corresponding detected envelope*

We will now discuss an important consideration of the selection of the RC time constant. Since the peaks of the AM waves are spaced at intervals equal to $1/f_c$ and the heights follow the message signal $s(t)$, the selection of RC is based on the following considerations:

(i) For the positive peaks increasing, the capacitor should hold the voltage between peaks – a long time constant is required.

(ii) For the peak amplitudes decreasing, the capacitor should discharge quickly to a value at least equal to the peak value of the next cycle in order to track the changes – a short RC is necessary. For too-large a time constant, the discharge cannot follow the falling edge of the envelope and few successive peaks may be overestimated by a diagonal envelope. This type of distortion is called diagonal distortion, and is shown in Fig. 3.37(a). For too-short a time constant, the capacitor discharges quickly to zero well before the positive-going-zero-crossing point of the next cycle and the output resembles a half-wave rectified one. The so-called rectifier distortion thus occurs.

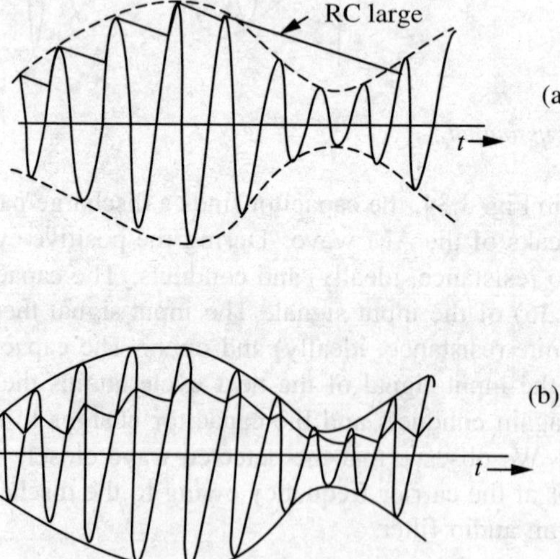

Fig. 3.37 *Detector distortion: (a) diagonal clipping and (b) rectifier clipping.*

Example 3.7

Determine the upper limit on RC time constant that removes the diagonal distortion in the detected wave.

Solution

We will show here that the RC time constant depends on the index of modulation m and the highest modulating frequency f_M. Let the AM wave be $(1 + m\cos\omega_m t)\cos\omega_c t$. Following Fig. 3.38, the amplitude of the AM wave at t_0 is $(1 + m\cos\omega_m t_0)$ and that at $t_0 + T$ is $[1 + m\cos\omega_m(t_0 + T)]$. Here T is taken as the time period of the carrier $(= 1/f_c)$. Since the envelope should track the changes in amplitudes,

$$A(t_0)e^{-T/RC} \leq A(t_0 + T) \tag{3.69}$$

or $\quad (1 + m\cos\omega_m t_0)e^{-T/RC} \leq 1 + m\cos\omega_m(t_0 + T)$

or $\quad (1 + m\cos\omega_m t_0)(1 - \frac{T}{RC}) \leq 1 + m\cos\omega_m t_0 \cos\omega_m T - m\sin\omega_m t_0 \sin\omega_m T$

(as $e^x \approx 1 + x$ for small x)

or $\quad 1 + m\cos\omega_m t_0 - \dfrac{T}{RC} - \dfrac{T}{RC} m\cos\omega_m t_0 \leq 1 + m\cos\omega_m t_0 - m\sin\omega_m t_0(\omega_m T)$

(since $\sin x \approx x$ and $\cos x \approx 1$)

or, $\quad \dfrac{1}{RC}(1 + m\cos\omega_m t_0) \geq m\omega_m \sin\omega_m t_0$

$$\text{or,} \quad RC \leq \frac{1 + m\cos\omega_m t_0}{m\omega_m \sin\omega_m t_0}, \quad \text{for all } t. \tag{3.70}$$

The worst case is when the right-hand-side is minimum; this is obtained by taking the derivative of the L.H.S and equating it to zero.

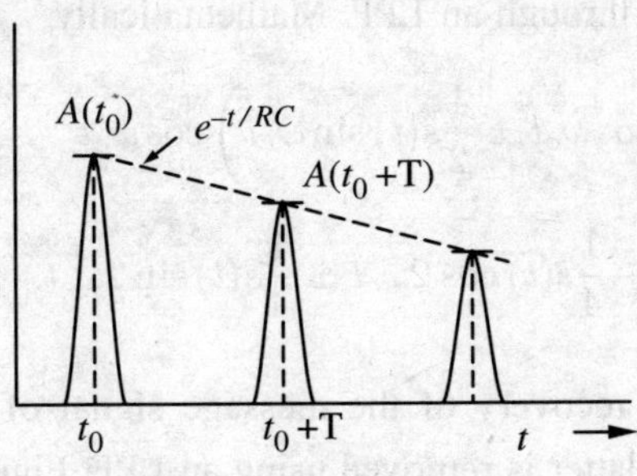

Fig. 3.38 *Capacitor discharge*

Let $y = \dfrac{1 + m\cos\omega_m t}{\sin\omega_m t}$.

Then, $\dfrac{dy}{dt} = \dfrac{\sin\omega_m t(-m\omega_m \sin\omega_m t) - (1 + m\cos\omega_m t)\omega_m \cos\cos\omega_m t}{\sin^2\omega_m t}$.

Equating $\dfrac{dy}{dt}$ to zero,

$-m\omega_m \sin^2\omega_m t - \omega_m \cos\omega_m t - m\omega_m \cos^2\omega_m t = 0,$

or, $\omega_m \cos\omega_m t + m\omega_m = 0$ or $\cos\omega_m t = -m.$

Then, $\sin\omega_m t = \sqrt{1-(-m)^2} = \sqrt{1-m^2}.$

So, (3.70) takes the form

$$RC \le \frac{1-m^2}{m\omega_m\sqrt{1-m^2}} = \left(\frac{\sqrt{1-m^2}}{m}\right)\frac{1}{\omega_m}. \tag{3.71}$$

For faithful detection, the highest modulation frequency f_M is given by

$$f_M = \frac{\sqrt{1-m^2}}{2\pi RCm}. \tag{3.72}$$

For 70.7% modulation ($m = 1/\sqrt{2}$), the value f_M is obtained as $f_M = 1/(2\pi RC)$.

3.7.5 Demodulation for SSB signals

SSB signals can be demodulated coherently in a synchronous demodulator, i.e., by multiplying it with the carrier $\cos\omega_c t$ and passing it through an LPF. Mathematically,

$$\begin{aligned} s_{\text{SSB}}(t)\cos\omega_c t &= \left(\frac{1}{2}s(t)\cos\omega_c t \pm \frac{1}{2}\hat{s}(t)\sin\omega_c t\right)\cos\omega_c t \\ &= \frac{1}{4}s(t) + \frac{1}{4}s(t)\cos 2\omega_c t \pm \frac{1}{4}\widehat{s}(t)\sin 2\omega_c t. \end{aligned} \tag{3.73}$$

The multiplication results in the recovery of the message signal of value $\dfrac{1}{4}s(t)$ and another SSB signal with a carrier at $2\omega_c$. The latter is removed using an LPF. Figure 3.39 shows the synchronous detector for SSB signals.

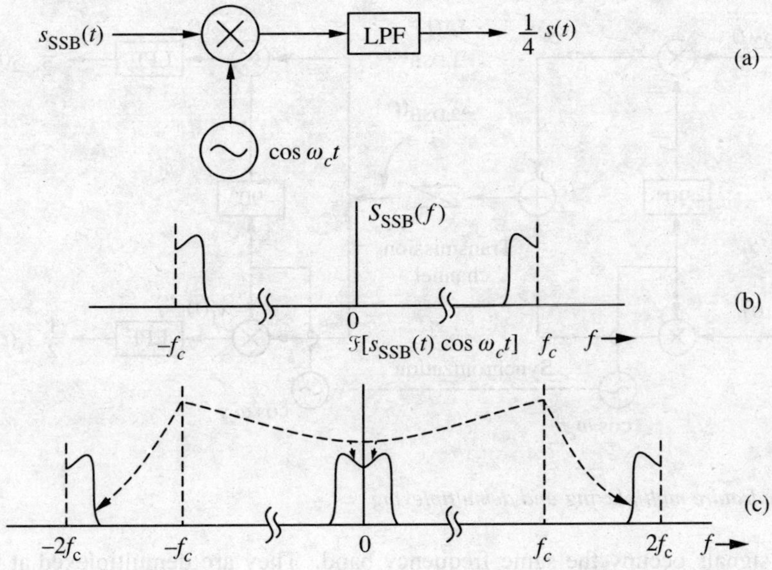

Fig. 3.39 *(a) Synchronous demodulator for SSB signals. (b) SSB spectra (lower sideband). (c) Spectra of the product* $s_{SSB}(t)\cos\omega_c t$.

3.8 Quadrature Amplitude Modulation

We know that DSB signals occupy twice the bandwidth of SSB signals. But DSB modulators and demodulators are simpler than SSB, which is a definite advantage; moreover, it is possible to send two message signals that overlap in time and frequency by employing DSB modulation. The resulting two DSB signals can occupy the same frequency band. This is accomplished by quadrature amplitude modulation (QAM), also called quadrature-carrier multiplexing.

In this scheme we generate two DSB signals for two different messages using carriers of the same frequency but in phase quadrature. Figure 3.40 explains the multiplexing of two DSB signals and their demultiplexing process. Consider two message signals $s_1(t)$ and $s_2(t)$, both being bandlimited to f_m. If $s_1(t)$ and $s_2(t)$ modulate the carriers $\cos\omega_c t$ and $\sin\omega_c t$ respectively, the transmitted QAM signal has the form

$$s_c(t) = s_1(t)\cos\omega_c t + s_2(t)\sin\omega_c t. \tag{3.74}$$

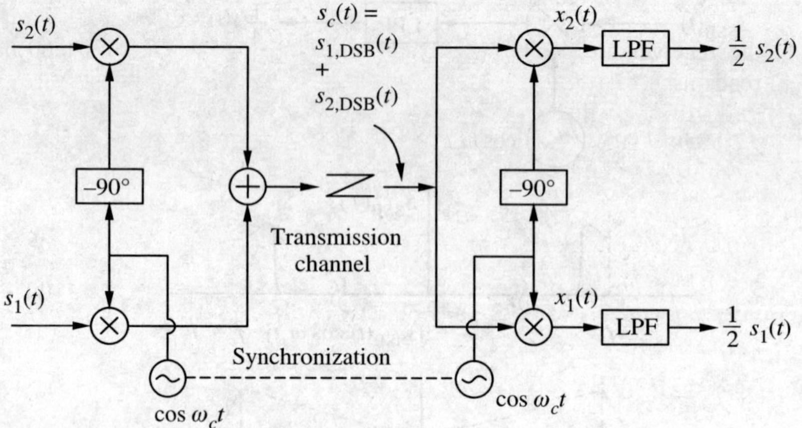

Fig. 3.40 *Quadrature multiplexing and demultiplexing*

The two DSB signals occupy the same frequency band. They are demultiplexed at the receiver by synchronous detector as shown in Fig. 3.40. The signal $x_1(t)$ at the input of the lower LPF is

$$\begin{aligned} x_1(t) &= s_c(t)\cos\omega_c t = (s_1(t)\cos\omega_c t + s_2(t)\sin\omega_c t)\cos\omega_c t \\ &= s_1(t)\cos^2\omega_c t + s_2(t)\sin\omega_c t\cos\omega_c t \\ &= \frac{1}{2}s_1(t) + \frac{1}{2}s_1(t)\cos 2\omega_c t + \frac{1}{2}s_2(t)\sin 2\omega_c t. \end{aligned} \tag{3.75}$$

The output of the filter is then $\frac{1}{2}s_1(t)$. The other two terms together represent quadrature carrier multiplexing at the carrier frequency $2\omega_c$ and are removed by the LPF.

The input of the upper LPF is

$$\begin{aligned} x_2(t) &= s_c(t)\sin\omega_c t \\ &= s_1(t)\cos\omega_c t\sin\omega_c t + s_2(t)\sin^2\omega_c t \\ &= \frac{1}{2}s_1(t)\sin 2\omega_c t - \frac{1}{2}s_2(t)\cos 2\omega_c t + \frac{1}{2}s_2(t). \end{aligned} \tag{3.76}$$

Hence the output of the filter is $\frac{1}{2}s_2(t)$.

The QAM involves perfect phase and frequency control at the receiver. To maintain this synchronisation, a pilot signal (sinusoidal) of reduced power may be sent outside the frequency band of the modulated signal. The phase and frequency of this signal correspond to the carrier at the transmitter. At the receiver the pilot signal is recovered by a tuned circuit.

Example 3.8

In the QAM scheme of Fig. 3.40, the oscillator in the receiver differs in phase from that of the oscillator in the transmitter. Show that the two signals at the receiver interfere with each other.

Solution

The carriers at the receiver are $\cos(\omega_c t + \theta)$ and $\sin(\omega_c t + \theta)$, where θ is the phase mismatch. Equation (3.75) reads as

$$x_1(t) = s_1(t)\cos\omega_c t\cos(\omega_c t + \theta) + s_2(t)\sin\omega_c t\cos(\omega_c t + \theta)$$

$$= s_1(t)[\cos^2\omega_c t.\cos\theta - \sin\omega_c t\cos\omega_c t.\sin\theta]$$

$$+s_2(t)[\sin\omega_c t\cos\omega_c t\cos\theta - \sin^2\omega_c t\sin\theta]$$

$$= \frac{1}{2}s_1(t)[\cos\theta + \cos 2\omega_c t\cos\theta - \sin 2\omega_c t\sin\theta]$$

$$+\frac{1}{2}s_2(t)[\sin 2\omega_c t\cos\theta - (1 - \cos 2\omega_c t)\sin\theta].$$

Hence output of the lower filter is $\frac{1}{2}s_1(t)\cos\theta - \frac{1}{2}s_2(t)\sin\theta$.

Equation (3.76) reads as

$$x_2(t) = s_1(t)\cos\omega_c t\sin(\omega_c t + \theta) + s_2(t)\sin\omega_c t\sin(\omega_c t + \theta)$$

$$= s_1(t)[\cos\omega_c t\sin\omega_c t\cos\theta + \cos^2\omega_c t\sin\theta]$$

$$+s_2(t)[\sin^2\omega_c t\cos\theta + \sin\omega_c t\cos\omega_c t\sin\theta]$$

$$= \frac{1}{2}s_1(t)[\sin 2\omega_c t\cos\theta + (1 + \cos 2\omega_c t)\sin\theta]$$

$$+\frac{1}{2}s_2(t)[(1 - \cos 2\omega_c t)\cos\theta + \sin 2\omega_c t\sin\theta].$$

The output of the upper filter is $\frac{1}{2}s_1(t)\sin\theta + \frac{1}{2}s_2(t)\cos\theta$.

Thus the message signals are not decoupled and hence one interferes with the other. A measure of interference is the ratio of the amplitude of the undesired term to that of the desired one. For the lower channel, the interference is $\sin\theta/\cos\theta = \tan\theta$. As the phase error θ approaches zero, the ratio $\tan\theta$ also approaches zero, which is expected. For a phase mismatch of $5^\circ, \tan\theta = 0.0875$ and the interference is $20\log_{10}(0.0875)$ dB, i.e., 21.16 dB below the desired signal.

3.9 Superheterodyne AM Receivers

A typical AM broadcast radio receiver must perform three basic operations :(i) station *separation*, i.e., tuning to select a signal and rejecting all other signals received with it, (ii) amplification to compensate for the signal loss in the media, and (iii) demodulation.

Tuning to a particular radio station requires a very sharp bandpass filter besides requiring a distinct centre frequency in order to pick it out from the other stations. Such bandpass characteristic is practically difficult to build. Fortunately, Edwin Armstrong devised the superheterodyne receiver in 1918 and this solved the problem. The superheterodyne or 'superhet' for short, is based on the multiplication of an incoming signal by a sinusoid, thereby shifting its frequency up and down depending on the frequency of the multiplier. Figure 3.41 illustrates the superhet AM receiver principle. The incoming signal $s_c(t)$ is first selected by a tunable radio-frequency (RF) bandpass filter. The next section is the RF amplifier that has a relatively broad bandwidth.

The mixer stage converts the RF to an intermediate frequency (IF). The mixer includes a tuneable local oscillator (LO) that generates sinusoids. The RF signal is heterodyned with the LO in a nonlinear device resulting in down-conversion of the RF output. The output of the mixer contains an infinite number of harmonics and the desired IF. The LO is tuned such that its frequency $f_{\rm LO}$ always maintains a constant deviation $f_{\rm IF}$ from the RF carrier f_c. Thus,

$$f_{\rm LO} = f_c + f_{\rm IF} \quad \text{or } f_c - f_{\rm IF}. \tag{3.77}$$

The LO can track either side of the RF. To select the upper side, $f_{\rm LO} = f_c + f_{\rm IF}$ and so $f_{\rm IF} = f_{\rm LO} - f_c$. Thus the RF upper side frequency is translated to the IF lower side frequency and vice versa. This is called sideband inversion.

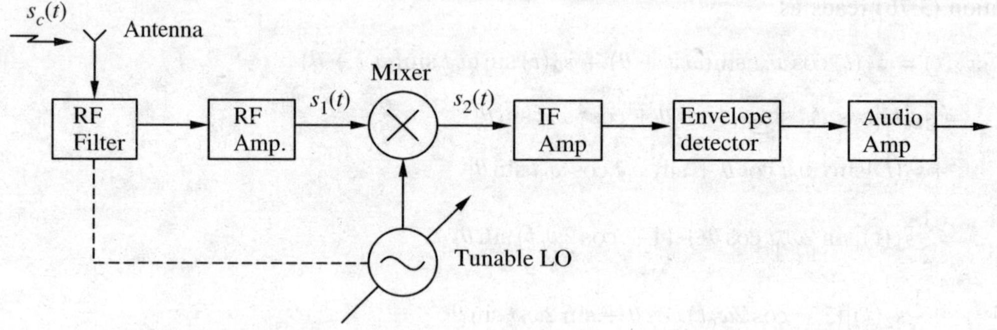

Fig. 3.41 *Superheterodyne AM receiver*

The IF signal is then amplified in IF amplifier, usually consisting of three stages, where maximum gain is provided to the IF signal. The receiver then demodulates the signal in an envelope detector and the detected signal is then amplified in an audio amplifier before being sent finally as the message signal to a loudspeaker.

The mixer unit may produce the same IF for two RF carriers. The undesired RF carrier may interfere with the desired RF. Let f_c be the desired RF carrier. We take the IF as $f_{\rm IF} = f_{\rm LO} - f_c$, that is, the local oscillator scans the upper side. Let the undesired RF carrier have the frequency $f_c' > f_c$ such that $f_{\rm IF} = f_c' - f_{\rm LO}$; the RF carrier f_c' is called the image frequency. The frequency difference between the image station and the desired station is $f_c' - f_c = 2f_{\rm IF}$. The image frequency is removed in the IF section by selecting the bandwidth $B_{\rm IF} < 2f_{\rm IF}$.

Commercial AM broadcast band specified by the Federal Communications Commissions (FCC) is 540–1600 kHz carrier frequency. The IF is set at 455 kHz with a bandwidth of 10 kHz.

The greatest benefit of the superhet receiver is that tuning to a particular station takes place at the front-end of the RF receiver. Since the RF from various radio stations are down-converted to a fixed IF (455 kHz), the circuitry following the IF amplifiers requires no adjustment at all. Secondly, a sharp bandpass filter of bandwidth 10 kHz is difficult to design if its centre frequency is high. But as the signal is down-converted to IF, one can implement sharp bandpass filters and thereby suppress adjacent channel interference.

Let us now determine the tuning range of the local oscillator. The choice $f_{\mathrm{LO}} = f_c + f_{\mathrm{IF}}$ is adopted for practical reasons as explained below. Suppose we want to tune to the station at $f_c = 540$ kHz. Since the IF is 455 kHz, the frequency of the LO must be either 540-455=85 kHz or 540+455=995 kHz. Similarly for a radio station at 1600 kHz, the LO frequency must be either $1600 - 455 = 1145$ kHz or $1600 + 455 = 2055$kHz. Thus for AM broadcast in the band $540 - 1600$ kHz, the LO must be tunable over either the range $995 < f_{\mathrm{LO}} < 2055$ kHz using $f_{\mathrm{LO}} = f_c + f_{\mathrm{IF}}$. For the first range, the ratio of the highest to lowest frequency of the LO is 1145:85=13:1 whereas that for the second range is 2055:995=2:1, which can readily be achieved. The reason why f_{LO} is taken higher than the incoming RF carrier is now obvious.

The superhet receiver has been fabricated in integrated circuits (IC). This includes a network called automatic gain control (AGC) which adjusts the overall gain of the receiver according to the received RF signal level. The voltage level from various stations depends on a number of factors: the distance of the radio station, the location of the transmitting antenna, its radiated power and power pattern, time of reception (daytime or night), characteristics of the transmitting media, and the others. The gain control is accomplished in a negative feedback mechanism. The average value of the audio signal is sensed and then fed back to the IF or RF stage, which modifies the Q-points thereby changing the gain.

Example 3.9

A superhet receiver is designed to receive signals from $540 - 1600$ kHz with $f_{\mathrm{IF}} = 455$ kHz. It is set to receive a 540 kHz signal. Assume that the LO has a significant third harmonic output. Find the possible carrier frequencies received if $f_{\mathrm{LO}} = f_c - f_{\mathrm{IF}}$ is used.

Solution

Since $540 < f_c < 1600$ kHz and $f_{\mathrm{IF}} = 455$kHz, so $995 < f_{LO} < 2055$ kHz.

For the station at $f_c = 540$ kHz, the image station is at $f_c' = f_c + 2f_{\mathrm{IF}} = 540 + 2 \times 455 = 1450$ kHz.

The third harmonic of f_{LO} is $3f_{\mathrm{LO}} = 3{\times}995 = 2985$ kHz. Thus, the RF signal at $f_{c1} = 3f_{\mathrm{LO}} - f_{\mathrm{IF}} = 2985 - 455 = 2530$ kHz. The corresponding image frequency is

$$f_{c1}' = f_{c1} + 2f_{\mathrm{IF}} = 2530 + 2 \times 455 = 3440 \text{ kHz.}$$

Thus, if the RF filter has a wide bandwidth, the receiver will receive signal from the stations at 540, 1450, 2530, and 3440 kHz.

Solved Numericals

1. Let the modulating signal be given by

   $s(t) = \cos(100\pi t) + 2\cos(300\pi t)$

Determine the time-domain expression for the upper side band and lower side band signals if the carrier frequency is $f_c = 1000\text{Hz}$

Solution

The USB and LSB signals are given by expression 3.47 and 3.48 and are written as

$$s_{\text{LSB}}(t) = \frac{1}{2}s(t)\cos\omega_c t + \frac{1}{2}\hat{s}(t)\sin\omega_c t$$

and $s_{\text{USB}}(t) = \frac{1}{2}s(t)\cos\omega_c t - \frac{1}{2}\hat{s}(t)\sin\omega_c t$

where $s(t)$ is the modulating signal and $s(t)$ is its Hilbert transform. We also know from Example 2.15 that the Hilbert transform of a sinusoid is simply the sinusoid phase-delayed by $\frac{\pi}{2}$. That is,

$$H\left[\cos\omega t\right] = \cos\left(\omega t - \frac{\pi}{2}\right) = \sin\omega t \text{ and}$$

$$H\left[\sin\omega t\right] = \sin\left(\omega t - \frac{\pi}{2}\right) = -\cos\omega t$$

So,

$$\hat{s}(t) = \sin(100\pi t) + 2sin(300\pi t)$$

$$s(t)\cos\omega_c t = [\cos(100\pi t) + \cos(300\pi t)]\cos(2000\pi t)$$

$$= \frac{1}{2}[\cos(2100\pi t) + \cos(1900\pi t)] + [\cos(2300\pi t) + \cos(1700\pi t)]$$

$$\hat{s}(t)\sin\omega_c t = [\sin(100\pi t) + 2\sin(300\pi t]\sin(2000\pi t)$$

$$= \frac{1}{2}[\cos(1900\pi t) - \cos(2100\pi t)] + [\cos(1700\pi t) - \cos(2300\pi t)]$$

So,

$$s_{LSB}(t) = \frac{1}{2}cos(1900\pi t) + \cos(1700\pi t)$$

and

$$s_{\text{USB}}(t) = \frac{1}{2}cos(2100\pi t) + \cos(2300\pi t)$$

2. An AM original is given by $s_{\text{AM}}(t) = [A + s(t)]\cos\omega_c t$ for a periodic triangular modulating signal $s(t)$ as shown in the adjoining figure. Derive the modulation efficiency η for indices of modulation m = 0.3, 0.5, 0.8 and 1.

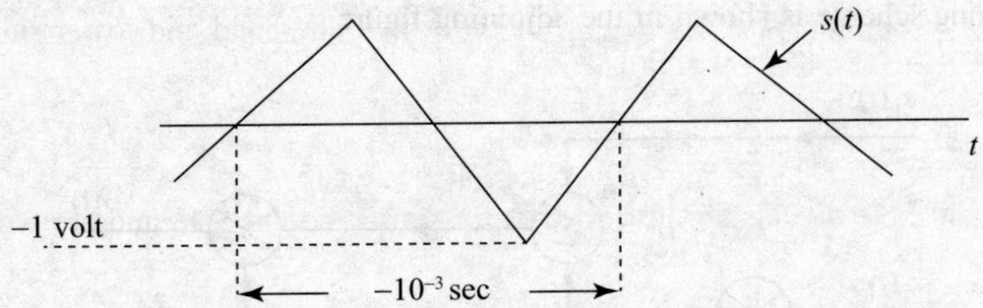

Solution

We find that

$$s_{\mathrm{AM}}(t) = \underbrace{A\cos\omega_c t}_{\text{carrier}} + \underbrace{s(t)\cos\omega_c t}_{\text{sidebands}}$$

The carrier power is $P_c = \frac{1}{2}A^2$ and the sideband is $P_s = \frac{1}{2}\overline{s^2(t)}$

The modulation efficiency η is

$$\eta = \frac{\text{Signal power}}{\text{Total power}} = \frac{\frac{1}{2}\overline{s^2(t)}}{\frac{1}{2}A^2 + \frac{1}{2}\overline{s^2(t)}} \times 100\% = \frac{\overline{s^2(t)}}{A^2 + \overline{s^2(t)}}$$

Similar to the case of tone modulation, we may assume the amplitude of $s(t)$ as μA where μ is modulation index (remember that $\frac{A_m}{A_c} = m$). Here we may take $mA = 1$ (amplitude of $s(t)$ is 1 volt), and

$$\overline{s^2(t)} = \frac{1}{T}\int_0^T s^2(t)dt = \frac{1}{T}\left(\frac{1}{2} \times \frac{T}{4} \times 1\right) \times 4 = \frac{1}{2}$$

Thus,

$$\eta = \frac{\frac{1}{2}}{\left(\frac{1}{m}\right)^2 + \frac{1}{2}} = \frac{\frac{1}{2}}{\frac{m^2+2}{2m^2}} = \frac{m^2}{2+m^2}$$

$$m = 0.3, \eta = \frac{(0.3)^2}{2+(0.3)^2} \times 100\% = \frac{0.09}{2.09} \times 100\% = 4.3\%$$

$$m = 0.5, \eta = \frac{(0.5)^2}{2+(0.5)^2} \times 100\% = \frac{0.25}{2.25} \times 100\% = 11.11\%$$

$$m = 0.8, \eta = \frac{(0.8)^2}{2+(0.8)^2} \times 100\% = \frac{0.64}{2.64} \times 100\% = 24.24\%$$

$$m = 1, \eta = \frac{1}{2+1} \times 100\% = 33.33\%$$

We note that as depth of modulation increases, the useful power (power in the sideband) increases and best condition is achieved for $m = 1$

3. A multiplying scheme is shown in the adjoining figure

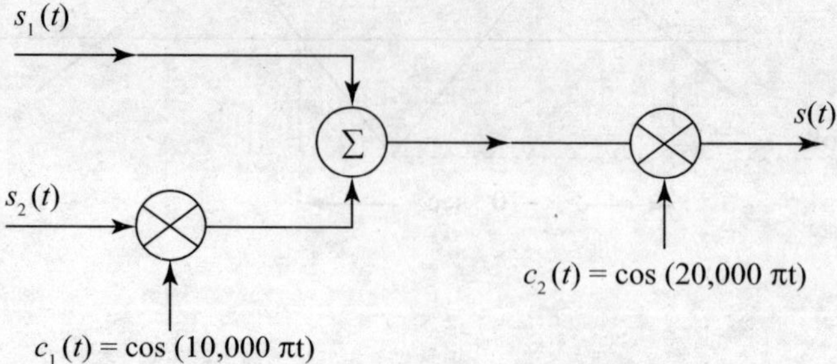

The signals $s_1(t)$ and $s_2(t)$ are modulating originals band-limited to 2000 Hz. The modulating signal $s(t)$ is transmitted over a channel and $c_1(t), c_2(t)$ are the carriers. Find the frequency bands over which the message signals lie and the capacity of the channel required.

Solution

Here, $s(t) = [s_1(t) + s_2(t)c_1(t)]\, c_2(t)$

$$= s_1(t)c_2(t) + s_2(t)c_1(t)c_2(t)$$

The first term $s_1(t)c_2(t)$ is a DSB and occupies the spectral range $f_{c2} - f_m$ to $f_{c2} + f_m$, Now, $f_{c2} = 10,000\text{Hz}$ and so $s_1(t)c_2(t)$ occupies the range from (10,000-2,000) Hz i.e., 8,000 to 12,000 Hz.

The term $s_2(t)c_1(t)$ is a DSB and occupies the spectral range from $f_{c1} - f_m$ to $f_{c1} + f_m$. Since $f_{c_1} = 5000\text{Hz}$, $s_2(t)c_1(t)$ occupies the range from $(5000 - 2000)$Hz to $(5000 + 2000)$Hz i.e., from 3000 to 7000 Hz. Now, $s_2(t)c_1(t)$ is a bandpass signal ($f_L = 3000\,\text{Hz}$, $f_H = 7000\,\text{Hz}$) which modulates the carrier $c_2(t)$ of $f_{c_2} = 10,000\,\text{Hz}$. The LSB will occupy the spectral range $(7000 - 3000)$Hz and the USB will occupy the spectral range $(13,000 - 17,000)$Hz.

So, the final modulated signal s(t) will contain the information signals in the following bands.

Signal $s_1(t)$ will occupy $8 - 12$ kHz range in $s(t)$

Signal $s_2(t)$ will occupy $7 - 3$ kHz and $13 - 17$ kHz range in $s(t)$.

The channel should support 17 kHz signal (channel bandwidth)

4. The signal $s(t) = \dfrac{1}{2}\cos(2\pi 50t) + \dfrac{1}{3}\cos(2\pi 100t)$ is applied to the product modulator of Fig. 3.5 (a) with the carrier frequency $f_c = 10\,\text{kHz}$. Assume the associated non-linearity be given as

$$v_o(t) = a_1 v_i + a_2 v_i^2$$

where $v_0(t)$ and $v_i(t)$ are, respectively, the output and input of the non-linear device (placed after the product modulator).

Assuming $A_c = 10$ volts and $m = 0.5$, find the centre frequency and bandwidth of the bandpass fitter that will produce a standard AM signal.

Solution

In this case, $v_0(t) = s(t) + A_c\cos(2\pi f_c t)$. Thus

$$v_0(t) = a_1\left[s(t) + A_c\cos(2\pi f_c t)\right] + a_2[s(t) + A_c\cos(2\pi f_c t)]^2$$

$$= a_1 s(t) + a_2 s^2(t) + a_2 A_c \cos^2(2\pi f_c t) + [a_1 A_c + 2a_2 A_c s(t)]\cos(2\pi f_c t)$$

$$= a_1 s(t) + a_2 s^2(t) + \frac{1}{2}a_2 A_c + \frac{1}{2}a_2 A_c \cos(4\pi f_c t) + \underbrace{a_1 A_c\left[1 + \frac{2a_2}{a_1}s(t)\right]\cos(2\pi f_c t)}_{\text{standard AM}}$$

The last term is the standard AM signal of the form $A\left[1 + ms(t)\right]\cos(2\pi f_c t)$. Possible spectrum of $v_0(t)$ will be similar to that of Fig 3.10.

The bandpass filter must have the centre frequency $f_c > 3W$ and bandwidth of $2W$ where W is the bandwidth of the base band $s(t)$.

In this case, highest frequency present in the modulating signal $s(t)$ is $f_m = 100\,\text{Hz}$ so that one can assume $W = 100\,\text{Hz}$. Since centre frequency must be at least three times W, that is, 300 Hz, and the carrier frequency is $f_c = 10\,\text{kH}_\text{z}$, one can safely take this as the centre frequency. However the width of the filter will be $2W = 200$ Hz.

In this case, $a_1 A_c =$ carrier amplitude and so $a_1 = 1$.
Also, $\dfrac{2a_2}{a_1} = m = 0.5$ and $a_2 = \dfrac{0.5}{2} = 0.25$.

5. A 400 W carrier is modulated to a depth of 75%. Calculate the total power transmitted in case of VSB modulation. Assume 20% of the other sideband is transmitted along with wanted sideband. Derive also the necessary power relation in VSB modulation in television signal.

Solution

The total power in the DSB wave is calculated as $P_t = P_c\left(1 + \dfrac{m^2}{2}\right)$, vide Eq. (3.21). The sideband power would be $P_c\dfrac{m^2}{2}$. If one of the side band is completely retained and a fraction k of the other sideband is retained, then the power in VSB wave would be

$$P_{VSB} = P_c\frac{m^2}{4} + k.P_c\frac{m^2}{4} \text{ (since each side band contains the power of } P_c\tfrac{m^2}{4})$$

$$= P_c\frac{m^2}{4}(1 + k)$$

This can also be derived from the envelope of VSB wave (vide Eq. 3.55).

In the given example, $P_c = 400$ W, $m = 0.75$ and $k = 0.2$.

Thus, total power in VSB is

$$P_{\text{VSB}} = 400 \times \frac{(3/4)}{4}(1+0.2) = 6.75\ \text{W}$$

The saving in power would be

$$0.8P_c\frac{m^2}{4} = 0.8 \times 400 \times \frac{(3/4)^2}{4}$$

$$= 45\text{W(ComparedtoDSB)}.$$

Exercises

Activity

3.1 Let $s(t) = \cos 2000\pi t$ be the baseband signal that modulates a carrier of frequency 10^6 Hz. Sketch the spectra of the AM signal with modulation index $m < 1$ and that of the DSB-SC signal. Identify the location where phase reversal occurs.

3.2 The message signal $s(t) = \dfrac{\sin 2\pi t}{t}$ amplitude-modulates a carrier of frequency 10Hz. Sketch the AM waveform and its spectrum.

3.3 The tuned circuit of the oscillator in an AM transmitter employs a 100 mH coil and a 1μF capacitor. If the oscillator output is modulated by an audio signal bandlimited to 10 kHz, what is the frequency range occupied by the sidebands?

Hint:

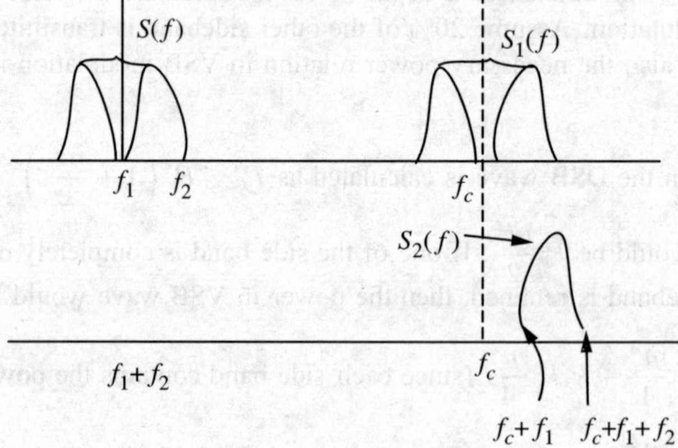

3.4 Let the message signal $s(t)$ be a periodic square wave as shown in Fig. P3.1. The signal amplitude-modulates the carrier $\cos 2\pi f_c t$, where $f_c >> 1/T$. Sketch the waveform when the modulation is AM for $m = 1$. Repeat for DSB case.

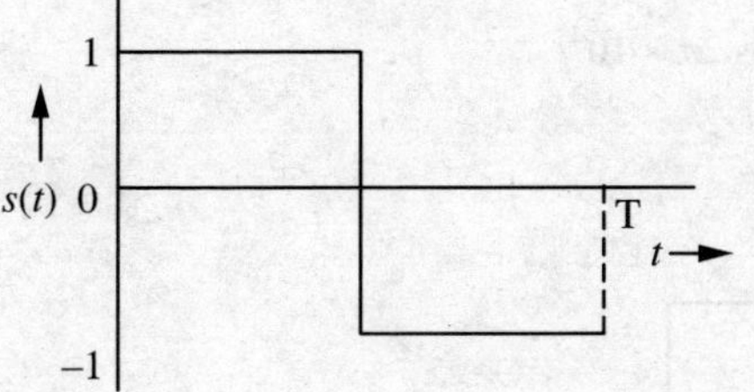

Fig. P3.1

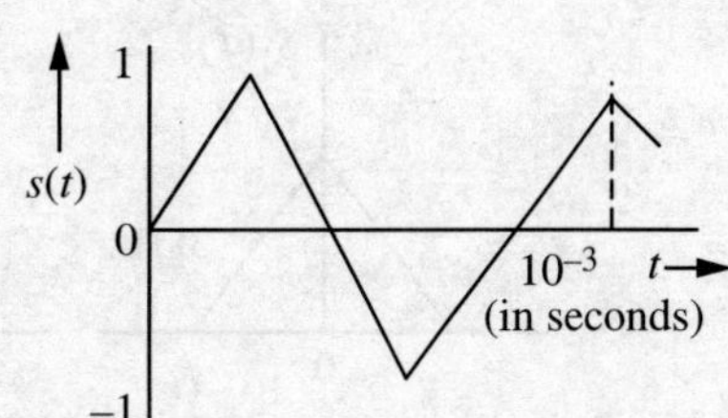

Fig.P3.2

3.5 Let the message signal $s(t)$ be a periodic triangle signal of Fig. P3.2. Sketch the AM signal $[1 + s(t)] \cos 2\pi t$. Find the total transmitted power and the sideband power. Also, calculate the power efficiency.

3.6 The system of Fig. P3.3 performs the scrambling operation. The message signal $s(t)$ is zero outside the interval $f_1 \leq |f| \leq f_2$. The filters have the same cutoff frequency $= f_c$. Assuming $f_c > f_2$, derive an expression for the scrambled output and sketch its spectrum.

3.7 Two message signals $s_1(t)$ and $s_2(t)$ both bandlimited to 3 kHz, are multiplexed in the scheme of Fig. P3.4. The multiplexed signal now modulates a carrier of frequency 100 kHz. Sketch the signal spectrum of the transmitted signal. What is the bandwidth of the channel? Design a demodulator that recovers $s_1(t)$ and $s_2(t)$.

3.8 An AM signal with carrier is given by $s_{\text{AM}}(t) = A_c[1 + ks(t)] \cos 2\pi f_c t$, where the message signal $s(t)$ is bandlimited to f_m. Find the output of the network shown in Fig. P3.5.

3.9 Find the envelope of the signal

$$s(t) = \cos 2\pi f_c t + \frac{1}{6} \cos 2\pi (f_c + f_m)t + \frac{1}{12} \cos 2\pi (f_c - f_m)t.$$

Determine the distortion produced by the quadrature component.

3.10 A superhet AM receiver is tuned to a radio station at $f_c = 850$ kHz. If the intermediate frequency is 455 kHz, find the frequency of the image station.

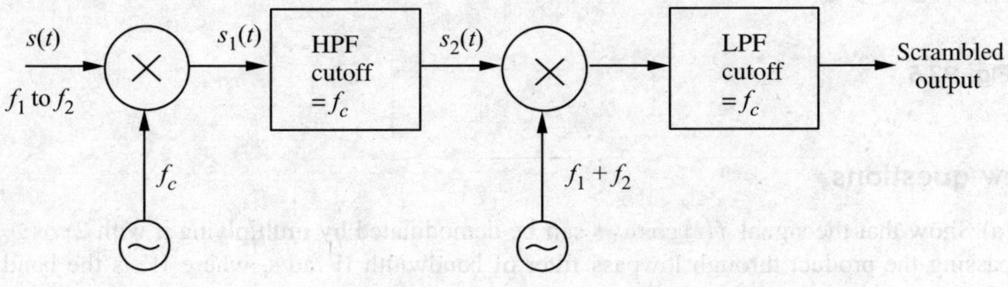

Fig. P3.3 *Scrambler*

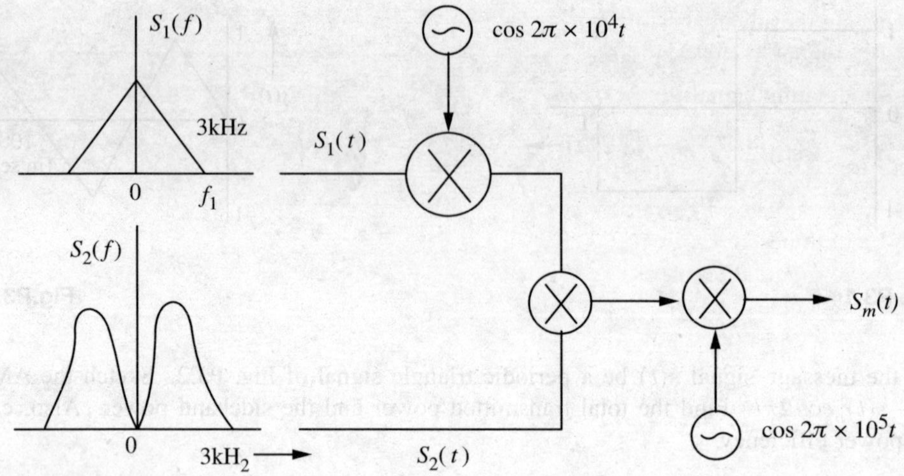

Fig. P3.4

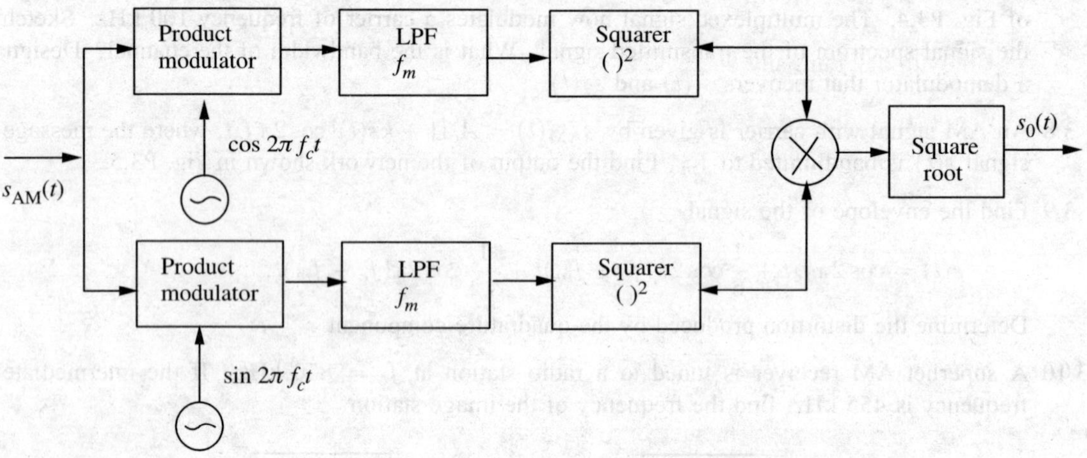

Fig. P3.5

Review questions

1. (a) Show that the signal $f(t) \cos \omega_0 t$ can be demodulated by multiplying it with $2 \cos \omega_c t$ and passing the product through lowpass filter of bandwidth W rad/s, where W is the bandwidth of $f(t)$. Assume that $W < W_e$.

   (b) The modulating signal is a single sinusoid given by $n(t) = 2 \cos 100\pi t$. Derive the expressions for full AM, DSB-SC, and SSB-SC if the carrier frequency and the amplitude are 15 kHz and 5 volts respectively. (K.U. 2000)

2. (a) What are the advantages of SSB modulation over standard AM? Describe phase-shift method for generating SSB modulation.
(b) A modulating signal given by

$$v_m = 2\sin(2\pi \times 500t) + 3\sin(2\pi \times 1100t) + 5\sin(2\pi \times 1300t),$$

amplitude-modulates a carrier given by $v_c = 10\sin(2\pi \times 10^6 t)$, where all amplitudes are in volts. Determine
(i) the total modulation index
(ii) the frequencies present in the modulated signal
(iii) the total transmitted power (K.U. 2001)

3. (a) With a neat circuit diagram, explain the principal of envelope detection of an amplitude modulated wave.
(b) What is meant by diagonal clipping in AM detection? Explain with an illustrative example. (KU 2001)

4. (a) The antenna current of an AM broadcast transmitter modulated to a depth of 40% by an audio sine wave is 11 A. It increases to 12 A as a result of instantaneous modulation by another audio sine wave. What is the modulation index due to the second wave?
(b) Show that for tone modulation the maximum efficiency that can be achieved is 33%. Explain with diagram, any one method of demodulating AM waves. (KU 2001)

5. (a) Why is modulation required?
(b) Define amplitude modulation and modulation index with a sketch of the amplitude modulated signal.
(c) What is the condition to be fulfilled for envelope detection of an amplitude modulated signal?
(d) Why is synchronous detection applied to detect an amplitude modulation signal? (WBUT 2003)

6. (a) For an AM signal, derive the carrier power and sideband power in terms of relevant parameters.
(b) A DSB-SC signal is demodulated using a carrier with a phase error of θ. Determine the expression of the output signal. (WBUT 2004)

7. (a) A sine wave carrier is amplitude modulated by square wave at 50 percent. Draw the waveform of the AM wave.
(b) An amplitude modulated signal has the form

$$x_c(t) = 10(1 + 0.5\cos 2000\pi t + 0.5\cos 4000\pi t)\cos 20,000\pi t.$$

(i) Sketch the amplitude spectrum of $x_c(t)$, showing the location of lines in the spectrum.
(ii) Find the normalised average power content of each spectral component.
(iii) Find the power efficiency. (WBUT 2004)

8. (a) Draw the circuit of a balanced modulator. Show that the modulator produces an output consisting of only sidebands with the carrier removed.
(b) A DSB-SC signal $m(t)\cos\omega_c t$ is demodulated using a locally-generated carrier $A\cos(\omega_c t+\theta)$ as shown in the following figure.

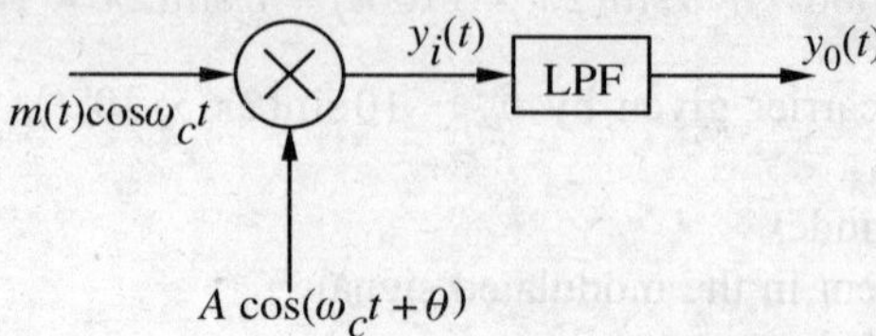

Show that this scheme can demodulate the signal as long as θ is a constant (i.e. not time-varying) and is not $\pi/2$. What is the output if $\theta=\pi/2$ and $\pi/4$? (WBUT 2004)

9. (a) Explain the generation of SSB-SC signal using selective filtering method.
(b) Compare the selective-filtering method with phase-shift method.
(c) Describe the signal squaring technique for carrier recovery. Explain why it cannot be used for SSB-SC detection.
(d) Explain the effects of frequency and phase error in the SSB-SC receiver. (WBUT 2005)

10. (a) Why is SSB modulation not suitable for video signals? Give an expression for the SSB-SC signal in the time domain, indicating the relationship between the inphase and the quadrature component.
(b) An amplitude-modulated signal is given as

$$\phi_{AM}(t)=10\cos(2\pi.10^6t)+5\cos(2\pi.10^6t)\cos(2\pi.10^3t)+2\cos(2\pi.10^6t)\cos(4\pi.10^3t).$$

Find the bandwidth and the total power of the signal. (RPE 2002)

11. Obtain and sketch the transfer function of a vestigial filter to obtain VSB-SC signals from DSB-SC signals. Also, sketch the spectrum of a VSB-SC signal. Find the impulse response of a $-90°$ phase-shift network. Mention its use in a modulator. (RPF 2004)

12. In a QAM system of Fig 3.38, the locally generated carrier suffers from a phase error of $(\Delta\phi)$ so that the receiver carriers are $\cos(w_ct+\Delta\phi)$ or $\sin(w_ct+\Delta\phi)$. Find the output signals $x_1(t)$ and $x_2(t)$ of the receiver branches.

13. A modulator consists of non-linear element with the nonlinearity given by $y(t)=aK^2x_1^2(t)-bx_1^2(t)$, where $x_1(t)=s(t)+A\cos\omega_ct$ and $x_2(t)=s(t)-A\cos\omega_ct$, with $A\cos w_ct$ being the carrier and $s(t)$ being the modulating signal. Verify that an appropriate choice of the constant K generates DSB signals.

14. An LSB signal is generated using Phase-shift method as in Fig. 3.18. If the input to this system is $\frac{1}{2}\hat{s}(t)$ instead of $\frac{1}{2}s(t)$, will this signal will be still an LSB signal with the same bandwidth equal to that of $s(t)$? Justify your answer.

15. The signal $s(t) = A_m \cos(2\pi f_m t)$ is applied to a VSB with Carrier modulator (Fig 3.19). Show that the resulting modulated signal will consist of the carrier and both USB and LSB components. Justify that the modulated signal will be truly a VSB signal.

Multiple-choice questions

1. An envelope detector is a
(a) synchronous detector (b) asynchronous detector
(c) coherent detector (d) product demodulator

2. An AM broadcast transmitter radiates 10 kW power when the index of modulation is 0.6. The carrier power is
(a) 8.47 kW (b) 7.69 kW (c) 5.00 kW (d) 9.17 kW

3. The rate of information transmission over a telecommunication channel depends on
(a) carrier frequency (b) channel bandwidth
(b) transmitter power (d) index of modulation

4. In TV broadcast, sound is modulated in
(a) VSB (b) FM (c) SSB (d) DSB-SC

5. Choose the connect answers.
(a) The bandwidth of AM signals depends on the modulating signal frequency.
(b) The bandwidth of AM signals depends on the modulation index.
(c) The upper and lower sidebands of a DSB-AM signal are equal in magnitude.
(d) The carrier power is half the sum of the powers in USB and LSB.

6. The choice of the RC product in an envelope detector consisting of a diode and an RC circuit is governed by:
(a) depth of modulation only
(b) both lowest and highest modulating frequencies
(c) depth of modulation and lowest modulating frequency
(d) depth of modulation and highest modulating frequency

7. The signal $(1 + m_1 \cos 4\pi t)\cos(2\pi \times 10^3 t)$ contains the frequency components (in Hz):
(a) 998, 1000, 1002 (b) 1000 and 2000 (c) dc, 2, and 1000 (d) none of the above

8. A 10 kW carrier is sinusoidally amplitude-modulated by two signals up to 30% and 40%. The total radiated power is
(a) 11.25 kW (b) 12.5 kW (c) 15 kW (d) none of the above

9. A superheterodyne receiver has IF of 455 kHz and is tuned to 1200 kHz. The image frequency is
(a) 1655 kHz (b) 745 kHz (c) 2110 kHz (d) 910 kHz

10. In an AM transmitter, the unmodulated output current is I_c. The modulated current I_t is

(a) $I_c\sqrt{1+m^2}$ (b) $I_c\sqrt{1+m}$ (c) $I_c\sqrt{1+m^2/2}$ (d) $I_c m/2$

11. The function of a pilot carrier in SSB system is to
   (a) provide increased BW (b) increase power consumption
   (c) improve noise immunity (d) make available carrier frequency at the receiver

12. A balanced modulator circuit is used to eliminate
   (a) carrier (b) USB (c) LSB (d) baseband

13. The AF range used in telephony is
   (a) 20 to 2000 Hz (b) 30 to 300 kHz (c) 300 to 3,000 Hz (d) 300 to 15 kHz

14. A linear diode detector utilises:
   (a) linear portion of the dynamic characteristic of a diode
   (b) square-law portion of the diode dynamic characteristics
   (c) rectification property of the diode
   (d) none of the above

15. In a homodyne receiver, the carrier frequency is obtained as follows
   (a) Locally generated carrier
   (b) Carrier recovered from the received signal
   (c) Pilot carrier in the transmitted signal
   (d) None

4 Exponential CW Modulation

4.1 Introduction

In the previous chapter we discussed linear CW modulation process in which the message signal modulated the amplitude of the carrier. The spectrum of the modulated signal was that of the message spectrum but translated on to a higher frequency of the carrier. Angle modulation on the other hand is a method where the message signal modulates the phase angle of the carrier. The two basic types of exponential modulation are phase modulation (PM) and frequency modulation (FM). Since the angular part of the carrier is varied in accordance with the baseband signal, the modulation is basically a nonlinear process. The spectrum of the modulated wave is not related in a simple way to the spectrum of the message signal. Also, the transmission bandwidth of exponential modulated signal is usually much larger than that of the amplitude modulated signal. In spite of these, angle modulation finds a variety of applications in commercial FM radio broadcasting, audio section of TV transmission, mobile radio communication, microwave and satellite communications, and so on. The advantages of angle modulation over amplitude modulation are noise reduction, improved system fidelity and efficient use of power. However, angle modulation requires wider bandwidth and utilises complex circuits in the transmitter and receiver sections.

4.2 Angle Modulation: Phase and Frequency

We start with a CW signal of constant amplitude and time-varying phase of the form

$$s_c(t) = A_c \cos[\omega_c t + \theta(t)] = A_c \cos \phi_c(t), \tag{4.1}$$

where A_c = peak carrier amplitude (volts)

ω_c = angular frequency of carrier in rad/s,

$\theta(t)$ = instantaneous phase angle.

The quantity $\phi_c(t) = \omega_c t + \theta(t)$ is the total instantaneous phase angle in which carrier reference phase is $\omega_c t$ and instantaneous phase deviation is $\theta(t)$. If the message information is embedded in $\phi_c(t)$, we have indeed accomplished angle or exponential modulation. The term exponential modulation stems from the fact that the message signal $s(t)$ and the modulated signal $s_c(t)$ are related through the exponential form as

$$s_c(t) = A_c \cos \phi_c(t) = A_c \text{Re}[e^{j\phi_c(t)}]. \tag{4.2}$$

The instantaneous frequency $f(t)$ of $s_c(t)$ is defined as the rate of change of the instantaneous phase. Thus, instantaneous frequency is expressed mathematically as

$$\begin{aligned} f(t) &= \frac{1}{2\pi}\frac{d}{dt}\phi_c(t) = \frac{1}{2\pi}\left[\frac{d}{dt}\{\omega_c t + \theta(t)\}\right] \\ &= f_c + \frac{1}{2\pi}\frac{d\theta}{dt}(t) \text{ Hz}. \end{aligned} \tag{4.3}$$

The second term, $\frac{d\theta}{dt}$, is also referred to as instantaneous frequency deviation, whereas the angle $\theta(t)$ is called the instantaneous phase deviation.

Note that the instantaneous frequency $f(t)$ is a time-varying property of $s_c(t)$ if there is exponential modulation. In the absence of modulation $d\theta/dt = 0$ and hence the instantaneous frequency is the same as the carrier frequency f_c.

In the case of frequency modulation (FM), we modulate the instantaneous frequency with the message signal $s(t)$. That is,

$$f(t) = f_c + k_f s(t), \text{ or equivalently, } \frac{d\theta}{dt} = 2\pi k_f s(t) \tag{4.4}$$

where k_f is the constant of proportionality which gives the deviation sensitivity. The quantity k_f has the unit of Hz/volt or $V^{-1}s^{-1}$. Note that $f(t)$ is varied around a positive value f_c rather than around zero. This makes it feasible for separation of various FM broadcast stations in the frequency domain. The restriction on k_f is that during the negative peak of $s(t)$ the instantaneous frequency should never become negative. We normally take the maximum frequency deviation to be much smaller than f_c so that the signal $s_c(t)$ can be treated as a bandpass signal.

Since the phase is the integration of instantaneous frequency, the phase is of the form

$$\phi(t) = 2\pi \int_{t_0}^{t} f(\tau)d\tau = 2\pi\left[f_c t + \int_0^t k_f s(\tau)d\tau\right] + \phi(t_0). \tag{4.5}$$

If t_0 is taken such that $\phi(t_0) = 0$, then

$$\phi(t) = 2\pi[f_c t + k_f \textstyle\int_0^t s(\tau)d\tau]. \tag{4.6}$$

The FM waveform is then

$$s_{\text{FM}}(t) = A_c \cos[\omega_c t + 2\pi k_f \textstyle\int_0^t s(\tau)d\tau]. \tag{4.7}$$

Phase modulation (PM) is another form of angle modulation in which the phase angle $\theta(t)$ is varied linearly with the message signal $s(t)$. That is,

$$\theta(t) = k_p s(t), \tag{4.8}$$

where the proportionality constant k_p defines the phase sensitivity of the modulator and has units of radian/volt. The PM signal is thus described by

$$s_{\text{PM}}(t) = A_c \cos[\omega_c t + k_p s(t)]. \tag{4.9}$$

A comparison of (4.7) and (4.9) reveals that there is no basic difference between FM and PM except that the message is direct in PM whereas it appears as integration in FM. Thus direct FM is indirect PM and vice versa. One can relate the two types by finding the instantaneous frequency. In (4.3), the instantaneous frequency is obtained as

$$f(t) = f_c + \frac{1}{2\pi} k_p \frac{d}{dt} s(t), \tag{4.10}$$

and that for the FM (stated in (4.4)) is

$$f(t) = f_c + k_f s(t), \text{ or } \omega(t) = 2\pi f(t) = \omega_c t + 2\pi k_f s(t) \tag{4.11}$$

The FM signal is thus described by

$$s_{\text{PM}}(t) = A_c \cos[\omega_c t + 2\pi k_f s(t)]. \tag{4.11a}$$

The above relations clearly indicate that an FM wave can be generated by first integrating the message signal $s(t)$ and then using the result as input to the phase modulator. Alternately, a differentiator followed by frequency modulation results in PM wave as shown in Fig. 4.1.

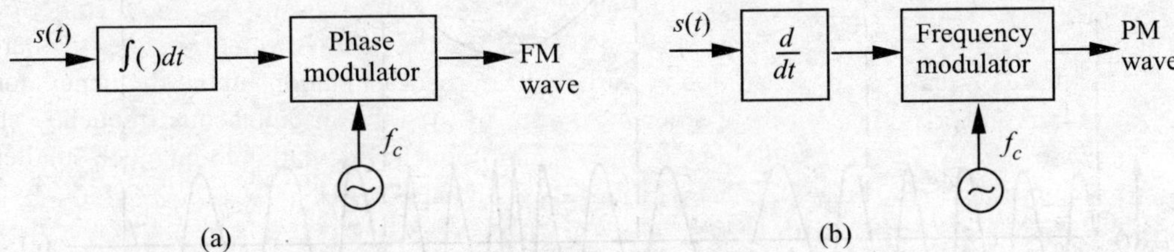

Fig. 4.1 *Illustrating the relations between PM and FM: (a) FM generation by phase modulator, (b) PM generation by frequency modulator.*

In the case of tone modulation (modulation signal is a single frequency sinusoidal) it is hardly possible to distinguish between PM and FM waves. Taking $s(t) = A_m \cos \omega_m t$, the expressions for PM and FM waves are given by

$$s_{\text{PM}}(t) = A_c \cos[\omega_c t + k_p A_m \cos \omega_m t] \tag{4.12a}$$

and

$$s_{\text{FM}}(t) = A_c \cos[\omega_c t + \frac{2\pi k_f A_m}{\omega_m} \sin \omega_m t] = A_c \cos[\omega_c t + \frac{k_f A_m}{f_m} \sin \omega_m t]. \tag{4.12b}$$

The arguments of the cosine terms in the above expressions consist of $\omega_c t$ plus the sinusoidal components. It is impossible to distinguish an FM waveform from a PM waveform without knowing the modulating signal. On a cathode ray oscilloscope, the two waveforms look identical. A closer look reveals that when $\sin \omega_m t$ in FM attains a maximum, $\cos \omega_m t$ in PM is zero; the maximum phase deviation in PM occurs at a quarter of a wave early or later than the FM wave. This is illustrated in Fig. 4.2 which shows the PM and FM waves for a sinusoidal message signal. Thus, unless one knows the modulating signal, it is impossible to distinguish an FM from a PM waveform. The distinction between exponential and linear modulations is now straightforward: the zero-crossing points of an exponential modulated wave are non-periodic while these takes place at regular intervals for linear modulation. Another difference is that the envelope of linear modulation depends on the message signal whereas for an FM or a PM wave it is a constant.

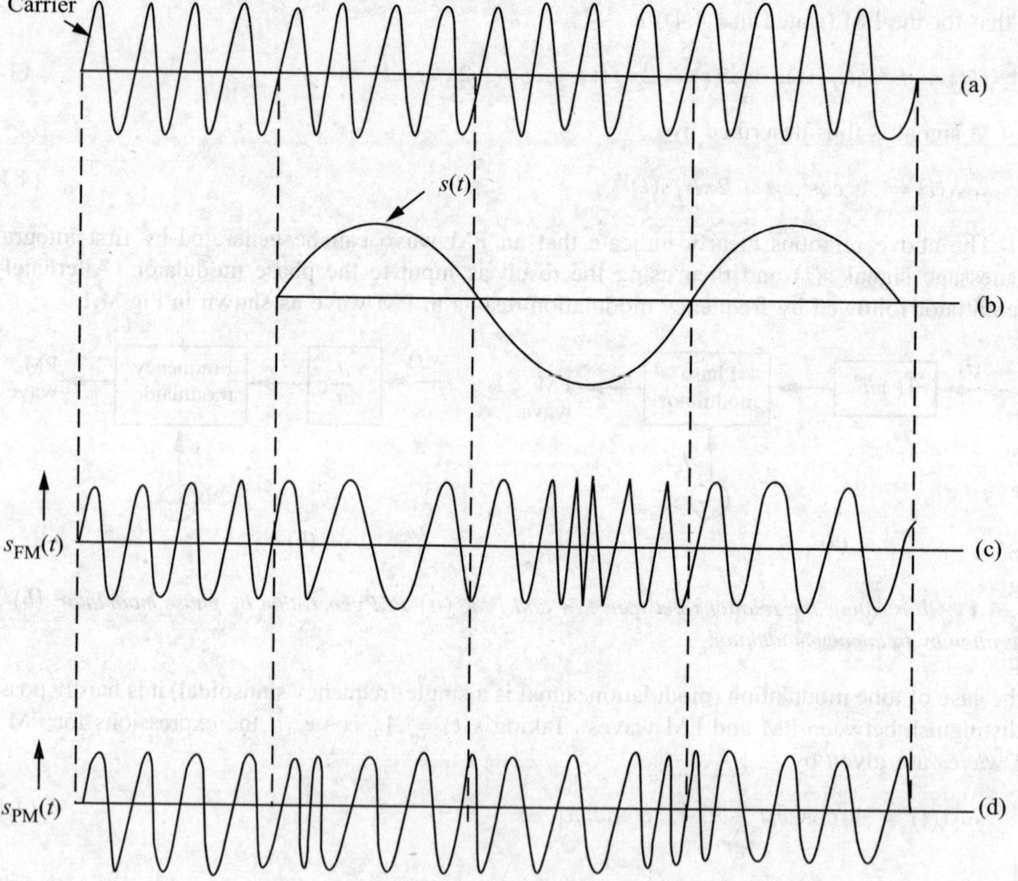

Fig. 4.2 *(a) Unmodulated sinusoidal carrier (b) modulating signal (c) FM wave (d) PM wave.*

Example 4.1
Find the instantaneous frequency of the function $s(t) = 5\cos[2\pi \times 10^4 t + \sin 2\pi \times 100t]$.

Solution
The total instantaneous phase $\phi(t)$ is given by

$$\phi(t) = 2\pi \times 10^4 t + \sin(2\pi \times 100t).$$

Thus, $f(t) = \dfrac{1}{2\pi}\dfrac{d\phi}{dt} = 10^4 + 100\cos(2\pi \times 100t)$. This is sketched in Fig. 4.3.

In summary, in the expression $s_c(t) = A_c \cos[\omega_c t + \theta(t)]$,

Instantaneous phase $= \omega_c t + \theta(t)$ radian.

Instantaneous phase deviation $= \theta(t)$ radian.

$$\text{Instantaneous frequency, } \omega(t) = \frac{d}{dt}\text{ (instantaneous phase)}$$

$$= \frac{d}{dt}(\omega_c t + \theta(t)) = \omega_c + \frac{d\theta}{dt}\text{rad/sec.}$$

Instantaneous frequency deviation,

$$= \frac{d\theta}{dt}\text{rad/sec}$$

$$= \frac{d\theta/dt\,\text{rad/ sec}}{2\pi\,\text{rad/cycle}} = \frac{1}{2\pi}\frac{d\theta}{dt}\frac{\text{cycle}}{\text{sec}}\text{ (Hz)}$$

Deviation Sensitivity

If one can take the phase modulated signal as given by Eq. (4.9), the instantaneous phase deviation is given by $\theta(t) = k_p s(t)$. Thus, sensitivity for phase deviation k_p has the unit of rad/V.

The frequency modulated signal as given by Eq. (4.11a), the instantaneous frequency deviation is given by $\dfrac{d\theta}{dt} = \dfrac{d}{dt}(2\pi k_f s(t)) = 2\pi k_f \dfrac{ds}{dt}$

Thus, the sensitivity for frequency deviation k_f has the unit

$$[k_f] = \frac{[d\theta/dt]}{[2\pi]\left[\dfrac{ds}{dt}\right]} = \left[\frac{d\theta/dt}{2\pi}\right]\cdot\frac{1}{\left[\dfrac{ds}{dt}\right]} = \text{cycle/Volt}$$

4.3 Frequency Modulation

Since the FM signal defined in (4.7) is a nonlinear function of the message signal $s(t)$, the spectrum of $s_{\text{FM}}(t)$ is not related in a simple manner to that of $s(t)$. If $s(t)$ were known, the Fourier transform of the FM waveform could be evaluated by using the integral property of the Fourier transform. In finding the range of frequencies occupied by the FM signal, we perform the analysis in the time domain for the simplest case of single-tone modulation.

Consider a message signal defined by $s(t) = A_m \cos 2\pi f_m t$. The instantaneous frequency of the FM wave is given by

$$f(t) = f_c + k_f A_m \cos 2\pi f_m t$$

$$= f_c + \Delta f \cos 2\pi f_m t, \tag{4.13}$$

where $\Delta f = k_f A_m$ is the maximum frequency deviation that occurs in the carrier frequency. The total phase angle $\phi(t)$ is then obtained as (vide (4.6))

$$\phi(t) = 2\pi \int_0^t f(\tau)d\tau = 2\pi \int_0^t [f_c + k_f s(\tau)]d\tau$$

$$= 2\pi f_c t + 2\pi k_f \int_0^t A_m \cos(2\pi f_m t)d\tau$$

$$= 2\pi f_c t + \frac{\Delta f}{f_m} sin 2\pi f_m t. \tag{4.14}$$

Equation (4.14) can also be derived as follows.

Keeping parity with the phase modulation as in Eq. (4.8), we can write $\frac{d\theta}{dt} = k_f s(t)$ where the sensitivity for frequency modulation has the unit $\frac{\text{rad/sec}}{\text{Volt}}$. Then $\theta(t) = \int k_f s(t)dt$ and hence,

$$\phi(t) = \omega_c t + \int_0^t k_f s(t)dt = \omega_c t + \frac{k_f A_m}{2\pi f_m} \sin 2\pi f_m t$$

$$= \omega_c t + \frac{\Delta f}{f_m} \sin 2\pi f_m t \tag{4.14a}$$

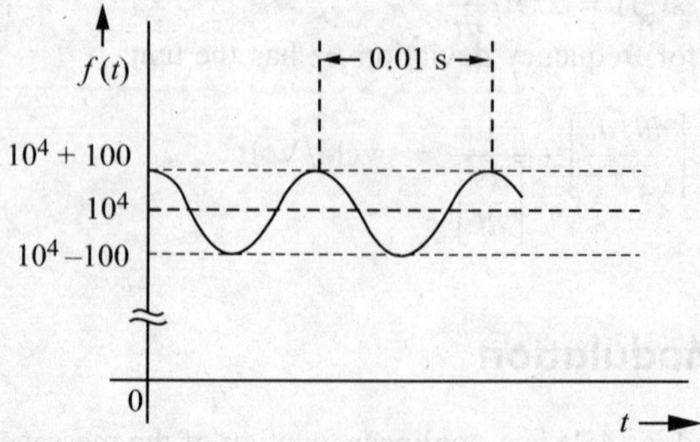

Fig. 4.3 *Instantaneous frequency*

We define the modulation index β as

$$\beta = \frac{\Delta f}{f_m} = \frac{k_f A_m}{f_m}. \tag{4.15}$$

The quantity β is a dimensionless quantity since k_f has the units of volt^{-1} second $^{-1}$. The FM signal is then given by

$$s_{\text{FM}}(t) = A_c \cos[2\pi f_c t + \beta \sin 2\pi f_m t]. \tag{4.16}$$

Although the instantaneous frequency $f(t)$ of (4.13) lies in the range $f_c \pm \Delta f$, the spectrum of FM signals does not lie in this range. Note that the modulation index is directly proportional to the amplitude of the modulating signal and inversely proportional to its frequency. The index of modulation for PM signal is $k_p A_m$ whereas that of FM is $\frac{k_p A_m}{f_m}$. Thus the modulation index of PM and FM are proportional to the amplitude of the modulating signal. Also index of modulation in PM is independent of f_m whereas it is inverse to f_m in FM.

Example 4.2

Sketch the FM waves for the modulating signal $s(t)$ as shown in Fig. 4.4 given that $k_f = 10^5$ rad/V.s and $f_c = 10$ MHz.

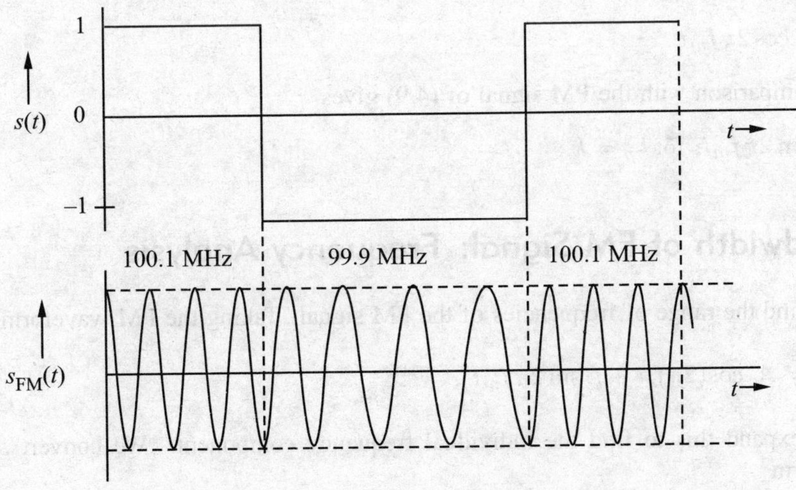

Fig. 4.4 *Angle modulated wave: (a) modulating wave, (b) FM wave.*

Solution

As k_f is in $\frac{\text{rad/sec}}{\text{V}}$, we must have

$\frac{d\theta}{dt} = k_f s(t)$ and hence

$$f(t) = f_c + \frac{k_f s(t)}{2\pi}$$
$$= f_c \pm \frac{k_f}{2\pi}$$

For FM, $f(t) = f_c \pm \frac{k_f}{2\pi}$ since $s(t)$ is either $+1$ Volt or -1 Volt).

$$= (10 \pm \frac{0.1}{2\pi}) \text{ MHz} = 9.984 \text{ MHz or } 10.016 \text{ MHz}.$$

Example 4.3

If the waveform $\cos(\omega_c t + k \sin \omega_m t)$ is a frequency modulated one, find the modulating signal. Also find the modulating signal if the carrier is phase modulated.

Solution

Comparing the given FM wave with that of (4.7), we see that

$$2\pi k_f \int_0^t s(\tau)d\tau = k \sin 2\pi f_m t$$

or, $\quad 2\pi k_f s(t) = 2\pi f_m k \cos 2\pi f_m t.$

So, the modulating signal $s(t)$ is given by

$$s(t) = k \cos 2\pi f_m t.$$

Similarly, a comparison with the PM signal of (4.9) gives

$$s(t) = \sin 2\pi f_m t, \text{ for } k_p = k.$$

4.4 Bandwidth of FM Signal: Frequency Analysis

We will now find the range of frequencies of the FM signal. Taking the FM waveform to be

$$s_{\text{FM}}(t) = A_c \cos(2\pi f_c t + \beta \sin 2\pi f_m t),$$

we will now expand this to find the individual frequency component. We convert $s_{\text{FM}}(t)$ to its exponential form

$$s_{\text{FM}}(t) = \text{Re}\,[A_c \exp(j2\pi f_c t) \exp(j\beta \sin 2\pi f_m t). \tag{4.17}$$

The term containing the message signal is

$$\exp(j\beta \sin 2\pi f_m t) = \cos(\beta \sin 2\pi f_m t) + j \sin(\beta \sin 2\pi f_m t), \tag{4.18}$$

the real part of which is $\cos(\beta \sin 2\pi f_m t)$. This is an even and periodic function with period $1/f_m$. We can expand the term in an exponential Fourier series as

$$e^{j\beta \sin 2\pi f_m t} = \sum_{n=-\infty}^{\infty} C_n e^{jn2\pi f_m t}, \tag{4.19}$$

where the series coefficients are given by the integral

$$C_n = \frac{1}{f_m} \int_{-1/2f_m}^{1/2f_m} \exp(j\beta \sin 2\pi f_m t) e^{-jn2\pi f_m t} dt. \tag{4.20}$$

The above integral cannot be evaluated in a closed form. The coefficients C_n are, of course, real and function of n and β. It is not a function of the modulating frequency f_m. The function C_n's are known as Bessel functions of the first kind and of order n, denoted by $J_n(\beta)$.

Now, (4.17) takes the form

$$s_{\text{FM}}(t) = \text{Re}\,[A_c e^{j2\pi f_c t} \sum_{n=-\infty}^{\infty} J_n(\beta) e^{jn2\pi f_m t}]$$

$$= \text{Re}\left[A_c \sum_{n=-\infty}^{\infty} J_n(\beta) e^{j2\pi(f_c + n f_m)t}\right].$$

Taking the real part,

$$s_{\text{FM}}(t) = A_c \sum_{n=-\infty}^{\infty} J_n(\beta) \cos 2\pi(f_c + n f_m)t. \tag{4.21}$$

Since for integer values of n, $J_{-n}(x) = (-1)^n J_n(x)$, (4.21) reduces to

$$\begin{aligned} s_{\text{FM}}(t) = A_c[&J_0(\beta)\cos\omega_c t + J_1(\beta)\{\cos 2\pi(f_c + f_m)t - \cos 2\pi(f_c - f_m)t\} \\ &+ J_2(\beta)\{\cos 2\pi(f_c + 2f_m)t + \cos 2\pi(f_c - 2f_m)t\} \\ &J_3(\beta)\{\cos 2\pi(f_c + 3f_m)t - \cos 2\pi(f_c - 3f_m)t\} + \cdots]. \end{aligned} \tag{4.22}$$

Note that the spectrum is composed of a carrier weighted by $J_0(\beta)$ and infinite number of sidebands situated symmetrically on both sides of the carrier with the frequency separations f_m, $2f_m$, $3f_m$, $\cdots$, etc. The magnitude spectrum is illustrated in Fig. 4.5 for positive frequencies. The amplitude of the nth sideband is proportional to $J_n(\beta)$.

The total power in $s_{FM}(t)$ can be calculated as follows.

Power in the carrier is $A_c^2 J_0^2(\beta)/2$, power in the first sideband component is $2A_c^2 J_1^2(\beta)/2 = A_c^2 J_1^2(\beta)$ and so on. The total power is thus

$$P = \frac{A_c^2}{2}[J_0^2(\beta) + 2J_1^2(\beta) + 2J_2^2(\beta) + \cdots] \tag{4.22a}$$

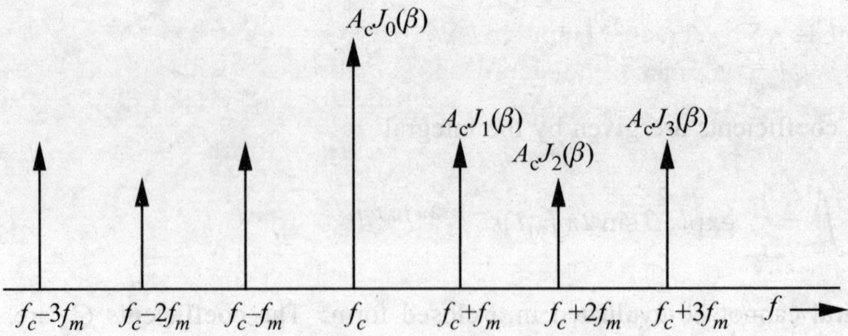

Fig. 4.5 *Line spectra of FM of equation (4.22) for sinusoidal modulation*

It appears from (4.22) that the bandwidth of FM signal is infinite. However, in practice, we consider only a reasonable number of sidebands that contain significant part of the energy. The quantum of energy outside this band is negligible for most applications and the resulting distortion is tolerable. Before we proceed further, we need to discuss the behaviour of the Bessel functions of order n with argument β.

Figure 4.6 shows Bessel functions $J_n(\beta)$ plotted as a function of β for values of $n = 0, 1$ and 2.

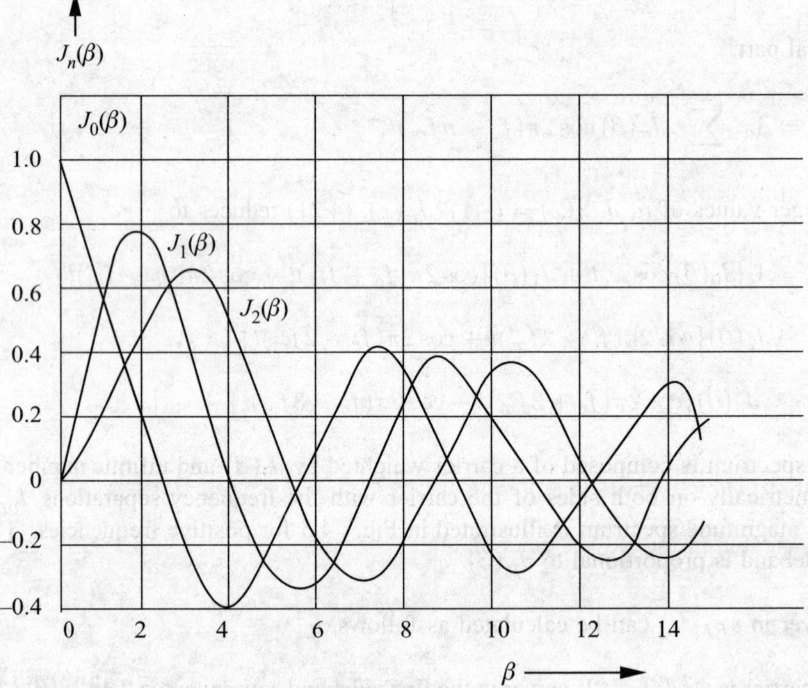

Fig. 4.6 *Bessel functions plotted as a function of β*

We see that $J_0(\beta)$ approaches unity and $J_1(\beta)$, $J_2(\beta)$ tend to zero for small β. At $\beta = 0$, $J_0(0) = 1$ and all other J_n's are zero. This indicates that if there is no modulation, only the carrier term is present, as expected. As β increases slightly, $J_0(\beta)$ decreases while both $J_1(\beta)$ and $J_2(\beta)$ increase. But the value of $J_0(\beta)$ continues to be significant in comparison with $J_1(\beta)$ and $J_2(\beta)$ which are still neglected. In fact, for a given n and $\beta << 1$, the Bessel function can be approximated by

$$J_n(\beta) \cong \frac{1}{n!}\left(\frac{\beta}{2}\right)^n \qquad n > 0. \tag{4.23}$$

In particular, for $n = 0$,

$$J_0(\beta) \simeq 1 - \left(\frac{\beta}{2}\right)^2. \tag{4.24}$$

By way of an example, let us take a small value of β equal to 0.2, say. Then, $J_0 = 0.99$, $J_1 = 0.1$, $J_2 = 0.005$, and higher-order Bessel functions are still smaller. The first carrier null occurs when $\beta = 2.4$ and $J_0(\beta) = 0$. It is then reasonable to consider J_0 and J_1 terms that include the carrier and a single pair of first sidebands located at $f_c \pm f_m$, giving a total bandwidth of $2f_m$. As β becomes larger, J_2 becomes significant and must be considered. In general, with the increase of the index β, higher-order terms must be considered. The selection rule can be stated as follows:

A sideband pair will not be considered unless it has an amplitude equal to or larger than one percent of the unmodulated carrier amplitude.

This rule gives the approximate bandwidth of the FM as a function of f_m and β. Since the power of a periodic signal depends on the square of the amplitude, this rule can also be stated in terms of power considerations. In approximating the bandwidth, we will consider those number of sidebands that contain 98% or more of the total transmitted energy. This will, of course, produce some amount of distortion in the received signal but tolerable for most purposes. In fact, for any value of β, only those J_n's need be considered which have significant nonzero values. Figure 4.7 shows the plot of $J_n(5)$ as a function of n. We can observe that for n larger than 5, the Bessel function asymptotically approaches zero. The power contained in the terms with $n = 0$ through 6 is

$$\begin{aligned} P &= \frac{1}{2}J_0^2(5) + J_1^2(5) + J_2^2(5) + J_3^2(5) + J_4^2(5) + J_5^2(5) + J_6^2(5) \\ &= \frac{1}{2}(-0.1776)^2 + (-0.3276)^2 + (0.0465)^2 + (0.3648)^2 + (0.3912)^2 \\ &\quad +(0.2611)^2 + (0.1310)^2 = 0.496. \end{aligned}$$

(The data is available in the Bessel function table.)

The total power in the FM signal is

$$\frac{1}{2}\left(J_0^2 + 2\sum_{n=1}^{\infty} J_n^2\right) = \frac{1}{2}[J_0^2 + 2J_1^2 + 2J_2^2 + \cdots] = \frac{1}{2},$$

since the value of the quantity within the square bracket is 1. Hence the power up to the six sidebands is 99% of the total transmitted power. We therefore conclude that for given β, the number of sidebands for 98% percent power transmission always occurs for $n > \beta + 1$. Hence for sinusoidal modulation, the approximate bandwidth of FM signal is

$$B = 2nf_m = 2(\beta + 1)f_m. \tag{4.25}$$

where n is the number of significant sidebands. This rule was proposed by John Carson and is called Carson's rule.

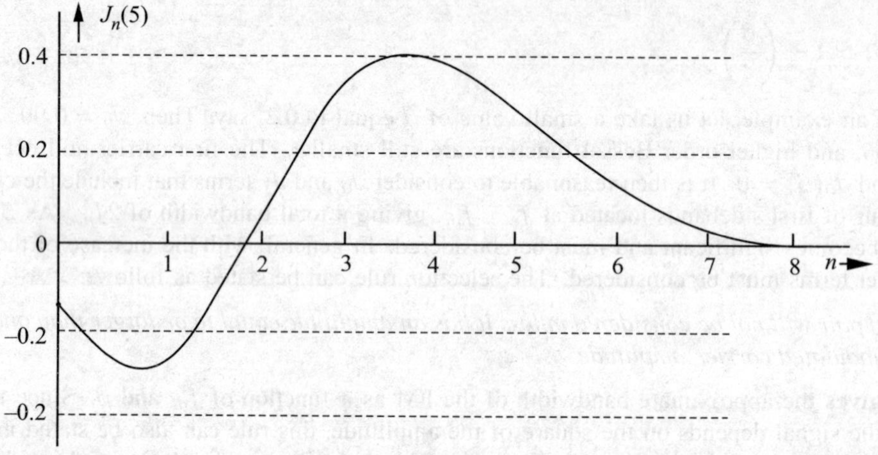

Fig. 4.7 *A qualitative plot of $J_n(5)$ as a function of n*

When β is small compared to unity, the bandwidth approximates to $B = 2f_m$. This is the case of a narrowband FM (NBFM), where only the first sideband pair is considered. The band of frequencies occupied by the NBFM is $f_c - f_m$ to $f_c + f_m$. For the other extreme case of $\beta >> 1$, the bandwidth is approximately $B = 2\beta f_m$. This is the nominal bandwidth of a wideband FM (WBFM).

Remember that the bandwidth relation (4.25) is stated for a sinusoidal modulation of a single frequency f_m. It can be generalised to the case of a sinusoidal information signal containing a continuum of frequencies up to a maximum of f_m; the maximum frequency deviation being $\Delta f = \beta f_m$, (4.25) generalises to

$$B = 2(\Delta f + f_m) = 2f_m(1 + \beta). \tag{4.26}$$

Example 4.4

A carrier of frequency 5 kHz and amplitude 5 volts is frequency modulated by a sinusoidal modulating wave of frequency 1 kHz and amplitude 100 volts with $k_f = 10$ Hz/V. Write down the expression for the FM wave. Find the approximate band of frequencies occupied by the FM.

Solution

Here the carrier wave $= 5\sin(2\pi \times 5 \times 10^3 t)$ volts and message signal, $s(t) = 100\sin(2\pi \times 10^3 t)$. Since $k_f = 10$ Hz/V, the FM signal is,

$$s_{\text{FM}}(t) = A_c \sin[\omega_c t + 2\pi k_f \int_0^t \sin(\tau) d\tau].$$
$$= A_c \sin[2\pi \times 5 \times 10^3 t - \frac{k_f \times 100}{10^3} \cos(2\pi \times 10^3 t)]$$

The maximum frequency deviation is

$$\Delta f = k_f A_m = 10 \times 100 = 10^3 \text{ Hz and index of modulation } \beta = \frac{\Delta f}{f_m} = \frac{10^3}{10^3} = 1$$

Thus, $BW = 2(\Delta f + f_m) = 2(10^3 + 10^3) = 4$ kHz.

The frequency range $= (f_c - \frac{1}{2}BW)$ to $(f_c + \frac{1}{2}BW) = 3$ to 7 kHz.

Example 4.5

A carrier of amplitude 5 volts and frequency 10 MHz is frequency-modulated by a sinusoidal signal of frequency 2 kHz. The maximum frequency deviation achieved is 75 kHz. The modulated signal attains its positive peak at $t = 0$. Write down the expression for the modulated carrier. Also, find the bandwidth of this waveform.

Solution

Since the modulated carrier attains positive peak at $t = 0$, so the modulation signal will be of the form $A_m \cos \omega_m t$ the FM signal can be put in the form

$$s_{\text{FM}}(t) = A_c \cos(\omega_c t + \beta \sin \omega_m t).$$

Here, $A_c = 5$ volts, $\omega_c = 2\pi \times 10^7$ Hz, $\omega_m = 2\pi \times 2 \times 10^3$ Hz, maximum frequency deviation is $\Delta f = \beta f_m$. Thus,

$$75 \times 10^3 = \beta \times 2 \times 10^3 \quad \Rightarrow \beta = 37.5.$$

Hence, $s_{\text{FM}}(t) = 5\cos(2\pi \times 10^7 t + 37.5 \sin 4\pi 10^3 t)$ volts.

The approximate bandwidth is given by

$$BW = 2(\Delta f + f_m)$$
$$= 2(75 \times 10^3 + 2 \times 10^3) = 154 \text{ kHz}.$$

The case is that of a WBFM since $\Delta f >> f_m$ or $\beta >> 1$.

4.5 Narrowband FM and PM Signals: Phasor Representation

The narrowband FM signal has been defined as a special case of FM for $\beta << 1$. To study the properties of the narrowband angle-modulated signals, we start with the angle modulated signal of the form

$$s_m(t) = A_c \cos[\omega_c t + km(t)], \tag{4.27}$$

where the message signal is $m(t)$. For phase modulation, the message signal directly relates $m(t)$ with the constant $k = k_p$. For frequency modulation, it is the integral of message signal $\int_0^t s(\tau)d\tau$ that relates $m(t)$ with the constant $k = 2\pi k_f$. Expanding (4.27) yields

$$s_m(t) = A_c \cos\omega_c t \cos[km(t)] - A_c \sin\omega_c t \sin[km(t)]. \tag{4.28}$$

If k is very small so that the angle $km(t)$ still remains small, then one can write $\cos[km(t)] \approx 1$ and $\sin[km(t)] \approx 0$. In doing so, (4.28) simplifies to

$$s_m(t) = A_c[\cos\omega_c t - km(t)\sin\omega_c t]. \tag{4.29}$$

This is the expression for the narrowband signals. The spectra of the signal is obtained by the Fourier transform of (4.29). Thus,

$$S_m(f) = \frac{A_c}{2}[\delta(f - f_c) + \delta(f + f_c)] - \frac{A_c k}{2j}[M(f - f_c) - M(f + f_c)], \tag{4.30}$$

where $M(f)$ is the spectrum of $m(t)$. For frequency modulation, $m(t) = \int_0^t s(\tau)d\tau$ and hence $M(f) = \frac{1}{j\omega}S(f) = \frac{1}{j2\pi f}S(f)$.

The Fourier transform of the NBFM is thus

$$\begin{aligned} S_{\text{NBFM}}(f) &= \frac{A_c}{2}[\delta(f - f_c) + \delta(f + f_c)] - \frac{A_c k}{2j}\frac{1}{j2\pi}\left[\frac{S(f - f_c)}{f - f_c} - \frac{S(f + f_c)}{f + f_c}\right] \\ &= \frac{A_c}{2}[\delta(f - f_c) + \delta(f + f_c)] + \frac{A_c k}{4\pi}\left[\frac{S(f - f_c)}{f - f_c} - \frac{S(f + f_c)}{f + f_c}\right]. \end{aligned} \tag{4.31}$$

Figure 4.8 shows the magnitudes of the Fourier transform of the PM and FM waves for a given message signal. Since we do not know the shape of the message spectra, we have given a representative plot of $S(f)$. For PM signals, $m(t) = s(t)$ and the spectrum looks like that of a DSB with carrier, as shown in Fig. 4.8(c). For FM signals, $M(f) = \frac{1}{j2\pi f}S(f)$ and the low-frequency parts are emphasised, as observed in Figs. 4.8(b) and (c).

We now present several phasor diagrams of angle-modulated waves. We start with a modulating signal of the form $s(t) = \cos\omega_m t$. The signal $m(t)$ is then

$$m(t) = \int_0^t s(\tau)d\tau = \frac{A_m}{2\pi f_m}\sin\omega_m t.$$

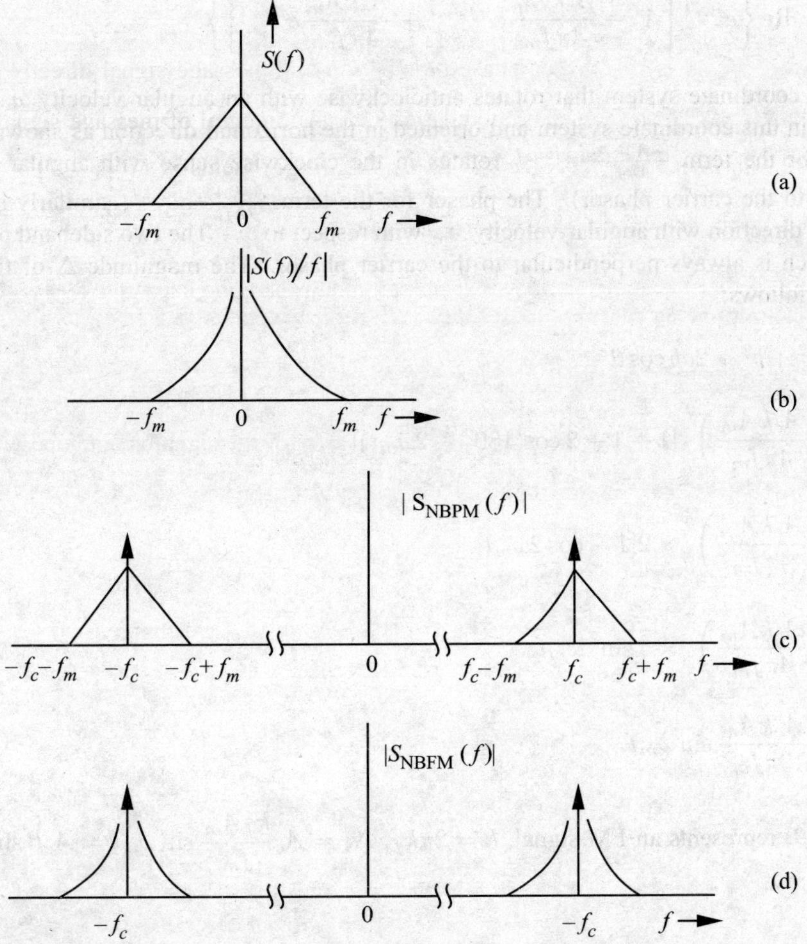

Fig. 4.8 *The magnitude spectrum of :* (a) $S(f)$ *of the message signal,* (b) *the form* $S(f)/f$, (c) *NBPM, (d) NBFM.*

Equation (4.29) becomes

$$s_m(t) = A_c[\cos\omega_c t - \frac{kA_m}{2\pi f_m}\sin\omega_m t\sin\omega_c t]$$

$$= A_c[\cos\omega_c t - \frac{1}{2}\left(\frac{kA_m}{2\pi f_m}\right)\cos(\omega_c - \omega_m)t + \frac{1}{2}\left(\frac{kA_m}{2\pi f_m}\right)\cos(\omega_c + \omega_m)t]$$

$$= \text{Re}\left\{A_c e^{j\omega_c t} - \frac{1}{2}\left(\frac{A_c A_m}{2\pi f_m}\right)e^{j(\omega_c-\omega_m)t} + \frac{1}{2}\left(\frac{A_c k A_m}{2\pi f_m}\right)e^{j(\omega_c+\omega_m)t}\right\}$$

$$= \text{Re}\left\{e^{j\omega_c t}\left[A_c - \frac{A_c k A_m}{4\pi f_m} e^{-j\omega_m t} + \frac{A_c k A_m}{4\pi f_m} e^{j\omega_m t}\right]\right\}. \tag{4.32}$$

We assume a coordinate system that rotates anticlockwise with an angular velocity ω_c. The carrier term is fixed in this coordinate system and oriented in the horizontal direction as shown in Fig. 4.9. The phasor for the term $\frac{-A_c k A_m}{4\pi f_m} e^{j\omega_m t}$ rotates in the clockwise sense with angular velocity ω_m (with respect to the carrier phasor). The phasor for the term $\frac{A_c k A_m}{4\pi f_m} e^{j\omega_m t}$, similarly rotates in the anticlockwise direction with angular velocity ω_m with respect to ω_c. The two sideband phasors result in a sum which is always perpendicular to the carrier phasor. The magnitude Δ of this phasor is calculated as follows:

$$\begin{aligned}
\Delta^2 &= a^2 + b^2 + 2ab\cos\theta \\
&= \left(\frac{A_c k A_m}{4\pi f_m}\right)^2 [1 + 1 + 2\cos(180^\circ - 2\omega_m t)] \\
&= \left(\frac{A_c k A_m}{4\pi f_m}\right)^2 \times 2[1 - \cos 2\omega_m t] \\
&= \left(\frac{A_c k A_m}{4\pi f_m}\right)^2 \times 4\sin^2 \omega_m t.
\end{aligned}$$

$$\text{Thus} \quad \Delta_1 = \frac{A_c k A_m}{2\pi f_m} \sin\omega_m t. \tag{4.33}$$

Note that (4.29) represents an FM signal, $k = 2\pi k_f$, $\Delta_1 = A_c \frac{k_f A_m}{f_m} \sin\omega_m t = A_c\beta \sin\omega_m t$.

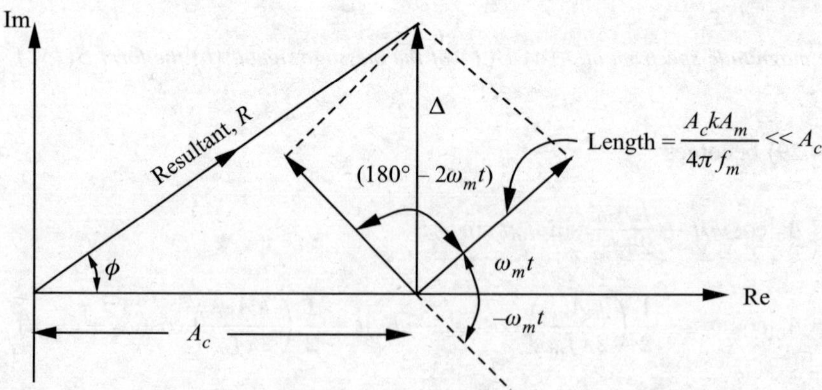

Fig. 4.9 *Phasor plot for NBFM signal*

The resultant phasor R attains its maximum value when $\Delta_1 = A_c\beta$ and minimum when $\Delta_1 = 0$. The phasor R is ahead of the carrier phasor by an angle ϕ given by

$$\tan\phi = \frac{\Delta_1}{A_c} = \frac{k_f A_m}{f_m} = \beta. \tag{4.34}$$

Since $\beta << 1$, the maximum value of ϕ is β.

Let us now compare the phasor diagram of the AM signal with that of the FM. For a sinusoidal message, the AM signal is

$$s_{\text{AM}}(t) = A_c[1 + ks(t)]\cos\omega_c t = A_c[\cos\omega_c t + ks(t)\cos\omega_c t]. \tag{4.35}$$

For a comparison of the time-domain representation, we rewrite the expression for NBFM:

$$s_{\text{NBFM}}(t) \approx A_c[\cos\omega_c t - km(t)\sin\omega_c t]. \tag{4.36}$$

The first term of NBFM is $\cos\omega_c t$ and the second term involves $\sin\omega_c t$, a quadrature component. But in AM both the first and second terms involve $\cos\omega_c t$, an in-phase relationship.

We now express (4.35) as the real part of a complex exponential with $s(t) = A_m\cos\omega_m t$.

$$\begin{aligned} s_{\text{AM}}(t) &= A_c[\cos\omega_c t + kA_m\cos\omega_c t\cos\omega_c t] \\ &= A_c\left[\cos\omega_c t + \frac{kA_m}{2}\{\cos(\omega_c+\omega_m)t + \cos(\omega_c-\omega_m)t\}\right] \\ &= A_c\,\text{Re}\left\{e^{j\omega_c t} + \frac{kA_m}{2}e^{j(\omega_c+\omega_m)t} + \frac{kA_m}{2}e^{j(\omega_c-\omega_m)}\right\} \\ &= A_c\,\text{Re}\left\{e^{j\omega_c t}\left[1 + \frac{kA_m}{2}e^{j\omega_m t} + \frac{kA_m}{2}e^{-j\omega_m t}\right]\right\}. \end{aligned} \tag{4.37}$$

Following the approach of NBFM, we now plot the phasor of (4.37) in Fig. 4.10. Note that the resultant phasor R is in phase with the carrier term. The magnitude of the resultant of the two sidebands phasors Δ is obtained as follows:

$$\begin{aligned} \Delta^2 &= \left(\frac{A_c k A_m}{2}\right)^2 [1 + 1 + 2\cos 2\omega_m t] \\ &= \left(\frac{kA_cA_m}{2}\right)^2 \times 4\cos^2\omega_m t, \end{aligned}$$

that is, $$\Delta = kA_cA_m\cos\omega_m t = A_c m\cos\omega_m t. \tag{4.38}$$

The resultant phasor is given by

$$R = A_c + mA_c\cos\omega_c t = A_c(1 + m\cos\omega_c t). \tag{4.39}$$

The resultant is in parallel to the carrier phasor with the magnitude varying between $A_c(1+m)$ and $A_c(1-m)$.

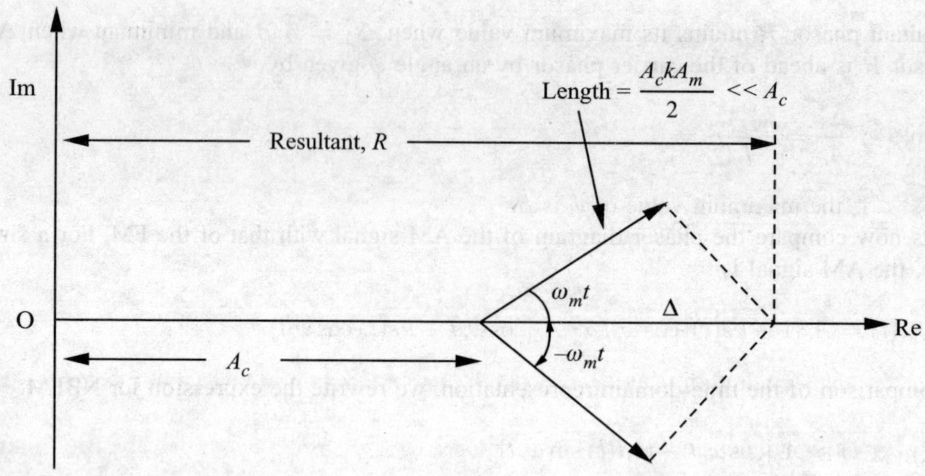

Fig. 4.10 *Phasor plot for an AM signal*

4.6 Frequency Modulation by Non-harmonic Modulating Waves

Consider an RF carrier of frequency ω_c angle modulated by two sinusoidal modulating waves of non-harmonic frequencies ω_1 and ω_2. The FM waveform is given by the equation

$$
\begin{aligned}
v(t) &= A_c \cos(\omega_c t + \beta_1 \sin\omega_1 t + \beta_2 \sin\omega_2 t) \\
&= A_c[\cos\omega_c t \cos(\beta_1 \sin\omega_1 t + \beta_2 \sin\omega_2 t) \\
&\quad - \sin\omega_c t \sin(\beta_1 \sin\omega_1 t + \beta_2 \sin\omega_2 t)],
\end{aligned} \tag{4.40}
$$

where β_1 and β_2 are the depths of modulation. Since for small angle x, $\cos x \approx 1 - x^2/2$ and $\sin x \approx x$,

$$
\begin{aligned}
v(t) &= A_c[\cos\omega_c t \left\{1 - \tfrac{1}{2}(\beta_1 \sin\omega_1 t + \beta_2 \sin\omega_2 t)^2\right\} \\
&\quad - \sin\omega_c t(\beta_1 \sin\omega_1 t + \beta_2 \sin\omega_2 t)],
\end{aligned} \tag{4.41}
$$

which on simplification leads to

$$
\begin{aligned}
v(t) &= A_c\left(1 - \frac{\beta_1^2 + \beta_2^2}{4}\right)\cos\omega_c t + \frac{\beta_1 A_c}{2}\left[\cos(\omega_c + \omega_1)t - \cos(\omega_c - \omega_1)t\right] \\
&\quad + \frac{\beta_2 A_c}{2}\left[\cos(\omega_c + \omega_2)t - \cos(\omega_c - \omega_2)t\right]
\end{aligned}
$$

$$+\frac{\beta_1^2 A_c}{8}\left[\cos(\omega_c + 2\omega_1)t + \cos(\omega_c - 2\omega_1)t\right]$$

$$+\frac{\beta_2^2 A_c}{8}\left[\cos(\omega_c + 2\omega_2)t + \cos(\omega_c - 2\omega_1)t\right]$$

$$+\frac{\beta_1\beta_2 A_c}{4}\left[\cos(\omega_c + \omega_1 + \omega_2)t + \cos(\omega_c - \omega_1 - \omega_2)t\right]$$

$$-\frac{\beta_1\beta_2 A_c}{4}\left[\cos(\omega_c + \omega_1 - \omega_2)t + \cos(\omega_c - \omega_1 + \omega_2)t\right]. \tag{4.42}$$

Thus we see that in addition to the carrier frequency, the spectrum consists of sidebands separated from the carrier at $\omega_1, \omega_2, 2\omega_1, 2\omega_2, \omega_1 + \omega_2, \omega_1 - \omega_2$. If higher-order terms in the approximations for $\cos x$ and $\sin x$ are considered, the sidebands are separated from the carrier frequency at multiples of $\omega_1, \omega_2, \omega_1 \pm \omega_2$.

If the depths of modulations β_1 and β_2 in (4.40) are sufficiently small such that $|\beta_1 \sin\omega_1 t + \beta_2 \sin\omega_2 t| << \pi/2$, then

$$v(t) \cong A_c[\cos\omega_c t - (\beta_1 \sin\omega_1 t + \beta_2 \sin\omega_2 t)\sin\omega_c t]. \tag{4.43}$$

4.7 Pulsed FM Signal

Frequency-modulated CW-radar utilises the linear FM pulse whose instantaneous frequency varies linearly with time. This linear FM is also called pulsed FM. Consider an RF pulse of rectangular envelope and duration τ_p as shown in Fig. 4.11. The instantaneous frequency $\omega(t)$ of the pulsed carrier is defined as

$$\omega(t) = \omega_0 + at, \tag{4.44}$$

where ω_0 is the unmodulated carrier frequency. The constant a determines the rate of linear frequency change and has the units of rad/s^2. Over the duration τ_p, the total frequency deviation is

$$2\omega_d = a\tau_p. \tag{4.45}$$

The total instantaneous phase $\theta(t)$ of the carrier is

$$\theta(t) = \int_0^t \omega(t)dt = \omega_0 t + \tfrac{1}{2}at^2. \tag{4.46}$$

The linear FM signal is then

$$\begin{aligned} s_{\text{LFM}}(t) &= A_0\cos(\omega_0 t + \tfrac{1}{2}at^2) \qquad -\tau_p/2 < t < \tau_p/2 \\ &= 0 \qquad\qquad\qquad\qquad\quad |t| > \tau_p. \end{aligned} \tag{4.47}$$

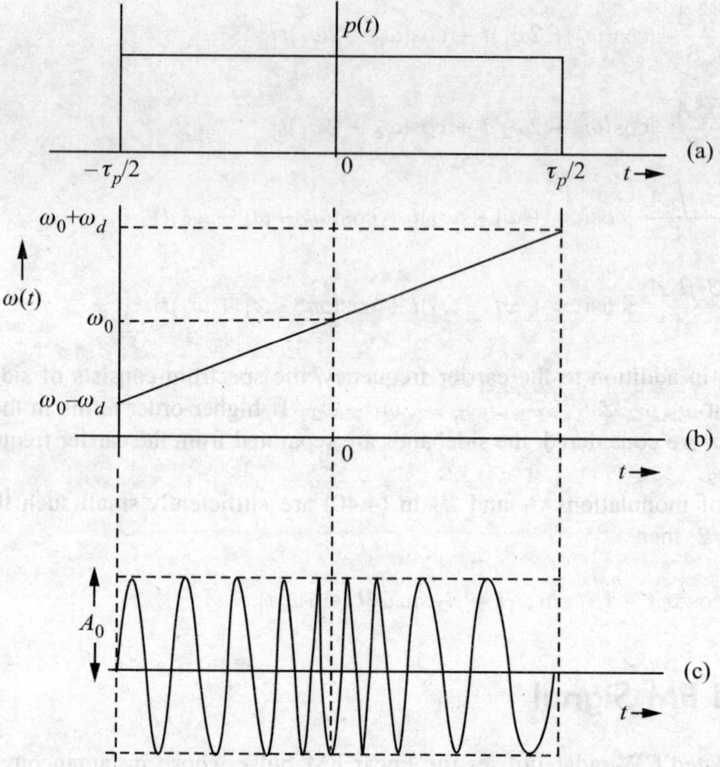

Fig. 4.11 *(a) RF pulse, (b) linear frequency variation, (c) linear FM pulse.*

The spectrum of $s_{\text{LFM}}(t)$ is obtained by the Fourier transform of (4.47). Thus,

$$S_{\text{LFM}}(f) = A_0 \int_{-\tau_p/2}^{\tau_p/2} \cos(\omega_0 t + \frac{1}{2}at^2)e^{-j\omega t}\,dt$$

$$= \frac{A_0}{2}\int_{-\tau_p/2}^{\tau_p/2} e^{j[(\omega-\omega_0)t-\frac{1}{2}at^2]}dt + \frac{A_0}{2}\int_{-\tau_p/2}^{\tau_p/2} e^{-j[(\omega+\omega_0)t+\frac{1}{2}at^2]}dt^2]dt. \quad (4.48)$$

The first integral gives a spike near the positive frequency $\omega = \omega_0$ and the second one gives the same near the negative frequency $\omega = -\omega_0$. To find the $S(f)$ in the region $\omega > 0$, the second integral is discarded as the two peaks are well separated and hence negligible overlap occurs. The spectrum for $\omega < 0$ may be derived in a similar fashion or by using the property of the Fourier transform.

Now, $\frac{1}{2}at^2 - (\omega-\omega_0)t = \frac{1}{2}a[t^2 - \frac{2}{a}(\omega-\omega_0)t] = \frac{1}{2}a\left[t - \frac{\omega-\omega_0}{a}\right]^2 - \left(\frac{\omega-\omega_0}{\sqrt{2a}}\right)^2.$

Let $\sqrt{\pi}x = \sqrt{a}\left(t - \dfrac{\omega - \omega_0}{a}\right)$ (4.49)

so that $\sqrt{\pi}dx = \sqrt{a}dt$. The first term of (4.48) becomes

$$S_1(f) = \frac{A_0}{2}\int_{-x_1}^{x_2} \exp\left[j\left\{\frac{1}{2}\pi x^2 - \frac{(\omega-\omega_0)^2}{2a}\right\}\right]\frac{\sqrt{\pi}}{\sqrt{a}}.dx$$

$$= \frac{A_0}{2}\sqrt{\pi a}\exp\left[-j\left(\frac{\omega-\omega_0}{\sqrt{2a}}\right)^2\right]\int_{-x_1}^{x_2} e^{j\pi x^2/2}dx, \tag{4.50}$$

where $x_1 = -\left[\dfrac{a\tau_p/2 + (\omega - \omega_0)}{\sqrt{\pi a}}\right]$ and $x_2 = -\left[\dfrac{a\tau_p/2 + (\omega - \omega_0)}{\sqrt{\pi a}}\right]$. (4.51)

Recall the Fourier integrals

$$c(x) = \textstyle\int_0^x \cos(\pi x^2/2)dx; \quad s(x) = \int_0^x \sin(\pi x^2/2)dx. \tag{4.52}$$

Then,

$$\int_{-x_1}^{x_2} e^{j\pi x^2/2}dx = \int_{-x_1}^{0}[\cos(\pi x^2/2) + j\sin(\pi x^2/2)]dx + \int_{0}^{x_2}[\cos(\pi x^2/2) + j\sin(\pi x^2/2)]dx$$

$$= C(x_1) + C(x_2) + j[S(x_1) + S(x_2). \tag{4.53}$$

Thus (4.50) reads as

$$S_1(f) = \frac{A_0}{2}\sqrt{\frac{\pi}{a}}\exp\left[-j\left(\frac{\omega-\omega_0}{\sqrt{2a}}\right)^2\right].\{C(x_1) + C(x_2) + j[S(x_1 + S(x_2)]\}. \tag{4.54}$$

The magnitude spectrum is

$$|S_1(f)| = \frac{A_0}{2}\sqrt{\frac{\pi}{a}}[\{C(x_1) + C(x_2)\}^2 + \{S(x_1) + S(x_2)\}^2] \tag{4.55}$$

and the phase spectrum is

$$\theta(\omega) = -\frac{(\omega-\omega_0)^2}{2a} + \arctan\left[\frac{S(x_1) + S(x_2)}{C(x_1) + C(x_2)}\right]. \tag{4.56}$$

The magnitude and phase spectra are obviously complicated functions of frequency. Define a dimensionless quantity B as the product of the frequency deviation and the pulse width.

That is, $B = 2\omega_d \tau_p = a\tau_p^2.$ (4.57)

$$\text{At } \omega = \omega_0, \quad -x_1 = \frac{a\tau_p}{2\sqrt{\pi a}} = \sqrt{\frac{a^2\tau_p^2}{4\pi a}} = \sqrt{\frac{a\tau_p^2}{4\pi}} = \sqrt{\frac{B}{4\pi}} = x_2.$$

For large $B(B >> 1$ is the case of practical interest),

$$C(x_1) = C(x_2) \approx 0.5 \text{ and } S(x_1) = S(x_2) \approx 0.5.$$

So $|S_1(\omega)| = A_0\sqrt{\pi/2a}$. Thus the magnitude spectrum approaches a rectangle shape with the spectral width close to $2\omega_d$, as shown in Fig. 4.12. The power spectral density of the signal is given by

$$G(f) = |S_1(\omega)|^2 = \frac{\pi A_0^2}{2a}, \text{ a constant.} \tag{4.58}$$

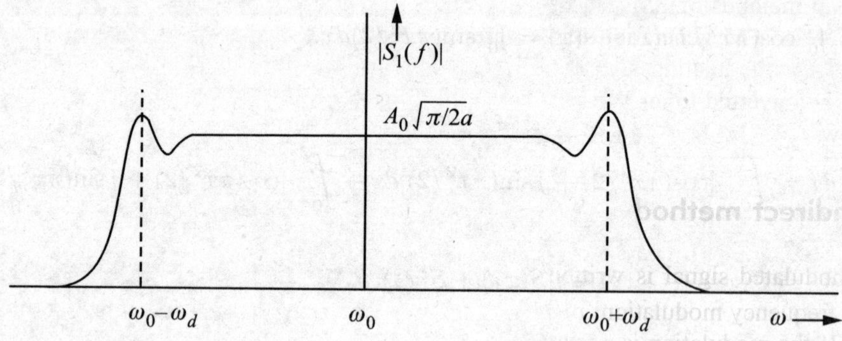

Fig. 4.12 *Spectrum of pulsed FM signal*

4.8 Amplitude–Frequency Modulated Signal (AM–FM)

The general form of a carrier which is modulated both in amplitude and frequency is given by

$$v_c(t) = A(t)\cos[\omega_c t + \theta(t)]. \tag{4.59}$$

Here the time-varying amplitude $A(t)$ varies in accordance with the message signal of frequency f_m. The phase $\theta(t)$ also varies with time following the message signal. For a sinusoidal message signal of the form $\cos\omega_m t$, the expression for the AM–FM signal can be written as

$$v_c(t) = A_c(1 + m\cos\omega_m t)\cos(\omega_c t + \beta\sin\omega_m t). \tag{4.60}$$

The quantities m and β denote the depths of modulation for amplitude and frequency, respectively. For small β, the FM part can be simplified as is done in NBFM. In doing so,

$$v_c(t) \approx A_c(1 + m\cos\omega_m t)[\cos\omega_c t - \beta\sin\omega_m t\sin\omega_c t]. \tag{4.61}$$

Example 4.6

Show that the signal $v_c(t)$ given by

$$v_c(t) = \cos(2\pi f_c t) + 0.02\cos[2\pi(f_c + f_m)t]$$

represents an AM-FM signal. Find the modulation indices m and β. Write an expression for the instantaneous frequency $f(t)$.

Solution

$$\begin{aligned} v_c(t) &= \cos(2\pi f_c t) + 0.02[\cos(2\pi f_c t)\cos(2\pi f_m t) - \sin(2\pi f_c t)\sin(2\pi f_m t)] \\ &= [1 + 0.02\cos(2\pi f_m t)]\cos(2\pi f_c t) - 0.02\sin(2\pi f_m t)\sin(2\pi f_c t). \end{aligned}$$

4.9 FM Generation

There are two methods of FM generation, namely, the direct method and the indirect method. The basic principle of FM generation in the direct method is to change the frequency of an electronic oscillator by the modulating signal. In the indirect method, the NBFM signal is first generated, which is then converted to an WBFM by the use of frequency multipliers. We will discuss both the methods now.

4.9.1 Indirect method

An angle modulated signal is written as $A_c\cos[\omega_c t + km(t)]$, where $m(t)$ relates the modulating signal. For frequency modulation, $m(t) = \int_0^t s(\tau)d\tau$, where $s(t)$ is the actual modulating signal and $k = 2\pi k_f$. If the modulation is narrowband, the waveform is approximated as

$$\begin{aligned} s_{\text{NBFM}}(t) &= A_c[\cos\omega_c t - 2\pi k_f m(t)\sin\omega_c t] \\ &= A_c[\cos\omega_c t - 2\pi k_f \int_0^t s(\tau)d\tau \cdot \sin\omega_c t]. \end{aligned}$$

The equation suggests the possible method of NBFM generator, whose block diagram representation is shown in Fig. 4.13. If we desire phase modulation rather than frequency, we simply remove the integration and replace the input $2\pi k_f m(t)$ by $k_p s(t)$.

The signals thus generated have strictly a narrow bandwidth since the maximum frequency deviation Δf is small compared to highest modulating frequency. The NBFM signal can be converted to an WBFM signal by the use of a frequency multiplier that increases the frequency deviation. Let us briefly discuss the operational principle of a frequency multiplier. Frequency multiplication is achieved by sending a periodic signal through a nonlinear device that generates harmonics, followed by bandpass filtering as shown in Fig. 4.14.

The input to the nonlinear device is a periodic signal, not necessarily sinusoidal. The output waveform has a similar fundamental frequency but is rich in higher-order harmonics. To attain an n-fold increase in frequency, the nonlinear device must offer nth order nonlinearity. The bandpass filter tuned to the nth harmonic selects the signal of frequency nf. We may multiply the frequency by any arbitrary integer n, at least in principle. But, in practice, the amplitudes decrease with increasing harmonic number. Normally, a multiplication factor of 2 to 5 is taken in a single stage. If it requires a 12-fold frequency multiplication, one can use a device offering 12th order nonlinearity. The same can be achieved by cascading two second-order devices and one third-order device.

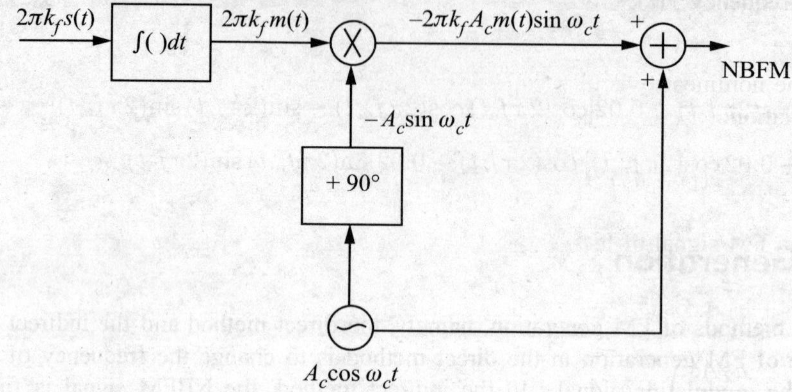

Fig. 4.13 *NBFM generator*

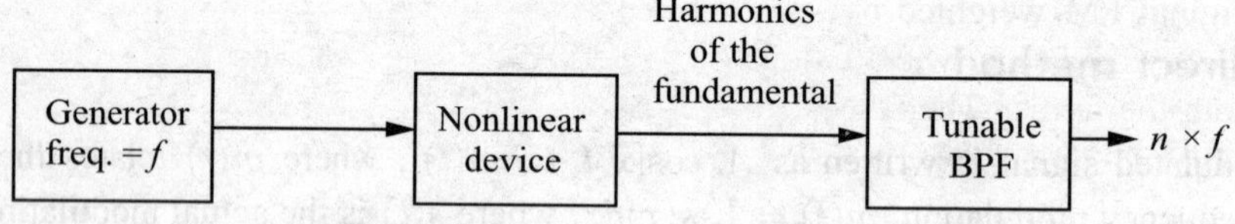

Fig. 4.14 *A frequency multiplier*

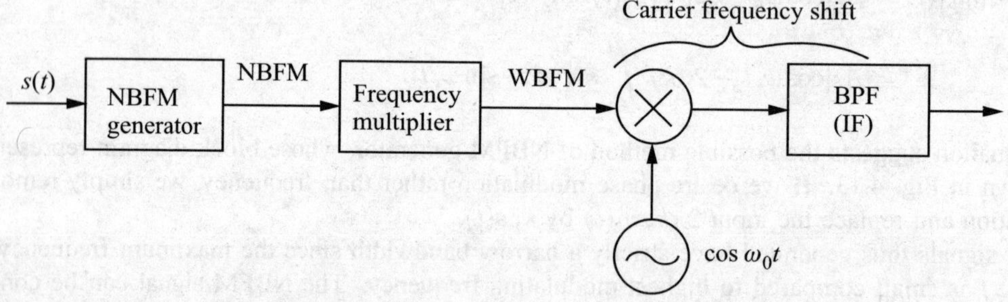

Fig. 4.15 *Block diagram of the modulator for WBFM*

Returning to the NBFM, let the signal be applied at the input of the frequency multiplier. Since the instantaneous frequency of the NBFM signal is $f(t) = f_c + k_f s(t)$, the resulting signal has the frequency $nf(t) = nf_c + nk_f s(t)$, where n is the multiplier. Both the carrier and the frequency deviation are multiplied by the same integer. Note that the modulation index is proportional to the frequency deviation for a given modulating frequency and hence this scheme increases the modulation index by factor n and the waveform is converted to a wideband signal. In practice, the higher carrier frequency is down-converted by heterodyning. The complete scheme for generating wideband FM is shown in Fig. 4.15.

Example 4.7
Show that the frequency multiplier, although a nonlinear device, does not distort the signal.

Solution
Suppose that the nonlinearity is of second order so that the multiplier is described by the following input–output relation:

$$y(t) = a_0 + a_1 x(t) + a_2 x^2(t).$$

When $x(t)$ is an FM signal of the form $x(t) = A\cos(\omega_c t + km(t))$, the output is

$$\begin{aligned} y(t) &= a_0 + a_1 A\cos[\omega_c t + km(t)] + a_2 A^2 \cos^2[\omega_c t + km(t)] \\ &= \left(a_0 + \frac{a_2 A^2}{2}\right) + a_1 A\cos[\omega_c t + km(t)] + \frac{a_2 A^2}{2}\cos[2\omega_c t + 2km(t)]. \end{aligned}$$

We observe that the output contains the dc term which can be filtered out by ac coupling. The second term is the original FM weighted by a_1. The last term is also an FM signal of carrier frequency $2\omega_c$ with frequency deviation twice that of the original FM wave. The message signal $m(t)$ remains unaltered in both the terms. The result can be extended to the case where the multiplier shows nth order nonlinearity.

Example of Armstrong FM system

In the Armstrong method, a relatively low frequency carrier is phase-shifted by 90° and fed to a balanced modulator (multiplier) (see Fig. 4.13), where it is mixed with the input modulating signal (f_m). The output is a narrowband FM with low index of modulation and hence low frequency deviation. Since small frequency deviations are possible in the basic Armstrong systems, extensive frequency multiplication and mixing are used to increase the deviation to the desired value (Fig. 4.16).

Suppose we want to design an Armstrong FM modulator to generate an FM signal of carrier frequency 96 MHz and frequency deviation $\Delta f = 75$ kHz using the indirect method. We first generate a NBFM signal with a carrier frequency $f_{c1} = 200$ kHz, as shown in Fig. 4.16. The frequency chosen is such that the stable crystal oscillator and the product modulator required for NBFM generation can easily be constructed (see Fig. 4.13 for NBFM generator). We select the frequency deviation $\Delta f = 25$ Hz at this stage based on the following consideration.

For commercial FM broadcast, the baseband spectrum ranges from 50 Hz to 15 kHz. The FM signal, $\cos(\omega_c t + \phi(t))$, turns into a narrowband FM provided the depth of modulation is low enough for $|\phi(t)| << \pi/2 = 1.57$. In the worst case we can take a value of ϕ as large as 0.5. For tone modulation, $\phi(t) = \beta\sin(2\pi f_m t)$, and so β can attain a maximum value of 0.5. Since $\beta = \frac{\Delta f}{f_m}$, $\Delta f = \beta f_m = 0.5 \times 50$ Hz= 25 Hz.

Since we require $\Delta f = 75$ kHz at the final output, the multiplication factor is $75000/25 = 3000$.

But, instead of multiplying directly, we multiply the frequency of NBFM in two stages: by 64 before heterodyning and by 48 after heterodyning. These values are selected based on the property that multiplication can be achieved using factors of 2 and 3, i.e., multiplication by 64 can be realised using

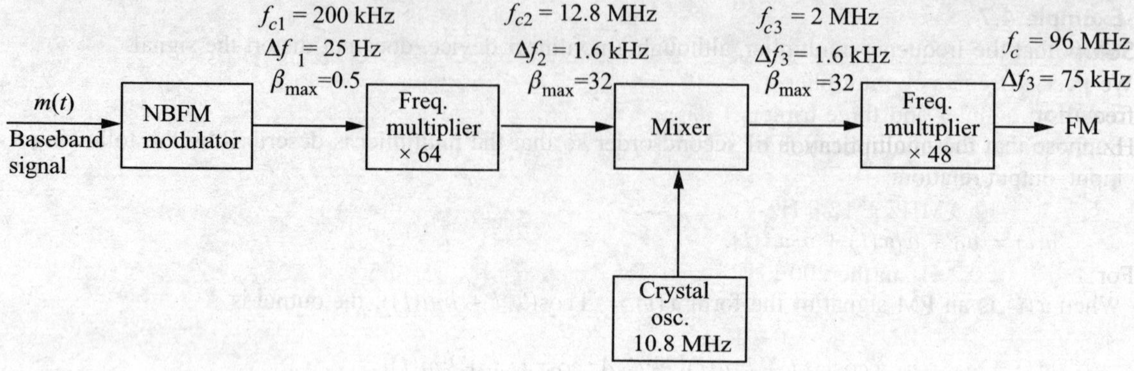

Fig. 4.16 *Indirect method of FM generation for Armstrong modulator*

a multiplier strip consisting of six doublers in cascade ($64 = 2^6$). The first multiplication results in a new carrier frequency $f_{c2} = 64 \times 200$ kHz $= 12.8$ MHz and a frequency deviation $\Delta f_2 = 64 \times 25$ Hz $= 1.6$ kHz. The frequency translation is achieved in a mixer with a carrier of frequency 10.8 MHz. The whole spectrum now shifts around an intermediate frequency of $f_{c3} = 12.8 - 10.8 = 2$ MHz. The frequency deviation is of course $\Delta f_3 = 1.6$ kHz, as it is not affected by the frequency-shifting. The FM signal is obtained by multiplying this signal a second time by 48. This gives the carrier frequency $f_c = 2$ MHz $\times 48 = 96$ MHz and a frequency deviation equal to 75 kHz. The crystal oscillator of carrier frequency 10.8 MHz is generated from the crystal oscillator of 200 kHz to avoid any phase mismatch.

Example 4.8

In the above FM modulator, if the oscillator at the mixer stage suffers a frequency drift of δf_{LO}, what will be the carrier frequency of the FM signal ?

Solution

The frequency of the oscillator is

$$f_L = 10.8 \text{ MHz} \pm \delta f_{LO} \text{ Hz}.$$

The output signal of the mixer has the carrier frequency

$$f_{c3} = 12.8\text{MHz} - (10.8 \text{ MHz} \pm \delta f_{LO} \text{ Hz}) = 2\text{MHz} \mp \delta f_{LO}.$$

Hence the FM carrier is at the frequency

$$f_c = 48 \times f_{c3} = 96 \text{ MHz} \pm 48 \times \delta f_{LO}.$$

To suppress the effect of this drift, the mixing signal is not taken from a separate oscillator but from the 200 kHz crystal oscillator through the use of frequency multipliers.

Example 4.9
Assume that the 10.8 MHz mixing signal is derived from the 200 kHz oscillator. If the 200 kHz oscillator drifts by 0.2 Hz (1 part per millions), find the drifts in the 10.8 MHz signal and the resulting FM signal.

Solution
We need a frequency multiplier of $\frac{10.8\text{MHz}}{200\text{kHz}} = 54 = 2 \times 3^3$. (This is obtained by the cascade of a frequency doubler and three triplers.) The carrier frequency of the NBFM is $f_{c1} = 200$ kHz ± 0.2 Hz. After the first multiplication by 64,

$$f_{c2} = 12.8 \text{ MHz} \pm 12.8 \text{ Hz}.$$

For a drift of 0.2 Hz in the 200 kHz signal, the frequency of the mixing signal is

$$f_{\text{LO}} = 10.8 \text{ MHz} \pm 54 \times 0.2 \text{ Hz} = 10.8 \text{ MHz} \pm 10.8 \text{ Hz}.$$

For the intermediate frequency after the mixer stage,

$$f_{c3} = (12.8 \text{ MHz} \ \pm \ 12.8 \text{ Hz}) - (10.8 \text{ MHz} \pm 10.8 \text{ Hz})$$
$$= 2 \text{ MHz} \pm 2 \text{ Hz}.$$

Hence the resulting FM signal drifts by $48 \times (2) = \pm 96$ Hz.

4.9.2 Direct method of FM generation

The direct way of FM generation is based on varying the output frequency of an oscillator in proportion to the baseband signal. The heart of this modulator is the oscillator. There are a wide variety of oscillators that accomplish this task, e.g., the Hartley or the Colpitt oscillator consisting of a tank circuit that produces sinusoidal waveform of frequency ω_c determined by the values of L and C (since $\omega_c = \frac{1}{\sqrt{LC}}$). The variation of the frequency of oscillation can be accomplished by varying any of the reactive parameters (L or C) of the tank circuit. If, for example, the capacitor variation is controlled by the baseband signal, an FM signal results. One way to accomplish this is with the use of semiconductor diodes that act as capacitors whose values vary with reverse bias voltage. Such capacitors are commonly called the varicaps.

Let the capacitance variation be of the form $C_m = \alpha s(t)$, where $s(t)$ is the baseband signal. The instantaneous frequency of the oscillator is then

$$\begin{aligned} \omega(t) &= \frac{1}{\sqrt{L[C+\alpha s(t)]}} = \frac{1}{\sqrt{LC}}\left[1+\frac{\alpha}{C}s(t)\right]^{-1/2} \\ &\approx \omega_c\left[1-\frac{\alpha}{2C}s(t)\right], \quad \text{since } \left|\frac{\alpha}{c}s(t)\right| << 1. \\ &= \omega_c - ks(t) \end{aligned} \tag{4.62}$$

where $k = +\dfrac{\alpha\omega_c}{2C}$. Thus FM signal is achieved. The instantaneous frequency depends on the instantaneous value of the modulating signal. The maximum capacitance deviation is $\Delta C = \alpha A_m$, where A_m is the signal amplitude. Thus,

$$\frac{\Delta C}{C} = \frac{\alpha A_m}{C} = \frac{2kA_m}{\omega_c} = \frac{2 \times 2\pi\Delta f}{2\pi f_c} = 2\left(\frac{\Delta f}{f_c}\right).$$

Since $\Delta f << f_c$, the fractional change in capacitance may be small compared to C.

There are other devices namely PIN diodes and FETs that can be used successfully to generate FM signals. Figure 4.17 shows a FET reactance modulator whose effective capacitance at the terminals (1,1') can be controlled by the modulating signal. If this capacitance variation be included in the tank circuit of an oscillator, we have accomplished the task of generating FM.

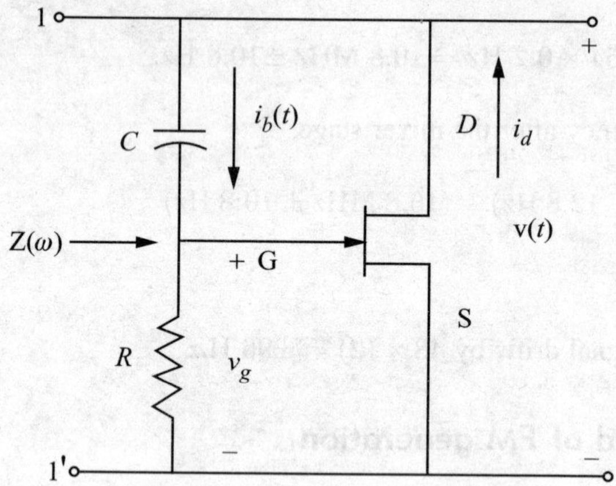

Fig. 4.17 *FET reactance modulator*

With the applied voltage $v(t)$, the bias current $i_b(t)$ is given by

$$i_b(t) = \frac{v(t)}{R - jX_C}, \tag{4.63}$$

where $X_c = 1/\omega C$.

The voltage at the gate of the FET is given by

$$v_g(t) = i_b(t)R = \frac{R}{R - jX_C}v(t). \tag{4.64}$$

The drain current $i_d(t)$ is then

$$i_d(t) = g_m v_g(t) = \frac{Rg_m}{R - jX_C}v(t), \tag{4.65}$$

where g_m is the FET's transconductance. The impedance $Z(\omega)$ at the terminals (1,1') between the drain and the ground is thus

$$Z(\omega) = \frac{v(t)}{i_d(t)} = \frac{R - jX_c}{g_m R} = \frac{1}{g_m}\left(1 - j\frac{X_c}{R}\right). \tag{4.66}$$

By the appropriate choice of C and the frequency of operation, we assume $X_c >> R$ and so

$$Z(\omega) \approx -j\frac{X_c}{g_m R} = -j\frac{1}{\omega RC g_m} = -\frac{j}{\omega C_m}, \tag{4.67}$$

where $C_m = g_m RC$ is the effective capacitance. Any modulating signal applied between the gate and the source of Fig. 4.17 produces a periodic variation of v_{gs}, the gate-to-source voltage, thereby making a proportional change in g_m. Thus the effective capacitance C_m can be varied with bias voltage.

An oscillator whose frequency of oscillation can be varied by the voltage of the modulating signal is called a voltage-controlled oscillator (VCO). The direct method of FM generation using VCO is attractive due to the fact that it produces sufficient frequency deviation and requires no frequency multiplication. The disadvantage of this method is that it has poor frequency stability. Of course, a crystal-oscillator may be used for carrier generation.

VCOs are now available in IC form. VCO-566 is one of the family that can be used for generating carrier frequencies less than 1 MHz.

Example 4.10

In the direct method of FM generation, the controlled capacitance C_m is related to the modulating signal $v_m(t)$ as

$$C_m = \frac{100}{\sqrt{1 + 2v_m}} \text{ pF.}$$

It is given that the capacitance C_0 of the resonant circuit is 200 pF. The inductance L is adjusted for resonance to occur at 5 MHz for $V_m = 4$ volts. If the modulating signal

$$v_m(t) = 4 + 0.05\sin(2\pi \times 10^3 t) \text{ volts,}$$

find the expression for the angle-modulated signal.

Solution

The carrier frequency is given by

$$f_c = \frac{1}{2\pi\sqrt{L(C_0 + C_m)}}.$$

Here, $C_0 + C_m = 200 + \dfrac{100}{\sqrt{1 + 2\{4 + 0.05\sin(2\pi \times 10^3 t)\}}}$ pF

$$\simeq \frac{700}{3}\left[1 - \frac{1}{1260}\sin(2\pi \times 10^3 t)\right] \text{ pF.}$$

Thus, $(C_0 + C_m)^{-1/2} = \sqrt{\frac{3}{700}}\left[1 + \frac{1}{2520}\sin(2\pi \times 10^3 t)\right] \times 10^6 (F^{-1/2})$.

For $v_m = 5$ volts, $f_c = 5$ MHz and $C_m = \frac{100}{3}$ pF,

$$5 \times 10^6 = \frac{1}{2\pi\sqrt{L\,(200 + \frac{100}{3}) \times 10^{-12}}} = \frac{10^6}{2\pi\sqrt{\frac{700}{3}L}} \text{ H. Thus } L = 4.342\ \mu\text{H}.$$

The instantaneous carrier frequency is given by

$$f(t) = \frac{1}{2\pi\sqrt{L}}\frac{1}{(C_0 + C_m)^{1/2}}$$

$$= \frac{1}{2\pi} \times 10\pi\sqrt{\frac{700}{3}} \times \sqrt{\frac{3}{700}}\left[1 + \frac{1}{2520}\sin(2\pi \times 10^3 t)\right] \times 10^6 \text{ Hz}$$

$$= 5\left[1 + \frac{1}{2520}\sin(2\pi \times 10^3 t)\right] \text{ MHz.}$$

The angle modulated signal is $v(t) = A_c \cos[5 \times 10^6 t + 1984 \sin(2\pi \times 10^3)t]$ volts.

4.10 FM Demodulators

In FM signals, the information resides in the instantaneous frequency. Recovery of the information signal can be done in two ways. In one, the demodulators employ discriminators (a frequency-selective network that yields an output proportional to the instantaneous frequency). A diode envelope detector following the discriminator detects the message signal. The frequency discriminator converts FM signals to AM signals and then demodulates the AM envelope using peak detector. The second method employs a phase-locked loop (PLL) that demodulates the message signal by matching the output of a local oscillator (VCO in this case) to the modulated carrier. We will discuss here both types of demodulators.

4.10.1 Detection based on discriminator

Consider the FM signal

$$s_{\text{FM}}(t) = A_c \cos[\omega_c t + 2\pi k_f m(t)]$$

where $m(t) = \int_0^t s(\tau)d\tau$, $s(t)$ being the message signal. The derivative of $s_{\text{FM}}(t)$ is given by

$$\frac{ds_{\mathrm{FM}}(t)}{dt} = -A_c[\omega_c + 2\pi k_f m(t)] \sin\left[\omega_c t + 2\pi k_f \int_0^t s(\tau)d\tau\right]. \tag{4.68}$$

The signal defined by (4.68) is an AM–FM waveform with amplitude $A_c[\omega_c + 2\pi k_f s(t)]$. Since the amplitude varies proportionately to the message signal, an envelope detector can be used to recover $s(t)$. Figure 4.18 shows such a scheme. For a phase modulated input, the envelope detector yields a signal that is the derivative of $s(t)$. In such a case an integrator must follow the detector.

A differentiator in the frequency domain is a network having transfer function of the form $H(f) = j2\pi f$. Its output is related to its input as $S_{\mathrm{out}}(f) = j2\pi f S_{\mathrm{in}}(f)$, i.e., the magnitude of the output is linearly proportional to the frequency of the input. We require such a linear transfer characteristic around the carrier frequency. A bandpass filter can be used if the operation is confined to a limited range of frequencies, as illustrated in Fig. 4.19.

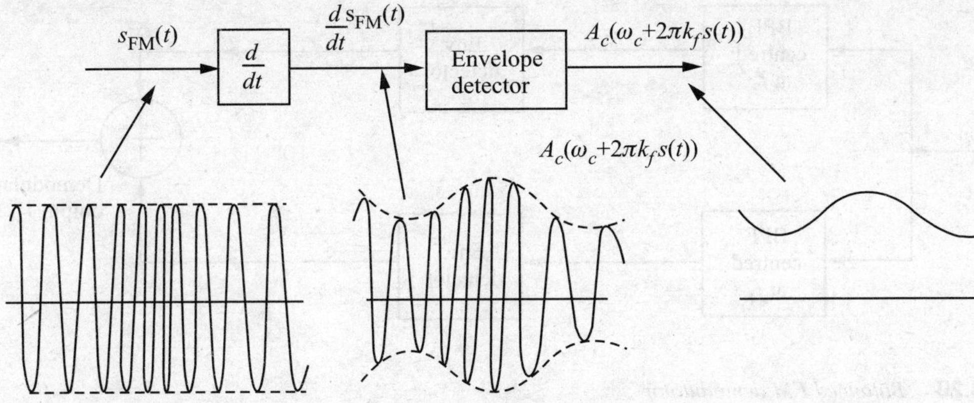

Fig. 4.18 *FM demodulation by differentiation followed by diode detection*

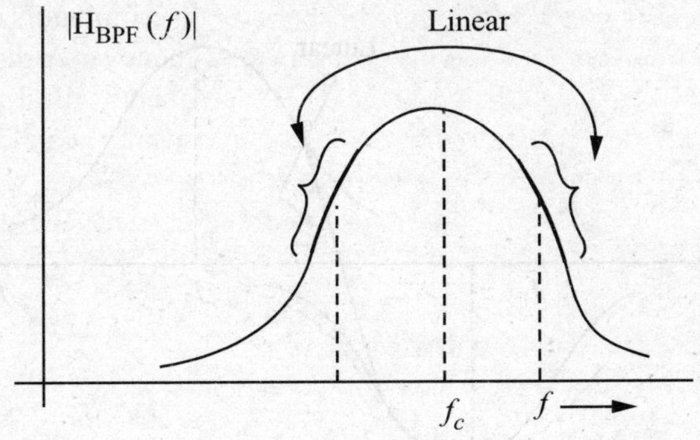

Fig. 4.19 *Bandpass filter showing linear portion*

The range of linear operation is required to be extended so as to accommodate sufficient number of sidebands. This is accomplished in a balanced demodulator as shown in Fig. 4.20. As we take the difference between two demodulator outputs, the overall transfer function of the bandpass filter discriminator is as shown in Fig. 4.21.

The bandpass filter consists of a tank circuit as shown in Fig. 4.22. For an FM input the discriminator produces an output voltage proportional to the input frequency. The magnitude of the output voltage attains a maximum for an input frequency equal to the resonant frequency (of either f_1 or f_2). The transfer characteristic of Fig. 4.21 can be realised in a circuit of Fig. 4.23. The upper half of the secondary winding and C_2 is tuned to f_2. The lower half and C_1 form the BPF centred at f_1. The precisely linear transfer characteristic of the filter's tail is a theoretical assumption. In practical filters the magnitude of the transfer function at the tail can be written as

$$|H(f)| = a_0 + a_1(f - f_c) + a_2(f - f_c)^2 + \cdots \tag{4.69}$$

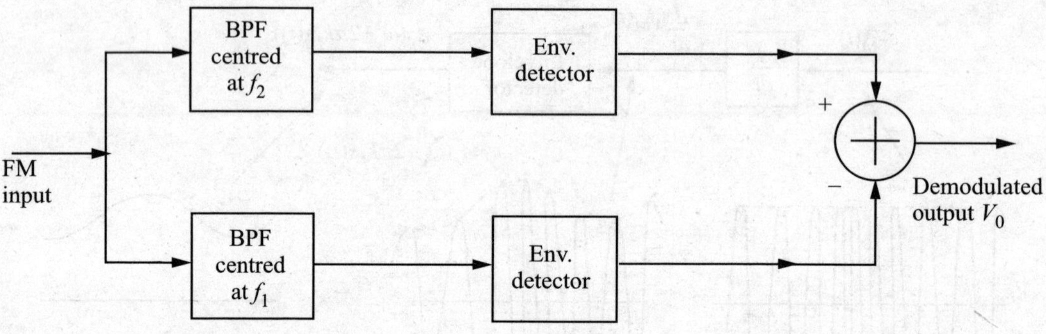

Fig. 4.20 *Balanced FM demodulator*

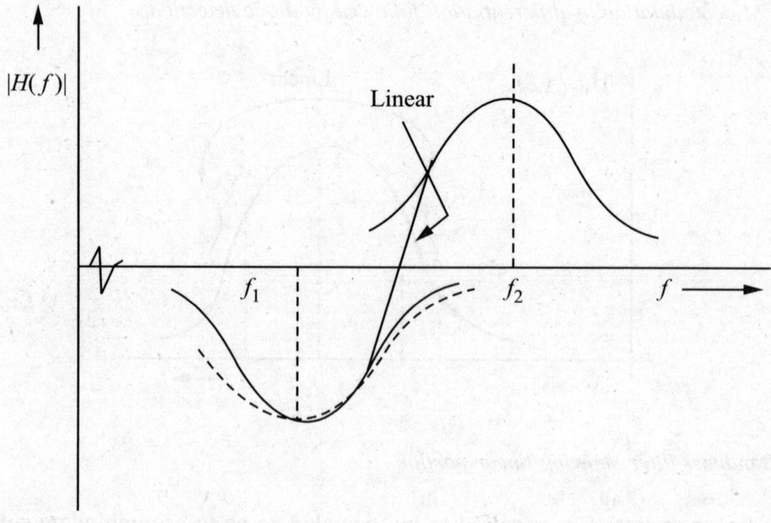

Fig. 4.21 *Bandpass filter discriminator*

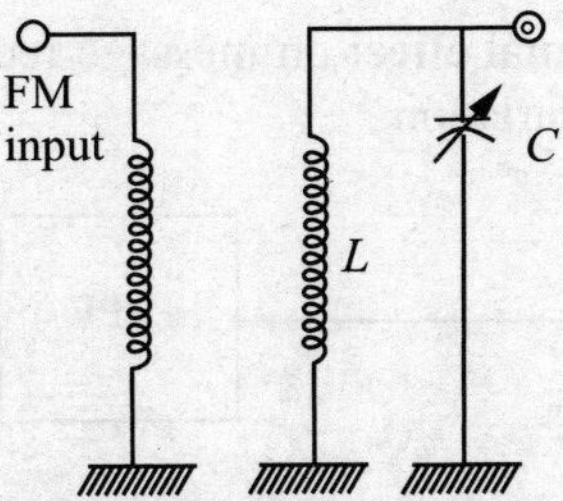

Fig. 4.22 *Possible frequency selective network (a) frequency-to-voltage converter*

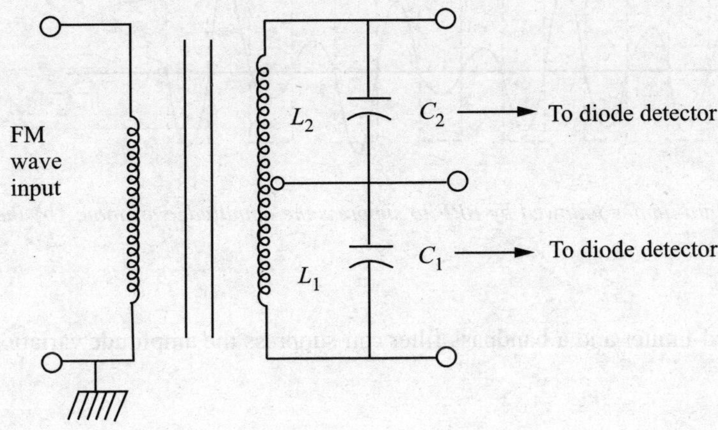

Fig. 4.23 *The circuit realising the transfer function of Fig. 4.21*

Assuming a second-order nonlinearity, the transfer function of BPF2 is

$$|H_2(f)| = a_0 + a_1(f - f_c) + a_2(f - f_c)^2 \tag{4.70}$$

and that of BPF1 is

$$|H_1(f)| = a_0 + a_1(f - f_c) + a_2(f - f_c)^2. \tag{4.71}$$

The difference signal at the output has the amplitude

$$\begin{aligned} V_0 &= V_i[|H_2(f)| - |H_1(f)|] \\ &= V_i \times 2a_1(f - f_c). \end{aligned} \tag{4.72}$$

So far, we have considered that the amplitude of the FM carrier remains constant. But in an FM system, channel noise, fading, etc. cause the amplitude A_c to vary. This leads to an additional term dA_c/dt in (4.69). Such amplitude variations can be removed by passing the signal through a hard-limiter followed by a bandpass filter as shown in Fig. 4.24. The limiting of the peaks of the envelope does not create any detrimental effect on message recovery as the message is impressed on the carrier in the form of frequency variation.

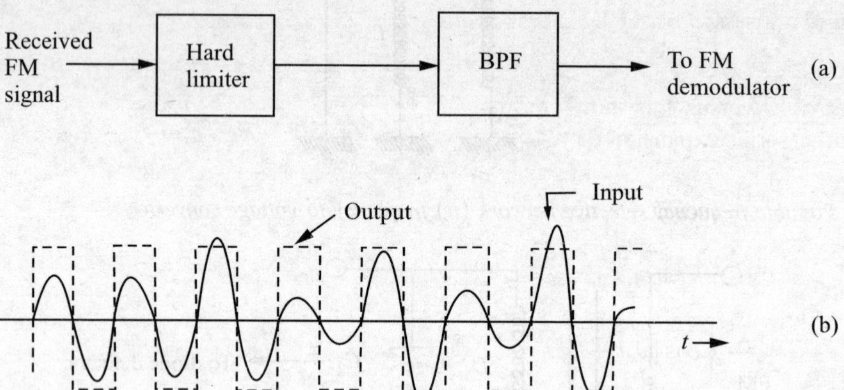

Fig. 4.24 *(a) Hard-limiter followed by BPF to suppress the amplitude variation; (b) the input and output of the hard-limiter.*

Example 4.11

Explain how a hard-limiter and a bandpass filter can suppress the amplitude variation in the received FM signal.

Solution

Let the received input be $v_i(t) = A_i(t)\cos(\phi_c t)$, where $\phi_c(t) = \omega_c t + 2\pi k_f(t)$ (see Fig. 4.25 (a)). The output of the hard-limiter is V or $-V$ depending on whether $v_i(t)$ is positive or negative. That is, irrespective of the magnitude of $v_i(t)$, the hard-limiter generates either $+V$ or $-V$. Hence a square wave results, the duration of which is determined by the zero-crossings of the input signal. Since the input signal is angle-modulated, the output is also angle-modulated.

The input–output relation of the hard-limiter can be written as

$$v_0(t) = V \quad \text{when } v_i(t) > 0$$
$$= -V \quad \text{when } v_i(t) < 0,$$

Fig. 4.25 *Illustrates the action of hard-limiter*

or equivalently,

$$v_O[\phi_c(t)] = \begin{cases} +V & \text{when } \cos\phi_c(t) \geq 0 \\ -V & \text{when } \cos\phi_c(t) < 0. \end{cases}$$

A plot of $v_O[\phi]$ versus ϕ is shown in Fig. 4.25 (b). Note that $v_O(\phi)$ is a periodic function with period 2π. Fourier series expansion of $v_O(\phi)$ gives

$$v_O(\phi) = \frac{4V}{\pi}(\cos\phi - \frac{1}{3}\cos 3\phi + \frac{1}{5}\cos 5\phi - \frac{1}{7}\cos 7\phi + \cdots).$$

Since $\phi = \omega_c t + 2\pi k_f s(t)$,

$$v_O(t) = \frac{4V}{\pi}\left\{\cos\left[\omega_c t + 2\pi k_f s(t)\right] - \frac{1}{3}\cos[3\omega_c t + 6\pi k_f s(t)] + \cdots\right\}.$$

Each term of the above equation represents an FM wave. We see that the hard-limiter output contains the original FM wave in addition to the FM waves of carrier frequencies $3\omega_c$, $5\omega_c$, etc. The bandpass filter tuned to ω_c passes only the original FM wave with constant amplitude.

4.10.2 Detection using phase-locked loops

The phase-locked loop (PLL) is a feedback system that can be used for FM demodulation, especially under high noise conditions when a conventional FM discriminator fails to yield satisfactory recovery of baseband signal. The PLL comprises two basic components namely a voltage-controlled oscillator (VCO) and a phase comparator (Fig. 4.26). A PLL FM demodulator does not require any tuned circuit. It automatically compensates for changes in the carrier frequency. Once frequency lock occurs, the VCO tracks the frequency changes in the input signal. When the PLL input is a deviated FM (frequency different from the natural frequency of VCO), the output of the phase comparator is proportional to the frequency deviation. This is the correction voltage to VCO. We have indeed demodulated the FM signal.

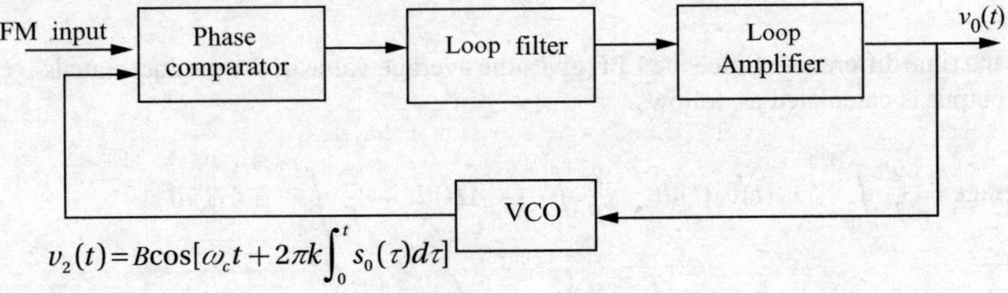

Fig. 4.26 *Block diagram of the PLL FM demodulator*

The simplest method of achieving phase comparison is using a product multiplier followed by a lowpass filter shown in Fig. 4.27. For the two input signals $v_1(t) = A\sin[\omega t + \theta_1(t)]$ and $v_2(t) = B\cos[\omega t + \theta_2(t)]$, the multiplier output is

$$x_1(t) = AB\sin[\omega t + \theta_1(t)]\cos[\omega t + \theta_2(t)]$$

$$= \frac{AB}{2}\{\sin[\theta_1(t) - \theta_2(t)] + \sin[2\omega t + \theta_1(t) + \theta_2(t)]\}. \tag{4.73}$$

The output of the LPF is

$$v_0(t) = \frac{AB}{2}\sin(\theta_1 - \theta_2) \tag{4.74}$$

provided that $\theta_1(t)$ and $\theta_2(t)$ have bandwidths less than 2ω. Since the instantaneous frequency of the VCO is given by

$$\omega_{\text{VCO}}(t) = \omega_c + kv_0(t), \tag{4.75}$$

the loop is locked when $\theta_1 = \theta_2$ and VCO oscillates at its free-running frequency ω_c.

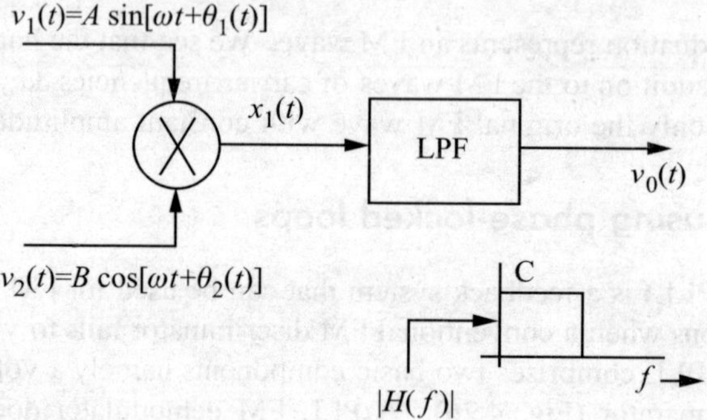

Fig. 4.27 *A simple phase comparator*

Example 4.12

Find and sketch the output of the phase comparator if the inputs to the phase comparator are square waveforms as shown in Fig. 4.28 (a).

Solution

Let t_d be the time difference. Since the LPF gives the average value of the product signals $v_1(t)$, $v_2(t)$, the final output is calculated as follows.

$$\text{Time average} = \frac{2}{T}\int_0^{T/2} v_1(t)v_2(t)dt = \frac{2}{T}\int_0^{t_d}(-AB)dt + \frac{2}{T}\int_{t_d}^{T/2}(AB)dt$$

$$= \frac{2}{T}(AB)\left[-t_d + \frac{T}{2} - t_d\right] = AB\left(1 - \frac{4t_d}{T}\right).$$

Thus the comparator output amplitude varies linearly with the time difference (or equivalently, the phase difference) as shown in Fig. 4.28(b).

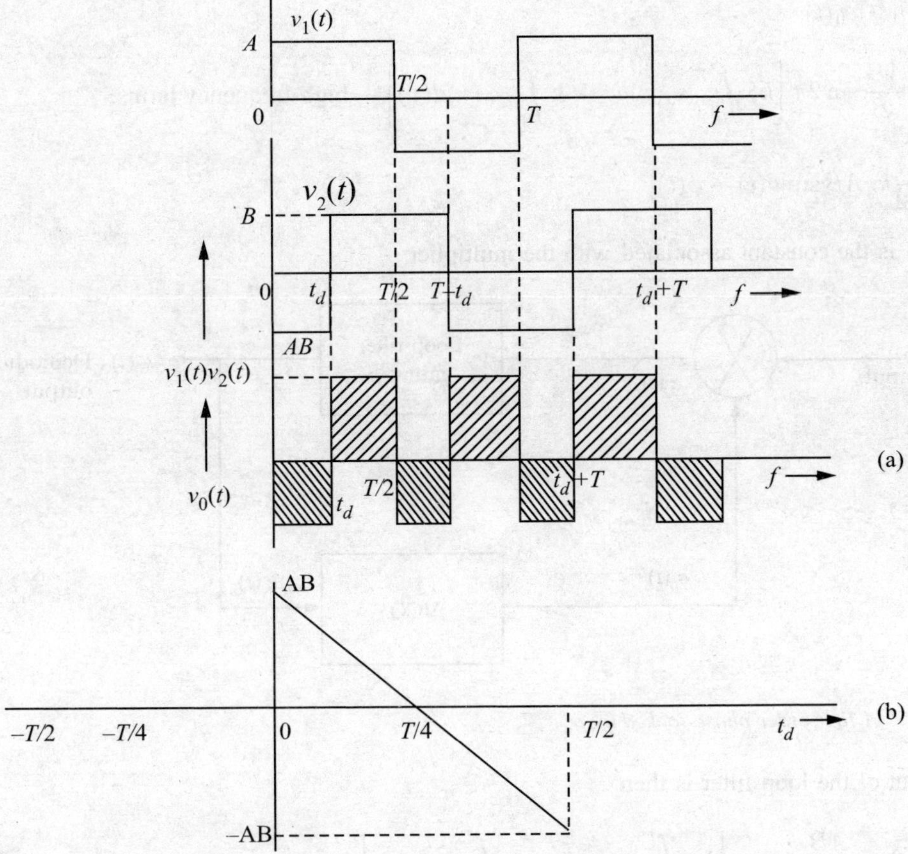

Fig. 4.28 *Illustrates phase comparison and average output.*

A phase comparator based FM demodulator is drawn in Fig. 4.29. Returning to the discussions on VCO, suppose the output of the VCO is a sinusoidal voltage of amplitude B and its free-running frequency is ω_c. The instantaneous frequency of the VCO is given by (4.75), where the constant k defines the measure of the sensitivity of the VCO. If one defines the VCO sensitivity as the rate of change of the instantaneous frequency with respect to the controlling voltage, then $k = \dfrac{d\omega_{\text{VCO}}}{dv_0(t)}$. Thus the VCO output is given by

$$e_0(t) = B\cos[\omega_c t + \phi(t)], \text{ where } \phi(t) = 2\pi k \int_0^t s_0(\tau)d\tau. \tag{4.76}$$

Let the FM signal be $v_1(t) = A\sin[\omega_c t + \theta(t)]$, where $\theta(t) = 2\pi k_f m(t)$ and $m(t) = \int_0^t s(\tau)d\tau$ and $s(t)$ is the message signal. The multiplier output is the error signal $e_d(t)$ given by

$$e_d(t) = v_1(t)e_0(t)$$

$$= \frac{AB}{2}\sin 2\pi\left[k_f\int_0^{t_v} s_0(\tau)d\tau - k\int_0^t s_0(\tau)d(\tau)\right] + \text{high-frequency terms.}$$

$$= \frac{1}{2}k_d AB\sin[\theta(t) - \phi(t)] \quad (4.77)$$

where k_d is the constant associated with the multiplier.

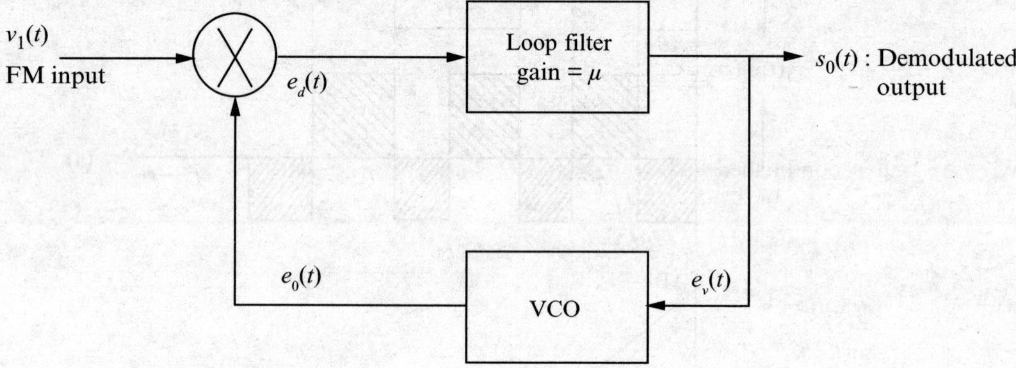

Fig. 4.29 *A first-order phase-locked loop*

The output of the loop filter is then

$$s_0(t) = \frac{AB}{2}\sin 2\pi\left[k_f\int_0^t s_0(\tau)d\tau - k\int_0^t s_0(\tau)d\tau\right].$$

$$= \frac{1}{2}\mu k_d AB\sin[\theta(t) - \phi(t)] \quad (4.78)$$

The phase difference $\beta(t)$ is given by

$$\beta(t) = \theta(t) - \phi(t) = 2\pi k_f\int_0^t s(\tau)d\tau - 2\pi k\int_0^t s_0(\tau)d\tau. \quad (4.79)$$

To perform transient analysis, we differentiate (4.79) to get

$$\frac{d\beta}{dt} = 2\pi k_f s(t) - 2\pi k s_0(t).$$

For a linear phase detector,

$$s_0(t) = \frac{1}{2}\mu k_d AB 2\pi\left[k_f\int_0^t s(\tau)d\tau - k\int_0^t s_0(\tau)d\tau\right], \quad (4.80)$$

which may be derived from (4.78) for small angles as for small phase error, $\theta(t) - \phi(t)$ is small and $\sin[\theta(t) - \phi(t)] \approx \theta(t) - \phi(t)$. Differentiating (4.80), we have

$$\frac{ds_0}{dt} = \pi\mu k_d AB[k_f s(t) - k s_0(t)] = \frac{1}{2}\mu k_d AB\frac{d\beta}{dt}. \tag{4.81}$$

This represents the relation between the frequency offset $\frac{d\beta}{dt}$ and the output voltage $s_0(t)$. The equilibrium steady-state is obtained by setting $\frac{ds_0}{dt} = 0$. Then,

$$k_f s(t) - k s_0(t) = 0$$

or, $$s(t) = \frac{k}{k_f} s_0(t). \tag{4.82}$$

We have thus accomplished FM demodulation. We rewrite (4.81) as

$$\frac{ds_0}{dt} + \pi\mu k_d ABk s_0(t) = \pi\mu k_d k_f ABs(t)$$

which is of the form

$$\frac{ds_0}{dt} + \frac{s_0}{\tau} = \frac{k_f}{k\tau}s(t), \tag{4.83}$$

where $\tau = \frac{1}{\pi\mu k_d kAB}$ is the time constant of the transient response.

The transfer function $H(s)$ of the PLL is obtained by taking the Laplace transform of (4.83) and then the finding the ratio of the output signal to the input signal. Thus,

$$sS_0(s) + \frac{1}{\tau}S_0(s) = \frac{k_f}{k_\tau}S(s)$$

or, $$H(s) \equiv \frac{S_0(s)}{S(s)} = \frac{k_f/(k\tau)}{(s + 1/\tau)}. \tag{4.84}$$

The loop transfer function $H(s)$ has a single pole at $s = -\frac{1}{\tau}$ and that is why the PLL is referred to as a first-order PLL. The PLL output is thus a distorted version of $v(t)$ (as if the signal has passed through a lowpass simple RC network of time constant τ). To reduce distortion, the cutoff frequency of the PLL is made much higher than the highest modulating frequency.

The unit impulse response is obtained by taking the inverse Laplace transform of Eq. (4.84). In doing so,

$$h(t) = \frac{k_f}{k} \cdot L^{-1}\left[\frac{1/\tau}{s + 1/\tau}\right] = \frac{k_f}{k} \cdot \frac{1}{\tau} \cdot e^{-t/\tau}u(t) \tag{4.84a}$$

where $\frac{k_f}{k\tau} = \pi\,\mu\,k_d\,AB.k_f = k_t$ is the loop gain of the PLL. As loop gain tends to infinity, we have $Lt\, h(t) \rightarrow \delta(t)$, Loop gain $\rightarrow \infty$ which is synonymous with the condition $\phi(t) \approx \theta(t)$ for large loop gain.

Example 4.13

Assume an input to the FM modulator is a unit-step function u(t). Find the demodulated output using a first-order PLL.

Solution

The FM signal in this case

$v_1(t) = A\, sin\, [\omega_c t + \theta(t)],$

where $\theta(t) = 2\pi k_f\, m(t) = 2\pi k_f \int\limits_0^t s(t) d(t)$, and $s(t) = u(t)$.

Now, for linear phase detector we have the relation:

$$\frac{ds_0(t)}{dt} = \frac{1}{2}\mu\, k_d\, AB \frac{d}{dt}\left[2\pi k_f s(t) - 2\pi k s_0(t)\right].$$

$$\frac{d\, s_0(t)}{dt} + \frac{s_0(t)}{\tau} = \frac{k_f}{k\tau} s(t)$$

$$\text{or, } s\, S_0(s) + \frac{1}{\tau} S_0(s) = \frac{k_f}{k\tau} . S(f)$$

Since $s(t) = u(t)$, we have, $S(s) = \dfrac{1}{s}$.

$$\text{So, } \left(s + \frac{1}{\tau}\right) S_0(s) = \frac{k_f}{k\tau} \cdot \frac{1}{s}$$

$$\text{or, } S_0(s) = \frac{k_f}{k\tau} \cdot \frac{1}{s} \frac{1}{s + 1/\tau}$$

$$= \frac{k_f}{k}\left(\frac{1}{s} - \frac{1}{s + 1/\tau}\right)$$

The demodulated output is given by

$$s_0(t) = \frac{k_f}{k}\left(1 - e^{-t/\tau}\right) u(t) \tag{4.84b}$$

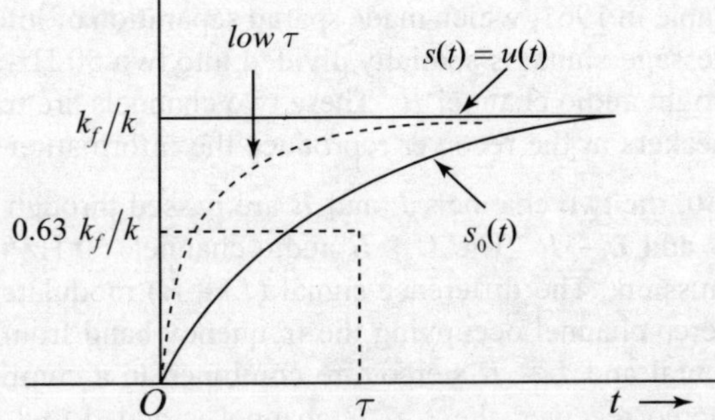

Fig. 4.29 *Demodulated output for a step input signal.*

and is shown in Fig. 4.29. The PLL output is thus a distorted version of the input signal. If the line constant of the loop (τ) is made small (that is, large loop gain), the demodulated output becomes closer to the information signal $u(t)$.

4.11 Stereophonic FM Broadcast: Transmission and Reception

The FM broadcast band extends from 88 MHz to 108 MHz (assigned by the FCC) with an adjacent channel separation of 200 kHz. The instantaneous maximum frequency deviation is limited by the FCC to 75 kHz. In monophonic commercial FM broadcasts, a single audio baseband signal is transmitted and the receiver reproduces the original sound via the loudspeaker. But in stereophonic FM broadcasts, two audio baseband signals are transmitted, and at the receiver the signals are applied to two independent loudspeakers. The microphones at the studio are physically separated as well. The speakers are also positioned some distance apart. This arrangement helps the listener to get a feel of the 'natural' sound; this is known as the stereophonic effect. For AM broadcasting, the maximum frequency of the information signal is set at 5 kHz in contrast to 15 kHz set for FM broadcasting. Inclusion of higher spectral components accounts for the improved tonal quality of FM reception (closer to the baseband) compared to AM. If one uses narrowband FM (commonly used in two-way wireless communications), the system requires a bandwidth of $2f_m = 30$ kHz although the bandwidth allocated to each FM station is much larger, i.e., 200 kHz. This avoids interference from the adjacent stations.

The monophonic FM broadcast is well established and there are millions of monophonic FM radio receivers. To facilitate their continued usability, the FCC has framed a rule that the stereophonic FM transmission should be compatible with the original monophone systems so that the old radio sets would continue to receive the monophonic version of the stereophonic transmission.

4.11.1 FM stereo

In commercial monaural FM broadcast systems, a single 50 Hz to 15 kHz audio channel makes up the entire information spectrum. This modulates the high frequency carrier and is transmitted through an FM communication channel with a bandwidth of 200 kHz. The receiver reproduces the information signal which appears to have originated from a point source. No spatial separation can be reproduced so that the sound looks like that from an extended source. Stereo FM transmission became commercially available in 1961, which made spatial separation of information signals possible. During transmission, the message signal is spatially divided into two 50 Hz audio channels, namely a left audio channel L and a right audio channel R. These two channels are transmitted simultaneously and independently. Two speakers at the receiver reproduce the information signal.

As shown in the Fig. 4.30, the two channels L and R are passed through a matrix which produces two audio channels, $L + R$ and $L - R$. The $L + R$ audio channel, 50 Hz to 15 kHz, modulates the carrier for monaural transmission. The difference signal $(L - R)$ modulates a subcarrier of 38 kHz and produces the $L - R$ stereo channel occupying the frequency band from 23 kHz–53 kHz. These two channels, $L + R$ monaural and $L - R$ stereo, are combined in a combining network and fed to the FM transmitter. In a stereo receiver, the $L - R$ channel is added to $L + R$ channel to produce the L channel; the difference produces the R-channel.

FM stereo basically multiplexes two independent audio channels. Consider the system shown in Fig. 4.31 that multiplexes two signals $s_1(t)$ and $s_2(t)$, each bandlimited to 15 kHz.

As shown in the figure, we amplitude modulate a carrier at 38 kHz with $s_2(t)$ and then add $s_1(t)$ to it. The combined signal is then $s_1(t) + s_2(t)\cos 4\pi f_c t$. Note that the individual spectrum of $s_1(t)$ and $s_2(t)$ are still preserved. The signal $s_2(t)\cos 4\pi f_c t$ occupies the frequency band $2f_c \pm f_m$, that is, from 23 to 53 kHz. If we frequency-modulate another carrier with this combined signal, the resulting NBFM wave uses only 106 kHz of the assigned 200 kHz.

Figure 4.32 depicts the FM stereo receiver for recovering the baseband signals $s_1(t)$ and $s_2(t)$. The signal $s_1(t)$ is easily recovered by a lowpass filter with cutoff frequency f_m = 15kHz. Although the spectrum of $s_2(t)$ is located at $2f_c$ = 38 kHz, we cannot recover $s_2(t)$ by a bandpass filter; it requires a synchronous demodulator. The reason for this is that for envelope detection to yield $s_2(t)$, the carrier frequency of the envelope must be higher than the frequency of the envelope.

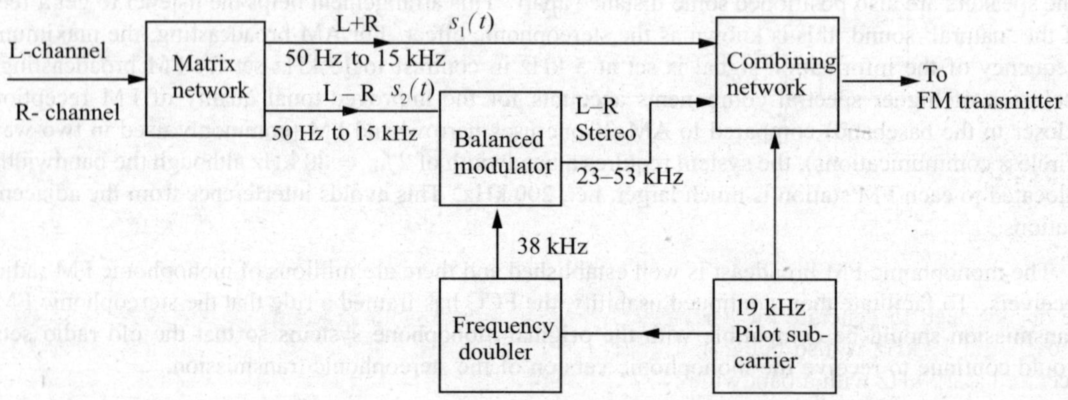

Fig. 4.30 *FM stereo transmitter*

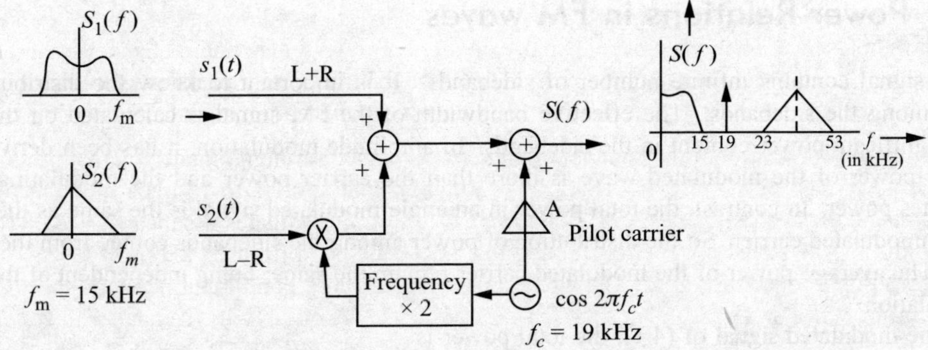

Fig. 4.31 *FM stereo and monophone transmitter*

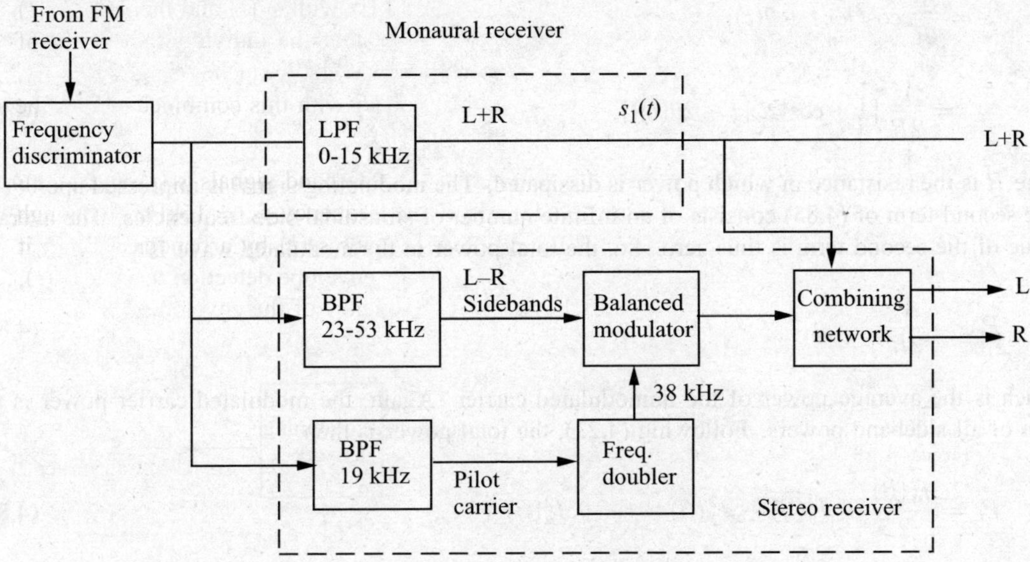

Fig. 4.32 *FM stereo and mono receiver*

We do not need to generate the carrier at 38 kHz in synchronous detection. Figure 4.30 shows that a carrier at 19 kHz is also being transmitted. This is separated at the receiver using a bandpass filter centred at 19 kHz with a bandwidth no more than 4 kHz. The carrier at 19 kHz is doubled and used to down-convert the spectrum so that an LPF can reconstruct the signal $s_2(t)$.

In the above discussions, $s_1(t)$ represents the $L + R$ channel, compatible with mono-receiver, and $s_2(t)$ represents the $L - R$ channel. These two are combined in a stereo decoder to produce L and R independently. These are then applied to the respective speakers.

4.12 Power Relations in FM waves

An FM signal contains infinite number of sidebands. It is important to know the distribution of power among the sidebands. The effective bandwidth of the FM signal is calculated on the basis of the significant power content in the sidebands. In amplitude modulation, it has been derived that the total power of the modulated wave is more than the carrier power and the modulating signal contributes power. In contrast, the total power in an angle-modulated signal is the same as the power of the unmodulated carrier. So the distribution of power among the sidebands comes from the carrier power. The average power of the modulated carrier remain the same, being independent of the depth of modulation.

For the modulated signal of (4.1), the total power is

$$P_t = s_c(t)/R$$

$$= \frac{A_c^2}{R}\cos^2[\omega_c t + \theta(t)]$$

$$= \frac{A_c^2}{2R}\{1 + \cos[2\omega_c t + 2\theta(t)]\}. \tag{4.85}$$

Here R is the resistance in which power is dissipated. The modulating signal is impressed upon $\theta(t)$. The second term of (4.85) consists of an infinite number of sinusoidal side frequencies. The average value of the second term is thus zero. So, the total power in the modulated wave is

$$P_{\text{av}} = \frac{A_c^2}{2R}, \tag{4.86}$$

which is the average power of the unmodulated carrier. Again, the modulated carrier power is the sum of all sideband powers. Following (4.22), the total power is then

$$P_t = \frac{J_0^2(\beta)}{2} + J_1^2(\beta) + J_2^2(\beta) + \ldots + J_n^2(\beta) + \ldots, \tag{4.87}$$

where the first term gives the modulated carrier power.

4.13 Frequency-Division Multiplexing

Frequency-division multiplexing (FDM) is the process of combining transmissions from several communication channels and transmitting them on one transmission facility (such as metallic or optical fibre cable, radio-frequency channel etc.). FDM is used extensively for combining many narrowband sources to form a wideband channel. Because of the bandwidth and power requirement considerations, single-sideband modulation is the most common application in the FDM systems.

The principle of combining several input messages in the FDM method is illustrated in Fig. 4.33, where N input channels $x_1(t), x_2(t), \ldots, x_N(t)$ modulate the subcarriers $f_{c1}, f_{c2}, \ldots, f_{cN}$ in SSB scheme. The individual channels are first band limited by the low-pass filters. The modulated signals are then summed in a linear manner to produce the baseband signal $x_b(t)$. The bandpass filter used after SSB modulation is to allow only one of the sidebands.

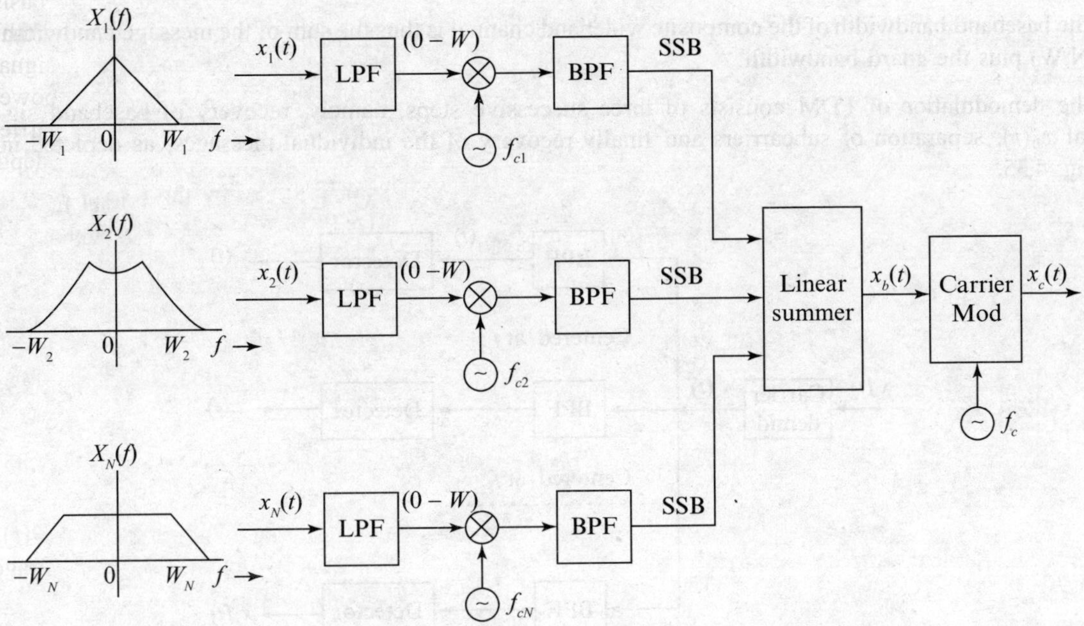

Fig. 4.33 *SSB FDM system block diagram*

The subcarrier frequencies $f_{c1}, f_{c2}, \ldots$ are properly chosen so that the composite wide spectra include non-overlapping spectra of individual channels, as in Fig.4.34. The baseband signal $x_b(t)$ thus formed can be transmitted directly or it can be used to modulate another carrier of frequency f_c, as shown in the figure to form $x_c(t)$. If the subcarrier frequencies are not separated enough, the tail of one message spectrum may couple with other adjacent message spectra. To avoid that, the modulated spectra are spaced out in frequency by guard bands. If we select the upper SSB modulation, the guard band frequency is calculated as follows: (vide Fig. 4.31)

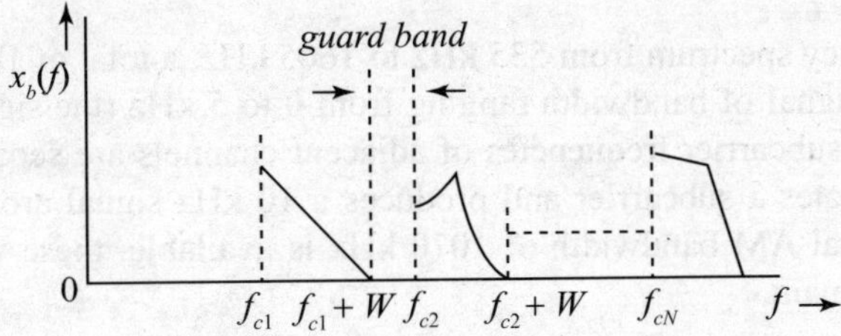

Fig. 4.34 *Baseband spectra of FDM*

U-SSB spectra of Message 1 range from f_{c1} to $f_{c1} + W$.

U-SSB spectra of Message 2 range from f_{c2} to $f_{c2} + W$.

So, guard band $= f_{c2} - (f_{c1} + W)$.

The baseband bandwidth of the composite wideband channel is thus the sum of the message bandwidth (N.W) plus the guard bandwidth.

The demodulation of FDM consists of three successive steps, namely, recovery of baseband signal $x_b(t)$, separation of subcarriers and finally recovery of the individual messages, as depicted in Fig. 4.35.

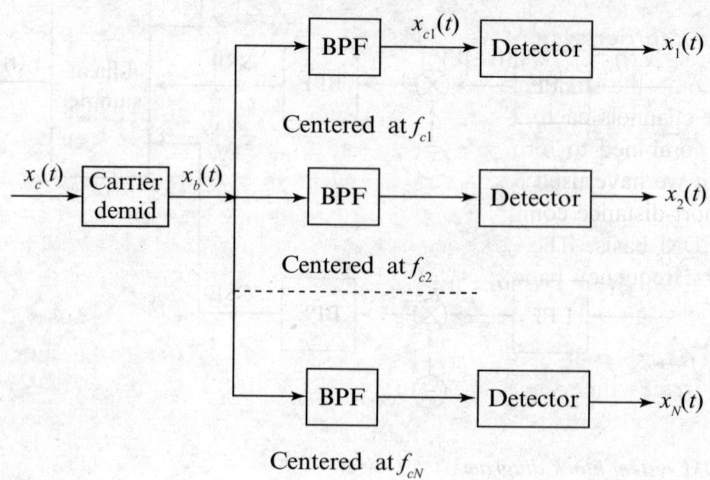

Fig. 4.35 *FDM Demodulation*

A familiar example of FDM system are the commercial AM and FM broadcast, TV broadcast, telephone and data communication system. We will illustrate here only two such examples, namely, the commercial AM broadcast and the AT & T type L_4 carrier system.

Commercial AM broadcast

This occupies the frequency spectrum from 535 kHz to 1605 kHz, a total of 1070 kHz bandwidth and accepts the information signal of bandwidth ranging from 0 to 5 kHz (the signal may be voice grade or musical signals). The subcarrier frequencies of adjacent channels are separated by 10 kHz. Each channel amplitude modulates a subcarrier and produces a 10 kHz signal around that subcarriers, as in Fig.(4.36). Since a total AM bandwidth of 1070 kHz is available, these will be $1070/10 = 107$ slots in the frequency domain.

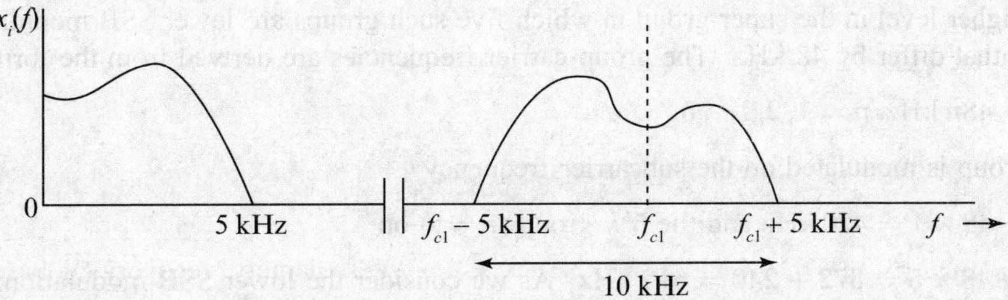

Fig. 4.36 *Channel bandwidth.*

The AT & T type L_4 carrier system

The system has become the standards for long-distance communication in the United States. This includes 3600 voice channels each of 4 kHz bandwidth, multiplexed together for transmission. The voice channels are combined to form groups; groups are combined to form super-groups etc. The digital hierarchy that we have used for T_1 carrier system multiplexes the signals on TDM basic and is an example of short-distance communication. A similar kind of hierarchy is also adopted here in L_4 system but on FDM basis. The voice band channel is actually band limited to approximately a 300 Hz to 3,000 Hz frequency band, but for practical design consideration, we consider a 4 kHz channel bandwidth.

The first level of FDM hierarchy is the formation of basic groups. The group multiplexes 12 input voice channels and stacks them on subcarriers 64, 68, . . . , 108 kHz using lower SSB modulation. The subcarrier frequencies for the channel bands are synthesised from a common oscillator and are of multiples of 4 kHz. These are determined by the formula

$$f_{c_n} = 112 - 4n\,\text{kHz}, n = 1, 2, \ldots, 12 \tag{4.88}$$

For $n = 12$, $f_c = 1124{\times}12 = 112 - 48 = 64$ kHz and we use the lower SSB. Thus, for channel no.12, the frequency band occupied is from 60 to 64 kHz. Channel no.1 has the subcarriers frequency $f_c = 112 - 4 = 108$ kHz and the spectrum occupies 104 to 108 kHz. In effect, the resulting group spectrum is 48 kHz wide and ranges from 60 kHz to 108 kHz, as shown in Fig. (4.37).

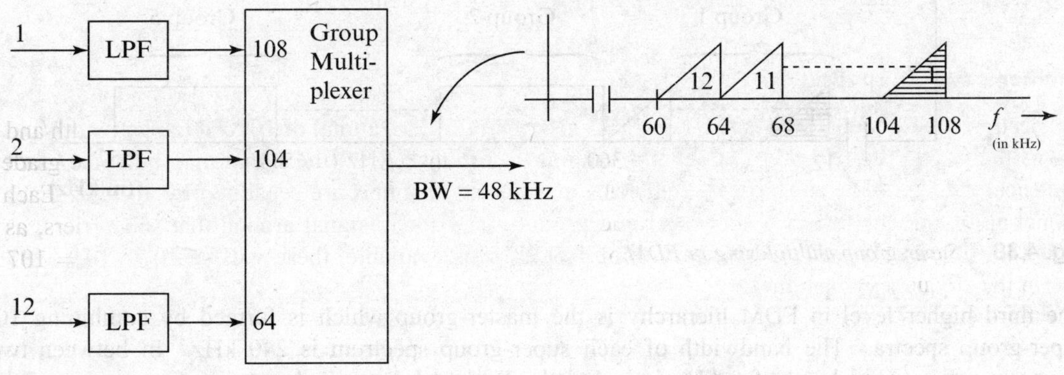

Fig. 4.37 *Group multiplexing*

The next higher level in the super-group in which five such groups are lower SSB modulated on the subcarriers that differ by 48 kHz. The group carrier frequencies are derived from the formula:

$$f_c = 372 + 48n\,\text{kHz}, n = 1, 2, \ldots, 5. \tag{4.89}$$

The first group is modulated on the subcarrier frequency

$f_c = 372 + 48 \times 1 = 420\,\text{kHz}$ and the $5^t h$ group (n =5) on

$f_c = 372 + 48 \times 5 = 372 + 240 = 612$ kHz. As we consider the lower SSB modulation, the range of the super-group spectrum can be calculated as follows:

Group 1: f_{out} = 420 kHz – (60 kHz to 108 kHz) = 312 kHz to 360 kHz

Group 2: f_{out} = 468 kHz – (60 kHz to 108 kHz) = 360 kHz to 408 kHz

Group 3: f_{out} = 516 kHz – (60 kHz to 108 kHz) = 408 kHz to 456 kHz

Group 4: f_{out} = 564 kHz – (60 kHz to 108 kHz) = 456 kHz to 504 kHz

Group 5: f_{out} = 612 kHz – (60 kHz to 108 kHz) = 504 kHz to 552 kHz

The super-group spectrum thus occupies the frequency range from 312 kHz to 552 kHz, as shown in Fig. (4.38). An interesting point in the figure is that the voice channel number 1 (104 – 108 kHz) appears is an upper SSB signal (312 – 316 kHz). We also note that the subcarrier for Group 1 (420 kHz) falls in the Group 3 spectrum, subcarrier for Group 2 (468 kHz) falls in the Group 4 spectrum.

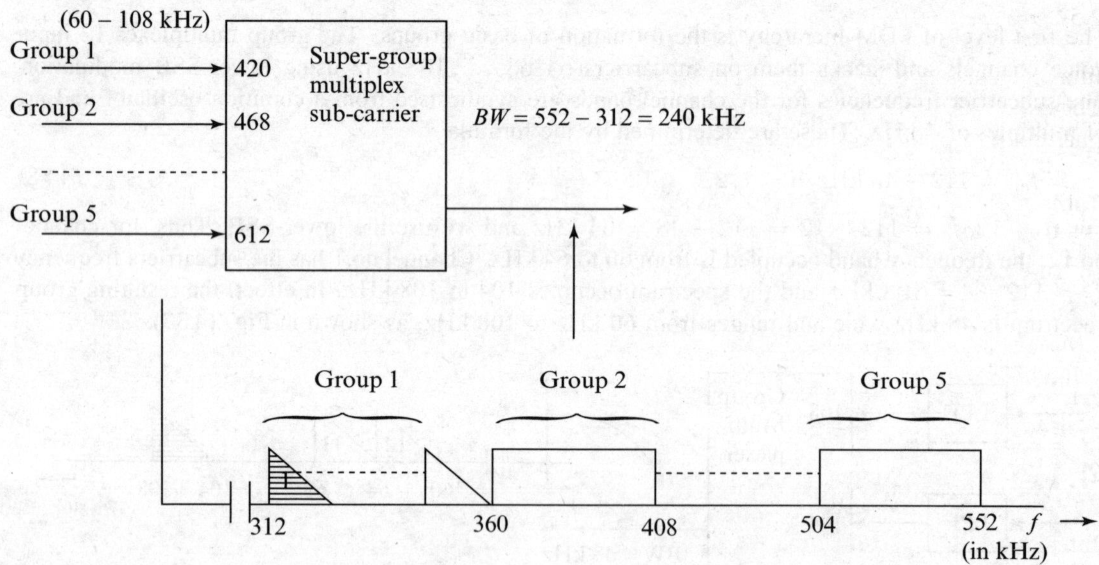

Fig. 4.38 *Super-group multiplexing in FDM*

The third higher level in FDM hierarchy is the master-group which is formed by combining 10 super-group spectra. The bandwidth of each super-group spectrum is 240 kHz. In between two super-groups, a guard band of 8 kHz is included. Table 4.1 lists all the carrier frequencies of the super-groups forming the master-group. Note that all the carrier frequencies are 248 kHz apart from the preceding frequencies except the carrier frequency of the super-group 7 which differs by 56 kHz. We now calculate the spectra of master-group as follows:

Table 4.1 Super-group carrier frequencies

Super-group	Carrier frequency in kHz
1	1116
2	1364
3	1612
4	1860
5	2108
6	2356
7	2652
8	2900
9	3148
10	3396

Super-group 1: $f_{out} = 1116$ kHz – (312 kHz to 552 kHz) = 564 kHz to 804 kHz

Super-group 10: $f_{out} = 3396$ kHz – (312 kHz to 552 kHz) = 2844 kHz to 3084 kHz

The spectra of master-group occupy the frequency band from 564 kHz to 3084 kHz, a bandwidth of 2.52 MHz.

In a similar fashion, 6 master-groups can be combined to form a bigger group, called a jumbo group. The final baseband spectra occupy the frequency range from 0.5 MHz to 17.5 MHz including the guard bands. Table 4.2 summarises the AT & T FDM hierarchy.

Table 4.2 AT & T FDM hierarchy: L_4 carrier spectrum.

	Frequency range	Bandwidth	No of voice band channel
Group	60–108 kHz	48 kHz	12
Super-group	312–552 kHz	240 kHz	60
Master-group	564–3084 kHz	2.52 MHz	600
Jumbo group	0.5–17.5 MHz	17 MHz	3600

4.14 Nonlinear Distortion and Interference in FM signals

Interference is the contamination of true signal by another similar signal. Severe interference may lead to unsuccessful recovery of true message signal. Since FM signal is of constant amplitude and as such, less affected by system nonlinearities. Assume a generalised nonlinear distortion dictated by the input $r(t)$ and output $s(t)$ relation

$$s(t) = a_0 + a_1 r(t) + a_2 r^2(t) + \cdots + a_n r^n(t) \tag{4.90}$$

Clearly, a_0 adds to the dc value while $a_1 r(t)$ is the signal amplification by gain a_1; the remaining terms are nonlinear distortion. For an angle-modulation, the signal is

$$s_c(t) = A_c \cos(\omega_c t + ?(t)) = A_c \cos\phi_c(t) \tag{4.91}$$

(vide Eq. (4.1)). The output of nonlinear system is given by

$$s(t) = a_0 + a_1 A_c \cos(\omega_c t + \theta(t)) + a_2 A_c^2 \cos^2(\omega_c t + \theta(t)) + \cdots + a_n A_c^n \cos^n(\omega_c t + \theta(t))$$

$$= c_0 + c_1 \cos(\omega_c t + \theta(t)) + c_2 \cos(2\omega_c t + 2\theta(t)) + \cdots + c_n \cos(n\omega_c t + n\theta(t)) \tag{4.92}$$

Since ω_c is large, each harmonic in the above equation s(t) is separable in the frequency domain. So, any bandpass filter centred at ω_c and with bandwidth of FM (or PM) signal, can recover the desired component $\cos(\omega_c t + \theta(t))$. Thus angle-modulated signals are immune to nonlinear distortion.

In discussing interference, it may be caused due to multipath propagation. Since FM is of constant amplitude, any amplitude change due to fading can be adjusted by automatic gain control. Of course, if the received signal contains two or more similar unwanted signals in the frequency band, it requires a little thought.

Let the receiver be turned to a particular carrier frequency ω_c. The received signal $r(t)$ consists of unmodulated carrier plus other interfering carrier of slightly different frequency. Thus,

$$r(t) = A_c \cos\omega_c t + A_i \cos((\omega_c + \omega_i)t + \phi_i) \tag{4.93}$$

where A_i is the amplitude of the interfering signal of frequency $(\omega_c + \omega_i)$ and ϕ_i is the relative phase. We may write the above equation as

$$r(t) = A_c \cos\omega_c t + A_i \cos\omega_c t \cos(\omega_i t + \phi_i) - A_i \sin\omega_c t \sin(\omega_i t + \phi_i)$$

$$= [A_c + A_i \cos(\omega_i t + \phi_i)] \cos\omega_c t - A_i \sin(\omega_i t + \phi_i)] \sin\omega_c t$$

Let $A_c + A_i \cos(\omega_i t + \phi_i) = A(t) \cos\theta(t)$

$$A_i \sin(\omega_i t + \phi_i) = A(t) \sin\theta(t)$$

so that

$$A^2(t) = A_c^2 + A_i^2 + 2A_c A_i \cos(\omega_i t + \varphi_i)$$

$$\left.\begin{aligned} &\text{or } A(t) = A_c \cdot \sqrt{1 + \rho^2 + 2\rho\cos(\omega_i t + \phi_i)} \\ &\text{and } \theta(t) = \tan^{-1}\frac{\rho\sin(\omega_i t + \phi_i)}{1 + \rho\cos(\omega_i t + \phi_i)} \end{aligned}\right\} \tag{4.94}$$

where $\rho = A_i/A_c$, a measure of interfering strength. Thus

$$r(t) = A(t)\cos(\omega_c t + \theta(t)) \tag{4.95}$$

in envelope-and-phase form. This shows that the interfering signal produces both amplitude and phase modulation. As a special case of low-level interference, $\rho = A_i/A_c \ll 1$, we have amplitude

$$A(t) \approx A_c(1 + \rho cos(\omega_i t + \varphi_i))$$

and phase

$$\theta\,(t) \approx \rho \sin(\omega_i t + \varphi_i)$$

Thus,

$$r\,(t) = [A_c\,(1 + \rho cos(w_i t + \phi_i))] \,\cdot cos(\omega_c t + \rho \sin(\omega_i t + \varphi_i)) \tag{4.96}$$

The above equation resembles Equation (4.60). The time-varying amplitude is with AM modulation, with modulation index ρ and the carrier undergoing phase or frequency modulation. On the other hand, if $\rho \gg 1$, then

$$\begin{aligned} A\,(t) &\approx A_c \cdot [\rho^2 + 2\rho cos(\omega_i t + \phi_i)]^{1/2} \\ &\approx A_c\rho\,[1 + \rho^{-1} cos(\omega_i t + \varphi_i)] = A_i\,[1 + \rho^{-1} cos(\omega_i t + \varphi_i)] \end{aligned}$$

$$\theta\,(t) \approx \omega_i t + \phi_i$$

So,

$$r(t) = A_i\,(1 + \rho^{-1}\,cos(\omega_i t + \varphi_i)) \cdot cos(\omega_c t + \omega_i t + \varphi_i) \tag{4.97}$$

Thus, the envelope looks like AM and phase of the carrier is shifted by ω_i in addition to ϕ_i.

Solved Numericals

1. In an FM system, the message signal is given by $s(t) = 2\cos(2\pi 500t)$ volts and the frequency deviation is $\Delta f = 5$ kHz. If the carrier is given by $v_c(t) = 5\cos(2\pi \times 10^6 t)$ volts, determine the following.

   (a) Index of modulation, and

   (b) The FM Waveform

   Repeat the problem if the amplitude of the modulating signal is increased to 3 volts.

   Solution

   By Eq. (4.16), we know the FM signal as

   $$s_{\text{FM}}(t) = A_c\text{cos}\,[2\pi\, f_c t + \beta \sin(2\pi f_m t)]$$

   where modulation index $\beta = \dfrac{\Delta f}{f_m}$ and $\Delta f = k_f A_m$ where k_f is a constant for FM.

   Here, $A_m = 2$ volts, $f_m = 500$ Hz, $f_c = 10^6$ Hz, $A_c = 5$ volts; $\Delta f = 5$ kHz

   (a) $\beta = \dfrac{\Delta f}{f_m} = \dfrac{5\,\text{kHZ}}{500\,\text{HZ}} = 10$

(b) $s_{FM}(t) = 5\cos(2\pi \cdot 10^6 t + 10\sin 2\pi \cdot 500t)$volts

Since $k_f = \dfrac{\Delta f}{A_m}$, we have $k_f = \dfrac{5 \times 10^3\text{Hz}}{2\text{volts}} = 2.5 \times 10^3\ \text{V}^{-1}\text{S}^{-1}$

Further, if amplitude A_m of the modulating signal is increased to 3 volts, $\Delta f = k_f A_m = 2.5 \times 10^3 \times 3 = 7.5$ kHz and so

$$\beta = \frac{7.5 \times 10^3}{500} = 15 \text{ and } s_{FM}(t) = 5\cos\left(2\pi 10^6 t + 15\sin(2\pi \cdot 500t)\right) \text{ Volts.}$$

2. In a PM system, the message signal is given by $s(t) = 2\cos(2\pi 500t)$ volts and the frequency deviation is $k_p A_m = 5$ kHz. If the carrier is given by $v_c(t) = 5\cos(2\pi \times 10^6 t)$ volts, answer the following.

(a) Index of modulation, and

(b) The PM waveform

Repeat the problem if the amplitude of the modulating signals is increased to 3 volts.

Solution

By Eq. (4.9), we know the PM signal as

$$s_{PM}(t) = A_c\cos(2\pi f_c t + k_p A_m \cos\omega_m t)$$

where modulation index k is $(k_p A_m)$, and k_p is a constant for PM.

Here $A_c = 5$ volts, $A_m = 2$ volts, $f_c = 10^6$ Hz

(a) The frequency deviation can be obtained from Eq. (4.10).

$$\Delta f = \frac{1}{2\pi} k_p \frac{d}{dt}[s(t)]\Big|_{\max} = \frac{1}{2\pi} k_p \frac{d}{dt}(A_m \cos 2\pi f_m t)\Big|_{\max}$$

$$= \frac{1}{2\pi} k_p \cdot (-A_m . 2\pi f_m) = k_p f_m . A_m \text{ (magnitude only)}$$

The constant k_p is obtained as

$$k_p = \frac{\Delta f}{f_m A_m} = \frac{5 \times 10^3\text{Hz}}{500 \times 2\,\text{Volt}\,Hz} = 5\text{V}^{-1}$$

and modulation index $k_p A_m = 5 \times 2 = 10$

(b) The PM waveform is given as

$s_{\text{PM}}(t) = 5\cos(2\pi \cdot 10^6 t + 10\,\cos 2\pi 500t)$ volts when the amplitude of the modulating signal is increased to 3 volts, then $k_p A_m = 5 \times 3 = 15 =$ modulation index.

The waveform is then

$$s_{\mathrm{PM}}(t) = 5\cos(2\pi \cdot 10^6 t + 15\cos 2\pi 500t) \text{ volts}$$

3. An angle modulated signal is described by

$$s_{FM}(t) = 5\cos(2\pi \times 10^6 t + 2\sin 2\pi 500t + 3\sin 2\pi 1000t) \text{ volts.}$$

Find

(a) Power dissipation

(b) Frequency and phase deviation

(c) Modulation index

(d) Bandwidth of the signal

Solution

(a) Since the carrier amplitude is $A_C = 5$ volts, normalised power dissipation will be

$$p = \frac{A_C{}^2}{2} = \frac{5^2}{2} = 12.5 \text{ watt.}$$

(b) To find frequency deviation, the instantaneous frequency is given by

$$f(t) = \frac{1}{2\pi}\left[2\pi \cdot 10^6 + 4\pi \times 500\cos(2\pi 500t) + 6\pi \times 1000\cos(2\pi 1000t)\right]$$

$$= \left[10^6 + 1000\cos(1000\pi t) + 3000\cos(2000\pi t)\right]$$

So, deviation from carrier frequency is

$$1000\cos(1000\pi t) + 3000\cos(2000\pi t)$$

Since these two sinusoids will add in phase at some time, the maximum frequency deviation would be

$$\Delta f = 1000 + 3000 = 4000 \text{ Hz}$$

The phase deviation can be calculated as follows

$$s_{FM}(t) = 5\cos(\omega_c t + \theta(t))$$

where the phase term $\theta(t) = 2\sin(2\pi 500t) + 3\sin(2\pi 1000t)$

Maximum value of this phase term is $2 + 3 = 5$ radian and this is the phase deviation $\Delta\theta$.

(c) The modulation index is

$$\beta = \frac{\Delta f}{f_h} = \frac{4000}{1000} = 4$$

(we have taken the highest modulating frequency)

(d) By Carson's formula (Eq. 4.26), signal bandwidth is $B = 2f_m(1+\beta) = 2\times 1000(1+4) =$ 10 kHz.

4. With reference to the FET reactance modulator of Fig.4.17, the effective capacitance of the FET is n-times larger than the gate-to-source impedance. Find the expression for the effective capacitance C_m. If the FET transconductance varies from 300 to 800 micro Siemen over the linear portion of the $g_m v_{gs}$ curve, find the minimum and maximum effective capacitance of the modulator. Assume that resistance R is one-tenth of the capacitive reactance and the resonant frequency is 100 MHz

Solution

By Eq. (4.67), we may write

$$X_c = \frac{1}{wC} = nR$$

Or, $\dfrac{1}{2\pi f_0 C} = nR$, thus $C = \dfrac{1}{2\pi f_0 nR}$.

Since $C_m = g_m RC$, we have controlled capacitance

$$C_m = \frac{g_m}{2\pi f_0 n}.$$

So, minimum effective capacitance

$$C_{m,\min} = \frac{300 \times 10^{-6}}{2\pi \times 100 \times 10^6 \times 10} = \frac{3}{20\pi} \times 10^{-12} = 0.0477\,\text{pF}.$$

Maximum effective capacitance

$$C_{m,\max} = \frac{800 \times 10^{-6}}{2\pi \times 100 \times 10^6 \times 10} = 0.1273\,\text{pF}.$$

5. If the FET reactance modulator is to offer a maximum frequency deviation of 75 kHz for the carrier frequency of $f_0 = 100$ MHz, find the fixed capacitance C and the inductance L of the tank circuit across which the FET is connected. Assume data of Example 4.

Solution

At the minimum effective capacitance $C_{m,\min}$, the frequency of oscillation is

$$f_{\max} = \frac{1}{2\pi\sqrt{L(C + C_{m,\min})}}$$

and that at the maximum effective capacitance $C_{m,\max}$

$$f_{\min} = \frac{1}{2\pi\sqrt{L(C + C_{m,\max})}}$$

Thus,

$$\frac{f^2{}_{\max}}{f^2{}_{\min}} = \frac{C + C_{m,\max}}{C + C_{m,\min}}$$

Or,

$$\frac{(f_{\max} - f_{\min})(f_{\max} + f_{\min})}{f^2_{\min}} = \left(\frac{C_{m,\max} - C_{m,\min}}{C + C_{m,\min}}\right)$$

Or,

$$\frac{2(\Delta f) \cdot 2f_0}{f^2_{\min}} \approx \frac{C_{m,\max} - C_{m,\min}}{C + C_{m,\min}}$$

(since $f_{\max} = f_0 + \Delta f, f_{\min} = f_0 - \Delta f$ and $f_{\min} \approx f_0$, assumed)

Or,

$$C + C_{m,\min} = \frac{(C_{m,\max} - C_{m,\min})f_0}{4\Delta f}$$

Or,

$$C = \frac{(C_{m,\max} - C_{m,\min})f_0}{4\Delta f} - C_{m,\max}$$

Substituting the values,

$$C = \frac{(0.1273 - 0.0477) \times 10^{-12} \times 100 \times 10^6}{4 \times 75 \times 10^3} - 0.0477 \times 10^{-12}$$

$$= 26.533\text{pF} - 0.0477\text{pF} = 26.48\text{pF}$$

Also, $f = \dfrac{1}{2\pi\sqrt{\text{LC}}} \approx \dfrac{1}{2\pi\sqrt{L(C + C_{av})}}$

(we assume the average value of the controlled capacitor C_m over the range, which is $\dfrac{C_{m,\max} + C_{m,\max}}{2} = 0.0875\,\text{pF}$)

So,

$$f = \frac{1}{2\pi\sqrt{\text{LC}}}, as\ C \gg C_{av}$$

Thus,

$$L = \frac{1}{4\pi^2 f_0^2 C} = \frac{1}{4\pi^2 \times (100 \times 10^6)^2 \times 26.48 \times 10^{-12}} = 9.56\ \mu\text{H}.$$

Exercises

Activity

4.1 An FM signal carrier amplitude of 2.5 volts has an instantaneous frequency $\omega(t)$ given by $f(t) = 10^8 + 10^4 \cos(2\pi \times 10^3 t)$. Find the modulation index. Write the expression for the modulated waveform.

4.2 An angle modulated signal is described by $s(t) = 10\cos(2\pi \times 10^6 t + 0.01 \sin 2\pi \times 10^3 t)$.

(a) What is the instantaneous frequency $f(t)$ of $s(t)$?
(b) Estimate the bandwidth of $s(t)$.
(c) Find the modulating signal if $s(t)$ is an FM signal.
(d) Repeat (c) if $s(t)$ is a phase modulated signal.

4.3 An angle modulated signal is described by $v(t) = 50\cos(2\pi \times 10^8 t + 20\cos 2\pi \times 10^3 t)$ volts.

(a) Find the modulating signal if the phase deviation constant $k_p = 100$ rad/volt.
(b) Find the modulating voltage if the frequency deviation constant $k_f = 10^6$ rad/volt second.

4.4 A single-tone FM signal has the frequency deviation $\Delta f = 75$ kHz. Find the frequencies of the modulating signal at which the carrier will be absent, given that $J_0(\beta) = 0$ for $\beta = 2.405, 5.520, 8.654$.

Hint: From (4.22) we find that the carrier frequency is absent for $J_0(\beta) = 0$. Here $\beta = \frac{\Delta f}{f_m}$ or, $f_m = \frac{\Delta f}{\beta}$; $f_1 = \dfrac{75\text{kHz}}{2.405} =$; $f_2 = \dfrac{75\text{kHz}}{5.520} =$, etc.

4.5 In FM, the carrier is sinusoidal-modulated by a modulating signal of frequency 3 kHz. What is the bandwidth occupied by the modulated wave if the frequency deviation is 5 kHz? The amplitude of the modulating signal is increased by three times and its frequency is also lowered by the same factor. What will be the new bandwidth?

4.6 The NBFM signal is given by (vide (4.36)) $s(t) = A_c[\cos\omega_c t - km(t)\sin\omega_c t]$. Find an expression for the instantaneous frequency.

4.7 A single tone modulating signal is applied to a frequency modulator and a phase modulator separately. The two output spectra are found identical. Identify the spectra with the following tests: (a) amplitude of the modulating signal is increased/decreased. (b) frequency of the modulating signal is increased/decreased.

4.8 Discuss the scheme for detecting FM signal in an envelope detector. (Hint: FM-to-AM conversion by time derivative)

4.9 Let $s_{PM}(t) = A_c \cos[\omega_c t + k_p\, s(t)]$ be the phase-modulated signal. If $s(t)$ has a jump discontinuity, then the phase term $k_p\, s(t)$ must be restricted to the range $(-\pi, \pi)$; otherwise, reception will be ambiguous. Discuss this effect by taking $s(t)$ to be a digital signal.

4.10 The narrowband FM (NBFM) signal is defined by Eq. (4.29).
$s_{NBFM}(t) = A_c[\cos\omega_c t - k\, s(t)\sin\omega_c t]$.
Explain the generation scheme for the same. Estimate the bandwidth of the NBFM signal.

Review questions

1. (a) What is pre-emphasis? Why is it used? Draw a typical pre-emphasis circuit and explain why de-emphasis must also be used .

   (b) A 20 MHz carrier is modulated by 1 kHz audio sine wave. If the carrier voltage is 5 V and the maximum deviation is 12.5 kHz, write the equations of this modulated wave for FM and PM. If the modulating frequency is now changed to 3 kHz, the other remaining constant, write the new equations for FM and PM.

   (c) The block diagram shown below is a typical narrowband FM system where $\sin \omega_c t$ is the carrier and $m(t) = \beta \sin \omega_m t$ is the modulating signal. What will be the expression and type of the output signal? (KU 2000)

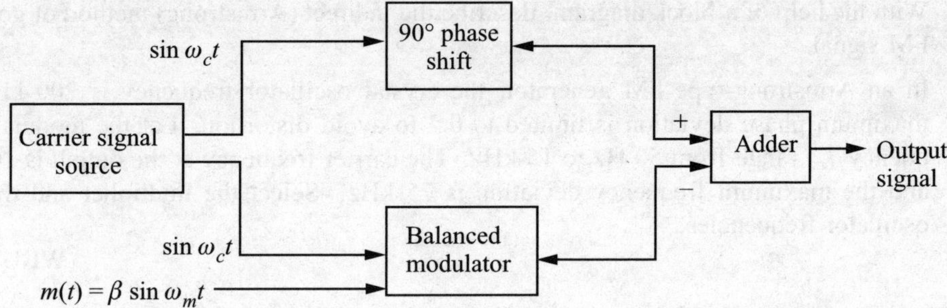

2. (a) Explain why modulation is needed for the purpose of electronic communications. (b) Derive the mathematical expression for a frequency and phase modulated waves. (c) What is the basic difference between frequency and phase modulation? (d) A 500 Hz modulating voltage fed into a PM generator produces a frequency deviation of 2.25 kHz. What is the modulation index? If the amplitude of the modulating voltage is kept constant but its frequency raised to 60 kHz, what is the new deviation? (KU 2001)

3. (a) Derive an expression for an FM signal.

   (b) Define modulation index for signal-tone FM signal. Under what conditions do FM signals become narrowband FM?

   (c) A 25 MHz carrier is modulated by a 400 Hz audio sine wave. If the carrier voltage is 4V and the maximum deviation is 10 kHz, write an expression for the modulated wave for (i) FM and (ii) PM. (KU 2001)

4. (a) What are the advantages of the frequency modulation over the amplitude modulation?

   (b) How are the frequency modulation and phase modulation related?

   (c) What is the advantage of wideband frequency-modulated signal over the narrowband frequency-modulated signal?

   (d) Obtain the peak frequency deviation (Δf) and modulation index for an FM modulator with a deviation sensitivity $k = 5$ kHz/V and a modulating signal $v_m(t) = 2\cos(2\pi \times 2000t)$. (WBUT 2003)

5. (a) Explain a method of FM generation. (b) Discuss exchange of bandwidth and SNR for a FM system.

6. (a) A tone-modulated signal is given by $x_c(t) = A\cos(\omega_c t + \beta \sin \omega_m t)$. When $\beta << 1$, we get narrowband (NB) angle modulation.
   (i) Find the spectrum of this NB angle-modulated signal.
   (ii) Compare the result with that of a tone-modulated AM signal.
   (iii) Discuss the similarities and differences by drawing their phasor representations.

   (b) Explain the differences between frequency modulation and phase modulation. Begin with the definitions of each type and the meaning of modulation index in each case.
   (WBUT 2004)

7. (a) With the help of a block diagram, describe the indirect (Armstrong) method of generating FM signal.

   (b) In an Armstrong-type FM generator, the crystal oscillator frequency is 200 kHz. The maximum phase deviation is limited to 0.2 to avoid distortion. Let the modulating frequency f_m range from 50 Hz to 15 kHz. The carrier frequency at the output is 108 MHz and the maximum frequency deviation is 75 kHz. Select the multiplier and the mixer oscillator frequencies.
   (WBUT 2004)

8. Obtain an expression for the spectrum of an angle-modulated signal. What is the difference between a standard AM signal and narrowband FM signal? An FM signal has sinusoidal modulation with frequency $f_m = 10$ kHz and modulation index $\beta = 3.0$. Find the transmission bandwidth of the FM signal. What is the percentage of the total FM signal? What is the percentage of the total FM signal power that lies within the Carson's rule bandwidth?
   (RPE 2004)

9. (a) Write the advantages and disadvantages of amplitude modulation over frequency modulation.

   (b) A carrier of frequency 50 kHz and amplitude 5 volts is frequency modulated by a sinusoidal modulating wave of frequency 1 kHz and amplitude 10 volts with $k_f = 10\,\text{Hz/V}$. Write the expression for the FM wave. Find the approximate band of frequencies occupied by FM.

   (c) Find the bandwidth of narrowband FM. Also draw the frequency spectra for the same.
   (MUST – Spring 2010)

10. What may be the possible reasons for FM reception to have higher received signal power than a comparable AM reception?

Multiple-choice questions

1. When the transmission bandwidth is doubled in FM, then the SNR is
   (a) doubled (b) improved four-fold (c) decreased by four-times (d) unaffected

2. An FM signal with a frequency deviation δ is passed through a mixer. The output of the mixer will have the frequency deviation

   (a) 5δ (b) $\delta/5$ (c) δ (d) indeterminate

3. In FM broadcasting, the maximum frequency deviation is usually

   (a) 10 kHz (b) 75 kHz (c) 200 kHz (d) 1.0 MHz

4. Varactor diode can be used for the generation of

   (a) FM (b) AM (c) PM (d) both AM and FM

5. Pre-emphasis in FM systems is used for

   (a) amplifying high frequency components of the modulating signal
   (b) amplifying low frequency components of the modulating signal
   (c) compressing the modulating signal
   (d) suppressing noise components

6. If the index of modulation of an FM signal is increased, the power in carrier component

   (a) increases (b) decreases (c) remains the same (d) nothing can be said

7. Two carrier components of frequencies 40 MHz and 80 MHz are frequency-modulated by a modulating frequency of 4 kHz in such a way that the bandwidths of the FM signals are the same. The peak frequency deviation in the two cases are

   (a) 1:1 (b) 1:2 (c) 1:4 (d) 2:1

8. An FM signal with index of modulation β is passed through a frequency-doubler network. The modulation index of the output signal is

   (a) β (b) 2β (c) 3β (d) 4β.

9. In FM, the modulation index β is related to the modulating frequency f_m as

   (a) $\beta \propto f_m$ (b) $\beta \propto (1/f_m)$ (c) $\beta \propto f_m^2$ (d) $\beta \propto (1/f_m^2)$

10. In FM broadcast, the maximum modulation frequency is restricted to

    (a) 5 kHz (b) 10 kHz (c) 15 kHz (d) 20 kHz

11. A 2.5 V, 500 Hz signal frequency modulates a carrier to cause a frequency deviation of 5 kHz. The modulation index is

    (a) 5 (b) 10 (c) 25 (d) 50

12 If the depth of modulation is doubled in FM, the output power

   (a) increases by a factor of $\sqrt{2}$ (b) increases by a factor of $\sqrt{3}$
   (c) doubles (d) remains at unmodulated value

13. One drawback of FM over AM is that

    (a) noise is very high for high modulation frequencies
    (b) larger bandwidth is required
    (c) higher modulating power is required
    (d) higher output power is required

14. Practical bandwidth of a very wideband FM approximately equals

    (a) f_m (b) $2f_m$ (c) f_d (d) $2f_d$.

15. The power distribution in FM sidebands is given by the Bessel function components $J_n(\beta)$. The number of significant sidebands ($\simeq 98\%$ or more power content) always occurs for

    (a) $n = \beta$ (b) $n > \beta$ (c) $n > \beta + 1$ (d) none

5 Theory of Probability and Random Process

5.1 Introduction

In radio communications systems, signal transmission is disturbed by the presence of undesired random signals called noise, generated by the random motion of the electrons in various physical structures. Being random in nature, noise is difficult to describe in exact mathematical terms. Such disturbing waves do not convey any information. The signal transmitted for any specific application is contaminated by noise from various sources. The receiver has, at its input the desired signal plus the undesired interfering noise components. This composite signal is again random in nature. The real challenge of a communications system is to separate the desired signal from the noise. It is of course possible to evaluate the quality of a radio link by means of probabilistic solutions. In this chapter we stress on the basic concepts of the theory of probability and apply it to describe random processes in order to manage the noise problems in the system.

5.2 Concept of Probability

A distinction of a random signal is that it is unpredictable; we cannot evaluate its instantaneous values in advance. However, we may define some of its properties with sufficient accuracy.

Before we develop a mathematical model for random signals, we must define some important terms.

1. *Experiment*: An experiment is a set of rules that govern an operation.
2. *Outcome*: The outcome is the result of the experiment being performed one at a time.
3. *Event*: Events are sets of possible outcomes of the experiment meeting some specifications.

An experiment may have several distinct outcomes. For example, consider the experiment of rolling a single dice. There are six possible outcomes namely, any one of the six faces of the dice may turn face upward. Rolling a single dice is an example of a 'random experiment' since its outcome cannot be predicted precisely. In the experiment under consideration, let the outcome corresponds to number 3 turning up. This is called an event, or more precisely, an elementary event since it is the same as any one of the possible six outcomes. If we specify event B as an even number turning up, it may result from a combination of three possible outcomes: number two, number four, and number six appear face-up.

We define the 'sample space' S as a collection of all the possible outcomes of an experiment; each outcome is an 'element' or 'sample point'. In the said example, the sample space consists of six elements $\{s_1, s_2, \cdots, s_6\}$, where s_1 is the sample point corresponding to the outcome 'number 1 turning face up' and so on. Figure 5.1 shows the sample space for the experiment of rolling a dice. The event $A = \{s_3\}$ and event $B = \{s_2, s_4, s_6\}$ are marked on the space. The complement event A^c of event A contains all sample points not included in A. In the example, $A^c = \{s_1, s_2, s_4, s_5, s_6\}$ and $B^c = \{s_1, s_3, s_5\}$.

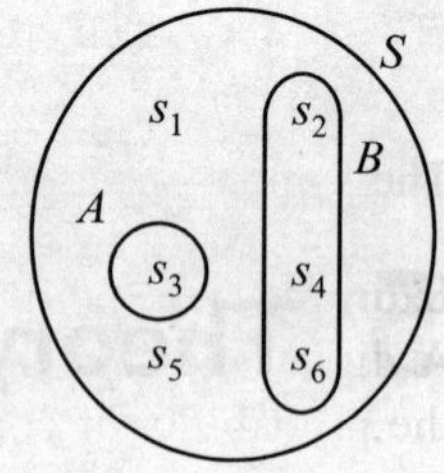

Fig. 5.1 *Sample space for the rolling of a dice*

We now try to explain the meaning of the probability of an event. To each event $A \in S$, a real number $P(A)$ may be assigned which we call the probability of that event. Suppose that a particular experiment is repeated N times and a particular event A occurs N_A times. The relative frequency of the event A is N_A/N which gives the fraction of times that the event A occurs. If the experiment is repeated very many times (N is large enough), this fraction approaches a fixed limit. This limiting value is called the probability $P(A)$ of the event A. Thus,

$$P(A) = \lim_{N \to \infty} \frac{N_A}{N} \tag{5.1}$$

The probability of an event satisfies the following rules:

1. The probability is non-negative and does not exceed unity : $0 \leq p(A) \leq 1$.

   If the event is impossible, $P(A) = 0$. Number 7 turning face-up, for example, is an impossible event. For an event being certain, $P = 1$.

2. The combination of all events of the sample space is a certain event:

   $$P(A_1) + P(A_2) + \cdots = \sum_{i=1}^{n} P(A_i) = 1.$$

3. Non-occurrence of an event A has the probability $P(A^c)$ given by

   $$P(A^c) = 1 - P(A).$$

Example 5.1

In the typical experiment of 'tossing a coin twice', list the outcomes. Sketch the sample space to show the events A= tail on the first toss and B= heads on both tosses.

Solution

Tossing a coin twice yields $2 \times 2 = 4$ distinct sequence. Denoting head and tail by the letters H and T, respectively, the sample space S is given by

$$S = \{HH, HT, TH, TT\}$$

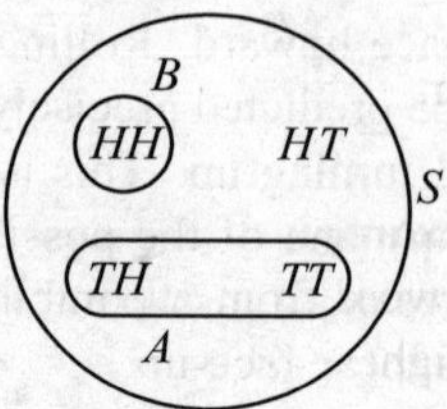

Event A includes the outcomes (TH, TT) and event B corresponds to the element (HH). Obviously,

$$P(HH) = P(HT) = P(TH) = P(TT) = \tfrac{1}{4}.$$

But $P(A) = P(TH, TT) = \frac{1}{4} + \frac{1}{4} = \frac{1}{2}$.

The sample space of the experiment is shown in the adjoining figure.

Example 5.2

A digital message is 100 symbols long. An average of 0.5% symbols are received in error. What is the probability of correct reception? How may symbols are correctly received?

Solution

The event A=erroneous reception.

$$P(A) = 0.5\% = \frac{5}{10 \times 100} = 5 \times 10^{-3}.$$

So, the probability of correct reception $P(A^c) = 1 - P(A) = 99.5\%$.

The number of symbols received correctly is $0.995 \times 1000 = 995$.

Suppose that we consider two different events A_1 and A_2 which are mutually exclusive or disjoint in the sense that the occurrence of one event does not depend on the other. The outcomes of A_1 and A_2 belong to the same sample space S. In a very large number N of repetitions of the experiment, let outcome A_1 occur N_1 times and let outcome A_2 occur N_2 times. The events A_1 and A_2 have their probabilities

$$P(A_1) = \lim_{N\to\infty} \frac{N_1}{N} \quad \text{and} \quad P(A_2) = \lim_{N\to\infty} \frac{N_2}{N}.$$

The probability of 'event A or event B' is

$$\begin{aligned} P(A_1 \text{ or } A_2) &= \frac{N_1 + N_2}{N} = \frac{N_1}{N} + \frac{N_2}{N}. \\ &= P(A_1) + P(A_2). \end{aligned} \tag{5.2}$$

The result may be extended to more than two mutually exclusive events, as

$$P(A_1 \text{ or } A_2 \text{ or } \cdots A_M) = \sum_{i=1}^{M} P(A_i). \tag{5.3}$$

If it happens that there are M possible outcomes each being equally likely, then,

$$MP(A_i) = 1$$

$$\text{or,} \quad P(A_i) = \frac{1}{M}. \tag{5.4}$$

Example 5.3

Two dice are thrown simultaneously. Find the probability that the sum of the numbers on the dice is eight.

Solution

The sample space consists of $6 \times 6 = 36$ sample points and all outcomes are equally likely. So the probability of each outcome is $1/36$. The 36 outcomes are: $(1,1),(1,2),\cdots,(1,6);(2,1),(2,2),\cdots$ $(2,6);(3,1),(3,2),\cdots,(3,6);(4,1),(4,2),\cdots,(4,6);(5,1),(5,2),\cdots,(5,6);$ $(6,1),(6,2),\cdots,(6,6)$. Out of these, a sum of 8 can be obtained by the following combinations: $(2,6),(3,5),(4,4),(5,3),(6,2)$, a total 5 in number.

Thus, $P(\text{sum of eight}) = 5 \times \dfrac{1}{36} = \dfrac{5}{36}$.

5.2.1 Conditional probability

There are many situations where the probability of one event is effected by the outcome of another event. As an example, consider the drawing of two balls in succession from six balls in a box, three of them white and three black. Let A denote the event that the first drawn ball is white. Then $P(A) = 1/2$. If we again replace the ball in the box and draw it again, then we are repeating the same experiment. Suppose we do not replace the ball drawn in the first experiment. If we draw a second ball now, we are performing a new experiment. Let B denote the event that the second ball drawn is again white. If the ball in the first trial was white, $P(B) = \frac{2}{5}$. If the first draw is not a white ball, the $P(B) = \frac{3}{5}$. We now thus arrive at the situation where the probability of event B depends on whether or not event A occurs. In other words, the outcome of the second event is conditional on the outcome of the first event. The probability of event B given that event A has occurred is called the conditional probability and is denoted as $P(B/A)$.

Suppose we perform an experiment N times in which n_1 trials favour event A. Of these n_1 trials, event B occurs n_2 times. We thus see that the pair of events A and B occur jointly n_2 times. Thus, the probability of joint events A and B is given by

$$\begin{aligned} P(A \text{ and } B) = P(AB) &= \lim_{N\to\infty} \frac{n_2}{N} \\ &= \lim_{N\to\infty} \left(\frac{n_1}{N}\right)\left(\frac{n_2}{n_1}\right) \\ &= \lim_{N\to\infty} \left(\frac{n_1}{N}\right) \lim_{n_1\to\infty} \left(\frac{n_2}{n_1}\right). \end{aligned} \tag{5.5}$$

On the right-hand side of (5.5), $\lim_{N\to\infty}(n_1/N) = P(A)$ and $\lim_{n_1\to\infty}(n_2/n_1) = P(B/A)$. Thus,

$$P(AB) = P(A)P(B/A)$$

or $$P(B/A) = \frac{P(AB)}{P(A)}, \qquad P(A) \neq 0 \tag{5.6a}$$

Similarly,

$$P(A/B) = \frac{P(AB)}{P(B)}, \quad P(B) \neq 0. \tag{5.6b}$$

The above results are called Bayes' theorem.

It follows from (5.6a) and (5.6b) that

$$P(A/B) = \frac{P(A)P(B/A)}{P(B)} \tag{5.6c}$$

$$\text{and } P(B/A) = \frac{P(B)P(A/B)}{P(A)} \tag{5.6d}$$

The events A and B are independent if the probability of A does not depend on the probability of B. In other words, event B is unconditional on the outcome of the event A.

Then, $P(B/A) = P(B)$ and $P(A/B) = P(A)$. That is, if event B is independent of event A, its probability is not affected by event A and vice versa.

Equation (5.6) yields

$$P(AB) = P(A)P(B). \tag{5.7}$$

The result can also be extended to any number of outcomes. Thus,

$$P(ABC \cdots) = P(A)P(B)P(C) \ldots \tag{5.8}$$

Note that when A and B are mutually exclusive, then event AB is empty and $P(AB) = 0$ (null). This means A and B cannot occur at the same time. So mutually exclusive and independent events are two different concepts.

Example 5.4

A coin is tossed twice. We define the events A and B as

Event A : head on the first toss; Event B : matching toss.

Find $P(A), P(B), P(AB)$, and $P(A/B)$.

Solution

The sample space consists of the elements $S = \{HH, HT, TH, TT\}$.

$A = \{HH, HT\}$ and $B = \{HH, TT\}$ as shown in the adjoining diagram.

So, $P(A) = P(B) = \frac{2}{4} = \frac{1}{2}$.

$$P(AB) = P\{HH\} = \tfrac{1}{4}.$$

So, $P(A/B) = \dfrac{P(AB)}{P(A)} = \dfrac{\frac{1}{4}}{\frac{1}{2}} = \dfrac{1}{2}.$

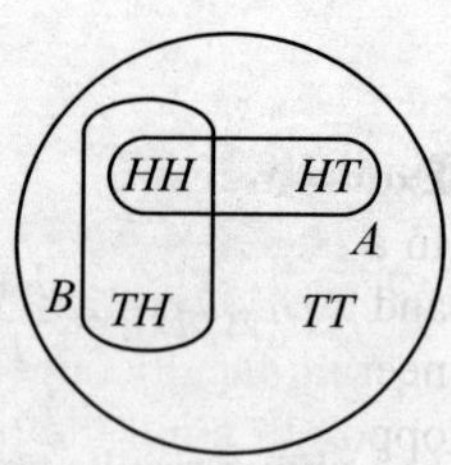

5.2.2 Total probability

The probability of an event A defined on a sample space S can be expressed in terms of conditional probabilities. Let us first define that there are n disjoint events $B_1, B_2, \ldots, B_n$ whose union equals the sample space S, as shown in the adjoining figure. That is,

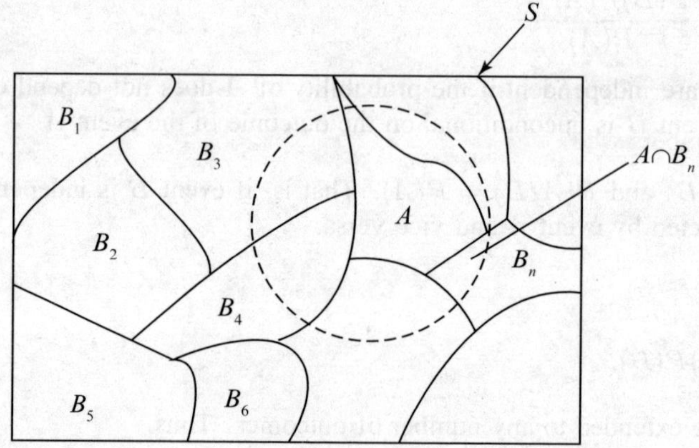

Figure: n-disjoint events $B_1, B_2, \ldots, B_n$ defined on the sample space S. The dotted curve specifies event A.

$B_1 \cup B_2 \cup B_3 \cdots \cup B_n = S$ and $B_i \cap B_j = \phi$ if $i \neq j$, where ϕ is a null set. Since B_i partitions the sample space S, we may write

$$\begin{aligned} A &= A \cap S = A \cap (B_1 \cup B_2 \cup \cdots \cup B_n) \\ &= (A \cap B_1) \cup (A \cap B_2) \cup \cdots \cup (A \cap B_n) \end{aligned}$$

Using the symbolic notation $A \cup B = A + B$ and $A \cap B = AB$, we may write

$$A = AB_1 + AB_2 + \cdots + AB_n$$

So, probability of event A is

$$P(A) = \sum_{i=1}^{n} P(B_i A) = \sum_{i=1}^{n} P(A/B_i) P(B_i)$$

So, the total probability is the sum of the conditional probability multiplied by the probability of the condition. This is the total probability theorem.

Example 5.4(a)

In a university, 50% of faculty members are from engineering, 40% are from Science-Commerce-Arts and 10% are from other divisions. They were asked about introducing college uniforms, 60% of engineering faculty opposed, 45% of Science-Commerce-Arts faculty opposed and 50% of other divisions opposed. Find the probability that a randomly selected faculty member opposes the introduction of uniforms.

Solution

Let $S =$ Total faculty members

$E =$ All engineering faculty members

$A =$ All Arts-Science-Commerce faculty members

$L =$ All law faculty members

and $N =$ Faculty members who opposed the decision

Then, $P(E) = 0.5$

$P(A) = 0.4$

$P(L) = 0.1$

Also, $P(N/E) = 0.6$

$P(N/A) = 0.45$

$P(N/L) = 0.5$

By total probability theorem,

$$P(N) = P(N/E) \cdot P(E) + P(N/A)P(A) + P(N/L)P(L)$$

$$= (0.6 \times 0.5) + (0.45 \times 0.4) + (0.5 \times 0.1)$$

$$= 0.3 + 0.18 + 0.05 = 0.485$$

5.3 Random Variable and Its Characteristics

We have observed that the outcome of an experiment may be real numbers (2, 3, etc. as in the case of rolling a dice) or it may be some non-numerical symbols (heads or tails as in the case of tossing a coin). For meaningful mathematical analysis, it is preferable to work with real numbers for each possible outcomes. Random variable is a rule or a mapping that assigns a number to each outcome. We could assign the number 0 with 'tail' and 1 with 'head' in the experiment of tossing a coin. Suppose that there are M outcomes $\{s_1, s_2, \cdots, s_m\}$. Let us assign real numbers $X(s_i)$ to each outcome. The symbol X is called the random variable (RV). Figure 5.2 depicts the mapping of the sample points of the dice-rolling experiment into real numbers on the real axis.

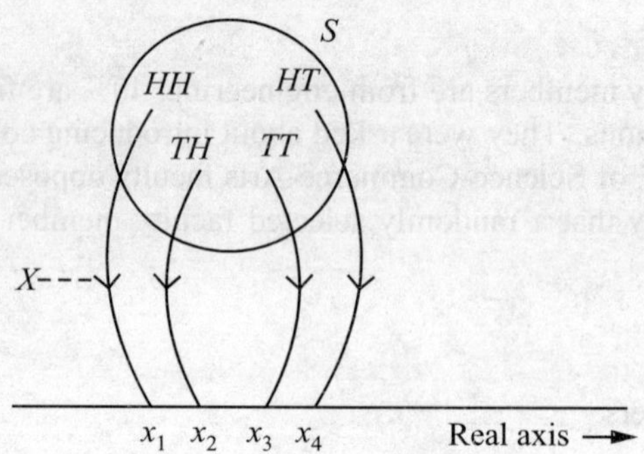

Fig. 5.2 *The mapping of sample points onto real numbers*

The numbers $x_1, x_2, x_3,$ and x_4 correspond to the outcomes HH, TH, TT and HT, respectively.

A random variable may be discrete or continuous. If the random variable X assumes only certain discrete values in any finite interval of x, then the RV is discrete. An example of a discrete RV is the sample points of the dice-rolling experiment in which $X = \{1, 2, \cdots, 6\}$ correspond to numbers one, two, etc. as the outcomes. In contrast, X is a continuous RV if it can take on all possible real numbers x in some specified interval. An experiment that measures the noise voltage across the terminals of some electronic device is an example of continuous random variable.

Example 5.5

Two regular dice are thrown. Suppose that the random variable $X(s_{ij})$ is the sum of the dots appearing on the dice, that is, $x = i + j$. Find the values taken by X and their probabilities.

Solution

The possible 36 outcomes are listed in Table 5.1.

Table 5.1 *Random variables and their probabilities of Example 5.5.*

RV X	Outcomes	$P_X(x)$
2	(1,1)	1/36
3	(1,2), (2,1)	2/36
4	(1,3), (2,2), (3,1)	3/36
5	(1,4), (2,3), (3,2), (4,1)	4/36
6	(1,5), (2,4), (3,3), (4,2), (5,1)	5/36
7	(1,6), (2,5), (3,4), (4,3), (5,2), (6,1)	6/36
8	(2,6), (3,5), (4,4), (5,3), (6,2)	5/36
9	(3,6), (4,5), (5,4), (6,3)	4/36
10	(4,6), (5,5), (6,4)	3/36
11	(5,6), (6,5)	2/36
12	(6,6)	1/36

5.3.1 Cumulative distribution function

The statistical behaviour of the random variable X can be described if we define a non-random function $F(x)$ which equals the probability that X can take any allowed values less than or equal to x on any given trial of the experiment. That is,

$$F(x) = P(X \leq x), \qquad -\infty < x < \infty. \tag{5.9}$$

The function $F(x)$ is called the cumulative distribution function (CDF) of the random variable X. For the continuous RV X, $F(x)$ is a smooth function. The CDF has the following properties.

$$0 \leq F(x) \leq 1 \tag{5.10a}$$

$$F(-\infty) = 0; F(\infty) = 1 \tag{5.10b}$$

$$F(x_1) \leq F(x_2) \text{ if } x_1 < x_2 \tag{5.10c}$$

The property in (5.10a) stems from the fact that $F(x)$ is after all a probability function. Since $F(-\infty) = P(X \leq -\infty)$ and $F(\infty) = P(X \leq \infty)$, the function $F(-\infty)$ includes no event at all and $F(\infty)$ includes all possible events, thereby proving property (5.10b). The property (5.10c) states that the CDF is a non-diminishing function with increasing argument. This is obvious since an increase in x can only add outcomes to the event.

In Example 5.5, we have defined a RV X as the sum $x = i + j$, where i and j denote, respectively, the number of dots appearing on the first and second dice. We now calculate the CDF $F(x)$ for $2 \leq x \leq 12$. From definition (vide (5.9)),

$$F(0) = F(1) = 0 \qquad \text{since it includes no possible events.}$$

$$F(2) = P(X \leq 2) = P(1) + P(2) = 0 + \frac{1}{36} = \frac{1}{36}.$$

$$F(3) = P(X \leq 3) = P(0) + P(1) + P(2) + P(3) = F(2) + P(3) = \frac{1}{36} + \frac{2}{36} = \frac{3}{36}.$$

In a similar way, $F(x)$ for other values of x can be calculated. Figure 5.3 shows the probability and CDF associated with the rolling of two dice.

5.3.2 Probability density function

The derivative of the distribution function

$$f(x) = \frac{dF}{dx} \tag{5.11}$$

is called the probability density function (PDF) of the random variable. It is obvious that

$$F(x_0) = \int_{-\infty}^{x_0} f(x)dx, \tag{5.12}$$

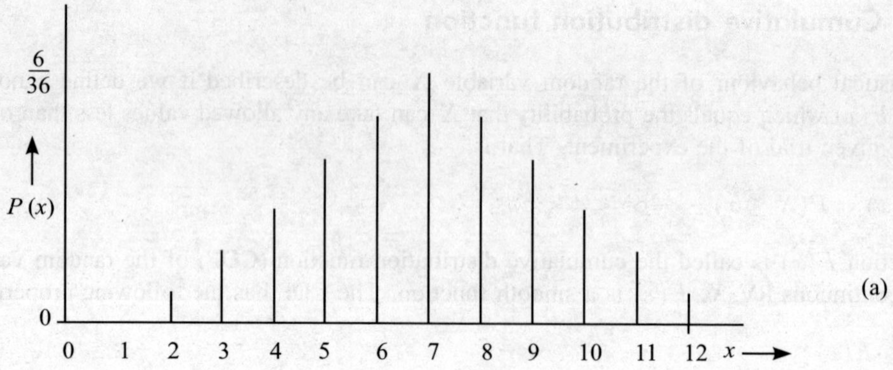

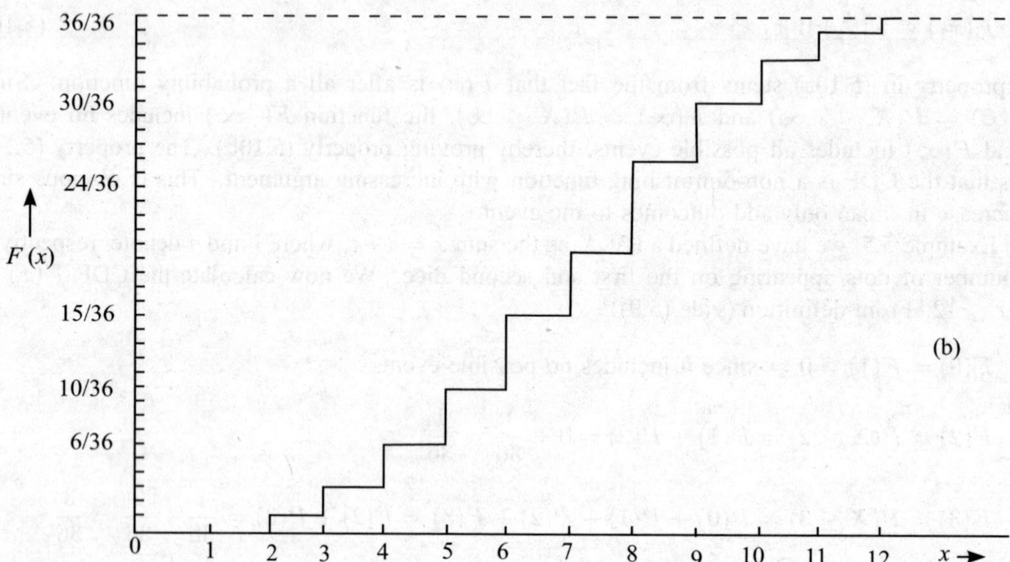

Fig. 5.3 *CDF and $P(x)$ of Example 5.5: (a) probability and (b) CDF of Example 5.5.*

$$\text{and } \int_{-\infty}^{\infty} f(x)dx = F(\infty) - F(-\infty) = 1 - 0 = 1. \tag{5.13}$$

Equation (5.13) is the condition of normality. Also,

$$f(x)dx = p(x < X \leq x + dx), \tag{5.14}$$

that is, the term $f(x)dx$ gives the probability of the random variable X falling within the interval $(x, x + dx)$.

For a continuous RV X, the PDF $f(x)$ is a smooth function. For a discrete RV, the CDF is a staircase (shown in Fig. 5.3) whose derivative is an impulse of strength equal to the amplitude of the step. Figure 5.4 displays the PDF $f(x)$ corresponding to the CDF of Fig. 5.3.

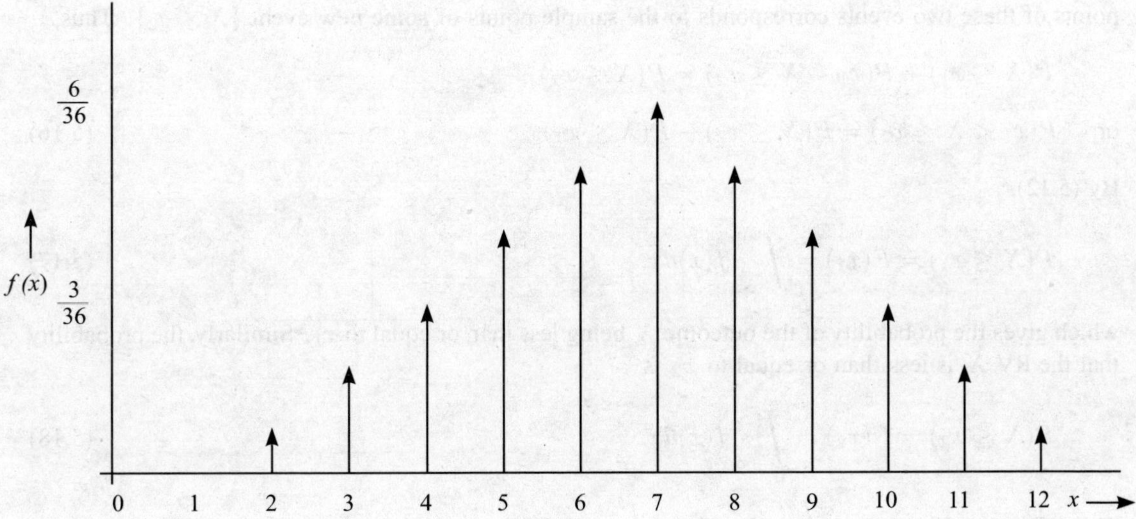

Fig. 5.4 *Probability density function $f(x) = dF/dx$, where $F(x)$ is given in Fig. 5.3.*

The shape of $f(x)$ of Fig. 5.4 is obvious since for discrete random variables $\{x_1, x_2, \cdots\}$ with probabilities $\{P_1, P_2, \cdots\}$,

$$f(x) = \sum_i P_i \delta(x - x_i). \tag{5.15}$$

Example 5.6

Explain the significance of the area $\int_{x_1}^{x_2} f(x)dx$ under the probability density function between two limits x_1 and x_2, as shown in Fig. 5.5.

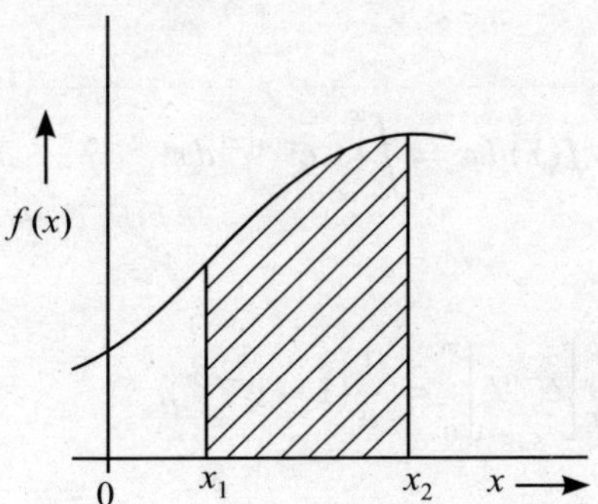

Fig. 5.5 *Illustrating the significance of the area*

Solution

The random variable $\{x_1 < X \le x_2\}$ defines an event. Consider two events defined by the RV $\{X \le x_1\}$ and $\{x_1 < X \le x_2\}$. These two events are mutually exclusive. The union of the sample points of these two events corresponds to the sample points of some new event $\{X \le x_2\}$. Thus,

$$P(X \le x_1) + P(x_1 < X \le x_2) = P(X \le x_2)$$

or $$P(x_1 < X \le x_2) = P(X \le x_2) - P(X \le x_1). \tag{5.16}$$

By (5.12),

$$P(X \le x_1) = F(x_1) = \int_{-\infty}^{x_1} f(x)dx, \tag{5.17}$$

which gives the probability of the outcome X being less than or equal to x_1. Similarly, the probability that the RV X is less than or equal to x_2 is

$$P(X \le x_2) = F(x_2) = \int_{-\infty}^{x_2} f(x)dx. \tag{5.18}$$

Thus, $$P(x_1 < X \le x_2) = \int_{-\infty}^{x_2} f(x)dx - \int_{-\infty}^{x_1} f(x)dx$$

$$= \int_{x_1}^{x_2} f(x)dx \tag{5.19}$$

Thus the area under the density function between two limits is the probability of X being between these two limits.

Example 5.7

The PDF of a random variable X is given by $f(x) = e^{-a|x|}$, where $a > 0$. Find (a) the CDF, (b) the number 'a' and (c) the probability that X lies between -1 and 2.

Solution

(a) The CDF is given by

$$F(x) = p(X \le x) = \int_{-\infty}^{x} f(x)dx = \int_{-\infty}^{x} e^{-a|x|}dx.$$

For $x \ge 0, |x| = x$ and so

$$F(x) = \int_0^x e^{-ax}dx = \frac{1}{-a}\left[e^{-ax}\right]_0^x = \frac{1}{a}(1 - e^{-ax}).$$

For $x < 0, |x| = -x$ and so

$$F(x) = \int_{-\infty}^{x} e^{ax}dx = \frac{1}{a}\left[e^{ax}\right]_{-\infty}^{x} = \frac{1}{a}e^{ax}.$$

(b) By the condition of normality (equation (5.13)),

$$\int_{-\infty}^{\infty} f(x)dx = \frac{1}{a}\left[e^{ax}\right]_{-\infty}^{0} + \frac{1}{a}\left[1 - e^{-ax}\right]_{0}^{\infty}$$

$$= \frac{1}{a}(1-0) + \frac{1}{a}[(1-0)-(1-1)] = \frac{2}{a} = 1.$$

So, $a = 2$.

(c) $P(-1 \le X \le 2) = P(X \le 2) - P(X \le -1) = F(2) - F(-1)$

$$= \frac{1}{a}(1 - e^{-2a}) - \frac{1}{a}e^{-a} = \frac{1}{a}(1 - e^{-2a} - e^{-a})$$

$$= \frac{1}{2}(1 - e^{-4} - e^{-2}).$$

Example 5.8

The CDF of a random variable X is given by

$$F(x) = \begin{cases} 0 & x \le 0 \\ ax^2 & 0 < x \le 5 \\ 25a & x > 5 \end{cases}$$

Evaluate the number 'a'. Also, find $P(X \le 2)$ and plot the corresponding PDF $f(x)$.

Solution

The function $F(x)$ is shown in the Fig. 5.6(a).

Since $f(x) = \dfrac{dF}{dx}$, so

$$f(x) = 2ax \qquad 0 < x < 5$$

$$= 0 \qquad \text{otherwise.}$$

The function $f(x)$ is plotted in the Fig. 5.6(b).

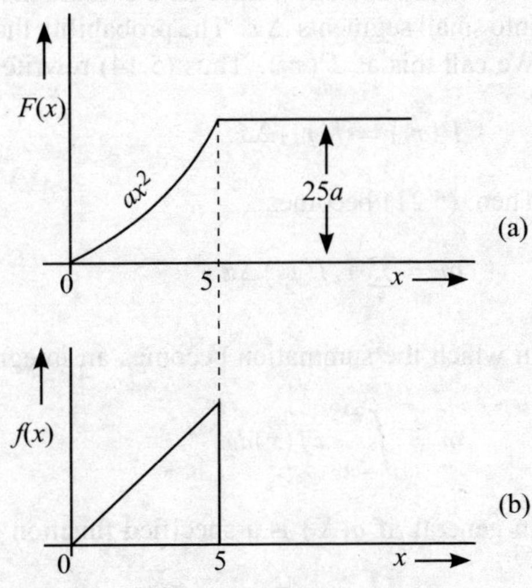

Fig. 5.6 *Plots of (a) $F(x)$ and (b) $f(x)$*

By the normalisation condition,

$$\int_{-\infty}^{\infty} f(x)dx = \int_{0}^{5} 2ax dx = a\left[x^2\right]_{0}^{5} = 25a = 1.$$

This gives $a = 0.04$.

Now, $P(X \le 2) = F(2) = a(2)^2 = 0.04 \times 4 = 0.16$.

5.3.3 Moments of random variables

For an efficient description of random signals, we need to define ensemble averages of the random variables. We first define the mean value of a random variable.

Consider that $x_1, x_2, \cdots$ are the possible values of a random variable X. Let $n_1, n_2, \cdots$ be the frequency of the respective outcomes. That is, the outcome $X = x_1$ occurs n_1 times, $X = x_2$ occurs n_2 times and so on. If the number of measurement of X be N, then the average of the RV is

$$\overline{X} = (x_1 n_1 + x_2 n_2 + \cdots)/N = \sum_{i=1}^{N} x_i \frac{n_i}{N}. \tag{5.20}$$

As $N \to \infty$, n_i/N becomes the probability $P(x_i)$ of x_i so that $\overline{X} = \sum_{i=1}^{N} x_i P(x_i)$. This average value is known as the mean, expectation, or first moment of X and is denoted by m or $E(X)$. Thus,

$$E(X) = \overline{X} = m = \sum_{i=1}^{N} x_i P(x_i). \tag{5.21}$$

To find the average values of a continuous random variable, we first divide the range of the variable into small segments Δx. The probability that X lies between x_i and $x_i + \Delta x$ is $P(x_i \le X \le x_i + \Delta x)$. We call this as $P(x_i)$. Thus (5.14) rewrites approximately as

$$P(x_i) = f(x_i)\Delta x. \tag{5.22}$$

Then, (5.21) becomes

$$m = \sum_i x_i f(x_i)\Delta x \tag{5.23}$$

in which the summation becomes an integral in the limit $\Delta x \to 0$. Thus,

$$m = \int_{-\infty}^{\infty} x f(x) dx. \tag{5.24}$$

In general, if $g(X)$ is a specified function of X, then its average or expectation is

$$\overline{g(X)} = E[g(X)] = \int_{-\infty}^{\infty} g(x) f(x) dx. \tag{5.25}$$

The above result is very important. The function $g(X)$ represents a new random variable. To find the average of $g(X)$, we simply integrate the function weighted by the PDF of X. We do not require the PDF of $g(x)$.

Statistical theory often uses the expectation of a power of the random variables called their 'moments'. The average value of x^n is referred to as the nth moment of the random variable.

$$\overline{x^n} = E[x^n] = \int_{\infty}^{\infty} x^n f(x) dx. \tag{5.26}$$

The simplest moment is the first moment for which $n = 1$ and by (5.24), this is the expected value of X. The second moment of the random variable is its 'mean square' value:

$$E[X^2] = \overline{X^2} = \int_{-\infty}^{\infty} x^2 f(x)dx. \tag{5.27}$$

Another important and useful statistical parameter of a random variable is its 'variance'. To explain its usefulness, consider an experiment which measures some physical parameter (output voltage of a dc generator, say) for many number of times. Suppose that the measured values are $x_1, x_2, x_3, \cdots$. From these data, we can estimate the mean 'm' of the measurement. What is the error in measurement? One might suggest that we find $(x_1 - m)$, $(x_2 - m)$, etc. and take the average. But the average $\overline{X - m}$ goes to zero since there are equal positive and negative contributions $(E(X - m) = E(X) - m = m - m = 0)$. We can define a 'central moment' which is the moment of the difference between X and m. The nth central moment of a random variable X is defined as

$$E[(X - m)^n] = \overline{(X - m)^n} = \int_{-\infty}^{\infty} (x - \overline{x})^n f(x). \tag{5.28}$$

The 'second central moment' is extremely useful and is called the 'variance' σ^2 of the random variable. Thus,

$$\begin{aligned}
\sigma^2 &= E[(X - m)^2] = E[X^2 + m^2 - 2mX] \\
&= E(x^2) + m^2 - 2mE(X) = \overline{X^2} + m^2 - 2m^2 \\
&= \overline{X^2} - m^2.
\end{aligned} \tag{5.29}$$

The quantity σ, called the standard deviation, is the root mean square (rms) value of $(X - m)$. It is a measure of the spread of the sample points about the sample mean. In particular, for a random variable having zero mean $(m = 0), \sigma^2 = E(X^2)$. A straightforward method to find the variance is to calculate the quantity 'mean of squares minus square of means'.

Example 5.9

A random variable X has a uniform density function $f(x)$ as shown in Fig. 5.7. Find (a) the distribution function $F(x)$, (b) the mean of X, and (c) the variance σ^2 of X.

Solution

(a) Following Fig. 5.7, the PDF will be

$$f(x) = \begin{cases} 0 & x < a \\ \dfrac{1}{b-a} & a \leq x \leq b \\ 0 & x > b \end{cases} \tag{5.30}$$

The CDF is obtained by integration.

$$F(x) = \int_{-\infty}^{x} f(x)dx$$

$$= \int_{a}^{x} \frac{1}{(b-a)}\, dx = \frac{x-a}{b-a},\ a \le x \le b.$$

$$\text{Thus, } F(x) = \begin{cases} 0 & x < a \\ \dfrac{1}{(b-a)}(x-a) & a \le x \le b \\ 1 & x > b. \end{cases} \tag{5.31}$$

Figure 5.8 shows the cumulative distribution function.

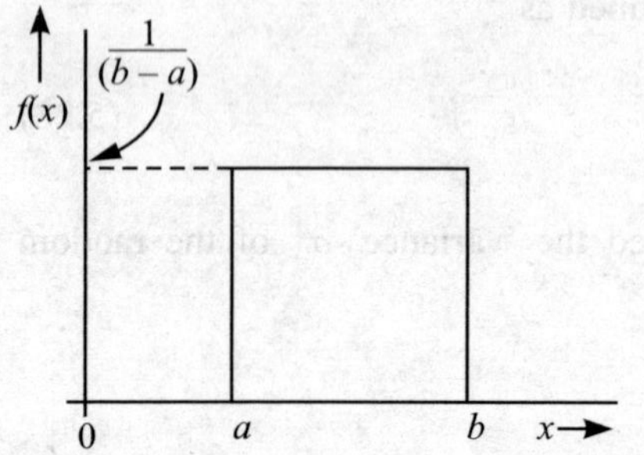

Fig. 5.7 *The uniform distribution function*

Fig. 5.8 *CDF for the uniform density*

(b) The mean, $m = E(X) = \int_{-\infty}^{\infty} x f(x)dx = \int_{a}^{b} x\left(\frac{1}{b-a}\right) dx$

$$= \frac{1}{(b-a)}\frac{1}{2}\left[x^2\right]_a^b = \frac{1}{2}(a+b).$$

(c) The variance, $\sigma^2 = E(X^2) - m^2 = \int_{-\infty}^{\infty} x^2 f(x)dx - m^2$

$$= \frac{1}{(b-a)}\frac{1}{3}\left[x^3\right]_a^b - \left[\frac{1}{2}(a+b)\right]^2$$

$$= \frac{1}{3}(b^2 + a^2 + ab) - \frac{1}{4}(a+b)^2$$

$$= \frac{1}{12}(4b^2 + 4a^2 + 4ab - 3a^2 - 3b^2 - 6ab)$$

$$= \frac{1}{12}(b^2 + a^2 - 2ab) = \frac{1}{12}(b-a)^2.$$

5.3.4 Function of a random variable

Let us consider that a random variable Y be connected to X by a unique functional relation $y = \phi(x)$. Suppose that the RV X has its density function $f_X(x)$. The density function $f_Y(y)$ of Y can be related to $f_X(x)$ in the following way.

Referring to Fig. 5.9, the event x corresponds to the event y and has the same probability. Thus $P(x \le X \le x + dx) = P(y \le Y \le y + dy)$. If the interval dx is very small, then, $f_X(x)dx = |f_Y(y)dy|$, where $f_X(x)$ and $f_Y(y)$ are the density functions for the random variables X and Y. Also, the interval dx corresponds to the interval $|dy| = |\phi(x)|dx$. Hence,

$$
\begin{aligned}
f_Y(y) &= f_X(x)\left|\frac{dx}{dy}\right| \\
&= f_X[\phi^{-1}(y)]\left|\frac{dx}{dy}\right|.
\end{aligned}
\tag{5.32}
$$

When the slope of the curve $y = \phi(x)$ is negative, we will take negative sign on the right side of (5.32). If the functional relation between X and Y is not unique and a number of functions exist,

$y = \phi_1(x_1),\ \ y = \phi_2(x_2)$ and so on; then (5.32) generalises to

$$
f_Y(y) = \sum_{i=1}^{N} f_X(x_i)\left|\frac{dx_i}{dy}\right|. \tag{5.33}
$$

For example, if $y = \phi(x) = x^2$ as in Fig. 5.10, then the event $\{1 < Y \le 4\}$ corresponds to either of the events $\{1 < X \le 2\}$ or $\{-2 \le X < -1\}$. In such case,

$$
f_Y(y) = f_X(x)\left|\frac{dx}{dy}\right|_{\text{at } x_1} + f_x(x)\left|\frac{dx}{dy}\right|_{\text{at } x_2}.
$$

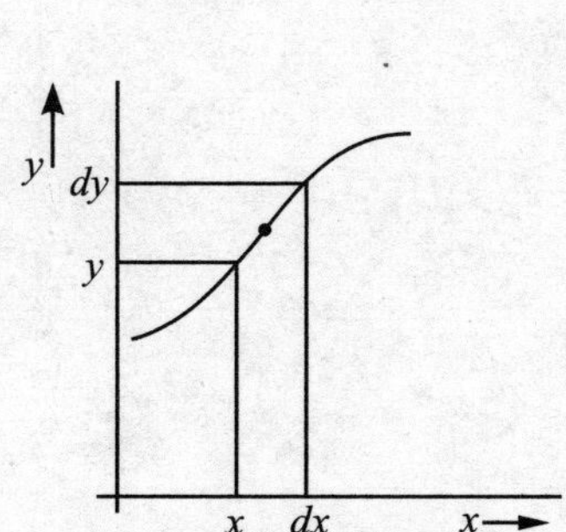

Fig. 5.9 *The functional relation $y = \phi(x)$; $\phi(x)$ has positive derivative.*

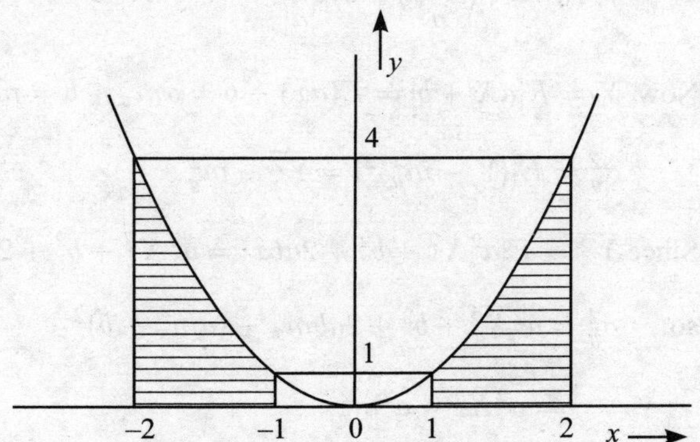

Fig. 5.10 *Functional relation $y = \phi(x) = x^2$*

Example 5.10

Consider the random process $x(t) = A\cos\omega t = A\cos\theta$, where θ is a random variable with a probability density

$$f(\theta) = \begin{cases} \dfrac{1}{2\pi} & 0 < \theta < 2\pi \\ 0 & \text{otherwise.} \end{cases}$$

Find the mean $\overline{x}$ and second moment $\overline{x^2}$.

Solution

The random variable X is function of another random variable Θ. Hence by (5.25),

$$\overline{x} = \int_0^{2\pi} x f(\theta) d\theta = \frac{1}{2\pi}\int_0^{2\pi} A\cos\theta d\theta = 0$$

and $\overline{x^2} = \int_0^{2\pi} x^2 f(\theta) d\theta = \dfrac{A^2}{2\pi}\int_0^{2\pi} \cos^2\theta d\theta = \dfrac{A^2}{2}.$

Example 5.11

The random variable X is linearly related to the random variable Y by $Y = aX + b$, where a and b are constants. If X has the mean m_x and variance σ_x^2, find the mean and variance of Y. Also, find the density function for Y in terms of that of X.

Solution

Since $\left.\dfrac{dx}{dy}\right| = \dfrac{1}{|a|}$, we have from (5.32) that

$$f_Y(y) = f_X[\frac{1}{a}(y-b)]\frac{1}{|a|}.$$

Now, $\overline{Y} = E[aX + b] = E(ax) + b = am_x + b = m_y.$

$$\sigma_y^2 = E[(Y - m_y)^2] = \overline{Y^2} - m_y^2.$$

Since $\overline{Y^2} = E[a^2X^2 + b^2 + 2abx] = a^2\overline{X^2} + b^2 + 2abm_x$,

so, $\sigma_y^2 = a^2\overline{X^2} + b^2 + 2abm_x - (am_x + b)^2$

$$= a^2\overline{X^2} - a^2m_x^2$$

$$= a^2(\overline{X^2} - m_x^2) = a^2\sigma_x^2.$$

That is, $\sigma_y^2 = a^2\sigma_x^2.$

5.3.5 The characteristic function of a random variable

The random variable can be described either by its probability density function or by its characteristic function. In some situations, characteristic function becomes convenient to find the density function of a random variable generated by some functional transform of another random variable. The characteristic function $C_X(\nu)$ of the random variable X is defined as the expectation of $e^{j\nu x}$. That is,

$$C_X(\nu) = E[e^{j\nu x}] = \int_{-\infty}^{\infty} f(x)e^{j\nu x}dx. \tag{5.34}$$

The variable ν is real and $4j = \sqrt{-1}$.

Note that $f(x)$ and $C_X(\nu)$ are related by the Fourier transform pair:

$$f(x) = \frac{1}{2\pi}\int_{-\infty}^{\infty} C_X(\nu)e^{-j\nu x}d\nu. \tag{5.35}$$

The characteristic function can also be used for finding the moments of random variables. Differentiating (5.34) with respect to ν,

$$\frac{d}{d\nu}C_X(\nu) = j\int_{-\infty}^{\infty} xf(x)e^{j\nu x}dx.$$

Evaluating the derivative at $\nu = 0$, we get

$$\frac{d}{d\nu}[C_X(\nu)]\bigg|_{\nu=0} = j\int_{-\infty}^{\infty} xf(x)dx = jE[X] = jm.$$

$$\text{or,}\quad m = E[X] = -j\frac{d}{d\nu}C_X(j\nu)\bigg|_{\nu=0}. \tag{5.36}$$

In other words, the mean of a random variable can be obtained by finding the derivative of the characteristic function at $\nu = 0$. Extending the result, the nth moment is found as

$$E[X^n] = (-j)^{-n}\frac{d^n}{d\nu^n}[C_X(\nu)]\bigg|_{\nu=0}. \tag{5.37}$$

Example 5.12
Find the characteristic function of a random variable X having the probability density function $f(x) = e^{-ax}, \quad a > 0$. Using the characteristic function, find the first and the second moments.

Solution
By (5.34), the characteristic function $C_X(j\nu)$ is

$$C_X(j\nu) = \int_0^{\infty} f(x)e^{j\nu x}dx = \int_0^{\infty} e^{-ax}e^{j\nu x}dx$$

$$= \int_0^\infty e^{-(a-iv)x} dx = \frac{1}{-(a-j\nu)} e^{-(a-j\nu)x}\Big|_{x=0}^{\infty} = \frac{1}{a-jv}.$$

The first moment is

$$E[X] = -j\frac{d}{d\nu}C_X(j\nu)\Big|_{\nu=0} = -j\frac{d}{d\nu}\left(\frac{1}{a-j\nu}\right)\Big|_{\nu=0}$$

$$= -j\left[\frac{1}{-(a-j\nu)^2}\right](-j)\Big|_{\nu=0} = \frac{1}{a^2}.$$

The second moment is

$$E[X^2] = (-j)^2\frac{d^2}{d\nu^2}C_X(j\nu)\Big|_{\nu=0} = \frac{d^2}{d\nu^2}C_X(j\nu) = \frac{2}{a^3}.$$

Example 5.13

Suppose that $y = A\cos x$, X is a random variable distributed uniformly over the interval $-\pi \le x \le \pi$ and A is constant. Find and sketch the density function of Y.

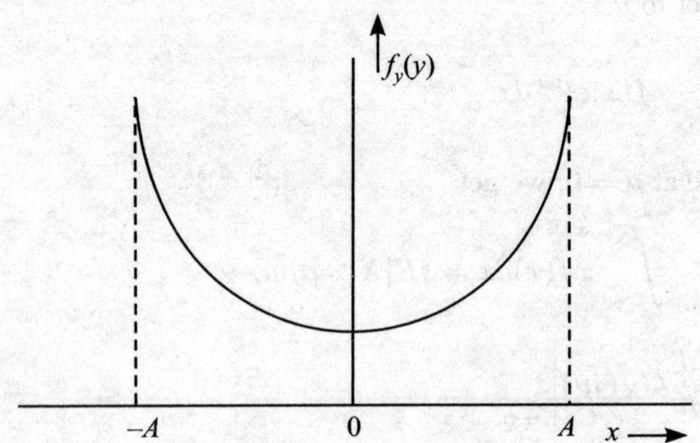

Fig. 5.11 *Plot of $f_Y(y)$ of Example 5.13*

Solution

Since $f_X(x) = \frac{1}{2\pi}$, $-\pi \le x \le \pi$, then the characteristic function of y is

$$C_Y(\nu) = \int_{-\pi}^{\pi} \frac{1}{2\pi}\exp[j\nu A\cos x]dx = J_0(\nu A),$$

where J_0 is the Bessel function of the first kind and of order zero. Then,

$$f_Y(y) = \frac{1}{2\pi}\int_{-\infty}^{\infty} C_y(\nu)e^{-j\nu y}d\nu = \frac{1}{2\pi}\int_{-\infty}^{\infty} J_0(\nu A)_e^{-j\nu y}d\nu$$

$$= \begin{cases} \frac{1}{\pi\sqrt{A^2-y^2}} & |y| \le A \\ 0 & |y| > A. \end{cases}$$

Figure 5.11 shows the plot of $f_Y(y)$ which is comparatively low as Y is passing through 0 and very high for Y approaching A.

5.3.6 Multidimensional random variables

Let us consider n random variables $\{X_1, X_2, \cdots, X_n\}$ in an n-dimensional space. These random variables may or may not be independent of one another. Similar to the case of a single random variable, we may define the distribution function of this n-dimensional RV as

$$F(x_1, x_2, \cdots, x_n) = P(X_1 \le x_1, X_2 \le x_2, \cdots, X_n \le x_n). \tag{5.38}$$

The probability density function of this variable is given by

$$f(x_1, x_2, \cdots, x_n)\, dx_1 dx_2 \cdots dx_n = P(x_1 \le X \le x_1 + dx_1, x_2 \le X_2 \le x_2 + dx_2, \cdots, x_n \le X_n \le x_n + dX_n). \tag{5.39}$$

The cumulative distribution function $\oint$ and the probability density functions are related by

$$F(x_1, x_2, \cdots, x_n) = \int_{-\infty}^{x_1} \int_{-\infty}^{x_2} \cdots \int_{-\infty}^{x_n} f(\lambda_1, \lambda_2, \cdots, \lambda_n) d\lambda_1 d\lambda_2 \cdots d\lambda_n. \tag{5.40}$$

If the random variables are statistically independent , their multivariate probability density can be represented as the product of the corresponding densities:

$$f(x_1, x_2, \cdots, x_n) = f_{X_1}(x_1) f_{X_2}(x_2) \cdots f_{X_n}(x_n). \tag{5.41}$$

We limit our discussions to the two-dimensional case. For two random variables X and Y, the probability that $x \le X \le x + dx$ and $y \le Y \le y + dy$ is given by

$$P(x \le X \le x + dx, y \le Y \le y + dy) = f_{XY}(x, y)\, dxdy. \tag{5.42}$$

For a single random variable X, the probability of the event $\{x \le X \le x + dx\}$ is given by $f_x(x)dx$ which is the area under the curve $f_X(x)$ as depicted in Fig. 5.12(a). But when two variables are involved, $f_{XY}(x, y)$ represents a surface and the probability of the joint event $\{x \le X \le x + dx, y \le \tau \le y + dy\}$ is given by (5.42) which encloses a volume as indicated in Fig. 5.12(b).

The probability of occurrence of the event $\{x_1 \le X \le x_2, y_1 \le Y \le y_2\}$ can be written as

$$P(x_1 \le X \le x_2,\ \ y_1 \le Y \le y_2) = \int_{x_1}^{x_2} \int_{y_1}^{y_2} f_{XY}(x, y) dydx. \tag{5.43}$$

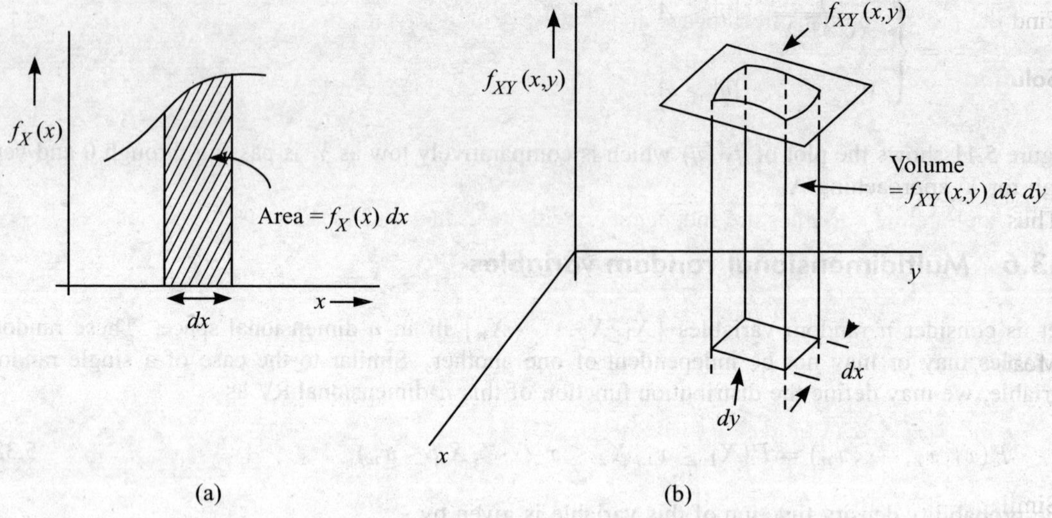

Fig. 5.12 *Density functions and probability representation for (a) single variable; (b) two variables.*

The cumulative distribution function is

$$F_{XY}(x, y) = P(X \leq x, Y \leq y) = \int_{-\infty}^{x} \int_{-\infty}^{y} f_{XY}(x, y) dy dx. \tag{5.44}$$

The first moments or the means of X and Y can be found from

$$\overline{X} = \iint\limits_{-\infty}^{\infty} x f_{XY}(x, y) dx dy, \tag{5.45}$$

$$\overline{Y} = \iint\limits_{-\infty}^{\infty} y f_{XY}(x, y) dx dy.$$

Finally, their variances are obtained as

$$\sigma_x^2 = \iint\limits_{-\infty}^{\infty} (X - \overline{X})^2 f_{XY}(x, y) dx dy,$$

$$\sigma_y^2 = \iint\limits_{-\infty}^{\infty} (Y - \overline{Y})^2 f_{XY}(x, y) dx dy. \tag{5.46}$$

Example 5.14

The joint probability density of the random variables X and Y is given by

$$f(x, y) = \frac{k^2}{\pi} \exp[-k^2(x^2 + y^2)], -\infty < x < \infty, -\infty < y < \infty.$$

Find the probability densities, means and variances of X and Y.

Solution

$$f(x, y) = \frac{k^2}{\pi} e^{-k^2(x^2+y^2)} = \frac{k}{\sqrt{\pi}} e^{-k^2x^2} . \frac{k}{\sqrt{\pi}} e^{-k^2y^2} = f_X(x) f_Y(y).$$

Thus the random variables are independent, with their density functions

$$f_X(x) = \frac{k}{\sqrt{\pi}} e^{-k^2x^2} \quad f_Y(y) = \frac{k}{\sqrt{\pi}} e^{-k^2y^2}.$$

Mean of X:

$$m_x = \int_{-\infty}^{\infty} x f_X(x) dx = \int_{-\infty}^{\infty} \frac{k}{\sqrt{\pi}} x e^{-k^2x^2} dx = 0.$$

Similarly, $m_y = 0$.

Variance of X:

$$\sigma_x^2 = \int_{-\infty}^{\infty} x^2 \frac{k}{\sqrt{\pi}} . e^{-k^2x^2} dx = 2 \int_0^{\infty} \frac{k}{\sqrt{\pi}} x^2 e^{-k^2x^2} dx.$$

Let $k^2x^2 = z$. Then $2k^2 x dx = dz$, or, $x dx = \frac{1}{2k^2} dz$ or, $x^2 dx = \frac{1}{2k^2} \frac{1}{k} z^{1/2} dz$.

$$\sigma_x^2 = 2 \int_0^{\infty} x^2 \frac{k}{\sqrt{\pi}} e^{-k^2x^2} dx = \int_0^{\infty} \frac{2k}{\sqrt{\pi}} e^{-z} \frac{1}{2k^3} z^{1/2} dz = \frac{1}{\sqrt{\pi} k^2} \int_0^{\infty} z^{1/2} e^{-z} dz$$

$$= \frac{1}{\sqrt{\pi} k^2} \Gamma(3/2) = \frac{1}{\sqrt{\pi} k^2} \frac{1}{2} \sqrt{\pi} = \frac{1}{2k^2}.$$

Similarly, $\sigma_y^2 = \dfrac{1}{2k^2}$.

Example 5.15
Suppose that $Z = X + Y$, where X and Y are random variables, statistically independent. Find the mean and variance of the random variable Z.

Solution
The means of X and Y are m_x and m_y, respectively, Using (5.45) for the mean of Z, we have,

$$m_z = \int_{-\infty}^{\infty} \int_{-\infty}^{\infty} (x + y) f(x, y) dx dy$$

$$= \int_{-\infty}^{\infty} \int_{\infty}^{\infty} x f(x + y) dx dy + \int_{-\infty}^{\infty} \int_{-\infty}^{\infty} y f(x + y) dx dy = m_x + m_y.$$

That is, the mean is the sum of the means, irrespective of whether or not the variables X and Y are independent. If X and Y are independent, then $f(x,y) = f(x)f(y)$ and

$$m_z = \int_{-\infty}^{\infty}\int_{\infty}^{\infty}(x+y)f(x)f(y)dxdy$$

$$= \int_{y=-\infty}^{\infty} f(y)dy \int_{-\infty}^{\infty} xf(x)dx + \int_{x=-\infty}^{\infty} f(x)dx \int_{-\infty}^{\infty} yf(y)dy$$

$$= m_x + m_y \quad \text{(using normality conditions } \textstyle\int_{-\infty}^{\infty} f(x)dx = \int_{-\infty}^{\infty} f(y)dy = 1\text{)}.$$

The variance σ_z^2 of z is given by

$$\sigma_z^2 = \overline{Z^2} - m_Z^2.$$

Now, $\overline{Z^2} = \overline{(X+Y)^2} = \displaystyle\int_{-\infty}^{\infty}\int_{-\infty}^{\infty}(x+y)^2 f(x,y)dxdy$

$$= \int_{-\infty}^{\infty} x^2 f(x)dx \int_{-\infty}^{\infty} f(y)dy + \int_{-\infty}^{\infty} y^2 f(y)dy \int_{-\infty}^{\infty} f(x)dx + 2\int_{-\infty}^{\infty} xf(x)dx \int_{-\infty}^{\infty} yf(y)dy$$

$$= \overline{X^2} + \overline{Y^2} + 2\overline{XY} = \overline{X^2} + \overline{Y}^2 + 2m_x m_y.$$

Therefore,

$$\sigma_z^2 = (\overline{X^2} + \overline{Y^2} + 2m_x m_y) - m_z^2$$

$$= (\overline{X^2} + \overline{Y^2} + 2m_x m_y) - (m_x^2 + m_y^2 + 2m_x m_y).$$

$$= (\overline{X^2} - m_x^2) + \left(\overline{Y^2} - m_y^2\right) = \sigma_x^2 + \sigma_y^2.$$

5.3.7 Chebyshev's inequality

Consider a random variable X with probability density function $f(x)$. Let m and σ denote, respectively, the mean and standard deviation of X. We have noted that the standard deviation gives a measure of the spread of the distribution function. We shall prove that for any number a, the probability that x differs from its mean m by more than a is within σ^2/a^2. This describes the Chebyshev's rule. That is,

$$P(|x-m| \geq a) \leq \sigma^2/a^2, \quad \text{regardless of } f(x). \tag{5.47}$$

To prove the inequality, we start with the standard deviation:

$$\sigma^2 = E\big[(x-m)^2\big] = \int_{-\infty}^{\infty} (x-m)^2 f(x)dx. \tag{5.48}$$

If we integrate over $x > (m+a)$, we get

$$\sigma^2 > \int_{m+a}^{\infty} (x-m)^2 f(x)dx \tag{5.49}$$

We replace $(x-m)$ by a; the right-hand side still decreases and we have

$$\sigma^2 > \int_{m+a}^{\infty} a^2 f(x)dx = a^2 \int_{m+a}^{\infty} f(x)dx,$$

that is,

$$\int_{m+a}^{\infty} f(x)dx < \sigma^2/a^2, \tag{5.50}$$

which says that the probability that x differs from m by more than a is less than σ^2/a^2.

5.4 Frequently Used Density Functions

The frequently encountered density functions in communications theory are the followings:

(a) Uniform distribution
(b) Binomial distribution
(c) Poisson distribution
(d) Gaussian or Normal distribution
(e) Rayleigh distribution
(f) Rice distribution

5.4.1 The uniform distribution

There are some processes in communications theory that follow uniform distribution of random variables. The PDF of a uniform distribution function can be expressed as

$$f(x) = \begin{cases} \dfrac{1}{a} & \dfrac{-a}{2} < x < \dfrac{a}{2} \\ 0 & \text{otherwise.} \end{cases} \tag{5.51}$$

The uniform (or rectangular) probability density function and its corresponding cumulative distribution are shown in Fig. 5.13. We have assumed that the random variable X takes on continuous values. We may compute the mean of the distribution by use of (5.24), which takes the form

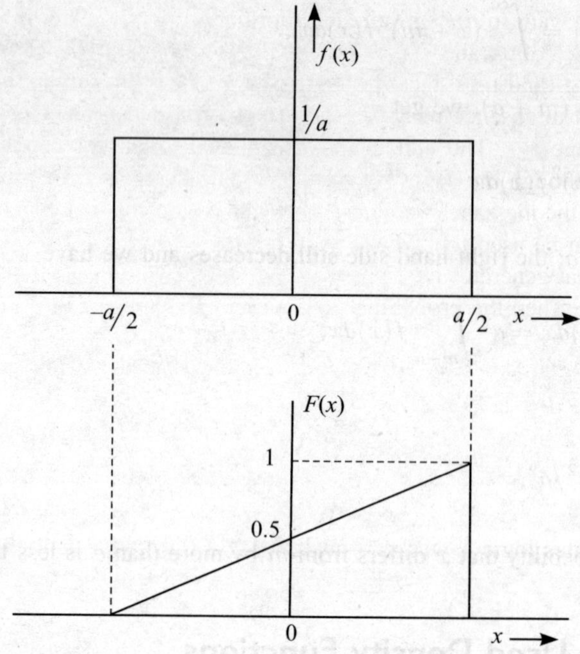

Fig. 5.13 *Uniform PDF and CDF*

$$m = \bar{X} = \int_{-a/2}^{a/2} x f(x) dx = \int_{-a/2}^{a/2} x \frac{1}{a} dx = \frac{1}{a}\left[\frac{1}{2}x^2\right]_{-a/2}^{a/2} = 0.$$

The variance or second central moment in this case is obtained by the use of (5.28). Thus,

$$\sigma^2 = E(X-m)^2 = \int_{-a/2}^{a/2} x^2 \left(\frac{1}{a}\right) dx = \frac{1}{a} \times \frac{1}{3}\left[x^3\right]_{-a/2}^{a/2}$$

$$= \frac{1}{3a}\left(\frac{a^3}{8} \times 2\right) = \frac{1}{12}a^2.$$

The power spectral density of white noise follows that of uniform PDF.

We can apply the uniform PDF in the analysis of error theory. In a variety of applications, the random variables may have discrete distribution, e.g., pulse amplitude modulation (PAM) and phase-shift keying (PSK) systems. These will be discussed later on.

5.4.2 The binomial distribution

The binomial distribution is applicable in communications experiments that have only two possible outcomes: one called a 'success' with the probability p, and the other a 'failure' with the probability

$q = 1 - p$. Suppose we want to transmit digital information over a channel. The digital information corresponds to a sequence of $0'$s and $1'$s. In other words, we have in this case two possible messages, message m_0 for the bit 0 and message m_1 for the bit 1. At the receiving end, the signal might be corrupted by noise. But in absence of noise, the message m_0 when received generates some voltage r_0 and m_1 some voltage r_1. The added noise may lead to false reception such as the received voltage being r_1 even though message m_0 was sent. The average error probability depends on the source probabilities for the messages m_0 and m_1 as well as the channel characteristics. The binomial distribution function can be used to estimate such errors.

Consider a source that generates two messages m_1 (logic 1) and m_0 (logic 0) with probabilities p and $1 - p$, respectively. Then the probability of a particular sequence of r ones and $(n - r)$ zeros is given by

$$P(r) = {}^nC_r p^r (1-p)^{n-r}, \tag{5.52}$$

where ${}^nC_r = n!/[r!(n-r)!]$ is the binomial coefficient.

Example 5.16

A digital source generates logic 1 with a probability $p = 0.6$. What is the probability that 8 out of 10 bits will be logic 1.

Solution

According to the problem, $p = 0.6, q = 1 - p = 0.4, r = 8$ and $n = 10$. So,

$$P(8) = {}^{10}C_8(0.6)^8(0.4)^2 = \frac{10!}{8!2!} \times (0.6)^8 \times (0.4)^2 = 0.121.$$

The mean or expectation of the binomial distribution is given by

$$\overline{X} = \frac{\sum_{r=0}^{n} rP(r)}{\sum_{r=0}^{n} P(r)} = \sum_{r=0}^{n} rP(r). \tag{5.53}$$

Note that the distribution is discrete and that is why the summation is adopted. In the above equation we have used $\sum_{r=0}^{n} P(r) = 1$. Now,

$$\overline{X} = \sum_{r=0}^{n} r\, {}^nC_r p^r (1-p)^{n-r} = \sum_{r=0}^{n} {}^nC_r r p^r q^{n-r}. \tag{5.54}$$

We treat p and q as independent variables and note that $p\dfrac{\partial}{\partial p}(p^r) = p.rp^{n-1} = rp^r$. Thus the series in (5.54) reads as

$$m = \overline{X} = p\frac{\partial}{\partial p} \sum_{r=0}^{n} {}^nC_r p^r q^{n-r}$$

$$= p\frac{\partial}{\partial p}(p+q)^n = np(p+q)^{n-1} = np \tag{5.55}$$

since $p+q=1$.

To evaluate the standard deviation of the binomial distribution, we first compute the second moment $E(X^2)$ about $x=0$ and make use of (5.29). Now,

$$E(X^2) = \overline{X}^2 = \sum_{r=0}^{n} r^2 P(r) = \sum_{r=0}^{n} r^{2\,n}C_r p^r q^{n-r} \tag{5.56}$$

$$= \left(p^2\frac{\partial^2}{\partial p^2} + p\frac{\partial}{\partial p}\right)\sum^{n} C_r p^r q^{n-r}$$

$$= p^2\frac{\partial^2}{\partial p^2}(p+q)^n + p\frac{\partial}{\partial p}(p+q)^n$$

$$= p^2 n(n-1) + p.n = np[1+p(n-1)]. \tag{5.57}$$

Then, $\sigma^2 = E(X^2) - m^2$

$$= np[1+p(n-1)] - (np)^2$$

$$= np - np^2 = np(1-p) = npq$$

or, $\sigma = (npq)^{1/2}$. (5.58)

5.4.3 The Poisson distribution

The statistical properties of the electron stream in electronic circuits follow Poisson distribution. In electron devices such as vacuum tubes and transistors, the current flow is discrete in nature owing to the movement of discrete charge carriers. For example, in vacuum diodes the electrons are emitted from the hot cathode at random times. The anode current shows peaks at the instants when electrons arrive.

Since electrons are emitted at statistically independent times, they can arrive at the anode singly, or in twos, threes, and so on. Let us denote the average number of electron reaching the anode per second by λ. Then the probability of an electron reaching the anode surface in the interval $(t, t+\Delta t)$ to within first order of smallness is $\lambda\delta t$.

Let us investigate the probability $P_0(t)$ that no electron arrives at the anode during the time interval $(0, t)$. In the interval $(0, t+\Delta t)$, this probability is the sum of the probabilities of two events: no electron arrives in the interval $(0, t)$ or in the interval $(t, t+\Delta t)$.

Thus we have

$$P_0(t+\Delta t) = P_0(t) + P_0(t)(1-\lambda\Delta t),$$

or, $$\frac{P_0(t+\Delta t) - P_0(t)}{\Delta t} = -\lambda P_0(t),$$

or in the limit $\Delta t \to 0$,

$$\frac{dp_0}{dt} = -\lambda P_0. \tag{5.59}$$

The solution of this equation is given by

$$P_0(t) = e^{-\lambda t} \tag{5.60}$$

with the initial condition that no one electron arrives at the anode, that is, $P_0(0) = 1$.

Let us now analyse the second process, that is, one electron arriving at the anode in the interval $(0, t)$. Let $P_1(t)$ designate this probability. In the interval $(0, t + \Delta t)$ this probability can be taken as the sum of the probabilities of two events:

(a) an electron arrives in the interval $(0, t)$ and no electron arrives in the interval $(t, t + \Delta t)$;

(b) no electron arrives in the interval $(0, t)$ but one electron in the interval $(t, t + \Delta t)$.

Thus,

$$P_1(t + \Delta t) = P_1(t)[1 - \lambda\Delta t] + P_0(t)\lambda\Delta t$$

or, $$\frac{P_1(t+\Delta t) - P_1(t)}{\Delta t} = -\lambda P_1(t) + \lambda P_0(t)$$

or, in the limit $\Delta t \to 0$,

$$\frac{dP_1}{dt} = -\lambda P_1 + \lambda P_0 \tag{5.61}$$

again with the initial condition $P_1(0) = 0$. By (5.60),

$$\frac{dP_1}{dt} + \lambda P_1 = \lambda e^{-\lambda t},$$

the solution of which is obtained as

$$P_1(t) = (\lambda t)e^{-\lambda t}. \tag{5.62}$$

Similarly, the probability $P_r(t)$ of exactly r electrons arriving at the anode in the interval $(0, t)$ is given by the differential equation

$$\frac{dP_r}{dt} = -\lambda P_r + \lambda P_{r-1} \tag{5.63}$$

whose solution is

$$P_r(t) = \frac{(\lambda t)^r}{r!} e^{-\lambda t}. \tag{5.64}$$

This is known as the Poisson distribution. This gives the probability of 'r' electrons (message) arriving in the time interval t seconds.

We note that

$$\sum_{r=0}^{\infty} P(r) = \sum_{r=0}^{\infty} \frac{(xt)^r}{r!}.e^{-\lambda t}$$

$$= e^{-\lambda t}[1 + \frac{\lambda t}{1!} + \frac{(xt)^2}{2!} + \cdots] = e^{-\lambda t}e^{\lambda t} = 1. \quad (5.65)$$

The mean $\overline{X}$ of the distribution is computed as

$$\overline{X} = m = \sum_{r=0}^{\infty} rP(r) = \sum_{r=0} r\frac{\lambda t)^r}{r!}e^{-\lambda t} = \sum \frac{(\lambda t)^r}{(r-1)!}e^{\lambda t}$$

$$= (\lambda t).e^{-\lambda t} \sum_{r=1}^{\infty} \frac{(\lambda t)^{r-1}}{(r-1)!} = \lambda t. \quad (5.66)$$

Since λ is defined as the average number of electrons arriving at the anode per second, the quantity λt on the right side of (5.66) gives the number of electrons arriving at the anode in time t.

To compute the standard deviation, let us first find $\overline{X^2}$.

$$\overline{X^2} = \sum_{r=0}^{\infty} r^2 \frac{(\lambda t)^r}{r!} e^{-\lambda t} = \sum_{r=0}^{\infty} [r(r-1) + r] \frac{(\lambda t)^r}{r!} e^{-\lambda t}$$

$$= \sum_{r=0}^{\infty} \frac{(\lambda t)^{r-2}}{(r-2)!} e^{-\lambda t} + \sum_{r=0}^{\infty} r.p(r)$$

$$= (\lambda t)^2 \sum_{r=2}^{\infty} \frac{(\lambda t)^{r-2}}{(r-2)!} e^{-\lambda t} + \lambda t$$

$$= \lambda t + (\lambda t)^2. \quad (5.67)$$

Thus, $\sigma^2 = \overline{X^2} - (\overline{X})^2$

$$= \lambda t + (\lambda t)^2 - (\lambda t)^2 = \lambda t$$

or, $\sigma = \sqrt{\lambda t}$. (5.68)

In the multiple-access digital transmission system, data are sent in packets. The data arrival statistics at the receiver is very often modelled as a Poisson process.

5.4.4 The Gaussian (normal) distribution

The most important probability distribution playing a fundamental part in communications is the normal or Gaussian distribution with the probability density function

$$f(x) = \frac{1}{\sqrt{2\pi\sigma^2}} e^{-(x-m)^2/2\sigma^2} \quad -\infty < x < \infty. \tag{5.69}$$

It is a continuous distribution defined by two parameters 'm' and 'σ'. We will show that m is the mean and σ is the standard deviation of the Gaussian distribution. Figure 5.14 shows the plot of the density function $f(x)$. The figure shows that though the continuous random variable X can take any value between $(-\infty, \infty)$, it is more likely that X is centred around the mean $\overline{x} = m$. The peak PDF is $\frac{1}{\sigma\sqrt{2\pi}}$ and the Gaussian PDF has even symmetry about its mean.

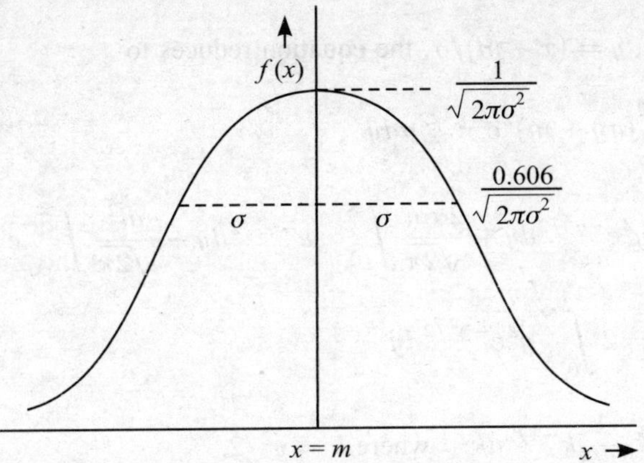

Fig. 5.14 *Plot of the normal or Gaussian density function*

We now compute the mean of the distribution. By definition,

$$E(X) = \overline{X} = \textstyle\int_{-\infty}^{\infty} x f(x) dx$$

$$= \frac{1}{\sqrt{2\pi\sigma^2}} \int_{-\infty}^{\infty} x e^{-(x-m)^2/2\sigma^2} dx. \tag{5.70}$$

We make the substitution

$$y = \frac{x-m}{\sigma}$$

so that $dx = \sigma dy$ and $x = (\sigma y + m)$. Equation (5.70) reduces to

$$\overline{X} = \frac{1}{\sqrt{2\pi\sigma^2}} \int_{-\infty}^{\infty} (\sigma y + m) e^{-y^2/2} . \sigma dy$$

$$= \frac{1}{\sqrt{2\pi}} \sigma \int_{-\infty}^{\infty} y e^{-y^2/2} dy + \frac{m}{\sqrt{2\pi}} \int_{-\infty}^{\infty} e^{-y^2/2} dy.$$

The first integral reduces to zero. Thus,

$$\overline{X} = \frac{m}{\sqrt{2\pi}} \int_{-\infty}^{\infty} e^{-y^2/2} dy = \frac{m}{\sqrt{2\pi}} \times \sqrt{2\pi} = m, \tag{5.71}$$

where $\int_{-\infty}^{\infty} e^{-y^2/2} dy = 2\sqrt{2}\Gamma(3/2) = \sqrt{2\pi}$, Γ represents the Gamma function. Thus it is verified that 'm' represents the mean or expected value of $f(x)$.

The standard derivation can be obtained as

$$\sigma^2 = \overline{X^2} - m^2$$

$$\overline{X^2} = E(X^2) = \frac{1}{\sqrt{2\pi}\sigma} \int_{-\infty}^{\infty} x^2 e^{-(x-m)/2\sigma^2} dx. \tag{5.72}$$

By a change of variable, $y = (x - m)/\sigma$, the equation reduces to

$$\overline{X^2} = \frac{1}{\sqrt{2\pi}\sigma} \int_{-\infty}^{\infty} (\sigma y + m)^2 e^{-y^2/2} \sigma dy$$

$$= \frac{\sigma^2}{\sqrt{2\pi}} \int_{-\infty}^{\infty} y^2 e^{-y^2/2} dy + \frac{2\sigma m}{\sqrt{2\pi}} \int_{-\infty}^{\infty} y e^{-y^2/2} dy + \frac{m^2}{\sqrt{2\pi}} \int_{-\infty}^{\infty} e^{-y^2/2} dy.$$

Now, $\int_{-\infty}^{\infty} y^2 e^{-y^2/2} dy = 2 \int_0^{\infty} y^2 e^{-y^2/2} dy$

$$= 2 \int_0^{\infty} 2k.e^{-k}.\frac{1}{\sqrt{2}} k^{-1/2} dk \quad \text{where } k = y^2/2.$$

$$= 2\sqrt{2} \int_0^{\infty} k^{\frac{3}{2}-1} e^{-k} dk = 2\sqrt{2}\Gamma(3/2) = 2\sqrt{2}\frac{1}{2}\sqrt{\pi} = \sqrt{2\pi}.$$

$$\int_{-\infty}^{\infty} y e^{-y^2/2} dy = 0 \quad \text{and} \quad \int_{-\infty}^{\infty} e^{-y^2/2} dy = \sqrt{2\pi}.$$

Thus $\overline{X^2} = \frac{\sigma^2}{\sqrt{2\pi}}\sqrt{2\pi} + \frac{m^2}{\sqrt{2\pi}}.\sqrt{2\pi} = \sigma^2 + m^2.$

So, $\overline{X^2} - m^2 = \sigma^2.$ (5.73)

The cumulative distribution function $F(x)$ of a Gaussian variable is

$$F(x) = \int_{-\infty}^{x} f(x) dx = \frac{1}{\sqrt{2\pi}\sigma} \int_{-\infty}^{x} e^{-(x-m)^2/2\sigma^2} dx.$$

With the change in variable $y = (x - m)/\sigma$, it reduces to the form

$$F(x) = \frac{1}{\sqrt{2\pi}} \int_{-\infty}^{(x-m)/\sigma} e^{-y^2/2} dy = Q[(x-m)/\sigma], \tag{5.74}$$

where the Q-function is defined by

$$Q(x) = \frac{1}{\sqrt{2\pi}} \int_{-\infty}^{x} e^{-u^2/2} du. \tag{5.75}$$

It is interesting to note that the density function $f(x)$ of Fig. 5.14 is maximum when $x = m$ and the maximum value is $\frac{1}{\sqrt{2\pi}\sigma}$. When $x = m \pm \sigma$, the value of $f(x)$ has fallen to $e^{-0.5} = 0.606$ of its peak value. It is of great interest in the theory of errors to know the probability that the deviation $y = x - m$ is within certain limits.

The Gaussian distribution is sometimes known as normal error law. The function can be used to estimate the probability of a measurement yielding an outcome which falls within a certain range about the mean of the random variable. In this context, one can relate the Q-function defined in (5.75) to the error function erf defined as

$$\text{erf}(x) = \frac{2}{\sqrt{\pi}} \int_0^x e^{-u^2} du. \tag{5.76}$$

From the definition of $\text{erf}(x)$, we can write

$$\left.\begin{aligned} \text{erf}(-x) &= -\text{erf}(x) \\ \text{erf}(0) &= 0 \\ \text{erf}(\infty) &= \frac{2}{\sqrt{\pi}} \int_0^\infty e^{-u^2} du = 1. \end{aligned}\right\} \tag{5.77}$$

The complementary error function $\text{erfc}(x)$ is defined as $\text{erfc}(x) = 1 - \text{erf}(x)$ and is given by

$$\text{erfc}(x) = \frac{2}{\sqrt{\pi}} \int_x^\infty e^{-u^2} du. \tag{5.78}$$

The error function and the Q-function cannot be evaluated in closed forms, i.e., the integrals are not integrable analytically. However, the functions can be tabulated using the method of numerical integration.

We now relate the error function to the Q-function.

$$\begin{aligned} Q(x) &= \frac{1}{\sqrt{2\pi}} \int_{-\infty}^{x} e^{-u^2/2} du = \frac{1}{\sqrt{2\pi}} \int_{-\infty}^{\infty} e^{-u^2/2} du - \frac{1}{\sqrt{2\pi}} \int_{x}^{\infty} e^{-u^2/2} du. \\ &= 1 - \frac{1}{\sqrt{2\pi}} \int_x^\infty e^{-u^2/2} du. \end{aligned} \tag{5.79}$$

To evaluate the integral on the right side we substitute $u^2/2 = k^2$ so that $du = \sqrt{2} dk$.

Thus, $Q(x) = 1 - \dfrac{\sqrt{2}}{\sqrt{2\pi}} \displaystyle\int_{x/\sqrt{2}}^{\infty} e^{-k^2}\, dk = 1 - \dfrac{1}{2}\dfrac{2}{\sqrt{\pi}} \displaystyle\int_{x/\sqrt{2}}^{\infty} e^{-u^2}\, du.$

Or, $$Q(x) = 1 - \frac{1}{2}\text{erfc}\left(\frac{x}{\sqrt{2}}\right). \tag{5.80}$$

Hence the cumulative distribution function

$$F(x) = Q\left[\left(\frac{x-m}{\sigma}\right)\right] = 1 - \frac{1}{2}\text{erfc}\left[\frac{x-m}{\sqrt{2}\sigma}\right]. \tag{5.81}$$

Figure 5.15 shows the plot of the function $F(x)$. The curve is monotonic, varying from zero to unity. Observe that at the mean $x = m$,

$$F(x) = 1 - \frac{1}{2}\text{erfc}(0) = 1 - \frac{1}{2} \times 1 = 0.5.$$

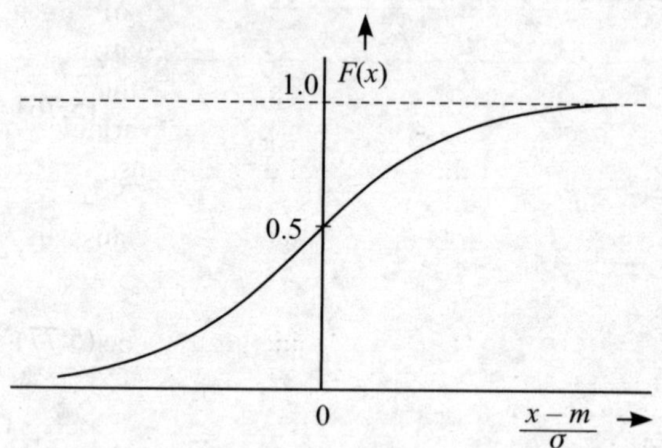

Fig. 5.15. *Illustrating the cumulative distribution function for Gaussian random variable*

Fig. 5.16. *Probability that X lies between x_1 and x_2*

In the calculation of probability of errors, the area under some part of the Gaussian distribution is needed. As an example, let it be required to calculate the probability that the random variable X lies between $X = x_1$ and $X = x_2$. The shaded area of Fig. 5.16 gives that probability which is given by

$$P(x_1 \le X \le x_2) = \int_{x_1}^{x_2} f(x)dx = \frac{1}{\sqrt{2\pi}\sigma} \int_{x_1}^{x_2} e^{-(x-m)^2/2\sigma^2} dx. \tag{5.82}$$

Let $\quad u = \dfrac{x-m}{\sigma}, \qquad dx = \sigma dy.$

Equation (5.82) becomes

$$P(x_1 \le X \le x_2) = \frac{1}{\sqrt{2\pi}} \int_{(x_1-m)/\sigma}^{(x_2-m)/\sigma} e^{-y^2/2} dy$$

$$= \frac{1}{\sqrt{2\pi}} \int_{-\sigma}^{(x_2-m)/\sigma} e^{-y^2/2} dy - \frac{1}{\sqrt{2\pi}} \int_{-\sigma}^{(x_1-m)/\sigma} e^{-y^2/2} dy$$

$$= F\left(\frac{x_2 - m}{\sigma}\right) - F\left(\frac{x_1 - m}{\sigma}\right). \quad \text{[by (5.17)]} \tag{5.83}$$

It is relevant here to revisit the Chebyshev's inequality (Section 5.3.7) which states that the probability that X differs from its mean m by more than 'a' is within σ^2/a^2 regardless of the form of the distribution function.

The Q-function and error functions are defined in several ways. All definitions are equally useful in finding the error probability. Gaussian distribution is frequently used and popular in various branches of engineering. This is because the errors in experimental measurements follow Gaussian distribution. Noise considerations in communications systems will be extremely difficult if the noise statistics are not Gaussian PDF. However, in situations where statistics are not Gaussian, one can approximate them as Gaussian without introducing significant error.

Many naturally occuring phenomena are Gaussian. These are produced by the summation of large number of distributed phenomenon; the individual distributions may not follow Gaussian statistics. Such statistical nature of natural phenomena is explained by the *central-limit* theorem. The theorem states that the probability distribution of the sum of n statistically independent random variables approach Gaussian distribution in the limit $n \to \infty$, irrespective of the individual distributions. That means, regardless of true distribution of X_i, the sum $\sum_{i=1}^{N} = X_i$ can be approximated by a Gaussian distribution with the mean $\sum_{i=1}^{N} = m_i$ and variance $\sum_{i=1}^{N} = \sigma_i^2$. Let us consider N independent random variables $X_1, X_2, \ldots X_N$ with means $m_1, m_2, \ldots, m_n$ and variances $\sigma_1^2, \sigma_2^2, \ldots, \sigma_N^2$. Such N random processes may or may not be Gaussian distribution. The probability density function of the random variable

$$Z = X_1 + X_2 + X_3 + \ldots + X_N$$

approaches a Gaussian PDF the mean of which is

$$m = m_1 + m_2 + \ldots + m_N,$$

and the variance σ^2 is given by

$$\sigma^2 = \sigma_1^2 + \sigma_2^2 + \ldots + \sigma_N^2$$

provided N is large enough.

For N random variables it will be difficult to find the PDF of Z on a piece of paper, but one can derive the same for two random variables; that is,

$$Z = X_1 + X_2 \tag{5.83a}$$

where X_1 and X_2 are both zero mean and independently Gaussian with PDFs given by

$$f_{X_1}(x) = \frac{1}{\sqrt{2\pi\sigma_1^2}} e^{-x^2/2\sigma_1^2}$$

and $$f_{X_2}(x) = \frac{1}{\sqrt{2\pi\sigma_2^2}} e^{-x^2/2\sigma_2^2}$$

The PDF of $Z = X_1 + X_2$ is given by the convolution integration

$$f_Z(x) = \int_{-\infty}^{\infty} f_{X_1}(x) f_{X_2}(z-x)\, dx \tag{5.83b}$$

$$= \frac{1}{2\pi\sigma_1\sigma_2} \int_{-\infty}^{\infty} e^{-x^2/2\sigma_1^2} . e^{-(z-x)^2/2\sigma_2^2} dx$$

To evaluate the integration, we find the term

$$\text{Exp}\left[-\frac{x^2}{2\sigma_1^2} - \frac{(z-x)^2}{2\sigma_2^2}\right]$$

Now,

$$\frac{x^2}{2\sigma_1^2} + \frac{(z-x)^2}{2\sigma_2^2} = x^2\left(\frac{1}{2\sigma_1^2} + \frac{1}{2\sigma_2^2}\right) - \frac{zx}{\sigma_2^2} + \frac{z^2}{2\sigma_2^2}$$

$$= \frac{\sigma_1^2 + \sigma_2^2}{2\sigma_1^2\sigma_2^2} x^2 - \frac{zx}{\sigma_2^2} + \frac{z^2}{2\sigma_2^2}$$

$$= \frac{\sigma_1^2 + \sigma_2^2}{2\sigma_1^2\sigma_2^2}\left[x^2 - \frac{zx \cdot 2\sigma_1^2}{\sigma_1^2 + \sigma_2^2}\right] + \frac{z^2}{2\sigma_2^2}$$

$$= \frac{\sigma_1^2 + \sigma_2^2}{2\sigma_1^2\sigma_2^2}\left[x^2 - 2x \cdot \frac{\sigma_1^2}{\sigma_1^2 + \sigma_2^2} \cdot z + \left(\frac{\sigma_1^2}{\sigma_1^2 + \sigma_2^2} z\right)^2 - \left(\frac{\sigma_1^2}{\sigma_1^2 + \sigma_2^2} z\right)^2\right] + \frac{z^2}{2\sigma_2^2}$$

$$= \left(\frac{\sigma_1^2 + \sigma_2^2}{2\sigma_1^2\sigma_2^2}\right)\left(x - \frac{\sigma_1^2}{\sigma_1^2 + \sigma_2^2} z\right)^2 + \frac{z^2}{2\sigma_2^2} - \frac{\sigma_1^2 z^2}{2(\sigma_1^2 + \sigma_2^2)\sigma_2^2}$$

$$= \left(\frac{\sigma_1^2 + \sigma_2^2}{2\sigma_1^2\sigma_2^2}\right)\left(x - \frac{\sigma_1^2}{\sigma_1^2 + \sigma_2^2} z\right)^2 + \frac{1}{2(\sigma_1^2 + \sigma_2^2)} \cdot z^2$$

Thus equation (5.83b) reduces to

$$f_Z(z) = \frac{1}{2\pi\sigma_1\sigma_2} e^{-\frac{z^2}{2(\sigma_1^2+\sigma_2^2)}} \int_{-\infty}^{\infty} \exp\left[-\frac{\sigma_1^2+\sigma_2^2}{2\sigma_1^2\sigma_2^2}\left(x - \frac{{\sigma_1}^2}{\sigma_1^2+\sigma_2^2} z\right)^2\right] dx$$

By changing the variable,

$$y = \frac{x - \frac{\sigma_1^2}{\sigma_1^2+\sigma_2^2} z}{\frac{\sigma_1\sigma_2}{\sqrt{\sigma_1^2+\sigma_2^2}}}, \text{ we may write}$$

$$f_Z(z) = \frac{1}{2\pi\sigma_1\sigma_2} e^{-\frac{z^2}{2(\sigma_1^2+\sigma_2^2)}} \cdot \int_{-\infty}^{\infty} e^{-y^2/2} dy \cdot \frac{\sigma_1\sigma_2}{\sqrt{\sigma_1^2+\sigma_2^2}}$$

$$= \frac{1}{2\pi\sqrt{(\sigma_1^2+\sigma_2^2)}} e^{-\frac{z^2}{2(\sigma_1^2+\sigma_2^2)}}$$

By changing $\sigma_{z^2} = \sigma_1^2 + \sigma_2^2$, we can write

$$f_Z(z) = \frac{1}{\sqrt{2\pi\sigma_z^2}} e^{-z^2/2\sigma_z^2} \tag{5.83c}$$

Thus $Z = X_1 + X_2$ is a Gaussian random variable with zero mean and variance given by

$$\sigma_z^2 = \sigma_1^2 + \sigma_2^2 \tag{5.83d}$$

Example 5.17

Find the probability that the Gaussian distributed random variable X is within the limit $m - k\sigma \leq X \leq m + k\sigma$, where k is a constant number. Express the result in terms of error function.

Solution

The required probability (see shaded portion in the adjoining figure) is given by

$$P(m - k\sigma \leq X \leq m + k\sigma) = \int_{m-k\sigma}^{m+k\sigma} f(x)dx$$

$$= \frac{1}{\sqrt{2\pi}\sigma} \int_{m-k\sigma}^{m+k\sigma} e^{-(x-m)^2/2\sigma^2} dx.$$

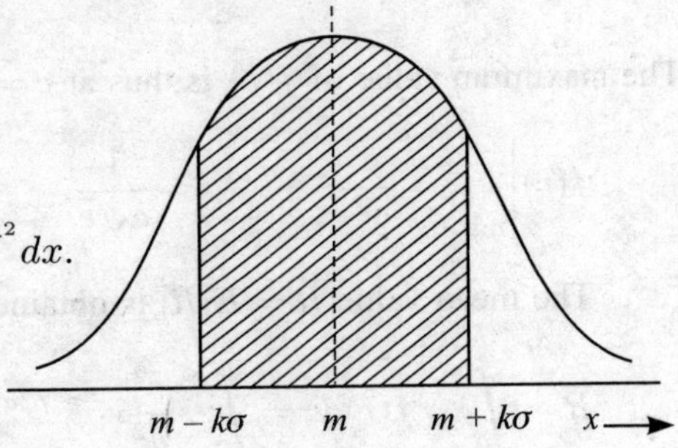

Let $u = \frac{x-m}{\sqrt{2}\sigma}$; $\quad dx = \sqrt{2}\sigma du.$

$$P(m - k\sigma \le X \le m + k\sigma) = \frac{\sqrt{2}.\sigma}{\sqrt{2\pi}\sigma}\int_{-k/\sqrt{2}}^{k/\sqrt{2}} e^{-u^2}du$$

$$= \frac{1}{\sqrt{\pi}}\int_{-k/\sqrt{2}}^{k/\sqrt{2}} e^{-u^2}du$$

$$= \frac{1}{\sqrt{\pi}}\int_{0}^{k/\sqrt{2}} e^{-u^2}du - \frac{1}{\sqrt{\pi}}\int_{0}^{-k/\sqrt{2}} e^{-u^2}du = \frac{1}{2}\text{erfc}\left(\frac{k}{\sqrt{2}}\right) - \frac{1}{2}\text{erfc}\left(-\frac{k}{\sqrt{2}}\right)$$

$$= \text{erf}\left(\frac{k}{\sqrt{2}}\right) \quad [\text{since } \text{erf}(-x) = -\text{erf}(x)].$$

5.4.5 The Rayleigh density function

The Rayleigh distribution function arises in various communications problems and is of great interest as it holds special relationship to the Gaussian distribution. Narrowband random processes have the envelopes with Rayleigh distribution. Also, mobile radio signals follow the Rayleigh distribution due to multipath components. Rayleigh distribution is frequently used to model multipath fading where no line-of-sight (LOS) direct path is present. The Rayleigh density function is defined as

$$f(r) = \begin{cases} \dfrac{r}{a^2}e^{-r^2/2a^2} & 0 \le r \le \infty \\ 0 & r < 0 \end{cases} \tag{5.84}$$

where r is the value assigned to the random variable R. Note that $f(r)$ is zero for $r < 0$ and is the most probable for some values of r as shown in Fig. 5.17. The most probable value of r is obtained by taking the first derivative of $f(r)$ and equating it to zero. Thus,

$$\frac{d}{dr}f(r) = \frac{r}{a^2}\left(-\frac{2r}{2a^2}\right)e^{-r^2/2a^2} + e^{-r^2/2a^2}\frac{1}{a^2} = 0.$$

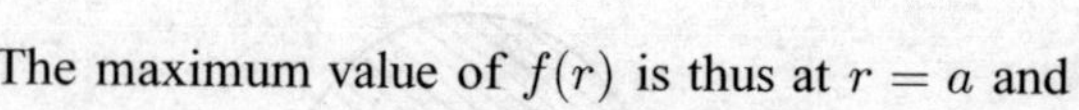

$$\Rightarrow r = a.$$

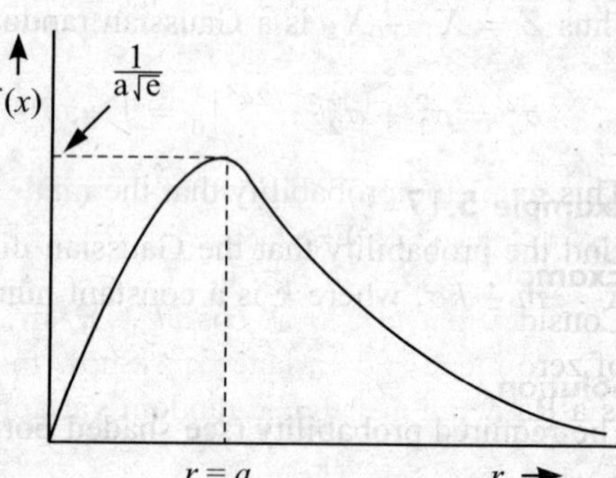

Fig. 5.17 *Plot of Rayleigh density function*

The maximum value of $f(r)$ is thus at $r = a$ and

$$f(r)\Big|_{\max} = \frac{a}{a^2}e^{-a^2/2a^2} = \frac{1}{a\sqrt{e}}.$$

The mean value $\overline{R} = E[R]$ is obtained as follows:

$$\overline{R} = \int_0^{\infty} rf(r)dr = \int_0^{\infty}\frac{r^2}{a^2}e^{-r^2/2a^2}dr.$$

By a change of variable

$$z = \frac{r^2}{2a^2}, \quad dr = \frac{1}{r}a^2dz = \frac{a^2dz}{\sqrt{2a^2z}} = \frac{1}{\sqrt{2}}az^{-1/2}dz$$

$$\overline{R} = \int_0^\infty 2ze^{-z}\frac{a}{\sqrt{2}}z^{-1/2}dz = \sqrt{2}a\int_0^\infty z^{\frac{3}{2}-1}e^{-z}dz = \sqrt{2}a\Gamma(3/2)$$

$$= \sqrt{2}a\frac{1}{2}\sqrt{\pi} = \sqrt{\frac{\pi}{2}}a. \tag{5.85}$$

Similarly, the mean-square value $\overline{R^2}$ can be calculated as $\overline{R}^2 = 2a^2$ so that the variance

$$\sigma^2 = \overline{R^2} - (\overline{R})^2 = 2a^2 - \frac{\pi}{2}a^2 = (2 - \pi/2)a^2. \tag{5.86}$$

Example 5.18

Find the distribution function $(F(x)$ for a Rayleigh distributed random variable defined in (5.84).

Solution

From the definition of the probability density,

$$F(x) = P(X \le x) = \int_{-\infty}^{x} f(x)dx$$

$$= \int_0^x \frac{x}{a^2}e^{-x^2/2a^2}dx = \int e^{-y}dy \quad \text{where } y = \frac{x^2}{2a^2}.$$

$$= -\left[e^{-x^2/2a^2}\right]_0^x = 1 - e^{-x^2/2a^2}.$$

This gives the probability that the random variable R does not exceed a specified value x.

Example 5.19

Consider that $v(t) = X\cos\omega t + Y\sin\omega t$ where X and Y are two independent Gaussian variables of zero mean and variance σ^2. Show that $v(t)$ may be expressed as $v(t) = R\cos(\omega t + \Theta)$, where R is a Rayleigh distributed random variable and Θ is a uniformly distributed random variable.

Solution

The PDFs of X and Y are given by

$$f(x) = \frac{1}{\sqrt{2\pi\sigma^2}}e^{-x^2/2\sigma^2} \text{ and } f(y) = \frac{1}{\sqrt{2\pi\sigma^2}}e^{-y^2/2\sigma^2}.$$

As X and Y are independent random variables, their joint probability density function is given by

$$f(x, y) = f(x)f(y) = \frac{1}{2\pi\sigma^2}e^{-(x^2+y^2)/2\sigma^2}. \tag{5.87}$$

The function $f(x, y)$ represents a bell-shaped surface in the $x - y$ plane. The probability of the joint event that X lies between x and $x + dx$ and Y lies between y and $y + dy$ is given by

$$P(x \leq X + dx, y \leq Y \leq y + dy) = f(x, y)dxdy$$

$$= \frac{1}{2\pi\sigma^2} e^{-(x^2+y^2)/2\sigma^2} dxdy. \tag{5.88}$$

This probability is measured by the volume above the surface area $dxdy$ and below the surface $f(x, y)$.

Transforming the Cartesian coordinates into polar (r, θ) form, the differential area $dxdy$ is represented by $rdrd\theta$ as shown in Fig. 5.18.

The volume above this area specifies the probability of the joint event that R lies between r and $r + dr$ and Θ lies between θ and $\theta + d\theta$. Here R and Θ denote the random variables corresponding to the envelope $r(t)$ and phase $\theta(t)$. Using the transformation

$$X = R\cos\theta, \qquad Y = R\sin\theta, \tag{5.89}$$

we have $R^2 = X^2 + Y^2$ and $\theta = \tan^{-1} Y/X$.

Thus, $P(x \leq X \leq x + dx, y \leq Y \leq y + dy) = P(r \leq R \leq r + dr, \theta \leq \Theta \leq \theta + d\theta)$

$$= \frac{e^{-r^2/2\sigma^2}}{2\pi\sigma^2} rdrd\theta = \frac{r}{2\pi\sigma^2} e^{-r^2/2\sigma^2} drd\theta = \frac{r}{\sigma^2} e^{-r^2/2\sigma^2} \left(\frac{d\theta}{2\pi}\right). \tag{5.90}$$

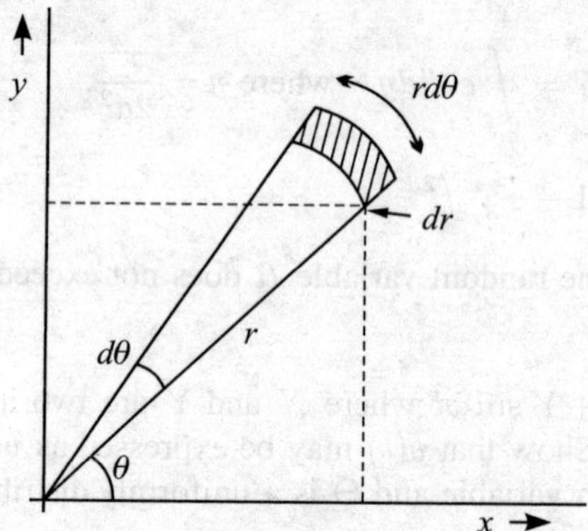

Fig. 5.18 *Illustrating the Cartesian and polar coordinate system transformation*

Hence the joint probability density function of R and Θ is

$$f(r, \theta) = \frac{r}{2\pi\sigma^2} e^{-r^2/2\sigma^2}, \tag{5.91}$$

which is independent of the phase θ. Expressing $f(r, \theta)$ as the product of $f_R(r)$ and $f_\Theta(\theta)$, we find that

$$f_R(r) = \begin{cases} \dfrac{r}{\sigma^2} e^{-r^2/2\sigma^2} & r \geq 0 \\ 0 & r < 0 \end{cases} \tag{5.92}$$

and

$$f_\Theta(\theta) = \begin{cases} \frac{1}{2\pi} & -\pi \leq \theta \leq 2\pi \\ 0 & \text{elsewhere.} \end{cases} \tag{5.93}$$

Hence the proof.

It reveals that the two independent Gaussian variables along two quadrature axes result in another random variable whose envelope is Rayleigh distributed and phase uniformly distributed. Although the two random variables have zero means, their envelope R has a nonzero positive mean. The peak PDF of Rayleigh distribution is at $r = \sigma$, whereas the peak PDF of Gaussian distribution occurs around the mean.

5.4.6 The Rice distribution

In communications engineering, the channels through which signals are transmitted behave as band-pass fillers. Along with the deterministic signal, narrowband noise may also appear at the output of the channel. It is important to investigate the envelope of the output wave for both strong and weak noise.

Consider that the deterministic signal $s(t)$ is of the form $s(t) = A\cos\omega_c t$ and the noise $n(t)$ is a narrowband process given by

$$n(t) = X(t)\cos\omega_c t - Y(t)\sin\omega_c t. \tag{5.94}$$

Here ω_c is a constant. Assume that the noise is Gaussian, of zero mean and variance σ^2. The joint process is given by

$$\begin{aligned} Z(t) &= s(t) + n(t) \\ &= (A + X)\cos\omega_c t - Y\sin\omega_c t. \end{aligned} \tag{5.95}$$

Define the transformation

$$A + X = R\cos\theta; \quad Y = R\sin\theta,$$

where R is the envelope and θ is the phase of the signal. Thus,

$$x(t) = r(t)\cos[\omega_c t + \theta(t)], \tag{5.96}$$

where the random amplitude of the envelope is

$$r(t) = \sqrt{\{A + X(t)\}^2 + Y^2(t)} = \sqrt{A^2 + X^2(t) + Y^2 + 2AX(t)} \tag{5.97}$$

and the random phase

$$\theta(t) = \tan^1 \frac{Y(t)}{A + X(t)}. \tag{5.98}$$

Since the noise is assumed to be Gaussian with zero mean and X and Y are independent, one can write

$$f(x, y) = \frac{e^{-(x^2+y^2)/2\sigma^2}}{2\pi\sigma^2}.$$

Thus in transformed coordinates, the joint probability density takes the form

$$f(r, \theta) = \frac{r}{2\pi\sigma^2} \exp\left[-\frac{(r\cos\theta - A)^2 + (r\sin\theta)^2}{2\sigma^2}\right]$$

$$= \frac{r}{2\pi\sigma^2} \exp\left[-\frac{A^2 + r^2 - 2Ar\cos\theta}{2\sigma^2}\right]. \tag{5.99}$$

Since R and Θ are dependent variables, we cannot express the joint PDF $f(r, \theta)$ as the product $f_R(r) f_\Theta(\theta)$ because of the presence of the product term $Ar\cos\theta$ in equation (5.99).

Since we are interested in the probability density function $f(r)$ of the envelope, we should integrate (5.99) over all possible values of θ. This gives

$$f(r) = \int_0^{2\pi} f(r, \theta) d\theta$$

$$= \frac{r}{2\pi\sigma^2} e^{-(A^2+r^2)/2\sigma^2} \int_0^{2\pi} e^{Ar\cos\theta/\sigma^2} d\theta.$$

The integral $\frac{1}{2\pi}\int_0^{2\pi} \exp\left(\frac{Ar}{\sigma^2}\cos\theta\right) d\theta$ is the modified Bessel function of zero order and is identified as $I_0\left(\frac{Ar}{\sigma^2}\right)$. Thus,

$$f(r) = \frac{r}{\sigma^2} \exp\left(-\frac{r^2 + A^2}{2\sigma^2}\right) I_0\left(\frac{Ar}{\sigma^2}\right). \tag{5.100}$$

The above relation expresses the Rayleigh distribution. The Rayleigh distribution reduces to the Gaussian distribution if $A = 0$ (signal is absent).

Equation (5.95) reveals that the new random variable Z comprises of a deterministic signal S of amplitude A and two independent Gaussian random variable X and Y of zero mean and variance σ^2. The deterministic signal is in phase with the random variable X. We can represent the sample functions along the two orthogonal axes of a Cartesian coordinate system. In one axis we represent the random variable $(S + X)$ which is Gaussian with variance σ^2 but with mean A. Along the other axis the random variable Y is represented. In Section 5.4.5, we had shown that two Gaussian random variables each of zero mean and identical variance at two orthogonal axes results in an envelope that follows Rayleigh distribution. But in the present case random variable at one axis has a nonzero mean due to deterministic signal and the joint distribution function of the envelope turns into a Rician one. Thus the relative strength of the deterministic signal S and the random signal X are the factors that decide the final PDF of the envelope.

We now define a parameter K, called the Rician factor, as the ratio of the deterministic signal power to the variance of the in-phase random signal. That is, $K = \frac{A^2/2}{\sigma^2} = \frac{A^2}{2\sigma^2}$. When $K = 0$, the deterministic signal is absent. This means there are two Gaussian random variables X and Y of zero means along the two orthogonal axes. The distribution boils down to Rayleigh distribution. On the other hand when K is large, the deterministic signal dominates over the random variables and the envelope follows Gaussian distribution. Figure 5.19 shows the Pician PDF with K as parameter.

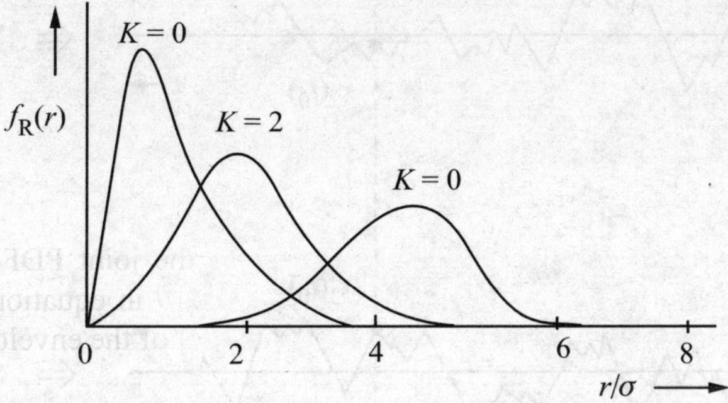

Fig. 5.19 *Rice distribution function*

5.5 Random Processes

When we deal with deterministic signals, there is no uncertainty about the values it assumes at any time t. However, in many situations, the deterministic model proves inadequate because the physical processes are influenced by too many factors varying in time.

A random process $X(t)$ is defined as a function having the property that at any time t the values it assumes are random variables. To find the statistical characteristics of the random process, we might perform either repeated measurement on the same random process, or make simultaneous measurements of a very large number of identical random processes.

A noise waveform generated by a noise voltage generator is an example of a random process. Let us perform a random experiment which consists of measuring the outputs of noise source. Each outcome of the experiment is associated with a sample point in the sample space. If we repeat the experiment of measuring the output voltage, the waveforms look like those of Fig. 5.20. Let every trial be subjected to the same set of conditions. The outcome of every observation in the experiment is a function of time over the observation interval $[0, T]$. The collection of all outcomes will then be a collection of functions over the interval $[0, T]$.

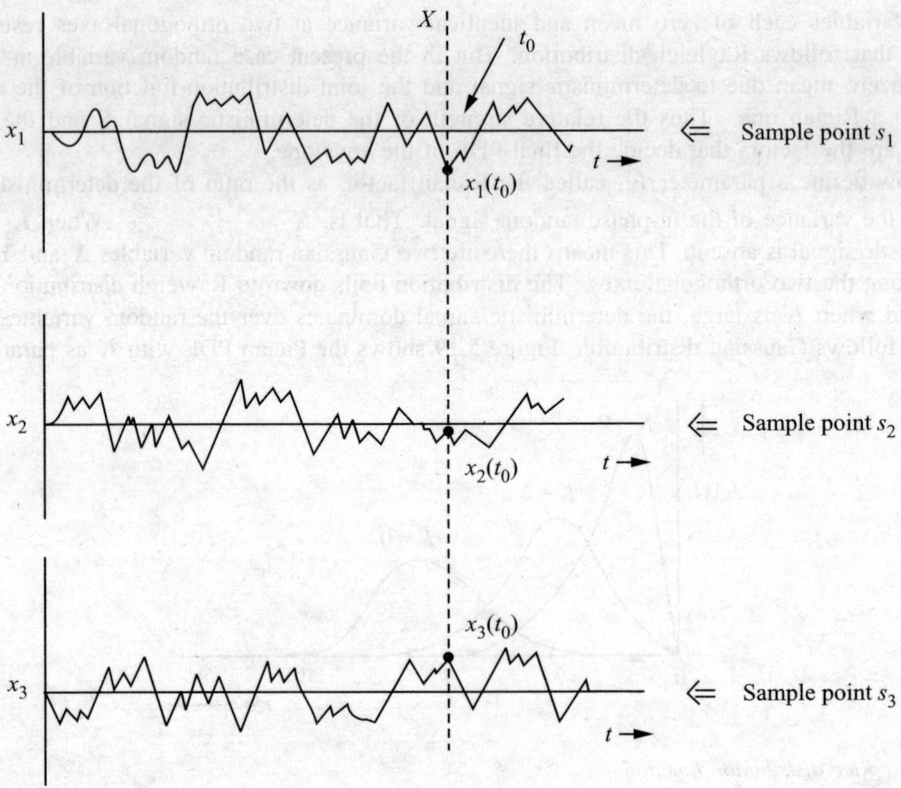

Fig. 5.20 *An ensemble of sample points realised in the random experiment*

Suppose at time $t = t_0$ we perform the random experiment many times. The measured noise voltages are $x_1(t_0), x_2(t_0), x_3(t_0)$, etc. These are the sample points of the random process X. In this way at another instant of time $t = t_1$ (say), we get a new set of sample points $x_1(t_1), x_2(t_1), x_3(t_1)$, etc. The collection of all these sets over all time instants constitute an ensemble of sample points.

We denote the collection of the outcomes by the symbol $X(t_0, s_i)$, where $0 \le t_0 \le T$ and i denotes the ith trial in a given set of measurements. For simplicity, we use the notation

$$X_i(t_0) = X(t_0, s_i). \tag{5.101}$$

For a given sample point s_i, the relation $X(t)$ is called the realisation of the random process. An infinite set of such realisations forms a statistical ensemble.

For a specific time $t_k, X(t_k, s)$ is a random variable $X(t_k)$, the value of which depends on the event s. For a specific event $s = s_i$, and a specific time $t_k, X(t_k, s_i)$ is a *number*.

5.5.1 Stationary random process

The statistical characteristics of a random process defined by the rule $X(t_0, s)$ depends on the time at which the process is initiated. If the statistical properties of the process remain unchanged with time (over the interval $[0, T]$), it is said to be a stationary process.

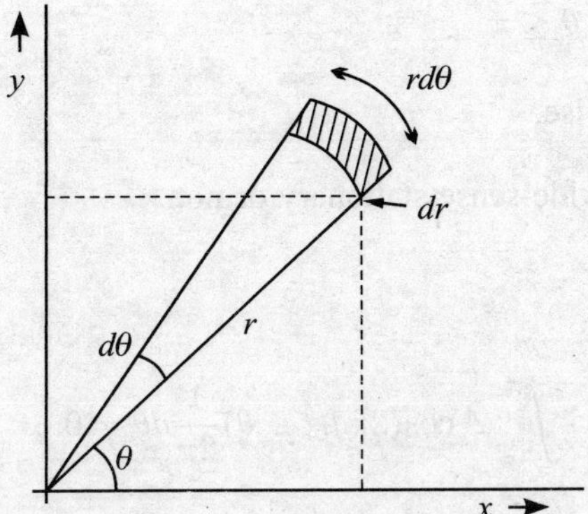

Fig. 5.21 *Illustrating the Cartesian and polar coordinate system transformation.*

To be precise, let a random process $X(t)$ be initiated at $t = -\infty$.

Over the observation time $[0, T]$, the process at times $t_1, t_2, t_3, \cdots$ gives the random variables $X(t_1), X(t_2), X(t_3), \cdots$, respectively. Then the probability density of this set of random variables is $f(x_1, x_2, x_3, \cdots, t_1, t_2, t_3, \cdots)$. Under a time shift τ of the observation times, the new set of random variables are $X(t_1+\tau), X(t_2+\tau), X(t_3+\tau), \cdots$. The probability density of this new set is $f(x_1, x_2, \cdots, t_1+\tau, t_2+\tau, t_3+\tau\cdots)$. The random process is *strictly stationary* if the probability density is invariant under a time shift. Thus, for the process be strictly stationary,

$$f(x_1, x_2, \cdots, t_1, t_2) = f(x_1, x_2, \cdots, t_1+\tau, t_2+\tau, \cdots). \tag{5.102}$$

That is, the statistics of a stationary random process is invariant to any translation in the time axis.

A random process is said to be *wide-sense stationary* if its mean and autocorrelation function do not change with a shift in time. Thus a process is said to be wide-sense stationary if it has a constant mean value

$$E[X(t)] = m_x = \text{constant} \tag{5.103}$$

and the autocorrection function satisfies

$$E[X(t)X(t+\tau)] = R_X(t_1, t_2) = R_X(t_1 - t_2) = R_X(\tau). \tag{5.104}$$

So, for a stationary process the autocorrelation function does not depend on the absolute time but on the difference $\tau = (t_1 - t_2)$.

Example 5.20

Consider a random process

$$X(\theta, t) = A\cos(2\pi f_c t + \theta),$$

where A and f_c are constants and Θ is a random variable with PDF

$$f_\Theta(\theta) = \begin{cases} \frac{1}{2}\pi & -\pi \le \theta \le \pi \\ 0 & \text{otherwise.} \end{cases}$$

Check whether the process is wide-sense stationary or not.

Solution

First we find the mean

$$m_X = \int_{-\infty}^{\infty} x f_\theta(\theta, t) d\theta = \int_{-\pi}^{\pi} A\cos(2\pi f_c t + \theta)\frac{1}{2\pi} d\theta = 0.$$

The autocorrelation function

$$R_X(\tau) = R_X(t_1 - t_2) = E[X(t_1)X(t_2)] = [A\cos(2\pi f_c t_1 + \theta) \times A\cos(2\pi f_c t_2 + \theta)]$$

$$= \frac{A^2}{2} E[\cos\{2\pi f_c(t_1 + t_2) + 2\theta\} + \cos\{2\pi f_c(t_1 - t_2)\}]$$

$$= \frac{A^2}{2}\int_{-\pi}^{\pi} \frac{1}{2\pi}\cos\{2\pi f_c(t_1 + t_2) + 2\theta\} d\theta + \frac{A^2}{2}\cos 2\pi f_c(t_1 - t_2)$$

$$= \frac{A^2}{2}\cos 2\pi f_c \tau, \quad \text{where } \tau = t_1 - t_2.$$

So, the mean of the random process is zero, and hence independent of time. The autocorrection function is again found to be independent of time. Hence the process is stationary in the wide sense.

Example 5.21

For the random process $X(\theta, t) = A\cos(2\pi f_c t + \theta)$, where A and f_c are constants and θ is a random variable with PDF

$$f_\Theta(\theta) = \begin{cases} \frac{1}{\pi} & |\theta| \le \pi/2 \\ 0 & \text{otherwise} \end{cases}$$

check for wide-sense stationarity.

Solution

$$\text{Mean, } m_X = \int_{-\pi/2}^{\pi/2} A\cos(2\pi f_c t + \theta)\frac{1}{\pi} d\theta = \frac{A}{\pi}\int_{-\pi/2}^{\pi/2} (\cos 2\pi f_c t \cos\theta - \sin 2\pi f_c t \sin\theta) d\theta$$

$$= \frac{A}{\pi}(\cos 2\pi f_c t)\sin\theta \Big|_{-\pi/2}^{\pi/2} + \frac{A}{\pi}\sin 2\pi f_c t \cos\theta \Big|_{-\pi/2}^{\pi/2}$$

$$= \frac{2A}{\pi} \cos 2\pi f_c t, \quad \text{which depends on time.}$$

Also, $R_X(t_1, t_2) = E[A^2 \cos(2\pi f_c t_1 + \theta) \cos(2\pi f_c t_2 + \theta)]$

$$= \frac{A^2}{2} E[\cos\{2\pi f_c(t_1 + t_2) + 2\theta\} + \cos 2\pi f_C(t_1 - t_2)]$$

$$= \frac{A^2}{2} E[\cos\{2\pi f_c(t_1 + t_2) + 2\theta\} + \cos 2\pi f_C(t_1 - t_2)]$$

$$= \frac{A^2}{2} \int_{-\pi/2}^{\pi/2} \frac{1}{\pi} \cos\{2\pi f_c(t_1 + t_2) + 2\theta] d\theta + \frac{A^2}{2} \cos 2\pi f_c(t_1 - t_2)$$

$$= \frac{A^2}{2\pi} \cos 2\pi f_c(t_1 + t_2) + \frac{A^2}{2} \cos 2\pi f_c \tau.$$

Thus, both mean and autocorrelation depend on time. Hence the process is not wide-sense stationary.

5.5.2 Ergodic process

A stationary process is said to be ergodic if the ensemble average is the same as the time average. Ensemble average refers to the expectation of a random process $X(t, s)$ at some fixed time t, that is, the average of all possible values of the sample functions $X(t_0, s_1)$, $X(t_0, s_2), X(t_0, s_3), \cdots$. On the other hand, time average gives the dc value of the random variable, defined as

$$< x(t) >= \lim_{T \to \infty} \frac{1}{T} \int_0^T x(t) dt, \tag{5.105}$$

over the observation time $[0, T]$. The time-averaging is performed on a single realisation $X(t)$ whose observation time approaches infinity.

The random process is ergodic if the following relation is satisfied:

$$E(X(t)] = m_X =< x(t) > \text{ when } T \to \infty. \tag{5.106}$$

Testing of ergodicity of a random process is usually difficult. We must test all higher order statistical parameters of the process and prove their equivalence to the respective time parameters. In practice, we normally come across the first two moments, that is, the mean and autocorrelation function. We can then say the process is ergodic in mean and in autocorrelation function. Since time averaging equals the ensemble average for ergodic process, one can relate the fundamental electrical parameters (dc value, rms value, average power, etc.) to some moments of the ergodic process as follows:

Moments of ergodic process	Electrical parameter
$E[X(t)]$	dc Value
$E^2[X(t)]$	Normalised power in dc component
$E[X^2(t)]$	Total average normalised power
$\sqrt{E[X^2(t)}$	rms Value

Example 5.22

For a sinusoidal signal $x(t,\theta) = A\cos(2\pi f_c t + \theta)$ with random phase having PDF

$$f_\Theta(\theta) = \begin{cases} \frac{1}{2\pi} & -\pi \leq \theta \leq \pi \\ 0 & \text{otherwise,} \end{cases}$$

find the time average of $x(t,\theta)$. Is the process ergodic in mean?

Solution

The time average is given by (5.105). Thus

$$< x(t) >= \lim_{T\to\infty} \frac{1}{T}\int_0^T A\cos(2\pi f_c t + \theta)dt$$

$$= \lim_{T\to\infty} \frac{A}{T}\left[\int_0^T (\cos 2\pi f_c t\cos\theta - \sin 2\pi f_c t\sin\theta)dt\right]$$

$$= \lim_{T\to\infty} \frac{A}{T}\left[\cos\theta \frac{\sin 2\pi f_c t}{2\pi f_c}\bigg|_0^T + \sin\theta \frac{\cos 2\pi f_c t}{2\pi f_c}\bigg|_0^T\right]$$

$$= \lim_{T\to\infty} \frac{A}{T}\left[\cos\theta \frac{\sin 2\pi f_c T}{2\pi f_c}\right]$$

$$= \lim_{T\to\infty} A\frac{\sin 2\pi f_c T}{2\pi f_c T}\cos\theta = 0.$$

So the time-average is zero.

5.5.3 Spectral Representation of Random Process

The realisation of random processes is explained in term of the cross and autocorrelation function in the time domain. An equally important realisation in the frequency domain is in terms of spectral analysis. For a single realisation (or sample) $x(t)$ of the random process $X(t)$, its spectral properties are determined by the Fourier transform pair

$$\left.\begin{aligned} X(\omega) &= \int_{-\infty}^{\infty} x(t)\, e^{-j\omega t} dt \\ x(t) &= \frac{1}{2\pi} \int_{-\infty}^{\infty} X(\omega)\, e^{j\omega t} d\omega = \int_{-\infty}^{\infty} X(f)\, e^{j2\pi f t} df \end{aligned}\right\} \quad \text{and} \tag{5.107}$$

(Note the abuse of symbol $X(t)$ which represents random process and $X(\omega)$ Fourier transform of $x(t)$).

To find the power spectrum, we consider the realisation of $x_\tau(t)$ over the small time section $-\tau \leq t \leq \tau$. That is,

$$x_\tau(t) = \begin{cases} x(t), & -\tau \leq t \leq \tau \\ 0 & \text{else where} \end{cases} \tag{5.108}$$

In other words, $x_\tau(t) = x(t)\pi(t/2\tau)$.

The Fourier transform of this truncated realisation is

$$X_\tau(\omega) = \int_{-\tau}^{\tau} x_\tau(t)\, e^{-j\omega t} dt = \int_{-\tau}^{\tau} x(t)\, e^{-j\omega t} dt \tag{5.109}$$

Over the time internal, the normalised energy ε_τ in $x(t)$ is

$$\varepsilon_\tau = \int_{-\tau}^{\tau} x^2{}_\tau(t)\, dt = \int_{-\tau}^{\tau} x^2(t)\, dt = \frac{1}{2\pi} \int_{-\infty}^{\infty} |X_\tau(\omega)|^2 d\omega \tag{5.110}$$

where in the last step, we have used Parseval's energy relation.

Average normalised power over the interval is

$$P_{\text{av}} = \frac{1}{2\tau}\varepsilon_\tau = \frac{1}{2\pi} \int_{-\infty}^{\infty} \frac{|X_\tau(\omega)|^2}{2\tau} d\omega \tag{5.111}$$

Obviously, the average power P_{XX} for the entire process can be obtained from above in the limit $2\tau \to \infty$. Thus

$$P_{XX} = \lim_{2\tau \to \infty} P_{av} = \frac{1}{2\pi} \int_{-\infty}^{\infty} \lim_{\tau \to \infty} \frac{E\left[\left|X_\tau(\omega)\right|^2\right]}{2\tau} d\omega \tag{5.112}$$

where E [.] denotes expectation for a given realisation over the entire duration $-\infty < t < \infty$, the expected value of $\dfrac{|X_\tau(\omega)|^2}{2\tau}$ gives the average power. We now define the power spectral density (PSD) $G_{XX}(\omega)$ of the random process $x(t)$ by

$$G_{XX}(\omega) = \lim_{\tau \to \infty} \frac{E\left[|X_\tau(\omega)|^2\right]}{2\tau} \tag{5.113}$$

Once PSD $G_{XX}(\omega)$ of $X(t)$ is known , the average power in $X(t)$ can be calculated by

$$P_{XX} = \frac{1}{2\pi} \int_{-\infty}^{\infty} G_{XX}(\omega) d\omega \tag{5.114}$$

Example 5.23

Find the PSD for the rectangular pulse sketched in Fig. 5.22.

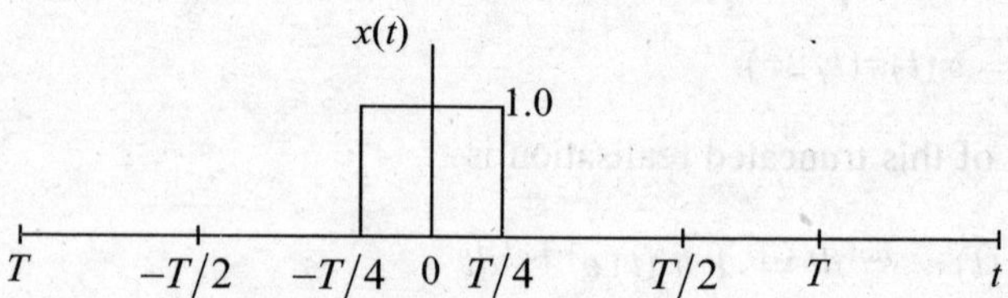

Fig. 5.22. *Rectangular pulse* $\pi\left({}^{t}/_{T/4}\right)$

In this case,

$$X_\tau(\omega) = \int_{-T/4}^{T/4} 1.e^{-j\omega t} dt = \frac{1}{-j2\pi f} e^{-j2\pi ft}\Big|_{-T/4}^{T/4} = \frac{-1}{j2\pi f}\left(e^{-j2\pi fT/4} - e^{-j2\pi fT/4}\right)$$

$$= \frac{e^{j\pi fT/2} - e^{-j\pi fT/2}}{2j\pi f} = \frac{T}{2}\frac{\sin(\pi fT/2)}{(\pi fT/2)} = \frac{T}{2}\sin \mathrm{c}(\pi fT/2)$$

so, $$\frac{|X_\tau(\omega)|^2}{2\tau} = \left(\frac{\tau}{2}\right)^2 \frac{\sin \mathrm{c}^2(fT/2)}{2\tau} = \frac{\tau}{8}\sin \mathrm{c}^2(fT/2)$$

Solved Numericals

1. A noisy binary channel has the following error probabilities:

   $p(y_0|x_1) = 0.15; p(y_1|x_0) = 0.25$

   where x_0, x_1 are the symbols transmitted from the source and y_0, y_1 are the symbols received at the source. We may assume $x_0 = 0, x_1 = 1, y_0 = 0$ and $y_1 = 1$. The probability of transmitting a O by the source is 0.4. Find the total error probability at the receiver and also the probabilities of receiving y_0 and y_1.

Solution

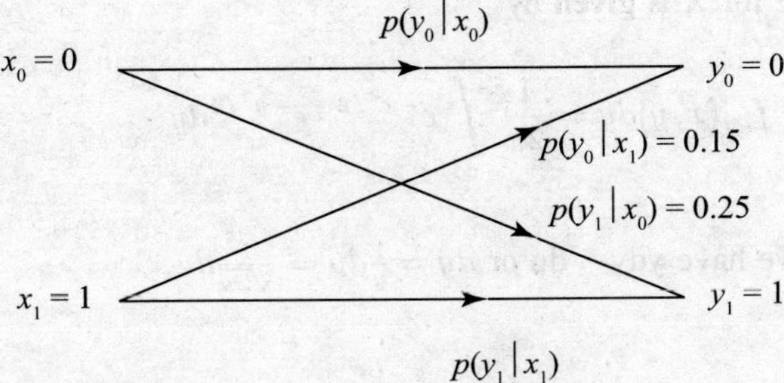

We have

$p(y_1|x_1) = 1 - p(y_0|x_1) = 1 - 0.15 = 0.85$

$p(y_0|x_0) = 1 - p(y_1|x_0) = 1 - 0.25 = 0.75$

Source symbol probabilities are $p(x_0) = 0.4$ and $p(x_1) = 1 - p(x_0) = 0.6$

Probability of receiving y_0 is

$$p(y_0) = p(y_0|x_0)p(x_0) + p(y_0|x_1)p(x_1)$$
$$= 0.75 \times 0.4 + 0.15 \times 0.6 = 0.3 + 0.09 = 0.39$$

Probability of receiving y_1 is

$$p(y_1) = p(y_1|x_0)\, p(x_0) + p(y_1|x_1)\, p(x_1)$$
$$= 0.25 \times 0.4 + 0.85 \times 0.6$$
$$= 0.1 + 0.51 = 0.61 = 1 - p(y_0)$$

The total error probability would be sum of the errors for x_0 transmission and for x_1 transmission. That is,

$$P_e = \text{p(error}\,|x_0\,\text{transmitted})p(x_0) + p(\text{error}\,|x_1\,\text{transmitted})p(x_1)$$
$$= 0.25 \times 0.4 + 0.15 \times 0.6 = 0.1 + 0.09 = 0.19.$$

2. The joint PDF of two random variables X and Y is given by

$$f_{XY}(x, y) = \frac{1}{2\pi} e^{-(x^2+y^2)/2}, -\infty < x < \infty, -\infty < y < \infty$$

Find the marginal PDFs for X and Y and hence show that the two random processes are independent. Also find the conditional PDFs.

Solution

The marginal PDF for X is given by

$$f_x(x) = \int_{-\infty}^{\infty} f_{xy}(x,y)dy = \frac{1}{2\pi}\int_{-\infty}^{\infty} e^{-x^2/2}\cdot e^{-y^2/2}dy$$

Letting $\frac{y^2}{2} = u$, we have ydy = du or $dy = \frac{1}{y}du = \frac{1}{\sqrt{2u}}du.$

Thus,

$$f_x(x) = \frac{1}{2\pi}\cdot e^{-x^2/2}\cdot 2\int_0^{\infty} e^{-u}\frac{1}{\sqrt{2}}u^{-1/2}du$$

$$= \frac{1}{\sqrt{2}\cdot\pi}e^{-x^2/2}\int_0^{\infty} e^{-u}u^{-1/2}dx$$

$$= \frac{1}{\sqrt{2\pi}}e^{-x^2/2}.\Gamma\left(\frac{1}{2}\right)$$

$$= \frac{e^{-x^2/2}}{\sqrt{2}\cdot\pi}\cdot\sqrt{\pi} = \frac{1}{\sqrt{2\pi}}e^{-x^2/2}.$$

Similarly, marginal PDF for Y is obtained as

$$f_y(y) = \frac{1}{\sqrt{2\pi}}e^{-y^2/2}$$

Now,

$$f_{xy}(x,y) = \frac{1}{2\pi}e^{-(x^2+y^2)/2}$$

$$= \frac{1}{\sqrt{2\pi}}e^{-x^2/2}\cdot\frac{1}{\sqrt{2\pi}}e^{-y^2/2} = f_x(x)\cdot f_y(y)$$

So, X and Y are independent.

Now by Eq. (5.6a and b), the conditional PDFs are

$$f_y(y/x) = \frac{f_{xy}(x,y)}{f_x(x)} = \frac{f_x(x)f_y(y)}{f_x(x)} = f_y(y)$$

and $f_x(x\,|y) = f_x(x)$

So conditional PDFs are the same as their marginal PDFs, as the variables are independent.

3. Random variables X and Y have joint PDF given by

$$f_{xy}(x, y) = \frac{1}{2\pi\sigma} \cdot e^{-(ax^2+by^2-2cxy)/2\sigma^2}$$

where $\sigma^2 = ab - c^2$. Find the marginal PDFs $f_x(x)$ and $f_y(y)$.

Also show that they are Gaussian and $\overline{x^2} = b, \overline{y^2} = a$ and $\overline{xy} = c$.

Solution

By definition, marginal PDF for X is

$$f_x(x) = \int_{-\infty}^{\infty} f_{xy}(x, y)dy = \frac{1}{2\pi\sigma} e^{-ax^2/2\sigma^2} \cdot \int_{-\infty}^{\infty} e^{-\frac{b}{2\sigma^2}y^2 + \frac{cx}{\sigma^2}} dy$$

We make use of the fact that

$$\int_{-\infty}^{\infty} e^{-ux^2+vx} dx = \sqrt{\frac{\pi}{u}} e^{v^2/4u}$$

Here $u = \dfrac{b}{2\sigma^2}, v = \dfrac{cx}{\sigma^2}$ and so

$$\int_{-\infty}^{\infty} e^{-\frac{b}{2\sigma^2}y^2 + \frac{cx}{\sigma^2}} dy = \sqrt{\frac{\pi}{b/2\sigma^2}} \cdot e^{\left(\frac{cx}{\sigma^2}\right)^2/4\left(\frac{b}{2\sigma^2}\right)}$$

$$= \sqrt{\frac{2\pi\sigma^2}{b}} . e^{\frac{c^2}{2b\sigma^2} . x^2}$$

So,

$$f_x(x) = \frac{1}{2\pi\sigma} \cdot e^{-\frac{a}{2\sigma^2}x^2} \cdot \sqrt{\frac{2\pi\sigma^2}{b}} e^{\frac{c^2}{2b\sigma^2}x^2}$$

$$= \frac{1}{\sqrt{2\pi b}} \cdot e^{\frac{c^2-ab}{2b\sigma^2}x^2}$$

$$= \frac{1}{\sqrt{2\pi b}} e^{-x^2/2b}, \text{ which is Gaussian with zero mean and variance } b.$$

Similarly,

$$f_y(y) = \int_{-\infty}^{\infty} f_{xy}(x, y)dx = \frac{1}{2\pi\sigma} e^{-\frac{b}{2\sigma^2}y^2} \int_{-\infty}^{\infty} e^{-\frac{a}{2\sigma^2}x^2 + \frac{2cy}{2\sigma^2}x^2} dx$$

Here, assuming $u = \dfrac{a}{2\sigma^2}$ and $v = \frac{2cy}{2\sigma^2}$, we have

$$\int_{-\infty}^{\infty} e^{-ux^2+vx}dx = \sqrt{\frac{\pi}{a/2\sigma^2}} \cdot e^{\left(\frac{2cy}{2\sigma^2}\right)^2/4\left(\frac{a}{2\sigma^2}\right)} = \sqrt{\frac{2\pi\sigma^2}{a}} \cdot e^{\frac{c^2}{2a\sigma^2}x^2}$$

and so

$$f_y(y) = \frac{1}{2\pi\sigma}e^{-\frac{b}{2\sigma^2}y^2} \cdot \sqrt{\frac{2\pi\sigma^2}{a}} \cdot e^{\frac{c^2}{2a\sigma^2}y^2}$$

$$= \frac{1}{\sqrt{2\pi a}}e^{-y^2/2a}, \text{ which is Gaussian of zero mean and variance } a.$$

Now, mean square of X is

$$\overline{x^2} = \int_{-\infty}^{\infty} x^2 f_x(x)dx = \int_{-\infty}^{\infty} x^2 \cdot \frac{1}{\sqrt{2\pi b}} \cdot e^{-x^2/2b}\,dx$$

Letting $x^2/2b = z$ so that $xdx = bdz$ or, $dx = \dfrac{b}{\sqrt{2bz}}dz$

So, $\overline{x^2} = \frac{2}{\sqrt{2\pi b}} \int\limits_0^{\infty} (2bz) \cdot e^{-z} \cdot \frac{b}{\sqrt{2b}} z^{-1/2}dz = \frac{2b}{\sqrt{\pi}} \int\limits_0^{\infty} e^{-z} \cdot z^{1/2}dz = \frac{2b}{\sqrt{\pi}}(3/2) = \frac{2b}{\sqrt{\pi}} \cdot \frac{\sqrt{\pi}}{2}$

$= b$

Similarly, it can be proved that $\overline{y^2} = a$.

Now, $\overline{xy} = \int\limits_{-\infty}^{\infty} \int\limits_{-\infty}^{\infty} (x, y) f_{xy}(x, y)dxdy.$

X and Y are not independent random variables, as $f_x(x).f_y(y) \neq f_{xy}(x, y)$, so we will have to take the joint PDF. Thus,

$$\overline{xy} = \int_{-\infty}^{\infty} \int_{-\infty}^{\infty} xy \cdot \frac{1}{2\pi\sigma} \cdot e^{-(ax^2+by^2-2cxy)/2\sigma^2}\,dxdy$$

or, $$\overline{xy} = \frac{1}{2\pi\sigma} \int_{-\infty}^{\infty} x \cdot e^{-ax^2/2\sigma^2}\,dx \int_{-\infty}^{\infty} y \cdot e^{-(by^2-2cxy)/2\sigma^2}\,dy.$$

4. For the random variable x satisfying Gaussian distribution

$$f_x(x) = \frac{1}{\sqrt{2\pi\sigma^2}}e^{-x^2/2\sigma^2},$$

find the cumulative distribution function (CDF) $F_X(x)$ of x.

Solution

We know that

$$F_X(x) = \int_{-\infty}^{x} f_x(x), dx \text{ (vide Eq. 5.12)}$$

Thus,

$$F_X(x) = \int_{-\infty}^{x} \frac{1}{\sqrt{2\pi\sigma^2}} e^{-y^2/2\sigma^2} dy, \text{ where the dummy variable y has been used.}$$

Let $\frac{y}{\sigma} = u$, then $dy = \sigma du$.

Then,

$$F_X(x) = \int_{-\infty}^{\infty} f_x(x)dx - \int_{x}^{\infty} \frac{1}{\sqrt{2\pi\sigma^2}} e^{-y^2/2\sigma^2} dy$$

$$= 1 - \frac{1}{\sqrt{2\pi\sigma^2}} \int_{x}^{\infty} e^{-u^2/2} \cdot \sigma du$$

$$= 1 - \frac{1}{\sqrt{2\pi}} \int_{x}^{\infty} e^{-u^2/2} du = 1 - Q(x).$$

where the Q-function is defined as $Q(x) = \frac{1}{\sqrt{2\pi}} \int_{x}^{\infty} e^{-u^2/2} du.$

5. Find the mean, second moment and standard deviation of the random variable X having PDF. $f_x(x) = ae^{-ax}u(x), a > 0.$

Solution

The mean is given by

$$\bar{x} = \int_{-\infty}^{\infty} x f_x(x)dx = \int_{0}^{\infty} ae^{-ax} \cdot xdx = a \int_{0}^{\infty} xe^{-ax} dx$$

$$= a \cdot \int_{0}^{\infty} \frac{z}{a} \cdot e^{-z} \frac{1}{a} dz = \frac{1}{a} \int_{0}^{\infty} ze^{-z} dz$$

$$= \frac{1}{a}\Gamma(2) = \frac{1}{a}$$

Second moment,

$$\overline{x^2} = \int_{-\infty}^{\infty} x^2 f_x(x)dx = \int_{0}^{\infty} x^2 \cdot a \cdot e^{-ax} dx$$

Let $ax = z$, then $x = z/a$ and $dx = \frac{1}{a}dz$

$$\therefore \quad \overline{x^2} = a \cdot \int_{0}^{\infty} \frac{z^2}{a^2} \cdot e^{-z} \cdot \frac{1}{a} dz$$

$$= \frac{1}{a^2} \int_{0}^{\infty} e^{-z} \cdot z^2 dz = \frac{1}{a^2}\Gamma(3) = \frac{2}{a^2}$$

Standard deviation

$$\sigma^2 = x^2 - (x)^2 = \frac{2}{a^2} - \left(\frac{1}{a}\right)^2 = \frac{1}{a^2}$$

Exercises

Activity

5.1 A card is drawn from a 52-cards deck. What is the probability of selecting at least 1 card larger than a 6?

5.2 An unbiased dice is rolled. Find the probability that the number appearing is greater than or equal to 3.

5.3 A box contains 3 white balls and 7 black balls. If 2 balls are drawn out in succession without replacing any, what is the probability that the two balls are white?

5.4 A dice is tossed and the number on the face turned up is n. If N be the random variable for the outcome, find $P(N \leq 4)$. Plot the probability $P(N \leq n)$ as a function of n.

5.5 Four coins are tossed simultaneously. Give a list of all the outcomes. Find the probabilities of the events: {3H, IT}, {all matches}, {one head only}.

5.6 Find the mean and variance of the random variable X whose probability density function is given by $f(x) = Ke^{-Kx}, \quad x > 0$ and $K > 0$.

5.7 Let two random variables X and Y be related by the linear transformation $Y = aX + b$. If Y be a Gaussian distributed random variable with mean m and variance σ^2, what is the mean and variance of X?.

5.8 A Gaussian random variable has a mean of 4. The probability $P(4 \leq X < 5) = 0.2$. Find the probability $P(3 < X < 4)$.

5.9 A zero mean Gaussian distributed random variable X has a variance of 9. Find x such that $P(X > x) = 10^{-2}$?

5.10 The joint probability density function of two random variables X and Y has the form

$$f(x, y) = \frac{1}{\pi a^2} e^{-(x^2+y^2)/a^2}.$$

Find the probability densities, first, and second moments of X and Y.

5.11 Consider the function

$$f(x) = \begin{cases} = ae^{-bx} & x > 0 \\ = 0 & x < 0 \end{cases}$$

Find the relation between a and b so that $f(x)$ is a density function. What is the mean of this density function?

5.12. Let A and B be two mutually exclusive events such that $P(AB) = P(A) \cdot P(B)$. Show that $P(AB^c) = P(A) - P(AB)$, where $P(X^c)$ stands for the probability of non-occurrence of the event X and is given as $P(X^c) = 1 - P(X)$.

5.13. The L-levels in a uniform quantisation process have the random values given by $A_m = (2m + 1 - L)d, m = 0, 1, \ldots, L-1$ and all levels are equally likely to occur. Find the mean, second moment and standard deviation of the discrete random variables.

5.14. We define a random variable X as $X = N^2$ where N is a number randomly selected in the range $-5 \leq N \leq 5$. Selection of any number in the range has equal probability. Find $P(x \leq 2), P(x \leq -2), P(-2 \leq x \leq 2)$. Plot the CDF of the random variable X.

5.15. In a certain setting, the voltage generated from a voltage generator is a random process $x(t)$ defined by $x(t) =$ at where 'a' is a uniformly distributed RV in the range $(-5, 5)$. (a) Find the PDF $f_X(x)$ and CDF $F_X(x)$. (b) Find $\overline{x(t)}$ and $\overline{x^2(t)}$.

Review questions

1. (a) Define probability and probability function.

   (b) A function is given by

$$f(x) = \begin{cases} c(x^2 + 1) & 1 < x < 3 \\ 0 & \text{otherwise} \end{cases}$$

   (i) Find the value of c for $f(x)$ to be a density function.
   (ii) Find the probability that x lies between 1 and 2.

(c) X and Y are two statistically independent random variable with corresponding mean value m_x and m_y and variances σ_x^2 and σ_y^2 respectively. Calculate the mean and variance of the sum of these two variables. (K.U. 2001)

2. (a) Define a random variable. Define CDF and PDF.

(b) Calculate the PDF of $z = \cos x$ where x is a uniformly distributed angle lying between 0 and 2π.

(c) Define mean, variance, and standard deviation. What is the physical significance of standard deviation?

(d) Write down the Chebyshev's inequality and explain briefly. (K.U. 2002)

3. (a) If X is Gaussian with $m = 5$ and $\sigma = 8$, calculate the probability $P(0 < x \leq 25)$, given $Q(0.5) - Q(2.5) = 0.30$.

(b) Define random process. Define stationary and ergodic process.

(c) $v(t) = x + 3t$, where the RV X has $\overline{X} = 0$, $\overline{x^2} = 5$. Find $\overline{v(t)}$, RV (t_1, t_2 and $\overline{v^2(t)}$). (K.U. 2002)

4. Define the conditions for a random process to be wide-sense stationary and ergodic. For which one of the above two cases are the conditions more stringent?

Let $r(t) = A_0 \cos \omega_0 t + n(t)$, where A_0 and ω_0 are constants. Assume that $n(t)$ is a wide-sense stationary random process with zero mean and an autocorrelation function $R_n(\tau)$.

(a) Find $\overline{r(t)}$ and determine whether $r(t)$ is a wide-sense stationary.

(b) Find $R_r(t_1, t_2)$, the autocorrelation function of $v(t)$. (RPE 2004)

5. (a) Binary data is transmitted over a noisy channel in a block of 16 bits. The probability that a received bit is in error due to channel noise is 0.01 and the error is random type.

Find (i) The mean or average error per block

(ii) The variance of the errors per block

(b) Let $Y = \sin(X)$ be a random variable with X uniformly distributed over $f_X(x) = \dfrac{1}{2\pi}, 0 \leq x \leq 2\pi$, and $f_X(x) = 0$, otherwise. Find $f_Y(y)$ and $F_y(y)$. Also sketch these functions. [MUST, Autumn 2009]

6. Let $n(t)$ be a sample function of wide sense stationary stochastic process with zero mean and is given as $n(t) = x(t)\cos(2\pi f_c t) - y(t) sin(2\pi f_c t)$.

Show that the autocorrelation functions of $x(t)$, $y(t)$ and $n(t)$ are related as

$R_{nn}(\tau) = R_{xx}(\tau)\text{Cos}(2\pi f_c t) - Ryy(\tau)\text{Sin}\,(2\pi f_c t)$. [MUST, Spring 2009]

7. (a) The error output of a delta modulator is uniformly distributed from -1.5 to 1.5 volts. Find the mean and variance of the error output.

(b) A random variable X has a PDF given by $f_X(x) = be^{-ax}, x > 0$ where 'a' is a positive constant. Find the value of 'b'.

(c) A certain random variable has the PDF given by

$$F_X(x) = \begin{cases} 0, & x \leq 0 \\ \beta x^2, & 0 < x \leq 10 \\ 100\beta, & x > 10 \end{cases}$$

Find β, corresponding PDF and the value of $P(5 < x < 7)$. [MUST, Spring 2014]

8. A continuous random variable X has a PDF $f_X(x) = 4x^3, 0 < x < 1$. Find α and β such that $p(x = \alpha) = P(x > \alpha)$ and $P(x > \beta) = 0.01$.

$$\left(Hint: P(x > \alpha) = 1 - P(x \leq \alpha) = 1 - F_X(x = \alpha) = 1 - \int_0^{\infty} f_X(x)dx, \text{etc} \right)$$

9. Let $f_X(x) = \begin{cases} 0, & x < -1 \\ A(1 - x^2)Cos\left(\frac{\pi x}{2}\right) & -1 \leq x < 1 \\ 0, & x > 1 \end{cases}$

Find the value of 'A' such that $f_X(x)$ truly represents a probability density function.

$$\left(Hint: \int_{-\infty}^{\infty} f_X(x)dx = 0 \right)$$

10. A continuous random variable X is Rayleigh distributed and has the PDF given by

$f_X(x) = \frac{2}{\beta}(x - \alpha)e^{(-x-\alpha)^2/b}, x > \alpha.$

Verify that $E(x) = \alpha + \sqrt{\pi\beta/_4}$ and $\sigma_x{}^2 = \beta\left(1 - \frac{\pi}{4}\right)$.

Multiple-choice questions

1. If $x(t) = \sin(\omega_0 t + \phi)$, ϕ is random, uniformly distributed in the range $-\pi/2 < \phi < \pi/2$. The mean of $x(t)$ is

   (a) zero (b) $2/\pi$ (c) $2/\pi \sin \omega_0 t$ (d) $2/\pi \cos \omega_0 t$

2. Let x and y are two independent Gaussian random variables. Each of x and y has zero mean and variance σ^2. The joint density function is defined as

   (a) $f(x, y) = f(x) + f(y)$ (b) $f(x, y) = f(x)f(y)$
   (c) $f(x, y) = f(x) - f(y)$ (d) $f(x, y) = f(x)/f(y)$

3. Two coins are flipped simultaneously. The probability of getting both tails is

   (a) 0 (b) 1 (c) 1/2 (d) 1/4

4. For two events A and B which are statistically independent and mutually exclusive, the probability $P(AB)$ is

   (a) zero (b) infinity (c) unity (d) 1/2

5. The Gaussian probability density is $f(x) = Ae^{(-x-m)^2/2\sigma^2}$. The normalisation constant A has the value

   (a) $\frac{1}{\sqrt{2\pi\sigma^2}}$ (b) $\frac{1}{\sqrt{2\pi\sigma^2}}$ (c) $\frac{1}{2\pi^2}$ (d) $\frac{1}{2\pi\sigma^2}$

6. Two dice are thrown simultaneously. The probability that the sum of the numbers is 7 is

   (a)1/7 (b) 1/6 (c) 7/36 (d) 1/36

7. For two independent random variables x and y, the means are denoted as $\overline{x}$ and $\overline{y}$. The the mean of product $\overline{xy}$ is given by

   (a) $\overline{xy} = \overline{x} + \overline{y}$ (b) $\overline{xy} = \overline{x + y}$ (c) $\overline{xy} = \overline{x}\ \overline{y}$ (d) $\overline{xy} = 0$

8. The zeroth order central moment of a random process is

   (a) 1 (b) 0 (c) infinity (d) none of these

9. The first order central moment of a random variable is

   (a) mean (b) square root of variance (c) 1 (d) 0

10. Standard deviation signifies

    (a) rms value of ac component (b) rms value of signal for $m_x \neq 0$
    (c) mean square value (d) square root of average normalised power

6 An Outline of the Theory of Noise

6.1 Introduction

The term 'noise' is used to designate any unwanted form of energy tending to interfere with signals in communications systems. Noise represents a basic limitation on the transmission, processing, or reception of signals. A knowledge of the characteristics of noise and its sources are obviously desirable to design an error-free communications systems. This chapter discusses the potential sources, mathematical model, and statistical characteristics of noise. It is important for an engineer to be conversant with the phenomena of noise as it is often possible to reduce or even suppress its effects by designing intelligent circuits.

6.2 Classification of Noise

There are several potential sources of noise in a communications system, both internal or external. The signal sources, the transmitter and the receiver systems themselves may produce noise. Electronic components, fluorescent lighting, electric motors, leakage from high-voltage transmission lines, etc. are examples of man-made noises. There are also non-man-made noises like the atmospheric noise, e.g., solar and cosmic noises. Internal noise is produced in electrical circuits due to the spontaneous fluctuations of currents and voltages. Thermal and shot noises are typical examples of this kind.

Atmospheric noise is produced by lighting discharge during thunderstorm and is in the form of electrical impulses. The spectrum of this noise falls off with frequency and is thus more severe at low frequencies. The cosmic noise (from the sun and distant stars) and the galactic noise (from quasars and pulsars) are always present in space and have a wide frequency band.

Shot noise originates from the random fluctuations of the electrons and holes and is present in all amplifying devices. This noise is encountered in thermionic devices (vacuum tube diode and triodes) and semiconductor bipolar junction transistors. Another kind of noise called the 'transit-time noise' also appears in transistors when the transit time of electrons between the electrodes becomes comparable to the reciprocal of input signal frequency. This effect is pronounced at very high frequency (VHF) range.

The most important class of noise in electronics circuits is the 'thermal', or 'resistor' or 'Johnson' noise originating from the random movement of conduction electrons in metals. The mechanism of this kind of noise is as follows. The conduction electrons being thermally agitated, collide with the positive ions of the metal and maintain the electrical neutrality of the resistor. However, the number of charges being high in metals, their erratic motion disturbs the uniformity of distribution of charges in the volume. The centres of positive and negative charges do not coincide, thus producing a voltage difference across the resistor terminals. The polarity of this voltage is unpredictable. This is the source of the so-called resistor noise.

At low audio frequencies, another kind of noise called the 'flicker noise' is felt in transistors. This is also called the $1/f$ noise as its strength varies inversely with frequency. The effect of this noise can be ignored above about 1 kHz.

Man-made noises may become severe in certain situations (particularly in industrial areas), the intensity of such noises often outstripping the signals from other sources. As industrial noise depends on factors like the quantum of automobile ignition, use of fluorescent tubes, electric motors, etc., it is difficult to analyse it on any analytical footing.

Superposition of noise on the signals may take several forms. Noise may be added to the signal, in which case it is termed additive noise. The noise may multiply the signal resulting in fading effect.

We may describe noise as an ergodic random process (discussed in the previous chapter), and hence the principle of statistical averages can be applied to study its effects on systems. In many communications systems of interest, the probability density of noise can be assumed to be Gaussian.

6.3 Thermal Noise

As discussed earlier, the polarity of the thermal noise voltage changes from time to time and so the mean voltage across the terminals over a finite time interval goes to zero. An effective way of representing the thermal noise is by its mean-square value.

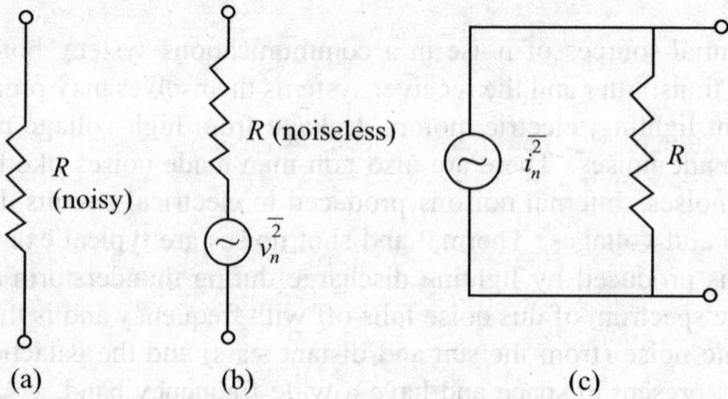

Fig. 6.1 *(a) An electrical resistor (noisy), (b) Thevenin equivalent circuit, (c) Norton equivalent circuit*

Consider a physical resistor R in thermal equilibrium with the surrounding at temperature T as shown in Fig. 6.1. It has been established that the mean square noise voltage across the terminals of the resistor is given by

$$\overline{v_n^2} = 4kTRdf \text{ volts}^2, \tag{6.1}$$

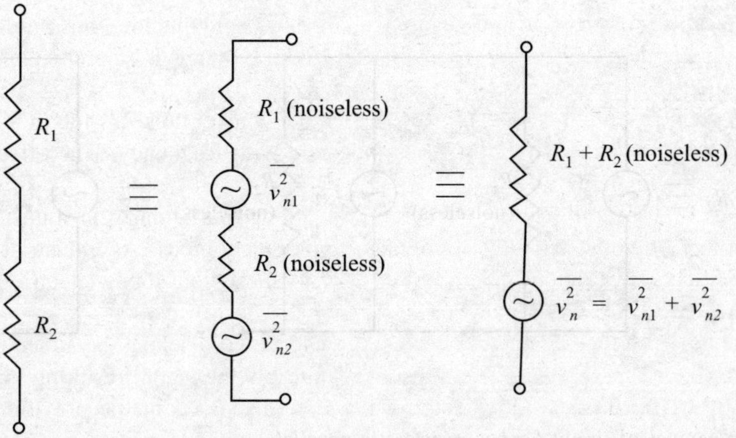

Fig. 6.2 *Thevenin equivalent circuit for the two resistors in series*

where k is the Boltzmann constant ($= 1.38 \times 10^{-23}$ J/K) and df is the bandwidth over which the noise is measured. We may model the noisy resistor by a Thevenin equivalent circuit consisting of a noise voltage generator of mean-square value $\overline{v_n^2}$ in series with a noiseless resistor R, as shown in Fig. 6.1(b). The Norton equivalent representation of the same resistor is shown in Fig. 6.1(c), which consists of a noise current generator of mean-square value $\overline{i_n^2}$ in parallel with a noise-free resistor. The mean-square value of the noise current is given by

$$\overline{i_n^2} = \frac{\overline{v_n^2}}{R^2} = \frac{4kTRdf}{R^2} = 4kTGdf \ \text{amps}^2 \tag{6.2}$$

where $G = 1/R$ is the conductance. The thermal noise follows the pattern of Gaussian distribution with zero mean.

When a number of resistors at the same temperature are connected in series, the effective mean-square noise voltage across the combination is obtained as the sum of all the mean squares (since they represent independent random processes). Thus,

$$\begin{aligned} \overline{v_n^2} &= \overline{v_{n_1}^2} + \overline{v_{n_2}^2} + \cdots \\ &= 4kT(R_1 + R_2 + \cdots)\, df \\ &= 4kTR_s df. \end{aligned} \tag{6.3}$$

where $R_s = R_1 + R_2 + \ldots$ is the total resistance.

So the effective noise resistor is the equivalent of the series combination. The addition of normalised power as in (6.3) accrues from the fact that the sum of independent Gaussian processes is itself a Gaussian process as dictated by the central limit theorem. The Thevenin equivalent circuit of the two resistors in series in depicted is Fig. 6.2. Similarly, the mean-square noise current for the resistors connected in parallel can be computed.

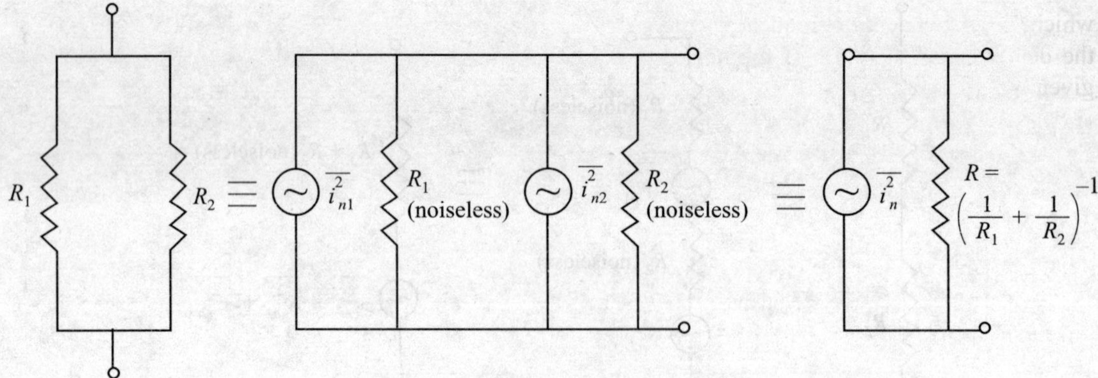

Fig. 6.3 *Norton equivalent circuit for two resistors in parallel*

$$\overline{i_n^2} = \overline{i_{n_1}^2} + \overline{i_{n_2}^2} + \cdots$$

$$= 4kT(G_1 + G_2 + \cdots)df. \tag{6.4}$$

Here $G_1, G_2, \ldots$ are the conductances of the resistors. The Norton equivalent circuit of the two resistors in parallel is shown in Fig. 6.3.

When two or more random processes are combined, the effective mean square value will be the sum of mean square of all the random processes. A proof of this statement is worked out in Section 6.13.

Example 6.1

Calculate the rms noise voltage at the input of an amplifier of bandwidth 5 MHz operating at 17°C. The equivalent input-noise resistor of the amplifier is 500Ω.

Solution

Here $R = 500\ \Omega$, $df = 5$ MHz, $T = (17 + 273)$ K = 290 K.

Now, $\overline{v_n^2} = 4kTRdf = 4 \times 1.38 \times 10^{-23}$ (JK^{-1}) 290 (K) 500(Ω) × 5 × 10^6 (Hz)

$$= 4.002 \times 10^{-11} \text{ volts}^2.$$

So, the rms noise voltage $v_n = \sqrt{\overline{v^2}} = 6.3\ \mu$V.

6.4 Power Spectral Density of Thermal Noise

The power spectral density (PSD) is defined as the mean power per unit frequency interval. The same for thermal noise is given by

$$G_n(f) = \frac{\overline{v_n^2}}{2df} = 2kTR, \tag{6.5}$$

which is independent of frequency. This truly defines the two-sided PSD because of the factor 2 in the denominator of (6.5). If the noise source contains reactive components, the two-sided PSD is given by

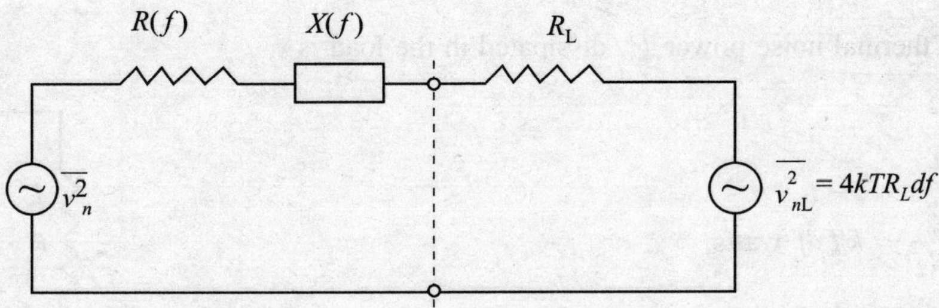

Fig. 6.4 *The noise source containing the reactive component is connected across the load resistor R_L*

$$G_n(f) = 2kTR(f), \tag{6.6}$$

where $R(f)$ denotes the real part of the impedance for which the PSD is sought. To prove this, consider the circuit shown in Fig. 6.4. The impedance $Z(f)$ has a resistive component $R(f)$ and a reactive component $X(f)$. Assuming all parts to be at thermal equilibrium at temperature T, the power delivered by $Z(f)$ to the load R_L must be the same as the power delivered by R_L to $Z(f)$. We represent the mean-square noise voltage of R_L to $Z(f)$. We represent the mean-square noise voltage of R_L by $\overline{V_{NL}^2}$. Thus,

$$\frac{\overline{v_n^2}}{(R(f)+R_L)^2+(X(f))^2}R_L = \frac{\overline{v_{nL}^2}}{(R(f)+R_L)^2+(X(f))^2}R(f)$$

or, $\overline{v_n^2} = \dfrac{R(f)}{R_L}.\overline{v_{nL}^2}$

$$= \frac{R(f)}{R_L}4kTR_L df.$$

$$= 4kTR(f)df.$$

Thus the two-sided PSD is $G_n(f) = \dfrac{\overline{v_n^2}}{2df} = 2kTR(f).$ (6.7)

Example 6.2

Consider that a resistor noise generator is terminated by a load resistance R_L. Find the available noise PSD.

Solution

Referring to Fig. 6.5, the rms noise voltage of the generator is $v_n = \sqrt{4kTRdf}$. So the RMS voltage at the output is $V = v_n/2$, assuming that $R = R_L$.

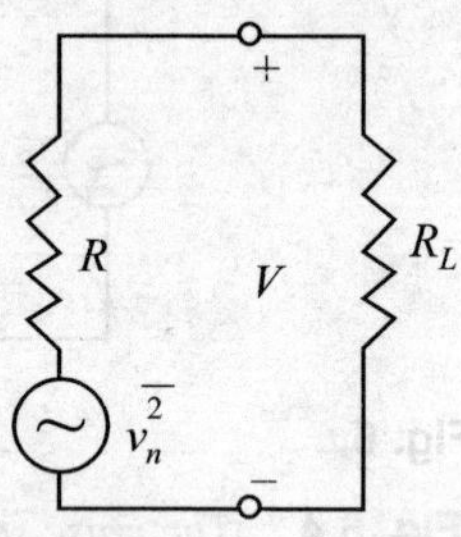

Fig. 6.5 *To calculate available noise PSD*

The maximum thermal noise power P_n dissipated in the load is

$$P_n = \frac{V^2}{R_L}$$

$$= \frac{v_n^2}{4R_L} = kTdf \text{ watts.}$$

The two-sided noise PSD is then

$$G_n(f) = \frac{P_n}{2df} = \frac{1}{2}kT.$$

An effective way to minimise the thermal noise is to cool the input circuits of the receivers to very low temperatures.

6.5 White Noise

White noise refers to that form of noise whose power spectrum is the same at all frequencies, that is,

$$G_n(f) = G_0 = \text{constant.}$$

The term 'white' is used in the sense that white light contains all the frequencies within the visible spectrum of electromagnetic radiation. But there is no physical noise source that gives a uniform PSD as required by the white noise. A reasonable expression for the PSD obtained from quantum mechanical treatment is given by

$$G_n(f) = \frac{hf/2}{e^{hf/(kT)} - 1}, \tag{6.8}$$

where h is Planck's constant and f is the frequency. In the frequency range for which $hf << kT$, (6.8) reduces to the classical limit of $\frac{1}{2}kT$.

$$G_n(f) \approx \frac{hf/2}{1 + \frac{hf}{kT} - 1} = \frac{1}{2}kT.$$

A plot of the available PSD of thermal noise as given in (6.8) is shown in Fig. 6.6. Note that at low frequencies $G_n(f)$ reduces to the classical limit of $\frac{1}{2}kT$.

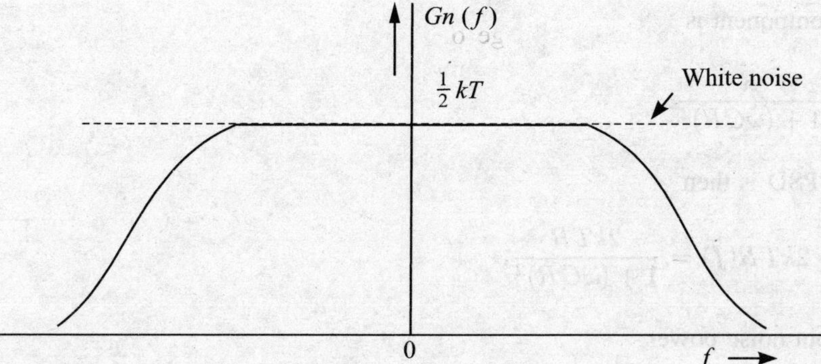

Fig. 6.6 *Power spectrum density of noise source as given in (6.8)*

The available noise power P_n can be obtained by integrating $G_n(f)$ over all frequencies. Thus,

$$P_n = \int_{-\infty}^{\infty} G_n(f)df = 2\int_{0}^{\infty} G_n(f)df. \tag{6.9}$$

For white noise, $G_n(f)$ is constant and (6.9) gives an indication of the infinitely large power in white noise.

Example 6.3

A resistor R is in thermal equilibrium with the surroundings at temperature T. A capacitor C is connected in parallel with the resistor. Derive the expression for the PSD of the noise voltage. Also, calculate the output noise power.

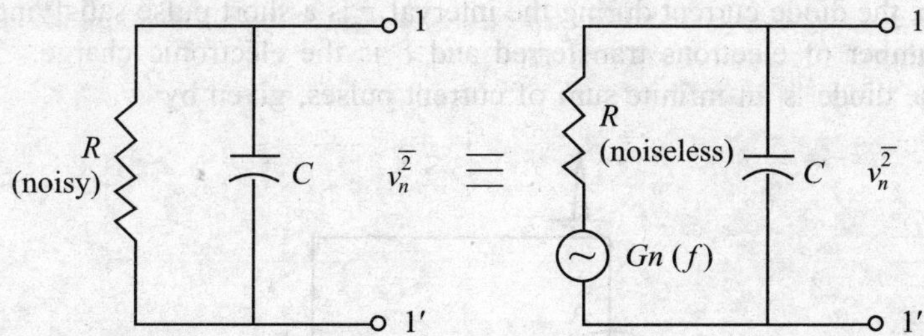

Fig. 6.7 *Figure of Example 6.3*

Solution

We replace the noisy resistor by a series combination of an ideal noise-free resistor R and a noise voltage generator as shown in Fig. 6.7. At the terminals $(1, 1')$ the impedance is

$$Z(f) = \frac{R\frac{1}{j\omega c}}{R + \frac{1}{j\omega C}} = \frac{R}{1 + j\omega cR} = \frac{R}{1 + (\omega CR)^2} - j\frac{\omega CR^2}{1 + (\omega CR)^2}.$$

The resistive component is

$$R(f) = \frac{R}{1 + (\omega CR)^2}.$$

The two-sided PSD is then

$$G_n(f) = 2kTR(f) = \frac{2kTR}{1 + (\omega CR)^2}.$$

Hence the output noise power

$$P_n = \int_{-\infty}^{\infty} G_n(f)df = 2kTR \int_{-\infty}^{\infty} \frac{df}{1 + (2\pi fCR)^2}$$

$$= \frac{kT}{\pi C}[\tan^{-1}(2\pi CR)]_{-\infty}^{\infty}$$

$$= \frac{kT}{C}.$$

6.6 Shot Noise in Electron Devices

In electron devices, shot noise arises because of the discrete nature of the current flow. For example, in a thermionic diode circuit, a current pulse is generated every time an electron from the cathode K reaches the anode A (refer to Fig. 6.8). The electrons are emitted at random times. If τ is the transit time of an electron, the diode current during the interval τ is a short pulse satisfying $\int_0^{\tau} i(t)dt = ne$, where n is the number of electrons transferred and e is the electronic charge. The total current flowing through the diode is an infinite sum of current pulses, given by

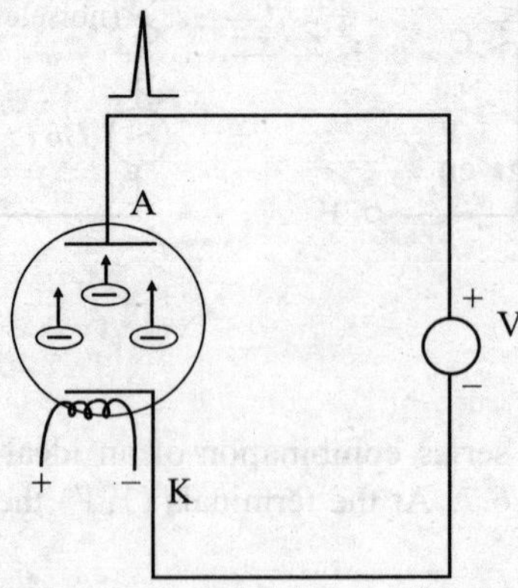

Fig. 6.8 *A vacuum diode showing the discrete nature of electron emission. The individual electron motion is statistically independent.*

$$X(t) = \sum_{n=-\infty}^{\infty} i(t - t_n), \tag{6.10}$$

where $i(t - t_n)$ denotes the current pulses generated at $t = t_n$. A plot of the pulsed diode current is shown in Fig. 6.9. The diode current does not remain constant, rather, it shows certain fluctuation. This kind of noise is called the shot noise. The shot noise has a Poisson distribution.

Let us now calculate the mean and variance of the diode current. Assuming x to be the number of electrons being collected at the anode over a time τ, the diode current is given by

$$i_\tau = \frac{xe}{\tau} = \frac{e}{\tau}x. \tag{6.11}$$

The mean of i_τ is given by

$$I_0 = \overline{i_\tau} = \frac{e}{\tau}\overline{x}$$

$$= \frac{e}{\tau}\sum_i x.\frac{(n\tau)^i}{i!}e^{-n\tau} = n\tau\frac{e}{\tau} = ne \tag{6.12}$$

(since $\overline{x} = \sum x f(x)$ and $f(x)$ follows Poisson distribution).

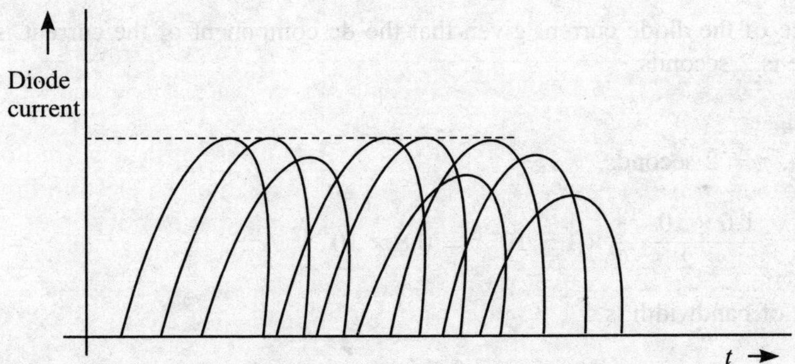

Fig. 6.9 *Pulse overlap representing diode current.*

The variance σ_i^2 of the current is given by

$$\sigma_i^2 = \left(\frac{e}{\tau}\right)^2 \sigma_x^2 = \frac{e}{\tau}I_0, \tag{6.13}$$

where $\sigma_x^2 = \overline{x^2} - (\overline{x})^2 = n\tau$. The ratio of the rms value to the mean value is

$$\frac{\sigma_i}{I_0} = \frac{\sqrt{(e/\tau)I_0}}{I_0} = \sqrt{\frac{e}{\tau}\frac{1}{I_0}}. \tag{6.14}$$

Thus the fluctuation of the current can be reduced by increasing the mean current I_0 and the time of observation τ. Again, τ is related to the maximum operating frequency as $f_{\max} = \frac{1}{2\tau}$. To process a signal over the observation time, the system must pass all frequencies up to f_{max}. Thus,

$$\sigma_i^2 = eI_0.2f_{max} = 2eI_0 f_{max}. \tag{6.15}$$

The variance per Hertz of bandwidth is

$$\overline{i^2} = \frac{\sigma_i^2}{f_{max}} = 2eI_0 \quad \text{A}^2/\text{Hz}. \tag{6.16}$$

Thus the device can be replaced by a noise current source producing white noise with a uniform spectrum. The above relation is called Schottky's equation.

Shot noise behaves in the same manner as thermal noise but has a different origin. Shot noise current is difficult to add to thermal noise voltage. A tractable method will be to derive a formula for an equivalent input-noise resistor representing the shot noise. This is a completely fictitious resistor assumed to be noiseless and preceding the device. The value of this resistance is chosen such that the same amount of noise appears at the output of the equivalent device as in the practical device. In this way, the noise current is replaced by an equivalent noise-generating resistor placed in series with the input electrode of the device. It is now simple to add the shot noise to the thermal noise.

Example 6.4

Find the variance of the diode current given that the dc component of the current is 1 mA and the observation time is 2 seconds.

Solution

Here $I_0 = 1$ mA, $\tau = 2$ seconds.

$$\sigma_i^2 = \frac{e}{\tau} I_0 = \frac{1.6 \times 10^{-19}}{2} \times 1 \times 10^{-3} = 0.8 \times 10^{-22} \text{ A}^2.$$

Variance per Hz of bandwidth is

$$\overline{i^2} = 2eI_0 = 2 \times 1.6 \times 10^{-19} \times 1 \times 10^{-3} = 3.2 \times 10^{-22} \quad \text{A}^2/\text{Hz}.$$

6.7 Noise Figure

Amplifiers and other active devices generate noise. At the output of such networks the message signal appears amplified, being corrupted by noise. A useful measure of the noise performance is furnished by the 'noise-figure'. The output noise is made up of two contributions, namely, the noise at the input and the device noise itself. Again, it is important to find the signal and noise at the same point for ensuring that the noise is not excessive. For this, the ratio of the signal-power to noise-power at the same point is very often used. The signal-to-noise power ratio (S/N ratio or SNR for short) deteriorates as the signal is transmitted through a device. This ratio may be used for comparing equipment while evaluating their performance.

Consider that any section of the receiver is modelled by linear, two-port device as shown in Fig. 6.10. Let $g(f)$ denote the available power gain of the two-port device defined as the ratio of the available signal power S_0 at the output of the device to the signal power S_i at the input. That is,

$g(f) = S_0/S_i$. We define the noise figure F of the two-port as the ratio of the S/N at the input to that at the output. Thus,

$$F = \frac{(S/N)_i}{(S/N)_0} = \frac{S_i/N_i}{S_0/N_0} = \frac{1}{g(f)}\frac{N_0}{N_i}, \tag{6.17}$$

where N_i and N_0 are the available noise powers at the input and output ports, respectively. The noise figure is commonly expressed in decibel unit for ideal receiver. If the device were noise-free, $S_i/N_i = S_0/N_0$; F would then be unity. Since the two-port generates some noise internally, $N_0 > N_i$ and $F > 1$. Please note that in (6.17) it is assumed that the input and output ports are matched to the source and the load respectively.

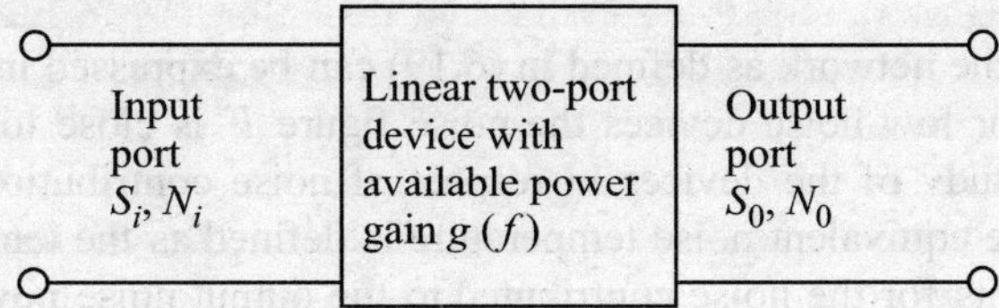

Fig. 6.10 *A two-port network of available power gain $g(f)$*

The two-port network can be modelled as a noise-free device of power gain $g(f)$ in series with an excess noise source that accounts for the noise generated internally in the device. Figure 6.11 shows such a noise-equivalent model. If is important here to note that the power gain $g(f)$ is not a characteristic of the two-port; rather, it depends on the source driving the network.

The noise ΔN generated in a two-port can be calculated as follows. Since the available noise power at the output is the available noise power at the input amplified by $g(f)$ plus the noise generated internal to the device, we have

$$N_0 = g(f)N_i + \Delta N \tag{6.18}$$

So, $F = \dfrac{1}{g(f)} \cdot \dfrac{N_0}{N_i} = \dfrac{1}{g(f)}\left[\dfrac{g(f)N_i + \Delta N}{N_i}\right] = 1 + \dfrac{\Delta N}{N_i g(f)}$

or, $$\Delta N = g(f)(F-1)N_i \tag{6.19}$$

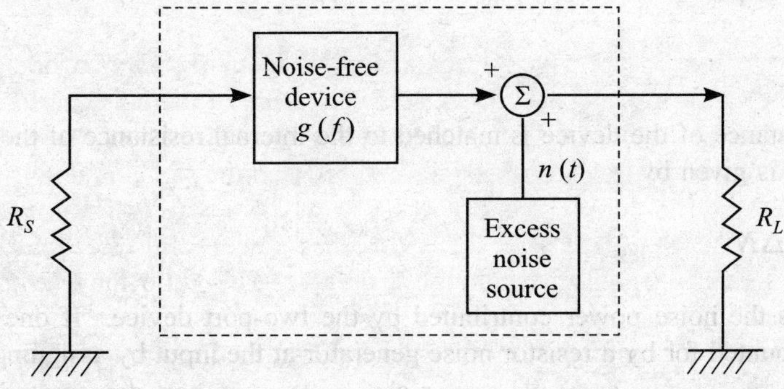

Fig. 6.11 *Noise-equivalent model for a real, two-port device*

Example 6.5
A two-port network is characterised by its noise figure $F = 0.5$ dB and its power gain $g = 10^{10}$. Calculate the noise power contributed by the two-port for an input noise of 1 μV.

Solution
Here $F = 0.5$ dB=1.122,

$$\Delta N = g(F-1)N_i = 10^{10}(1.122-1)\times(10^{-6})^2\text{watts} = 0.122\times10^{-2}\text{ watts}= 1.22\text{ mW}.$$

6.8 Equivalent Noise Temperature

The noise contributed by the network as defined in (6.19) can be expressed in terms of the equivalent noise temperature, T_e. For low-noise devices the noise figure F is close to unity and it is difficult to make a comparative study of the devices in respect of noise contribution. The parameter T_e is helpful in such cases. The equivalent noise temperature is defined as the temperature at the input of the network which accounts for the noise contributed to the output noise power.

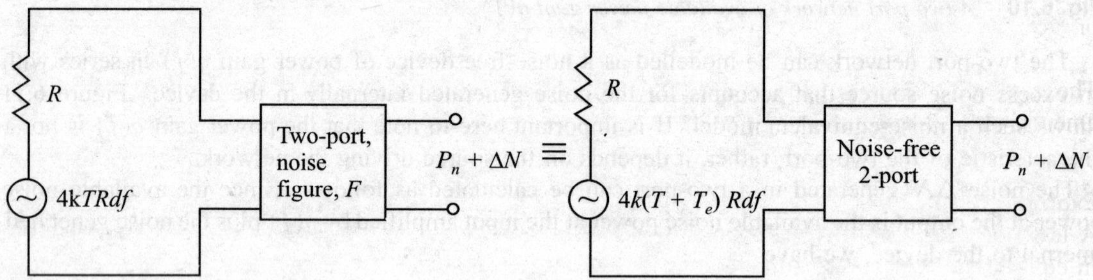

Fig. 6.12 *A two-port device driven by a noise generator. The network is made noise-free by the inclusion of the temperature term T_e to the resistor.*

Consider a thermal noise voltage generator driving a linear two-port device as shown in Fig. 6.12. The input resistance of the two-port device is matched to the source resistance. The available noise power at the device input is

$$N_i = kTdf \tag{6.20}$$

since the input resistance of the device is matched to the internal resistance of the source. The total output noise power is given by

$$N_0 = gN_i + \Delta N,$$

where ΔN denotes the noise power contributed by the two-port device. If one assumes that this excess noise is accounted for by a resistor noise generator at the input by assigning a temperature T_e to it, then

$$\Delta N = kT_e df g(f). \tag{6.22}$$

Thus,

$$N_O = gN_i + kT_e df g(f)$$

$$= gkTdf + gkT_e df$$

$$= gk(T + T_e)df. \tag{6.23}$$

Hence the noise figure

$$F = \frac{N_O}{gN_i}$$

$$= \frac{gk(T + T_e)df}{gkTdf} = 1 + \frac{T_e}{T}$$

or $\quad T_e = T(F - 1).$ (6.24)

The equivalent noise temperature need not be the physical temperature of any part of the system. Once noise figure is known, the equivalent noise temperature can be calculated from (6.24).

Example 6.6.
A low-noise RF amplifier has a noise figure $F = 2.3$ dB and power gain $g = 40$ dB. The system is in equilibrium with room temperature $T = 290$ K. Find the effective input noise temperature of the receiver.

Solution
Here, $\quad F = 2.3\text{dB} = 1.7, \quad g = 40\text{dB} = 10^4.$

So, $\quad T_e = T(F - 1) = 290(1.7 - 1) = 203$ K.

6.9 Noise Bandwidth

The noise bandwidth of a filter is defined as the bandwidth of an ideal filter which passes the same amount of noise power as does the real filter under consideration. Consider a bandpass filter of transfer function $H(f)$, symmetrical about the centre frequency f_0, as depicted in Fig. 6.13. Assume that the filter is driven by a noise source of power spectral density $G_n(f)$. The PSD of the noise appearing at the output of the filter is then $G_n(f)|H(f)|^2$. The noise power at the output is given by

$$N_O = \int_{-\infty}^{\infty} G_n(f)|H(f)|^2 df. \tag{6.25}$$

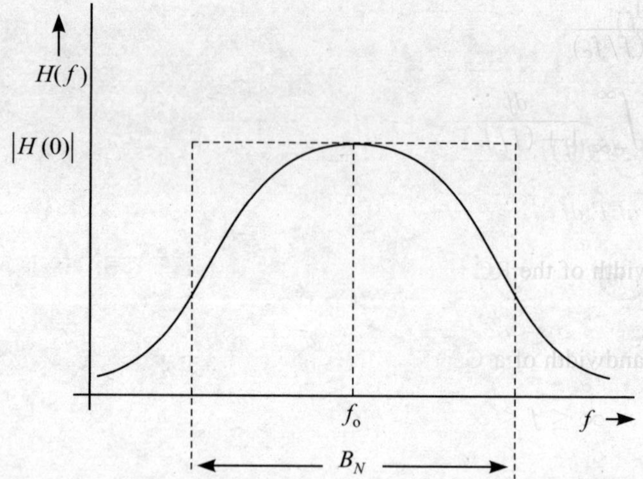

Fig. 6.13 *Bandpass filter characteristics. The noise bandwidth is also depicted.*

Assume an ideal filter with a rectangular passband (shown dotted in Fig. 6.13) and width B_N such that the same noise power is available at the output. This gives

$$N_0 = \int_{-\infty}^{\infty} G_n(f)|H(f)|^2 df = G_n(f)|H(0)|2B_N$$

or, $$B_N = \frac{1}{2|H(0)|G_n(f)} \int_{-\infty}^{\infty} G_n(f)|H(f)|^2 df. \tag{6.26}$$

The factor 2 signifies the contribution from the negative frequency components. As a special case, if the filter be driven by a white noise source of $G_n(f) = \frac{1}{2}kT$, then the noise bandwidth becomes

$$B_N = \frac{1}{2|H(0)|} \int_{-\infty}^{\infty} |H(f)|^2 df = \frac{1}{|H(0)|} \int_{-0}^{\infty} |H(f)|^2 df. \tag{6.27}$$

Example 6.7

Derive the noise bandwidth of the first-order RC lowpass filter of 3 dB cutoff frequency f_c.

Solution

The first-order low-pass RC filter is shown in the adjoining diagram. The transfer function of the filter is $H(f) = \dfrac{1}{1 + j2\pi fCR}$.

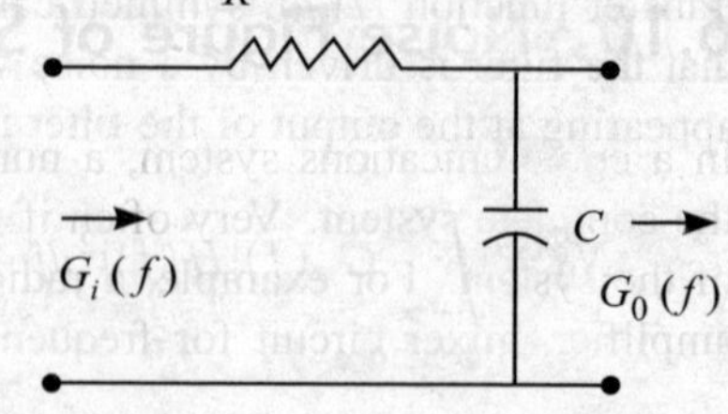

At 3dB points, $|H(f)| = \frac{1}{\sqrt{2}}|H(0)|$, implying

$$1 + (2\pi fCR)^2 = 2 \Rightarrow f_c = \frac{1}{2\pi CR}.$$

Thus, $H(f) = \dfrac{1}{1 + j(f/f_c)}$.

Hence, $B_N = \dfrac{1}{2 \times 1} \displaystyle\int_{-\infty}^{\infty} \frac{df}{1 + (f/f_c)^2} = \frac{f_c^2}{2} \int_{-\infty}^{\infty} \frac{df}{f^2 + f_c^2} = \frac{\pi}{2} f_c.$

So, $\quad B_N = \dfrac{\pi}{2} f_c.$

Thus the noise bandwidth of the RC filter is $\pi/2 = 1.57$ times the 3 dB bandwidth f_c.

Example 6.8

Calculate the noise bandwidth of a Gaussian filter having the transfer function

$$\mathcal{H}(f) = e^{-k^2 f^2}, -\infty \leq f \leq \infty.$$

Solution

By the (6.27)

$$B_N = \frac{1}{2\left|H(0)\right|} \int_{-\infty}^{\infty} \left|H(f)\right|^2 df = \int_0^{\infty} e^{-2k^2 f^2} df \quad \text{(since } |H(0)| = 1\text{)}.$$

Let $2k^2 f^2 = x$ so that $f = \dfrac{x^{1/2}}{\sqrt{2}k}$ and $df = \dfrac{x^{-1/2}}{2\sqrt{2}k} dx$.

Thus, $B_N = \dfrac{1}{2\sqrt{2}k} \displaystyle\int_0^{\infty} e^{-x}\, x^{-1/2} dx = \frac{1}{2\sqrt{2}k} \Gamma\left(\frac{1}{2}\right) = \sqrt{\frac{\pi}{2}} \times \frac{1}{2k}.$

The 3 dB cutoff frequency is given by

$$e^{-k^2 f_c^2} = \frac{1}{\sqrt{2}} H(0) = \frac{1}{\sqrt{2}}.$$

So, $(kf_c)^2 = \frac{1}{2} \ln 2$ or, $f_c = \sqrt{\dfrac{1}{2k^2} \ln 2}$.

Therefore, $\dfrac{B_N}{f_c} = \sqrt{\dfrac{\pi}{2}} / \ln 2$.

6.10 Noise Figure of Several Two-port Networks in Cascade

In a communications system, a number of two-port networks are assumed to be cascaded to form the complete system. Very often if is required to find the overall noise temperature and noise figure of the system. For example, a radio earth-station includes a receiving antenna, front-end low-noise amplifier, mixer circuit for frequency-down conversion, IF amplifier, etc. If the power gains and the temperatures of the elements be known, one can compute the noise figure of the whole system. Consider three two-port networks of noise figures F_1, F_2, and F_3 and power gains g_1, g_2, and g_3 connected in cascade as shown in Fig. 6.14. It is assumed that the networks are matched and the noise figure F_2 and F_3 are defined for an input noise power N_i. The noise power contributed by any device is $(F-1)N_i$ and appears at its output as $g(F-1)N_i$. The total output noise power consists of:

(a) The term $N_i g_1 g_2 g_3$, as the input noise N_i gets multiplied by the gains of individual stages
(b) The noise generated by the first network which appears at the output, being multiplied by $g_1 g_2 g_3$. This term is $N_i(F_1-1)g_1 g_2 g_3$.
(c) The noise generated by the second port that appears at the output as $(F_2-1)N_i g_2 g_3$.
(d) The noise generated by the third network that appears at the output as $(F_3-1)N_i g_3$.

Combining the above, total output noise power

$$N_O = N_i g_1 g_2 g_3 + N_i(F_1-1)g_1 g_2 g_3 + N_i(F_2-1)g_2 g_3 + N_i(F_3-1)q_3. \tag{6.28}$$

We now define the overall noise figure F as the ratio of the output noise power to the noise power when all the networks are assumed noiseless. Thus

$$F = \frac{N_O}{N_i g_1 g_2 g_3} = 1 + (F_1-1) + \frac{(F_2-1)}{g_1} + \frac{(F_3-1)}{g_1 g_2}$$

$$= F_1 + \frac{F_2-1}{g_1} + \frac{F_3-1}{g_1 g_2}. \tag{6.29}$$

Fig. 6.14 *Three noisy two-ports connected in cascade*

The above result can readily be extended to the cascade connection of n two-ports as

$$F = F_1 + \frac{F_2-1}{g_1} + \frac{F_3-1}{g_1 g_2} + \cdots + \frac{F_{n-1}}{g_1 g_2 \cdots g_{n-1}}. \tag{6.30}$$

If $T_1, T_2, \cdots, T_n$ be the equivalent noise temperatures of the individual stages, then by (6.24) we can write the expression for the overall equivalent noise temperature T_e as

$$T_e = T_1 + \frac{T_2}{g_1} + \frac{T_3}{g_1 g_2} + \cdots + \frac{T_n}{g_1 g_2 \cdots g_{n-1}}. \tag{6.31}$$

It is observed from (6.30) and (6.31) that the overall F and T_e are dominated by the first stage, provided the gain of the first stage is sufficiently high.

Example 6.9

A microwave receiving system consists of a cascade of a maser amplifier, a travelling wave tube, and an IF amplifier. The system is in equilibrium with a temperature of 4K. The power gains and noise figures are indicated in the diagram. Find the overall noise figure and the equivalent noise temperature.

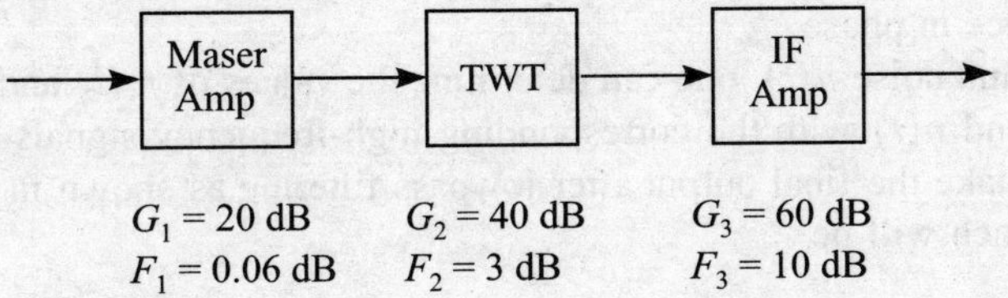

Solution

For the first block, $g_1 = 20$ dB $= 100$; $F_1 = 0.06$ dB$= 1.0139$.

For the second block, $g_2 = 40$ dB $= 10^4$; $F_2 = 3$ dB$= 2$.

For the third block, $g_3 = 60$ dB $= 10^6$; $F_3 = 10$ dB $= 10$.

By (6.30),

$$F = 1.0139 + \frac{2-1}{100} + \frac{10-1}{100 \times 10^6} = 1.0239 \approx F_1.$$

Now, $T_1 = (F_1 - 1)T_0 = (1.0139 - 1) \times 4K = 0.0556K$

$$T_2 = (F_2 - 1)T_0 = (2 - 1) \times 4\text{K} = 4\text{K}$$

$$T_3 = (F_3 - 1)T_0 = (10 - 1) \times 4\text{K} = 36\text{ K}.$$

By (6.31), $T_e = 0.0556 + \dfrac{4}{100} + \dfrac{36}{100 \times 10^4} = 0.0956$ K.

6.11 Narrowband Noise

In this section we will discuss a special class of noise called the narrowband noise that appears at the output of frequency-selective circuits. Normally, the receiver section of a communications system incorporates some means of processing the received signal in order to make the signal free from noise. If the processing unit be a narrowband filter having a passband large enough to pass the modulated signal, then the output of the filter is the signal plus the admissible noise. The noise that is allowed to pass is narrowband in the sense that its spectrum is nonzero only within the frequency bands of a width that equals the bandwidth of the filter. If the midband frequency be ω_c and the bandwidth be

BW, then for narrowband noise, $BW/\omega_c << 1$ must be satisfied.

The most general representation of a narrowband noise is the following form

$$n(t) = n_i(t)\cos\omega_0 t - n_q(t)\sin\omega_0 t, \tag{6.32}$$

where $n_i(t)$ is called the 'in-phase component' and $n_q(t)$ is called the 'quadrature-phase component' of a narrowband signal $n(t)$. The function $n_i(t)$ and $n_q(t)$ are low-frequency functions as they change slightly over the period $T = 2\pi/\omega_0$. Since $n_i(t)$ is a low-frequency function, the spectrum of $n_i(t)\cos\omega_0 t$ will be concentrated near the points $\pm\omega_0$. The same is true for the term $n_q(t)\sin\omega_0 t$ except for the difference in phase.

Given the narrowband noise $n(t)$, one can determine the values of $n_i(t)$ and $n_q(t)$ as follows. We multiply the narrowband $n(t)$ with the corresponding high-frequency signals ($\sin\omega_0 t$ or $\cos\omega_0 t$ as the case may be) and take the final output after lowpass filtering as shown in Fig. 6.15. The output noise in the upper branch will be

$$\begin{aligned} n_0(t) &= n(t)\cos\omega_0 t \\ &= [n_i(t)\cos\omega_0 t - n_q(t)\sin\omega_0 t]\cos\omega_0 t \\ &= \frac{n_i(t)}{2} + \frac{1}{2}n_i(t)\cos 2\omega_0 t - \frac{1}{2}n_q(t)\sin 2\omega_0 t. \end{aligned} \tag{6.33}$$

The lowpass filter suppresses the $2\omega_0$-frequency components allowing $\frac{1}{2}n_i(t)$. Similarly, the lower branch of Fig. 6.15 will extract $\frac{1}{2}n_q(t)$.

The analysis the spectrum of the narrowband noise can be tackled using the concept of the complex envelope. Let us define the 'complex envelop' of the narrowband noise by the complex low-frequency function

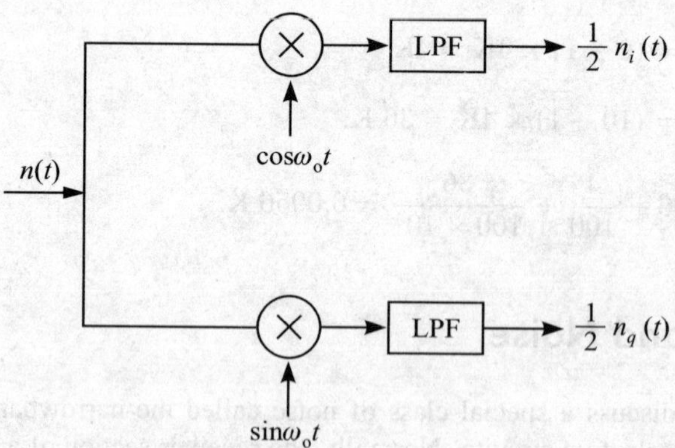

Fig. 6.15 *Extraction of in-phase and quadrature phase components*

$$\tilde{A}(t) = n_i(t) + jn_q(t). \tag{6.34}$$

In terms of the complex envelope, the narrowband noise $n(t)$ is expressed as

$$n(t) = n_i(t)\text{Re}\,[e^{j\omega_0 t}] - n_q(t)\text{Im}\,[e^{j\omega_0 t}]$$
$$= \text{Re}\,[\tilde{A}(t)e^{j\omega_0 t}]. \tag{6.35}$$

The complex envelope can be expressed in exponential from as $\tilde{A}(t) = A(t)e^{j\phi(t)}$, where $A(t)$ is a real, non-negative function of time called the physical envelope, and $\phi(t)$ is a slowly-varying initial phase of the narrowband process.

The physical envelop $A(t)$ is given by

$$A(t) = \sqrt{n_i^2(t) + n_q^2(t)}. \tag{6.36}$$

If $S_n(\omega)$ denotes the spectrum of the complex envelope of $n(t)$, then the spectrum $G_n(\omega)$ of the noise $n(t)$ is obtained as follows:

$$G_n(\omega) = \int_{-\infty}^{\infty} n(t)e^{-j\omega t}dt = \int_{-\infty}^{\infty} \text{Re}[\tilde{A}(t)e^{j\omega_0 t}]e^{-j\omega t}dt$$
$$= \int_{-\infty}^{\infty} \left[\frac{1}{2}\tilde{A}(t)e^{j\omega_0 t} + \frac{1}{2}\tilde{A}^*(t)e^{-j\omega_0 t}\right] e^{-j\omega t}$$
$$= \frac{1}{2}\int_{-\infty}^{\infty} \tilde{A}(t)e^{-j(\omega-\omega_0)t}dt + \frac{1}{2}\int_{-\infty}^{\infty} \tilde{A}^*(t)e^{-j(\omega+\omega_0)t}dt$$
$$= \frac{1}{2}S_n(\omega-\omega_0) + \frac{1}{2}S_n^*(-\omega-\omega_0). \tag{6.37}$$

Equation (6.37) states that the spectrum of $n(t)$ is obtained by translating the spectrum of its complex envelope from the surrounding of $\omega = 0$ to $\omega = \pm\omega_0$.

Example 6.10

Show that the in-phase and quadrature-phase amplitudes of the narrowband noise $n(t)$ are connected through

$$n_i(t) = n(t)\cos\omega_0 t + \widehat{n}(t)\sin\omega_0 t,$$
$$n_q(t) = \widehat{n}(t)\cos\omega_0 t - n(t)\sin\omega_0 t,$$

where $\widehat{n}(t)$ denotes the Hilbert transform of $n(t)$.

Solution

By definition,

$$\widehat{n}(t) = H[n(t)] = n(t) * \frac{1}{\pi t} = \frac{1}{\pi}\int_{-\infty}^{\infty} \frac{n(\tau)}{t-\tau}d\tau. \tag{6.38}$$

Expanding $n_i(t)$ and $n_q(t)$ in a Taylor series about the point $\tau = t$ and substituting in (6.38), we have

$$\widehat{n}(t) = \frac{1}{\pi}\int_{-\infty}^{\infty}\frac{d\tau}{(t-\tau)}\left[n_i(t) + n_i'(t)(t-\tau) + \frac{1}{2}n_i''(t)(t-\tau)^2 + \cdots\right]\cos\omega_0\tau$$

$$-\frac{1}{\pi}\int_{-\infty}^{\infty}\frac{d\tau}{(t-\tau)}\left[n_q(t) + n_q'(t)(t-\tau) + \frac{1}{2}n_q''(t)(1-\tau)^2 + \cdots\right]\sin\omega_0\tau. \tag{6.39}$$

In (6.38), the higher-order derivatives $n_i'(t), n_q'(t), n_i''(t), n_q''(t)$ etc. are negligibly small and can be omitted. In doing so,

$$\widehat{n}(t) = n_i(t)\frac{1}{\pi}\int_{-\infty}^{\infty}\frac{d\tau}{(t-\tau)}\cos\omega_0\tau - n_q(t)\frac{1}{\pi}\int_{-\infty}^{\infty}\frac{d\tau}{(t-\tau)}\sin\omega_0\tau$$

$$= n_i(t)\sin\omega_0 t + n_q(t)\cos\omega_0 t. \tag{6.40}$$

Also , for the original noise we have

$$n(t) = n_i(t)\cos\omega_0 t - n_q(t)\sin\omega_0 t. \tag{6.41}$$

Solving for n_i and n_q, we get the results.

6.12 Correlation Theory of Noise

There is a close correspondence between the correlation and the spectral properties of noise. In general, we can state that a random process in the time domain is related to another random process in the frequency domain. Precisely, the power spectrum of a noise of zero mean is the Fourier transform of its autocorrelation function. That is,

$$G_n(f) = \int_{-\infty}^{\infty} R_n(\tau)e^{-j2\pi f\tau}d\tau, \tag{6.42}$$

where $R_n(\tau)$ signifies the mean power. As an example, the white noise is specified by its power spectrum $G_n(f) = \frac{1}{2}kT$, which is independent of frequency. The autocorrelation function of white noise is then

$$R_n(\tau) = \mathcal{F}^{-1}[G_n(f)] = \frac{1}{2}kT\frac{1}{2\pi}\int_{-\infty}^{\infty} e^{j\omega\tau}dw = \frac{1}{2}kT\delta(\tau). \tag{6.43}$$

The white noise is thus a delta-correlated random process.

6.12.1 Correlation function of narrowband process

The autocorrelation function is given by

$$R_n(\tau) = \frac{1}{2\pi}\int_{-\infty}^{\infty} G_n(f)e^{j2\pi f\tau}d\omega. \tag{6.44}$$

Both $R_n(\tau)$ and $G_n(f)$ are even functions. Equation (6.44) then reads as

$$R_n(\tau) = \frac{1}{\pi}\int_0^{\infty} G_n(f)\cos(2\pi f\tau)d\omega, \tag{6.45}$$

where $G_n(f)$ is the one-sided power spectral density of the random process. The power spectral density should have a sharply defined peak around a certain nonzero frequency ω_0, say. By a change of variable $\omega = \omega_0 + \Omega$, equation (6.45) reads as

$$R_n(\tau) = \frac{1}{\pi}\int_{-\omega_0}^{\infty} G_n(\omega_0 + \Omega)\cos(\omega_0 + \Omega)\tau d\Omega. \tag{6.46}$$

The change of variable implies that the power spectrum of the process has been shifted from ω_0 to zero frequency. Since for a narrowband process the power spectrum is vanishingly small at frequencies close to zero, we can safely replace the lower limit of integration in (6.46) by $-\infty$ without committing a large error. Thus,

$$\begin{aligned} R_n(\tau) &= \frac{1}{\pi}\int_{-\infty}^{\infty} G_n(\omega_0 + \Omega)\cos(\omega_0 + \Omega)\tau d\tau \\ &= \frac{1}{\pi}G_n(\omega_0 + \omega)\cos\omega_0\tau\cos\omega\tau d\omega - \frac{1}{\pi}\int_{-\infty}^{\infty} G_n(\omega_0 + \Omega)\sin\Omega\tau d\Omega \\ &= a(\tau)\cos\omega_0\tau - b(\tau)\sin\omega_0\tau, \end{aligned} \tag{6.47}$$

where $a(\tau) = \frac{1}{\pi}\int_{-\infty}^{\infty} G_n(\omega_0 + \Omega)\cos\Omega\tau d\Omega$ and $b(\tau) = \dfrac{1}{\pi}\displaystyle\int_{-\infty}^{\infty} G_n(\omega_0 + \omega)\sin\Omega\tau d\Omega.$

The functions $a(\tau)$ and $b(\tau)$ are slowly-varying functions. It can be proved that $a(\tau)$ is even and $b(\tau)$ is odd.

For the case when $G_n(\omega)$ is symmetric about its centre frequency ω_0, $b(\tau) = 0$ and

$$R_n(\tau) = a(\tau)\cos\omega_0\tau. \tag{6.48}$$

On setting $\tau = 0$ in (6.48), $R_n(0) = a(0) = \sigma_n^2$.

The above mathematical background will be helpful for establishing the correlation properties of the in-phase and quadrature amplitudes of the random process $n(t)$ defined in (6.32). The conjugate $\widehat{n}(t)$ of the random process $n(t)$ is defined in (6.40). Rewriting,

$$n(t) = n_i(t)\cos\omega_0 t - n_q(t)\sin\omega_0 t$$

and $\widehat{n}(t) = n_i(t)\sin\omega_0 t + n_q(t)\cos\omega_0 t.$

So, $n_i(t) = n(t)\cos\omega_0 t + \widehat{n}(t)\sin\omega_0 t$ (6.49)

and $n_q(t) = -n(t)\sin\omega_0 t + \widehat{n}(t)\cos\omega_0 t.$ (6.50)

If we assume that the original random process $n(t)$ is a Gaussian process with zero mean $E[n(t)] = 0$, then the conjugate process $\widehat{n}(t)$ will again be Gaussian with zero mean $E[\widehat{n}(t)] = 0$. It follows that the processes $n_i(t)$ and $n_q(t)$ are also Gaussian with $E[n_i(t)] = E[n_q(t)] = 0$.

The autocorrelation function of the process $n_i(t)$ is given by

$$R_{n_i}(\tau) = \int_{-\infty}^{\infty} n_i(t)n_i(t-\tau)dt = \int_{-\infty}^{\infty} [n(t)\cos\omega_0 t + \widehat{n}(t)\sin\omega_0 t] \times$$

$$[n(t-\tau)\cos\omega_0(t-\tau) + \widehat{n}(t-\tau)\sin\omega_0(t-\tau)]dt$$

$$= R_n(\tau)\cos\omega_0\tau + R_{n,\widehat{n}}(\tau)\sin\omega_0\tau. \tag{6.51}$$

In the above equation, $R_n(\tau)$ is given by (6.47). The cross-correlation function $R_{n,\widehat{n}}(\tau)$ is derived as follows:

$$R_{n,\widehat{n}}(\tau) = \int_{-\infty}^{\infty} n(t)\widehat{n}(t-\tau)dt = \int_{-\infty}^{\infty} n(t)\left[\frac{1}{\pi}\int_{-\infty}^{\infty} \frac{n(t')}{t-\tau-t'}dt'\right]dt$$

$$= \frac{1}{\pi}\int_{-\infty}^{\infty} \frac{R_n(t'-t)}{\tau-(t'-t)}dt' = \frac{1}{\pi}\int_{-\infty}^{\infty} \frac{R_n(\zeta)}{\tau-\zeta}d\zeta\ , \tag{6.52}$$

which is nothing but the Hilbert transform of the autocorrelation function of the process $n(t)$. Now, by (6.47),

$$R_n(\tau) = a(\tau)\cos\omega_0\tau - b(\tau)\sin\omega_0\tau.$$

On Hilbert transformation,

$$R_{n,\widehat{n}}(\tau) = H[R_n(\tau)] = a(\tau)\sin\omega_0\tau + b(\tau)\cos\omega_0\tau. \tag{6.53}$$

Substituting (6.47) and (6.53) into (6.51), we have

$$R_{n_i}(\tau) = a(\tau)\cos^2\omega_0\tau - b(\tau)\sin\omega_0\tau\cos\omega_0\tau + a(\tau)\sin^2\omega_0\tau + b(\tau)\sin\omega_0\tau\cos\omega_0\tau$$

$$= a(\tau). \tag{6.54}$$

In a similar manner, it is proved that

$$R_{n_q}(\tau) = a(\tau) \tag{6.55}$$

and $\quad R_{n_i,n_q}(\tau) = b(\tau).$ $\hfill (6.56)$

Setting $\tau = 0$ in (6.54) and (6.55) yields,

$$R_{n_i}(0) = R_{n_q}(0) = a(0),$$

that is, $\sigma^2_{n_i} = \sigma^2_{n_q} = \sigma^2_n.$ (6.57)

Thus it is proved that the powers (variances) of the in-phase and quadrature amplitudes are equal to the power (variance) of the original narrowband process $n(t)$.

Example 6.11

A white Gaussian noise of zero mean has the power spectral density $\eta/2$. It is applied to an ideal lowpass filter of bandwidth B. Find the autocorrelation function of the bandlimited noise.

Solution

The power spectral density $G_n(f)$ is given by

$$G_n(f) = \begin{cases} \eta/2 & -B < f < B \\ =0 & \text{otherwise.} \end{cases} \tag{6.58}$$

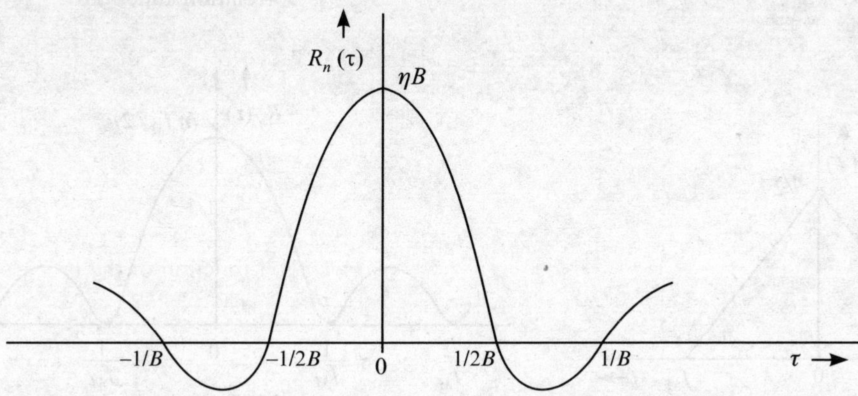

Fig. 6.16 *Autocorrelation function of Example 6.11*

The autocorrelation function

$$R_n(\tau) = \mathcal{F}^{-1}[G_n(f)] = \int_{-\infty}^{\infty} G_n(f)e^{j2\pi f\tau}\,df$$

$$= \frac{n}{2}\int_{-B}^{B} e^{j2\pi f\tau}\,df = \frac{n}{2}\frac{1}{j2\pi f\tau}\left[e^{j2\pi f\tau}\right]_{B}^{B}$$

$$= \frac{\eta}{j4\pi\tau}\left(e^{j2\pi B\tau} - e^{-j2\pi B\tau}\right) = \frac{\eta}{2\pi\tau}\sin(2\pi B\tau)$$

$$= \eta B\left[\frac{\sin 2\pi B\tau}{2\pi B\tau}\right]. \tag{6.59}$$

Figure 6.16 shows the plot of $R_n(\tau)$. The variance σ^2 or mean power of the noise is $R_n(0) = \sigma^2 = \eta B$.

Example 6.12

The power spectral density of a noise $n(t)$ is shown in Fig. 6.17. Find and sketch the autocorrelation function of $n(t)$. Find also the mean power.

Solution

$$\text{Here } G_n(f) = \frac{\eta}{2}\left(1 - \frac{|f|}{f_M}\right), \qquad |f| \le f_M \tag{6.60}$$

$$= 0, \qquad |f| > f_M$$

The autocorrelation function

$$R_n(\tau) = \int_{-\infty}^{\infty} G_n(f) e^{j2\pi f\tau}\, d\tau$$

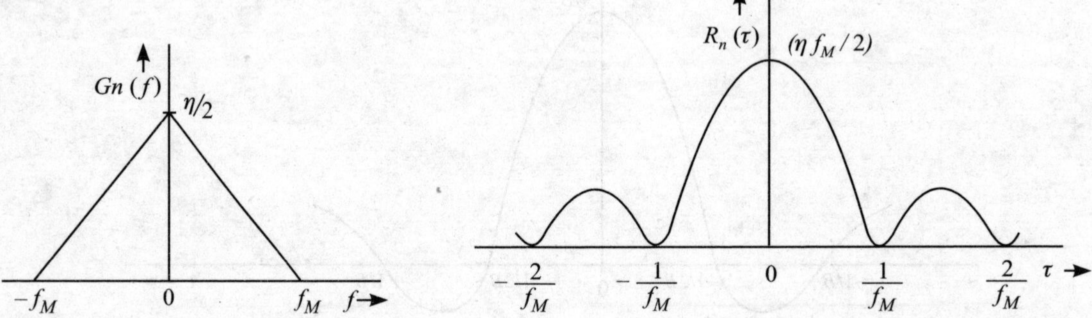

Fig. 6.17 *PSD of a noise $n(t)$*

Fig. 6.18 *Autocorrelation function of n(t)*

$$= \frac{n}{2}\int_{-f_M}^{0}\left(1 + \frac{f}{f_M}\right) e^{j2\pi f\tau}\, d\tau + \frac{n}{2}\int_{0}^{f_M}\left(1 - \frac{f}{f_M}\right) e^{j2\pi f\tau}\, df.$$

Integrating, we have

$$R_n(\tau) = \frac{\eta f_M}{2}\left(\frac{\sin \pi f_M \tau}{\pi f_M \tau}\right)^2$$

$$= \frac{\eta f_M}{2}\,\text{sinc}^2(f_M\tau). \tag{6.61}$$

The plot is shown in Fig. 6.18.

Example 6.13

A network consists of two RC sections is decoupled by an ideal amplifier of voltage gain A. The network is driven by a white noise source of power spectral density $n/2$. Find the autocorrelation function of the output signal.

Solution

The magnitude square of the transfer function of the network of Fig. 6.19 is given by

$$\left|H(f)\right|^2 = \left|\frac{1/j\omega C_1}{R_1 + 1/j\omega C_1}\right|^2 A^2 \left|\frac{1/j\omega C_2}{R_2 + 1/j\omega C_2}\right|^2$$

$$= \frac{A^2}{(1+\omega^2\tau_1^2)(1+\omega^2\tau_2^2)},$$

where $\tau_1 = R_1C_1$ and $\tau_2 = R_2C_2$. The PSD of the output noise is

$$G_{no}(f) = G_{ni}(f)\left|H(f)\right|^2 = \frac{A^2\eta/2}{(1+\omega^2\tau_1^2)(1+\omega^2\tau_2^2)}.$$

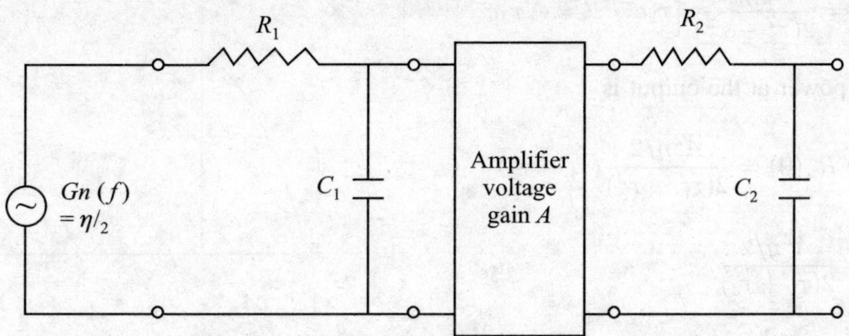

Fig. 6.19 *Network of Example 6.13*

Hence the autocorrelation function

$$R_n(\tau) = A^2\eta/2 \int_{-\infty}^{\infty} \frac{e^{j2\pi f\tau}}{(1+\omega^2\tau_1^2)(1+\omega^2\tau_2^2)}df.$$

The integral is calculated using Cauchy's residue theorem. The integrand

$$I(f) = \frac{e^{j2\pi f\tau}}{(1+\omega^2\tau_1^2)(1+\omega^2\tau_2^2)}$$ has four poles at $\pm j/(2\pi\tau_1)$ and $\pm j/(2\pi\tau_2)$.

The residue of $I(f)$ at $f = j/(2\pi\tau_1)$ is

$$\left.\frac{e^{j2\pi f\tau}}{\frac{d}{df}\left[(1+4\pi^2 f^2\tau_1^2)(1+4\pi^2 f^2\tau_2^2)\right]}\right|_{f=\frac{j}{2\pi\tau_1}} = \frac{\tau_1 e^{-\tau/\tau_1}}{4\pi j(\tau_1^2-\tau_2^2)}.$$

Similarly, the residue at $f = j/(2\pi\tau_2)$ is $\dfrac{-T_2 e^{-j\tau/\tau_2}}{4\pi j(\tau_1^2 - \tau_2^2)}$.

Hence,

$$R_n(\tau) = A^2\eta/2(2\pi j)\left[\frac{\tau_1 e^{-\tau/\tau_1}}{4\pi j(\tau_1^2 - \tau_2^2)} - \frac{\tau_2 e^{-\tau/\tau_2}}{4\pi j(\tau_1^2 - \tau_2^2)}\right]$$

$$= \frac{A^2\eta/2}{2(\tau_1^2 - \tau_2^2)}\left[\tau_1 e^{-\tau/\tau_1} - \tau_2 e^{-\tau/\tau_2}\right] \tag{6.62}$$

for $\tau > 0$.

In the same way $R_n(\tau)$ for $\tau < 0$ can be calculated. The results will be the same as the above since $R_n(\tau) = R_n(-\tau)$. Thus,

$$R_n(\tau) = \frac{A^2\eta/2}{2(\tau_1^2 - \tau_2^2)}\left[\tau_1 e^{-|\tau|/\tau_1} - \tau_2 e^{-|\tau|/\tau_2}\right]. \tag{6.63}$$

The average power at the output is

$$\bar{P}_{\text{avg}} = R_n(0) = \frac{A^2\eta/2}{2(\tau_1^2 - \tau_2^2)}(\tau_1 - \tau_2)$$

$$= \frac{A^2\eta/2}{2(\tau_1 + \tau_2)}.$$

6.13 Autocorrelation of Sum of Two Random Processes

In deriving Equation (6.3) or (6.4) we assumed the noise sources as independent random processes, and thus the noise voltage or current added in mean square sense. A proof of this is now presented here.

Two stationary random processes $X(t)$ and $Y(t)$, when added, results in another stationary random process $Z(t) = X(t) + Y(t)$. By the autocorrelation function (vide Eq. 5.104), the autocorrelation of $Z(t)$ is obtained as

$$R_Z(\tau) = E[Z(t)Z(t+\tau)] = E[\{X(t) + Y(t)\}\{X(t+\tau) + Y(t+\tau)\}]$$

$$= E[X(t)X(t+\tau) + Y(t)Y(t+\tau) + X(t)Y(t+\tau) + Y(t)X(t+\tau)]$$

$$= E[X(t)X(t+\tau)] + E[Y(t)Y(t+\tau)] + E[X(t)Y(t+\tau)] + E[Y(t)X(t+\tau)]$$

$$= R_X(\tau) + R_Y(\tau) + R_{XY}(\tau) + R_{YX}(\tau) \tag{6.64}$$

where $R_{XY}(\tau)$ and $R_{YX}(\tau)$ are two cross-correlation terms, given as

$$R_{XY}(\tau) = E[X(t)Y(t+\tau)] \tag{6.65}$$

When the two processes on $X(t)$ and $Y(t)$ are uncorrelated, their cross- correlation should be the product of their means. That is, for uncorrelated $X(t)$ and $Y(t)$,

$$R_{XY}(\tau) = E[X(t)Y(t+\tau)] = E[X(t)]E[Y(t+\tau)] = \overline{X(t)}.\overline{Y(t)}$$

$$= m_x.m_y \tag{6.66}$$

For all t and τ in the above, $m_x = \overline{X(t)}$ and $m_y = \overline{Y(t)}$ are the means of $X(t)$ and $Y(t)$, respectively. Then Eq.(6.64) can be written as

$$R_Z(\tau) = R_X(\tau) + R_Y(\tau) + 2m_x m_y \tag{6.67}$$

If we further assume that either of m_x or m_y or both are zero mean, that is $m_x = 0$ or $m_y = 0$ or both $m_x, m_y = 0$, then

$$R_Z(\tau) = R_X(\tau) + R_Y(\tau) \tag{6.68}$$

Consequently, the power spectral density of $Z(t)$ is

$$G_Z(f) = G_X(f) + G_Y(f) \tag{6.69}$$

$$= \overline{X^2(t)} + \overline{Y^2(t)}$$

The above notion has been applied in the calculation of mean square noise voltage or current (Eqs (6.3) and (6.4))

6.14 Filtering of Stationary Process in an LTI System

Filtering is frequently used to suppress some frequency band and to allow the desired frequency band to pass through the system. This finds application in removing out-of-band interference and noise components in the desired signal. This section studies the effects of filtering by an LTI system to stationary random processes.

Let an LTI system with an impulse response $h(t)$ be subjected to a stationary random process $X(t)$, as shown in fig 6.21

$$x(t) \longrightarrow \boxed{h(t)} \longrightarrow y(t) = x(t) * h(t)$$

Fig.6.21 *A random input $x(t)$ to an LTI system outputs another random signal $y(t)$*

Let $x(t)$ be a sample function of the stationary stochastic process $X(t)$. The quantity $y(t)$ is also a sample of the output stochastic process $Y(t)$. The output of the LTI filter is

$$y(t) = x(t) * h(t) = \int_{-\infty}^{\infty} x(\tau)\, h(t-\tau)\, d\tau = \int_{-\infty}^{\infty} h(\tau)\, x(t-\tau) d\tau \tag{6.70}$$

The mean value or expectation of $y(t)$ is given by

$$m_y = E\left[y\left(t\right)\right]Z$$

$$= \int_{-\infty}^{\infty} h\left(\tau\right) E\left[x\left(t-\tau\right)\right] d\tau = m_x \int_{-\infty}^{\infty} h\left(\tau\right) d\tau = m_x H\left(j0\right) \tag{6.71}$$

where $H\left(j\omega\right) = \int_{-\infty}^{\infty} h\left(\tau\right) e^{-j\omega\tau} d\tau$ is the frequency response of the filter and $H\left(j0\right) = \int_{-\infty}^{\infty} h\left(\tau\right) d\tau$ is the $H\left(j\omega\right)$ evaluated at $\omega = 0$. Thus, the mean value of the output process is a constant.

Now, the output autocorrelation function is

$$R_y\left(\tau\right) = E\left[y\left(t\right) y\left(t+\tau\right)\right] \tag{6.72}$$

We note that, by equation (6.70),

$$y\left(t\right) = \int_{-\infty}^{\infty} h\left(\tau\right) x\left(t-\tau\right) d\tau = \int_{-\infty}^{\infty} h\left(\alpha\right) x(t-\alpha) d\tau$$

and

$$y\left(t+\tau\right) = \int_{-\infty}^{\infty} h\left(\beta\right) x\left(t+\tau-\beta\right) d\beta.$$

Thus,

$$R_y\left(\tau\right) = E\left[\int_{-\infty}^{\infty} h\left(\alpha\right) x\left(t-\alpha\right) d\alpha \int_{-\infty}^{\infty} h\left(\beta\right) x\left(t+\tau-\beta\right) d\beta\right]$$

$$= \int_{-\infty}^{\infty}\int_{-\infty}^{\infty} h\left(\alpha\right) h\left(\beta\right) E\left[x\left(t-\alpha\right) x\left(t+\tau-\beta\right)\right] d\alpha d\beta$$

$$= \int_{-\infty}^{\infty}\int_{-\infty}^{\infty} h\left(\alpha\right) h\left(\beta\right) R_x\left(\tau+\alpha-\beta\right) d\alpha d\beta \tag{6.73}$$

[The derivation of E [ ] in the above equation is as follows:

Let $t-\alpha = t'$, then $t+\tau-\beta = t'+\alpha+\tau-\beta = t'+\left(\tau+\alpha-\beta\right)$.

Thus, $E\left[x\left(t-\alpha\right) x\left(t+\tau-\beta\right)\right] = E\left[x\left(t'\right) x\left(t'+\left(\tau+\alpha-\beta\right)\right)\right] = R_x\left(\tau+\alpha-\beta\right)$

Now, taking the Fourier transform on both sides of Eq. (6.73), we get

$$G_y\left(\omega\right) = \mathfrak{F}\left[R_y\left(\tau\right)\right] = \int_{\tau=-\infty}^{\infty}\left[\int_{-\infty}^{\infty}\int_{-\infty}^{\infty} h\left(\alpha\right) h\left(\beta\right) R_x\left(\tau+\alpha-\beta\right) d\alpha d\beta\right] e^{-j\omega\tau} d\tau$$

$$= \int_{-\infty}^{\infty}\int_{-\infty}^{\infty} h(\alpha)\, h(\beta)\, d\alpha d\beta \int_{\tau=-\infty}^{\infty} R_x(\tau+\alpha-\beta) e^{-j\omega\tau} d\tau$$

Since, $\int_{\tau=-\infty}^{\infty} R_x(\tau+\alpha-\beta) e^{-j\omega\tau} d\tau = \int_{\tau'=-\infty}^{\infty} R_x(\tau')\, e^{-j\omega(\tau'-\alpha+\beta)} d\tau'$

$$= e^{-j\omega(-\alpha+\beta)} \int_{\tau'=-\infty}^{\infty} R_x(\tau') e^{-j\omega\tau'} d\tau'$$

$$= G_x(\omega)\, e^{-j\omega(-\alpha+\beta)}$$

We may write,

$$G_y(\omega) = \int_{-\infty}^{\infty}\int_{-\infty}^{\infty} h(\alpha)\, h(\beta)\, G_x(\omega)\, e^{-j\omega(-\alpha+\beta)} d\alpha d\beta$$

$$= G_x(\omega) \int_{-\infty}^{\infty} h(\alpha) e^{j\omega\alpha} d\alpha \int_{-\infty}^{\infty} h(\beta)\, e^{-j\omega\beta} d\beta$$

$$= G_x(\omega) \left[\int_{-\infty}^{\infty} h^*(\alpha)\, e^{-j\omega\alpha} d\alpha\right]^* \left[\int_{-\infty}^{\infty} h(\beta) e^{-j\omega\beta} d\beta\right]$$

$$= G_x(\omega)\, H^*(j\omega)\, H(j\omega)$$

$$= G_x(\omega)\, |H(j\omega)|^2 \tag{6.74}$$

Thus, we arrive at the important relation

$$G_y(\omega) = G_x(\omega) . |H(j\omega)|^2$$

stating that the output PSD is the input PSD multiplied by the power gain $|H(j\omega)|^2$ of the LTI system. The total output power can be obtained by integrating the above relation. That is,

$$P_{out} = \int_{-\infty}^{\infty} G_y(\omega)\, d\omega = \int_{-\infty}^{\infty} G_x(\omega) |H(j\omega)|^2 d\omega \tag{6.75}$$

Example 6.14

A noise with $G_{ni}(f) = \eta/2$ is filtered by a low-pass process to over $-B < f < B$ and then passed through a differentiator. Find the output noise power.

Solution

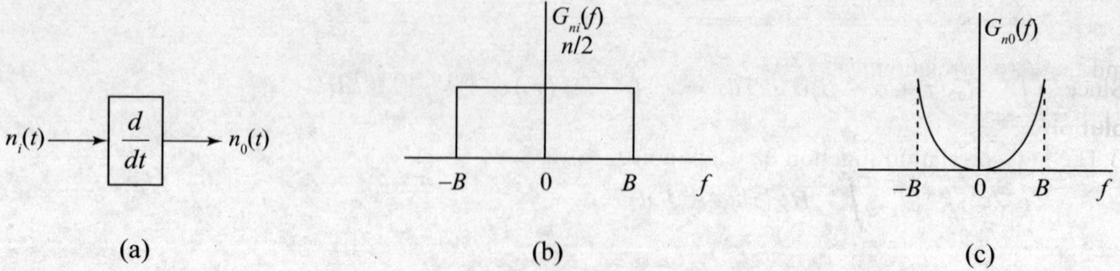

Fig 6.22 *(a) LTI system being a differentiator; (b) Input noise PSD after low-pass filtering; (c) output noise PSD*

The transfer function of the differentiator is

$$H(j\omega) = j2\pi f$$

Thus, $G_{no}(f) = G_{ni}(f)\,|H(jf)|^2 = \frac{\eta}{2}4\pi f^2 = 2\pi\eta f^2$

So, output noise power

$$N_0 = \int_{-B}^{B} 2\pi\eta f^2 df = 2\pi\eta\frac{f^3}{3}\int_{-B}^{B} = \frac{2\pi\eta}{3}.2B^3 = \frac{4\pi\eta}{3}B^3$$

Example 6.15

A white noise is modelled as a stationary random process $N(t)$ when power spectral density $G_n(f)$ is given by

$$G_n(f) = \frac{N_0}{2} = \text{Constant}, \; -\infty < f < \infty$$

(a) Find its autocorrelation function.

(b) When the white noise is passed through the low-pass filter, the band limited white noise has the power spectrum in the low-pass model as

$$G_{LP}(f) = \begin{cases} \frac{N_0}{2} & f_u < f < fu \\ 0, & \text{elsewhere} \end{cases} \tag{6.76}$$

Find the autocorrelation function.

(c) In the bandpass model, let the power spectral density of white noise be

$$G_{Bp}(f) = \begin{cases} \frac{N_0}{2}, & f_0 - \frac{f_u}{2} < |f| < f_0 + \frac{f_u}{2} \\ 0, & \text{elsewhere} \end{cases} \quad (6.77)$$

Find the autocorrelation.

Solution

(a) The autocorrelation function of white noise is given by,

$$R_n(\tau) = F^{-1}[G_n(f)]$$

$$= \int_{-\infty}^{\infty} \frac{N_0}{2} e^{j2\pi f\tau} df = \frac{N_0}{2}\delta(\tau) \quad (6.78)$$

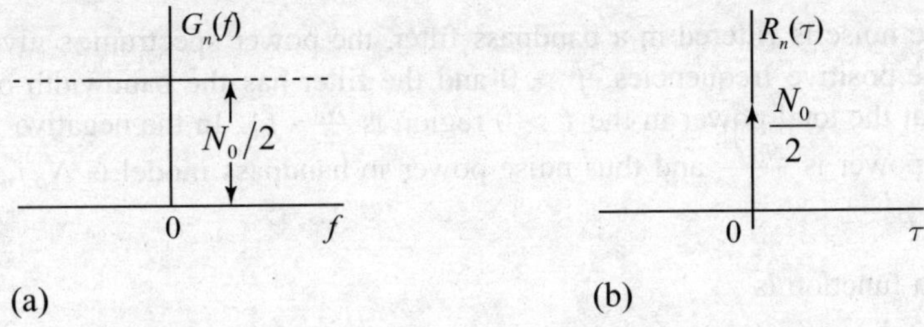

Fig 6.23 *(a) Power spectral density of white noise (b) Delta- correlated autocorrelation function.*

The power spectrum and autocorrelation function are shown in Fig. 6.23

(b) The autocorrelation function of a band limited white noise in a low-pass model is derived as follows:

$$R_{LP}(\tau) = \text{Inverse Fourier Transform of } G_{LP}(f)$$

$$= \int_{-f_u}^{f_u} \frac{N_0}{2} e^{j2\pi f\tau} df$$

$$= \frac{N_0}{2}\frac{1}{j2\pi\tau}\left[e^{j2\pi f_u\tau} - e^{-j2\pi f_u\tau}\right]$$

$$= \frac{N_0}{2}\frac{\sin(2\pi f_u\tau)}{2\pi f_u\tau}.2f_u = N_0 f_u \sin c\,(2f_u\tau) \quad (6.79)$$

The autocorrelation is thus a sinc function (which is nothing but the impulse response of the low-pass filter through which the white noise has been filtered out).

The power spectrum and autocorrelation function of a white noise in a low-pass model are sketched in Fig.6.24

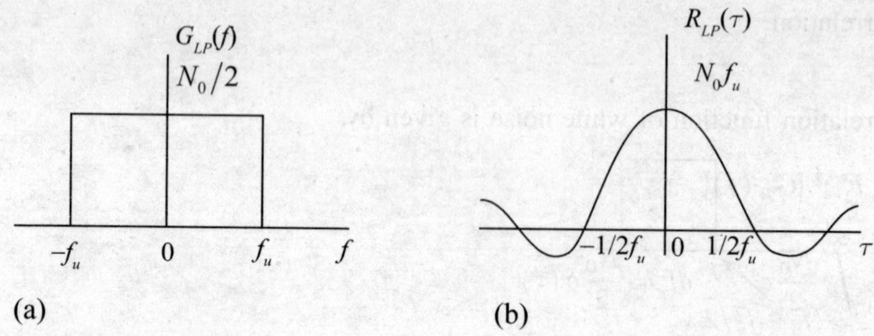

Fig 6.24 *(a) Power spectral density of white noise in a low-pass model (b) The autocorrelation function.*

(c) When the white noise is filtered in a bandpass filter, the power spectrum is given by Eq.(6.77). We note that in the positive frequencies, $f > 0$ and the filter has the bandwidth of f_u around the frequency f_o so that the total power in the $f > 0$ region is $\frac{N_0}{2} \cdot f_u$. In the negative frequency region $(f < 0)$, the total power is $\frac{N_0 f_u}{2}$ and thus noise power in bandpass model is $N_o f_u$, the same as in low-pass model.

The autocorrelation function is

$$R_{BP}(\tau) = \int_{-\infty}^{\infty} G_{BP}(f) e^{j2\pi f\tau} df$$

$$= \int_{-f_0-f_u/2}^{-f_0+f_u/2} \frac{N_0}{2} e^{j2\pi f\tau} df + \int_{f_0-f_u/2}^{f_0+f_u/2} \frac{N_0}{2} e^{j2\pi f\tau} df$$

$$= \frac{N_0}{2} \cdot \frac{1}{j2\pi\tau} \cdot \left[e^{j2\pi\tau(-f_0+f_u/2)} - e^{j2\pi\tau(-f_0-f_u/2)} \right]$$

$$+ \frac{N_0}{2} \cdot \frac{1}{j2\pi\tau} \left[e^{j2\pi\tau(f_0+f_u/2)} - e^{j2\pi\tau(f_0-f_u/2)} \right]$$

$$= \frac{N_0}{2\pi\tau} \cdot \left[e^{-j2\pi f_0\tau} \left(\frac{e^{j\pi f_u\tau} - e^{-j\pi f_u\tau}}{2j} \right) + e^{j2\pi f_0\tau} \left(\frac{e^{j\pi f_u/2} - e^{-j\pi f_u\tau}}{2j} \right) \right]$$

$$= \frac{N_0}{\pi\tau} \sin(\pi f_u\tau) \left(\frac{e^{j2\pi f_0\tau} + e^{-j2\pi f_0\tau}}{2} \right)$$

$$= N_0 f_u \cdot \text{sin c}\,(f\pi\tau) \cdot \cos(2\pi f_0\tau) \tag{6.80}$$

The autocorrelation function has the sinc-type envelope and oscillates at the bandpass centre frequency of f_o. Figure 6.25 depicts the variation.

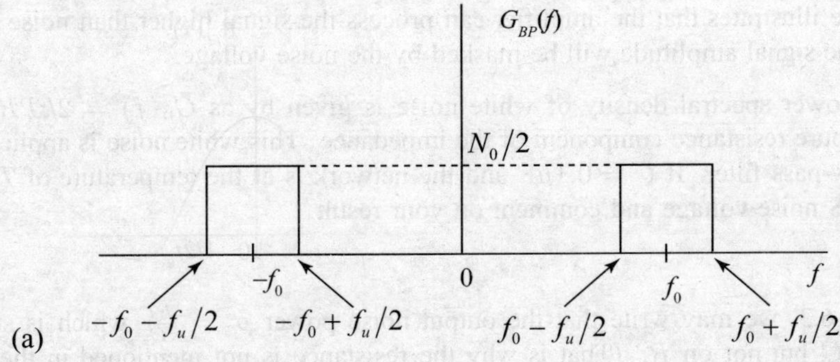

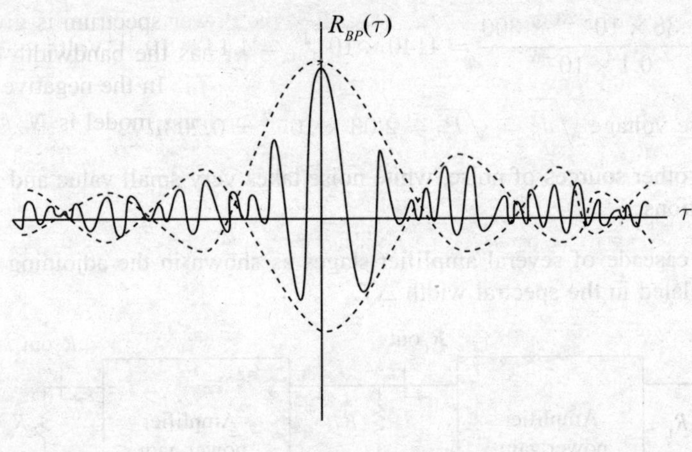

Fig. 6.25 *(a) Band limited noise power spectral density in bandpass model (b) Autocorrelation function.*

Solved Numericals

1. An amplifier has an input impedance of 1kΩ at an operating temperature of 27°C. Calculate from Eq. 6.1 the r.m.s. noise voltage across this resistance (at the input of the amplifier) if the noise comes from the frequency components in the range from 10 to 20 MHz.

Solution

By Eq. (6.1),

$$v_n = \sqrt{4kTR\Delta f}$$

$$= \sqrt{4 \times 1.38 \times 10^{-23} \times (27 + 273) \times 1 \times 10^3 \times (20 - 10) \times 10^6}$$

$$= \sqrt{4 \times 1.38 \times 300 \times 10 \times 10^{-14}}$$
$$= 128.68 \times 10^{-7}\,\text{volt} = 12.87\mu\text{V}$$

This example illustrates that the amplifier can process the signal higher than noise amplitude; otherwise, the signal amplitude will be masked by the noise voltage.

2. Two-sided power spectral density of white noise is given by as $G_n(f) = 2kTR(f)$ where $R(f)$ is the pure resistance component of the impedance. This white noise is applied to a RC-noiseless low-pass filter. If $C = 0.1\mu$F and the network is at the temperature of $T = 300\,$K, find the RMS noise voltage and comment on your result.

Solution

By example 6.3, we may write that the output noise power $p_n = \frac{kT}{C}$ which is surprisingly depends on C but not on R. (That is why the resistance is not mentioned in the problem). Substituting the values,

$$P_n = \frac{1.38 \times 10^{-23} \times 300}{0.1 \times 10^{-6}} = 4140 \times 10^{-17} = 4.14 \times 10^{-14}\text{Volt}^2$$

So RMS noise voltage $\sqrt{\overline{v_n^2}} = \sqrt{P_n} = 2.03 \times 10^{-7} = 0.203\mu$V.

Compared to other sources of noise, white noise takes very small value and remains unnoticed in most situations.

3. Consider the cascade of several amplifier stages as shown in the adjoining figure. The noise is to be calculated in the spectral width Δf.

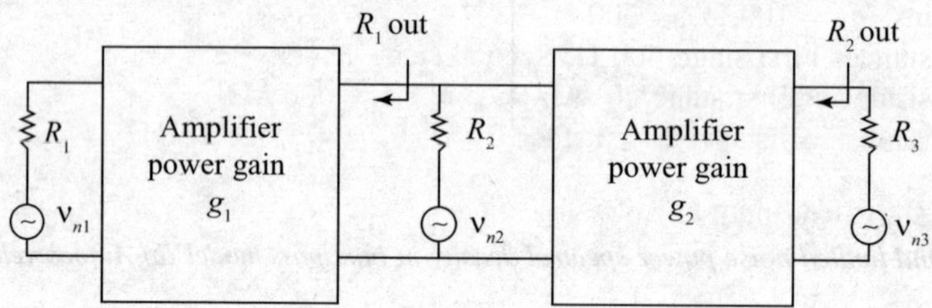

Find the equivalent noise resistance at the input of the first stage.

Solution

RMS noise voltage at the output due to R_3 alone is $v_{n3} = \sqrt{4kTR_3\Delta f}$ and noise power $= 4kTR_3\Delta f = \overline{v_{n3}^2}$

The same noise power can be thought of being originated from the input resistance of value R_3' (and multiplied by the power gain g_3).

Thus,

$$v_{n3}' = \sqrt{4kTR'_3\Delta F}$$

So, $4kTR_3\Delta f = g_2.4kTR_3'\Delta f$

or, $R_3' = \dfrac{R_3}{g_2}$

This shows that if resistance at the output is transformed to the input, the equivalent resistance is reduced by the power gain of the stage.

So, equivalent noise resistance at the input of Stage 2 is

$$R_{2,eq} = R_2 + R_3' = R_2 + \frac{R_3}{g_2}.$$

By the same analogy, if this resistance is transformed back to the input of the first stage, its transformed noise resistance would be

$$R_2' = \frac{R_{2,eq}}{g_1} = \frac{R_2 + {R_3}/{g_2}}{g_1} = \frac{R_2}{g_1} + \frac{R_3}{g_1 g_2}$$

So, equivalent input resistance at the first stage would be

$$R_{eq} = R_1 + R_2' = R_1 + \frac{R_2}{g_1} + \frac{R_3}{g_1 g_2}.$$

Note that the resistance R_3 at the output stage produces the same noise effect as a $\frac{R_3}{g_1 g_2}$ resistor at the input.

4. In the above example of a two-stage amplifier in cascade, assume the following values.

Power gains: $g_1 = 100$, $g_2 = 500$
Input resistance: First stage 500 Ω; Second stage 75 $k\Omega$
Output resistance: First stage 20 $k\Omega$; Second stage 500 $k\Omega$
Equivalent noise resistance: First stage 2 $k\Omega$; Second stage 10 $k\Omega$,

Find the equivalent input noise resistance (at the input of the first stage).

Solution

In the example, we will have to first identify and find the values of R_1, R_2 and R_3; then apply the above formula.

Now, $R_1 =$ Sum of the input resistance and equivalent noise resistance

$$= 500\Omega + 2\text{k}\Omega = 2.5\text{k}\Omega$$

$R_2 =$ Parallel equivalent of output resistance of the first stage and input resistor of the second stage + Equivalent noise resistor of the second stage.

$$= 20\text{k}\Omega//75\text{k}\Omega + 10\text{k}\Omega = \frac{20 \times 75}{20 + 75} + 10 = 25.79\ \text{k}\Omega$$

$R_3 =$ Output resistance of 2nd stage $= 500\ k\Omega$

For applying the formula

$$R_{eq} = R_1 + \frac{R_2}{g_1} + \frac{R_3}{g_1 g_2} = 2.5 + \frac{25.79}{100} + \frac{500}{100 \times 500}$$

$$= (2.5 + 0.2579 + 0.01)\text{k}\Omega = 2.768 \text{ k}\Omega$$

We observe that 500 kΩ output resistor has the same noise effect as a 10Ω resistor transformed at the input.

5. The RF amplifier of a receiver section has the input resistance of 1000 Ω and equivalent noise resistance of 1500 Ω. The receiver load resistance is 100 $k\Omega$. Assume power gain of the amplifier as 500.

   (a) Find the equivalent input noise resistance R_{eq}.

   (b) If the receiver is operating at a temperature of 27°C and having a bandwidth of 1 MHz, calculate the equivalent noise voltage at the input of this receiver.

   (c) If the receiver is connected to an antenna with an input resistance of 75Ω, find the noise figure.

Solution

We identify the resistors R_1 and R_2 as follows.

$$R_1 = (1000 + 1500)\Omega = 2500\Omega$$

$$R_2 = \text{Load resistance} = 100k\Omega$$

So,

(a) $R_{eq} = R_1 + \frac{R_2}{g_1} = 2500 + \frac{100\times10^3}{500} = 2500 + 200 = 2.7\text{k}\Omega.$

(b) Equivalent noise voltage (rms)

$$V_n = \sqrt{4\text{k}TR_{\text{eq}}\Delta f} = \sqrt{4 \times 1.38 \times 10^{-23} \times (27 + 273) \times 2700 \times 10^6}$$

$$= \sqrt{4 \times 1.38 \times 300 \times 2700 \times 10^{-17}} = 66.867 \times 10^{-7} \text{ volt} = 6.69\,\mu\text{V}$$

(c) Receiver noise figure

$$F = 1 + \frac{R_{eq}}{R_a}, \text{ where } R_a = \text{input resistance of the antenna}$$

$$= 1 + \frac{2700}{75} = 37 = 15.6\,\text{dB}$$

Exercises

Activity

6.1 Derive the power spectral density of the noise voltage $v_n(t)$ of Fig. P6.1 . Find also the total output noise power.

6.2 The power spectral density of noise is given by $G_n(f) = \dfrac{a}{a^2 + 4\pi^2 f^2}$. Calculate the autocorrelation function.

6.3 Given the following autocorrelation function $R_n(\tau)$, find the power spectral density.

$$R_n(\tau) = \begin{cases} a^2(1 - |\tau|/t_0) & |\tau| \leq t_0 \\ 0 & |\tau| > t_0 \end{cases}$$

6.4 A transistor amplifier is modelled by h-parameters as shown in Fig. P6.2. Find the available power gain. Ans: $g = \dfrac{h_{fe}^2 R_s}{h_{oe}(R_s + h_{ie})^2}$

6.5 A series RLC circuit is in equilibrium with a temperature T. Develop an expression for the noise bandwidth and the variance of the noise voltage across the inductor. The circuit has a 3 dB bandwidth $B << f_0$, where f_0 is the series resonant frequency.

6.6 Calculate the noise bandwidth of a parallel RLC filter having a 3 dB bandwidth B.

6.7 A 75Ω resistor is kept at a temperature of 300 K. Calculate the rms value of the voltage across its terminals and the maximum available noise power delivered to a matched load. Assume a bandwidth of 1 MHz for the measuring instrument.

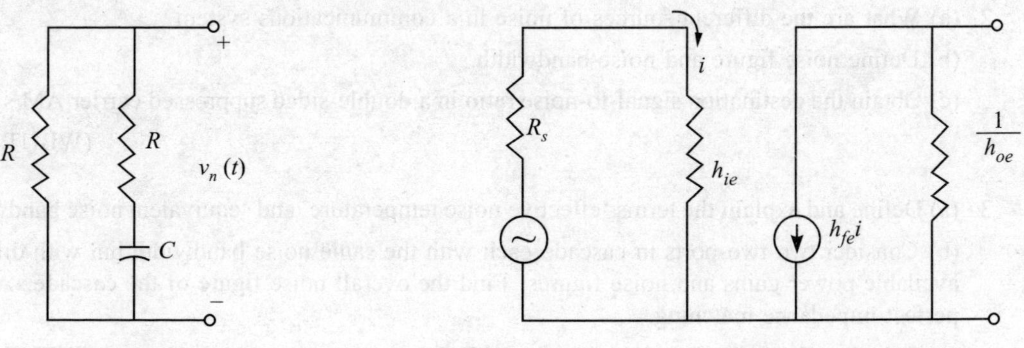

Fig. P6.1 **Fig. P6.2**

6.8 The measured noise figure of the amplifier circuit of Fig. P6.3 is $F = 2.5$ and rms voltmeter reads 10 volts when the setup is in equilibrium with 300 K. When the resistor R is cooled down to 25 K, what will be the reading of the voltmeter?

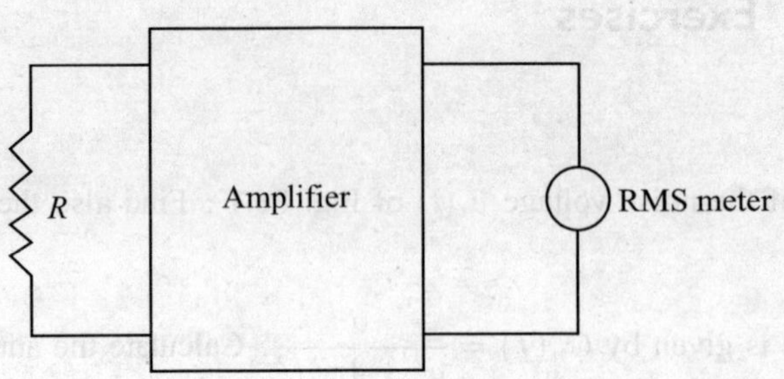

Fig. P6.3

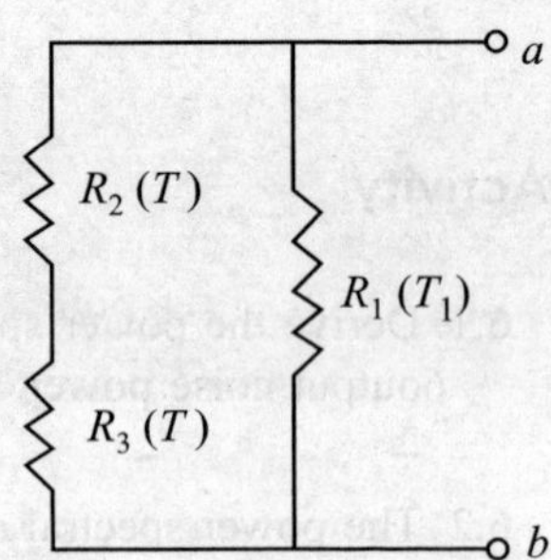

Fig. P6.4

6.9 In Fig. P6.4, resistors R_2 and R_3 are at temperature T, while R_1 is at temperature T_1. Calculate the rms voltage at the terminals (a, b).

6.10 An antenna has a noise temperature $T_a = 4$ K. A receiver follows the antenna. The specifications of the receiver are: equivalent noise temperature $T_e = 100$ K, mid-band gain $g_a = 10^{10}$(power gain), bandwidth = 10 MHz. Find the noise bandwidth B_N.

Review questions

1. (a) What is thermal noise? Give the voltage-generator and current-generator equivalent circuit for thermal noise and find out their rms noise voltage and current respectively.

   (b) Two resistors 30 kΩ and 50 kΩ are at room temperature of 30°C. For a BW of 2 MHz, determine the thermal noise voltage of (i) each resistor (ii) two resistors in series and (iii) two resistors in parallel.

   (c) For two amplifiers in cascade, derive the expressions for the overall noise factor and noise temperature. (KU 2001)

2. (a) What are the different sources of noise in a communications system?

   (b) Define noise figure and noise bandwidth.

   (c) Obtain the destination signal-to-noise ratio in a double-sided suppressed carrier AM system. (WBUT 2003)

3. (a) Define and explain the terms 'effective noise temperature' and 'equivalent noise bandwidth'.

   (b) Consider two two-ports in cascade each with the same noise bandwidth but with different available power gains and noise figures. Find the overall noise figure of the cascade. Assume perfect impedance matching.

   (c) Calculate the SNR at the output of a DSB-SC system. (WBUT 2004)

4. What is 'white noise'? Give an example. Obtain the thermal noise power across a resistor at a temperature T within a bandwidth B. (RPE 2002)

5. (a) A random signal $x(t)$ is passed through an LTI system with impulse response $h(t)$ for $-\infty < t < \infty$. The output of the system is a random sample $y(t)$. Derive the relation between power density spectra of the input and output processes.

   (b) An input noise with PSD $Gn_{\text{i}}(t) = k$ is applied to a differentiator. Find and sketch the PSD of the output noise. [MUST, Spring 2011]

6. (a) Define noise temperature.

   (b) The input terminals of a cascade of two-port network of noise figures $F_1, F_2, \ldots$, and power gains $G_1, G_2, \ldots$ are excited by an input noise power N_i, Derive the overall equivalent noise temperature.

7. The power spectral density of a baseband noise $n(t)$ is given by $G_n(f) = \dfrac{2}{(1+\pi^2 f^2)^2}$. Find the r.m.s bandwidth.

   (Hint: RHS bandwidth is defined by $B_{\text{rms}} = \sqrt{\dfrac{\int_{-\infty}^{\infty} \omega^2 G_n(\omega) d\omega}{\int_{-\infty}^{\infty} G_n(\omega) d\omega}}$

8. A random noise has the autocorrelation function given by

   $$R_n(\tau) = \begin{cases} 1 - |\tau|, & |\tau| \leq 1 \\ 0, & \text{otherwise} \end{cases}$$

   Find the power spectral density of the noise,

   (Hint: PSD $G_{\text{n}}f = \int_{-\infty}^{\infty} R_{\text{n}}(\tau) e^{-i\omega t} dt$)

9. Two resistors R_1 and R_2 at the temperature T_1 and T_2 are in parallel configuration. Find the noise temperature of the combined resistors.

   (Hint: mean square currents $\overline{i_{\text{n}}^2} = \overline{i_1^2} + \overline{i_2^2}$ (by Eq. 6.4) and $\overline{i_{\text{n}}^2} = 4kT/R \cdot df$.)

10. An amplifier has three stages in cascade. The effective input noise temperature of the stages are $T_1 = 100\,\text{K}, T_2 = 500\,\text{K}$ and $T_3 = 1000\,\text{K}$. The available power gain of the second stage is $g_2 = 10$. Find the power gain g_1 of the first stage to have an effective input noise temperature of the amplifier of 400 K.

    (Hint: Use Eq.6.31, $T_{\text{e}} = T_1 + \frac{T_2}{g_1} + \frac{T_3}{g_1 g_2}$)

Multiple-choice questions

1. The autocorrelation function of white noise is

   (a) A constant (b) Gaussian (c) Delta function (d) $\exp(-|\tau|)$

2. The spectral density of white noise is

   (a) Uniform (b) Gaussian (c) Poisson (d) Exponential

3. The PDF of the envelope of narrowband noise is

(a) Gaussian (b) Rayleigh (c) Rician (d) Binomial

4. The energy content of atmospheric noise varies as

   (a) f (b) f^2 (c) $1/f$ (d) $1/f^2$

5. Cosmic noise is caused by

   (a) Lightning discharge (b) Solar eruption

   (c) Distant stars (d) Industrial electrical discharge

6. Thermal noise power equals

   (a) kTB (b) kTB^2 (c) kT/B (d) kT/B^2

7. Shot noise current in a diode is proportional to

   (a) I_p^2 (b) I_p (c) $\sqrt{I_p}$ (d) I/I_p

8. If R is doubled, the thermal noise power generated in it

   (a) Becomes halve the previous value (b) Doubles

   (c) Remains unchanged (d) None of the above.

9. Noise figure of a system is defined as

   (a) S_O/N_O (b) S_i/N_i (c) $\frac{S_O.N_O}{S_i/N_i}$ (d) $\frac{S_i/N_i}{S_O/N_O}$

10. The equivalent noise temperature T_e of a system varies

    (a) Linearly with T_a (b) Linearly with T_a^2,

    (c) Inversely with T_a (d) Inversely with T_a^2

    where T_a = the ambient temperature.

11. The noise that interferes very much with HF transmission is

    (a) White (b) Transit time (c) Flicker (d) Shot

12. Consider the following parameters.

    (1) Input noise voltage (2) Equivalent noise resistance

    (3) Noise temperature (4) Noise figure

    Select the correct combinations that is helpful in comparing the receiver noise performance.

    (a) 1, 2 and 3 (b) 1, 3 and 4 (c) 2, 3 and 4 (d) 1, 2 and 4.

13. Which of the following statement is false regarding white noise?

    (a) It is a Gaussian noise

    (b) Theoretically, it has infinite power

    (c) It is a thermal noise

    (d) Its frequency spectrum contains all frequency components but in unequal proportion

14. Zero-mean Gaussian noise process of variance σ_n^2 is applied to a half wave rectifier. The mean-squared value of the rectified output will be

    (a) Zero (b) σ_n^2 (c) $\sigma_n^2/\sqrt{2}$ (d) $\sigma_n^2/2$.

15. The noise bandwidth of an RC low-pass filter is

    (a) $\frac{1}{5RC}$ (b) $\frac{1}{4RC}$ (c) $\frac{1}{3RC}$ (d) $\frac{1}{2RC}$

7 Noise in AM and FM Systems

7.1 Introduction

In the process of communicating an information signal from the source to the destination, noise gets incorporated into the signal in various ways. It may arise in the transmitter, the communication channel or the receiver. Noise corrupts the signal either additively or multiplicatively. Let us first discuss the noise occuring in the transmitter. The modulator section in the transmitter is made up of electronic devices which generate thermal noise. The signal $s(t)$ may be corrupted by some noise before he modulation process itself. The modulated output $s_{\mathrm{AM}}(t)$ is then of the form $[s(t) + n_1(t)] \cos \omega_c t$.

The noise $n_1(t)$ adds to the signal $s(t)$ and is called the additive noise. The most significant contribution to noise occurs during the transmission of the signal from the transmitter to the receiver. The characteristics of a channel vary with time owing to the turbulence in the medium. Also, the signal propagation through the medium (air, for example) is severely affected due to multipath propagation (reflected signal from say hills, may interfere with the primary signal). The signal received at the receiving antenna is of the form

$$As_{\mathrm{AM}}(t) + As_{\mathrm{AM}}(t)n_2(t) = A[1 + n_2(t)]s_{\mathrm{AM}}(t),$$

where A is the attenuation factor in the medium and $n_2(t)$ the multiplicative noise. In addition to these, the modulated signal experiences additive noise during propagation. This noise may originate from overhead high-tension transmission lines, sudden lightning in the atmosphere, etc. Representing this kind of noise as $n_3(t)$, the signal received is of the form

$$A[1 + n_2(t)].[s(t) + n_1(t)] \cos \omega_c t + n_3(t).$$

In addition to the above noise terms, the electronic devices and the wires in the receiver generate thermal and shot noises which behave as additive noise and may be included in $n_3(t)$. It is worth mentioning that all the noise terms are not equally significant. The most important additive noise is the transmission noise. Multiplicative noises become significant at very high frequencies — the television signal, for example, experiences this kind of noise.

The modulated signals are bandpass in nature, that is, the signal spectrum is centred on some high-frequency carrier. So, in order to study the effects of noise in the received signal, it is essential to concentrate on the bandpass noise voltages. Since the baseband signal spans over a frequency band from zero to f_M, we may consider the noise spectrum to have a bandwidth of $2f_M$ centred on carrier frequency f_c. The quadrature representation of bandpass noise spectrum can be written as

$$x(t) = x_c(t)\cos\omega_c t + x_s(t)\sin\omega_c t$$

where $x_c(t)$ are $x_s(t)$ are the low-frequency equivalent in phase and quadrature components of $x(t)$. The spectrum of $x(t)$ in the bandpass spectrum is shown in the Figure 7.1 as $G_x(f)$. The power spectral densities $G_{x_c}(f)$ and $G_{x_s}(f)$ are equal and are given as

$$G_{x_c}(f) = G_{x_s}(f) = \begin{cases} G_x(f+f_c), & |f| < f_M \\ 0, & |f| > f_M \end{cases}$$

and is plotted in Fig. 7.1.

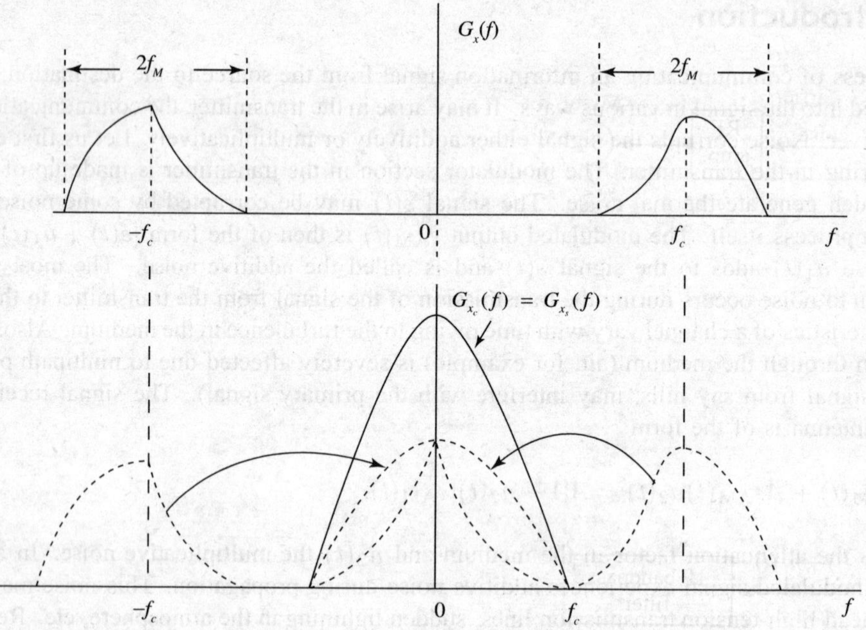

Fig. 7.1 *Power spectra: (a) bandpass signal; (b) quadrature components merged into low-pass region.*

7.2 AM Receiver

We are interested in the performance of the AM receiver in the presence of noise. With this aim, let us briefly discuss the AM receiving system. This will help to analyse the detrimental effects of noise arising in various parts of the receiver. Figure 7.1 shows an AM receiving system. The receiving antenna picks up the modulated RF carrier plus the noise. After some filtering, the received signal is amplified in an RF amplifier and down-converted in frequency by multiplying it with a sinusoidal signal from the local oscillator (LO). This results in a modulated intermediate frequency (IF) carrier which is applied to an IF amplifier. The output of the IF amplifier is applied to the bandpass IF carrier filter. After synchronous demodulation, the baseband filter recovers the original signal. The detection is essentially coherent in the sense that a synchronous source is needed to perform demodulation.

7.3 Noise in Coherent Detection

We will study in this section the analysis of signal power and noise power at various point of the receiving system for both the DSB and SSB signals. The receiver of Fig. 7.2(a) can receive and demodulate both DSB or SSB, with or without carrier by modifying the demodulator section and selecting proper bandwidth of the IF filter. Hence we will concentrate on the portions of the receiver beginning with the IF bandpass filter (shown in Fig. 7.2(b)). In the subsequent discussions, we will assume that the noise is white, Gaussian-distributed, having two-sided power spectral density $\eta/2$.

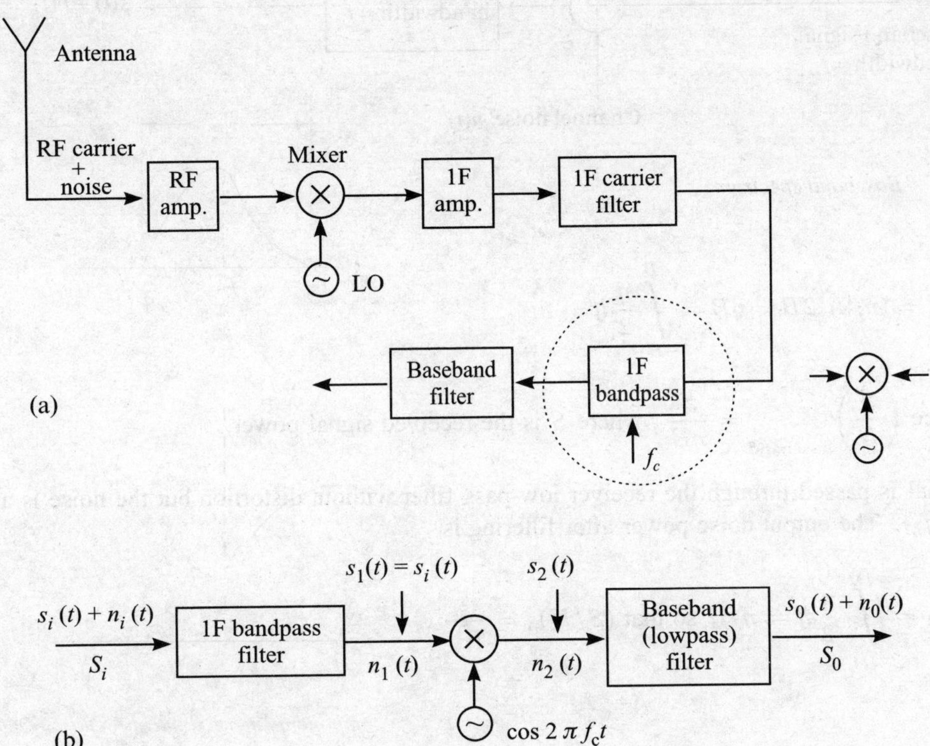

Fig. 7.2 (a) *An AM receiver.* (b) *Synchronous demodulation*

This is instructive to compare the output signal-to-noise ratio $(S/N)_0$ for different bandpass systems. For that, we need common measurement criteria for the receiver input. We first consider a baseband signal (without modulation) of spectral bandwidth f_M over which the noise is assumed to have a spectral density of $\eta/2$ over a bandwidth of B Hz. The baseband signal plus this noise is present at the input of the receiver, as shown in Fig. 7.3. The noise power at the input of the receiver is spread over the bandwidth B Hz (this is the input noise power N_i to the receiver) and is given by

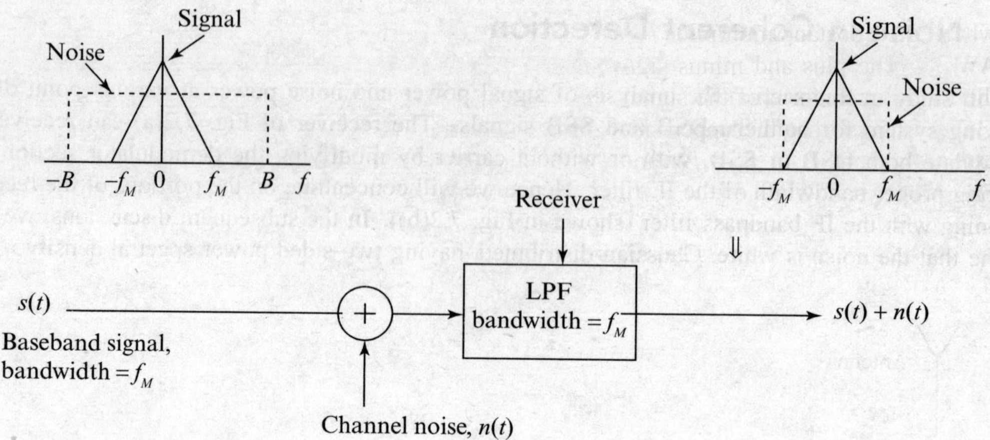

Fig. 7.3 *Baseband spectrum*

$$N_i = (\eta/2)\,.2B = \eta B = \int_{-B}^{B} \frac{\eta}{2} df$$

and hence $\left(\dfrac{S}{N}\right)_{\text{baseband}} = \dfrac{S}{\eta B}$, where S is the received signal power .

The signal is passed through the receiver low-pass filter without distortion but the noise is arrested beyond f_M. The output noise power after filtering is

$$N_0 = \int_{-f_M}^{f_M} \frac{\eta}{2} df = \eta f_M \text{ so that } (S/N)_0 = \frac{S}{\eta f_M} .$$

Thus,

$\dfrac{(S/N)_0}{(S/N)_{\text{baseband}}} = \dfrac{S/(\eta f_M)}{S/(\eta B)} = \dfrac{B}{f_M}$. This indicates that the low-pass filter enhances the signal-to-noise power ratio by the factor (B/f_M).

7.3.1 Single-sideband (SSB) reception

We begin with the single-sideband suppressed carrier (SSB-SC) transmission. The waveform at the input to the receiver is [vide (3.49)]

$$r(t) = k\left[\frac{1}{2}s(t)\cos 2\pi f_c t \pm \frac{1}{2}\widehat{s}(t)\sin 2\pi f_c t\right] + n(t), \tag{7.1}$$

where the constant multiplier k accounts for the attenuation during transmission and $n(t)$ is the AWGN. The plus and minus signs signify the upper and lower sidebands, respectively and $\hat{s}(t)$ is the Hilbert transform of the signal $s(t)$.

We assume that the upper sideband is being used so that the bandpass filter has the passband extending from f_c to $f_c + f_M$ and the passband of the baseband extends from 0 to f_M as shown in Fig. 7.4.

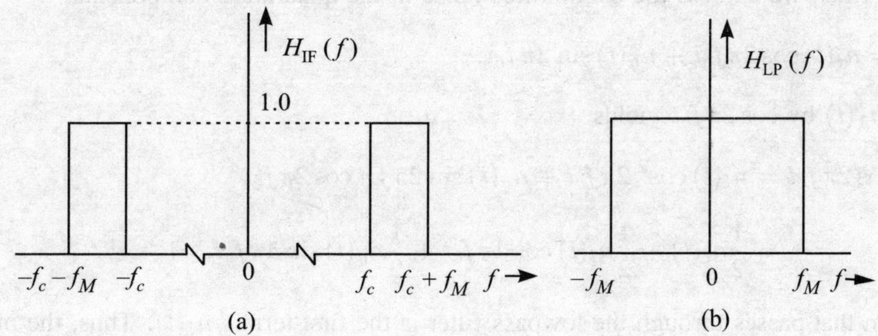

Fig. 7.4 *(a) Passband of IF bandpass filter; (b) passband of the lowpass filter.*

The received signal is the signal portion of $r(t)$ of (7.1) and can be written as

$$s_i(t) = A\cos 2\pi(f_c + f_m)t, \quad f_m \leq f_M; \tag{7.2}$$

the amplitude A includes all losses.

The input signal power is

$$S_i = \frac{A^2}{2}. \tag{7.3}$$

For a carrier signal of $\cos 2\pi f_c t$ from the LO, the output of the multiplier is

$$\begin{aligned} s_2(t) &= s_1(t)\cos 2\pi f_c t = A\cos[2\pi(f_c + f_m)t]\cos 2\pi f_c t \\ &= \frac{A}{2}[\cos 2\pi(2f_c + f_m)t + \cos 2\pi f_m t]. \end{aligned} \tag{7.4}$$

The lowpass filter allows only the difference-frequency signal to pass through so that

$$s_0(t) = \frac{A}{2}\cos 2\pi f_m t. \tag{7.5}$$

The output signal power is then

$$S_0 = \frac{1}{2}(A/2)^2 = \frac{A^2}{8}. \tag{7.6}$$

Thus the output-to-input signal power ratio is $S_0/S_i = 1/4$.

Now we turn our attention to the calculation of noise power at the input and output.

The input noise power N_i is $\eta/2$ multiplied by the bandwidth of the bandpass filter. That is,

$$N_i = \left(\frac{\eta}{2}\right)(2f_M) = \eta f_M, \tag{7.7}$$

the factor 2 accounting for both negative and positive frequencies. To find the noise at the output of the baseband filter, we express the bandlimited noise in the quadrature components.

$$n_1(t) = n_i(t)\cos 2\pi f_c t + n_q(t)\sin 2\pi f_c t. \tag{7.8}$$

Multiplying $n_1(t)$ by $\cos 2\pi f_c t$ yields

$$\begin{aligned} n_1(t)\cos 2\pi f_c t &= n_i(t)\cos^2 2\pi f_c t + n_q(t)\sin 2\pi f_c t \cos 2\pi f_c t \\ &= \frac{1}{2}n_i(t) + \frac{1}{2}n_i(t)\cos 4\pi f_c t + \frac{1}{2}n_q(t)\sin 4\pi f_c t. \end{aligned} \tag{7.9}$$

The only term that passes through the lowpass filter is the first term $\frac{1}{2}n_i(t)$. Thus, the output noise is

$$n_0(t) = \tfrac{1}{2}n_i(t). \tag{7.10}$$

Now, applying the modulation theorem, the spectral densities of $n_i(t)$ and $n_q(t)$ are obtained as

$$Gn_i(f) = Gn_q(f) = Gn_1(f - f_c) + Gn_1(f + f_c), \text{ for } f_c - f_M < |f| < f_c + f_M. \tag{7.11}$$

Thus for $0 \le f \le f_M$, $Gn_1(f + f_c) = \eta/2$ and $Gn_1(f - f_c) = 0$.

Figure 7.5 shows the quadrature spectral densities.

Thus, the narrowband noise as represented by Eq. (7.8) has the power spectral density given by

$$Gn_i(f) = Gn_q(f) = \begin{cases} \eta/2, & |f| < f_M \\ 0, & |f| > f_M \end{cases}$$

It follows from Fig. 7.5 that mean square values

$$\overline{n_1^2(t)} = \frac{\eta}{2}\cdot(2f_M) = \eta f_M$$

Similarly,

$$\overline{n_i^2(t)} = \overline{n_q^2(t)} = \eta f_M.$$

Since the output noise amplitude is reduced by 2 (vide Eg. 7.10), its power is reduced by 4. The spectral density $Gn_0(f)$ of output noise is $Gn_0(f) = \frac{1}{4}Gn_i(f) = \frac{1}{4}(\eta/2) = \eta/8$. Thus, output noise power,

$$N_0 = \left(\frac{\eta}{8}\right)(2f_M) = \frac{\eta f_M}{4}. \tag{7.12}$$

Finally, the signal-to-noise power ratio at the input of the detector is

$$\left(\frac{S}{N}\right)_i = \frac{S_i}{N_i} = \frac{A^2/2}{\eta f_M} = \frac{A^2}{2\eta f_M}, \tag{7.13}$$

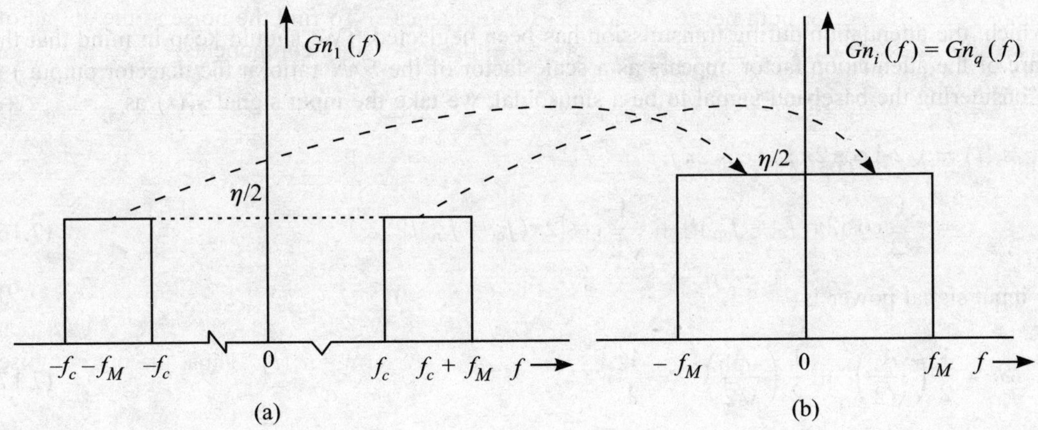

Fig. 7.5 *Quadrature expansion of $n_1(t)$ (a) PSD of noise $n_1(t)$; (b) PSD of noise components $n_i(t)$ and $n_q(t)$.*

and the same at the output is

$$(S/N)_0 = \frac{S_0}{N_0} = \frac{A^2/8}{\eta f_M/4} = \frac{A^2}{2\eta f_M}. \tag{7.14}$$

From (7.13) and (7.14) we observe that the (S/N) ratio at the detector output is the same as that at the input.

Example 7.1

An information signal $s(t) = 5\cos 2000\pi t$ is transmitted using SSB-SC and detected in a synchronous demodulator. The signal is accompanied by noise with uniform power spectral density 10^{-4} watts/Hz. Find the (S/N) ratio at the output.

Solution

We assume that the signal that is detected is the same as that transmitted. For the given signal, $A = 5$ volts, $f_M = 1000$, Hz.

Also, $\eta/2 = 10^{-4}$ watts/Hz.

So, the (S/N) ratio is found using (7.14) as

$$(S/N)_0 = \frac{5^2}{2 \times (2 \times 10^{-4}) \times 1000} = 62.5 = 18 \text{ dB}.$$

7.3.2 Double-sideband-suppressed carrier (DSB-SC) reception

In this case, the waveform at the input of the detector is

$$r(t) = s(t)\cos 2\pi f_c t + n(t), \tag{7.15}$$

in which, the attenuation during transmission has been neglected. (We should keep in mind that the square of the attenuation factor appears as a scale factor of the S/N ratio at the detector output.)

Considering the baseband signal to be a sinusoidal, we take the input signal $s_i(t)$ as

$$\begin{aligned} s_i(t) &= \sqrt{2}A\cos 2\pi f_m t\cos 2\pi f_c t \\ &= \frac{A}{\sqrt{2}}\cos[2\pi(f_c + f_m)t] + \frac{A}{\sqrt{2}}\cos[2\pi(f_c - f_m)t]. \end{aligned} \tag{7.16}$$

The input signal power is

$$S_i = \frac{1}{2}\left(\frac{A}{\sqrt{2}}\right)^2 + \frac{1}{2}\left(\frac{A}{\sqrt{2}}\right)^2 = \frac{A^2}{2}. \tag{7.17}$$

The amplitude of the input signal is chosen as $\sqrt{2}A$, which gives the same input power $A^2/2$ as in the SSB case and a direct comparison with SSB can be possible. In the demodulator (vide Fig. 7.2), $s_i(t)$ is multiplied by $\cos 2\pi f_c t$ to give

$$\begin{aligned} s_2(t) &= s_1(t)\cos 2\pi f_c t = s_i(t)\cos 2\pi f_c t \\ &= \frac{A}{\sqrt{2}}\{\cos 2\pi(f_c + f_m)t + \cos 2\pi(f_c - f_m)t\}\cos 2\pi f_c t \\ &= \frac{A}{2\sqrt{2}}\{\cos 2\pi(2f_c + f_m)t + \cos 2\pi(2f_c - f_m)t + 2\cos 2\pi f_m t\}. \end{aligned} \tag{7.18}$$

At the output of the baseband filter,

$$s_O(t) = \frac{A}{2\sqrt{2}}2\cos 2\pi f_m t = \frac{A}{\sqrt{2}}\cos 2\pi f_m t. \tag{7.19}$$

The output signal power is then,

$$S_O = \frac{1}{2}\left(\frac{A}{\sqrt{2}}\right)^2 = \frac{A^2}{4}. \tag{7.20}$$

The output signal lower is related to the input signal lower by $S_O/S_i = 1/2$. Observe that the output signal power is increase by a factor of 2 as compared to the SSB case (vide equation.(7.6)). This is due to the contribution from both the sidebands.

Let us now compute the average noise power at the detector input and output. The average power of the input noise is its PSD times the bandwidth. That is,

$$N_i = \frac{\eta}{2} \times 2(2f_M) = 2\eta f_M. \tag{7.21}$$

Here the noise is assumed to have uniform spectral density of $\eta/2$. The total bandwidth of the filter is $4f_M$ including both the positive and negative frequencies. The noise power at the detector output is obtained as

$$N_0 = \frac{1}{4}Gn_1(f)(4f_M) = \frac{\eta}{8} \times 4f_M = \frac{\eta f_M}{2}. \tag{7.22}$$

The result of Eq. (7.21) can be better understood if one considers the double-sideband of bandpass noise $n(t)$ in quadrature components as (vide Eq. (7.8))

$$n(t) = n_i(t)\cos\omega_c t + n_q(t)\sin\omega_c t.$$

In the case of SSB reception, we have considered the bandpass noise to occupy the bandwidth of f_M (either from f_c to $f_c + f_M$, $-f_c$ to $-f_c - f_M$ or from $f_c - f_M$ to f_c, $-f_c + f_M$ to $-f_c$). But in the case of DSB, noise power spectral density $G_n(f)$ is $\eta/2$, of bandwidth 2 f_M and centered at $\pm f_c$, as shown in Fig. 7.6.

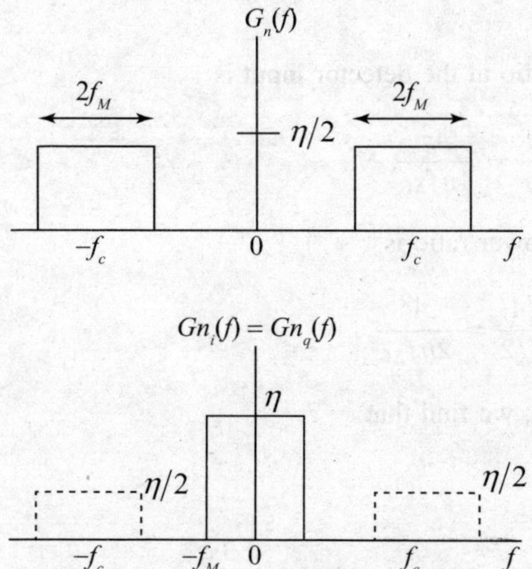

Fig. 7.6 *(a) Power spectral density of bandpass white noise; (b) The same for the quadrature components.*

The equivalent low-pass components $n_i(t)$ and $n_q(t)$ have the power spectral densities given by

$$G_{n_i}(f) = G_{n_q}(f) = \begin{cases} G_n(f+f_c) + G_n(f-f_c), & |f| \le f_M \\ 0, & |f| > f_M \end{cases} \tag{7.22a}$$

$$\text{where } G_n(f) = \begin{cases} \eta/2, & f_c - f_M < f < f_c + f_M \text{ and } -f_c - f_M < -f_c < -f_c + f_M \\ 0 & \text{elsewhere} \end{cases} \tag{7.22b}$$

It is thus seen that

$$G_{n_i}(f) = G_{n_q}(f) = \begin{cases} \eta, & |f| \leq f_M \\ 0, & |f| > f_M \end{cases} \tag{7.22c}$$

So, the average noise power is given by

$$\overline{n^2(t)} = 2 \int_{f_c - f_M}^{f_c + f_M} \frac{\eta}{2} df = 2 \cdot \frac{\eta}{2} \cdot 2f_M = 2\eta f_M.$$

same as Eq. (7.21). The factor 2 in the integration appears due to two-sided spectral density. Similarly, the mean square noise in quadrature components are given by

$$\overline{\eta_i^2(t)} = \overline{\eta_q^2(t)} = 2 \int_0^B \eta \, df = 2\eta B$$

Thus, $\overline{\eta^2} = \overline{\eta_i^2} = \overline{\eta_q^2} = 2\eta B$.

The signal-to-noise power ratio at the detector input is

$$(S/N)_i = \frac{S_i}{N_i} = \frac{A^2/2}{2\eta f_M} = \frac{A^2}{4\eta f_M}. \tag{7.23}$$

The output signal-to-noise power ratio is

$$(S/N)_0 = \frac{S_0}{N_0} = \frac{A^2/4}{\eta f_M/2} = \frac{A^2}{2\eta f_M}. \tag{7.24}$$

Comparing (7.23) and (7.24), we find that

$$\left(\frac{S}{N}\right)_0 = 2\left(\frac{S}{N}\right)_i. \tag{7.25}$$

A point of interest in connection with the DSB demodulation is that the demodulation process has doubled the signal-to-noise ratio. The result can be interpreted as follows. In the DSB case, the upper and lower sidebands are related to each other in the sense that one sideband can be obtained from the other. Since the two signal are in phase, the addition implies that the amplitude doubles causing a 4-times increase in power. On the other hand, the noises in the two bands are uncorrelated although having he same frequency. The resultant noise power in this case is the sum of the mean-square values and so the noise power doubles. The combined effect is an overall improvement in the output S/N ratio by a factor of 2.

Example 7.2

For an arbitrary modulating signal $v_m(t)$, show that in the case of DSB-SC reception, the output signal power is related to the input signal power by $S_0/S_i = 1/2$.

Solution

The input received signal is

$$s_i(t) = v_m(t)\cos v_c t.$$

The power of $s_i(t)$ is thus

$$S_i = \overline{s_i^2(t)} = \overline{v_m^2(t)\cos^2\omega_c t} = \frac{1}{2}\overline{v_m^2(t)} + \frac{1}{2}\overline{v_m^2(t)\cos 2\omega_c t}.$$

As $v_M(t)$ is assumed bandlimited to $f_m, v_m^2(t)$ is bandlimited to $2f_M$. The term $v_m^2(t)\cos 2\omega_c t$ consists of sinusoidal waves in the frequency range $2f_c \pm 2f_M$. So the average value of $v_m^2(t)\cos 2\omega_c t$ is zero (being sinusoidal). So,

$$S_i = \overline{s_i^2(t)} = \tfrac{1}{2}\overline{v_m^2(t)}.$$

In the course of demodulation, the signal $s_i(t)$ is first multiplied by $\cos\omega_c t$, then passed through the baseband filter. Thus,

$$s_i(t)\cos\omega_c t = v_m(t)\cos^2\omega_c t = \frac{1}{2}v_m(t) + \frac{1}{2}v_m(t)\cos 2\omega_c t.$$

The baseband filter accepts only $\frac{1}{2}v_m(t)$. Thus the output is

$$s_0(t) = \tfrac{1}{2}v_m(t).$$

The output signal power is

$$S_0 = \overline{s_0^2(t)} = (\overline{\frac{1}{2}v_m(t))^2} = \frac{1}{4}\overline{v_m^2(t)}.$$

Hence,

$$\frac{S_0}{S_i} = \frac{\frac{1}{4}\overline{v_m^2(t)}}{\frac{1}{2}\overline{v_m^2(t)}} = \frac{1}{2}.$$

7.3.3 Double-sideband with transmitted carrier

For the case of suppressed carrier transmission discussed above, the receiving circuitry needs to regenerate the carrier component for coherent detection. On the other hand, an AM waveform with transmitted carrier provides the reference carrier signal $\cos\omega_c t$ which can be extracted from the received signal. Perfect matching of phase and frequency for the synchronous demodulation can be obtained. However, the transmitted carrier component increases the total input signal power but not the total output signal power.

We assume the received signal of the form

$$s_i(t) = A\cos 2\pi f_c t + s(t)\cos 2\pi f_c t,$$

where $s(t)$ is the baseband modulating signal. The carrier power is $A^2/2$. The power associated with the sideband components is $\overline{s^2(t)\cos^2 2\pi f_c t} = \frac{1}{2}\overline{s^2(t)}$. So the input signal power is

$$S_i = \frac{A^2}{2} + \frac{1}{2}\overline{s^2(t)}. \tag{7.26}$$

In the demodulator, signal $s_i(t)$ is multiplied by $\cos 2\pi f_c t$. Thus

$$\begin{aligned} s_i(t)\cos 2\pi f_c t &= [A\cos 2\pi f_c t + s(t)\cos 2\pi f_c t]\cos 2\pi f_c t \\ &= \frac{A}{2} + \frac{A}{2}\cos 4\pi f_c t + \frac{1}{2}s(t) + \frac{1}{2}s(t)\cos 4\pi f_c t. \end{aligned}$$

The output of the baseband filter is

$$s_0(t) = \tfrac{1}{2}s(t).$$

The output signal power is

$$S_0 = \overline{\left(\frac{1}{2}s(t)\right)^2} = \frac{1}{4}\overline{s^2(t)}. \tag{7.27}$$

The average noise power at the detector input and output are given by (7.21) and (7.22) that is, $N_i = 2\eta f_M$ and $N_0 = \eta f_M/2$.

The signal-to-noise power ratio at the input of the detector is

$$\left(\frac{S}{N}\right)_i = \frac{S_i}{N_i} = \frac{(A^2 + \overline{s^2(t)})}{4\eta f_M}.$$

The same at the detector output is

$$\left(\frac{S}{N}\right)_0 = \frac{S_0}{N_0} = \frac{\frac{1}{4}\overline{s^2(t)}}{\eta f_M/2} = \frac{\overline{s^2(t)}}{2\eta f_M}.$$

So,
$$\frac{(S/N)_0}{(S/N)_i} = \frac{\overline{s^2(t)}}{2\eta f_M}\,\frac{4\eta f_M}{A^2 + \overline{s^2(t)}} = \frac{2\overline{s^2(t)}}{A^2 + \overline{s^2(t)}}. \tag{7.28}$$

For a comparison with the baseband system,

$$(S/N)_i = (S/N)_{\text{baseband}} = \frac{S_i}{\eta f_M} = \frac{A^2 + \overline{s^2(t)}}{2\eta f_M}$$

and hence

$$\frac{(S/N)_0}{(S/N)_{\text{baseband}}} = \frac{\overline{s^2(t)}}{A^2 + \overline{s^2(t)}}$$

Example 7.3

For the DSB with sinusoidal modulation to the depth 'm', find the relation between S/N-ratios at the detector input and output.

Solution

Assume the modulating signal $s(t) = A_m \cos \omega_m t$. The AM waveform is then

$$v(t) = A \cos \omega_c t + A_m \cos \omega_m t \cos \omega_c t$$

$$= A(1 + m \cos \omega_m t) \cos \omega_c t,$$

where $\frac{A_m}{A} = m$ is the depth of modulation.

Thus, $\overline{s^2(t)} = A_m^2/2 = m^2 A^2/2$.

Hence,

$$\frac{(S/N)_0}{(S/N)_i} = \frac{2m^2 A^2/2}{A^2 + m^2 A^2/2} = \frac{2m^2}{2 + m^2}.$$

For a 100% modulation, $m = 1$ and $\frac{(S/N)_0}{(S/N)_i} = \frac{2}{2+1} = \frac{2}{3}$.

7.4 Noise in Incoherent Detection

We now consider the effect of noise in incoherent detection, where the signal is transmitted with the carrier. The carrier amplitude is large enough so that it can serve as a reference signal for incoherent detection. The carrier increases the input signal power; it contributes nothing to the output since this signal does not appear at the output of the detector. To simplify the analysis, assume that the received signal plus the noise is

$$r(t) = s_i(t) + n(t) = [\epsilon + s(t)] \cos 2\pi f_c t + n(t), \text{ where } \epsilon \text{ is the carrier amplitude}$$

so that the signal part is

$$s_i(t) = [\epsilon + s(t)] \cos 2\pi f_c t = \epsilon[1 + m(t)] \cos 2\pi f_c t, \text{ where } m(t) = \frac{s(t)}{\epsilon}.$$

Figure 7.7 shows a scheme for obtaining the reference carrier signal for demodulation.

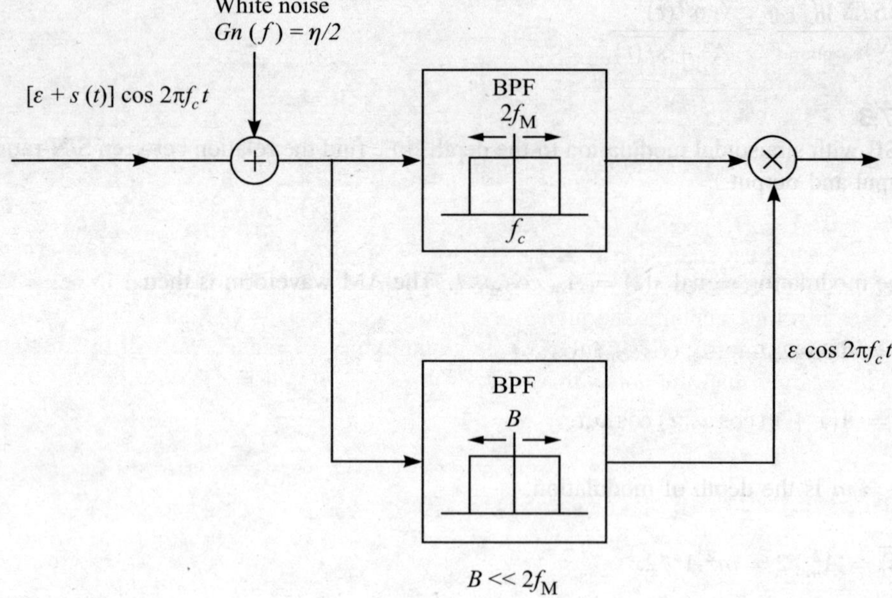

Fig. 7.7 *Recovery of reference carrier in incoherent demodulation*

Expanding the input noise into quadrature form and combining it with the signal, the signal pulse noise at the demodulator input is

$$s_i(t) + n(t) = [\epsilon + s(t)] \cos 2\pi f_c t + n_i(t) \cos 2\pi f_c t - n_q(t) \sin 2\pi f_c t$$

$$= [\epsilon + s(t) + n_i(t)] \cos 2\pi f_c t - n_q(t) \sin 2\pi f_c t. \quad (7.29)$$

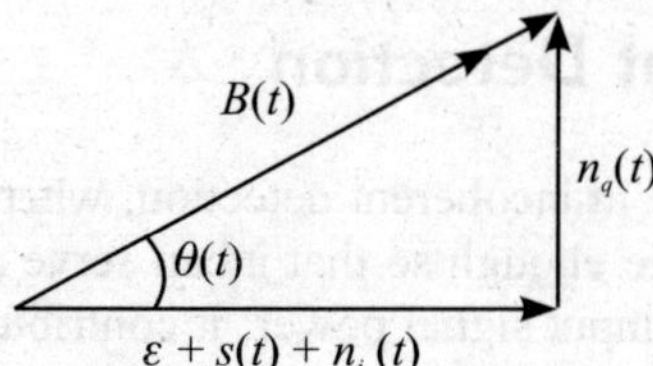

Fig. 7.8 *Phasor diagram of (7.28)*

In the phasor diagram, the first term is represented by a phasor of amplitude $[\epsilon + s(t) + n_i(t)]$ and the second term is represented by a phasor perpendicular to the first and of amplitude $n_q(t)$. Figure 7.8 illustrates this, with the abscissa representing the cosine term. The phasor sum $B(t)$ gives the envelope of composite received signal plus noise $s_i(t) + n(t)$ and is given by

$$B(t) = \left[\{\epsilon + s(t) + n_i(t)\}^2 + n_q^2(t)\right]^{1/2} \quad (7.30)$$

and the angle $\theta(t)$ is given by

$$\theta(t) = \tan^{-1}\left[\frac{n_q(t)}{\epsilon + s(t) + n_i(t)}\right]. \tag{7.31}$$

In terms of $B(t)$ and $\theta(t)$, (7.29) takes the form

$$s_i(t) + n(t) = B(t)\cos[2\pi f_c t + \theta(t)]. \tag{7.32}$$

The envelope $B(t)$ contains the message information. This can be detected in a diode envelope detector. The envelope contains the higher order noise components and cross products of signal and noise. One cannot simplify (7.30) unless some assumptions are made. We will first assume that the signal is much larger than the noise, that is, $|n_i(t)| << |s(t)|$ and $|n_q(t)| << |s(t)|$.

In that case,

$$\begin{aligned} B(t) &= [\{\epsilon + s(t)\}^2 + n_i^2(t) + 2\{\epsilon + s(t)\}n_i(t) + n_q^2(t)]^{1/2} \\ &\approx [\{\epsilon + s(t)\}^2 + 2\{\epsilon + s(t)\}n_i(t)]^{1/2} = [\epsilon + s(t)]\left\{1 + \frac{2n_i(t)}{\epsilon + s(t)}\right\}^{1/2} \\ &\approx (\epsilon + s(t))\left[1 + \frac{n_i(t)}{\epsilon + s(t)}\right] = \epsilon + s(t) + n_i(t) = \epsilon + \epsilon m(t) + n_i(t). \end{aligned} \tag{7.33}$$

We now calculate the signal-to-noise ratio at the input and output of the detector. The signal power at the input of the detector is

$$S_i = \frac{1}{2}\overline{s^2(t)} + \frac{\epsilon^2}{2} = \frac{1}{2}\left[\epsilon^2 + \overline{s^2(t)}\right] = \frac{\epsilon^2}{2}(1 + \overline{m^2(t)}), \tag{7.34}$$

where we assume that $\overline{m(t)} = 0$.

The noise power at the input to the detector is

$$N_i = \frac{\eta}{2}2(2f_M) = 2\eta f_M.$$

So the input signal-to-noise power ratio is

$$\left(\frac{S}{N}\right)_i = \frac{S_i}{N_i} = \frac{\frac{1}{2}[\overline{s^2(t)} + \epsilon^2]}{2\eta f_M} = \frac{\overline{s^2(t)} + \epsilon^2}{4\eta f_M}. \tag{7.35}$$

The output signal power is measured after the baseband filter. Since $s(t)$ is the detected modulation and $x_i(t)$ is the detected noise, neglecting the dc term, we have

$$S_0 = \overline{s^2(t)} \quad \text{and} \quad N_0 = 2\eta f_M.$$

$$\left(\frac{S}{N}\right)_0 = \frac{S_0}{N_0} = \frac{\overline{s^2(t)}}{2\eta f_M}. \tag{7.36}$$

Thus, combining Equations (7.35) and (7.36)

$$\frac{(S/N)_0}{(S/N)_i} = \frac{2\overline{s^2(t)}}{\epsilon^2 + \overline{s^2(t)}} = \frac{2\overline{m^2(t)}}{1 + \overline{m^2(t)}}. \tag{7.37}$$

Considering the information signal $s(t)$ of (7.27) a pure sinusoid $s(t) = \cos 2\pi f_M t$, the minimum value of ϵ is 1 to make the quantity $[\epsilon + s(t)]$ always positive.

Since $\overline{s^2(t)} = \frac{1}{2}$, (7.37) reduces to

$$\frac{(S/N)_0}{(S/N)_i} = \frac{2 \times \frac{1}{2}}{1 + \frac{1}{2}} = \frac{2}{3},$$

that is, for incoherent detection, $\left(\frac{S}{N}\right)_0 = \frac{2}{3}\left(\frac{S}{N}\right)_i$; for a 100% modulation. (7.38)

Threshold

We have seen that for low-noise condition, output S/N ratio is related linearly to the input S/N ratio. As the input signal-to-noise ratio decreases, a point is reached where the dependence approaches a quadratic relationship. When the noise vector is larger than the signal vector, the envelope $B(t)$ of (7.30) modifies as follows:

$$B(t) = [n_i^2(t) + n_q^2(t) + 2\{\epsilon + s(t)\}n_i(t) + \{\epsilon + s(t)\}^2]^{1/2}$$

$$= [n_i^2(t) + n_q^2(t)]^{1/2}\left[1 + \frac{2\{\epsilon + s(t)\}n_i(t)}{n_i^2 + n_q^2}\right]^{1/2}$$

$$\approx [n_i^2(t) + n_q^2(t)]^{1/2}\left[1 + \frac{\{\epsilon + s(t)\}n_i(t)}{n_i^2(t) + n_q^2(t)}\right]$$

$$= \sqrt{n_i^2(t) + n_q^2(t)} + [\epsilon + s(t)]\cos\theta(t), \tag{7.39}$$

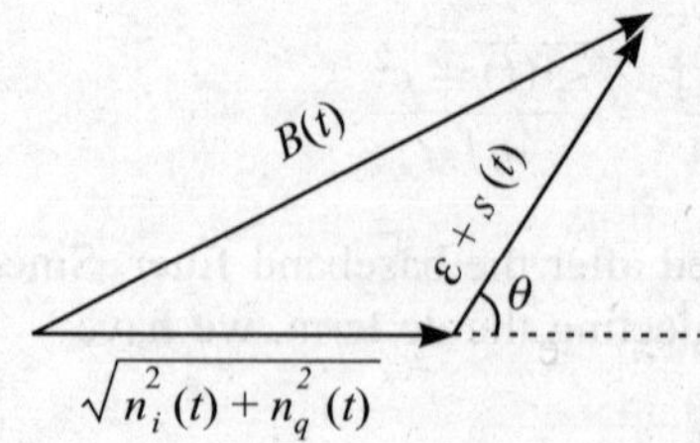

Fig. 7.9 *Phasor diagram for low-input S/N ratio*

where $\theta(t)$ is the angle between the signal and the noise. The corresponding phasor diagram is shown in Fig. 7.9. The angle $\theta(t)$ is given by

$$\tan\theta(t) = \frac{n_q(t)}{n_i(t)}. \tag{7.40}$$

It is thus seen that in the expression for the envelope, the signal $s(t)$ is multiplied by $\cos\theta(t)$, which is a random number. So, the signal cannot be recovered from the envelope.

It is imperative to compare expression (7.39) with (7.33); both give the amplitude of the envelope $B(t)$. In the low-noise case (high signal-to-noise ratio), the noise is additive in nature as evident from (7.33). And in this case, the output signal-to-noise-power ratio is linearly related to the detector input signal-to-noise-power ratio (vide (7.38)). On the other-hand, when the noise is large (low signal-to-noise ratio), the noise is both additive and multiplicative, as seen in (7.3). In this case, the signal-to-noise-power ratio at the output decreases more rapidly than at the input. The dependence approaches a quadratic relationship.

7.5 Noise in FM Systems

Noise adds to the FM signal in the path from the transmitter to the receiver. Since FM is a nonlinear modulation, the noise calculation in this system is more complicated than in AM systems. It can be proved that in FM systems the improvement in signal-to-noise ratio at the receiver can be made by sacrificing the bandwidth. To analyse a noisy FM system, we start with an FM receiver system as shown in Fig. 7.10.

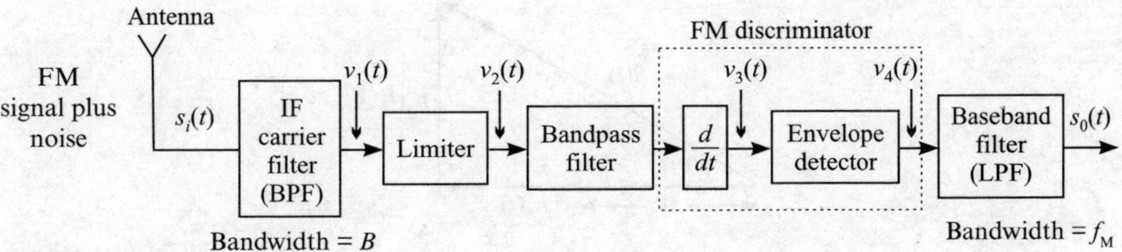

Fig. 7.10 *An FM receiver to demodulate FM signal.*

In an FM system, the information signal varies only the frequency of the carrier, not its amplitude. So any amplitude variation in the FM wave must be due to the additive noise alone. The FM receiver, therefore, clips the amplitude of the received waveform in a 'limiter' circuit. This provides constant amplitude square waveform. However, the clipping operation adds higher harmonics; these are eliminated in the bandpass filter following the limiter circuit. The signal is then applied to a frequency discriminator consisting of a differentiating network followed by an envelope detector. The constant-amplitude FM signal appears at the output of the differentiator with amplitude variation. The amplitude varies precisely as the instantaneous carrier frequency. The envelope detector recovers the baseband signal. Note that in the Fig. 7.8 we have excluded the heterodyning and IF strip as these do not modify the signal-to-noise ratio.

7.5.1 Output signal-to-noise ratio

Let us assume that a noise-free FM signal received at the IF carrier filter of Fig. 7.8 is

$$s_i(t) = A\cos[\omega_c t + 2\pi k_f \int_0^t s(\tau)d\tau]. \tag{7.41}$$

The bandpass filter passes this signal to the limiter circuit. The limiter circuit limits the amplitude of the waveform to and within a constant amplitude A_L. The bandwidth of this filter is sufficient to accept significant sideband components (as given by Carson's rule vide (4.26)). This of course eliminates the out-of-band noise. Assume that the input noise is white noise of two-sided power spectral density $\eta/2$. Both the signal and this bandlimited noise appear at the input to the limiter circuit. We consider the case of a large carrier-to-signal ratio, that is, the amplitude A is much larger than $s(t)$. In the bandlimited quadrature representation of noise, the carrier plus the noise at the limiter input is

$$\begin{aligned} v_1(t) &= A\cos 2\pi f_c t + n_i(t)\cos 2\pi f_c t - n_q(t)\sin 2\pi f_c t \\ &= [A + n_i(t)]\cos 2\pi f_c t - n_q(t)\sin 2\pi f_c t. \end{aligned} \tag{7.42}$$

Figure 7.11 illustrates the phasor diagram of the signal and the noise. In terms of envelope $B(t)$ and phase $\theta(t)$, the signal $v_1(t)$ can be written as

$$v_1(t) = B(t)\cos(2\pi f_c t + \theta(t)), \tag{7.43}$$

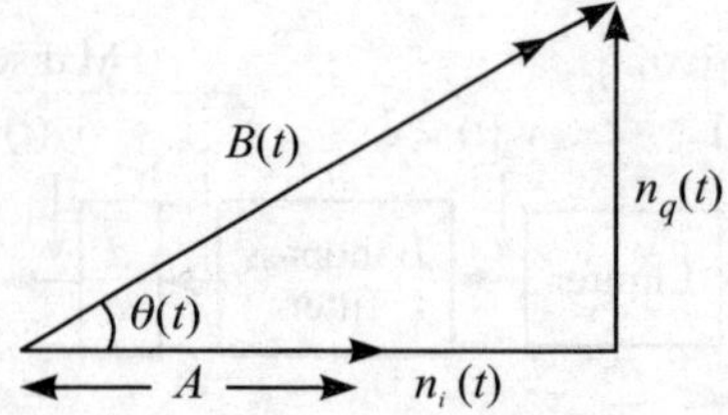

Fig. 7.11 *Phasor diagram of the FM signal and the noise*

where the envelope is

$$B(t) = \sqrt{[A + n_i(t)]^2 + n_q^2(t)} \tag{7.44}$$

and the phase

$$\theta(t) = \tan^{-1}\left(\frac{n_q(t)}{A + n_i(t)}\right). \tag{7.45}$$

Since the limiter provides constant-amplitude A_L waveform, the limiter output can be written as

$$v_2(t) = A_L \cos(2\pi f_c t + \theta(t)) + \text{harmonics}. \tag{7.46}$$

The bandpass filter following the limiter rejects the higher harmonics and passes only the fundamental component. Now, taking the derivative of $v_2(t)$ with respect to time,

$$\frac{dv_2}{dt}(t) = -A_L \sin(2\pi f_c t + \theta(t)) \times [2\pi f_c + \frac{d\theta}{dt}(t)]. \tag{7.47}$$

Also,

$$\frac{d\theta(t)}{dt} = \frac{[A + n_i(t)]\frac{dn_q}{dt}(t) - n_q(t)\frac{dn_i}{dt}(t)}{[A + n_i(t)]^2} \times \frac{1}{1 + \left[\dfrac{n_q(t)}{A + n_i(t)}\right]^2} \tag{7.48}$$

Under the assumption of high-input carrier-to-noise ratio, $A >> |n_i(t)|$ or $|n_q(t)$. Thus,

$$\frac{d\theta}{dt}(t) \approx \frac{1}{A}\frac{dn_q(t)}{dt}. \tag{7.49}$$

So, the differentiator output is

$$v_3(t) = \frac{dv_2}{dt}(t) = -A_L \left[2\pi f_c + \frac{1}{A} \cdot \frac{dn_q(t)}{dt}\right] \sin[2\pi f_c t + \theta(t)]. \tag{7.50}$$

The envelope detector detects the envelope part of (7.50). Eliminating the constant term arising from the carrier frequency, the noise at the input to the baseband filter is v_4, given by

$$v_4(t) = \frac{A_L}{A}\frac{dn_q(t)}{dt}. \tag{7.51}$$

Note that the signal part $s(t)$ is absent. This is obvious because we have started with a pure carrier and added noise to it.

To find the output noise power, we require the power spectral density of $n_q(t)$. Since we have assumed the white noise $n(t)$ of PSD equal to $\eta/2$, the same for the quadrature component $n_q(t)$ is η for the frequencies $-B/2$ to $B/2$, where B is the bandwidth of the receiver input filter (IF carrier filter). The power spectral density of $\frac{dn_q(t)}{dt}$ is $(2\pi f)^2$ multiplied by the PSD of $n_q(t)$. Thus, the noise $v_4(t)$ of (7.51) has the power spectral density given by

$$Gv_4(f) = |j2\pi f|^2 \left(\frac{A_L}{A}\right)^2 \eta = 4\pi^2 f^2 \eta \left(\frac{A_L}{A}\right)^2, \quad |f| < B/2. \tag{7.52}$$

This is plotted in Fig. 7.12. The PSD of the noise out of the FM detector has a parabolic shape.

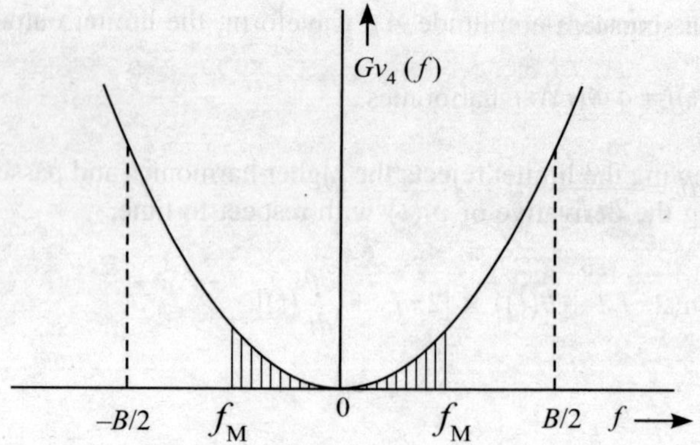

Fig. 7.12 *Variation of PSD of the output noise with frequency*

The bandwidth occupied by wideband FM signal can be many times f_M (highest modulating frequency). But the final baseband filter cuts off the components above f_M. So, output noise power N_0 is the filtered noise given by

$$N_0 = \int_{-f_M}^{f_M} Gv_4(f)df = 2\int_0^{f_M} 4\pi^2 f^2 \eta \left(\frac{A_L}{A}\right)^2 df = \frac{8\pi^2 \eta A_L^2}{A^2} \times \frac{1}{3} f_M^3$$

$$= \frac{8\pi^2}{3} \frac{\eta A_L^2}{A^2} f_M^3. \tag{7.53}$$

In the similar way, the signal component that appears at the output of the discriminator can be calculated. At the limiter input, $s_1(t) = s_i(t)$. The signal $s_2(t)$ at the limiter output is given by

$$s_2(t) = A_L \cos[2\pi f_c t + 2\pi k_f \int_0^t s(t)dt], \tag{7.54}$$

The differentiator output is

$$s_3(t) = \frac{ds_2}{dt} = -A_L[2\pi f_c + 2\pi k_f s(t)] \sin\left(\omega_c t + 2\pi k_f \int s(t)dt\right). \tag{7.55}$$

Again, removing the dc term, the output of the envelope detector is

$$s_4(t) = A_L 2\pi k_f s(t). \tag{7.56}$$

Thus the output signal power S_0 is

$$S_0 = A_L^2 (2\pi k_f)^2 \overline{s^2(t)}. \tag{7.57}$$

The output signal-to-noise ratio is then

$$\left(\frac{S}{N}\right)_0 = \frac{S_0}{N_0} = \frac{4\pi^2 k_f^2 A_L^2 \overline{s^2(t)}}{(8\pi^2/3)\eta(A_L^2/A^2) f_M^3} = \frac{3A^2 k_f^2 \overline{s^2(t)}}{2\eta f_M^3}. \tag{7.58}$$

Now, for the case of a sinusoidal information signal,

$$s(t) = a\cos 2\pi f_m t$$

so that $\displaystyle\int_0^t s(t)dt = \frac{a}{2\pi f_m}\sin 2\pi f_m t.$

The input signal $s_i(t)$ then becomes

$$\begin{aligned} s_i(t) &= A\cos(2\pi f_c t + 2\pi k_f \frac{a}{2\pi f_m}\sin 2\pi f_m t) \\ &= A\cos(2\pi f_c t + \frac{ak_f}{f_m}\sin 2\pi f_m t) \\ &= A\cos(2\pi f_c t + \beta \sin 2\pi f_m t), \end{aligned} \tag{7.59}$$

where $\beta = \dfrac{\Delta f}{f_m} = \dfrac{ak_f}{f_m}$. For the highest modulating signal frequency, $\beta = ak_f/f_M$.

The input signal power is $S_i = A^2/2$. The received noise power over the FM bandwidth of $B = 2\beta f_M$ is $N_i = \frac{\eta}{2}\times 2\beta f_M = \eta\beta f_M$.

Thus the input signal-to-noise power is

$$\left(\frac{S}{N}\right)_i = \frac{S_i}{N_i} = \frac{A^2}{2\eta\beta f_M} \tag{7.60}$$

So, $\dfrac{(S/N)_0}{(S/N)_i} = \dfrac{3A^2k_f^2a^2/2}{2\eta f_M^3} \times \dfrac{2\eta\beta f_M}{A^2}$

$$= \frac{3}{2}\frac{a^2k_f^2}{f_M^2}\beta = \frac{3}{2}\beta^3.$$

Thus, $(S/N)_0 = \dfrac{3}{2}\beta^3(S/N)_i.$ (7.61)

The calculation shows that an improvement in the signal-to-noise ratio is dependent on the square of β. Thus at higher β, the (S/N) improvement becomes more pronounced, but this is achieved at the expense of greater bandwidth. This leads to a trade-off decision in the system design. When the bandwidth of the systems becomes large, the system admits more noise power and the performance of the system degrades. Thus the system exhibits a 'threshold'. Such is indeed the case of wideband FM but not observed in narrowband FM.

It may be mentioned here that input noise power is $N_i = \eta\beta f_M$, which is proportional to β. Thus the output signal-to-noise ratio effectively improves according to the square of β. In other words,

$$\left(\frac{S}{N}\right)_0 = \frac{3}{2}\beta^3\frac{S_i}{\eta f_M} = \frac{3}{2}\beta^2\frac{S_i}{\eta\beta f_M} = \frac{3}{2}\beta^2\left(\frac{S}{N}\right)_{\text{base band}}$$

7.5.2 Threshold effect in FM

We have derived (7.61) with the assumption of small noise, that is, large carrier-to-noise ratio. $(S/N)_i \gg 1$ implies that the signal is above threshold. But if the reverse is true, noise gets larger than the signal, the receiver locks onto the noise and suppresses the signal. In this section, we shall discuss the mechanism for determining the threshold and show that FM noise performance can be substantially better than the baseband performance.

In FM, we have the result vide (7.60) that

$$(S/N)_0 = \frac{3}{2}\beta^2 (S/N)_i = \frac{3}{2}\beta^2 \left(\frac{S_i}{\eta f_M}\right),$$

or in decibel unit,

$$(S/N)_0\Big|_{\text{dB}} = \left(\frac{S_i}{\eta f_M}\right)_i\Big|_{\text{dB}} + 10\log_{10}\frac{3}{2}\beta^2. \tag{7.62}$$

Figure 7.13 gives the plots of output (S/N) ratio against the input (S/N) ratio for various β-values. The plots are straight lines having y-intercepts of $10\log_{10}(\frac{3}{2}\beta^2)$. For each value of β, as $(S/N)_i$ decreases, a point is reached when $(S/N)_0$ declines much more sharply than $(S/N)_i$. We define the threshold value of $(S/N)_i$ as the value at which $(S/N)_0$ falls 1 dB below the value that would be if the curve is extended. Observe that for higher β, the threshold is also higher. The performance can be improved further by using de-emphasis network. A threshold point may be observed seeing the noise output of the FM discriminator on a CRO. At high $(S/N)_i$ ratio (the straight-line portion), the CRO displays random variation of thermal-type noise, as shown in Fig. 7.14(a). As $(S/N)_i$ decreases (bending of the curve), the noise characteristic changes dramatically as indicated in Fig. 7.14(b). In this case, superimposed on the thermal-type noise, a pulse-type noise appears.

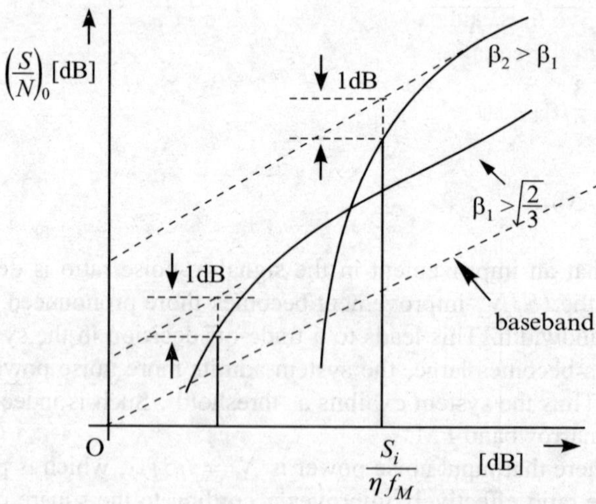

Fig. 7.13 *Illustrating threshold effect in FM. The output S/N characteristic showing the threshold effect is derived in Eq. (7.88).*

This impulsive noise (also called spike) is associated with a very large energy compared to the background noise (although the spike occurs rarely) and hence it increases the total noise output significantly thereby causing a threshold. In FM broadcast receiver, we would probably hear a click each time a spike appears.

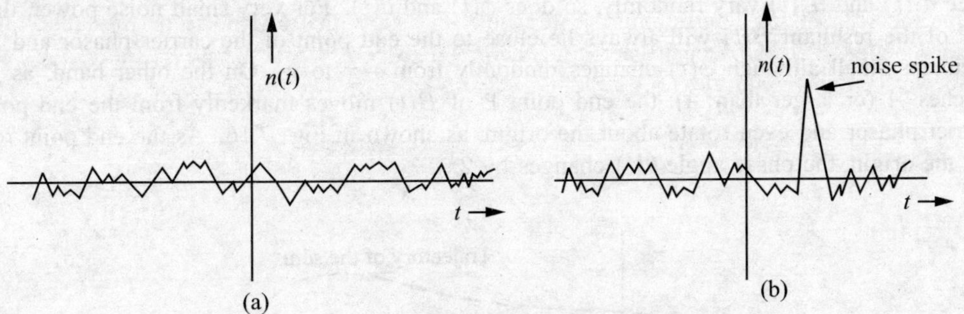

Fig. 7.14 *(a) Thermal noise at FM discriminator output; (b) noise spike indicating the onset of threshold.*

The occurrence of a spike can easily be assessed from the phasor diagram. We rewrite here (7.39) which gives the signal plus the noise at the input to the demodulator.

$$\begin{aligned} v_1(t) &= [A + n_i(t)] \cos \omega_c t - n_q(t) \sin \omega_c t \\ &= \text{Re}\left\{ [A + n_i(t)] e^{j\omega_c t} + n_q(t) e^{j(\omega_c t + \pi/2)} \right\} \\ &= \text{Re}\left\{ e^{j\omega_c t} [A + n_i(t) + jn_q(t)] \right\}. \end{aligned} \tag{7.63}$$

The phasor diagram of $v_1(t)$ is shown in Fig. 7.15. The signal amplitude phasor A adds to the noise amplitude phasor $z(t)$ to give the resultant sum phasor $B(t)$. The signal vector, the noise vector and the sum are related through the equation

$$\begin{aligned} A + n_i(t) + jn_q(t) &= A + z(t)e^{j\phi(t)} \\ &= B(t)e^{j\theta(t)}, \end{aligned} \tag{7.64}$$

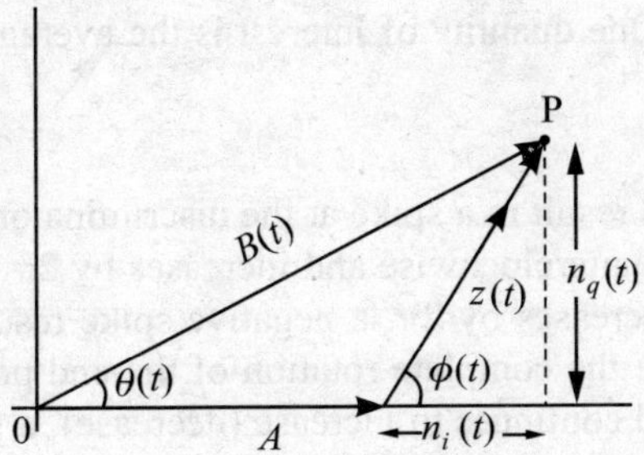

Fig. 7.15 *Phasor diagram of the signal plus noise*

where noise amplitude $z(t) = \sqrt{n_i^2 + n_q^2}$; random phase angle of noise, $\phi(t) = \tan^{-1} \dfrac{n_q(t)}{n_i(t)}$; the sum $B(t) = \sqrt{[A + n_i(t)]^2 + n_q^2(t)}$ and $\theta(t) = \tan^{-1} \dfrac{n_q(t)}{A + n_i(t)}$.

Since $n_i(t)$ and $n_q(t)$ vary randomly, so does $\phi(t)$ and $\theta(t)$. For very small noise power, the end point P of the resultant $B(t)$ will always lie close to the end point of the carrier phasor and angle $\theta(t)$ remains small although $\phi(t)$ changes randomly from $-\pi$ to π. On the other hand, as $|z(t)|$ approaches A (or larger than A), the end point P of $B(t)$ moves markedly from the end point of the carrier phasor and even rotate about the origin, as shown in Fig 7.16. As the end point rotates around the origin, the phase angle $\theta(t)$ changes by 2π.

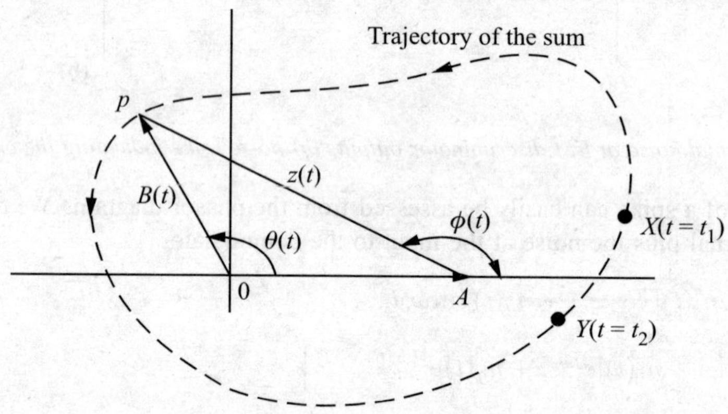

Fig. 7.16 *Locus of the sum vector* $B(t)$ *as* $|z(t)| \cong A$

At the point X at $t = t_1, \theta(t)$ is close to 0° but at the point Y at $t = t_2$, $\theta(t)$ attains 360° (approx). The noise capture takes effect in the interval (t_1, t_2) when $\theta(t)$ changes by 2π radian. Since the output of the discriminator is proportional to $\dfrac{d\theta}{dt}(t)$, a sharp spike of strength 2π appears there. Figure 7.17 shows the variation of $\theta(t)$ and consequent spike generation. Owing to the randomness, the frequency of occurrence of the spikes cannot be characterised deterministically, rather we will have to adopt a probabilistic approach. One quantity of interest is the average number of spikes per second.

Mean time between spikes

Let us identify the conditions that result in a spike at the discriminator output. We have argued that if the phase angle $\theta(t)$ is rotated counterclockwise and increases by 2π, a spike results. Similarly if the rotation is clockwise and $\theta(t)$ decreases by 2π, a negative spike results. However, to ensure a spike output, it is not essential to chase the complete rotation of the end point of $B(t)$. We assume that if $\theta(t)$ passes through π radians and continues to increase (decrease), a positive (negative) spike results. As shown in Fig. 7.18, the phase angle $\theta(t)$ of π radian corresponds to the zero-crossing point P of the locus of the end point of $B(t)$. For the locus to cross the point P, we must have $n_i(t) < -A$ and

$n_q(t)$ must have zero-crossing ($n_q(t) = 0$ and $\dfrac{dn_q}{dt}(t) < 0$). Since $n_i(t)$ and $n_q(t)$ are independent, the probability of this joint event is the product of their individual probabilities.

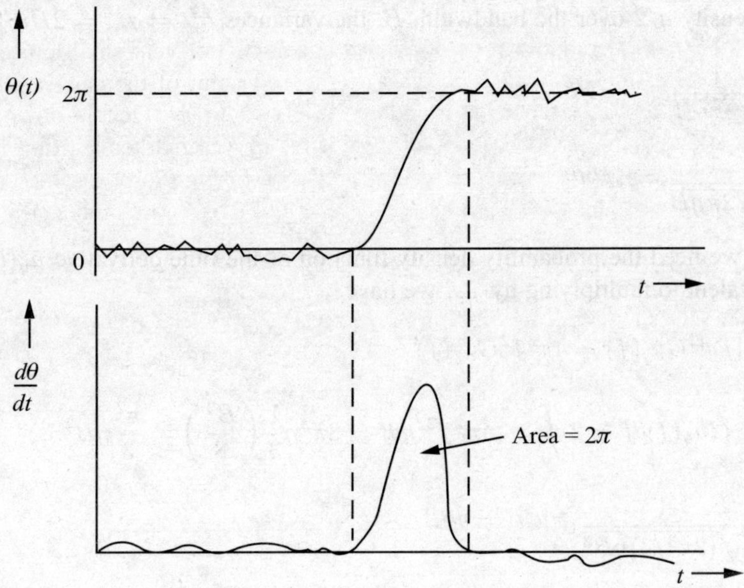

Fig. 7.17 *Illustrating the generation of impulsive noise as the phase angle $\theta(t)$ changes by 2π radians*

To be specific, for the event 'a positive spike generation', the following conditions must be satisfied simultaneously at time $t = t_1$ (say).

$$n_i(t_1) < -A; \qquad n_q(t_1) = 0; \qquad \dot{n}_q(t) = \frac{dn_q}{dt} < 0. \tag{7.65}$$

Fig. 7.18 *Phasor diagram using quadrature noise components*

For the negative spike at $t = t_2$, we have

$$n_i(t_2) < -A; ; \qquad n_q(t_2) = 0; ; \qquad \dot{n}_q(t) = \frac{dn_q}{dt}(t) > 0. \tag{7.66}$$

We assume $n_i(t)$ and $n_q(t)$ are Gaussian random process with zero mean. Since the noise $n(t)$ has a power spectral density $\eta/2$ over the bandwidth B, the variances $\sigma_{n_i}^2 = \sigma_{n_q}^2 = 2B\eta/2 = \eta B$. Thus,

$$f(n_i) = \frac{1}{\sqrt{2\pi\eta B}} e^{-n_i^2/2\eta B} \tag{7.67}$$

$$\text{and} \quad f(n_q) = \frac{1}{\sqrt{2\pi\eta B}} e^{-n_q^2/2\eta B}. \tag{7.68}$$

At this moment, we need the probability density function of the time derivative $\dot{n}_q(t)$. Since differentiation is equivalent to multiplying by $j\omega$, we have

$$G_{\dot{n}_q}(f) = |j\omega|^2 G_{n_q}(f) = 4\pi^2 f^2 G_{n_q}(f). \tag{7.69}$$

$$\text{So, } \sigma_{\dot{n}_q}^2 = \int_{-B/2}^{B/2} G\dot{n}_q(f)df = 2\int_0^{B/2} 4\pi^2 f^2 .\eta df = 8\pi^2\eta \frac{1}{3}\left(\frac{B^3}{8}\right) = \frac{\pi^2}{3}\eta B^3. \tag{7.70}$$

$$\text{Thus, } f(\dot{n}_q) = \frac{1}{\sqrt{(2\pi^3/3)\eta B^3}} e^{-\dot{n}_q^2/[\frac{2\pi^2}{3}\eta B^3]}. \tag{7.71}$$

Let us now calculate the probability P_1 of the occurrence of a positive spike over an interval Δt at t_1. This is,

$$P_1 = P[x_i(t_1) < -A, n_q(t_1) = 0, \dot{n}_q(t_1) < 0]$$

$$= \int_{n_i}\int_{n_q}\int_{\dot{n}_q} f(n_i, n_q, \dot{n}_q) dn_i \, dn_q \, d\dot{n}_q, \tag{7.72}$$

where $f(n_i, n_q, \dot{n}_q)$ is the joint density function and given by

$$f(n_i, n_q, \dot{n}_q) = \frac{e^{-n_i^2/2\eta B}}{\sqrt{2\pi\eta B}} \; \frac{e^{-n_q^2/2\eta B}}{\sqrt{2\pi\eta B}} \; \frac{e^{-\dot{n}_q^2/[2\pi^2 B^3\eta/3]}}{\sqrt{2\pi^3 B^3\eta/3}}. \tag{7.73}$$

$$\text{Thus, } P_1 = \int_{-\alpha}^{-A} dn_i \int_{-\Delta n_q/2}^{\Delta n_q/2} dn_q \int_{-\infty}^{0} d\dot{n}_q . f(n_i, n_q, \dot{n}_q). \tag{7.74}$$

The term $\int_{-\infty}^{-A} df(n_i)dn_i$ is the probability that $n_i(t_1) < -A$. The term $\int_{-\Delta n_q/2}^{\Delta n_q/2} f(n_q)dn_q$ gives the probability that $n_q(t_1)$ passes through 0. The term $\int_{-\infty}^{0} f(\dot{n}_q)d\dot{n}_q$ gives the probability that $\dot{n}_q(t_1) < 0$.

Imposing top priority to the condition $n_q(t_1) = 0$, we first evaluate

$$\int_{-\frac{\Delta}{2}n_q}^{\Delta n_q/2} f(n_q)dn_q = f(0)\Delta n_q, \tag{7.75}$$

where $f(0) = \dfrac{1}{\sqrt{2\pi\eta B}}$ and Δn_q is the change in n_q over the interval Δt which is $\Delta n_q = \dfrac{dn_q}{dt}\Delta t = \dot{n}_q\Delta t$. Thus (7.74) becomes

$$\begin{aligned} P_1 &= \int_{-\infty}^{-A} dn_i f(n_i) \int_{-\infty}^{0} d\dot{n}_q f(n\dot{n}_q)\, \frac{\dot{n}_q\Delta t}{\sqrt{2\pi\eta B}} \\ &= \int_{\infty}^{-A} dn_i f(n_i). \frac{\Delta t}{\sqrt{2\pi\eta B}} \int_{-\infty}^{0} d\dot{n}_q f(\dot{n}_q)\dot{n}_q. \end{aligned} \tag{7.76}$$

Next we compute

$$\int_{-\infty}^{0} d\dot{n}_q f(\dot{n}_q)\dot{n}_q = \int_{-\infty}^{0} d\dot{n}_q \dot{n}_q \frac{e^{-\dot{n}_q^2/[2\pi^2 B^s \eta/3]}}{\sqrt{\frac{2}{3}\pi^3 B^3 \eta}}.$$

By a change of variable

$$x = \frac{\dot{n}_q^2}{2\pi^2 B^3 \eta/3}; \qquad dx = \frac{\dot{n}_q d\dot{n}_q}{\pi^2 B^3 \eta/3},$$

we have

$$\begin{aligned} \int_{-\infty}^{0} d\dot{n}_q f(\dot{n}_q)\dot{n}_q &= \int_{\infty}^{0} \frac{\pi^2 B^3 \eta}{3} dx \frac{e^{-x}}{\sqrt{\frac{2}{3}\pi^3 B^3 \eta}} \\ &= \sqrt{\frac{\pi B^3 \eta}{6}}. \int_{\infty}^{0} e^{-x} dx \\ &= -\sqrt{\frac{\pi B^3 \eta}{6}}. \end{aligned}$$

Equation (7.76) becomes

$$\begin{aligned} P_1 &= \int_{-\infty}^{-A} dn_i f(n_i) \frac{\Delta t}{\sqrt{2\pi\eta B}} \sqrt{\frac{\pi B^3 \eta}{6}} \\ &= \int_{-\infty}^{-A} dn_i \frac{\Delta t}{\sqrt{2\pi\eta B}} \sqrt{\frac{\pi B^3 \eta}{6}}\, \frac{e^{-n_i^2/2\eta B}}{\sqrt{2\pi\eta B}} \\ &= \frac{B\Delta t}{\sqrt{12}} \int_{-\infty}^{-A} dn_i \frac{e^{-n_i^2/2\eta B}}{\sqrt{2\pi\eta B}}. \end{aligned} \tag{7.77}$$

By a change of variable

$$x = \frac{-n_i}{\sqrt{2\eta B}}; \quad dn_i = -\sqrt{2\eta B}dx,$$

$$P_1 = \frac{B\Delta t}{2\sqrt{3}} \int_{\infty}^{\frac{A}{\sqrt{2\eta B}}} \left(-\frac{1}{\sqrt{\pi}}\right) e^{-x^2} dx = \frac{B\Delta t}{4\sqrt{3}} \frac{2}{\sqrt{\pi}} \int_{\frac{A}{\sqrt{2\eta B}}}^{\infty} e^{-x^2} dx$$

$$= \frac{B\Delta t}{4\sqrt{3}} \operatorname{erfc}\sqrt{\frac{A^2}{2\eta B}}. \tag{7.78}$$

Thus, the total number of spikes per second is

$$N_1 = \frac{P_1}{\Delta t} = \frac{B}{4\sqrt{3}} \operatorname{erfc}\sqrt{\frac{A^2}{2\eta B}}. \tag{7.79}$$

In a similarly way, the total number of negative spikes per second is

$$N_2 = \frac{B}{4\sqrt{3}} \operatorname{erfc}\sqrt{\frac{A^2}{2\eta B}} \tag{7.80}$$

so that on the average the total number of spikes occuring per second is

$$2N_1 = \frac{B}{2\sqrt{3}} \operatorname{erfc}\sqrt{\frac{A^2}{2\eta B}}. \tag{7.81}$$

The mean time T_s between the spikes is given by the reciprocal of the average number of spikes occurring per second. That is,

$$T_s = \frac{2\sqrt{3}}{B \operatorname{erfc}\sqrt{\frac{A^2}{2\eta B}}}. \tag{7.82}$$

Our next aim is to find the output signal-to-noise ratio. The output noise power, N_s, due to the spike is obtained by integrating the power spectral density $G_s(f)$ of the spike over the frequency band of interest. The power spectral density of the spikes is given by

$$G_s(f) = \frac{1}{T_s}|P(j\omega)|^2, \tag{7.83}$$

where $P(j\omega)$ is the Fourier transform of the spike. As we are interested only in the range $-f_M$ to f_M, the quantity $|P(j\omega)|$ remains almost constant, equal to its value $|P(0)|$. We know that the output noise spike is $A_L \frac{d\theta}{dt}$ [see (7.50) & (7.51)]. As the area under $\frac{d\theta}{dt}$ is 2π, we have

$$|P(0)| = 2\pi A_L. \tag{7.84}$$

Thus, from (7.83),

$$G_s(f) = \frac{1}{T_s} 4\pi^2 A_L^2 \tag{7.85}$$

and the output noise power due to spike

$$N_s = \frac{4\pi^2 A_L^2}{T_s} 2f_M. \tag{7.86}$$

Substituting T_s in (7.86), we get

$$\begin{aligned} N_s &= \frac{4\pi^2 A_L^2 2f_M B}{2\sqrt{3}} \operatorname{erfc}\sqrt{\frac{A^2}{2\eta B}} \\ &= \frac{4\pi^2 A_L^2 B f_M}{\sqrt{3}} \operatorname{erfc}\sqrt{\frac{A^2}{2\eta B}}. \end{aligned} \tag{7.87}$$

The total output noise power is $N_T = N_S + N_O$, where N_O is the Gaussian noise given by (7.53). Thus the output signal-to-noise ratio is

$$\begin{aligned} \left(\frac{S}{N}\right)_0 &= \frac{S_O}{N_s + N_O} = \frac{4\pi^2 k_f^2 A_L^2 \overline{s^2(t)}}{\left[\frac{4\pi^2 A_L^2 B f_M}{\sqrt{3}} \operatorname{erf}\sqrt{\frac{A^2}{2\eta B}} + \frac{8\pi^2}{3}\frac{\eta A_L^2}{A^2} f_M^3\right]} \\ &= \frac{\frac{3k_f^2 \overline{s^2(t)}}{f_M^2}\left[\frac{S_i}{\eta f_M}\right]}{1 + \left(\frac{\sqrt{3}B}{f_M}\right)\left(\frac{S_i}{\eta f_M}\right)\operatorname{erfc}\sqrt{\frac{S_i}{\eta B}}}. \end{aligned} \tag{7.88}$$

In the above equation, S_O is obtained from (7.57), where $S_i = A^2/2$ has been used. The threshold is evident from (7.88). At very high carrier-to-noise ratio (CNR) $\left(\frac{S_i}{\eta B}\right)$, the erfc function approaches zero and $(S/N)_O$ varies linearly with the CNR. As CNR decreases, the denominator increases and $(S/N)_O$ decreases at a high rate. The onset of threshold happens when the carrier power is 10 times the channel noise power. Thus, a threshold occurs when

$$\eta B = \frac{1}{10} \cdot \frac{A^2}{2}. \tag{7.89}$$

Example 7.4

The total output signal-to-noise ratio of an FM discriminator is given by (7.88). A threshold is defined as the value of $Si/\eta f_M$ at which $(S/N)_O$ decreases by 1 dB from the value

$$\frac{S_O}{N_O} = \frac{3k_f^2 \overline{s^2(t)}}{f_M^2}\left(\frac{S_i}{\eta f_M}\right).$$

Show that threshold is reached when

$$0.26 = \frac{\sqrt{3}B}{f_M}\frac{S_i}{\eta f_M} \operatorname{erfc}\sqrt{\frac{S_i}{\eta B}}.$$

Solution

Since $(S/N)_O = \frac{x}{y}$, where x and y are the quantities in the numerator and denominator of (7.88), we have

$$\left(\frac{S}{N}\right)_O\bigg|_{\text{dB}} = 10\log_{10}^{x} - 10\log_{10}^{y}.$$

By definition of threshold, $(S/N)_O|_{\text{dB}}$ decreases by 1 dB from $10\log_{10}^{x}$. This occurs when $10\log_{10}^{y} = 1$, that is $y = 10^{0.1} = 1.2589 \approx 1.26$. Hence the result.

Example 7.5

A carrier $A\cos 2\pi f_c t$ is accompanied by bandlimited noise having quadrature components $n_c(t)$ and $n_s(t)$. Assuming

$$n_c(t) = \frac{A}{2}\sin[2\pi(f_c + f)t],$$

$$n_s(t) = \frac{A}{2}[\cos 2\pi(f_c + f)t - 1],$$

give the expression for the resultant phasor $B(t)$. Also, plot the phasor $B(t)$ in the phasor diagram.

Solution

$$\begin{aligned}\text{Here, } n(t) &= n_c(t)\cos 2\pi f_c t - n_s(t)\sin 2\pi f_c t \\ &= \frac{A}{2}[\sin 2\pi(f_c + f)t]\cos 2\pi f_c t - \frac{A}{2}[\cos 2\pi(f_c + f)t - 1]\sin 2\pi f_c t\end{aligned}$$

So, the signal plus the noise is

$$v(t) = A\cos 2\pi f_c t + n(t) = A[1 + \frac{1}{2}\sin 2\pi(f_c + f)t]\cos 2\pi f_c t.$$

7.5.3 Pre-emphasis and de-emphasis filters: Output S/N improvement with pre-emphasis

It is possible to improve the output (S/N) ratio by deliberate distortion of the transmitted signal and correspondingly compensating at the receiver. We recall that the noise output at the FM receiver is not white [see (7.52)], rather, it varies as f^2 even through the input noise is white channel noise. So, high information frequencies are badly affected by noise than are the lower frequencies. In this case we can boost (emphasise) the higher signal frequencies and de-emphasise the lower frequencies in a manner that keeps the total power constant. Since the signal has relatively less power at high frequencies (speech has less power spectral density above about 3 kHz), this pre-emphasis will require only a small increment in the transmitted power. At the receiver, the high-frequency components are de-emphasised to undo the signal pre-emphasis at the transmitter. Because the channel noise is added after the transmitter, it is not pre-emphasised but de-emphasised at the receiver. Thus, the de-emphasis filter at the receiver restores the signal power and simultaneously reduces the noise power, improving the signal-to-noise ratio. An FM system with a pre-emphasis filter $H_p(\omega)$ at the transmitter and de-emphasis filter $H_d(\omega)$ at the receiver is shown in Fig. 7.19. The composite filter transfer function of the two filters is unity so that the deliberate distortion at the transmitter is compensated at the receiver. The pre-emphasis filter is placed before FM modulator and the de-emphasis filter is placed before the baseband filter, as shown in the figure. Since $H_d(\omega)H_p(\omega) = 1$, any modification introduced in the information signal by the first filter is undone by the second filter.

We have discussed that the power spectral density of the output noise of an FM discriminator varies squarely with the frequency (see Fig. 7.12). But from commercial viewpoint, the exact function shown in Fig. 7.12 is not used. De-emphasis is performed by a simple RC lowpass filter of Fig. 7.20(a). The inverse filter is used at the transmitting station. One such network is as shown in Fig. 7.20(b).

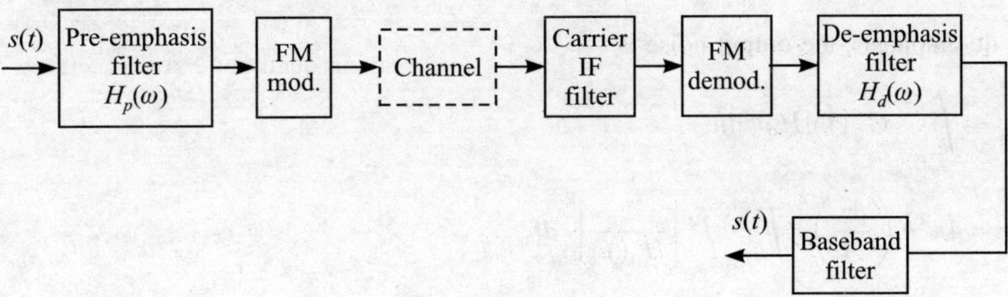

Fig. 7.19 *Pre-emphasis and de-emphasis filters in the FM systems*

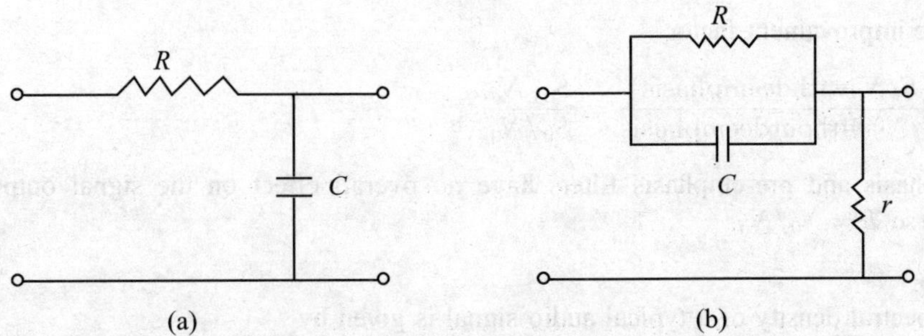

Fig. 7.20 *(a) De-emphasis filter, (b) pre-emphasis filter.*

The transfer function $H_d(f)$ of the de-emphasis network is given by

$$H_d(f) = \frac{1}{1 + jf/f_1}, \tag{7.90}$$

where $f_1 = \frac{1}{2\pi RC}$, defines the 3 dB cutoff frequency. The pre-emphasis network has the transfer function $H_p(f)$ given by

$$H_p(f) = \frac{r}{R}\left(1 + j\frac{f}{f_1}\right). \tag{7.91}$$

In deriving this result, we have assumed that $r << R$ and $r << \frac{1}{\omega C}$ at the baseband signal. Since

$H_p(f)H_d(f) = \frac{r}{R}$, a constant, no net distortion is introduced in the signal. Thus the signal itself is unaffected in the overall process. To estimate the (S/N) ratio improvement using pre-emphasis – de-emphasis networks, it is reasonable to consider only the noise. We will investigate the ratio of the noise outputs without and with de-emphasis, and if this ratio exceeds unity we have indeed attained an improvement in the S/N ratio.

The output noise spectral density without de-emphasis was derived in (7.52) and is re-written here.

$$N_O = 4\pi^2\eta\left(\frac{A_L}{A}\right)^2\int_{-f_M}^{F_M} f^2\,df. \tag{7.92}$$

With the de-emphasis, the output noise is

$$N_d = \int_{-f_M}^{f_M} G_n(f)|H_d|^2 df$$

$$= 4\pi^2\eta\left(\frac{A_L}{A}\right)^2\int_{-f_M}^{f_M} f^2\left|\frac{1}{H_p(f)}\right|^2 df. \tag{7.93}$$

$$\text{Thus,}\quad \frac{N_O}{N_d} = \frac{\int_{-f_M}^{f_M} f^2 df}{\int_{-f_M}^{f_M} f^2/|H_p(f)|^2 df} = \frac{f_M^3/3}{\int_0^{f_M}\frac{f^2}{|H_p(f)|^2}df}. \tag{7.94}$$

We define the improvement factor

$$I = \frac{(S/N)\text{withdeemphasis}}{(S/N)\text{withoutdeemphasis}} = \frac{S_d/N_d}{S_O/N_O}$$

Since deemphasis and pre-emphasis filters have no overall effect on the signal output, we have $S_d = S_O$ and so $I = N_O/N_d$.

Example 7.6

The power spectral density of a typical audio signal is given by

$$G(f) = \frac{G_O}{1+(f/f_1)^2},\quad |f| \le f_M$$

$$= 0 \quad \text{otherwise}$$

The pre-emphasis network has the transfer function

$$H_p(f) = C\left(1 + j\frac{f}{f_1}\right).$$

Find the value of C and the signal improvement factor.

Solution

The value of C is calculated on the basis that the total power of the baseband signal is unchanged by employing the pre-emphasis network. Thus,

$$\text{Power} = \int_{-f_M}^{f_M} G(f)df = \int_{-f_M}^{f_M} G(f)|H_p(f)|^2 df$$

or $$\int_{-f_M}^{f_M} \frac{G_0}{1+(f/f_1)^2} df = \int_{-f_M}^{f_M} G_0 C^2 df$$

or $$\frac{2f_1^2}{f_1}\left[\tan^{-1}\left(\frac{f}{f_1}\right)\right]_0^{f_M} = G_0 C^2 2 f_M$$

or $$C^2 = \frac{f_1}{f_M}\tan^{-1}(f_M/f_1).$$

The improvement factor is obtained from (7.94).

$$\int_0^{f_M} \frac{f^2}{|H_p(f)|^2} df = \int_0^{f_M} \frac{f^2}{C^2(1+f^2/f_1^2)} df$$

$$= \frac{f_1^2}{C^2}\int_0^{f_M} \frac{f^2}{f_1^2+f^2} df$$

$$= \frac{f_1^2}{C^2}\left[\int_0^{f_M} df - \frac{f_1^2}{f_1}\tan^{-1}\left[\frac{f}{f_1}\right]_0^{f_M}\right]$$

$$= \frac{f_1^2}{C^2}[f_M - f_1\tan^{-1}(f_M/f_1)]$$

$$= \frac{f_1^2 f_M[1-\frac{f_1}{f_M}\tan^{-1}(f_M/f_1)]}{(f_1/f_M)\tan^{-1}(f_M/f_1)}.$$

Thus, the improvement factor, $I = \dfrac{f_M^3\tan^{-1}(f_M/f_1)}{f_1 f_M^2[1-\frac{f_1}{f_M}\tan^{-1}\frac{f_M}{f_1}]}$

$$= \frac{\tan^{-1}(f_M/f_1)}{2(f_1/f_M)[1-(f_1/f_M)\tan^{-1}(f_M/f_1)]}.$$

By adjusting the value of C one may make $f_M/f >> 1$ and this is indeed the case because the information signal is basically concentrated near the low frequency range. In doing so, $\tan^{-1}(f_M/f_1) \approx \pi/2$ and the improvement factor $= \frac{\pi/2}{3(f_1/f_M)} \approx \frac{\pi}{6}\left(\frac{f_M}{f_1}\right)$.

For academic interest, suppose that in FM broadcasting $f_1 = 2$ kHz for a 75 μs emphasis, the 3 dB frequency of the receiver deemphasis is

$$f_1 = \frac{1}{2\pi \times \mu s} = 2.12 \text{ kHz}$$

and $f_M = 15$ kHz. Then, $f_M/f_1 = 7.5$, and the improvement factor $= \frac{\pi}{6} \times 7.5 = 3.92 = 5.9$ dB.

For voice communication, output S/N ratio is typically 40 dB. An improvement of nearly 6 dB corresponds to roughly 15% improvement.

FM System with de-emphasis

Incorporating de-emphasis at the receiver, the noise power output of the receiver becomes

$$N_0 = \int_{-f_M}^{f_M} 4\pi^2\eta\left(\frac{A_L}{A}\right)^2 \cdot f^2 \cdot \frac{1}{1+(f/f_1)^2}df = 4\pi^2\eta\left(\frac{A_L}{A}\right)^2 f_1^2 \cdot 2\int_0^{f_M}\frac{f^2}{f_1^2+f^2}df$$

$$= 8\pi^2\eta\left(\frac{A_L}{A}\right)^2 \cdot f_1^2\left[\int_0^{f_M} df - \frac{f_1^2}{f_1}\tan^{-1}\frac{f}{f_1}\Bigg|_0^{f_M}\right]$$

$$= 8\pi^2\eta f_1^2\left(\frac{A_L}{A}\right)^2\left[f_M - f_1\tan^{-1}\frac{f_M}{f_1}\right] = 8\pi^2\eta\left(\frac{A_L}{A}\right)^2 \cdot f_1^3\left[\frac{f_M}{f_1} - \tan^{-1}\frac{f_M}{f_1}\right] \quad (7.95)$$

In practical case $f_M/f_1 >> 1$ so that $\tan^{-1} f_M/f_1 = \pi/2$. Thus

$$N_0 = 8\pi^2\eta\left(\frac{A_L}{A}\right)^2 f_M f_1^2, \quad (7.96)$$

as $\frac{f_M}{f_1} >> \pi/2$

Solved Numericals

1. The bandpass noise can be represented by Eq. (6.32)

$$n(t) = n_i(t)\text{Cos}\omega_0 t - n_q(t)\sin\omega_0 t$$

$$= A(t)Cos\left(\omega_0 t + \phi_n(t)\right)$$

where $A(t)Cos\phi_n(t) = n_i(t)$ and $A(t)Sin\phi_n(t) = n_q(t)$ so that

$$A^2(t) = n_i^2(t) + n_q^2(t) \text{ and } \phi_n(t) = \tan^{-1}\left[\frac{n_q(t)}{n_i(t)}\right].$$

The PDF of the envelope $A_n = \sqrt{n_i^2 + n_q^2}$ follows Rayleigh distribution with

$$f_{A_n}(A_n) = \frac{A_n}{\sigma_n^2}e^{-A_n{}^2/2\sigma^2}u(A_n),\ u(x) \text{ is the unit-step function.}$$

Given that $\sigma_n^2 = 1\mu$W, find the mean and rms value of the envelope across a 1-Ω resistor. Also calculate the probability that this envelope exceeds its mean value.

Solution

The mean of the envelope is obtained as

$$\overline{A_n} = \int_{A_n=0}^{\infty} A_n f_{An}(A_n) dA_n = \int_0^{\infty} \frac{A_n^2}{\sigma_n^2} e^{-A_n^2/2\sigma_n^2} dA_n$$

Let $\dfrac{A_n^2}{2\sigma_n^2} = x$ so that $A_n dA_n = \sigma_n^2 dx$ or, $dA_n = \dfrac{\sigma_n^2}{\sqrt{2\sigma_n^2 x}} dx$

Thus,

$$\overline{A_n} = \int_0^{\infty} (2x) \cdot e^{-x} \cdot \frac{{\sigma_n}^2}{\sqrt{2{\sigma_n}^2}} \cdot x^{-1/2} dx$$

$$= \sqrt{2} \cdot \sigma_n \int_0^{\infty} e^{-x} x^{1/2} dx = \sqrt{2} \cdot \sigma_n \cdot \Gamma\left(3/2\right)$$

$$= \sqrt{2}.\sigma_n \left(\frac{1}{2}\sqrt{\pi}\right) = \sqrt{\frac{\pi \sigma_n^2}{2}} = \sqrt{\frac{\pi \times (10^{-6})}{2}} = 1.25 \text{ millivolt.}$$

RMS value of the noise envelope is given by

$$\overline{A_n^2} = \int_0^{\infty} A_n^2 f_{A_n}(A_n) dA_n = \int_0^{\infty} \frac{A_n^3}{{\sigma_n}^2} e^{-A_n^2/2\sigma_n^2} dA_n$$

Taking $A_n^2/2\sigma_n^2 = x$, we may write,

$$\overline{A_n^2} = \int_0^{\infty} \frac{A_n^2}{\sigma_n^2} e^{-A_n^2/2\sigma_n^2} \cdot (A_n dA_n)$$

$$= \int_0^{\infty} (2x) e^{-x} \cdot \sigma_n^2 dx = 2\sigma_n^2 \int_0^{\infty} x e^{-x} dx$$

$$= 2\sigma_n^2 \Gamma(2) = 2\sigma_n^2 \text{ (since } \Gamma(2) = 1)$$

Thus, rms noise envelope

$$A_n = \sqrt{2\sigma_n^2} = \sqrt{2 \times 1 \times 10^{-6}} = 1.41 \text{ millivolt.}$$

Finally, we evaluate the probability that the envelope exceeds its mean value, that is, $P(A_n > \bar{A}_n)$. Now,

$$P(A_n > \bar{A}_n) = \int_{\bar{A}_n}^{\infty} f_{A_n}(A_n)dA_n = \int_{\bar{A}_n}^{\infty} \frac{A_n}{\sigma_n^2} e^{-A_n{}^2/2\sigma_n^2} dA_n.$$

By letting $A_n^2/2\sigma_n^2 = x$, we may write,

$$P(A_n > \bar{A}_n) = \int_{y}^{\infty} \frac{1}{\sigma_n^2} e^{-x} \cdot \sigma_n^2 dx$$

$$= \int_{y}^{\infty} e^{-x} dx = \left[\frac{e^{-x}}{-1}\right]_y^{\infty} = -e^{-\infty} + e^{-y}$$

$$= e^{-y}, \text{ where } y = (\bar{A}_n)^2 \Big/ 2\sigma_n^2 = \tfrac{\pi\sigma_n^2}{2} \Big/ 2\sigma_n^2 = \tfrac{\pi}{4}$$

$$= e^{-\pi/4}$$

$$= 0.456.$$

2. A realistic expression for the PSD of noise obtained from quantum mechanical treatment is given by Eq. (68) as

$$G_n(f) = \frac{hf/2}{e^{hf/kT} - 1},$$

where h is the Planck constant, k is the Boltzmann Constant.

T is the ambient temperature and f is the frequency. Show that even at room temperature, this PDF can be considered with white noise in practical communication system.

Solution

$G_n(f)$ attains its maximum value $kT/2 = 2.12 \times 10^{-21}$J at room temperature at $f = 0$.

$$\left[G_n(f=0) = \underset{f\to 0}{\text{Lt}} \left[\frac{hf/2}{e^{hf/kT} - 1}\right] = \underset{f\to 0}{\text{Lt}} \left[\frac{h/2}{\frac{h}{kT} e^{hf/kT}}\right] = \frac{h/2}{h/kT} = \frac{kT}{2}\right]$$

Now, at room temperature, let us calculate $G_n(f)$ at the frequency of $f = 10^{12}$ Hz (say, a reasonable frequency for communication).

$$\frac{hf}{kT} = \frac{6.62 \times 10^{-34} \times 10^{12}}{1.38 \times 10^{-23} \times 300} = \frac{6.62}{1.38 \times 30} = 1.59903$$

$$hf/2 = \frac{6.62 \times 10^{-34} \times 10^{12}}{2} = 3.31 \times 10^{-22} \text{ J},$$

$$kT = 1.38 \times 10^{-23} \times 300 \text{ J} = 4.14 \times 10^{-21} \text{ J.}$$

$$\text{So, } G_n(f) = \frac{3.31 \times 10^{-22}}{e^{1.59903} - 1} = \frac{3.31 \times 10^{-22}}{0.17339} = 19.089 \times 10^{-22} \text{J/Hz}$$

or, in terms of $kT/2$, this figure is

$$G_n(f) = \frac{19.089 \times 10^{-22}}{\left(\frac{4.14}{2}\right) \times 10^{-21}} = 0.92 \times \left(\frac{kT}{2}\right)$$

Thus, we conclude that the PSD of noise drops to 92% its maximum value at about f $\approx 10^{12}$ Hz at room temperature. Hence, this spectrum can be safely considered as identical to that of white noise for most practical purposes.

3. Let $n(t)$be a bandpass noise centred at f_c. We define the process $y(t) = 2n(t)\text{Cos}(2\pi f_c t + \theta)$. Show that $y(t)$ consists of both the low-pass and the bandpass components.

Solution

The bandpass noise is represented by

$$n(t) = n_i(t)Cos(2\pi f_c t) - n_q(t)\sin(2\pi f_c t)$$

Thus,

$$y(t) = 2\left[n_i(t)Cos(2\pi f_c t) - n_q(t)sin(2\pi f_c t)\right] Cos(2\pi f_c t + \theta)$$

$$= 2\left[n_i(t)Cos\omega_c t - n_q(t)Sin\omega_c t\right](Cos\omega_c t\text{Cos}\theta - Sin\omega_c t\text{Sin}\theta)$$

$$= 2n_i(t)Cos\theta\cdot\text{Cos}^2\omega_c t + 2n_q(t)\text{Sin}\theta\cdot\text{Sin}^2\omega_c t - 2\left[n_i(t)\text{Sin}\theta + n_q(t)\text{Cos}\theta\right] Sin\omega_c tCos\omega_c t$$

$$= n_i(t)Cos\theta\left[1 + \text{Cos}2\omega_c t\right] + n_q(t)sin\theta\left[1 - \text{Cos}2\omega_c t\right]$$

$$- \left[n_i(t)sin\theta + n_q(t)\text{Cos}\theta\right] \cdot \sin 2\omega_c t$$

$$= \left[n_i(t)Cos\theta + n_q(t)Sin\theta\right] + \left[n_i(t)Cos\theta - n_q(t)Sin\theta\right] \text{Cos}(2\omega_c\text{t})$$

$$- \left[n_i(t)Sin\theta + n_q(t)Cos\theta\right] \sin 2\omega_c t$$

So, $y(t)$ consists of:

(i) low-frequency components around $f = 0$ and

(ii) bandpass components around $f = 2f_c$.

4. A DSB with transmitted carrier is applied to an ideal envelope detector including a DC block. Find the output $y(t)$ of the detector. Comment on the threshold effect in presence of noise.

Solution

The threshold effect concerns with the loss of message signal at low signal-to-noise ratio. This effect can be minimised when the carrier amplitude is much higher than the rms noise.

As in Section 7.4, Eq. (7.29), we may write the received signal plus noise at the detector as

$$r(t) = [A_c + s(t)]\, \text{Cos}\omega_c t + [n_i(t) Cos\omega_c t - n_q(t) \sin \omega_c t]$$

where A_c is the carrier amplitude, the first term is the DSB with carrier and the second term is the bandpass noise .

Rewriting,

$$r(t) = [A_c + s(t) + n_i(t)]\, Cos\omega_c t - n_q(t) sin\omega_c t$$

and the resultant envelope B(t) and phase $\phi(t)$ are (vide Eqs. 7.30 and 7.31)

$$B(t) = \sqrt{[A_c + s(t) + n_i(t)]^2 + [n_q(t)]^2}$$

and $\phi(t) = \tan^{-1}\left[\dfrac{n_q(t)}{A_c + s(t) + n_i(t)}\right].$

Assuming $|s(t)| \gg |n_i(t)|\, or\, |n_q(t)|$, the envelope can be approximated as

$$B(t) \approx A_c + s(t) + n_i(t) \text{ (vide Eq. 7.33)}$$

The mean of $B(t)$ is $\overline{B(t)} = A_c$. So, suppressing the DC, the envelope detector will detect the envelope $B_d(t)$ as

$$B_d(t) = B(t) - \overline{B(t)}$$

$$= s(t) + n_i(t).$$

Since the signal component has power $\overline{s^2(t)}$ and noise component has power $\overline{n_i^2(t)} = \sigma_i^2 = \sigma_n^2$, the threshold effect is minimised when $\overline{s^2(t)} \big/ \overline{n_i^2(t)} \gg 1$; that is, when S/N-ratio is high.

5. Representing the bandpass noise as

$$n(t) = n_i(t) Cos\omega_c t - n_q(t) Sin\omega_c t,$$

Find the autocorrelation function $R_{nn}(\tau)$ of $n(t)$, assuming $n(t)$ be a stationary random process.

Solution

By definition,

$R_{nn}(\tau) = E\left[n(t)n(t-\tau)\right]$,where E [·] stands for expectation operator.

$$E\left\{\left[n_i(t)\text{Cos}\omega_c t - n_q(t)\sin\omega_c t\right]\left[n_i(t-\tau)\text{Cos}\omega_c(t-\tau) - n_q(t-\tau)Sin\omega_c(t-\tau)\right]\right\}$$

$$= E\left[n_i(t)n_i\left(t-\tau\right)\right]\text{Cos}\omega_c t\cdot\text{Cos}\omega_c\left(t-\tau\right) + E\left[n_q(t)n_q\left(t-\tau\right)\right]\text{Sin}\omega_c t\cdot\text{Sin}\omega_c\left(t-\tau\right)$$
$$- E\left[n_i(t)n_q\left(t-\tau\right)\right]\text{Cos}\omega_c t\cdot\text{Sin}\omega_c\left(t-\tau\right) - E\left[n_q(t)n_i\left(t-\tau\right)\right]\text{Sin}\omega_c t\cdot\text{Cos}\omega_c\left(t-\tau\right)$$

$$= R_{n_in_i}(\tau)\text{Cos}\omega_c t\text{Cos}\omega_c(t-\tau) + R_{n_qn_q}(\tau)Sin\omega_c t\text{Sin}\omega_c(t-\tau)$$

$$- R_{n_in_q}(\tau)\text{Cos}\omega_c t\sin\omega_c(t-\tau) - R_{n_qn_i}(\tau)\text{Sin}\omega_c tCos\omega_c(t-z)$$

(where $E[x(t)x(t-\tau] = R_{xx}(\tau), E[x(t)y(t-\tau] = R_{xy}(\tau))$

$$= \frac{1}{2}R_{n_in_i}(\tau)\cdot\left[(1+\text{Cos}2\omega_c t)Cos\omega_c\tau + \text{Sin}2\omega_c t\cdot\text{Sin}w_c\tau\right]$$

$$+\frac{1}{2}R_{n_qn_q}(\tau)\cdot\left[(1-\text{Cos}2\omega_c t)Cos\omega_c\tau - \text{Sin}2\omega_c t\cdot\text{Sin}w_c\tau\right]$$

$$-\frac{1}{2}R_{n_in_q}(\tau)\cdot\left[\text{Sin}2\omega_c tCos\omega_c\tau - (1+\text{Cos}2\omega_c t)\text{Sin}w_c\tau\right]$$

$$-\frac{1}{2}R_{n_qn_i}(\tau)\cdot\left[\text{Sin}2\omega_c tCos\omega_c\tau + (1-\text{Cos}2\omega_c t)\text{Sin}w_c\tau\right]$$

$$= \frac{1}{2}\left[R_{n_in_i}(\tau) + R_{n_qn_q}(\tau)\right]\text{Cos}\omega_c\tau + \frac{1}{2}\left[R_{n_in_q}(\tau) - R_{n_qn_i}(\tau)\right]\text{Sin}\omega_c\tau$$

$$+\frac{1}{2}\left[R_{n_in_i}(\tau) - R_{n_qn_q}(\tau)\right]\text{Cos}2\omega_c t.\text{Cos}\omega_c\tau$$

$$+\frac{1}{2}\left[R_{n_in_q}(\tau) + R_{n_qn_q}(\tau)\right]\text{Cos}2\omega_c t.Sin\omega_c\tau$$

$$+\frac{1}{2}\left[R_{n_in_i}(\tau) - R_{n_qn_q}(\tau)\right]\text{Sin}2\omega_c t.Sin\omega_c\tau$$

$$-\frac{1}{2}\left[R_{n_in_q}(\tau) + R_{n_qn_i}(\tau)\right]\text{Sin}2\omega_c t.Cos\omega_c\tau$$

Since $n(t)$ is stationary, $R_{nn}(\tau)$ must be independent of time

So,

$R_{n_in_i}(\tau) = R_{n_qn_q}(\tau)$ and $R_{n_in_q}(\tau) = -R_{n_qn_i}(\tau)$

and

$$R_{nn}(\tau) = R_{n_in_i}(\tau)\text{Cos}\omega_c t - R_{n_qn_i}(\tau)\sin\omega_c t$$

Exercises

Activity

7.1 A baseband signal is transmitted using SSB. The power spectral density of the signal is as shown in Fig. P.7.1. If white Gaussian noise of power spectral density $\eta/2$ adds to the SSB signal, find the output signal-to-noise ratio.

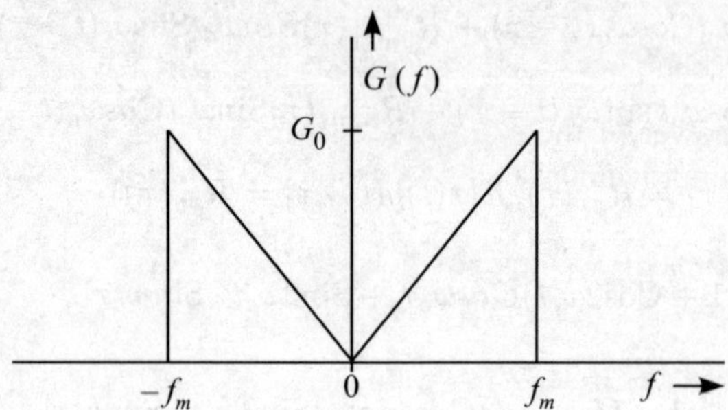

Fig. P.7.1

7.2 A signal $s(t) = \cos 1000\pi t + 5\cos 2000\pi t$ volts is transmitted employing SSB modulation. Noise of PSD 10^{-4} watt/Hz adds to the signal during transmission. Find the signal-to-noise ratio at the output of the receiver.

7.3 An information signal $s(t) = 5\cos 2\pi \times 1000t$ frequency modulates a carrier $100\cos 2\pi \times 10^6 t$ to depth $k_f = 100$ Hz/V. A noise of PSD 10^{-3} is added during transmission. Find the signal-power to noise-power ratio at the output of the FM receiver.

7.4 An FM signal $s_{FM}(t) = 10\cos(2\pi \times 10^5 t + 0.01\cos 2\pi \times 50t)$ is received along with AWGN of PSD $\eta/2 = 0.05$. Find the average number of clicks per second in the output of the detector.

7.5 A received AM signal $s_{AM}(t) = a[1 + V_m(t)]\cos 2\pi f_c t$ is accompanied by AWGN of PSD $\eta/2$. The power spectral density of $v_m(t)$ is given by

$$G(f) = \begin{cases} \dfrac{G_0}{F_M}|f| & |f| < f_M \\ 0 & \text{otherwise.} \end{cases}$$

Calculate the output signal-to-noise ratio.

7.6 A bandpass noise is given by $n(t) = n_i(t)\,Cos\omega_c t + n_q(t)Sin\omega_c t$, has the variance σ_n^2. Show that the variance of the in-phase and quadrature components, namely, $\sigma_{n_i}^2$ and $\sigma_{n_q}^2$ will also be equal to σ_n^2.

7.7 If the bandpass noise in 7.6 above follows the Gaussian distribution with zero mean and is applied to an envelope detector resulting in the envelope $A(t) = \sqrt{n_{\mathrm{i}}^2(t) + n_{\mathrm{q}}^2(t)}$, (vide Eq. 6.36). We define the random variable $y(t) = A^2(t)$. Find $\overline{y}, \overline{y^2}$.

7.8 A baseband signal $s(t)$ with PSD $G_s(f) = \dfrac{G_0}{1 + af^2}, |f| \leq f_M$ is transmitted with pre-emphasis filtering given by

$$H_{\mathrm{p}}(f) = K(1 + jaf)$$

Assuming the total power of the signal remains unchanged with pre-emphasis filter, find the value of K and the improvement factor.

7.9 The PSD of a filtered bandpass noise is given by

$$G_{\mathrm{n}}(f) = \frac{\eta}{2}, \quad \begin{array}{c} f_c - B_T/2 < f < f_c + B_T/2 \\ -f_c - B_T/2 < f < -f_c + B_T/2 \end{array}$$

Find the autocorrelation function $R_n(\tau)$ and $\hat{R}_n(\tau)$ where $\hat{R}_n(\tau)$ is the Hilbert transform of $R_n(\tau)$.

(Hint: Take inverse Fourier transforms of $G_n(f)$ and $(-jsgn(f))G_n(f)$.)

7.10 (a) Explain the effect of random noise at an FM receiver, which includes amplitude limiter (hard limiter). In this connection, explain the concept of noise triangle.

(b) Explain in detail if the carrier is modulated by a single noise frequency.

Review questions

1. Obtain the destination signal-to-noise ratio for an AM + C system for the small noise case. What is the effect of increased noise on this parameter? [RPE,CU 2002]

2. Why is pre-emphasis used in FM broadcast transmitter? Sketch a typical pre-emphasis network. [R.P.E., C.U. 2002]

3. A baseband signal $f(t) = 10\cos 2000\pi t$ is transmitted using SSB-SC added with noise with PSD 10^{-4} W/Hz. Find the S_0/N_0 at the output of the receiver considering that the SSB-SC signal is received without any detectable attenuation. [R:P.E., CU 2004]

4. Derive the signal-to-noise ratio at the output of the demodulator of a DSB-SC receiver. [R.P.E., CU 2004]

5. Show that $\dfrac{\gamma_{FM}}{\gamma_{AM}} = \dfrac{9}{2}\beta^2$, where γ is the figure of merit in the respective modulation technique. [Ber.U. 2001]

6. (a) Compare effect of noise in DSB-SC, DSB with carrier SSB, SSB-SC, SSB-SC.

(b) A carrier of amplitude 10 mV at f_c is 50 percent amplitude-modulated by a sinusoidal waveform of frequency 750 Hz. It is accompanied by thermal noise of two-sided power spectral density $\eta/2 > 10^3$ watt/Hz. The signal plus noise is passed through the filter of Fig. P.7.2 below. The signal is demodulated by multiplication with a local carrier of amplitude 1 volt.

(i) Find the output signal power. (ii) Find the output noise power.

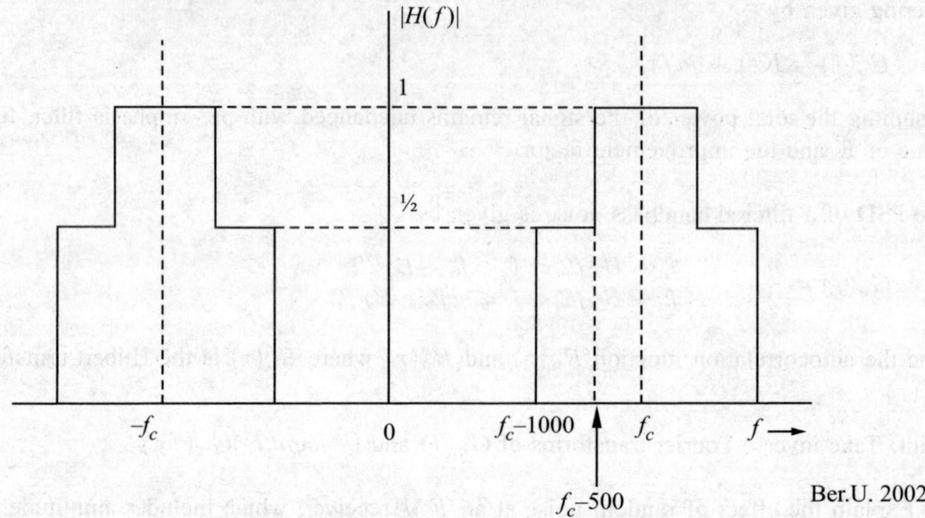

Ber.U. 2002

Fig. P.7.2

7. In an FM system, the allowable maximum frequency deviation is 3.5 kHz and the modulating frequency ranges from 300 Hz to 3000 Hz. Find the total bandwidth required by the FM system. Assume that any sideband with an amplitude of 1% or less relative to the de component may be ignored. (Hint: Use Bessel function table)

8. In an AM system, the message signal s(t) is band-limited to 4kHz with a triangular amplitude distribution given by the following figure.

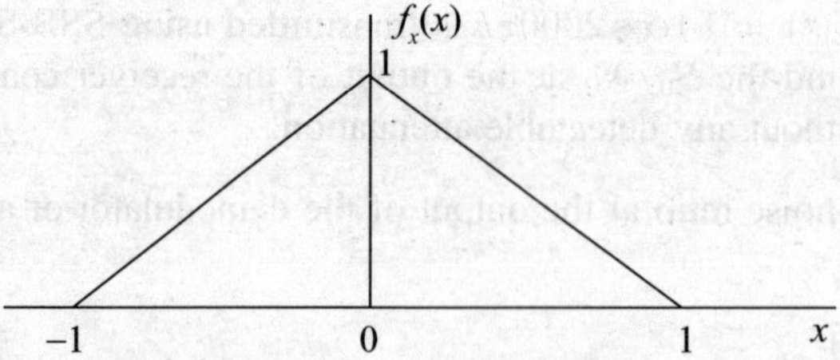

The signal is demodulated in an envelope detector. Find the minimum value of carrier amplitude A_c that will ensure an (S/N)-ratio of at least 40 dB. Assume the noise is thermal noise, additive and have power spectral density $\eta/2 = 10^{-12}$W/Hz.

(Hint: $S_{\text{AM}}(t) = A_{\text{C}}\left[1 + s(t)\right] \text{Cos}\omega_c t$, calculate signal power $P_s = \int_{-\infty}^{\infty} s^2(t) f(s) ds$,, Input signal power $= \frac{1}{2} E\left[A_c{}^2 (1 + s(t))^2\right]$ output signal power, $S_{\text{O}} = \frac{1}{2} A_c^2 P_s$, $SNR = \frac{P_s}{1 + P_s}\gamma$, where $\gamma = \frac{S_i}{\eta B}$, etc)

9. The power spectral density of a typical music signal is given by

$$G(f) = \begin{cases} \dfrac{1}{1 + (f/f_1)^2}, |f| < 15\,\text{kHz} \\ 0, \text{otherwise} \end{cases} \quad \text{and } f_1 = 2 \text{ kHz}.$$

The frequency response of the pre-emphasis filter is given by

$$H_p(f) = k_p \cdot (1 + jf/f_1),$$

Find the change in modulation level with and without pre-emphasis. Assume a fixed modulating signal power. (Hint: Find signal power with and without pre-emphasis and obtain the value K_p, as in Example 7.6. Find the improvement factor.)

10. An AM signal is required to have an (S/N) ratio of at least 40 dB at the receiver. If the transmission loss in the communication channel be 60 dB, find the required transmission power when the modulation is (a) SSB and (b) AM with $\mu = 1$. It is given that the background noise is thermal noise with $\eta/2 = 10^{-12}$W/Hz, channel bandwidth of 8 kHz and normalised average signal power of $\frac{1}{2}$.

Multiple-choice questions

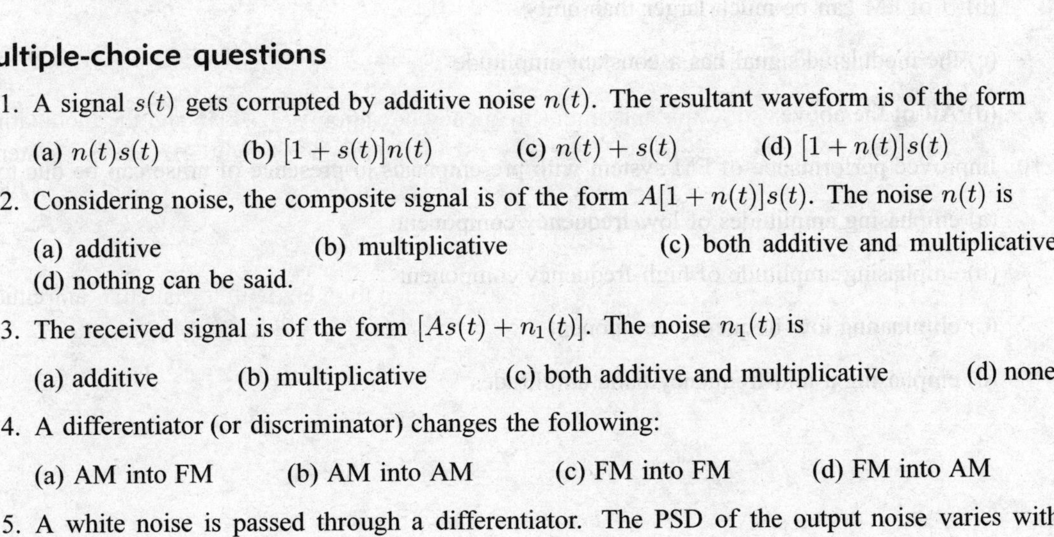

1. A signal $s(t)$ gets corrupted by additive noise $n(t)$. The resultant waveform is of the form

   (a) $n(t)s(t)$ (b) $[1 + s(t)]n(t)$ (c) $n(t) + s(t)$ (d) $[1 + n(t)]s(t)$

2. Considering noise, the composite signal is of the form $A[1 + n(t)]s(t)$. The noise $n(t)$ is

   (a) additive (b) multiplicative (c) both additive and multiplicative (d) nothing can be said.

3. The received signal is of the form $[As(t) + n_1(t)]$. The noise $n_1(t)$ is

   (a) additive (b) multiplicative (c) both additive and multiplicative (d) none

4. A differentiator (or discriminator) changes the following:

   (a) AM into FM (b) AM into AM (c) FM into FM (d) FM into AM

5. A white noise is passed through a differentiator. The PSD of the output noise varies with frequency as

   (a) f^3 (b) f^2 (c) $1/f$ (d) $1/f^3$

6. In an wideband FM, the output signal-to-noise ratio is approximately proportional to

   (a) β (b) β^2 (c) β^3 (d) $1/\beta$

7. Study the following statements.

   (1) DSB-SC has a bandwidth $2f_m$
   (2) DSB with transmitted carrier has a bandwidth $2f_m$
   (3) DSB-SC and DSBTC both are 100% efficient.

   Choose the correct alternative.

   (a) (1) & (2) (b) (1) & (3) (c) (2) & (3) (d) (1), (2), & (3).

8. In the spectrum of FM wave of index of modulation, β, find out the inappropriate statement.

   (a) number of sidebands depends on β

   (b) amplitude of any sideband depends on β

   (c) f_c is not present for large β

   (d) power is distributed in all sidebands

   (e) f_c must be present in the spectrum

9. Performance improvement of FM over AM in the presence of noise can be accounted for by the fact that

   (a) noise is frequency-dependent for FM

   (b) β of FM can be much larger than unity

   (c) the modulated signal has a constant amplitude

   (d) All of the above.

10. Improved performance of FM system with pre-emphasis in presence of noise can be due to

    (a) emphasing amplitudes of low-frequency component

    (b) emphasing amplitude of high-frequency component

    (c) eliminating low-frequency component

    (d) emphasing a mid-frequency band amplitudes

8 Pulse Modulation

8.1 Introduction

In chapters 3 and 4 we described CW modulation for transmitting analog information. Pulse modulation is a system where continuous waveforms of the signals are sampled at regular intervals. If one can have sufficiently close spacing between the sample points, the original waveform may be reconstructed from the samples. Usually, the pulses have quite short duration compared to the time between them. Thus the transmitter need not supply power continuously. This is the potential advantage of pulse modulation systems since the designer has the option of selecting devices that operate only on a pulsed basis. The message transmitted in pulse modulation remains off in between two pulses and this makes the transmission of sampled pulses of other signals possible. This process is called time-division multiplexing (TDM).

Pulse modulation can be either analog or digital. In analog pulse modulation, either the pulse amplitude, pulse position or the pulse width may be varied in accordance with the modulating signal. Pulse-code and pulse delta modulation systems are digital in the sense that the sample amplitudes may be some discrete values within the signal range. The digital or coded pulse modulation is increasingly used because of some of the advantages that will become apparent as we will go along.

8.2 Sampling Theorem

We will see shortly that if the sampling is done rapidly, the process preserves the spectrum of the message and the signal may be reconstructed from these samples.

Let $x(t)$ be a message signal that is bandlimited to f_M, which requires that the message spectrum $X(f)$

$$X(f) = 0, \quad \text{for} \quad |f| > f_M. \tag{8.1}$$

The sampling theorem states that one can uniquely determine the signal $x(t)$ from its samples if the samples are taken at regular intervals separated by $T_s \leq \frac{1}{2f_M}$. The time T_s is called the sampling interval. To study the sampling operation, assume that the message signal $x(t)$ is applied to a multiplier of Fig. 8.1, the input is an infinite train of pulses $p(t)$ of width τ, period T_s and amplitude unity. The figure shows that the signal is sampled when the pulse occurs, and at all other times the product is zero. Representing $p(t)$ in the exponential Fourier series, we have

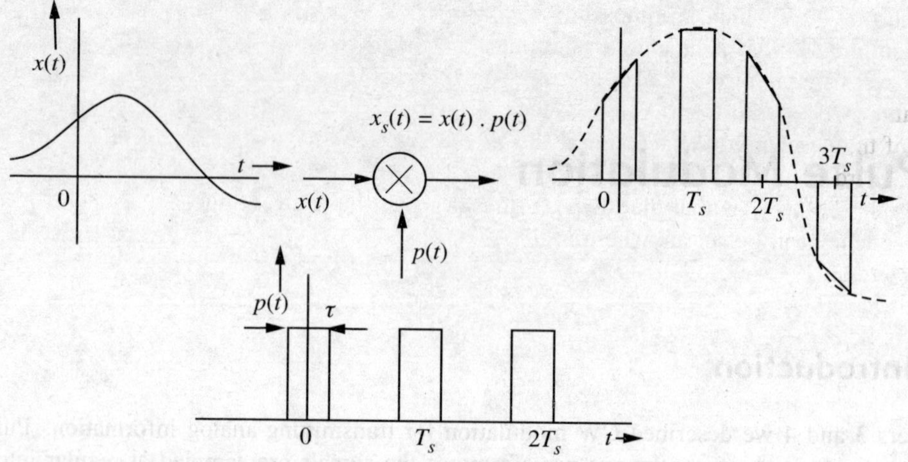

Fig. 8.1 *Sampling or multiplication operation*

$$p(t) = \sum_{n=-\infty}^{\infty} C_n e^{j2\pi n f_s t} = C_0 + \sum_{n=1}^{\infty} 2C_n \cos 2\pi n f_s t, \tag{8.2}$$

where the series coefficient $C_n = f_s \tau$ sinc $n f_s \tau$ and $f_s = 1/T_s$. Thus the product $x_s(t) = x(t)p(t)$ is

$$x_s(t) = C_0 x(t) + 2C_1 x(t) \cos 2\pi f_s t + 2C_2 x(t) \cos 4\pi f_s t + \ldots\ldots \tag{8.3}$$

The first term is the message signal multiplied by a constant factor C_0. The higher terms corresponds to the double-sideband-suppressed carrier (DSB-SC) signals around the frequencies $f_s, 2f_s$, etc. Except the multiplier, $2C_n$, the spectra of the DSB-SC signals are similar to the message spectra $X(f)$.

Taking the Fourier transform of both sides of (8.3), we obtain the output spectrum as

$$X_s(f) = C_0 X(f) + C_1[X(f - f_s) + X(f + f_s)] + C_2[X(f - 2f_s) + X(f + 2f_s)] + \ldots \tag{8.4}$$

Figure 8.2 plots the spectra of the message signal as well as the sampled signal. The spectrum of the message signal extends from 0 to f_M, as it is band-limited to f_M. The spectrum of the second term of (8.4) extends from $f_s - f_M$ to $f_s + f_M$. If the product signal $x(t)p(t)$ is filtered by an ideal lowpass filter having a cutoff at f_M, the filtered output will be proportional to $x(t)$ provided that the spectrum about the frequency f_s does not overlap (extends into) the message spectrum. This is necessary for exact recovery of the message signal. To ensure that, we require

$$f_s - f_M \geq f_M$$

or $$f_s \geq 2f_M. \tag{8.4a}$$

The minimum sampling frequency $f_s = 2f_M$ is called the *Nyquist rate* and this minimum rate of sampling just prevents spectral overlapping (Fig. 8.2(c)). This requires that the LPF should have a cutoff sharp at f_M, which is impossible in practice. Another problem associated with the Nyquist rate of sampling is that if an initial sampling happens to be at the zero-crossing point of a signal, all the successive samples will also be zero. When $f_s < 2f_M$, the spectra overlaps (Fig. 8.2(d)) and $X(f)$ cannot be filtered from $X_s(f)$. When $f_s > 2f_M$, there is a gap called *guard band* between the end of the baseband message signal and the start of the lower-side spectrum of DSB-SC around f_s. The guard band thus extends from f_M to $f_s - f_M$ (Fig. 8.2(b)). The potential advantage of sampling at $f_s > 2f_M$ is that the lowpass filter need not have an infinitely sharp cutoff. Rather, its cutoff frequency can be set anywhere in this guard band.

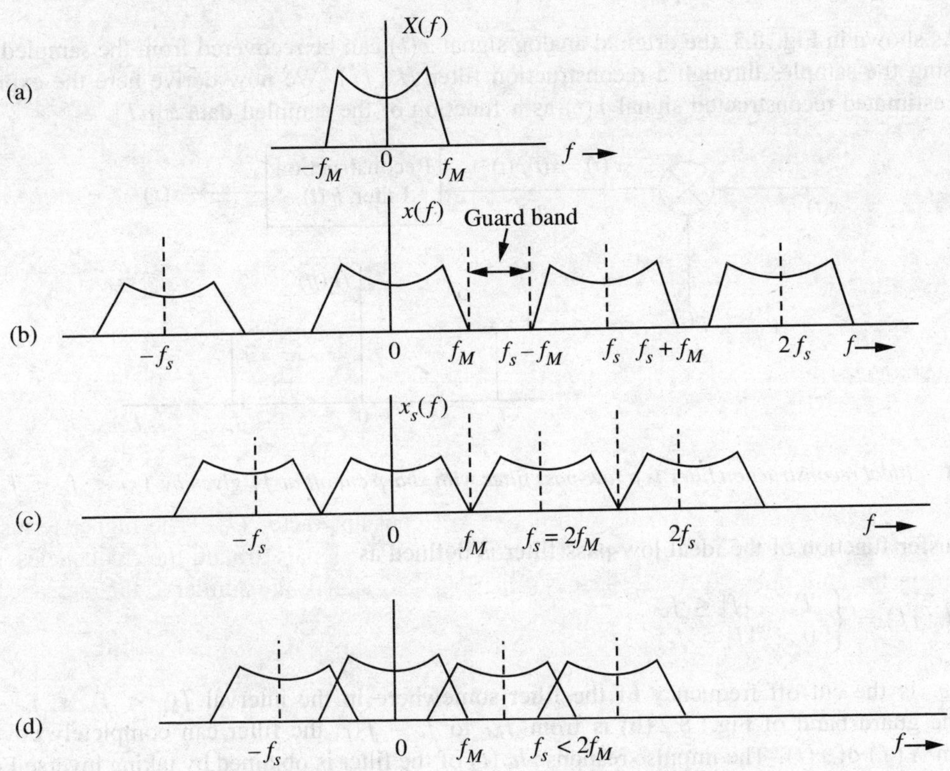

Fig. 8.2 *Magnitude spectra for sampling* (a) *message spectrum;* (b), (c), *and* (d) *are the sampled spectra for* $f_s > 2f_M$, $f_s = 2f_M$, *and* $f_s < 2f_M$, *respectively.*

Example 8.1

The signal $s(t) = \cos 2\pi t + 0.5 \cos 4\pi t$ undergoes instantaneous sampling process. Find the maximum allowable interval T_s between samples.

Solution

Here, the highest modulating frequency $f_M = 2$ Hz and so sampling frequency $f_s(\text{min}) = 2f_M = 4$ samples per second.

$$\text{So, } T_s(\max) = \frac{1}{f_s(\min)} = \frac{1}{4} \text{ seconds} = 0.25 \text{ seconds.}$$

8.2.1 Signal reconstruction

There are situations when the only information available about a signal is a set of sample values taken at regular instants over some time interval. From this sampled values we may have to reconstruct the signal. We may arrive at some reasonable approximation of the signal from the sampled values by a process called 'interpolation'. An ideal lowpass filter can be employed to interpolate between samples.

As shown in Fig. 8.3, the original analog signal $x(t)$ can be recovered from the sampled values by passing the samples through a reconstruction filter $H_r(jf)$. We now derive here the expression for the estimated reconstructed signal $\hat{x}(t)$ as a function of the sampled data $x[nT]$.

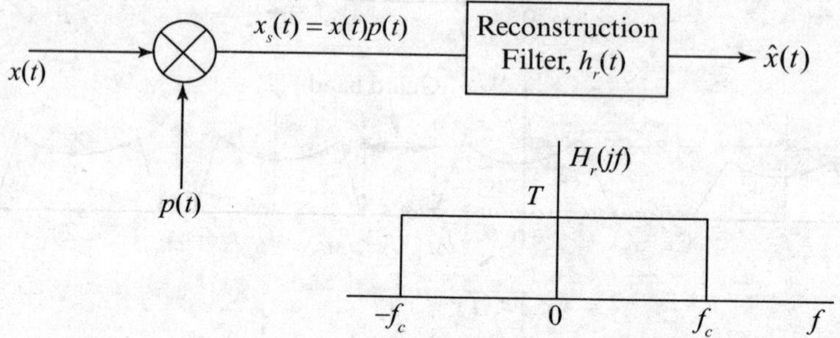

Fig. 8.3 *Ideal reconstruction filter is a low-pass filter with sharp cut off at f_c given by $f_M < f_c < f_s - f_M$*

The transfer function of the ideal low-pass filter is defined as

$$H_r(jf) = \begin{cases} T_s, & |f| \le f_c \\ 0, & |f| > f_c \end{cases} \tag{8.5}$$

where ω_c is the cut off frequency of the filter somewhere in the interval $f_M < f_c < f_s - f_M$. Since the guard band of Fig. 8.2.(b) is from f_M to $f_s - f_M$, the filter can completely cover the spectrum $X(f)$ of $x(t)$. The impulse response $h_r(t)$ of the filter is obtained by taking inverse Fourier transform, that is,

$$h_r(t) = F^{-1}\left[H_r(jf)\right] = \int_{-\infty}^{\infty} H_r(jf)e^{j2\pi ft}df = T_s \int_{-f_c}^{f_c} e^{j2\pi ft}df$$

$$= \frac{T_s}{j2\pi t}e^{j2\pi ft}\Big|_{-f_c}^{f_c} = \frac{T_s}{j2\pi t}\left(e^{j2\pi f_c t} - e^{-j2\pi f_c t}\right) = \frac{\text{T}\sin(2\pi f_c t)}{\pi t}$$

$$= \frac{\sin(2\pi f_c t)}{f_s \cdot \pi t} \tag{8.5a}$$

The output $\hat{x}(t)$ can be obtained by the convolution of $x_s(t)$ with the impulse function $h_r(t)$.Thus,

$$\hat{x}(t) = x_s(t) * h_r(t) = \sum_{n=-\infty}^{\infty} x[nT_s]\, h_r(t - nT_s)$$

or, if one assumes that $f_c = f_s/2 = \dfrac{1}{2T_s}$, then

$$h_r(t - nT) = \frac{\sin 2\pi f_c (t - nT_s)}{f_s \pi (t - nT_s)} = \frac{\sin \frac{2\pi}{2T_s}(t - nT_s)}{\frac{1}{T_s}\pi (t - nT_s)}$$

$$= \frac{\sin[\pi (t - nT_s)/T_s]}{\pi (t - nT_s)/T_s}$$

$$\text{Thus, } \hat{x}(t) = \sum_{n=-\infty}^{\infty} x[nT_s] \cdot \frac{\sin[\pi (t - nT_s)/T_s]}{\pi (t - nT_s)/T_s} \tag{8.5b}$$

One can expand the above expression giving

$$\hat{x}(t) = \cdots\cdots + x[-T] \cdot \frac{\sin \frac{\pi}{T_s}(t + T_s)}{\frac{\pi}{T_s}(t + T_s)} + x[0]\frac{\sin \frac{\pi}{T_s} t}{\frac{\pi}{T_s} \cdot t} + x[1] \cdot \frac{\sin \frac{\pi}{T_s}(t - T_s)}{\frac{\pi}{T_s}(t - T_s)} + \cdots$$

So, the reconstruction is by interpolation, as shown in Fig. 8.4.

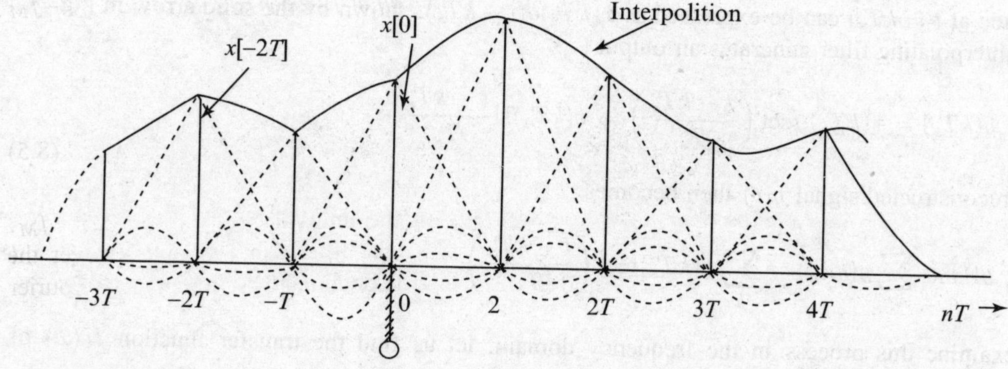

Fig. 8.4 *Interpolation by sinc-type interpolating function.*

The interpolating scheme is referred to as ideal interpolation in which case, the tops of the consecutive samples are joined by the curve obtained by summing the tails of all the preceding and following samples. Such reconstruction is ideal as the impulse response of the filter is non-causal and hence unrealisable. Hence, practical interpolation schemes are required. These schemes are explained below.

We can reconstruct the signal using a zero-order hold filter or a first-order hold filter, as explained below. The zero-order hold is a poor approximation of the ideal lowpass filter, resulting in a staircase interpolation as shown in Fig. 8.5.

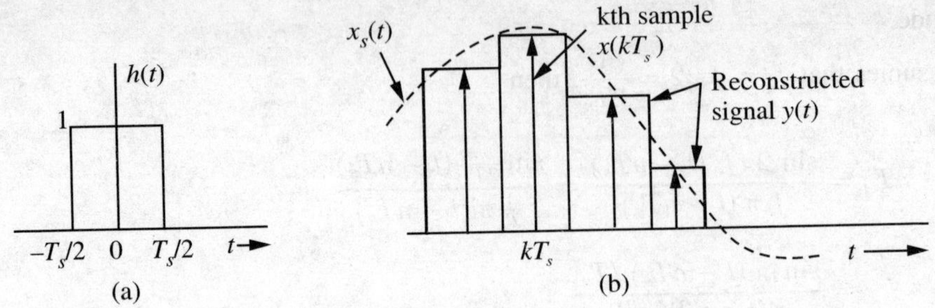

Fig. 8.5 *Signal reconstruction by interpolation. (a) Impulse response of ideal interpolator, (b) reconstructed signal.*

In the zero-order hold, the filter's impulse response is a rectangular pulse of unit height, centred at the origin, and of width T_s. Thus,

$$h(t) = \text{rect}(t/T_s) = \Pi(t/T_s). \tag{8.5}$$

Since each sample generates a pulse of height equal to the strength of the sample, the kth sample (located at $t = kT_s$) can be expressed as $x(kT_s)\delta(t - kT_s)$, shown by the solid arrow in Fig. 8.5(b). The interpolating filter generates an output

$$y(kT_s) = x(kT_s).\text{rect}\left(\frac{t - kT_s}{T_s}\right) = x(kT_s)\Pi\frac{t - kT_s}{T_s}. \tag{8.6}$$

The reconstructed signal $y(t)$ then becomes

$$y(t) = \sum_k y(kT_s) = \sum_k x(kT_s)\text{rect}\left(\frac{t - kT_s}{T_s}\right). \tag{8.7}$$

To examine this process in the frequency domain, let us find the transfer function $H(\omega)$ of the zero-order hold filter.

$$H(\omega) = \mathcal{F}[h(t)] = \int_{-T_s/2}^{0} 1 \times e^{-j\omega t}dt + \int_{0}^{T_s/2} 1 \times e^{-j\omega t}dt$$

$$= \frac{1}{-j\omega}\left[e^{-j\omega t}\right]_{-T/2}^{0} + \frac{1}{-j\omega}\left[e^{-j\omega t}\right]_{0}^{T_s/2}$$

$$= -\frac{1}{j2\pi f}\left[1 - e^{j\pi f T_s}\right] - \frac{1}{j2\pi f}\left[e^{-j\pi f T_s} - 1\right]$$

$$= \frac{1}{j2\pi f}\left[e^{j\pi f T_s} - e^{-j\pi f T_s}\right] = \frac{T_s \sin(\pi f T_s)}{\pi f T_s}.$$

Thus, $H(f) = T_s \operatorname{sinc}(fT_s)$. (8.8)

The magnitude transfer function of the zero-order hold filter is plotted in Fig. 8.6.

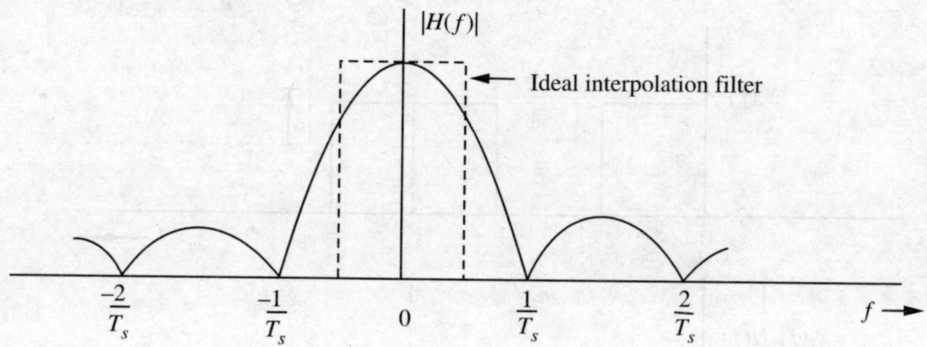

Fig. 8.6 *Deviation of zero-order hold filter from the ideal lowpass filter characteristics*

It is observed from the figure that the sinc-function nature of the transfer function of the zero-order filter distorts the signal. This distortion is pronounced at the higher modulating frequencies. This effect is known as *aperture effect*. Figure 8.7 illustrates how the signal is distorted as a result of this aperture effect for a hypothetical signal having uniform spectrum. Note that the spectrum is modified according to the sinc function.That is, the zero-order hold attenuates the unwanted replicas centered at $f_s, 2f_s,$ etc.

The distortion can be minimised if the first zero-crossing of the sinc-function occurs at large frequency given by

$$\pi f T_s = \pi \quad \text{or} \quad f = \tfrac{1}{T_s} = f_s.$$

So, a higher sampling rate reduces this error. If one tolerates the distortion at this stage, an appropriate inverse filter ($H(f) \propto \frac{x}{\sin x}$), called an equaliser, can be designed to compensate this error.

Example 8.2

Figure 8.8 shows a zero-order hold filter. Find its transfer function $H(f)$.

Solution

An integrator has the transfer function $\frac{1}{j2\pi f}$ and the delay network has the transfer function $e^{-j2\pi f T_s}$. So,

$$H(f) = \frac{1}{j2\pi f}\left(1 - e^{-j2\pi f T_s}\right) = \frac{1}{j2\pi f}[e^{j\pi f T_s} - e^{-j\pi f T_s}]e^{-j\pi f T_s}.$$

$$= T_s \ \operatorname{sinc}\ (\pi f T_s).e^{-j\pi f T_s}.$$

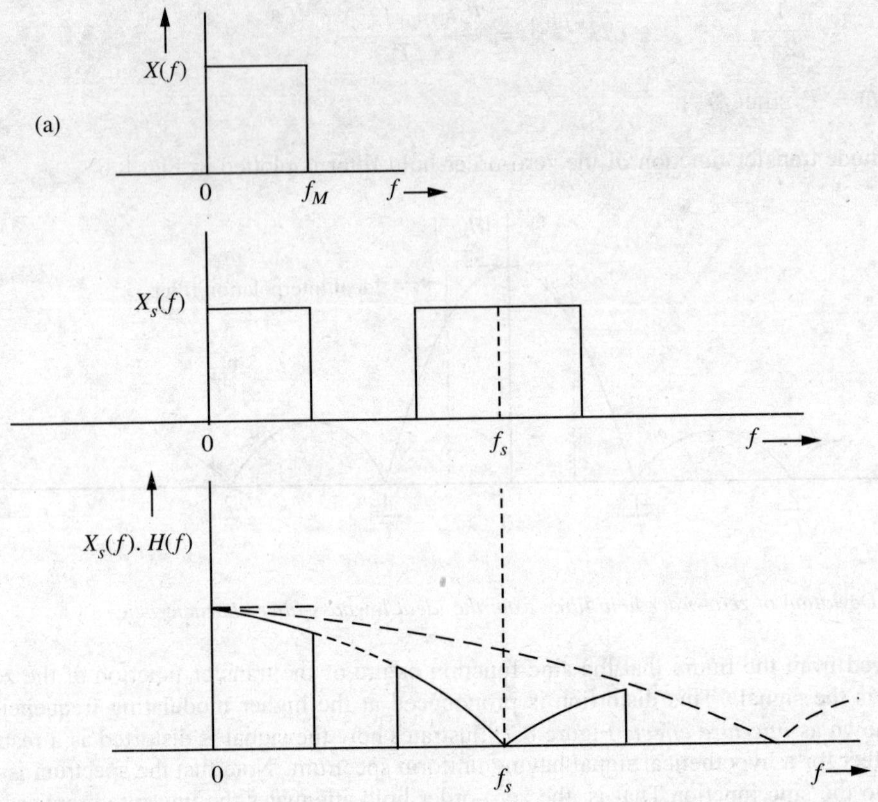

Fig. 8.7 *Aperture effect in signal reconstruction by interpolation. (a) Spectrum of a hypothetical signal, assumed uniform spectrum up to f_M; (b) spectrum of sampled signal; (c) spectrum of reproduced signal.*

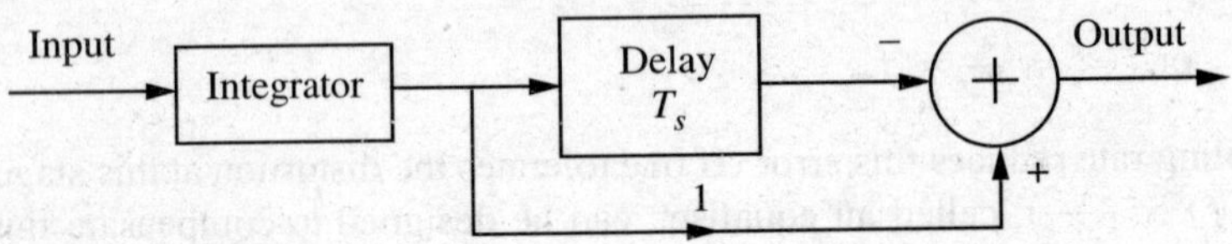

Fig. 8.8 *A zero-order hold filter*

So, $|H(f)| = T_s$ sinc $(\pi f T_s)$.

The first-order hold filter performs linear interpolation. The impulse response of the lowpass filter is a triangular pulse of unit height, centred at the origin, and of width $2T_s$. That is,

$$h(t) = \bigwedge\left(\frac{t}{2T_s}\right). \tag{8.9}$$

For the kth sample impulse $x(kT_s)\delta(t - kT_s)$, the filter output is $x(kT_s)\bigwedge\left(\frac{t-kT_s}{2T_s}\right)$. The reconstructed signal then becomes

$$y(t) = \sum_k x(kT_s) \bigwedge \left(\frac{t - kT_s}{2T_s} \right). \tag{8.10}$$

In this case the sample tops are connected by straight line segments, as illustrated in Fig. 8.9.

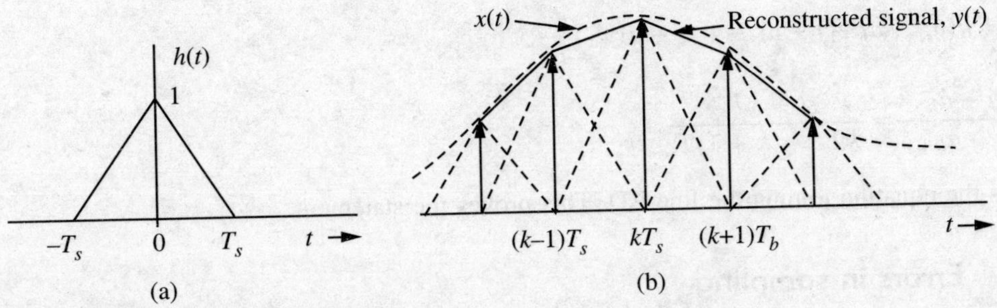

Fig. 8.9 *Signal reconstruction using first-order hold filter*

Example 8.3

For the linear interpolation scheme, prove that the sample tops are connected by line segments.

Solution

Consider two samples of strengths a_k and a_{k+1} at the instants kT_s and $(k+1)T_s$, as shown in Fig. 8.10. For any time $t(kT_s \leq t \leq (k+1)T_s)$, the result is obtained by superposing the contribution from a_k and a_{k+1}. The line BC (see Fig. 8.10) varies as

$$\text{BC:} \quad y = a_k \left[(k+1) - \frac{t}{T_s} \right].$$

Similarly,

$$\text{AD:} \quad y = a_{k+1} \left[\frac{t}{T_s} - k \right].$$

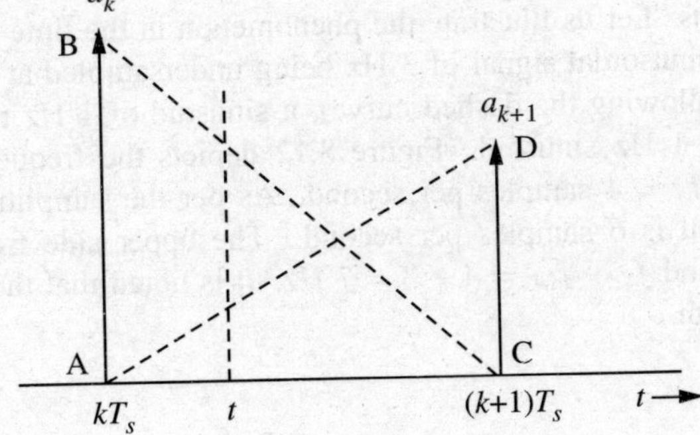

Fig. 8.10 *Linear interpolation*

At any time t, the resultant is obtained by adding the two amplitudes. Thus,

$$y = a_k(k+1) - a_{k+1}k - \left(\frac{a_k}{T_s} - \frac{a_{k+1}}{T_s}\right)t$$

or, $$y = k(a_k - a_{k+1}) + a_k - (a_k - a_{k+1})\frac{t}{T_s}$$

or, $$\frac{y - a_k}{a_k - a_{k+1}} = \frac{t - kT_s}{kT_s - (k+1)T_s},$$

which is the equation joining the line BD. This proves the statement.

8.2.2 Errors in sampling

The sampling theorem indicates that the signal $x(t)$ can be recovered from its samples if the sampling is performed at a sufficiently high rate. The interpolating schemes (zero-order and first-order) distort the spectra of $x(t)$ in the form of aperture effect, which can be minimised with the proper design of an inverse filter. However, sampling in the real world differs from the ideal sampling in three major aspects. The sampled values consists of discrete amplitude values because of quantisation in the signal levels (as will be discussed later in the digital communications systems). This process of rounding off of the sampled values leads to 'round-off' errors (the other name of it is quantisation noise, to be discussed in detail in connection with pulse-code modulation). Secondly, the messages to be sampled are 'time-limited' signals, so that the sampling is practised over a finite time. But sampling theorem requires that samples be taken for all times and that every samples be used to reconstruct the original signal. Practical sampling is observed for a limited time. This incorporates an error known as 'truncation error'. Since the spectra of a time-limited signal cannot be band-limited, the signal may be enriched with significant high-frequency components. If the sampling rate is not high enough, this will lead to 'aliasing error', where the high-frequency signal components suffer a downward frequency translation and appear disguised in the lower frequency spectra. To minimise this effect it is a normal practice to use a lowpass filter (called anti-aliasing filter) before sampling. However the cutoff frequency of the filter should be properly selected so that the signal spectra beyond the cutoff convey negligible information. We will view the aliasing effect in both time and frequency domains. Let us illustrate the phenomenon in the time domain with reference to Fig. 8.11, which shows a sinusoidal signal of 3 Hz being undersampled at 4 samples per second. If one chases the samples (following the dashed curve), a sinusoid of 1 Hz results. Thus the sinusoid of 3 Hz folds back into a 1 Hz sinusoid. Figure 8.12 depicts the frequency-domain picture of a 3 Hz sinusoid sampled at $f_s = 4$ samples per second. As per the sampling theorem, the minimum sampling rate for the signal is 6 samples per second. The upper side frequency signal occurs at $f_s - f_M = 4 - 3 = 1$ Hz and $f_s + f_M = 4 + 3 = 7$ Hz. It is noted that the low frequency lies well within the message spectrum.

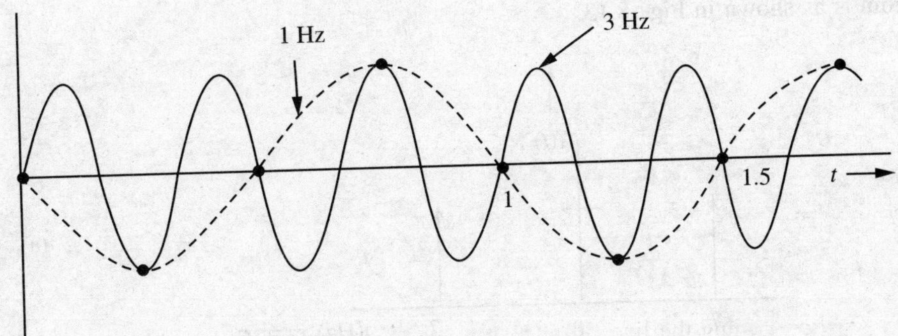

Fig. 8.11 *Aliasing effect in time domain*

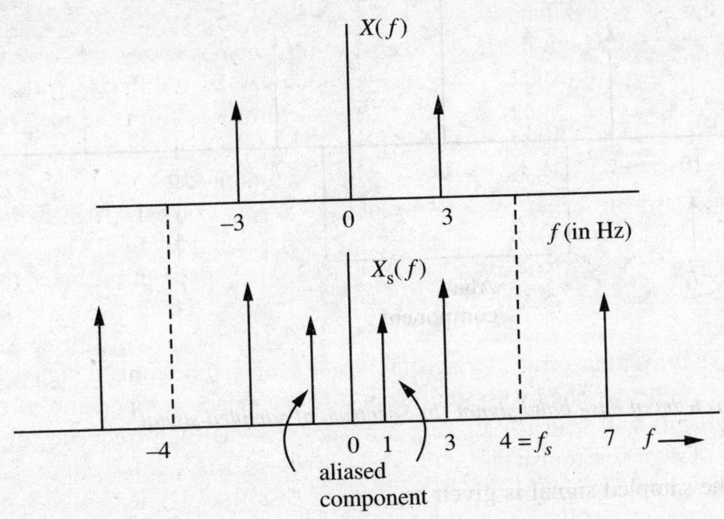

Fig. 8.12 *Aliasing error in frequency domain*

Example 8.4

The signal $x(t) = \cos(14\pi t)$ is sampled at $f_s = 10$ Hz. Explain the effect of aliasing in frequency domain.

Solution

$x(t) = \cos(14\pi t) = \cos(2\pi \times 7t)$, so frequency $f = 7$ Hz. Taking the Fourier transform,

$$X(f) = \pi\,[\delta(f-7) + \delta(f+7)]$$

The spectrum is as shown in Fig. 8.13.

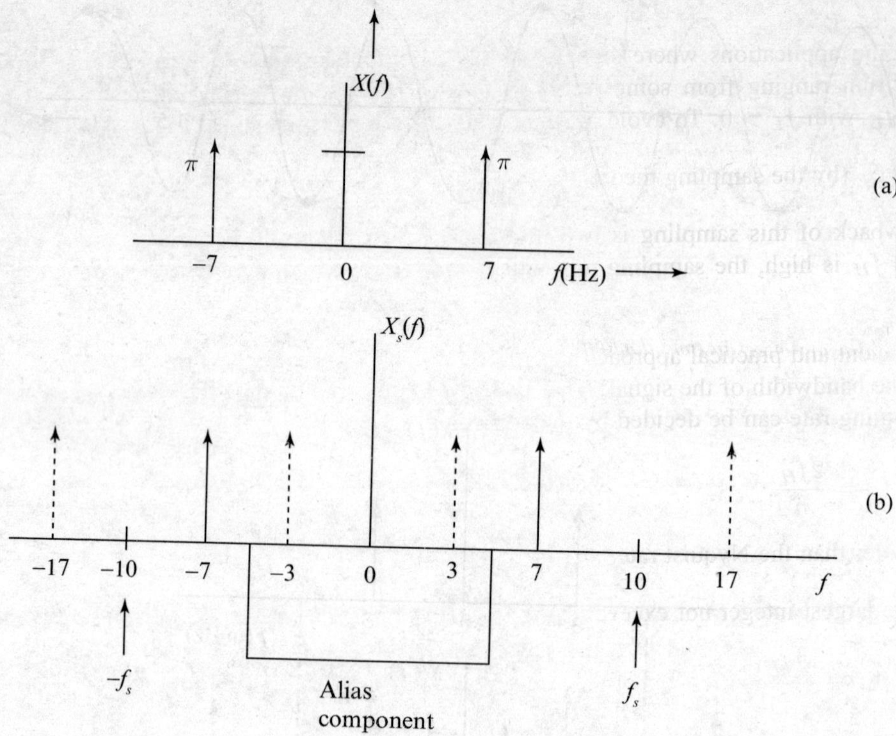

Fig. 8.13 *(a) Spectrum of base band signal (b) Spectrum of sampled signal.*

The spectrum of the sampled signal is given by

$$X_s(f) = \sum_{k=-\infty}^{\infty} X(jf - jkf_s).$$

Since the sampling frequency $f_s = 10$ Hz (it is under-sampled), we have

$$X_s(f) = \dots\dots + X(f + f_s) + X(f) + X(f - f_s) + \dots.$$

$X(f)$ consists of two delta functions at $+7$ Hz and -7 Hz

$X(f - f_s)$ consists of two delta functions at $10 + 7$ or 17 Hz and $10 - 7 = -3$ Hz

$X(f + f_s)$ consists of two delta functions at $-10 + 7$ or -3 Hz and $-10 - 7 = -17$ Hz

Thus, within the main spectrum $X(f)$, there are two components at ± 3 Hz appearing from the adjacent spectra $X(f \pm f_s)$. These are the aliased components.

8.2.3 Bandpass sampling

There are some applications where signals are band-limited and such bandpass signal $s_b(t)$ may have a spectrum ranging from some lower frequency f_L to some upper frequency f_H , that is, $f_L < |f| < f_H$ with $f_L > 0$. To avoid aliasing, one can sample such signal at the rate

$$f_s \geq 2f_H \text{ (by the sampling theorem)}$$

But the drawback of this sampling is two-fold: one is that there will be spectral gap. The other is that when f_H is high, the sampling rate will also be high, which may be objectionable in some applications.

An efficient and practical approach is as follows. Suppose the highest frequency is an integral multiple of the bandwidth of the signal. That is, if $f_B = f_H - f_L$ be the bandwidth and $f_H = nf_B$, then the sampling rate can be decided by the condition

$$f_s = 2f_B = \frac{2f_H}{n}, \tag{8.10a}$$

which is smaller than the Nyquist rate.

Here, n is the largest integer not exceeding f_H/f_B. Bandpass sampling theorem can be written as

$$\frac{2f_H}{n} \leq f_s \leq \frac{2f_L}{n-1} \tag{8.10b}$$

A special case occurs when the band edge frequencies f_L and f_H are integral multiples of the signal bandwidth f_B. In such cases, the signal can be sampled at the theoretical minimum rate of 2 f_B without aliasing. That is,

$$f_s(\min) = 2f_B \tag{8.10c}$$

when f_H/f_B and f_L/f_B are integers. This condition is called integer-positioned bandpass sampling. It may be mentioned that higher sampled rates are not usable unless $n < 1$.

As an example, let the frequency edges of the bandpass signal be $f_L = 25$ kHz and $f_H = 30$ kHz so that $f_B = f_H - f_L = 5$ kHz. Thus $n = \dfrac{f_H}{f_B} = \dfrac{30}{5} = 6$ so, $f_s = \dfrac{2f_H}{n} = \dfrac{2 \times 30}{5} = 12$ kHz

Since $\dfrac{f_H}{f_B}$ or $\dfrac{f_L}{f_B}$ are both integers,

$$f_s\,(\min) = 2f_B = 2 \times 5 = 10 \text{ kHz}.$$

From the sampled values, it is also possible to recover $x(t)$ by using bandpass filtering. Consider the bandpass signal

$$x(t) = 10\cos\left(2\pi \times 5 \times 10^3 t\right) \cdot \cos\left(2\pi \times 20 \times 10^3 t\right)$$

$$= 5\cos(2\pi \times 15 \times 10^3 t) + 5\cos(2\pi \times 25 \times 10^3 t)$$

The lower frequency is $f_L = 15$ kHz and upper frequency is $f_H = 25$ kHz. So, by the low-pass uniform sampling theorem, the minimum sampling rate should be

$$f_s = 2f_M = 2f_H = 50 \text{ kHz}.$$

On the other hand, the bandpass spectrum width is $f_B = f_H - f_L = 10$ kHz and so, $n = \dfrac{f_H}{f_B} = \dfrac{25}{10} = 2.5$. We take $n = 2$. So, based on bandpass sampling theorem, the sampling rate is $f_s = \dfrac{2f_H}{n} = \dfrac{2 \times 25}{2} = 25$ kHz.

Extensions of bandwidth of the signal for achieving alias-free bandpass under sampling

We have discussed earlier that in case of integer-positioned bandpass sampling, one can sample a narrowband high-frequency signal at the minimum sampling rate of $2f_B$ Hz and still avoid errors due to aliasing. However, when f_H/f_B or f_L/f_B are not integers, there must be aliasing. In such cases, one can avoid aliasing by extending the bandwidth of the narrowband signal or the centre frequency f_c such that n becomes an integer. In doing so, the situation becomes similar to the integer-position case. The extension of bandwidth can be done in two ways: by decreasing the lower frequency or by increasing the upper frequency.

Case 1: Decreasing the lower frequency

Let f_L be decreased to $f_1 < f_L$ such that the modified bandwidth is $f'_B = (f_H - f_1)$ and $n = \dfrac{f_H}{f'_B} = \dfrac{f_H}{f_H - f_1}$

This gives $nf_H - nf_1 = nf_H$

$$\text{or, } f_1 = \left(\frac{n-1}{n}\right) f_H \le f_L \tag{8.10d}$$

Case 2: Increasing the higher frequency

Let f_H be increased to $f_2 > f_H$ such that the modified bandwidth is $f'_B = f_2 - f_L$ and $n = \dfrac{f_L}{f'_B} = \dfrac{f_L}{f_2 - f_L}$.

This gives $nf_2 - nf_L = f_L$

$$\text{or, } f_2 = \left(\frac{n+1}{n}\right) f_L \geq f_H \tag{8.10e}$$

As an example, let us consider a narrow bandpass signal with $f_L = 30$ kHz, $f_H = 34$ kHz and bandwidth $f_B = 4$ kHz. In this case,

$$n = \frac{f_H}{F_B} = \frac{34}{4} = 8.5.$$

We take $n = 8$ (lowest integer). So f_B should be increased by decreasing f_L. By using $f_1 = \left(\frac{n-1}{n}\right) f_H$, we have

$$f_1 = \left(\frac{8-1}{8}\right) \times 34 = \frac{7 \times 17}{4} = 29.75\,\text{kHz}$$

Now, the bandwidth is $f'_B = f_H - f_1 = 34 - 29.75 = 4.25$ kHz

Hence, sampling rate, $f_s = 2f'_B = 8.5\,\text{kHz}$

On the other hand,

$$n' = \frac{f_L}{f_B} = \frac{30}{4} = 7.5$$

We take $n' = 7$ (lowest integer). So, f_B should be increased by increasing f_H. By using $f_2 = \left(\frac{n+1}{n}\right) f_L$, we have

$$f_2 = \left(\frac{7+1}{7}\right) \times 30 = 34.29\,\text{kHz}.$$

Now bandwidth is $f'_B = f_2 - f_L = 34.29 - 30 = 4.29$ kHz

Hence, sampling rate $f'_s = 2f'_B = 8.58\,\text{kHz}$

8.3 Pulse-Amplitude Modulation

Pulse-amplitude modulation (PAM) is a technique in which the amplitude of a pulse train is modulated in accordance with the sample values of the message waveform. Figure 8.14 shows that the amplitude of the pulse varies in direct proportion to the sample values of the message waveform $x(t)$.

Here the sampled waveform consists of a pulse train of duration τ and separated by T_s. Observe that the top of the pulses preserves the shape of the message signal. This scheme of sampling is referred to as 'natural sampling'. However, we seldom encounter such PAM signals, instead, flat-top pulses of Fig. 8.15 are frequently employed. As the figure depicts, a flat-topped pulse train consists of constant amplitude pulses within the pulse duration τ. Flat-top sampling can easily be obtained by employing sample-and-hold (S/H) electronic circuitry.

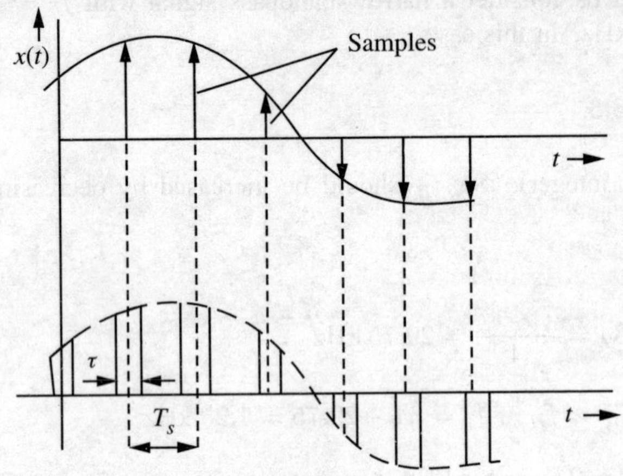

Fig. 8.14 *Natural sampling of a baseband signal $x(t)$*

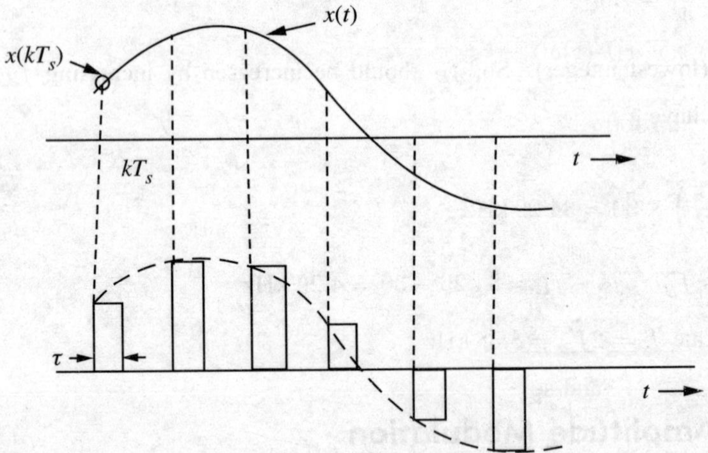

Fig. 8.15 *Flat-top sampling*

Consequently, the electronic circuitry should broaden the instantaneously sampled signal $x(kT_s)$ into a pulse of duration τ. The network shown in Fig. 8.16 performs the operation.

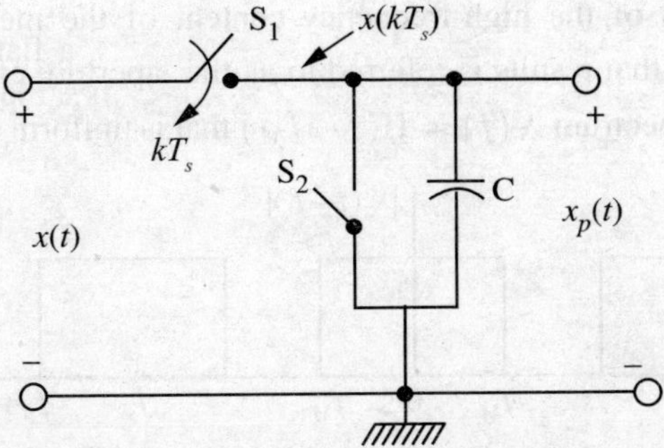

Fig. 8.16 *A simple sample-and-hold circuit to broaden the pulse width*

The sampling switch S_1 briefly closes and the capacitor C charges up to the sample value $x(kT_s)$. It holds the voltage until discharged by the discharge switch S_2. The network then generates

$$x_p(t) = \sum_k x(kT_s)p(t - kT_s), \tag{8.11}$$

a sequence of sampled wave.

Since $p(t - kT_s) = p(t) * \delta(t - kT_s)$, we may write (8.11) as

$$\begin{aligned} x_p(t) &= \sum_k x(kT_s)p(t) * \delta(t - kT_s) \\ &= p(t) * \sum_k x(kT_s)\delta(t - kT_s). \end{aligned} \tag{8.12}$$

Taking its Fourier transform,

$$X_p(f) = P(f)\left[\frac{1}{T_s}\sum_n X(f - nf_s)\right], \tag{8.13}$$

where $P(f)$ is the Fourier transform of $p(t)$ and $f_s = 1/T_s$ is the sampling rate.

Specifically, for a pulse $p(t)$ of unit amplitude and extending from $t = -\tau/2$ to $\tau/2$,

$P(f) = \tau\dfrac{\sin(\pi f\tau)}{(\pi f\tau)}$, and so

$$X_p(f) = \frac{\tau}{T_s}\left(\frac{\sin \pi f\tau}{\pi f\tau}\right)\sum_n X(f - nf_s). \tag{8.14}$$

Thus, in the message frequency range 0 to f_M, the Fourier transform of the flat-topped signal is given by

$$X_p(f) = \frac{\tau}{T_s}\left(\frac{\sin \pi f\tau}{\pi f\tau}\right)X(f), \qquad 0 \le f \le f_M. \tag{8.15}$$

The result is the attenuation of the high-frequency content of the message spectrum by a factor $\frac{\tau}{T_s}\left(\frac{\sin \pi f \tau}{\pi f \tau}\right)$. The distortion that results is referred to as the 'aperture effect' distortion, as indicated in Fig. 8.17 for an idealised spectrum $X(f) = \Pi(f/2f_M)$ that is uniform over the message bandwidth.

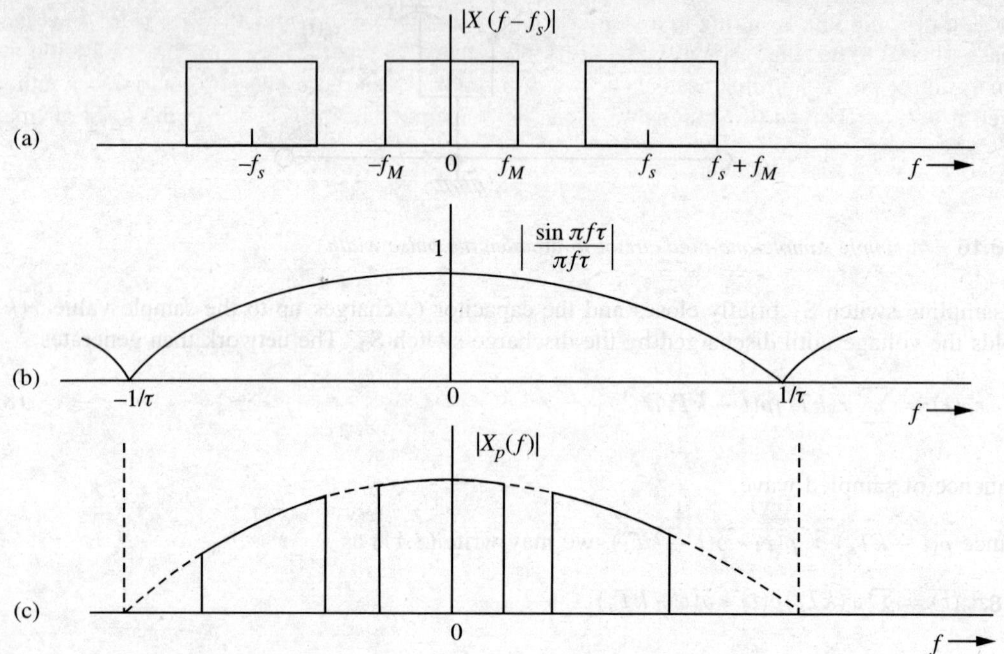

Fig. 8.17 *(a) Idealised spectrum for* $X(f) = \Pi(f/2f_M)$. *(b) The plot of distortion factor* $\frac{\sin \pi f \tau}{\pi f \tau}$. *(c) Aperture effect, causing the spectrum* $X_p(f)$ *to decrease at higher frequencies.*

The aperture effect can be minimised if the first zero-crossing point of the distortion term corresponds to a frequency much higher than f_M. The first zero-crossing occurs at a frequency f_0 given by $\pi f_0 \tau = \pi$ or $f_0 = \frac{1}{\tau}$. Consequently, the aperture effect is minimum if $f_0 = 1/\tau >> f_M$ or $\tau << \frac{1}{f_M}$ and so $\tau/T_s << 1$. Note that, as $\tau \to 0$ (instantaneous sampling), the distortion gets removed. But too low a value of τ reduces the amplitude of the signal (vide (8.14)) and hence a trade-off between the acceptable signal strength and the distortion is sought. An attempt to correct the aperture effect is through the use of an 'equaliser' network which has a transfer function of the form $\frac{\pi f \tau}{\sin(\pi f \tau)}$. The equaliser in cascade with the reconstruction filter will provide a flat transfer characteristic.

A significant point about a PAM signal is that its spectrum extends from dc to several harmonics of f_s. This rules out the possibility of efficient PAM signal transmission through frequency multiplexing. It is evident that only one signal can be sent at a time. However, we seldom encounter a single-channel system with PAM; a different form of multiplexing called time-division multiplexing is suited to analog pulse modulated signals.[1]

[1] The transmission bandwidth B_T of PAM signal is based on the fact that the channel should preserve the shape of the pulse. For a pulse of duration $\tau << T_s \leq \frac{1}{2f_M}$, the pulse resolution requires that $B_T \geq \frac{1}{2\tau} >> f_M$.

In practical applications, the PAM signal is not transmitted directly, rather, we often modulate a carrier wave by PAM signal by amplitude or frequency modulation.

We now discuss the demodulation of PAM signals. A sample-and-hold circuit could be used to recover the baseband signal (see Fig. 8.18). The sample value of the baseband signal is held for the period of sampling, resulting in a staircase approximation of the original waveform. This is then lowpass-filtered to get the original waveform. The receiver must perform sampling operation at the original sample points and this needs synchronisation. The required timing information can either be derived from the received PAM signal or it can be transmitted to the receiver in the form of 'marker pulses'.

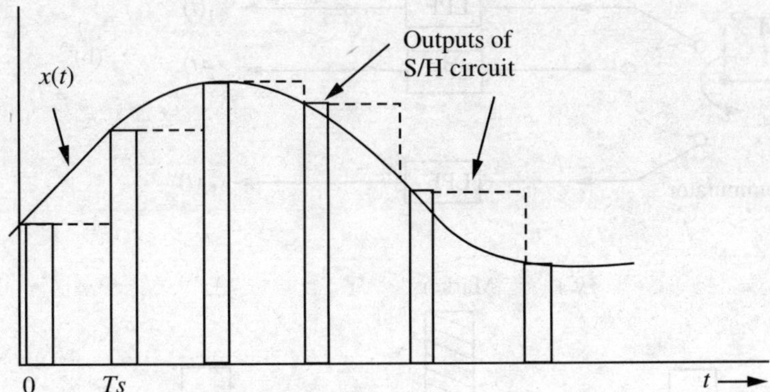

Fig. 8.18 *Illustrating the PAM signal demodulation using S/H circuit*

8.4 Time-Division Multiplexing

Time-division multiplexing (TDM) is the process of transmitting a number of signals from different sources over the same communication channel on a time-sharing basis. This is possible in pulse communication since the sample waveform is 'off' most of the time and samples from other signals can be interspersed into a single waveform. Figure 8.19 illustrates this process of multiplexing N different signals. At the transmitting end, the commutator switch rotates and samples each signal sequentially. As a result, the output of Fig. 8.19(a) is a PAM signal consisting of periodically interspersed samples of various inputs. The width τ of each sample pulse is determined by the contact time of the commutator arm with the signal terminals. The LPFs before the commutator switch band-limit the signals to avoid aliasing error.

The rotary switch at the receiver (called the decommutator) separates the samples of a given signal $x_i(t)$ from the other samples. The signals are lowpass-filtered to reconstruct the individual messages. The decommutator switch at the receiver must rotate in 'synchronism' with the rotary switch at the transmitter. This is called 'frame synchronisation' and requires a great effort. A popular method of synchronisation is to sacrifice one time slot (i.e., channel) per frame (per revolution of the rotary switch) and send a distinct synchronising signal (called marker pulse) in its place as illustrated in Fig. 8.19(c).

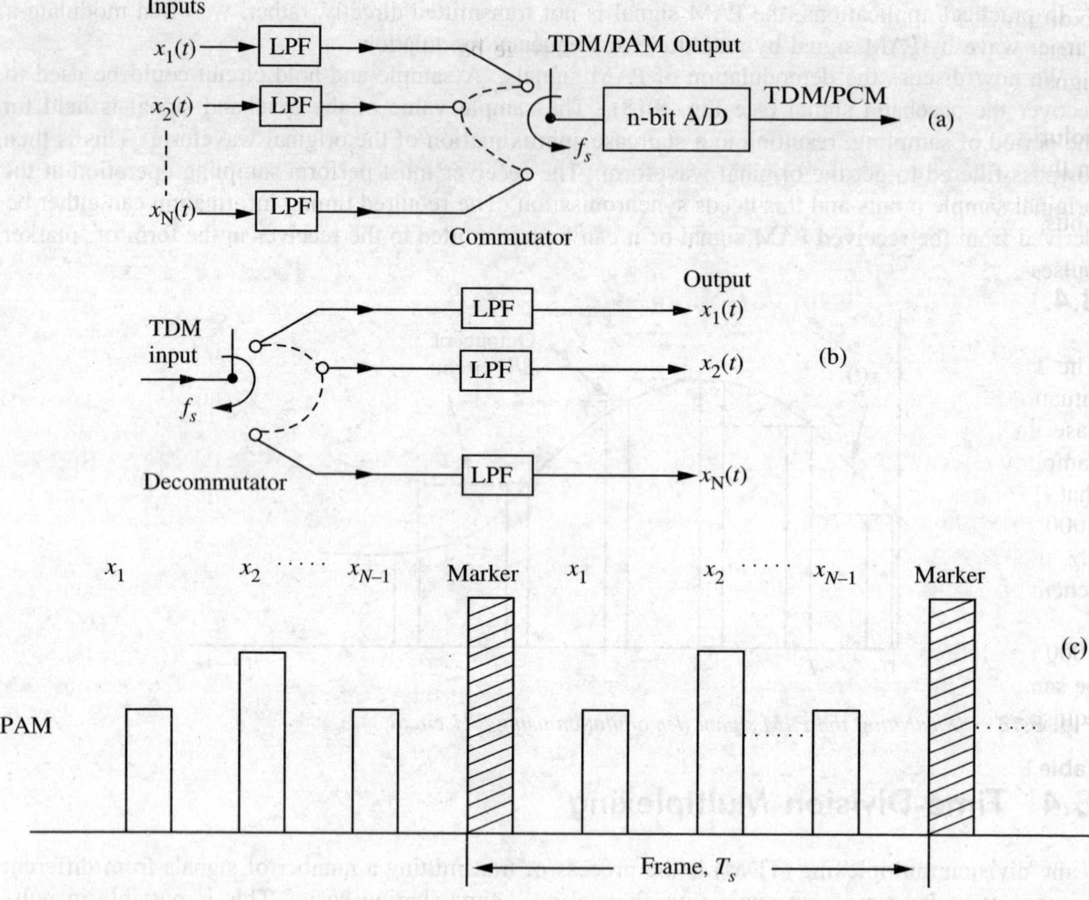

Fig. 8.19 *Bock diagram of time-division multiplexing system: (a) transmitter, (b) receiver. (c) synchronising pulses.*

The number of channels that can be multiplexed this way is based on how fast the rotary switch can rotate. If f_M be the highest message bandwidth of N inputs, the commutator arm must make at least $2f_M$ revolutions in one second. That is, the commutator speed $f_s \geq 2f_M$. The frame time T_s defined as the pulse-to-pulse spacing between adjacent samples of a single channel is given as $T_s = \dfrac{1}{f_s} \leq \dfrac{1}{2f_M}$. Since there are N channels, the distance between two consecutive pulses is

$$\frac{T_s}{N} = \frac{1}{Nf_s} \leq \frac{1}{2Nf_M}.$$

The signalling rate, r, is then given by

$$r = Nf_s \geq 2Nf_M. \tag{8.16}$$

Example 8.5

Consider that 5 input channels have the same message bandwidth $f_M = 1500$ Hz. Final the TDM signalling rate.

Solution

In this case, $f_s = 2 \times 1500 \text{ Hz} = 3000$ Hz.

Thus, signalling rate $r = Nf_s = 5 \times 3000 \text{ Hz} = 15$ kHz.

8.4.1 TDM of dissimilar channels

The TDM system described so far is restricted to identical sampling rate of various channels. Some situations demand the multiplexing of signals of different sampling rates. The TDM system in this case includes 'submultiplexers' in addition to the main multiplexer. The technique requires that all sampling rates must be multiples of same basic rate. We illustrate this with an example. Suppose that five input signals are to be sent using TDM/PAM. Let the maximum frequency of the signals be 2000, 500, 500, 250, and 200 Hz. If one uses a 5-channel multiplexer with $f_s = 2 \times 2000 = 4000$ Hz, the signalling rate would be $r = Nf_s = 5 \times 4000\text{Hz} = 20$ kHz. Table 8.1 illustrates an efficient scheme using a 3-channel main multiplexer with $f_s = 1000$ Hz and a 2-channel submultiplexer with $f_s = \frac{1}{2} \times 1000 = 500$ Hz. At the basic sampling rate of 1000 Hz, the commutator arm rotates at 1000 revolutions per second, giving one sample per 1 ms. The highest-rate signal, $x_1(t)$, requires to be sampled four times during each 1 millisecond rotation of the commutator arm and appears in four equispaced slots on the data frame.

Table 8.1 *Main multiplexer and sub-multiplexer*

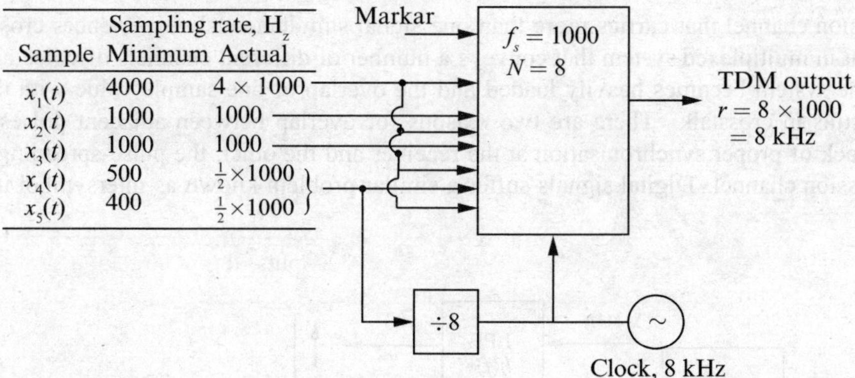

Sample	Sampling rate, H_z Minimum	Actual
$x_1(t)$	4000	4×1000
$x_2(t)$	1000	1000
$x_3(t)$	1000	1000
$x_4(t)$	500	$\frac{1}{2} \times 1000$
$x_5(t)$	400	$\frac{1}{2} \times 1000$

The channels sampled at less than 1000 Hz need not occupy slot in every frame. For example, the two lowest-rate signals $x_4(t)$ and $x_5(t)$ need be sampled once every 2 rotations of the commutator arm. The overall signalling rate is then, $r = 8 \times 1000 = 8000$ Hz.

Suppose we have three sources with the following specifications.

Source 1: Analog, 2 kHz bandwidth

Source 2: Analog, 4 kHz bandwidth

Source 3: Analog, 2 kHz bandwidth

To satisfy the Nyquist rate of sampling, the sources 1, 2 and 3 need to be sampled at 4, 8 and 4 kHz respectively. This can be accomplished by rotating the commutator at $t_1 = 4$ kHz and sampling the source 2 twice at each revolution as shown in Fig. 8.20.

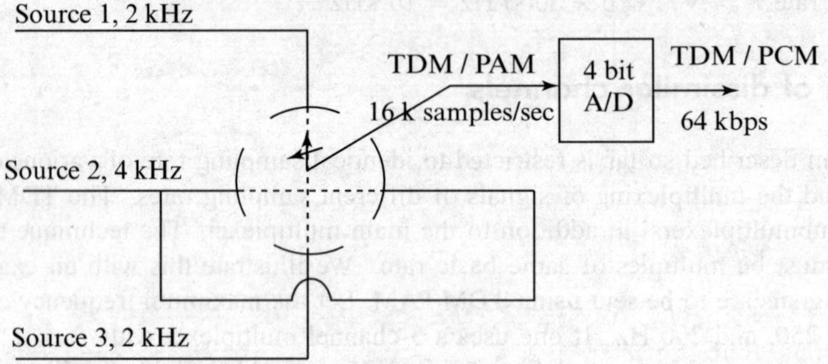

Fig. 8.20 Data frame TDM with dissimilar channels.

This produces a $4 + 8 + 4 = 16$ kilo samples per second TDM/PAM signal. For a 4-bit PCM word, the A/D output comes out at 64 kbps

8.4.2 Crosstalk and intersymbol interference

A transmission channel that carries more than one signal simultaneously experiences crosstalk. This is prominent in multiplexed system that conveys a number of different channels of voice and data. In this case, the system becomes heavily loaded and the overlap of one sample value with the adjacent samples results in crosstalk. There are two reasons for overlap between adjacent pulses. The first one is the lack of proper synchronisation at the receiver and the other, the pulse-spreading caused by the transmission channel. Digital signals suffer a similar problem known as intersymbol interference (ISI).

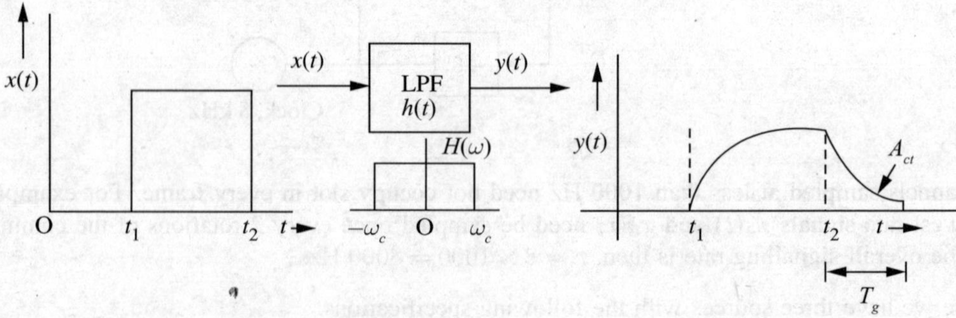

Fig. 8.21 *Crosstalk in lowpass channel*

For a quantitative measure of the pulse overlap, let us assume that the channel has an exponential transient response, that is, it acts like a first-order lowpass filter. A perfect square wave will thus broaden as shown in Fig. 8.21.

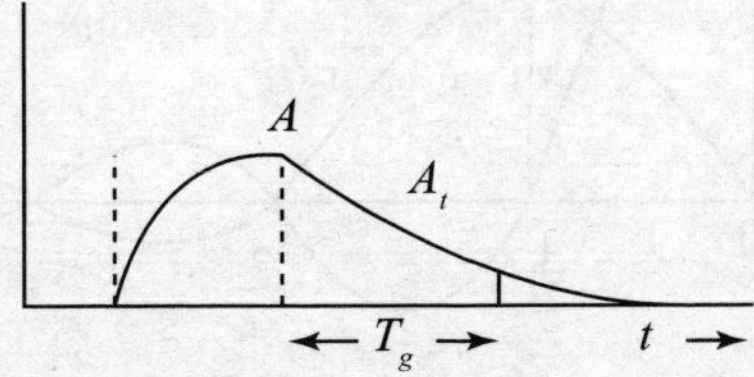

The response to the rectangular pulse decreases exponentially. We define the guard time T_g as the minimum pulse spacing so that the pulse tail decays to a value no larger than A_t given by $A_t = A_e^{-2\pi BT_g}$, where B is the 3 dB bandwidth of the LPF. In the meantime the next pulse arrives. The ratio (A_t/A) is a measure of the crosstalk reduction factor and its value in dB unit is $10log_{10}(A_t/A)^2 \approx -54.5BT_g$. A crosstalk measure below 30 dB demands $T_g > 1/2B$.

Here T_g = Grand time
= Minimum pulse spacing.

To analyse the shape of the pulse at the output, assume a constant transfer function $H(\omega) = 1$ of the channel over the frequency range $-\omega_c$ to ω_c. The impulse response of the ideal lowpass filter is given by

$$h(t) = \int_{-\omega_c}^{\omega_c} H(f)e^{j2\pi ft}df$$
$$= 2f_c\left(\frac{\sin 2\pi f_c t}{2\pi f_c t}\right). \tag{8.17}$$

If one considers an ideal impulse sampling, the channel shapes each impulse into a $\frac{\sin t}{t}$ type of pulse. The sample pulse strength I_1 at $t = 0$ appears as

$$s_1(t) = \int_{-\infty}^{\infty} I_1\delta(t)h(t-\tau)d\tau$$
$$= 2I_1 f_c \frac{\sin 2\pi f_c t}{2\pi f_c t}. \tag{8.18}$$

Thus the received waveform clearly extends into the adjacent time slots, as depicted in Fig. 8.22. The first zero-crossing occurs when $\omega_c t = \pi$, that is, at $t = \frac{1}{2f_c}$. If the samples are taken at the Nyquist rate f_s, the spread pulse will be zero at all adjacent sample points. Figure 8.22 also shows the plot of the spread pulse $s_2(t)$ sampled at $t = \frac{1}{2f_c}$, where

$$s_2(t) = 2I_2 f_c \frac{\sin 2\pi f_c(t - \frac{1}{2f_c})}{2\pi f_c(t - \frac{1}{2f_c})}. \tag{8.19}$$

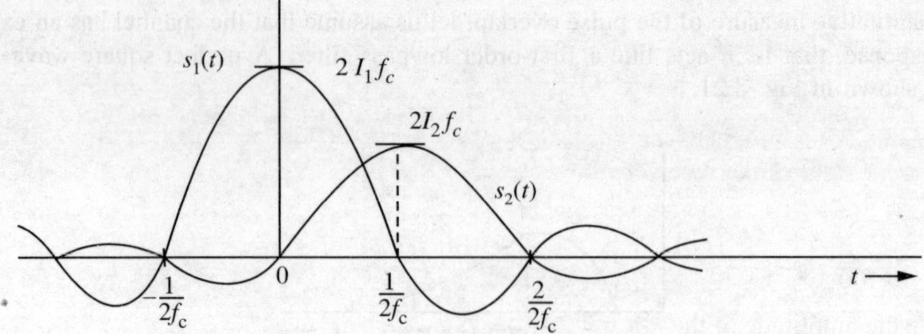

Fig. 8.22 *Pulse spreading of two adjacent pulses sampled at* $t = 0$ *and* $t = 1/2f_c$

Thus to avoid crosstalk, the samples are separated by interval $1/2f_c$ where f_c is the bandwidth of the given channel. However, the samples of a given signal should be separated no more than $\frac{1}{2f_M}$. For N-channel multiplexing, the figure is $\frac{1}{2Nf_M}$. So $\frac{1}{2Nf_M} = \frac{1}{2f_c}$, giving $f_c = Nf_M$, the theoretical minimum bandwidth of TDM systems.

The pulse spreading can be decreased by increasing the transmission bandwidth. But the transmission channel imposes bandwidth limitation and consequently, rectangular pulses get severely distorted during transmission. The other option is to shape that sample pulses. We have shown that when the channel transfer function $H(f)$ is rectangular, its impulse response $h(t)$ is of the form $\sin t/t$. Even though $h(t)$ has long tails, it passes through zero amplitude at the instant $t = \frac{1}{2f_c}$ when $h(t - \frac{1}{2fc})$ is to be sampled (vide Fig. 8.22). Thus, assuming sampling time is perfect, there will be no channel-induced degradation and hence no ISI. Such a $(\sin t)/t$ shaped pulse is called 'Nyquist pulse'. But such ideal filters are not realisable because of the sharp corners on the frequency spectrum. One frequently used transfer function is the raised cosine filter, whose spectrum has a smooth roll-off following a sinusoidal curve.

8.4.3 Pulse shaping

A time-limited rectangular pulse cannot be band-limited. If such pulses are transmitted through a band-limited channel, part of their spectra are suppressed leading to ISI. On the contrary, if one uses bandlimited pulses, these can readily be transmitted through a bandlimited channel. But, since bandlimited pulses can not be time-limited, each pulse extends into adjacent pulses causing ISI. Thus, there seems no way out of the problem of ISI. One can resolve the problem of ISI by giving proper 'shape' to the pulses such that at the correct sampling time, the spread-pulse amplitude reduces to zero, thereby eliminating ISI.

We are indebted to Nyquist who investigated the problem of pulse shaping and suggested a method to overcome the problem. A Nyquist pulse is of the form $(\sin t)/t$ multiplied by some time function and this pulse shape satisfies the condition of zero ISI at the sampling points. This pulse shaping is accomplished with a Nyquist filter. A Nyquist filter is a network whose frequency transfer function is a rectangular function convolved with any real and even-frequency function. A common Nyquist filter is the raised-cosine filter, which is discussed here.

Let us consider that $p(t)$ is the pulse shape used to transmit messages and that

$$p(t) = \begin{cases} 1 & t = 0 \\ 0 & t = kT_b, \quad k = \pm 1, \pm 2, \ldots \end{cases} \tag{8.20}$$

where T_b is the separation between pulses and $T_b = 1/r$, r being the signalling rate. The pulse that appears at the output of the receiver is

$$y(t) = \sum_k a_k p(t - kT_b), \tag{8.21}$$

where a_k is the amplitude of the kth pulse, which we require to detect at $t = kT_b$. The time function

$$p(t) = \text{ sinc } (rt) = \text{ sinc } \left(\frac{t}{T_b} \right) \tag{8.22}$$

satisfies the condition stipulated in (8.20). Furthermore, let us justify that this pulse shape fulfills the bandwidth requirement. The Fourier transform of (8.22) gives

$$P(f) = \frac{1}{r} \Pi \left(\frac{f}{r} \right), \tag{8.23}$$

which has the bandwidth of $r/2$. Since a system with bandwidth B Hz can support a maximum transmission rate of $2B$ bits/s, the pulse in (8.22) can be transmitted at the rate of r pulse per second with zero ISI.

The sinc function imposes serious problems although it satisfies (8.20). This is due to the fact that it decays too slowly at a maximum rate of $1/t$. If the data rate at the transmitter or the sampling rate at the receiver deviates slightly from r bits/s, the pulse amplitudes will not be zero at the positions of other pulses. The cumulative effect of all the amplitudes will add up to a large value, causing ISI. Any error in sampling time at the receiver (called time jitter), will again lead to erroneous detection.

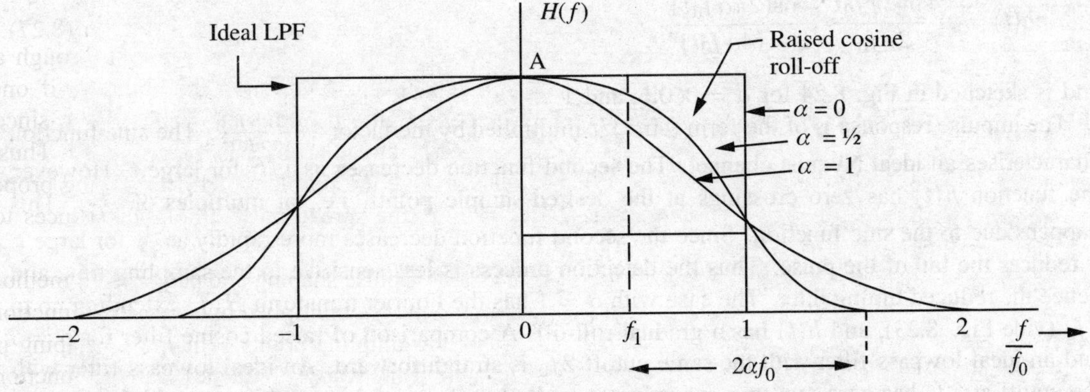

Fig. 8.23 *Frequency response of raised cosine filter.* $\alpha = 0$ *is the ideal LPF.*

The above practical difficulties can be overcome by selecting a pulse shape that satisfies (8.20) but decays faster than $1/t$. A desirable choice is the 'raised-cosine spectrum'. This is similar to ideal lowpass characteristics having flat response, except that the roll-off portion follows a sinusoidal curve, as shown in Fig. 8.23. This can be expressed as

$$H(f) = \begin{cases} A & \text{for } \; 0 < |f| < f_1 \\ \frac{A}{2}\left[1 - \sin \frac{\pi}{2\alpha f_0}(|f| - f_0)\right] & \text{for } \; f_1 \leq |f| < 2f_0 - f_1 \\ 0 & \text{for } \; |f| \geq 2f_0 - f_1. \end{cases} \tag{8.24}$$

Here, the constant α determines the width of the flat portion of the curve and it lies between 0 and 1. If $\alpha = 0$, the curve follows an ideal lowpass characteristics. The other extreme value $\alpha = 1$ is the case where the flat portion reduces to zero and the filter characteristics falls off right at $f = 2f_0$.

The frequency parameter f_1 characterises the raised-cosine roll-off and frequency f_0 describes the minimum Nyquist bandwidth for the rectangular spectrum, the -6 dB bandwidth (or half amplitude) of the raised cosine spectrum. These two are related as

$$\alpha = \frac{f_0 - f_1}{f_0} = 1 - \frac{f_1}{f_0}, \tag{8.25}$$

where α is the roll-off factor, and $0 \leq \alpha \leq 1$. It gives the 'excess bandwidth' over Nyquist minimum reduced by the filter -6 dB bandwidth. It indicates the steepness of the filter roll-off.

The bandwidth of $H(f)$ is

$$B_T = f_0 + \alpha f_0 = (1 + \alpha) f_0. \tag{8.26}$$

The corresponding time function or the raised-cosine impulse response $h(t)$ can be obtained by the inverse Fourier transform of $H(f)$. This is

$$h(t) = A \frac{\sin 2\pi f_0 t}{2\pi f_0} \frac{\cos(2\pi\alpha f_0 t)}{1 - (4\alpha f_0 t)^2} \tag{8.27}$$

and is sketched in Fig. 8.24 for $\alpha = 0, 0.5$, and 1.

The impulse response is of the form $(\sin t)/t$ multiplied by the factor $\frac{\cos 2\pi\alpha f_0 t}{1-(4\alpha f_0)^2}$. The sinc function characterises an ideal Nyquist channel. The second function decreases as $1/t^2$ for large t. However, the function $h(t)$ has zero crossings at the desired sample points, i.e., at multiples of $\frac{1}{2f_0}$. This happens due to the sinc function. Since the second function decreases more rapidly as $\frac{1}{t^2}$ for large t, it reduces the tail of the pulse. Thus the detection process is less sensitive to the sampling time and hence the reduced timing jitter. The case with $\alpha = 1$ has the Fourier transform $H(f)$ extending up to $2f_0$ (vide Fig. 8.23), and $h(t)$ has a gradual roll-off. A comparison of raised-cosine filter for $\alpha = 1$ and an ideal lowpass filter with the same cutoff $2f_0$ is straightforward. An ideal lowpass filter with the cutoff at $2f_0$ has zero crossing occurring at multiples of $\frac{1}{4f_0}$ and hence has two additional zeros in its time response. However, beyond $t = 1/2f_0$, the zeros of lowpass filter coincide with those of the raised cosine filter.

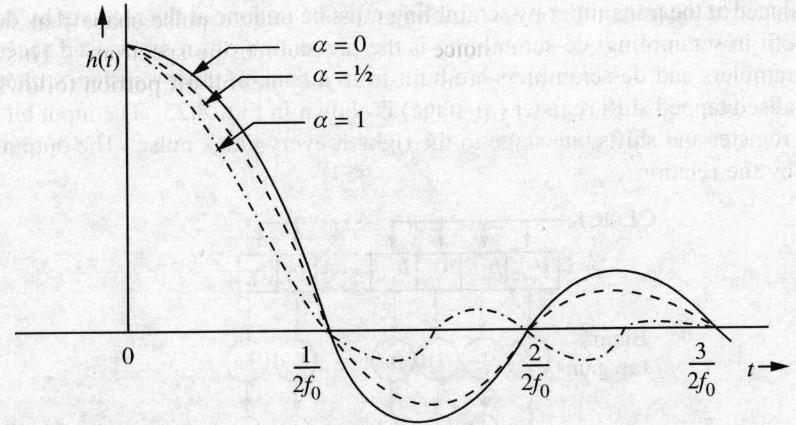

Fig. 8.24 *Raised cosine time response for different roll-off factors*

An interesting case of the maximum roll-off with $\alpha = 1$ and hence $f_1 = 0$ reduces (8.24) to

$$H(f) = \begin{cases} \frac{A}{2}\left[1 + \cos\left(\frac{\pi f}{2f_0}\right)\right] & 0 < f < 2f_0 \\ 0 & f > 2f_0. \end{cases} \tag{8.28}$$

This is known as 'full-cosine roll-off' characteristics. The corresponding impulse response can be obtained as

$$h(t) = \frac{\text{sinc}(4f_0 t)}{1 - (4f_0 t)^2}. \tag{8.29}$$

It is interesting to note that at $4f_0 t = \pm 1$, i.e. at $t = \pm \frac{1}{4f_0} = \pm \frac{1}{2}T_b$, $h(t) = \frac{1}{2}$. That is, the pulsewidth at its half amplitude is exactly the bit duration $2T_b$. There are additional zero-crossings at the instants $t = \pm\frac{3}{2}T_b, \pm\frac{5}{2}T_b, \ldots$ besides the usual zero-crossings at the sampling instants $t = \pm T_b, \pm 2T_b, \ldots$. This property of the raised-cosine time impulse is useful for extracting the synchronising signal at the receiver.

8.4.4 Scrambling

In Section 8.4.2, we have discussed that crosstalk and ISI produces lack of proper synchronisation at the receiver. Pulse shaping is one technique for extracting synchronisation signal at the receiver, as discussed in 8.4.3. When a long string of like bits (zero's or one's) appear at the receiver, this may product lack of receiver synchronisation. Also, in the transmitted message bit stream, if there be periodic bit patterns, this is equivalent to undesirable discrete frequency components in the power spectrum. Scrambling is an operation that tends the randomise the transmitted data. Thus, when a long string of 0s in the data is randomised, this operation can be made helpful in synchronising. The distortion introduced at the transmitter by scrambling must be undone at the receiver by de-scrambling. The added benefit in scrambling/ de-scrambling is the prevention of un-authorised access to the data stream. The scramblers and de-scramblers are built from a bank of shift-registers with adjustable tap gains. A generalised tapped shift register (n-stage) is shown in Fig. 8.25. The input bit sequence b_n enters the shift register and shifts one stage to the right at every clock pulse. The output bit sequence b'_n is obtained by the relation

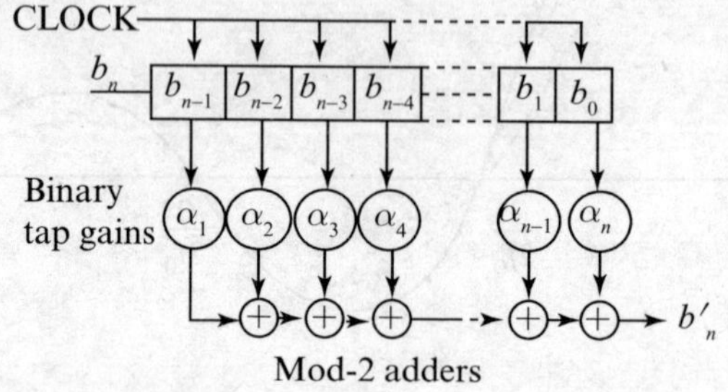

Fig. 8.25 *n-stage tapped shift register*

$$b'_n = \alpha_1 b_{n-1} + \alpha_2 b_{n-2} + \cdots + \alpha_{n-1} b_1 + \alpha_n b_0 \tag{8.29a}$$

Where α'_is are the binary tap gains (either 0s or 1s). The combining is performed through modulo-2 addition.

We now illustrate here, a binary scrambler and de-scrambler as shown in Fig. 8.26.

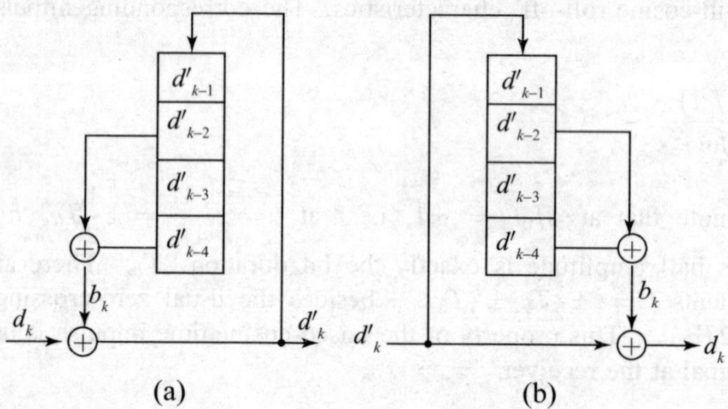

Fig. 8.26 *4-stage binary (a) scrambler (b) de-scrambler.*

We have considered here the tap gains as ($\alpha_1, \alpha_2, \alpha_3, \alpha_4) = (0, 1, 0, 1)$; zero corresponds to the case of no-connection to the binary odder. Observe that the tap gain connections at the de-scrambler are matched to the scrambler network.

The input message sequence is d_k and the scrambled sequence is d'_k given by

$$d'_k = d_k \oplus b_k = d_k \oplus \left(d'_{k-2} \oplus d'_{k-4}\right) \tag{8.29b}$$

Again, the input to the de-scrambler is d'_k and the output of the de-scrambler comes out

$$d'_k \oplus b_k = d'_k \oplus (d'_{k-2} \oplus d'_{k-4})$$

$$= \{d_k \oplus (d'_{k-2} \oplus d'_{k-4})\} \oplus (d'_{k-2} \oplus d'_{k-4})$$

$$= d_k, \text{ since } x \oplus x = 0 \text{ and}$$

$$x \oplus 0 = x.$$

Example 8.6

Let the input binary bit stream

$$d_k = 1010000010111101$$

be fed to the scrambler. Find the scrambled output. Given that MSB is feeding first to the scrambler.

Solution

We have shown the shift register contents $d'_{k-1}\, d'_{k-2}\, d'_{k-3}\, d'_{k-4}$ initialised to 0000. The sum b_k is obtained by $d'_{k-2} \oplus d'_{k-4}$. The output bit d'_k is obtained by $b_k \oplus d_k$. Finally, the bit d'_k is shifted to the shift register d'_{k-1} in the next clock cycle. The previous contents in d'_{k-1} shifts to d'_{k-2}, d'_{k-2} to d'_{k-3}, and d'_{k-3} to d'_{k-4}. This continues until the end of the input bit shifts in d'_{k-1}. In the scrambling operation, the 5 consecutive zeros in d_k have been eliminated in d'_k.

Scrambling

Register contents	d'_{k-1}	0	1	0	0	0	1	0	1	0	1	0	1	1	1	0	0	0
	d'_{k-2}	0	0	1	0	0	0	1	0	1	0	1	0	1	1	1	0	
	d'_{k-3}	0	0	0	1	0	0	0	1	0	1	0	1	0	1	1	1	
	d'_{k-4}	0	0	0	0	1	0	0	0	1	0	1	0	1	0	1	1	
	b_k	0	0	1	0	1	0	1	0	0	0	0	0	0	1	0	1	
Input →	d_k	1	0	1	0	0	0	0	0	1	0	1	1	1	1	0	1	
Output ←	d'_k	1	0	0	0	1	0	1	0	1	0	1	1	1	0	0	0	

DeScrambling

	d'_{k-1}	0	1	0	0	0	1	0	1	0	1	0	1	1	1	0	0	1
	d'_{k-2}	0	0	1	0	0	0	1	0	1	0	1	0	1	1	1	0	
	d'_{k-3}	0	0	0	1	0	0	0	1	0	1	0	1	0	1	1	1	
	d'_{k-4}	0	0	0	0	1	0	0	0	1	0	1	0	1	0	1	1	
	b_k	0	0	1	0	1	0	1	0	0	0	0	0	0	1	0	1	
Input →	d_k	1	0	0	0	1	0	1	0	1	0	1	1	1	0	0	0	
Output ←	d'_k	1	0	1	0	0	0	0	0	1	0	1	1	1	1	0	1	

Thus,

$$\{d_k\} = \{1\ 0\ 1\ 0\ 0\ 0\ 0\ 0\ 1\ 0\ 1\ 1\ 1\ 1\ 0\ 1\}$$
$$\{d'_k\} = \{1\ \ 0\ 0\ 0\ 1\ 0\ 1\ 0\ 1\ 0\ 1\ 1\ 1\ 0\ 0\ 0\}$$

Following the de-scrambling operation, the scrambled data sequence d'_k converts into d_k again.

8.5 Pulse-Time Modulation

In the case of PAM, the samples of a message modulate the amplitude of a carrier that is a periodic train of pulses. The sample values can also modulate the time parameters of the carrier. Such a process is named as 'pulse-time modulation (shown in Fig. 8.27). The width or position of pulses varies in accordance with the instantaneous sample values of the information signal $x(t)$, while the amplitude of the pulses are constant.

8.5.1 Pulse-width modulation

In pulse-width modulation (PWM), the width of each pulse varies in accordance with the sample values of the message $x(t)$. The larger (smaller) the sample value is, the wider (narrower) is the duration of the pulse. The variation follows the expression

$$\tau_k = \tau_0[1 + \mu x(kT_s)], \tag{8.30}$$

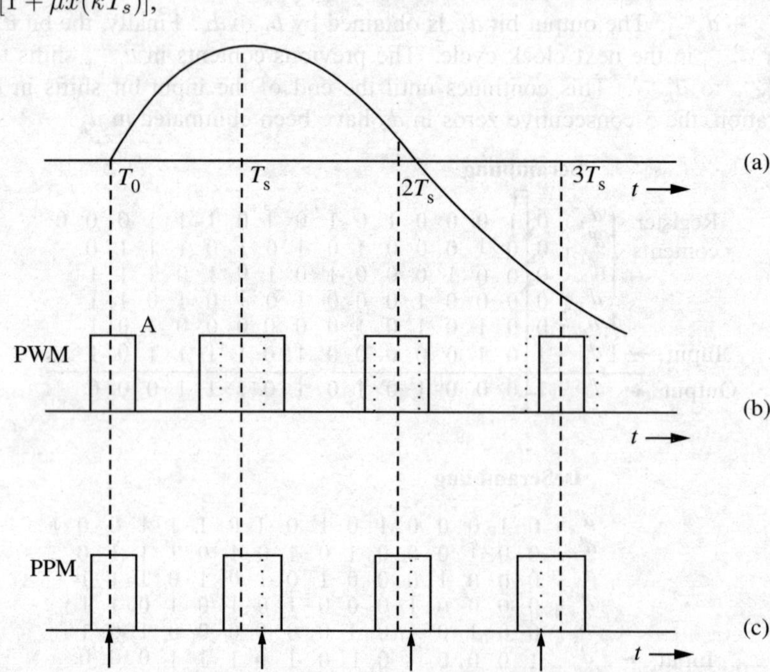

Fig. 8.27 *Pulse-time modulation: (a) message $x(t)$, (b) PWM, (c) PPM.*

where τ_k is the width of the pulse obtained for the sample value $x(kT_s)$ at $t = kT_s$ and τ_0 is that of the unmodulated pulse train. As the sample amplitude $x(kT_s)$ increases, the width τ_k increases and hence the transmitted power also increases. It follows from (8.30) that PWM is a nonlinear form of modulation.

The spectral analysis of the PWM waveform is rather complex. However, an approximate spectral analysis is possible if one assume a slowly-varying message so that adjacent pulses have nearly the same width. This assumption is met with easily if the signal is sampled at a fast rate compared to the Nyquist rate. If one takes rectangular pulses of amplitude A centred at $t = kT_s$ and repeating every T_s seconds, its Fourier series expansions becomes

$$x_s(t) = a_0 + \sum_{n=1}^{\infty} a_n \cos n\omega_s t, \tag{8.31}$$

where the sine component is absent because of the evenness of the pulse train. The constants a_0 and a_n are evaluated as follows.

$$a_0 = \frac{1}{T_s}\int_{-\tau_k/2}^{\tau_k/2} A dt = \frac{A\tau_k}{T_s} = Af_s\tau_0[1 + \mu x(t)] \tag{8.32}$$

$$\text{and } a_n = \frac{2}{T_s}\int_{-\tau_k/2}^{\tau_k/2} A\cos(n\omega_s t)dt = \frac{2A}{n\pi}\sin[n\pi f_s\tau_0[1 + \mu x(t)]]. \tag{8.33}$$

$$\text{Thus, } x_s(t) = Af_s\tau_0[1 + \mu x(t)] + \sum_{n=1}^{\infty}\frac{2A}{n\pi}\sin[n\phi(t)]\cos n\omega_s t, \tag{8.34}$$

where $\phi(t) = \pi f_s\tau_0[1 + \mu x(t)]$. A careful examination of (8.34) reveals that the PWM waveform contains the information signal $x(t)$, a dc component ($= Af_s\tau_0$) and angle-modulated waves at the harmonics of f_s. When $\tau_0 << T_s$, the angle-modulated spectra do not overlap with the message band and hence PWM signal can be demodulated using a lowpass filter after dc blocking.

8.5.2 Pulse-position modulation

In pulse-position modulation (PPM), the position of each pulse varies in direct proportion to the sample values of $x(t)$. The larger the sample value, the more the pulse deviates from its unmodulated position. The pulse deviates to the right for positive $x(t)$ and to the left for negative $x(t)$ (vide Fig. 8.27(c)). We thus see that the PPM pulses are non-uniformly spaced. Corresponding to the sampling instant $t = kT_s$, the position of the kth pulse is at $t = t_k$, given by

$$t_k = kT_s + \alpha x(kT_s). \tag{8.35}$$

The constant 'α' determines the displacement from the sample point $t = kT_s$. The PPM signal power is uniform (since pulse amplitude and width are constant) unlike PWM where the signal power is a function of the message signal.

A PPM signal can be derived from a PWM signal. In PWM, it is the width of the pulse that contains the message whereas the message lies in the position of the pulse in PPM signal. Thus, any method that generates pulses at the leading (or trailing) edge of the PWM wave produces a PPM signal. Applying PWM signal to a differentiator results in positive and negative pulses at the leading and trailing edges, respectively. If we generate pulses of fixed duration at the location of negative pulses, the result is PPM signal. This is depicted in Fig. 8.28 for a sawtooth-type message signal.

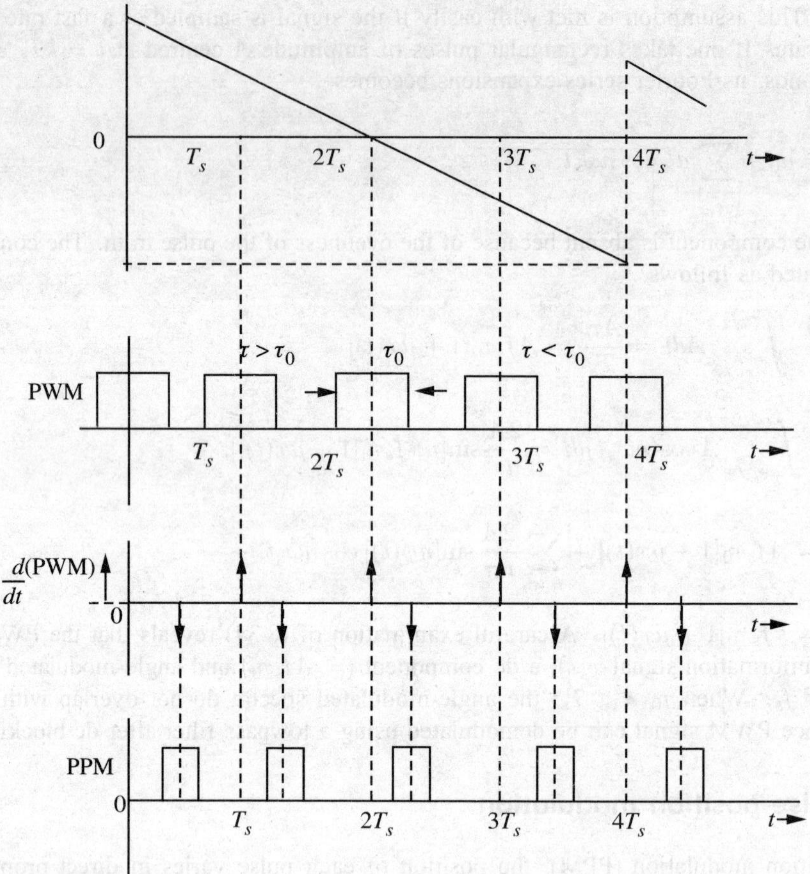

Fig. 8.28 *PPM from PWM*

8.5.3 Modulator and demodulator for PWM and PPM

Figure 8.29 shows the block diagram and typical waveforms of the PWM and PPM modulator. It consists of a sample-and-hold (S/H) circuit, a comparator, a sawtooth-wave generator and a monostable pulse generator. The sampling takes place T_s second apart and the time-period of the sawtooth-wave is also T_s second.

The S/H circuit at the input provides uniform sampling at T_s intervals. For the times during which the sample values are higher than the sawtooth wave, the comparator gives a positive constant; it is zero otherwise. This results in a PWM signal. The PPM signal is obtained by putting the PWM signal to a monostable pulse generator that is triggered on the trailing edges of its input pulses. This result in pulses of fixed duration and is position-modulated. Our illustration assumes that the message signal is scaled so as to lie within the sawtooth-wave range. Demodulation is performed by integrating the PWM over its width. Since the pulse height is constant, the integral output is a ramp

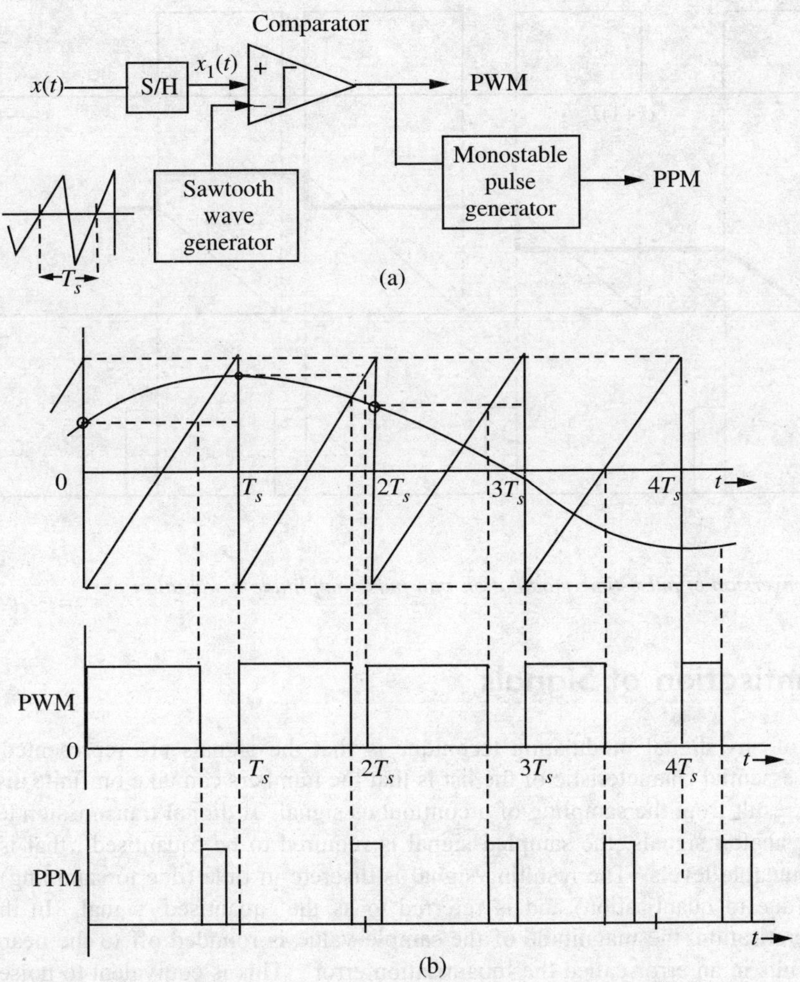

Fig. 8.29 *(a) Modulator for PWM and PPM, (b) waveforms.*

function starting at kT_s and ending at the trailing edge of the PWM pulse as shown in Fig. 8.30. The integrator output is sampled and held at its final value resulting in PAM signal. The lower and upper limits for the integration can be extracted from the leading and trailing edges of the PWM pulse. But the PPM pulse provides only the upper limit and so its demodulation needs a synchronisation signal to know the start of the integration.

Since the off-times between either PWM or PPM pulses are non-uniform, it creates complication in the multiplexed systems. In TDM systems we cannot simply insert pulses in the spaces; rather sufficient spaces must be maintained to avoid interaction. Consequently, less number of channels can be multiplexed.

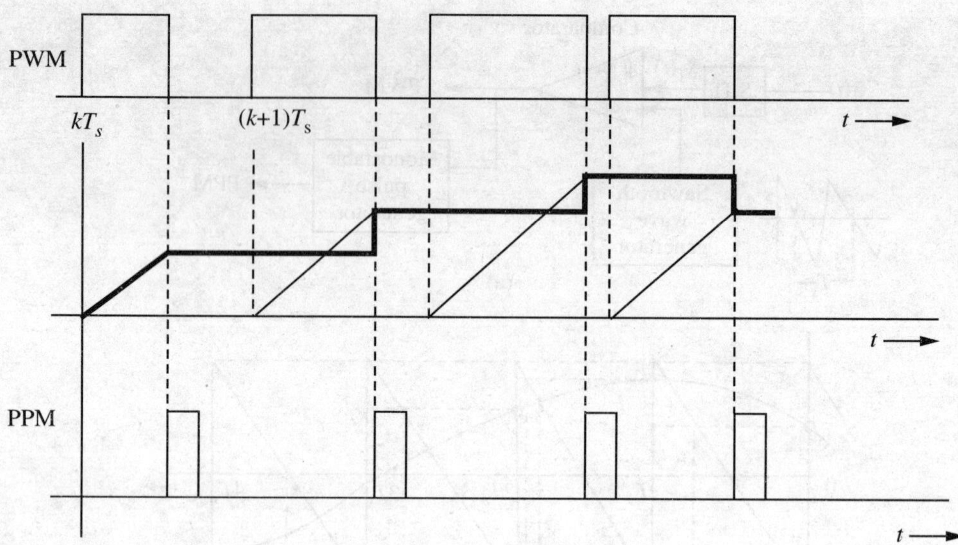

Fig. 8.30 *Conversion of pulse-time modulation into pulse-amplitude modulation*

8.6 Quantisation of Signals

A distinction of any digital modulation technique is that the signals are represented by a list of numbers. The essential characteristic of the list is that the numbers can take on finite discrete values. The list might result from the sampling of a continuous signal. If digital transmission techniques are to be used for analog signals, the sampled signal is required to be 'quantised', that is, represented by a set of countable levels. The resulting signal is discrete in time (due to sampling) and discrete in amplitude (due to quantisation) and is referred to as the 'quantised' signal. In the process of amplitude discretisation, the magnitude of the sample value is rounded off to the nearest quantised level. This results in an error called the 'quantisation error'. This is equivalent to noise that adds to or subtracts from the actual sampled values and is also called as 'quantisation noise'.

A straightforward way to illustrate the process of quantisation is by taking an example of analog-to-digital (A/D) converter, called ADC in brief. Figure 8.31 shows how a linear ramp voltage is quantised.

We have considered here the 'uniform quantisation' where the signal range is divided into uniform regions. At the sampling instant corresponding to the point A, the quantised signal is higher than the analog input (shown dotted). The reverse is true for the point B. Thus, quantisation noise defined by $q(t) = s(t) - s_q(t)$ is positive at A and negative at B. However, in no case the quantisation noise exceeds $\Delta/2$, as illustrated in Fig. 8.31 (b). The quantisation noise power may be obtained as follows.

Let the analog message signal vary between V_L and V_H. Uniform quantisation subdivides the range $(V_H - V_L)$ into M equal steps so that the height of each step is $\Delta = (V_H - V_L)/M$. We

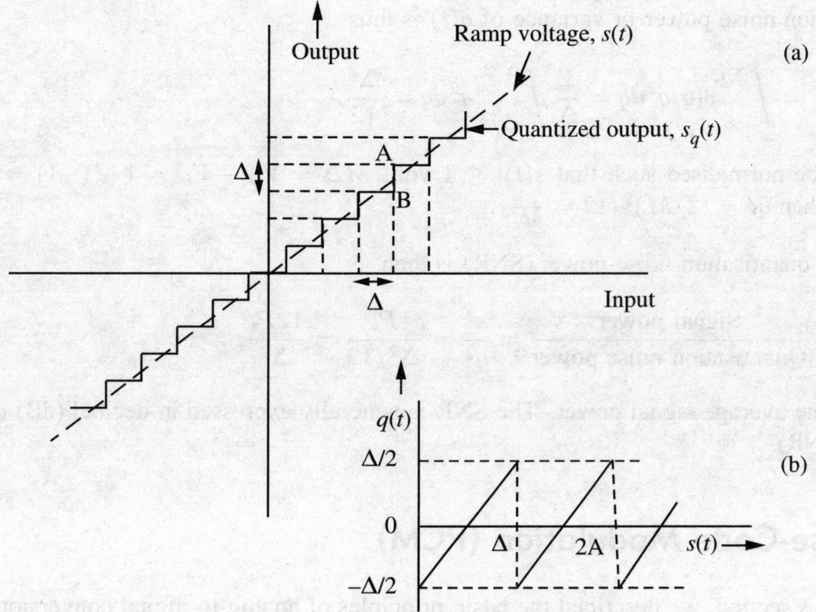

Fig. 8.31 *Quantisation: (a) staircase quantised signal for a linear ramp, (b) quantisation noise.*

locate the quantisation levels at $s_0, s_1, s_2, \ldots$ at the centre of each interval, as shown in Fig. 8.32. The samples $s(nT_s)$ of the signal will be assigned a quantisation level to which it is the closest. The quantisation noise is then given by

$$q(nT_s) = s(nT_s) - s_q(nT_s). \tag{8.36}$$

Assume that the quantisation noise has zero-mean uniform probability density function $p(q)$ over the range $-\Delta/2 \le q \le \Delta/2$. Figure 8.32(b) shows the PDF of $p(q)$.

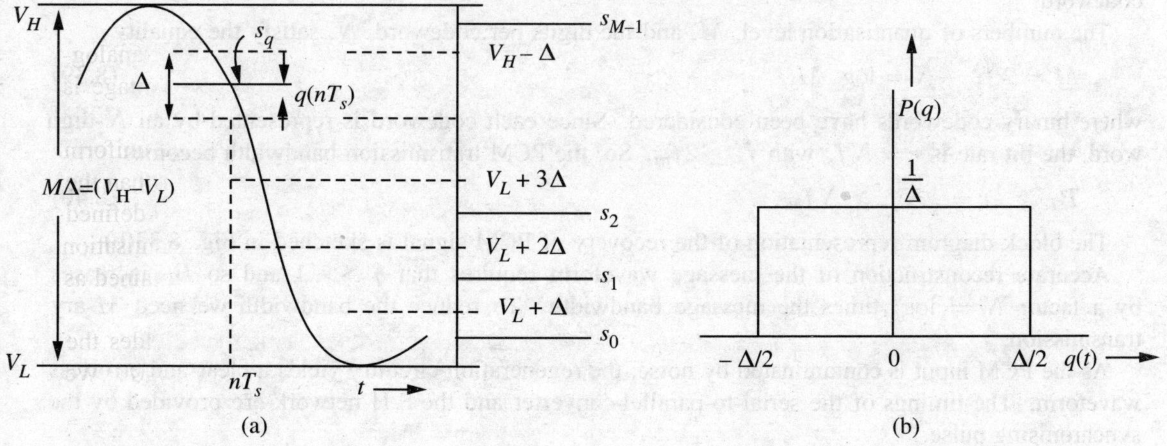

Fig. 8.32 *(a) Signal quantisation, (b) Uniform PDF of quantisation noise.*

The quantisation noise power or variance of $q(t)$ is thus

$$\sigma_q^2 = \overline{q^2} = \int_{-\infty}^{\infty} p(q)q^2 dq = \frac{1}{\Delta}\int_{-\Delta/2}^{\Delta/2} q^2 dq = \frac{\Delta^2}{12}. \tag{8.37}$$

If the signal be normalised such that $s(t)| \leq 1$ volt, $M\Delta = V_H - V_L = 1 - (-1) = 2$ volts and $\Delta = 2/M$. Then $\overline{q^2} = (2/M)^2/12 = \frac{1}{3M^2}$.

The signal-to-quantisation-noise-power (SNR) is then

$$\text{SNR} = \frac{\text{Signal power}}{\text{Quantisation noise power}} = \frac{\bar{s^2}}{\bar{q^2}} = \frac{P_s}{\Delta^2/12} = \frac{12P_s}{\Delta^2}, \tag{8.38}$$

where P_s is the average signal power. The SNR is generally expressed in decibel (dB) or SNR (dB) $= 10\log_{10}(\text{SNR})$.

8.7 Pulse-Code Modulation (PCM)

In the previous section, we described the basic principles of analog-to-digital conversion techniques. In the coded pulse modulation, the quantised levels are represented by code numbers, which are then converted to binary digits. These digits are transmitted in the form of electrical pulses.

Figure 8.33(a) shows the block diagram of a PCM generator. The analog input signal $s(t)$ is band-limited to f_M in a lowpass filter and sampled at the rate of $f_s \geq 2f_M$. The sampled amplitude $s(nT_s)$ is then rounded off to the nearest quantisation level $s_q(nT_s)$ in a quantiser. The encoder assigns N-bit codewords to the quantised samples. Successive codewords are read out serially by a parallel-to-serial converter.

A great advantage of coded communication is its improved error correction facility provided by regenerative repeaters. If the noise voltage is less than one-half of the quantisation step size, the repeater restores the signal to the corresponding quantisation level and retransmits the refreshed codeword.

The numbers of quantisation level, M, and the digits per codeword, N, satisfy the equality

$$M = 2^N, \quad N = \log_2 M, \tag{8.39}$$

where binary codewords have been considered. Since each codeword is represented by an N-digit word, the bit rate is $r = Nf_s$ with $f_s \geq 2f_m$. So, the PCM transmission bandwidth becomes

$$B_T \geq \tfrac{1}{2}r = \tfrac{1}{2}Nf_s \geq Nf_M. \tag{8.40}$$

The block diagram representation of the recovery of PCM signal is sketched in Fig. 8.33(b).

Accurate reconstruction of the message waveform requires that $q >> 1$ and so B_T increases by a factor $N = \log_M^q$times the message bandwidth. To reduce the bandwidth we need M-ary transmission.

As the PCM input is contaminated by noise, the regeneration circuitry yields a clear and errorless waveform. The timings of the serial-to-parallel converter and the S/H network are provided by the synchronising pulse.

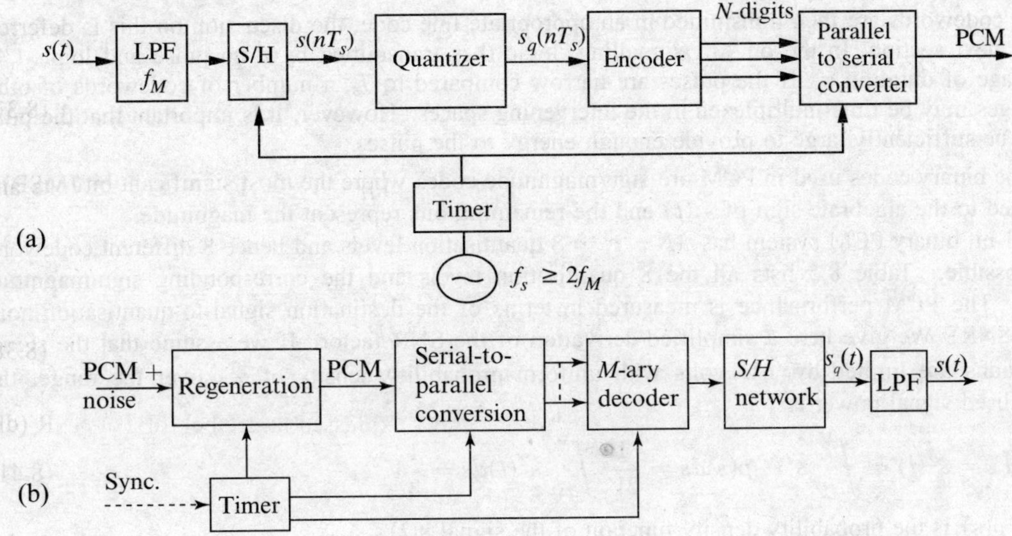

Fig. 8.33 (a) *PCM generator* (b) *PCM receiver*

We illustrate in Fig. 8.34 the encoding of PCM codewords for $N = 2$ bits. We require here $N^2 = 2^2 = 4$ quantisation levels (s_0, s_1, s_2, and s_3 shown dotted). We assign binary codes 00, 01, 10, and 11 for the quantisation levels s_0, s_1, s_2, and s_3, respectively. The signal sample lying between V_2 and V_1 is assigned to the appropriate quantisation level s_0 with the corresponding code word 00, and so on. At the first sampling point $t = 0$, the sample value lies between V_2 and V_3, with the closest quantisation level s_2. So it is encoded as 10. Following this, the message signal of Fig. 8.33 may be encoded as 00 11 11 11 10 01 . . .

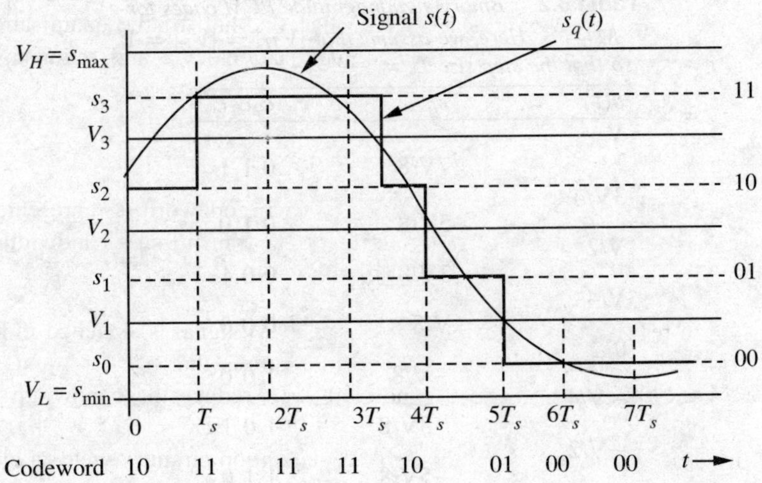

Fig. 8.34 *A 2-bit PCM encoding*

These codewords are then transmitted in an appropriate line code; the discussion on this is deferred to the next section. In the on–off signalling, logic 0 is transmitted by a no pulse and logic 1 by a voltage of duration τ . If the pulses are narrow compared to T_s, a number of codewords of other messages may be time-multiplexed in the intervening spaces. However, it is important that the pulse width be sufficiently large to provide enough energy to the pulses.

The binary codes used in PCM are sign/magnitude codes where the most significant bit (MSB) is assigned to the algebraic sign of $s_q(t)$ and the remaining bits represent the magnitude.

A 3-bit binary PCM system has $M = 2^3 = 8$ quantisation levels and hence 8 different codewords are possible. Table 8.2 lists all the 8 quantisation levels and the corresponding sign/magnitude codes. The PCM performance is measured in terms of the destination signal-to-quantisation-noise ratio (SNR). We give here a simplified derivation of the SNR factor. If we assume that the signal excursions are limited by $\pm V$ volts with uniform probability density of $\frac{1}{2V}$ over the range, the normalised signal power is

$$P_s = \overline{s^2}(t) = \int_{-V}^{V} s^2(t)p(s)ds = \frac{1}{2V}\int_{-V}^{V} s^2(t)ds = \frac{1}{3}V^2. \tag{8.41}$$

where $p(s)$ is the probability density function of the signal $s(t)$.

Thus the destination SNR is given by

$$\text{SNR} = \frac{\overline{s^2}}{\overline{q^2}} = \frac{\frac{1}{3}V^2}{\Delta^2/12} = M^2 = 2^{2N}, \tag{8.42}$$

where we have used $\Delta = 2V/M$.

In dB,

$$\text{SNR (dB)} = 10\log_{10} 2^{2N} = 6N, \tag{8.43}$$

which shows that for every 1 bit added to the codewords, the SNR increases by 6 dB.

Table 8.2 *Binary sign/magnitude PCM codes for $M = 8$. Here, we assume that $|V_H| = |V_L| = V$ so that the step size $\Delta = \frac{2V}{8} = \frac{V}{4}$.*

$s(t)$	s_q	$c_2c_1c_0$
V		
	7V/8	0 1 1
3V/4		
	5V/8	0 1 0
V/2		
	3V/8	0 0 1
V/4		
	V/8	0 0 0
0		
	−V/8	1 0 0
−V/4		
	−3V/8	1 0 1
−V/2		
	−5V/8	1 1 0
−3V/4		
	−7V/8	1 1 1
−V		

The number of bits per codeword in the PCM encoder is mostly determined by the 'resolution' and the 'dynamic range'. Resolution measures the quality of the converter. It is the ratio of the LSB increment to the maximum output. For an N-bit codeword,

$$\text{Resolution} = 1/(2^N - 1).$$

The smaller the step size, the better is the resolution and more close is the quantised sample to the actual signal. The term 'dynamic range' is defined as the ratio of the largest to the smallest magnitudes of the signal that can be decoded by a digital-to-analog (D/A) converter. For acceptable voice transmission, a value of at least 30 dB is required for SNR at the destination point. Consider an 8-level PCM system which encodes a message of maximum voltage 7 volts. The maximum magnitude of the voltage has the binary code (natural code) 111 and the minimum decodable magnitude has the binary code 001. The resolution is $1/(2^3 - 1) = 1/7$ corresponding to 1 volt and dynamic range is $7/1 = 7 = 16.9$ dB. When a smaller resolution is required, say 0.5 V, the signal amplitude must be reduced by the factor of 2 to maintain the same dynamic range.

8.8 Non-uniform Quantisation

In some cases it becomes advantageous to adopt quantisation with non-uniform step sizes. This is particularly helpful when the signal amplitude remains small over an appreciable time as in the case of a musical signal. Suppose that we have a musical signal whose voltage varies between +2 and −2 volts. During the soft part of the music the signal strength may not exceed 0.5 volts for a long time. Using a 3-bit uniform PCM, the step size is $4/8 = 0.5$ volt. All sample values in between 0 and 0.5 volt will be encoded as 000 (sign/magnitude). Similarly, all sample values within 0 and −0.5 volt will be encoded as 100. Thus the codewords 000 or 100 will be repeated many times although the signal has significant variations (may have large slopes) within ±0.5 volts. This means that the PCM coder cannot resolve such undulations at low amplitudes. Musical quality is obviously lost to a great extent. In order to protect the quality at low amplitude, it is desirable to use small quantisation steps at the lower signal levels and larger step size at the higher signal levels. In other words, the resolution should be improved at the lower amplitudes at the expense of that at the higher amplitudes; this improves the dynamic range. The added advantage is the reduction in the quantisation noise at reduced step size. We know that the maximum quantising error is half the step size. For a 128 levels PCM system, the maximum error is $\frac{1}{2} \times \frac{V_P}{128} = \frac{V_P}{256}$, where V_P is the peak of the message signal. The figure is quite small in practical systems. Now suppose that the signal spends a considerable time at the low levels, $\frac{V_P}{64}$, say. The quantisation error in the same system would be as large as $\frac{V_P}{256} \div \frac{V_P}{64} = \frac{1}{4}$ of the peak of this small signal. Non-uniform quantisation can be implemented by using a tapered transfer characteristic, which we will discuss now.

To start with, assume that the signal samples can be represented by the probability density function $f(s)$ as shown in Fig. 8.35. The representative PDF has even symmetry. The non-uniform quantisation is also indicated. The levels $\pm s_1, \pm s_2, \ldots \pm s_{M/2}$ are closely spaced near the centre when $f(s)$ is maximum and $|s(t)|$ is small most of the time. These are gradually dispersed as $f(s)$ decreases and $|s(t)|$ becomes large. In the signal range a_i to b_i, the quantisation level is s_i. Our aim is to find out the optimum location of s_i that could produce the minimum quantisation noise. In the said range the quantisation error $q_i = s_i - s$ and mean-square quantisation error is given by

$$\overline{q_i^2} = \int_{a_i}^{b_i} (s - s_i)^2 f(s) ds. \tag{8.44}$$

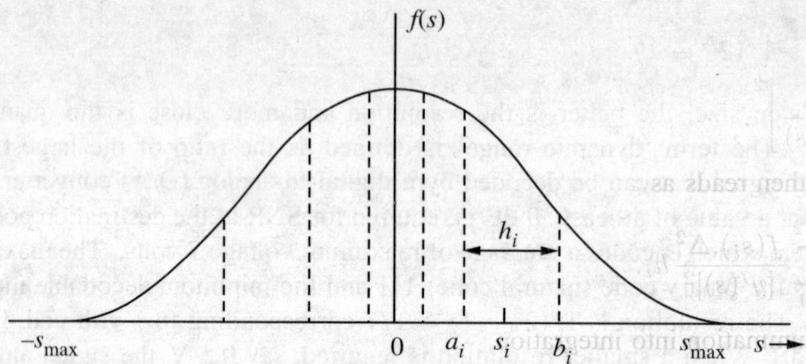

Fig. 8.35 *Probability density function with non-uniform quantisation*

Summing $\overline{q_i^2}$ over such $M/2$ intervals,

$$\overline{q^2} = 2\sum_{i=1}^{M/2} \overline{q_i^2} = 2\sum_{i=1}^{M/2} \int_{a_i}^{bi} (s - s_i)^2 f(s) ds. \tag{8.45}$$

The probability density can be assumed to be a constant over the width $h_i = (b_i - a_i)$ provided h_i is small enough. The best choice of the quantisation level is in the middle of each interval, that is, $s_i = (a_i + b_i)/2$. Equation (8.45) then becomes

$$\overline{q^2} = 2\sum_{i=1}^{M/2} f(s_i) \int_{s_i - h_i/2}^{s_i + h_i/2} (s - s_i)^2 ds = 2\sum_{i=1}^{M/2} f(s_i) \frac{1}{12} h_i^3 = \frac{1}{6} \sum_{i=1}^{M/2} f(s_i) h_i^3. \tag{8.46}$$

For uniform quantisation, $f(s_i) = \frac{1}{M\Delta}$ and $h_i = \Delta$ and (8.46) reduces to (8.37).

In an attempt to optimise the PCM performance, we should have to find the best locations of the quantisation levels. This can be obtained by differentiating (8.45) with respect to s_i and equating it to zero. This gives

$$\int_{a_i}^{b_i} (s - s_i) f(s) ds = 0, \qquad \text{for all } i. \tag{8.47}$$

Therefore, such optimisation needs the knowledge of PDF $f(s)$. In the range $a_i < s < b_i$, the quantisation level is the centre of gravity of $f(s)$ as dictated by (8.47). Towards this goal, the most common practice is to use signal compression followed by uniform quantisation. Here, the signal is first compressed in a nonlinear device and then uniformly quantised. The distortion introduced by the process of compressing is undone at the reconstruction by an 'expander' having complementary nonlinearity. This process is referred to in the text as 'companding', a name derived from the words 'compression and expanding'. Figure 8.36 illustrates a compressor curve $y(s)$ plotted against the signal $s(t)$. An analog input $s(t)$ at the compressor yields the compressed output $y(s)$. If the output is uniformly quantised, the result is non-uniform quantisation of $s(t)$ with increasing steps for higher signal levels.

Now, the slope of the curve is given by $y'(s) = \frac{dy}{ds}$. If the step size Δ is small over the portion AB, the slope of the straight line AB can be approximated as $\frac{dy}{ds}$ at $s = s_i$.

Thus, $y'(s) = \frac{dy}{ds}|_{s=s_i} = \frac{\Delta}{h_i}$

or $$h_i^2 = \left[\frac{\Delta}{y'(s)}\right]^2. \tag{8.48}$$

Equation (8.46) then reads as

$$\overline{q^2} = \frac{1}{6}\sum_{i=1}^{M/2} \frac{f(s_i).\Delta^2}{[y'(s)]^2} h_i. \tag{8.49}$$

Converting the summation into integration,

$$\overline{q^2} = \frac{\Delta^2}{6}\int_0^1 \frac{f(s_i)}{[y'(s)]^2} ds. \tag{8.50}$$

The signal-power-to-quantisation-noise-power then becomes

$$SNR = \frac{\overline{s^2}}{\overline{q^2}} = \frac{P_s}{\Delta^2/12} \Bigg/ \left[2\int_0^1 \frac{f(s)}{[y'(s)]} ds\right]. \tag{8.51}$$

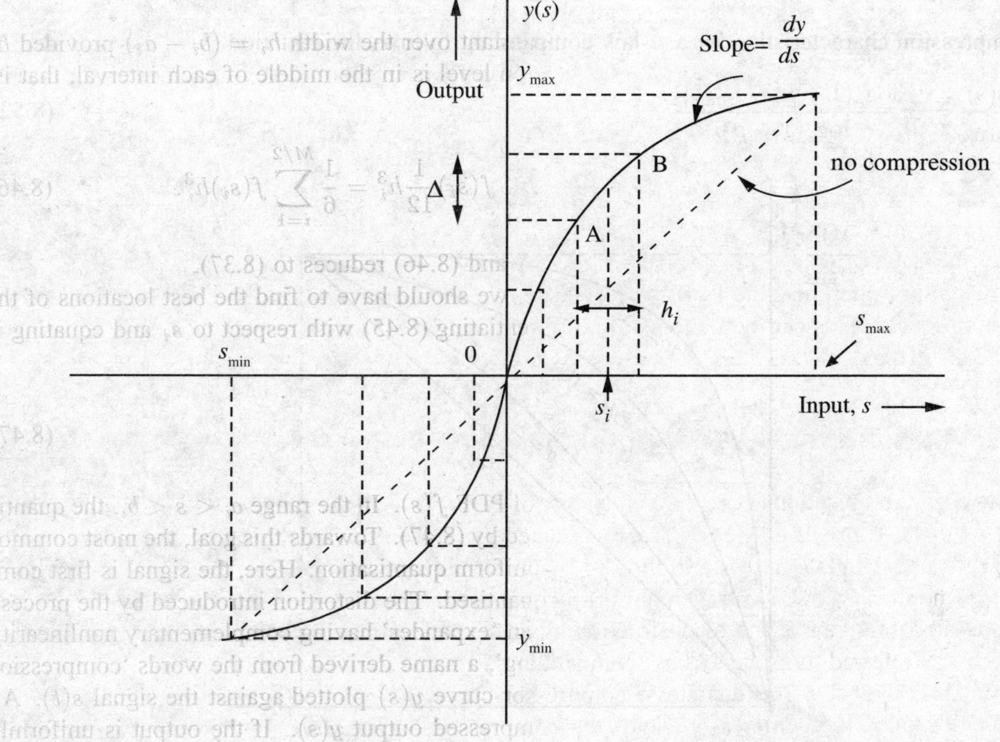

Fig. 8.36 *Compression characteristic of $y(s)$*

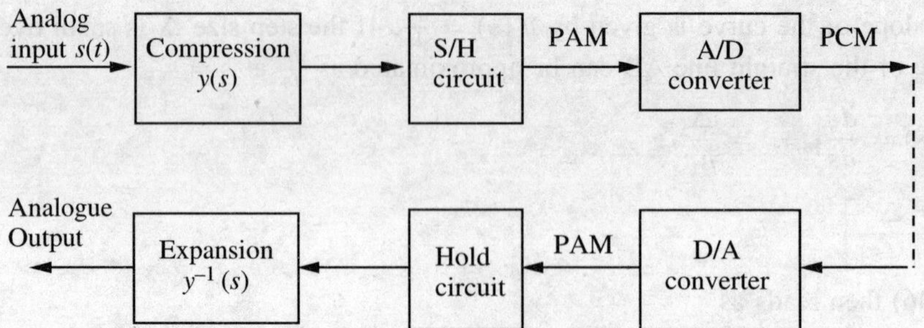

Fig. 8.37 *Model of PCM system with companding*

The factor $2\int_0^1 \frac{f(s)}{[y'(s)]^2}ds$ accounts for the compression. Improvement in PCM performance is observed when $2\int_0^1 \frac{f(s)}{[y'(s)]^2}ds < 1$.

Figure 8.37 illustrates a PCM system with companding. The two methods of analog companding are 'μ-law companding' and 'A- law companding'. We describe here both the compressor characteristics.

μ-law companding

The compression characteristics for a μ-law companding follows the equation

$$\frac{y(s)}{y_{\max}} = \pm\frac{\log_e(1+\mu|s/s_{max}|)}{\log_e(1+\mu)} \tag{8.52}$$

Fig. 8.38 *A μ-law compression characteristic curve*

and is plotted in Fig. 8.38. Here the positive sign applies for $s(t) > 0$ and the negative sign for $s(t) < 0$. The input $s(t)$ and output $y(s)$ are normalised by their maximum value $s_{\max}$ and $y_{\max}$ respectively. The parameter μ defines the degree of compression. A standard value of μ is 255.

Example 8.7

Find the expansion characteristic curve corresponding to (8.52).

Solution

To find the inverse of (8.52), we will have to find out the explicit expression of $s(t)$. The function $y(s)$ is uniformly quantised to generate s_q. Considering the case $s(t) > 0$, we have

$$\frac{y(s)}{y_{\max}}\ln(1+\mu) = \ln\left(1+\mu\left|\frac{s}{s_{\max}}\right|\right)$$

$$\text{or,}\qquad 1+\mu\left|\frac{s}{s_{\max}}\right| = \exp\left[\frac{y(s)}{y_{\max}}\ln(1+\mu)\right] = \exp\left[\ln(1+\mu)^{y(s)/y_{\max}}\right] = (1+\mu)^{\frac{y(s)}{y_{\max}}}.$$

$$\text{or,}\qquad \frac{s}{s_{\max}} = \frac{1}{\mu}\left[(1+\mu)^{y(s)/y_{\max}} - 1\right].$$

Since $y(s)/y_{\max}$ is the input to the expander and $s/s_{\max}$ is its output, we rewrite the above as

$$s = \frac{s_{\max}}{\mu}\left[(1+\mu)^{\frac{y(s)}{y_{\max}}} - 1\right].$$

Thus,

$$y^{-1}(s_q) = \frac{s_{\max}}{\mu}\left[(1+\mu)^{|s_q|s_{\max}|} - 1\right].$$

This is the expansion function. We find that $y(y^{-1}(s)) = s$.

A value of $\mu = 0$ corresponds to the linear curve indicating no compression.

In implementing the μ-law curve, the tapering portion is approximated by piecewise linear segments as shown in Fig. 8.39.

We illustrate here a μ-255 curve for a message signal $s(t) > 0$. The compressed output (i.e. the vertical axis) is divided into 8 equal intervals (called segments). In each segment, the curve is represented by a straight-line segment. For a 12-bit-to-8bit digital compression, the slope of a segment is one-half of the previous segment. In other words, each interval on the horizontal axis is double the previous interval. This implies that the resolution of any one segment is twice that of the proceeding segment. All the input segments have the same resolution.

Out of the 8 bits in the compressed code, the MSB represents the sign bit, following 3 bits are used to identify the segment and the rest 4 bits refer to the quantisation level within the segment. With reference to Fig. 8.39, an 8-bit PCM codeword of 01011101, the meaning of the 8 bits are as follows:

A 0-bit at the MSB indicates that the signal is positive and hence the compression curve in the first quadrant is used. The next three bits 101 signifies that the signal lies in the segment number 5. The last four bits, 1101, refer to signal encoding.

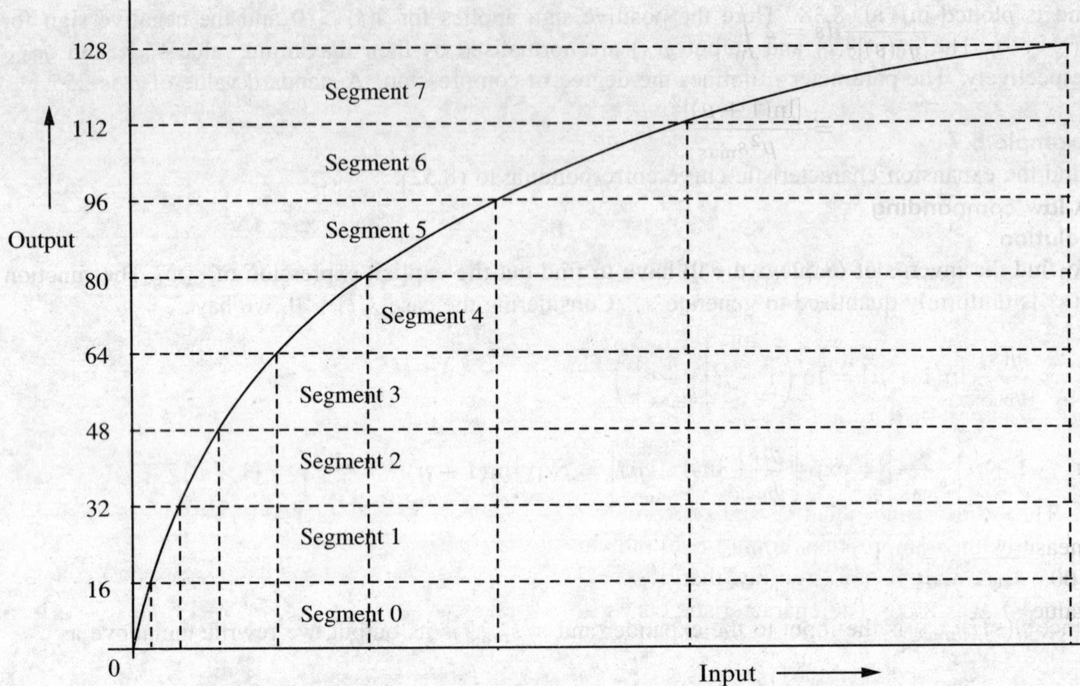

Fig. 8.39 *A piecewise linear approximation of a μ-255 logarithmic compressor*

For a given μ, the slope of the compression curve is given by

$$\frac{dy}{ds} = y'(s) = \frac{d\left(\frac{y(s)}{y_{\max}}\right)}{d\left(\frac{s(t)}{s_{\max}}\right)} = \frac{1}{\ln(1+\mu)} \frac{\mu}{[1+\mu|s/s_{\max}|]}. \tag{8.53}$$

Obviously, the curve is not linear nor logarithmic. Since μ is a large number, we consider first the case of small signal so that $\mu|\frac{s}{s_{\max}}| << 1$. Equation (8.53) yields $y'(s) \approx \frac{\mu}{\log(1+\mu)}$, a constant, and the curve is linear. However, at higher input levels $\mu|s/s_{\max}| >> 1$ and so

$$y'(s) \approx \frac{1}{s} \frac{s_{\max}}{\ln(1+\mu)}$$

indicating a logarithmic curve.

Example 8.8

Find the factor $2\int_0^1 \frac{f(s)}{[y'(s)]^2} ds$ for the μ-law compression curve.

Solution

Since $y'(s)$ is given by (8.53), we have,

$$2\int_0^{s_{\max}} \frac{f(s)}{[y'(s)]^2} ds = 2\int_0^{s_{\max}} \frac{1}{2s_{\max}} \cdot \left[\frac{\ln(1+\mu)}{\mu}\right]^2 [1+\mu|s/s_{\max}|]^2\, ds$$

$$= \frac{[\ln(1+\mu)]^2}{\mu^2 s_{\max}} \int_0^{s_{\max}} \left[1+\mu\frac{s}{s_{\max}}\right]^2 ds = \frac{[\ln(1+\mu)]^2}{\mu^2}(1+\mu+\mu^2/3)$$

A-law companding

An A-law compander employs the following compression characteristic equation.

$$\frac{y(s)}{y_{\max}} = \pm\frac{A|s/s_{\max}|}{1+\log A}, \quad \text{for } 0 \le \frac{s}{s_{\max}} \le 1/A$$

$$= \pm\frac{1+\log\left(A|\frac{s}{s_{\max}}|\right)}{1+\log A}, \quad \text{for} \frac{1}{A} \le \frac{s}{s_{\max}} \le 1. \tag{8.54}$$

The positive and negative signs are for $s(t) > 0$ and $s(t) < 0$, respectively. The parameter A measures the compression. Figure 8.40 plots the corresponding compressor curve for $A = 1$, 2 and 100. The value $A = 1$ signifies that $y(s) = s(t)$ and there is no compression. A commonly used value of A is 87.6. The characteristic curve can be explained as follows.

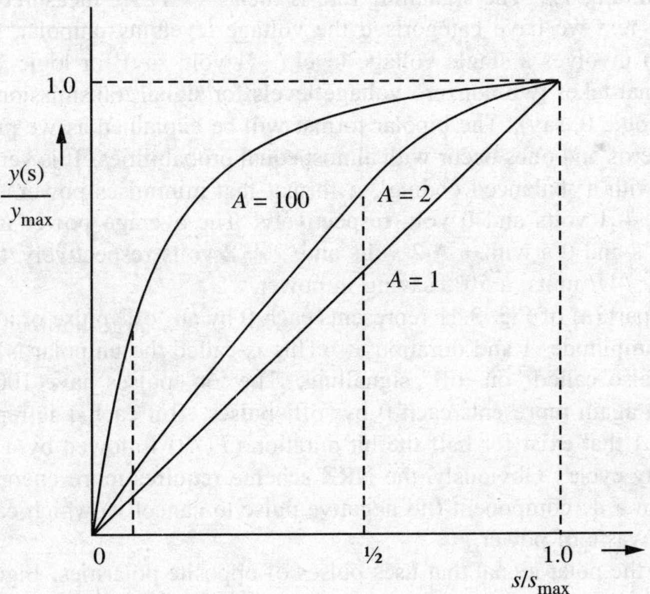

Fig. 8.40 *A-law compression characteristics*

In the range $0 \le \frac{s}{s_{\max}} \le \frac{1}{A}$ the curve is linear and in the range $\frac{1}{A} \le \frac{s}{s_{\max}} \le 1$ the curve is logarithmic. The nonlinearity starts at an early stage for higher A. To illustrate this, take the value $A = 2$ so that the curve is linear up to $s/s_{\max} \le 1/2$ with a slope $2/(1+\log 2)$. Beyond $\frac{s}{s_{\max}} = \frac{1}{2}$, the curve is logarithmic.

8.9 Line Coding

In the process of sampling and quantising, a message signal is translated into a set of discrete numbers. However, a basic requirement of signal transmission is that the signal must be in electrical form. So we require an encoding process that converts the discrete values into codewords which are then represented in some electrical format. There are several factors to be considered while selecting a line coding format, e.g., the transmission bandwidth, error detection capabilities, line voltage levels, etc. The important line codes for the electrical representation of data stream are listed below.

1. Unipolar non-return-to-zero (NRZ) signalling
2. Unipolar return-to-zero (RZ) signalling
3. Polar NRZ signalling
4. Polar RZ signalling
5. Bipolar NRZ signalling
6. Bipolar RZ signalling
7. Manchester code
8. Polar quaternary NRZ signalling

Each code has its merits and demerits which we will discuss later. Let us look at the waveforms of the various codes (shown in Fig. 8.41) for a binary message 01001011.

We have depicted the codes with rectangular pulses for clarity. The time allotted to one bit is known as the bit duration, T_b. The signalling rate is then $r = 1/T_b$, measured in bits per seconds (bps or b/s). In this text we have categorised the voltage levels as unipolar, polar, or bipolar. A unipolar transmission involves a single voltage level ($+A$ volt, say) for logic 1 and a zero volt for logic 0. A polar format takes two nonzero voltage levels for signal transmission ($+A$ volts for logic 1 and $-A$ volts for logic 0, say). The bipolar format will be explained as we proceed.

In binary codes, zeros and ones occur with almost equal probabilities. It is very effective to encode the binary symbols with a 'balanced coding', a format that minimises power. Suppose we encode the 1's and 0's with $+A$ volts and 0 volt, respectively. The average power is $A^2/2R$ units. But if one encodes the 1's and 0's with + A/2 volts and $-A/2$ volts respectively, the average power is $2 \times (A/2)^2/2R = A^2/4R$ units, a 50% saving in power.

The waveform in part (a) of Fig. 8.41 represents each 0 by an 'off' pulse or a no pulse and each 1 by an 'on' pulse of amplitude A and duration T_b. This is called the unipolar NRZ format. Because of its structure it is also called "on–off" signalling. The 'on' pulses have 100% duty cycle. The waveform in part (b) again represents each 0 by 'off' pulses. But each 1 is represented by an 'on' pulses of amplitude A that exist for half the bit duration ($T_b/2$) followed by a return to zero. The pulses have 50% duty cycle. Obviously, the NRZ scheme requires more energy. Both the RZ and NRZ formats result in a dc component (no negative pulse to cancel it) which carries no meaningful information and is a waste of power.

A better format is the polar signal that uses pulses of opposite polarities. Figure 8.41 (c) portrays the polar NRZ format that represents each 0 by pulses of amplitude $-\frac{A}{2}$ and duration T_b and each 1 by pulses of amplitude $A/2$ and duration T_b. Its dc component is zero if the 1's and 0's are in equal proportion in the data. Polar RZ format again uses two voltage levels: $-A/2$ to represent each 0 and $+A/2$ for each 1. But here the pulses exist for $T_b/2$ followed by a return to zero, as shown in Fig. 8.41 (d). Here also the dc component will be zero if the 1s and 0s are in equal proportion in the data stream.

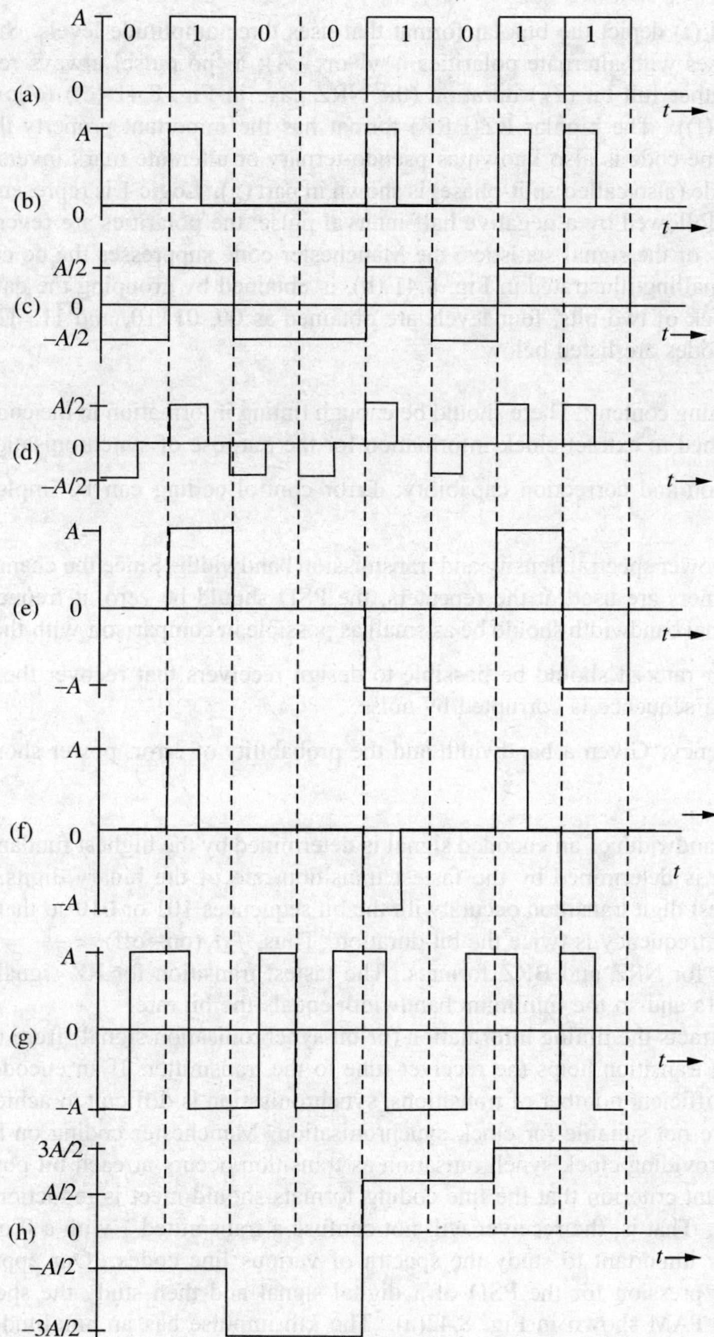

Fig. 8.41 *Line codes (a) unipolar NRZ; (b) unipolar RZ; (c) polar NRZ; (d) polar RZ; (e) bipolar NRZ; (f) bipolar RZ; (g) Manchester code; (h) polar quaternary*

Figure 8.41 (e) and (f) depict the bipolar format that uses three amplitude levels. Successive 1s are represented by pulses with alternate polarities ($+A$ or $-A$); a 'no pulse' always represents a zero. The pulses have either full bit (T_b) duration (the NRZ case in Fig. 8.41 (e)) followed by a return to zero (Fig. 8.41 (f)). The bipolar RZ(BRZ) format has the important property that it has no dc component. This line code is also known as pseudo-ternary or alternate mark inversion (AMI).
The Manchester code (also called split-phase) is shown in part (g). Logic 1 is represented by a positive half-interval pulse followed by a negative half-interval pulse; the polarities are reversed to represent logic 0. Regardless of the signal statistics, the Manchester code suppresses the dc component.
The quaternary signalling illustrated in Fig. 8.41 (h), is obtained by grouping the data bits in blocks of two. With a block of two bits, four levels are obtained as 00, 01, 10, and 11. The few desirable properties of line codes are listed below.

- Adequate timing content: There should be enough timing information in the code so that system can be designed to extract clock information for the purpose of synchronisation.
- Error detection and correction capability: Error control coding can be implemented into the line code.
- Favourable power spectral density and transmission bandwidth: Since the channel is ac-coupled and transformers are used at the repeaters, the PSD should be zero at frequencies near zero. Also, the signal bandwidth should be as small as possible in comparison with the channel bandwidth.
- Low bit error rate: It should be possible to design receivers that recover the data even when the input data sequence is corrupted by noise.
- Power efficiency: Given a bandwidth and the probability of error, power should be as low as possible.

The transmission bandwidth of an encoded signal is determined by the highest fundamental frequency f_M, which in turn is determined by the fastest transition rate of the binary digits. In the on–off signalling, the fastest digit transition occurs with the bit sequences 101 or 010 so that the time period of the fundamental frequency is twice the bit duration. Thus, f_M (on–off) $= \frac{1}{2T_b} = \frac{1}{2}\times$ the bit rate. Similar is the case for NRZ and BRZ formats. The fastest transition for RZ signalling takes place during successive 1s and so the minimum bandwidth equals the bit rate.

The receiver extracts the timing information (or bit synchronisation signal) from the encoded data stream. The signal transition helps the receiver tune to the transmitter. If an encoded signal format does not provide sufficient number of transitions, synchronisation is difficult to achieve. The on–off or NRZ formats are not suitable for clock synchronisation. Manchester coding on the other hand is very effective in providing clock synchronisation as transition occurs at each bit position.

Another important criterion that the line coding formats should meet is reduction in bit errors in the received signal. That is, the receiver will not confuse a transmitted 1 with a 0 or vice versa.

It is particularly important to study the spectra of various line codes. Our approach will be to derive a general expression for the PSD of a digital signal and then study the specific cases. We start with a digital PAM shown in Fig. 8.42(a). The kth impulse has an amplitude a_k and so on. Corresponding to the data we construct a pulse train $y(t)$ shown in Fig. 8.42(b) represented by

$$y(t) = \sum_{n=-\infty}^{\infty} a_n p(t - nT_b), \tag{8.55}$$

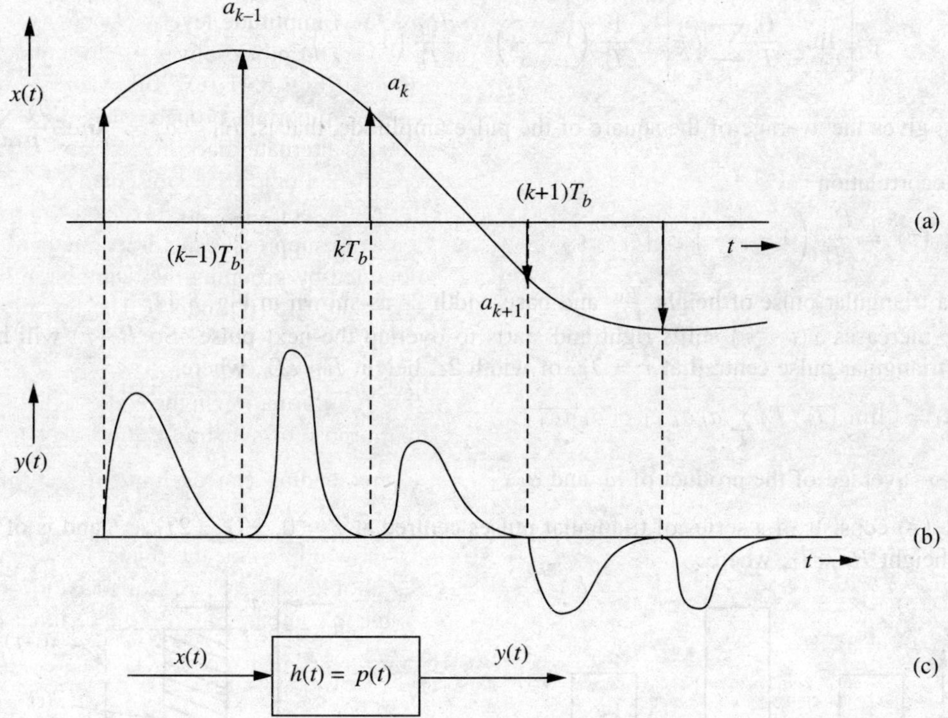

Fig. 8.42 *(a) A digital PAM signal; (b) pulse train; (c) block representing the conversion of a digital PAM into pulse train.*

where $p(t)$ is the basic pulse shape and T_b is the pulse duration. For binary signalling, the data transmission rate $r = 1/T_b$. The value of a_k is random. For example, for unipolar NRZ data, $a_k = +A$ for binary 1 and $a_k = 0$ for binary 0.

Suppose that $S_x(f)$ and $S_y(f)$ are the Fourier transforms of $x(t)$ and $y(t)$. These two are related through the equation

$$S_y(f) = |P(f)|^2 S_x(f), \tag{8.56}$$

where $P(f)$ is the Fourier transform of the pulse shape $p(t)$. Since an impulse function is the limiting form of rectangular pulse of width $\epsilon \to 0$, the strength of the kth impulse is $\epsilon h_k = a_k$, where h_k is the height of the kth impulse. Figure 8.43 shows the rectangular pulse train $\widehat{x}(t)$ for the impulse of Fig. 8.42(a). The autocorrelation is given by

$$\begin{aligned} R_{\widehat{x}}(\tau) &= \lim_{T\to\infty} \frac{1}{T} \int_{-T/2}^{T/2} \widehat{x}(t)\widehat{x}(t-\tau)dt \\ &= \lim_{T\to\infty} \frac{1}{T} \sum_k h_k^2(\epsilon - \tau) = \lim_{T\to\infty} \frac{1}{T} \sum_k a_k^2 \left(\frac{\epsilon - \tau}{\epsilon^2}\right) \end{aligned}$$

$$= \left[\lim_{T\to\infty} \frac{T_b}{T} \sum_k a_k^2 \right] \cdot \frac{1}{\epsilon T_b}\left(1 - \frac{\tau}{\epsilon}\right) = \frac{R_0}{\epsilon T_b}\left(1 - \frac{\tau}{\epsilon}\right). \tag{8.57}$$

Here R_0 gives the average of the square of the pulse amplitude, that is, $R_0 = \overline{a_k^2} = \lim_{T\to\infty} \frac{T_b}{T} \sum_k a_k^2$.

The autocorrelation

$$R_{\widehat{x}}(\tau) = \frac{R_0}{\epsilon T_b}\left(1 - \frac{|\tau|}{\epsilon}\right), \quad |\tau| < \epsilon$$

is thus a triangular pulse of height $\frac{R_0}{\epsilon T_b}$ and base width 2ϵ as shown in Fig. 8.44.

As τ increases $\widehat{x}(t-\tau)$ shifts right and starts to overlap the next pulse. So $R_{\widehat{x}}(\tau)$ will have a second triangular pulse centred at $\tau = T_b$, of width 2ϵ, height $R_1/\epsilon T_b$, where,

$$R_1 = \lim_{T\to\infty} (T_b/T) \sum_k a_k a_{k+1} = \overline{a_k a_{k+1}}$$

$$= \text{average of the product of } a_k \text{ and } a_{k+1}.$$

Thus $R_{\widehat{x}}(\tau)$ consists of a series of triangular pulses centred at $\tau = 0, \pm T_b, \pm 2T_b, \ldots$ and is of width 2ϵ and height $R_n/\epsilon T_b$, where

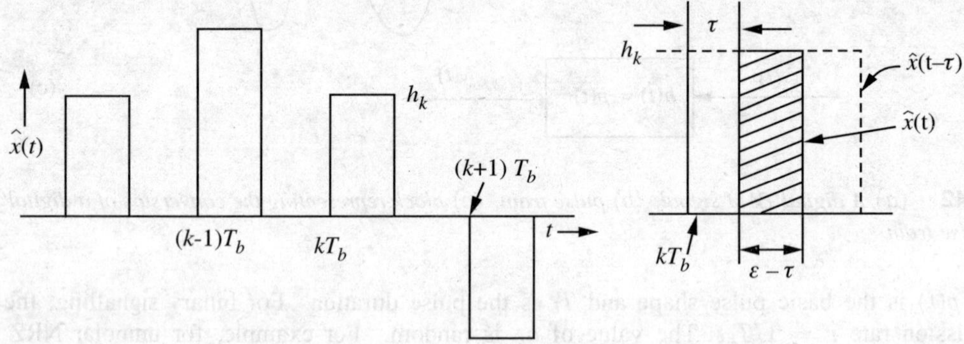

Fig. 8.43 *Rectangular pulse train of finite width ϵ, height being h_k at the kth instant.*

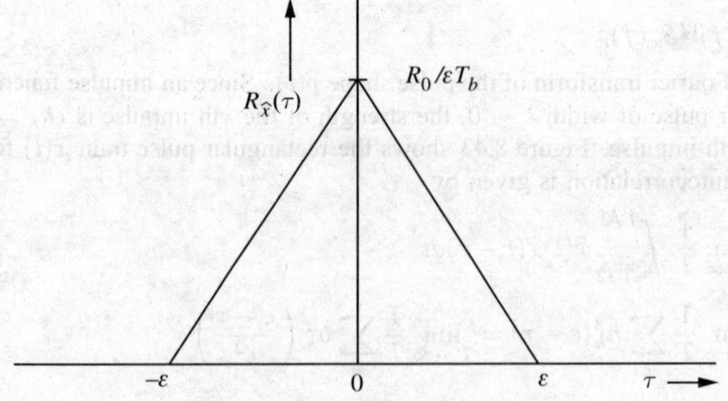

Fig. 8.44 *Autocorrelation of a rectangular pulse*

$$R_n = \lim_{T\to\infty} \frac{T_b}{T} \sum_k a_k a_{a+n} = \overline{a_k a_{k+n}}.$$

Now the derivation of $R_x(\tau)$ from $R_{\widehat{x}}(\tau)$ is straightforward. In the limit $\epsilon \to 0$, the height h_k becomes unbounded in such a way that the area is finite. The area of the nth pulse is $\frac{1}{2}2\epsilon\frac{R_n}{\epsilon T_b} = \frac{R_n}{T_b}$. So, $R_x(\tau)$ has the shape of a series of impulses of strength R_n/T_b whose general analytic expression is given by

$$R_x(\tau) = \frac{1}{T_b} \sum_{n=-\infty}^{\infty} R_n \delta(\tau - nT_b). \tag{8.58}$$

The power spectral density of $x(t)$ is then

$$\begin{aligned} S_x(f) = \mathcal{F}[R_x(t)] &= \int_{-\infty}^{\infty} \left[\frac{1}{T_b} \sum_{n=-\infty}^{\infty} R_n \delta(t - nT_b) \right] e^{-j2\pi f t} dt \\ &= \frac{1}{T_b} \sum_{n=-\infty}^{\infty} R_n e^{-j2\pi n f T_b} = \frac{1}{T_b} R_0 + \frac{1}{T_b} \sum_{\substack{n=-\infty \\ n\neq 0}}^{\infty} R_n e^{-j2\pi n f T_b} \\ &= \frac{1}{T_b} [R_0 + 2 \sum_{n=1}^{\infty} R_n \cos(2\pi n f T_b)]. \end{aligned} \tag{8.59}$$

So the expression for the PSD of the digital signal $y(t)$ is

$$S_y(f) = \frac{|P(f)|^2}{T_b} \left[R_0 + 2 \sum_{n=1}^{\infty} R_n \cos(2\pi n f T_b) \right] \tag{8.60}$$

Using above equation, we will now evaluate the PSDs of various line codes of Fig. 8.41.

Unipolar NRZ

Here the possible levels for the a's are $a_k = +A$ for binary 1 and $a_k = 0$ for binary 0; a_k is equally likely to be 1 or 0. Out of the N pulses, $N/2$ pulses are for $a_k = A$ and $N/2$ pulses for $a_k = 0$. By (8.57) $R_0 = \lim_{T\to\infty} \frac{T_b}{T} \sum_k a_k^2$. As there are N bits, $T/T_b = N$ and as $T \to \infty$, $N \to \infty$. Thus,

$$\begin{aligned} R_0 &= \lim_{N\to\infty} \frac{1}{N} \sum_k a_k^2 \\ &= \lim_{N\to\infty} \frac{1}{N} \left[\frac{N}{2} \times A^2 + \frac{N}{2} \times 0 \right] = \frac{A^2}{2}. \end{aligned}$$

The product values $a_k a_{k+n}$ are $A \times A$, $A \times 0$, $0 \times A$, and 0×0, that is, $A^2, 0, 0, 0$. They all occur with a probability of 1/4. Thus,

$$R_n = \lim_{N\to\infty} \frac{1}{N} \sum_k a_k a_{k+n}$$

$$= \lim_{N\to\infty} \frac{1}{N}\left[\frac{N}{4} \times A^2 + 3.\frac{N}{4} \times 0\right] = \frac{A^2}{4}.$$

So, $S_x(f) = \dfrac{1}{T_b}\left[\dfrac{A^2}{4} + \dfrac{A^2}{4}\displaystyle\sum_{n=-\infty}^{\infty} e^{-j2\pi n f T_b}\right]$, $n = 0$ **included in the summation.**

But, $\displaystyle\sum_{n=-\infty}^{\infty} e^{-j2\pi n f T_b} = \frac{2\pi}{T_b}\sum_n \delta(f - n/T_b)$.

Thus, $$S_x(f) = \frac{A^2}{4T_b}\left[1 + \frac{\pi}{T_b}\sum_n \delta\left(f - \frac{n}{T_b}\right)\right]. \tag{8.61}$$

The PSD for unipolar NRZ line code is given by

$$P(f)_{\text{UPNRZ}} = \int_0^{T_b} e^{-j2\pi f t}dt = \frac{1}{-j2\pi f}\left[e^{-j2\pi f t}\right]_0^{T_b}$$

$$= \int_0^{T_b} e^{-j2\pi f t}dt = \frac{1}{-j2\pi f}\left[e^{-j2\pi f t}\right]_0^{T_b}$$

$$= \frac{1}{-j2\pi f}[e^{-j2\pi f T_b} - 1]$$

$$= \frac{1}{-j2\pi f}\left(e^{-j\pi f T_b} - e^{j\pi f T_b}\right)e^{-j\pi f T_b}$$

$$= \frac{1}{\pi f}\sin(\pi f T_b)e^{-j\pi f T_b} = T_b\frac{\sin(\pi f T_b}{\pi f T_b}e^{-j\pi f T_b}.$$

Thus, $$|P(f)_{\text{UPNRZ}}|^2 = T_b^2\left(\frac{\sin \pi f T_b}{\pi f T_b}\right)^2. \tag{8.62}$$

So, $S_y(f) = \dfrac{A^2 T_b}{4}\left(\dfrac{\sin \pi f T_b}{\pi f T_b}\right)^2\left[1 + \dfrac{\pi}{T_b}\delta\left(f - \dfrac{n}{T_b}\right)\right]$ and since $\left(\dfrac{\sin \pi f T_b}{\pi f T_b}\right) = 0$ at $f = n/T_b$, this becomes

$$S_y(f) = \frac{A^2 T_b}{4}\left(\frac{\sin \pi f T_b}{\pi f T_b}\right)^2\left[1 + \frac{\pi}{T_b}\delta(f)\right]. \tag{8.63}$$

If the amplitude A is selected such that the normalised average power is unity, then $A = \sqrt{2}$. Figure 8.45 depicts the plot of this $S_y(f)$ where $R = \frac{1}{T_b}$ is the data bit rate. The spectrum is not approaching zero near dc and hence dc- coupled circuits are needed. Note the strength of the impulse at $f = 0$ is $\pi/4 = 0.78$.

Unipolar RZ

The autocorrelation for unipolar data has already been calculated as $R_0 = A^2/2$ and $R_n = A^2/4$. But for RZ scheme (halfwidth rectangular pulse), the pulse duration is $T_b/2$ instead of T_b. Thus for RZ signalling,

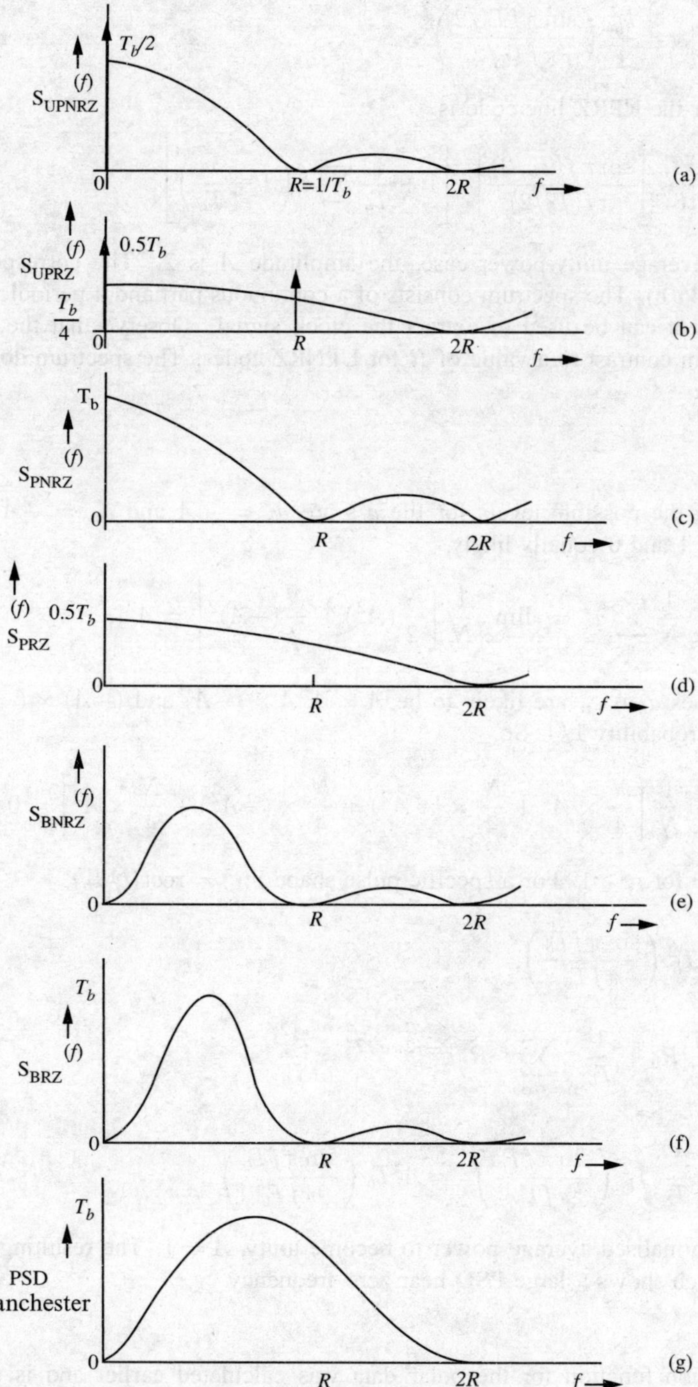

Fig. 8.45 *Power spectra of line codes: (a) unipolar NRZ signal, (b) unipolar RZ signal, (c) polar NRZ signal, (d) polar RZ signal, (e) bipolar NRZ signal, (f) bipolar RZ signal, (g) Manchester coded signal.*

$$|P(f)_{\text{UPRZ}}|^2 = \frac{T_b^2}{4}\left(\frac{\sin \pi f T_b/2}{\pi f T_b/2}\right)^2. \tag{8.64}$$

Then the PSD for the UPRZ line code is

$$S_y(f) = \frac{A^2 T_b}{16}\left[\frac{\sin \pi f(T_b/2)}{\pi f(T_b/2)}\right]^2 \left[1 + \frac{\pi}{T_b}\sum_n \delta\left(f - \frac{n}{T_b}\right)\right]. \tag{8.65}$$

For normalised average unity-power case, the amplitude A is 2. The corresponding spectrum is plotted in Fig. 8.45(b). The spectrum consists of a continuous part and a periodic discrete part. This periodic component can be used to extract the clock signal. Observe that the first null occurs at $f = 2/T_b = 2R$ in contrast to a value of R for UPNRZ codes. The spectrum does not pass through zero for $f = 0$.

Polar NRZ

In this signalling the possible levels for the a's are $a_k = +A$ and $a_k = -A$ so that $a_k^2 = A^2$. Assuming binary 1 and 0 equally likely,

$$R_0 = \lim_{N\to\infty} \frac{1}{N}\sum_k a_k^2 = \lim_{N\to\infty} \frac{1}{N}\left[\frac{N}{2}(A^2) + \frac{N}{2}(-A)^2\right] = A^2.$$

The product values $a_k a_{k+n}$ are likely to be $A \times A$, $A \times (-A)$ and $(-A) \times (-A)$, i.e., A^2, $-A^2$, $-A^2$, A^2, with probability 1/4. So,

$$R_1 = \lim_{N\to\infty} \frac{1}{N}\left[\frac{N}{4}\times A^2 + \frac{N}{4}\times(-A^2) + \frac{N}{4}\times(-A^2) + \frac{N}{4}\times A^2\right] = 0$$

and also, $R_n = 0$ for $n \geq 1$. For a specific pulse shape $p(t) = \text{rect}(t/T_b)$,

$$|P(f)|^2 = T_b^2\left(\frac{\sin \pi f T_b}{\pi f T_b}\right)^2.$$

Now, $S_x(f) = \dfrac{1}{T_b}R_0 + \dfrac{1}{T_b}\displaystyle\sum_{\substack{n=-\infty \\ n\neq 0}}^{\infty} R_n e^{-j2\pi n f T_b} = \dfrac{A^2}{T_b}.$

So, $$S_y(f) = \frac{A^2}{T_b}T_b^2\left(\frac{\sin \pi f T_b}{\pi f T_b}\right)^2 = A^2 T_b\left(\frac{\sin \pi f T_b}{\pi f T_b}\right)^2. \tag{8.66}$$

Again, the for normalised average power to become unity, $A = 1$. The resulting PSD is sketched in Fig. 8.45(c), which shows a large PSD near zero frequency.

Polar RZ

The autocorrelation function for the polar data was calculated earlier and is given as $R_0 = A^2$ and $R_n = 0$. For RZ signalling the pulse width is $T_b/2$ instead of T_b. Hence the pulse spectrum squared is

$$|P(f)|^2 = \frac{T_b^2}{4}\left(\frac{\sin \pi f T_b/2}{\pi f T_b/2}\right)^2.$$

Thus, $$S_y(f) = \frac{A^2 T_b}{4}\left(\frac{\sin \pi f T_b/2}{\pi f T_b/2}\right)^2 \tag{8.67}$$

Here $A = \sqrt{2}$ makes the average normalised power unity. The PSD for this unity-power case is shown in Fig. 8.45(d). We find that polar RZ line codes have large PSD near dc.

Bipolar NRZ

Here the possible values of $a_k' s$ are $a_k = \pm A$ and $a_k = 0$. Binary 1s are transmitted by alternating $+A$ and $-A$ and binary 0 is transmitted by a 'no pulse' ($a_k = 0$). With consecutive pulses alternating, we can avoid here the dc wander and thus cause a dc null in the PSD. The product terms $a_k a_k$ are A^2 and 0 with respective probabilities of $1/2$. Thus,

$$R_0 = \lim_{N\to\infty} \frac{1}{N}\sum_k a_k^2$$

$$= \lim_{N\to\infty} \frac{1}{N}\left[\frac{N}{2}(A^2) + \frac{N}{2}(0)\right] = \frac{A^2}{2}.$$

For the products $a_k a_{k+1}$, we find that there are four possible, equally likely sequence of two bits: 11, 10, 01 and 00 with possible products $-A^2, 0, 0,$ and 0. Out of the N bit sequences, $\frac{3N}{4}$ combinations have $a_k a_{k+1} = 0$ and $\frac{N}{4}$ combination have nonzero $a_k a_{k+1}$. The bit sequence 11 can only be encoded by two consecutive pulses of opposite polarities for which $a_k a_{k+1} = -A^2$. Consequently,

$$R_1 = \lim_{N\to\infty} \frac{1}{N}\left[\frac{N}{4}(-A^2) + \frac{3N}{4}(0)\right] = -\frac{A^2}{4}.$$

The products $a_k a_{k+2}$ have eight possible configuration 111, 101, 110, 100, 011, 010, 001, and 000. The last six combinations have either the first bit or the last bit or both bits zero, giving $a_k a_{k+2} = 0$. For the combination 111, $a_k a_{k+2} = A^2$ and for the combination 101, $a_k a_{k+2} = -A^2$. Thus, on the average, $a_k a_{k+2} = A^2$ for $\frac{N}{8}$ terms; it equals $-A^2$ for $\frac{N}{8}$ terms; and it equals zero for $\frac{3N}{4}$ term. So,

$$R_2 = \lim_{N\to\infty} \frac{1}{N}\left[\frac{N}{8}(A^2) + \frac{N}{8}\left(-A^2\right) + \frac{3N}{4}(0)\right] = 0.$$

In general, $R_n = 0$ for $n > 1$.

So, by (8.60) we have,

$$S_y(f) = \frac{|P(f)|^2}{T_b}(R_0 + 2R_1 \cos 2\pi f T_b)$$

$$= \frac{A^2|P(f)|^2}{2T_b}(1 - \cos 2\pi f T_b) = \frac{A^2|P(f)|^2}{T_b}\sin^2(\pi f T_b). \tag{8.68}$$

Now, $|P(f)|^2 = T_b^2\left(\frac{\sin \pi f T_b}{\pi f T_b}\right)^2.$

So, $$S_y(f) = A^2 T_b\left(\frac{\sin \pi f T_b}{\pi f T_b}\right)^2 \sin^2(\pi f T_b). \tag{8.69}$$

The PSD for bipolar signalling is depicted in Fig. 8.45 (e). We note that $S_y(f) = 0$ for $f = 0$ regardless of $P(f)$ and hence bipolar signalling has a spectral null at dc, desirable for ac coupling. Also, the bandwidth is not excessive. Bipolar signalling also has single error correction capability since an error will violate the alternate pulse rule.

Bipolar RZ

The derivation of PSD in the bipolar RZ code is straightforward. For the half-width pulse we replace T_b by $T_b/2$ in the expression for $|P(f)|^2$. In doing so,

$$S_y(f) = \frac{A^2 T_b}{4}\left(\frac{\sin \pi f T_b/2}{\pi f T_b/2}\right)^2 \sin^2(\pi f T_b) \tag{8.70}$$

and is sketched in Fig. 8.45 (f). For unity average normalised power, $A = 2$.

Manchester coding

Manchester code uses the pulse shape as shown in Fig. 8.46. The code has zero dc level on a bit-by-bit basis. The pulse shape has the general analytic expression

$$p(t) = \Pi\left(\frac{t - T_b/4}{T_b/2}\right) - \Pi\left(\frac{t - 3T_b/4}{T_b/2}\right).$$

The resulting pulse spectrum can be obtained by taking its Fourier transform. The result is

$$P(f) = \frac{j\pi f T_b^2}{2} e^{-j\pi f T_b}\left(\frac{\sin \pi f T_b/2}{\pi f T_b/2}\right)^2. \tag{8.71}$$

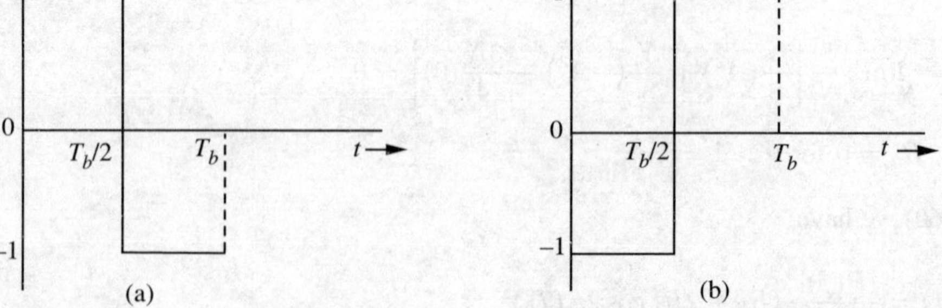

Fig. 8.46 *Manchester code. (a) Binary 1 is represented by a positive pulse of duration $T_b/2$ followed by a negative pulse of duration $T_b/2$. (b) Binary 0 is represented by a negative pulse followed by a position pulse, each of duration $T_b/2$.*

For Manchester coding, $a_k^2 = A^2$ and so $R_0 = A^2$. The product $a_k a_{k+1}$ may be $+A^2$ or $-A^2$ so that $R_1 = \lim_{N\to\infty} \frac{1}{N}\left[\frac{N}{2}A^2 + \frac{N}{2}(-A^2)\right] = 0$. Similarly, R_2, R_3, etc. are all zero. Thus,

$$S_y(f) = \frac{|P(f)|^2}{T_b}\left[R_0 + 2\sum_n R_n \cos n\omega T_b\right] \qquad = \frac{|P(f)|^2}{T_b}A^2$$

$$= A^2 T_b \left(\frac{\sin \pi f T_b/2}{\pi f T_b/2}\right)^2 \sin^2(\pi f T_b/2). \tag{8.72}$$

The average power is normalised to unity for $A = 1$. The spectrum is shown in Fig. 8.45 (g).

8.10 Delta Modulation

Delta modulation (DM) is basically a single-bit PCM. Instead of sending each sample value by its PCM code, a DM system sends the difference between a sample value and the previous sample value. The algorithm is as follows. If the current sample value is larger than the previous one, a logic 1 is transmitted; logic 0 is transmitted when the reverse is true. In DM the incoming message signal is oversampled at a rate much higher than the Nyquist rate so that consecutive samples are highly 'correlated'. This ensures that the difference between the consecutive sample values is less than the sample itself and hence the 'dynamic range' of the samples to be encoded is reduced. At the Nyquist rate of sampling the sample difference happens to be the peak-to-peak signal range (this occurs when a particular sample point falls at the signal maximum, the next sample point will be at the signal minimum). Due to oversampling, the sample value changes by a small amount from the previous sample and this very often allows a reasonable guess of the next sample value. In other words, one can 'predict' the sample with some tolerance (called the prediction error). The difference between the message input and its prediction is quantised to two levels: $\pm\Delta$. This is the prediction error that is transmitted. A similar prediction circuit at the receiver reconstructs the waveform by recombining the incoming prediction error with its own predicted value.

To describe delta modulation, let $s(t)$ be the message waveform sampled at the rate $f_s = \frac{1}{T_s}$. Let us designate by $s(k)$ the sample value at the instant $t = kT_s$, that is, $s(k) = s(kT_s)$, where $k = 0, \pm 1, \pm 2, \ldots$. Since the sample values are highly correlated, a reasonable prediction of $s(k)$ would be the previous sample value $s(k-1)$. Thus the prediction of $s(k)$ is $\widehat{s}_q(k)$, given as

$$\widehat{s}_q(k) = s(k-1). \tag{8.73}$$

The prediction error $e(k)$ defined as the difference between the actual and predicted values is given by

$$e(k) = s(k) - \widehat{s}_q(k) = s(k) - s(k-1). \tag{8.74}$$

This error may be positive or negative. If the predicted value falls below the actual value, we increase it by Δ. If the reverse be true, it is decreased by $-\Delta$. We may then formulate the DM signal as

$$e_q(k) = [\text{sgn}\{e(k)\}]\Delta, \tag{8.75}$$

where $e_q(k)$ is the quantised prediction error. Figure 8.47 illustrates $s(t)$, $s_q(t)$, and $e_q(t)$.

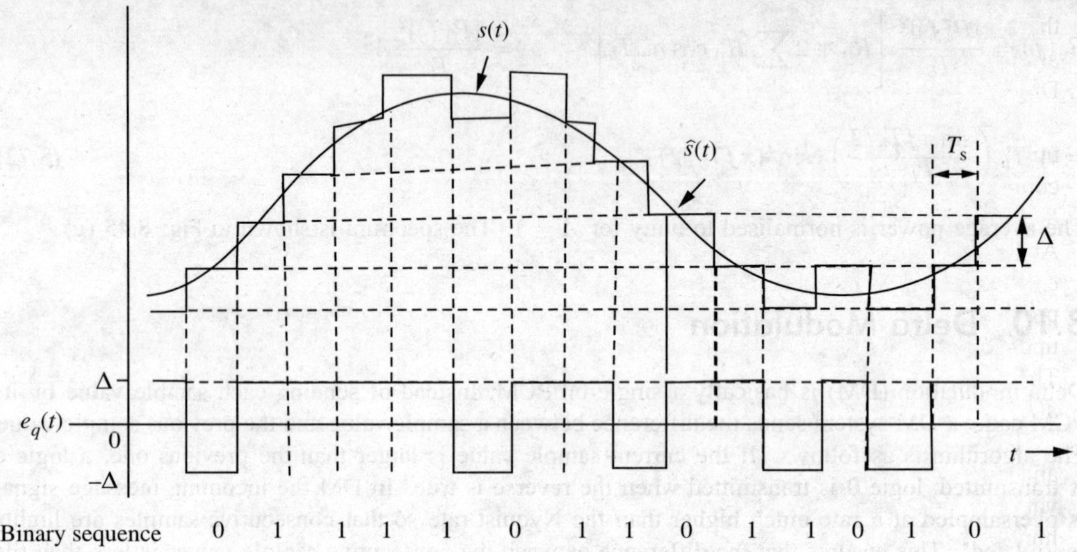

Fig. 8.47 *Delta modulation. It illustrates the signal $s(t)$, quantised signal $\hat{s}_q(t)$ and quantised prediction $e_q(t)$.*

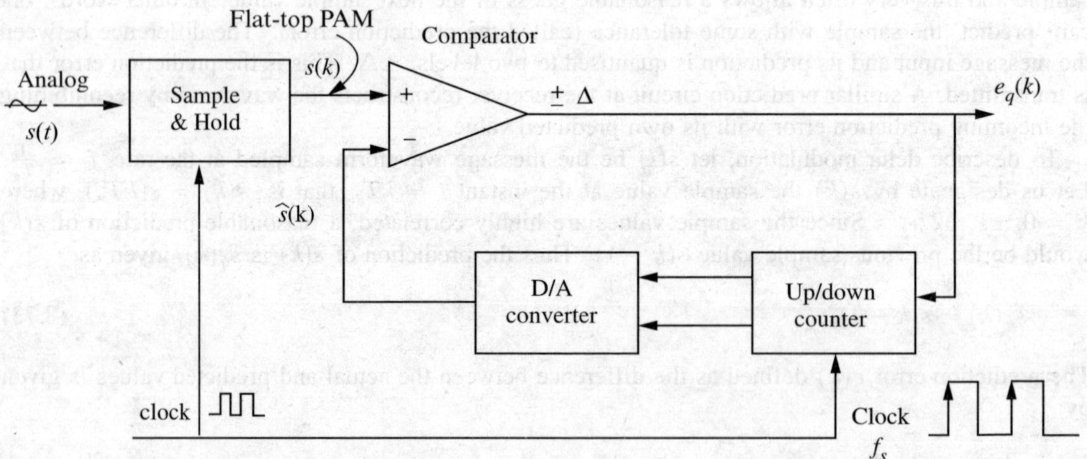

Fig. 8.48 *Delta modulator transmitter*

Figure 8.48 shows a DM transmitter that consists of a comparator, an up/down counter and a digital-to-analog converter. The comparator serves as a one-bit quantiser with two output values $+\Delta$ and $-\Delta$ following (8.75). The up/down counter is clocked at the sampling rate f_s.The D/A converter and the up/down counter in the feedback path produce the prediction $\widehat{s}(k)$ from $e_q(k)$. It is apparent that the rate of information transmission is simply the sampling rate f_s. Since the sampled input $s(k)$ is encoded as a single pulse of height $+\Delta$ or$-\Delta$, the DM is also called a one-bit PCM. The DM signal is a polar signal.

The DM transmitter starts with an input sample $s(k)$ and an arbitrary initial prediction $\widehat{s}(k)$. The up/down counter is initialised to zero. This is updated at the active edge of the clock pulses. It starts counting up or down depending on whether the prediction error $e_q(k)$ is positive or negative.

Initially, when the counter is reset to zero, the D/A converter outputs a 0 volt to the comparator. At the starting, $k = 0$ for which $\widehat{s}(0) = 0$ so that $s(0) > \widehat{s}(0)$ for a positive sample input. Then $e_q(0) = +\Delta$. At the next clock pulse the counter is incremented to a count 1. The D/A converter produces a voltage $\widehat{s}(1)$ corresponding to the magnitude of the minimum step size Δ. Continual updating at each clock edge causes $\widehat{s}(k)$ to increase by Δ until $\widehat{s}(k)$ exceeds $s(k)$ and $e_q(k) = -\Delta$. The counter then starts counting down. The process repeats so that the updated prediction $\widehat{s}(k) = s(k-1) + e_q(k)$ remains within $\pm\Delta$ of the input sample $s(k)$.

The delta modulation is subject to two types of quantisation errors, namely, the 'granular noise' and the 'slope-overload distortion'. If the message signal $s(t)$ has a relatively flat segment (slowly varying with time), the prediction $\widehat{s}(t)$ swings up and down about $s(t)$ as shown in Fig. 8.49. This hunting behaviour of $\widehat{s}(t)$ is the source of granular noise $e(t)$, which is bounded by $|e(t)| \leq \Delta$. It is noted from Fig. 8.49 that at the start-up, where $\widehat{s}(t)$ is a poor approximation of $s(t)$, there are few sampling periods that have been elapsed. A reasonable approximation of $s(t)$ may be obtained at the receiver by an intelligent choice of the step size Δ and the sampling interval T_s.

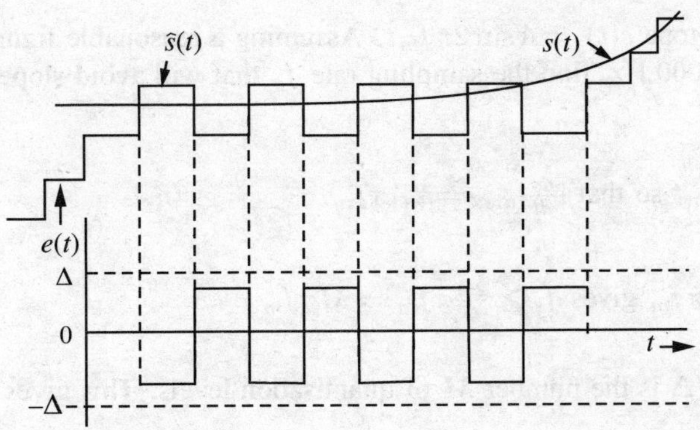

Fig. 8.49 *Granular noise*

The DM system that maintains constants values of the parameters Δ and T_s is referred to as 'linear DM' or LDM. The slope of the staircase approximation in LDM is Δ/T_s. To follow a faster change in $s(t)$, either the sampling rate $\frac{1}{T_s}$ or the step size Δ will need to be increased. But an increased Δ yields more quantisation noise and increased sampling rate increases the transmission bandwidth of the DM waveform.

When the message signal $s(t)$ changes too rapidly, $\widehat{s}(t)$ cannot keep up with it and the error $s(t) - \widehat{s}(t)$ becomes progressively larger as shown in Fig. 8.50. This inaccuracy is described as 'slope-overload distortion'. For the staircase approximation $\widehat{s}(t)$ to follow the fastest change in $s(t)$, we require that

$$\frac{\Delta}{T_s} \geq |\frac{ds}{dt}|_{\max}. \tag{8.76}$$

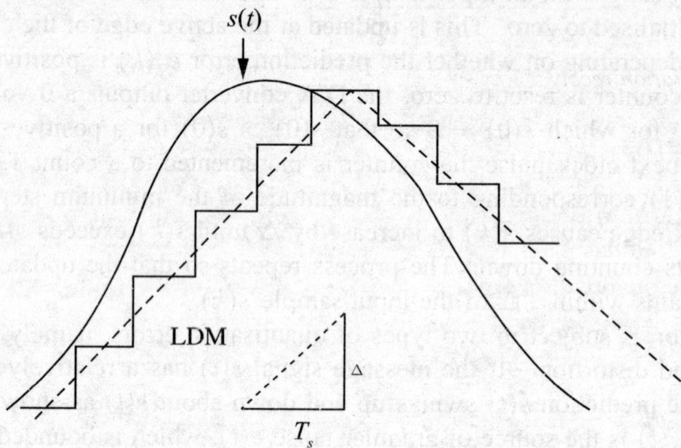

Fig. 8.50 *Slope-overload distortion*

Example 8.9

Consider a modulating tone $s(t) = A \sin 2\pi f_m t$. Assuming a reasonable figure of $M = 256$ quantisation level and f_m= 1000 Hz, find the sampling rate f_s that will avoid slope-overload distortion.

Solution

$$\frac{ds}{dt} = A\, 2\pi f_m \cos 2\pi f_m t \text{ so that } |\tfrac{ds}{dt}|_{\max} = 2A\pi f_m.$$

$$\frac{\Delta}{T_s} = \Delta f_s \geq 2A\pi f_m \text{ gives } f_s \geq \tfrac{2A}{\Delta}\pi f_m = M\pi f_m$$

where the quantity $2A/\Delta$ is the number M of quantisation levels. This gives

$$f_s \geq 256\pi \times 1000 \approx 800 \text{ kHz}.$$

Delta modulation receiver

Figure 8.51 shows the block diagram of a delta modulation receiver. As the logic 1s and 0s are received, the up/down counter is incremented or decremented at the active clock edge. The output of the DAC in the docoder is the PAM signal. The lowpass filtering reconstructs the analog signal.

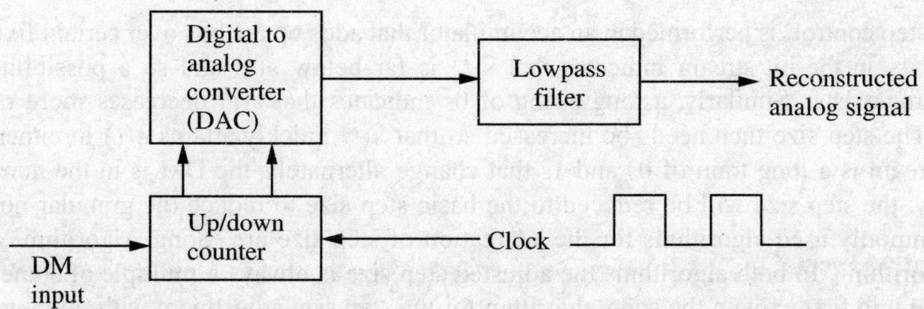

Fig. 8.51 *Delta modulation receiver*

Noise in DM systems

The DM systems suffer from two classes of noise; one, the granular noise and the other, the slope overload noise. The granular kind of noise is assumed uniformly distributed over the frequency interval $|f| \le f_s$. Since quantising noise power is $\overline{v_n^2} = \frac{\Delta^2}{12}$, the power spectral density (PSD) of this noise is $G_n(f) = \overline{v_n^2}\frac{2}{f_s} = \frac{\Delta^2}{6f_s}$. So, the granular noise power in the band of analog signal $|f| < f_M$ is

$$N = \int_{-f_M}^{f_M} G_n(f)\, df = \frac{\Delta^2 f_M}{3f_s} \tag{8.76a}$$

Since we know that $f_s \ge \frac{2\pi f_M A}{\Delta}$, and using equality sign,

$$N = \left(\frac{2\pi f_M A}{f_s}\right)^2 \cdot \frac{f_M}{3f_s} = \frac{4\pi^2 A^2 \cdot f_m^2 f_M}{3f_s^3}$$

Also, the signal power is $S = \langle s^2(t)\rangle = A^2/2$. The signal-to-quantisation noise ratio for the DM system with a sinusoid input is

$$\left(\frac{S}{N}\right)_{\text{out}} = \frac{3}{8\pi^2} \cdot \frac{f_s^3}{f_m^2 f_M} \tag{8.76b}$$

This is restrictive only for the sinusoidal type of input.

8.10.1 Adaptive delta modulation

The LDM scheme suffers from granular noise and slope overload distortion. If the successive prediction errors $e_q(k)$ are of opposite polarity, the DM system is on a hunting mode. The reconstructed wave at the receiver will be reasonably close to $s(t)$ if Δ and T_s are made sufficiently small. On the contrary, if successive errors are of the same polarity, the delta modulator is in its slope overload mode. In this case it may be advantageous to increase the step size. Thus reducing the step size in an attempt to suppress granular noise will enhance slope-overload distortion.

An adaptive delta modulation (ADM) system is a DM system that includes a specially designed circuitry to provide variable step sizes. This reduces the slope overload effects without increasing the granular noise significantly. When the message signal $s(t)$ changes rapidly, tending to produce slope overload, the step size becomes progressively larger so that $\widehat{s}(t)$ catches up with $s(t)$ in a rather short time. The 'step control' is performed in an accumulator that adds up the bits over certain fixed periods. A string of 1s in the bit stream indicates that $\widehat{s}(t)$ is far below $s(t)$ and so a possibility of slope overload error exists. Similarly, a long string of 0s indicates that $s(t)$ decreases more rapidly than does $\widehat{s}(t)$. The step size then needs be increased so that $\widehat{s}(t)$ quickly attains $s(t)$ in either direction. If the bit stream is a long train of 0s and 1s that change alternately, the DM is in the hunting mode. Accordingly, the step size will be reduced to the basic step size to reduce the granular noise.

Two commonly used algorithms for the adaptation of step size are 'Song Algorithm' and 'Space Shuttle Algorithm'. In both algorithms the adjusted step size is always a multiple of some basic step size Δ_0. We will first explain the song algorithm for the step size adjustment with reference to a step input signal $s(t)$ shown in Fig. 8.52. Up to $t = 3T_s$, $s(t)$ is constant and so $\widehat{s}(t)$ oscillates about $s(t)$ with the basic step size Δ_0 and at a frequency one-half that of the clock. At $t = 5T_s$, $s(t) > \widehat{s}(t)$ and the accumulator increases the step size by Δ_0. At $t = 6T_s$, $\widehat{s}(t)$ is again less than $s(t)$; here the total step size is $3\Delta_0$. However, at $t = 7T_s$ the predicted value $\widehat{s}(t)$ exceeds $s(t)$. The accumulator decreases the step size by Δ_0 from $4\Delta_0$. Prior to $9T_s$, $s(t) > \widehat{s}(t)$ and so at the arrival of the next clock the step size becomes $2\Delta_0$ and so on. Thus the step control algorithm can be stated as follows.

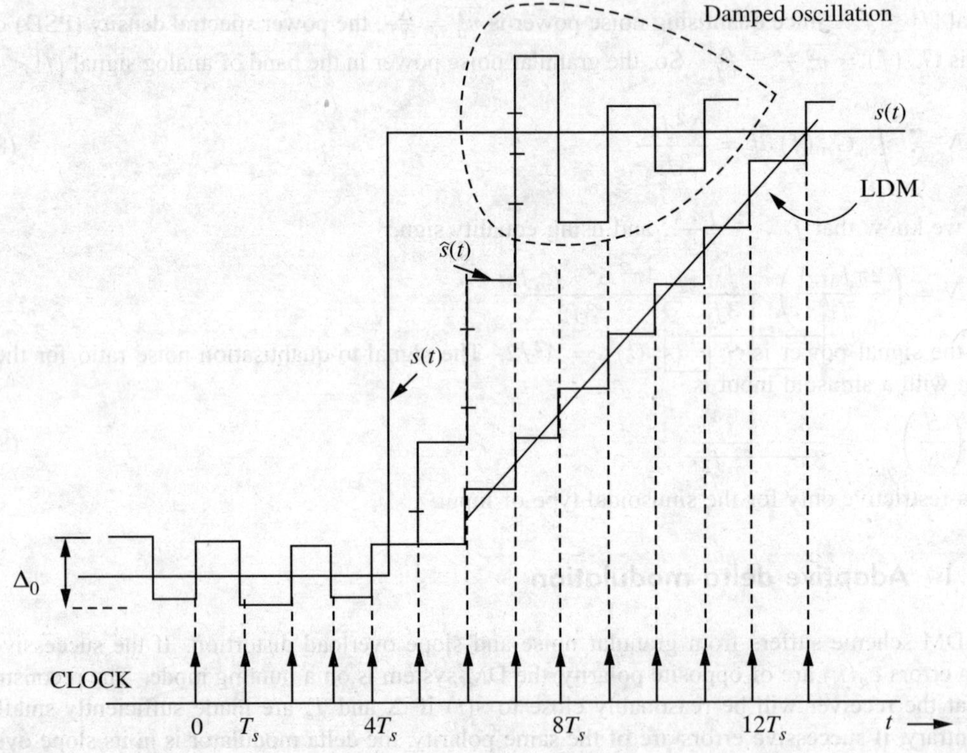

Fig. 8.52 *Illustrating song algorithm*

1. The signal $\widehat{s}(t)$ will jump in the positive direction if $s(t) > \widehat{s}(t)$ and in the negative direction if $s(t) < \widehat{s}(t)$.

2. If the jumps at two consecutive clock pulses are in the same direction, the step size will increase by Δ_0. It will decrease by Δ_0 if the jumps are in the opposite directions.

The processor thus needs the information about two jumps. Let us denote by $d(n)$ the discrepancy between $s(t)$ and $\widehat{s}(t)$ prior to the nth clock edge and set $d(n) = 1$ for $s(t) > \widehat{s}(t)$ and $d(n) = -1$ for $s(t) < \widehat{s}(t)$. The step-controller algorithm can be written as

$$\Delta(n) = |\Delta(n-1)|d(n) + \Delta_0 d(n-1), \tag{8.77}$$

where $\Delta(n)$ is the step at the nth clock. It is observed in Fig. 8.52 that $\widehat{s}(t)$ exceeds $s(t)$ in a rather short time $(= 7T_s)$ as compared to $13T_s$ in the LDM case.

The song algorithm suffers from undue damped oscillations whenever two consecutive bits disagree. This oscillation will eventually decay after some time. This problem can be overcome by modifying the algorithm slightly as in the case of the space shuttle algorithm. When the two consecutive bits are the same, the step size in the space shuttle algorithm increases by the basic size Δ_0, same as in the song algorithm. But when they disagree the step size immediately comes to the basic size Δ_0. In song algorithm the step size decreases to Δ_0 at the rate Δ_0 at every active clock edge. Figure 8.53 illustrates the space shuttle algorithm for the step input function.

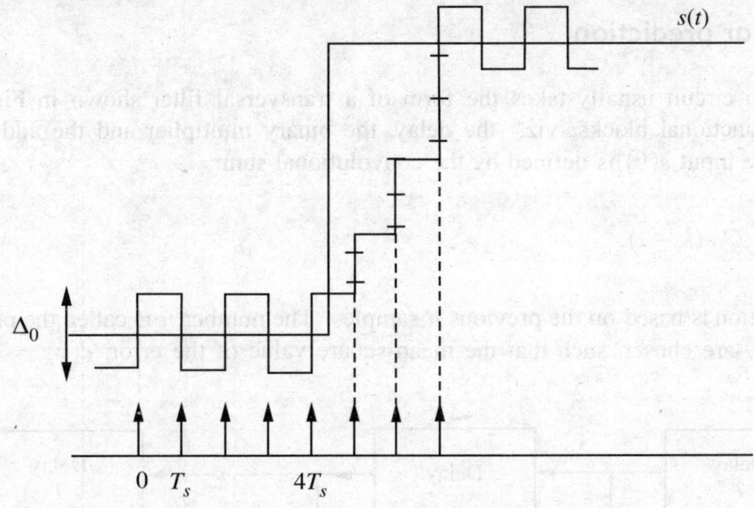

Fig. 8.53 *Illustrating space shuttle algorithm for step size controller*

8.11 Differential PCM

A differential PCM (called DPCM) is designed specifically to reduce the signalling rate of a typical PCM-encoded waveform. In conventional PCM there are often successive samples taken in which there is little difference between the amplitudes of the consecutive samples. Thus identical PCM codes are transmitted. A DPCM scheme removes such transmission of redundant codes. With DPCM, the difference in amplitudes of successive samples is encoded and transmitted rather than the actual samples. As the range of the sample difference is lesser than the range of individual samples, relatively lesser bits are needed for encoding by DPCM than PCM system.

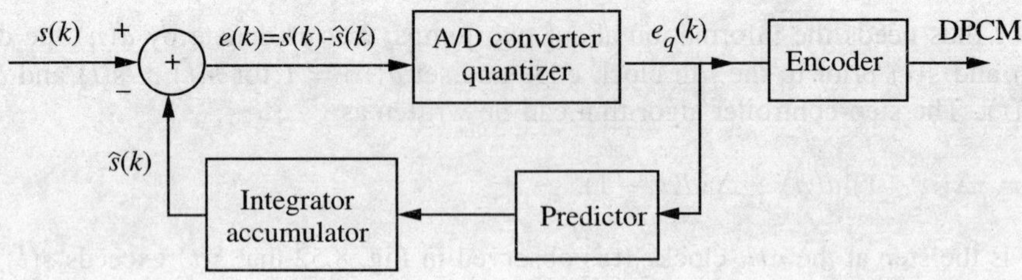

Fig. 8.54 *DPCM transmitter*

Figure 8.54 shows a simplified block diagram of a DPCM transmitter. The analog input signal $s(t)$ is bandlimited and sampled to yield the PAM signal $s(k)$. This is then compared with the preceding accumulated signal $\widehat{s}(k)$. The output of the subtractor is $e(k)$, which is the difference between two signals. This difference is PCM encoded and quantised in ADC/quantiser and is then transmitted. The ADC operation here is the same as in PCM but typically uses less number of bits per sample.

8.11.1 Linear prediction

Linear prediction circuit usually takes the form of a transversal filter shown in Fig. 8.55. This involves three functional blocks, viz. the delay, the binary multiplier and the adder. The linear predication of the input $s(k)$ is defined by the convolutional sum

$$\widehat{s}(k) = \sum_{i=1}^{n} C_i s(k-i), \tag{8.78}$$

where the prediction is based on the previous n samples. The number n is called the prediction order. The tap gains C_i are chosen such that the mean-square value of the error, $e(k) = s(k) - \widehat{s}(k)$, is minimum.

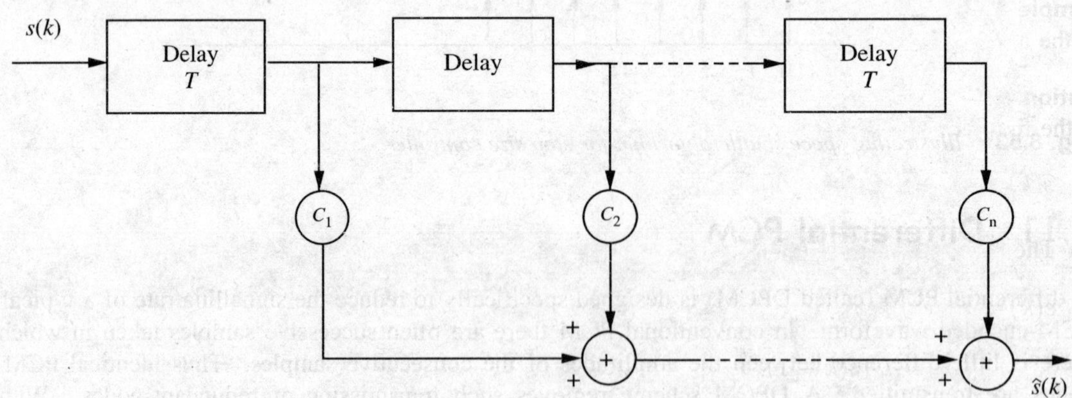

Fig. 8.55 *Linear prediction circuit*

Now, the mean square error,

$$\text{mse} = E[e^2(k)]$$

$$= E\left[\left\{s(k) - \sum_{i=1}^{n} C_i s(k-i)\right\}^2\right]$$

$$= E[s^2(k)] - 2E[s(k)\sum_{i=1}^{n} C_i s(k-i)] + E[\{\sum_{i=1}^{n} C_i s(k-i)\}^2]$$

$$= E[s^2(k)] - 2\sum_{i=1}^{n} C_i E[s(k)s(k-i)] + \sum_{i=1}^{n}\sum_{j=1}^{n} C_i C_j E[s(k-i)s(k-j)]. \tag{8.79}$$

If one assumes that $s(t)$ belongs to the stationary process $S(t)$ having zero mean and variance σ^2, then $\sigma^2 = E[s^2(k)] - (E[s(k)])^2 = E[s^2(k)]$, as $E[s(k)] = 0$.

The autocorrelation function is

$$R(i) = E[s(k)s(k-i)].$$

Thus, (8.79) can be rewritten as

$$\text{mse} = \sigma^2 - 2\sum_{i=1}^{n} C_i R(i) + \sum_{i=1}^{n}\sum_{j=1}^{n} C_i C_j R(i-j). \tag{8.80}$$

To optimise the performance, one should take the derivative of the mse with respect to the tap gain and set it equal to zero. The following example illustrates one such method.

Example 8.10

For the a first-order predictor, find the tap gain that makes $\widehat{s}(t)$ close to $s(t)$.

Solution

For the first-order predictor, we consider only the previous one sample. We may thus write (8.78) as

$$\widehat{s}(k) = C_1 s(k-1).$$

$s(k)$ → T_s → C_1 → $s(k) = C_1\, s(k-1)$

The prediction error is then

$$e(k) = s(k) - \widehat{s}(k)$$

$$= s(k) - C_1 s(k-1).$$

The mean-square error is

$$\text{mse} = \overline{e^2(k)} = E[\{s(k) - C_1 s(k-1)\}]^2$$

$$= E[s^2(k)] - 2C_1E[s(k)s(k-1)] + C_1^2E[s^2(k-1)]$$

$$= (1 + C_1^2)R(0) - 2C_1R(1).$$

Here $R(0) = E[s^2(k)] = E[s^2(k-1)]$ is the autocorrelation function. The quantity $R(1)$ refers to the autocorrelation of the process $s(t)$ for a lag of T_s seconds. Differentiating the expression for mse with respect to C_1 and equating it to zero, we have

$$C_1R(0) = R(1) \Rightarrow C_1 = R(1)/R(0).$$

Thus, by taking the predictor multiplier as $R(1)/R(0)$, the prediction $\widehat{s}(t)$ can be made close to $s(t)$.

8.12 Digital Multiplexing

In Section 8.4 we explained the concept of time-division multiplexing of a number of signals from different analog sources. All such signals were sampled sequentially at a common sampling rate. The advancements in digital electronics revolution has made possible the multiplexing of digital signals at different bit rates.

Digital signals may originate from audio, video or graphic information sources. The data bits from such different sources are merged using a 'bit-by-bit' interleaving procedure into one signal and applied to a high-speed transmission line. The data bits are successfully separated (or demultiplexed) at the receiver and delivered into respective destinations.

Two categories of multiplexing patterns have been adopted for digital data transmission. These are AT&T hierarchy and CCIT hierarchy. Both hierarchies basically time-division multiplex low-rate data with a much higher-rate data stream. The general feature of the hierarchies is that it starts at 64 kbps standard voice PCM unit. This incoming bit stream is referred to as 'digital signal level zero (DS0)'. Figure 8.56 shows the digital multiplexing hierarchy. The AT&T digital TDM hierarchy is described here. The DS0 single channel uses an 8 kHz sample rate and an 8-bit PCM code. The PCM line speed is

$$\text{Line speed} = \frac{8000 \text{ samples}}{\text{second}} \times \frac{8 \text{ bits}}{\text{sample}}$$

$$= 64{,}000 \frac{\text{bits}}{\text{second}} = 64 \text{ kbps}.$$

The first level combines twenty-four DSO input bit stream and produces the output signal referred to as 'digital signal level 1 (DS1)'. The DS1 is at a bit rate of 1.544 Mbps and is delivered on the second level via T_1 system. The second level hierarchy mixes four DS1 bit streams and produces 'digital signal level 2 (DS2)' at the bit rate of 6.312 Mbps. This is also called the T_2 signal. The third level multiplexer combines seven DS2 bit stream to produce 'digital signal level 3 (DS3)' at 44.736 Mbps. Six such DS3 bit streams (also called T_3 lines) are multiplexed at the fourth level that generates 'digital signal level 4 (DS4)' or T_4 lines at 274.176 Mbps. The parameters for AT&T multiplexing hierarchy are tabulated in Table 8.3. We note that the bit rate of the output signal at any one of these four levels exceeds the sum of the incoming bit rate. This allows for control bits and stuff bits needed for synchronisation and also provides a steady output rate.

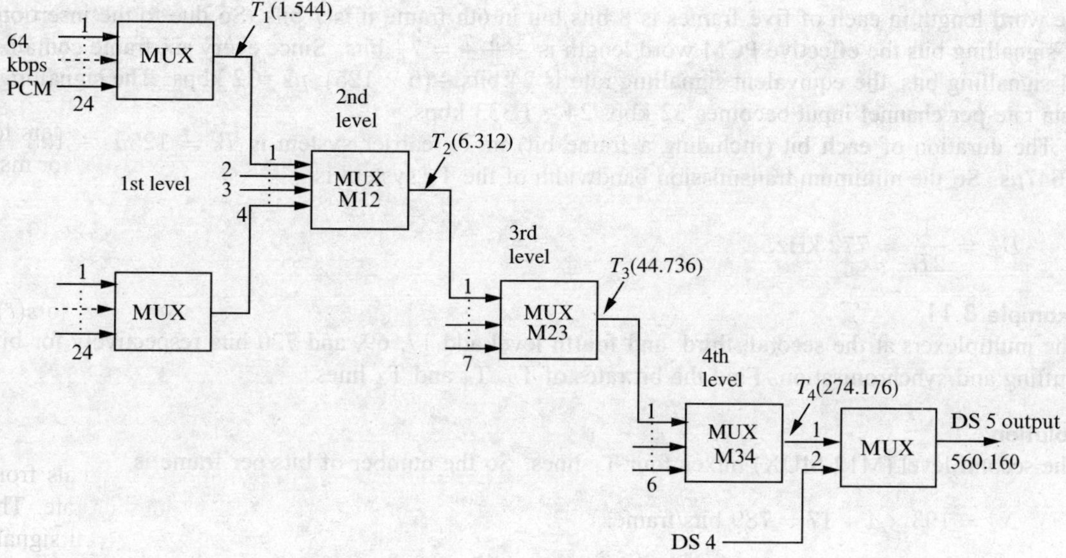

Fig. 8.56 *Digital multiplexing hierarchy (AT&T standard)*

Table 8.3 AT&T multiplexing

Level	No. of input lines	Output bit rate, Mbps	No. of channels
1st level	24	1.544	24
2nd level	4	6.312	96
3rd level	7	44.736	672
4th level	6	274.176	4032

The control bits are the supervisory bits required to send some signalling information. In telephone systems these additional bits provide information like dial pulses, telephone hook on–off signals, etc. Small variations of input data rates are adjusted in the multiplexer by a technique known as 'bit stuffing'. The incoming digital data streams are stuffed with sufficient number of additional bits in order to increase the bit rate to match the master clock. The demultiplexer identifies the stuffed bits and removes them from the signal. We will consider here the first level synchronous multiplexer known as the T_1 system.

The T_1 system

The first level of multiplexing hierarchy recombines 24 voice band channels sequentially sampled at 8 kHz and uses μ-255 companding. The resulting TDM–PAM signal is encoded into an 8-bit PCM with interleaved words. Each voice-band channel has a bandwidth of 300 Hz to 3000 Hz. Each frame contains one sample from each of the 24 channels plus a single framing bit, a total of 8000 samples per second, i.e., the frame duration is 1/8000 second = 125 μs. So the T_1 bit rate is 193 bits/125 μs = 1,544 Mbps. The signalling information is provided as follows. In every sixth frame the least significant bit of each channel word is robbed and a signalling bit is inserted there. That is, the word length in each of five frames is 8 bits but in 6th frame it is 7 bits. So due to the insertion of signalling bits the effective PCM word length is $\frac{5\times8+7}{6} = 7\frac{5}{6}$ bits. Since every six frame contains 24 signalling bits, the equivalent signalling rate is 24 bits $\div(6 \times 125)$ μs =32 kbps. The signalling data rate per channel input becomes 32 kbps/24 = 1.333 kbps.

The duration of each bit (including a frame bit) in T_1 carrier system is $T_b = 125\mu s \div 193 = 0.647\mu s$. So the minimum transmission bandwidth of the T_1 system is

$$B_T = \frac{1}{2T_b} = 772 \text{ kHz}.$$

Example 8.11

The multiplexers at the second, third, and fourth level add 17, 69, and 720 bits respectively for bit stuffing and synchronisation. Find the bit rates of T_2, T_3 and T_4 lines.

Solution

The second level (M12-MUX) mixes four T_1 lines. So the number of bits per frame is

$$N_2 = 193 \times 4 + 17 = 789 \text{ bits/frame}.$$

The T_2-line bit rate

$$r_b(T_2) = 789\frac{\text{bits}}{\text{frame}} \times 8000\frac{\text{frames}}{\text{second}} = 6.312 \text{ Mbps}.$$

The third level (M23-MUX) mixes seven T_2 lines. Hence the number of bits per frame is

$$N_3 = 789 \times 7 + 69 = 5592 \text{ bits/frame}.$$

So the T_3-line bit rate

$$r_b(T_3) = 5592 \times 8000 = 44.736 \text{ Mbps}.$$

The fourth level (M34-MUX) mixes six T_3 lines. So the number of bits per frame is

$$N_4 = 5592 \times 6 + 720 = 34,272 \text{ bits/frame}.$$

Hence the T_4-line bit rate

$$r_b(T_4) = 34,272 \times 8000 = 274.176 \text{ Mbps}.$$

Essentially, T_1 carrier systems were designed to combine PCM and TDM techniques for short-haul communication of voice-grade telephone signals or data on a 64 kbps channel. It uses BPRZ–AMI encoding with regenerative repeaters. Regenerative repeaters should be placed at regular intervals to combat the power loss of digital signals while propagating down a metallic cable. When a channel is idle or transmitting a long strings of consecutive 0s, the T_1 carriers are susceptible to losing clock synchronisation. A monitoring is thus needed to ensure that long strings of consecutive 0s are not allowed for transmission through the channel. Early T_1 carrier systems made provisions for checking the presence of a PCM sample 8-bit word without at least one nonzero bit or 15 consecutive logic 0s in the data. When either of these occurred, a logic 1 is substituted in some appropriate bit position. In the worst-case, an original DS1 signal d_1 = 1000 0000 0000 0001 contains 14 consecutive 0*s* and hence needs no substitution and is transmitted without any alteration. But the DS1 signal d_2 = 1000 0000 0000 0000 contains a string of 15 consecutive 0s and hence a 1 is substituted in the second LSB position. The modified DS1 signal is d_2' = 1000 0000 0000 0010. This of course introduces an encoding error but is permitted in view of synchronisation. The choice of the bit position is not the LSB since the LSB of every sixth frame is the signalling bit.

Modern T_1 carrier systems use binary eight zero substitution (B8ZS) format. Whenever eight consecutive 0s are encountered, a special pattern is substituted for the 0s. The pattern may be either $+-0-+000$ or $-+0+-000$. Being bipolar in nature the + or − sign represents logic 1 and a 0 indicate a logic 0 condition. Observe that the pattern violates the bipolar nature of signalling. Whenever the receiver detects this violation in the substituted pattern it re-inserts the eight 0s back into the data signal.

T_1 carrier systems employ word interleaving for multiplexing data from various terminals in the time domain. Higher speed TDM systems and DM systems use bit interleaving.

In AT & T terminology, the transmission medium for the multiplier levels depends on the DS level. For DSO and DSI levels, the signal is transmitted over one pair of twisted wire. For higher bit rate levels, i.e., DS2 onwards, the transmission media may be co-axial cable, microwave radio or optical fibre.

The CCITT TDM standard for digital multiplexing hierarchy is illustrated in Fig. 8.57.

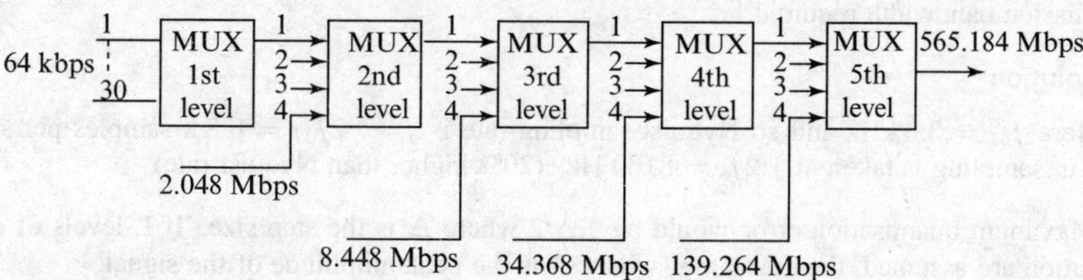

Fig. 8.57 *Digital multiplexing hierarchy (CCITT standard)*

The first level of multiplexing combines 30 DSO input channels generating a bit rate by 2.048 Mbps line speed.

$$\text{Line speed} = \frac{8\,\text{bits}}{\text{sample}} \times \frac{8000\,\text{sample}}{\text{sec}} = 64\,\text{kbps for DSO.}$$

Since these are 30-channel inputs, we have $8 \times 30 = 240$ bits of data plus 16 bits for frame synchronisation, a total of 256 bits per frame. Data rate is then

$$\text{Data rate} = \frac{256\,\text{bits}}{\text{frame}} \times \frac{8000\,\text{frames}}{\text{sec}} = 2.048\,\text{Mbps}.$$

Subsequent data rates at different levels of multiplexing can be calculated by considering the stuff bits into account.

TDM standards for CCITT hierarchy is tabulated in Table 8.4

Table 8.4 CCITT multiplexing

Level	No. of input times	Bit rate (Mbps)	No. of channels
1	30	2.048	30
2	4	8.448	120
3	4	34.368	480
4	4	139.264	1920
5	4	565.148	7680

Solved Numericals

1. An audio signal is band-limited to 3.4 kHz and is sampled at 20% higher than the Nyquist rate. The samples are quantised and binary encoded. Allowed quantisation error is 1% of the peak signal amplitude.

   (a) Find the minimum channel bandwidth required for transmission of the binary data.
   (b) If 24 such audio channels are multi-plexed on time sharing basis, find the minimum transmission bandwidth required.

Solution

Here $f_M = 3.4\,\text{kHz}$ and so Nyquist sampling rate is $f_s = 2f_M = 6.8\,\text{k}$ samples per second. But sampling is taken at $1.2f_s = 8160$ Hz. (20% higher than Nyquist rate)

Maximum quantisation error would be $\pm\Delta/2$ where Δ is the step size. If L-levels of quantisation are assumed, then $\Delta = \frac{2A}{L}$, where A is the peak amplitude of the signal.

Since quantisation error is 1% of peak amplitude, we have

$$\frac{\Delta}{2} = \frac{A}{L} = \frac{1}{100}A \text{ or } L = 100$$

For binary encoding, L must be in power of 2. So we take $L = 128$ (next higher value) for which n-bits per sample can be assigned. Since $2^n = 128$, we have $n = 7$ bits per sample.

So, the transmission rate would be

$$B = 8160\frac{\text{sample}}{\text{sec}} \times 7\frac{\text{bits}}{\text{sample}} = 57{,}120 \text{ bits per second.}$$

Since binary channel has spectral efficiency of 2 bits/Hz of bandwidth, the minimum channel bandwidth required is

$$C = \frac{B}{2} = 57120\frac{\text{bits}}{\text{second}} \div 2\frac{\text{bits}}{\text{Hz}} = 28.56\,\text{kHz}$$

(b) For 24 voice channels, the total bit rate would be $24\,B = 24 \times 57120 = 1.37 \times 10^6$ bits/sec and hence transmission bandwidth required $B_T = 0.685\,\text{Mbits/second}$.

2. For recording of music on a compact disc (CD), the signals are sampled at 44.1 kHz, quantised at $L = 65,536$ levels. Find the number of bits per sample required and minimum transmission bandwidth. The musical signal has the bandwidth of 15 kHz. Find also the signal-to-quantisation noise ratio if peak value of the signal is one volt.

Solution

Here $f_M = 15$ kHz, so minimum sampling rate required is $2f_m = 30$ kHz, but actual sampling rate used is 44.1 kHz (approximately 50% more than Nyquist rate).

Since these are $L = 65,536$ levels which is 2^{16}, the number of encoded bits per sample is 16 bits.

Number of bits produced per second would be

$$B = 44.1 \times 10^3 \frac{\text{samples}}{\text{sec}} \times 16 \frac{\text{bits}}{\text{sample}} = 705.6 \times 10^3 \text{ bits/sec}$$

So the minimum transmission bandwidth B_T required is

$$B_T = \frac{B}{2} = 352.8 \times 10^3 \text{Hz} \approx 353 \text{ kHz}$$

Since the signal peak is 1 V, the quantisation step size would be $\Delta = \frac{2}{L} = \frac{2}{2^{16}} = 2^{-15}$ *volt*.

Thus, quantisation noise power

$$N_q = \Delta^2/12 = \frac{2^{-30}}{12} \text{ watt} = 7.761 \times 10^{-11} \text{ watt}$$

Normalised signal power is $S = \frac{1}{2}A^2 = 0.5$ watt.

So, signal-to-quantisation noise power ratio is

$$\frac{S}{N_q} = \frac{0.5}{7.761 \times 10^{-11}} = 6.442 \times 10^9 = 98 \text{ dB}$$

3. In PCM, the signal-to-quantisation ratio in dB is given by (Eq.8.43) $\text{SNR(dB)} = 6N$, where N is the number of bits per sample. The above formula is valid when the signal has uniform distribution over its entire amplitude range of $\pm$ A volts. Derive the SNR(dB) in case the signal distribution is not specified.

Solution

Following the similar treatment as in Section 8.7, we may write,

step size, $\Delta = \frac{2A}{L} = \frac{2A}{2^N}$

quantisation noise power, $Nq = \frac{\Delta^2}{12} = \left(\frac{2A}{2^N}\right)^2 \Big/ 12 = \frac{A^2}{3 \times 2^{2N}}$

If the signal is $s(t)$, then signal power, $S = \overline{s^2(t)}$. Thus

$$\text{SNR} = \frac{\overline{s^2(t)}}{Nq} = \frac{\overline{s^2(t)} \cdot 3 \times 2^{2N}}{A^2} = \frac{3}{A^2} \cdot \overline{s^2(t)} \cdot 2^{2N}$$

So,

$$\text{SNR(dB)} = 10\log_{10}\left[\frac{3}{A^2}\overline{s^2(t)}.2^{2N}\right]$$

$$= 10\log_{10}\left(\frac{3\overline{s^2(t)}}{A^2}\right) + 2\text{N}\log_{10}{}^{2}$$

$$= \alpha + 6\text{N}$$

where $\alpha = 10\log_{10}\left(\frac{3\overline{s^2(t)}}{A^2}\right)$ can be found by knowing $s(t)$. Once again, the SNR increases by 6 dB per bit added to the coded word.

4. In a certain observation, the signal-to- quantisation noise ratio was calculated as 30 dB using 8-bit PCM. Let us suppose that the SNR is increased to a reasonable value of 42dB by increasing the quantisation level L. Find the change in the transmission bandwidth requirement.

Solution

Since $SNR(dB) = \alpha + 6N_1$

we have $\alpha + 6 \times 8 = 30$

or, $\alpha = -18 \text{ dB} = 10\log_{10}\left(\frac{3\overline{s^2(t)}}{A^2}\right)$

So, $\text{SNR}(dB) = -18 + 6N$

To achieve an SNR of 42 dB, let the number of bits per sample be N_2. Then,

$$42 = -18 + 6\,N_1$$

or, $N_2 = 10\,\text{bits}$.

So, quantisation level has to be increased.

Now, for quantisation level L, we have $\text{N} = \log_2^L$, if signal bandwidth be B Hz, then transmission bandwidth $B_T = \text{BN}$ so that

$$N = (B_T/B)$$

So,

$$N_2 - N_1 = \frac{(B_T)_2 - (B_T)_1}{B}$$

or, $\frac{N_2 - N_1}{N_1} = \frac{(B_T)_2 - (B_T)_1}{(B_T)_1} = \frac{10-8}{8} = \frac{1}{4}$

Thus 25% excess transmission bandwidth will be required.

5. Derive the expression for the signal-to-quantisation noise ratio for delta modulation when slope-overload noise dominates.

Solution

The delta modulated signal is given by (eq.8. 75)

$$e_q(k) = \{sgn\,[e(k)]\}\,\Delta$$

The transmission bandwidth requirement would be

$$B_T \geq r_b/2 = fs/2$$

Slope overload prevails if (Eq. 8.76)

$$\frac{\Delta}{T_s} = \Delta \cdot f_s \geq \left|\frac{ds(t)}{dt}\right|_{\max}$$

The granular noise $e(t)$ is bounded by $|e(t)| \leq \Delta$. We may assume uniform amplitude distribution of the granular noise over the range from $-\Delta < e(t) < \Delta$. Thus mean square error due to granular noise would be

$$\overline{e^2} = \int_{-\Delta}^{\Delta} \frac{1}{2\Delta} \cdot e^2 de$$

$$= \frac{1}{2\Delta} \cdot \left[\frac{e^3}{3}\right]_{\Delta}^{\Delta} = \frac{1}{6\Delta}(2\Delta^3) = \Delta^2/3.$$

The power spectral density of $e(t)$ is expected to be almost flat over the range $|f| \leq f_s$ and so we may write

$$G_e(f) = \frac{\overline{e^2}}{2f_s},\ |f| \leq f_s$$

So, the average power of the granular noise at the filtered output will be

$$N_{gr} = \int_{-fM}^{fM} G_e(f)df = \frac{\overline{e^2}}{2f_s} \cdot 2f_M = \frac{f_M}{f_s}\frac{\Delta^2}{3}.$$

where f_M is the message bandwidth.

Thus signal-to-granular noise power ratio is given by

$$(S/N)_{gr} = \frac{P_s}{N_{gr}} = \frac{3f_s}{\Delta^2 \cdot f_M} \cdot P_s.$$

Exercises

Activity

8.1 Find the Nyquist rate and Nyquist interval for the following signals.

(a) $s(t) = \text{sinc}(2t)$; (b) $s(t) = \sin 2\pi t$; (c) $s(t) = \sin^2 2\pi t$;

(d) $s(t) = \sin 2\pi t + 2sin^2 2\pi t$; (e) $s(t) = \cos^3(2\pi 10^3 t)$.

8.2 Figure P8.1 shows the spectrum $S(f)$ of a message signal $s(t)$. It is sampled at $f_s = 2.5f_m$ using rectangular pulses of unit amplitude and width $\tau = 0.1/f_s$. Find the spectrum of the PAM signal.

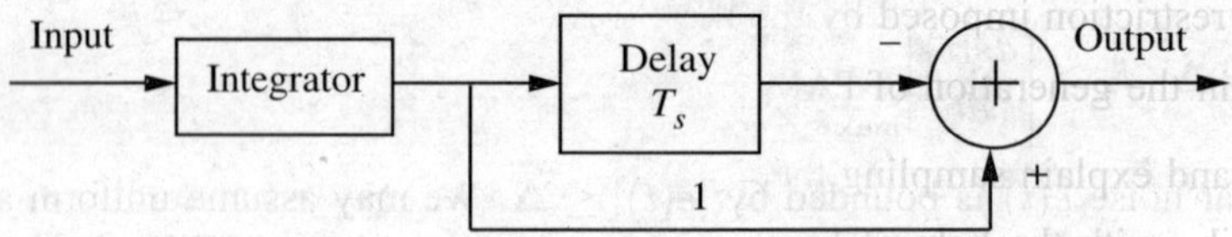

Fig. P8.1

8.3 If the spectral range of a bandpass signal extends from 20 to 30 kHz, find the range of the sampling frequency f_s so that the signal may be reconstructed.

8.4 Five different message signals each with a bandwidth of 10 kHz are sampled uniformly and time-division multiplexed. Find the minimum bandwidth required for the TDM-PAM signal.

8.5 The signal $s(t) = \cos 2\pi t$ is sampled every 0.4 seconds. The signal is reconstructed using a lowpass filter. Find the truncation error.

8.6 Consider that the signal $s(t) = \cos 2\pi t$ is quantised into 16 levels. Find the mean-square quantisation noise power.

8.7 A PCM-TDM system multiplexes 32 voice channels (each of bandwidth 0–4 kHz), encodes with an 8-bit PCM code and uses UPNRZ line coding. Find (a) line speed in bits per second and (b) minimum Nyquist bandwidth.

8.8 A PCM carrier system comprises of two DSO channels and are time-division multiplexed. A sampling rate of 8 kHz is used and the data is encoded in an 8-bit PCM. Find the line speed of the multiplexer.

8.9 The reconstruction of signal from its samples $\{x(kT)\}$ is done by the interpolation scheme

$$y(t) = \sum_{k=0}^{\infty} x(kt)h_r(t - kT),$$

where $h_r(t)$ is the interpolation filter. For infinite-order interpolation, $h_{\mathrm{r}}(t) = \sin\frac{(\pi fT)}{\pi fT}$; for-first-order interpolation, $h_{\mathrm{r}}(t) = \pi(t/T)$; and for second order (or linear) interpolation, $h_r(t) = \wedge(t/2T)$. Sketch $y(t)$ to interpret the reconstruction operation for the samples $\{x(k)\} = \{x(0), x(T), x(2T), \ldots\}$.

8.10 A pulse modulated wave can be generated by sampling the signal $s(t) = \text{Cos}2\pi(10)t$ at $t = kT_s$ where T_s is the sampling interval given by $T_s = \frac{1}{f_s}$ and $f_s = 4f_m$. Sketch the PPM signal with pulse width $\tau_0 = 0.2T_s$. (Hint: Position of k-th pulse is at $t_k = kT_s + \alpha x(kT_s)$ (by Eq.8.35) Here, $f_m = 10$ Hz, $f_s = 40$ Hz,$T_s = \frac{1}{40}$ sec, and in one cycle of 0.1 sec, there will be 4 samples)

Review questions

1. (a) Explain the difference between analog modulation and pulse modulation. Name and define the four types of modulation.

   (b) Name and state the basic theorem governing pulse modulation and indicate what happens if the restriction imposed by the theorem is violated.

   (c) Explain the generation of PAM signals. (KU 2000)

2. (a) State and explain sampling theorem.

   (b) Describe with the help of block diagram how a speech signal from a single channel is converted into a pulse code modulated signal.

   (c) Defined PAM, PWM and PPM with the necessary waveforms of the modulated signals. (KU 2001)

3. (a) State and prove the sampling theorem.

   (b) Deduce the expression for the quantisation error of a PCM mid-riser quantiser. Deduce the SNR as well.

   (c) Describe PAM, PWM and PPM for a communications system. (KU 2001)

4. (a) Show that companding improves PCM performance by reducing the quantisation noise. Illustrate the variation of $(S/N)_D$ vs. S_x in dB for μ-law companding. Hence point out the usefulness of companding.

   (b) A voice signal having $W = 3$ kHz and $S_x = 1/4$ watt is to be transmitted via M-ary PCM. Determine the values of M, γ and f_s such that $(S/N)_D \geq 40$ dB if $B_T = 16$ kHz. (KU 2001)

5. (a) Describe with a block diagram the function of a PCM transmitter and a receiver.

   (b) Show that the signal-to-noise ratio at the destination for PCM system is $(S/N)_D = 3q^2 S_x$, where all the terms have their usual meaning.

   (c) A TV signal has a BW of 4.5 MHz. This signal is sampled, quantised and binary coded to obtain a PCM signal.

      (i) Determine the sampling rate if the signal is to be sampled at a rate 20% above the Nyquist rate.

      (ii) If the samples are quantised into 1024 levels, determine the number of binary pulses required to encode each sample.

      (iii) Determine the binary pulse rate (bits/s) of the binary coded signal and the minimum BW required to transmit this signal. (KU 2002)

6. (a) What is slope-overload distortion in delta modulation? Explain a scheme that overcomes this distortion.

   A function $s(t) = A\sin(2\pi f_m t)$ is applied to a delta modulator of step size S and sampling period T_s. What is the maximum amplitude of the signal that avoids the slope-overload distortion?

   (b) Explain the need for non-uniform quantisation in PCM systems. Describe the μ-law compression. What determines the polarity of the samples in μ-255 companding system? (KU 2002)

7. (a) With a suitable block diagram, explain the principle of pulse code modulation (PCM).

   (b) What is companding? Discuss the two laws of companding.

   (c) An audio output of a microphone $m(t)$ swings between $\pm V$ volts and is characterised by a uniform probability density function. If the signal is sampled and uniformly quantised by an R-bit quantiser, derive the expression for the output signal-to-noise ratio in dB in terms of R.

8. (a) Describe the frequency-division multiplexing (FDM) with a neat diagram.

   (b) A commercial AM broadcast occupies the spectrum from 535 kHz 1605 kHz with a sub-carrier spacing of 10 kHz. How many stations can be incorporated in that band?

   (c) Explain the AT & T FDM hierarchy for an L $_4$ carrier system. (MUST, Spring 2010)

9. (a) Why is it desirable to have non-uniform quantisation? Show that the quantisation noise σ_q for M-level non-uniform quantisation is given by

   $\sigma_{\rm q}^2 \approx \frac{1}{6}\sum_{i=1}^{M/2} f_{\rm x}(x_{\rm i})h_{\rm i}^3$ where $f_X(x)$ is the PDF of the message and h_i is the step height at the i-th level.

   (b) What is the benefit of using companding? Give an example of A-law companding for voice PCM with analytical details. (MUST, Spring 2011)

10. (a) Describe the concept of time-division multiplexing (TDM) with a neat diagram.

    (b) A T_1- carrier system multiplexes 24 voice band channels with a bandwidth 4 KHz. Each channel is sampled at 8000 samples per second and encoded in 8-bit PCM. A single framing bit is added at each frame. Find the line speed in bits per second. (MUST, Spring 2012)

Multiple-choice questions

1. A data signal having frequencies from dc to 2 kHz is transmitted on a 4-bit PCM. The minimum bandwidth of the carrier channel should be

   (a) 2kHz (b) 4kHz (c) 8kHz (d) 16kHz

2. In PCM systems the signal-to-quantisation noise ratio depends on

   (a) sampling rate (b) message bandwidth
   (c) number of quantisation levels (d) size of the transmission system

3. The demodulator for PAM signal is a

   (a) differentiator followed by an LPF
   (b) clipper circuit followed by an LPF
   (c) Schmitt trigger followed by an LPF
   (d) LPF alone

4. The Nyquist sampling rate for a signal $f(t) = 10\cos(50\pi t)\cos^2(150\pi t)$ in samples per second is

   (a) 150 (b) 200 (c) 300 (d) 600

5. A message signal is bandlimited to 5 kHz. It is sampled at the Nyquist rate, quantised using 64-levels and finally encoded in binary form. The transmission rate would be

   (a) 60 kbps (b) 64 kbps (c) 128 kbps (d) 32 kbps

6. quantising noise can be reduced by increasing

   (a) bandwidth (b) sampling rate (c) number of quantisation levels (d) all of these

7. In delta modulation systems, the granular noise occurs when the

   (a) pulse amplitude decrease
   (b) modulating signal has fast variation
   (c) system is in noisy environment
   (d) modulating signal remains constant

8. In the PCM system with uniform quantisation, increasing the number of bits from 8 to 9 will reduce the quantisation noise power by a factor of

   (a) 9 (b) 8 (c) 4 (d) 2

9. The PCM signal can be generated by amplitude modulating the following signal.

   (a) PWM (b) PPM (c) PDM (d) PAM

10. A signal of $f_{\max} = 10$ kHz is sampled at the Nyquist rate. The time interval between the samples is

    (a) 50 μs (b) 10 μs (c) 5μs (d) 100 μs.

9 Information Theory and Coding

9.1 Introduction

In the design of a communications system, the focus is on finding the best method of transmitting information from the source to the receiver. Information, in the broadest sense, is defined as the knowledge or intelligence carried by a 'message' signal. This, for instance could be a telephone conversation, radio or television broadcast, graphics over a fax machine or the signals of data-processing and computing devices. When the output of the source is an analog signal, the source is referred to as analog source. In contrast, sources like computer storage devices, etc produce discrete outputs and hence they are discrete sources.

The major concern of information theory is to develop a mathematical model for signals and then find the best method of separating the information content of the signal from the ever-present noise. This requires an efficient description of the information source and the information-carrying medium (called the information channel). In this chapter, we shall develop a mathematical measure and an intuitive understanding of information, followed by techniques of source and channel coding. The latter govern the processing and transmission of information from the source and its reconstruction at the destination.

9.2 Information Measure

Any information source, be it analogue or digital, is random in nature. If it were not so and had we known it exactly before its transmission, there would be no reason to transmit it. The amount of information in a message or an event, and hence its measure, is strongly related to the probability of occurrence of that message or event.

We then recognise that a known message provides no information. If one event is more probable than other, observation of that event is less informative. Let us first develop an intuitive understanding of information measure before we formulate it mathematically. Consider the following three sentences:

(a) Tomorrow, the sun will rise in the East.
(b) One of my guests will visit us in the next three hours.
(c) There will be a tornado in Jaipur city.

The above three sentences do not carry equal amount of information. The first one provides virtually no information at all! The fact that the sun rises in the East is a certainty and everyone know it. The second sentence carries some information, since we are not absolutely certain that the guest will visit us. However, there is some chance that the guest may visit us. The third forecast provides a great deal of information as tornado is unexpected and rare in Jaipur, as confirmed by the past statistics of the meteorology department. So, what we see is that a message with a low probability of occurrence seems to carry more information than the one with a high probability of occurrence.

Let a digital source generate a message with a probability of occurrence p. The information I conveyed by the message is defined as

$$I = \log_2 \left(\frac{1}{p}\right) \text{ bit} \tag{9.1}$$

$$= -\log_2 p \text{ bit}.$$

This is called the self-information of the event.

Note that a certain event for which $p = 1$ provides no information. Although the probability p is a dimensionless pure number, the unit 'bit' is assigned to I. In fact, the units of I are determined by the base of the logarithm. When base 2 logarithm is used, the units are in 'bits'. If base 10 is used, the units are in 'Hartley' or 'decit' and when the base is 'e', the units are in 'nat' (natural units). Base 2 logarithm is very convenient when binary PCM is used for signal modulation. The unit 'bit' is also used to denote the unit of binary data. So take care not to confuse between the two. Since $0 \leq p \leq 1$, $I \geq 0$, the information is 'non-negative'.

When required, one can convert information in bit into information measured in Hartley or in nat. For example,

$$I = \log_2^{1/p} \text{ bits} = \log_{10}^{1/p} \text{ Hartley}$$

or $\log_{10}^{1/p}$ Hartley of information $\equiv \log_2^{1/p}$bits of information.

If one assumes $p = 1/10$, then

$$\log_{10} 10 \text{ Hartley of information} \equiv \log_2 10 \text{ bits of information}$$

or, 1 Hartley of information $= 3.32$ bits of information.

Similarly, the relation between bit of information and nat of information can be established as follows:

$$I = \log_e 1/p \text{ nat of information} = \log_2 1/p \text{ bit of information.}$$

If we assume $p = 1/e$, then

$$\log_e e \text{ nat of information} = \log_2 e \text{ bits of information}$$

or 1 nat of information $= \frac{\log_{10} e}{\log_{10} 2} = 1.44$ bits of information.

Thus, 1 Hartley of information $= 3.32$ bits of information

1 nat of information $= 1.44$ bits of information.

When a source generates a number of independent messages, m_1, m_2, $\cdots$ with the respective probabilities $p_1, p_2, \cdots$ and the receiver correctly identifies individual messages, the information measure in the composite message is

$$I = \log_2 \frac{1}{p_1} + \log_2 \frac{1}{p_2} + \cdots$$

$$= \sum_i \log_2 \frac{1}{p_i} = I_1 + I_2 + \cdots, \tag{9.2}$$

where $p_1 + p_2 + \cdots = 1$. The information of a given message differs from the others since their probabilities are different. Suppose that on a sunny day a person makes a phone call to the meteorological office to enquire about the weather report of the city. He receives two reports 'there is daylight here' and 'there is a chance of rain in the next six hours', with their respective probabilities 3/4 and 1/4. The first report conveys $\log_2 4/3 = 0.42$ bit of information while the latter report $\log_2 4 = 2$ bits of information. We consequently enquire of the 'average information' of the source.

Suppose that a source generates a sequence of L independent messages over a long period of time. For very large L, one can expect that message 1 with probability p_1 occurs $p_1 L$ times, message 2 occurs $p_2 L$ times and so on. The information content of the first message is $p_1 L \log_2 1/p_1$, that of the second message is $p_2 L \log_2 1/p_2$ and so on. The complete information provided by the sequence is

$$I = p_1 L \log_2 1/p_1 + p_2 L \log_2 1/p_2 + \cdots. \tag{9.3}$$

We define the 'average information' per message as

$$I_{\text{av}} = \frac{I}{L} = p_1 \log_2 \frac{1}{p_1} + p_2 \log_2 \frac{1}{P_2} + \cdots$$

$$= \sum_i p_i \log_2 \frac{1}{p_i} \text{ bits/message} \tag{9.4}$$

and is referred to as the 'entropy' of the source. Thus the source entropy H is

$$H = \sum_i p_i \log_2 \frac{1}{p_i} \text{ bits/message} \tag{9.5}$$

If the digital source emits only a single message, the message carries no information since $p = 1$ for the message and so $H = 0$. Again, as $p \to 0$, $\lim_{p \to 0} p \log_2 \frac{1}{p} = 0$. Thus the average information provided by the source is always zero whether the message is certain or completely uncertain.

The simplest discrete source is a binary source that emits a binary sequence of the form $1011010001 \cdots$. Let us find the dependence of entropy H on the chances of occurrence of the two messages with probabilities p and $(1-p)$. The source entropy is

$$H = p \log_2 1/p + (1-p) \log_2 \frac{1}{(1-p)}. \tag{9.6}$$

In order to find the maximum source entropy, we differentiate H with respect to p and set it equal to zero. Now,

$$\frac{dH}{dp} = p \times p\left(-\frac{1}{p^2}\right) \times \log_2 e + \log_2 \frac{1}{p}$$

$$+ (1-p)(1-p)\frac{1}{(1-p)^2} \times \log_2 e - \log_2 \frac{1}{(1-p)}$$

$$= \log_2 \frac{1}{p} - \log_2 \frac{1}{1-p} = 0.$$

So, $p = 1 - p$, giving $p = 1/2$. Thus, the source entropy is maximum when it sends two equally likely messages and the maximum value is

$$H_{\max} = \frac{1}{2}\log_2 2 + \frac{1}{2}\log_2 2 = 1 \text{ bit/message.} \tag{9.7}$$

A plot of H as a function of the probability p is depicted in Fig. 9.1. Note that H is zero for both $p = 0$ and $p = 1$.

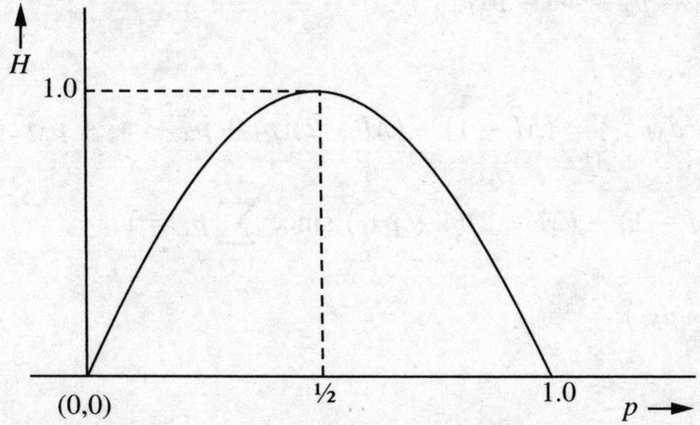

Fig. 9.1 *Average information or source entropy H for two messages as a function of probability p*

This result can be extended to M number of independent messages sent by the source. Consider that a digital source sends M messages, all uncorrelated. Our aim is to show that the source entropy attains a maximum value if all the messages are equiprobable, that is, the source has no bias to generate a particular set of messages more often than the others. To start with, assume that $p_1, p_2, \cdots, p_M$ are the respective probabilities of the messages. The source entropy is then

$$H = p_1 \log_2 \frac{1}{p_1} + p_2 \log_2 \frac{1}{p_2} + \cdots + p_M \log_2 \frac{1}{p_M} \tag{9.8}$$

with $p_1 + p_2 + \cdots + p_M = 1$. The partial derivatives $\partial H/\partial p_1, \partial H/\partial p_2, \cdots, \partial H/\partial p_M$ are equated to zero to give a set of $(M-1)$ equations. The first partial derivative when equated to zero gives

$$\frac{\partial H}{\partial p_1} = \left[p_1 \times p_1\left(-\frac{1}{p_1^2}\right)\log_2 e + \log_2 \frac{1}{p_1}\right]$$

$$+\left[(1-p_1-p_2-\cdots-p_{M-1})^2\frac{1}{(1-p_1-p_2-\cdots-p_{M-1})^2}\log_2 e\right.$$

$$\left.-\log_2\frac{1}{(1-p_1-p_2-\cdots-p_{M-1})}\right]=0$$

or $\log_2\frac{1}{p_1}=\log_2\frac{1}{(1-p_1-p_2-\cdots-p_{M-1})}$

or $p_1=1-p_1-p_2-p_3-\cdots-p_{M-1}$.

In this way, the $(M-1)$ equations are obtained as:

$$\begin{aligned}2p_1&=1-p_2-p_3-\cdots-p_{M-1}\\2p_2&=1-p_1-p_3-\cdots-p_{M-1}\\&\cdots\\2p_{M-1}&=1-p_1-p_2-\cdots-p_{M-2}\end{aligned}\tag{9.9}$$

Adding all the equations,

$$2(p_1+p_2+\cdots+p_{M-1})=(M-1)-(M-2)(p_1+p_2+\cdots+p_{M-1})$$

or, $2(1-p_M)=(M-1)-(M-2)(1-p_M)$ since $\sum_{i=1}^{M}p_i=1$

or, $M-1=M(1-p_M)$.

That is, $p_M=\frac{1}{M}$.

Thus, $p_M=\frac{1}{M}$ is the condition that gives $\partial H/\partial p_i=0$ for all $i=1,2,\cdots,M-1$ with p_M being expressed in terms of all other p's . In the same way it can be proved that H is maximum when all the messages have equal probabilities, that is, $p_1=p_2=\cdots=p_M=\frac{1}{M}$ and $H_{\max}$ is given by

$$H_{\max}=M\times\frac{1}{M}\log_2 M=\log_2 M\text{ bits/message.}\tag{9.10}$$

Source information rate

If the source generates r messages/second, the source information rate R is defined as

$$R=\frac{H}{T}=rH\text{ bit/second}\tag{9.11}$$

and is equivalent to the average information per second. In (9.11), T represents the time required to send a message $(=1/r)$.

We have defined the information measure in terms of digital sources, not for the continuous sources.

The simple reason for this is that communications is inherently a discrete process; any analog source can be viewed as an approximations of the discrete sources.

Example 9.1

A digital source sends messages of 8 digit word-length. Each digit in the word may take on two levels (binary). Assume that the probability of sending any of the two levels is equal and all the 8-digit words are equi-probable. Find the source information and entropy.

Solution

Here, 2^8 different words can be obtained. Since each level is equally likely, all different words are equally likely. Thus, for the ith word,

$$p_i = \frac{1}{2^8} = \left(\frac{1}{2}\right)^8$$

and $I_i = \log_2 \frac{1}{(1/2)^8} = 8$ bits.

The average information is $H = 8$ bits/message, as the information content in any message is equal to that in any other possible message.

Example 9.2

A digital source sends 0 V and -1 V levels with a probability of 0.2 each and $+2$ V and $+3$ V levels with a probability of 0.3 each. Find the average information of the source.

Solution

The average information of the source is

$$H = \sum_i p_i \log_2 \frac{1}{p_i} = \left(0.2 \log_2 \frac{1}{0.2}\right) \times 2 + \left(0.3 \log_2 \frac{1}{0.3}\right) \times 2$$

$= 1.97$ bits/message. [Note also that $\sum P_i = 2 \times 0.2 + 2 \times 0.3 = 1$.]

Example 9.3

A source transmits four messages with probabilities $1/8, 1/8, 1/4$, and $1/2$. Find the average information. What will be the average information if the source sends four messages with equal probability?

Solution

The average information is given by

$$H = \left(\frac{1}{8} \log_2 8\right) \times 2 + \frac{1}{4} \log_2 4 + \frac{1}{2} \log_2 2$$

$$= \frac{3}{4} + \frac{1}{2} + \frac{1}{2} = 1.75 \text{ bits/message.}$$

For the messages to be equally likely, H is maximum and is given by

$$H_{\max} = \left(\frac{1}{4}\log_2 4\right) \times 4 = 2 \text{ bits/message} \qquad [\text{since } p = \tfrac{1}{4}].$$

Example 9.4

A seven-segment display emits a 0 with a probability 0.25; a 1 with a probability of 0.20; a 2 with a probability of 0.15; a 3, 4, 5, and 6 with a probability of 0.09 each; 7 and 8 with probabilities of 0.015 each; and a 9 with a probability of 0.01. Find the average information of the source. Compare your result with that when the display emits all the digits with equal probability.

Solution

The average information is

$$H = \sum p \log \frac{1}{p} = 0.25 \log_2 \frac{1}{0.25} + 0.20 \log_2 \frac{1}{0.20} + 0.15 \log_2 \frac{1}{0.15}$$

$$+ \left(0.09 \log_2 \frac{1}{0.09}\right) \times 4 + \left(0.015 \log_2 \frac{1}{0.015}\right) \times 2 + 0.01 \log_2 \frac{1}{0.01}$$

$$= 0.5 + 0.464 + 0.410 + 1.250 + 0.181 + 0.066 = 2.871 \text{ bits/message.}$$

If the likelihood of all the digits glowing is equal,

$$H = \frac{1}{10}\log_2 10 = 0.332 \text{ bits/message.}$$

9.2.1 Differential entropy of continuous random variables

In the previous section, we considered the sources generating random variables that are discrete in amplitude. We can extend the concept of mutual information from discrete random variables to continuous random variables.

Consider a continuous random variable X with the probability density function $p(x)$. By analogy with equation (9.5), we define

$$h(x) = \int_{-\infty}^{\infty} p(x) \log_2 \frac{1}{p(x)} dx \tag{9.12}$$

We call $h(x)$ as the differential entropy of X, to distinguish it from the absolute entropy. Although defined by the mathematical equation, the above quantity has no physical meaning and it does not make sense to measure the randomness of X. The self information content of a continuous source is infinite.

Example 9.5

A voltage source generates a voltage x which is a random variable uniformly distributed over the interval $-1 < x < 1$. The signal is passed through an amplifier of gain A. The output y is also a random variable, uniformly distributed over $-A < y < A$. Find the entropies $H(x)$ and $H(y)$.

Solution

The probability density function of x is

$$\begin{aligned} p(x) &= \tfrac{1}{2} \qquad |x| < 1 \\ &= 0 \qquad \text{otherwise.} \end{aligned}$$

Applying equation (9.12), we get

$$h(x) = \int_{-1}^{1} \frac{1}{2} \log_2 2dx = \frac{1}{2}\Big[x\Big]_{1}^{1} = 1 \text{ bit/message.}$$

The probability density function of the amplifier output is

$$p(y) = \begin{cases} \frac{1}{A} & |y| < A \\ 0 & \text{otherwise} \end{cases}$$

Hence, $h(y) = \displaystyle\int_{-A}^{A} \frac{1}{A} \log_2 Adx = 2\log_2 A$ bits/message.

In particular, for an amplifier gain of $A = 2$,

$$h(y) = 2\log_2 2 = 2 \text{ bits.}$$

Note that $h(y)$ can be negative if $A < 1$.

9.2.2 Maximum entropy calculation for a given continuous random variable with fixed average power

The differential entropy given by Eq. (9.12) provides the average amount of information for a sample of $x(t)$. The probability density function $p(x)$ of $x(t)$ follows the condition $\int_{-\infty}^{\infty} p(x)dx = 1$. We are interested to compute the maximum value of source entropy under the constraint of fixed average power of $x(t)$:

This is, $\int_{-\infty}^{\infty} x^2 p(x)dx = \overline{x^2} = \sigma^2$, a constant (9.12a)

Stated in another way, we seek the PDF $p(x)$ which maximises $h(x)$. The integrand of Eq.(9.12) is a function of x and $p = p(x)$ and so we may write

$$F(x, p) = p(x) \log \frac{1}{p(x)} \tag{9.12b}$$

Thus, the problem boils down to optimise the integral

$$I = \int_a^b F(x,p)dx \tag{9.12c}$$

subject to the constraint

$$\int_a^b F_k(x,p)dx = \lambda_k \tag{9.12d}$$

where λ_k are the constants. The integral I becomes extremum when p satisfies the equation

$$\frac{\partial F}{\partial p} + \sum_k c_k \frac{\partial F_k}{\partial p} = 0 \tag{9.12e}$$

where c_k are Lagrange's undetermined multipliers. In the present case, there are two constraints namely

$$\int_{-\infty}^{\infty} p(x)dx = 1 \text{ and } \int_{-\infty}^{\infty} x^2 p(x)dx = \overline{x^2} = \sigma^2 \tag{9.12f}$$

So, we consider two unknown constants c_1 and c_2. The relation becomes

$$\frac{\partial F}{\partial p} + c_1 \frac{\partial F_1(x,p)}{\partial p} + c_2 \frac{\partial F_2(x,p)}{\partial p} = 0$$

where $F_1(x,p) = p$ and $F_2(x,p) = x^2 p$. Of course $F = p \log \frac{1}{p}$.

Substituting these values of F, F_1 and F_2.

$$\frac{\partial}{\partial p}\left(p \log \frac{1}{p}\right) + c_1 \frac{\partial}{\partial p}(p) + c_2 \frac{\partial}{\partial p}(x^2 p) = 0$$

or, $-\frac{1}{\ln 2}(1 + \ln p) + c_1 + c_2 x^2 = 0$

or, $(1 + \ln p) + \lambda_1 + \lambda_2 x^2 = 0$

where $\lambda_1 = -c_1 \ln 2$ and $\lambda_2 = -c_2 \ln 2$

Solving for p, we have

$$p = e^{(\lambda_1 - 1)} e^{\lambda_2 x^2} \tag{9.12g}$$

We may use the constraint equations to find the multipliers

$$1 = \int_{-\infty}^{\infty} p\,dx = e^{\lambda_1 - 1} \int_{-\infty}^{\infty} e^{\lambda_2 x^2} dx = 2e^{\lambda_1 - 1} \int_0^{\infty} e^{\lambda_2 x^2} dx$$

Let $\lambda_2 x^2 = -y$ so that $dx = \frac{1}{2\sqrt{-\lambda_2}} y^{-1/2} dy$. Thus,

$$\int_0^\infty e^{\lambda_2 x^2} dx = \int_0^\infty \frac{e^{-y}}{2\sqrt{-\lambda_2}} \cdot y^{-1/2} dy = \frac{1}{2\sqrt{-\lambda_2}} \cdot \Gamma({}^1\!/_2) = \frac{\sqrt{\pi}}{2\sqrt{-\lambda_2}}. \text{ So,}$$

$$1 = 2.\, e^{\lambda_1 - 1} \frac{\sqrt{\pi}}{2\sqrt{-\lambda_2}} = \sqrt{\frac{\pi}{-\lambda_2}}.e^{\lambda_1 - 1}.$$

By using the other constraint $\int_{-\infty}^{\infty} x^2 p dx = \sigma^2$, we have

$$\sigma^2 = 2.\, e^{\lambda_1 - 1} \int_0^\infty x^2 e^{\lambda_2 x^2} dx = 2e^{\lambda_1 - 1} \int_0^\infty \frac{y}{(-\lambda_2)} e^{-y} \frac{1}{2\sqrt{-\lambda_2}} y^{-1/2} dy$$

$$= \frac{e^{\lambda_1 - 1}}{(-\lambda_2)\sqrt{-\lambda_2}} \cdot \int_0^\infty y^{1/2} . e^{-y} dy = \frac{e^{\lambda_1 - 1}}{(-\lambda_2)\sqrt{-\lambda_2}} \cdot \Gamma({}^3\!/_2) = \frac{\sqrt{\pi}}{2 \cdot \sqrt{-\lambda_2}\,(-\lambda_2)} e^{\lambda_1 - 1}$$

Thus, we have

$$\sigma^2 = \frac{\sqrt{\pi}}{2\,(-\lambda_2)\sqrt{-\lambda_2}} e^{\lambda_1 - 1}$$

and

$$1 = \frac{\sqrt{\pi}}{\sqrt{-\lambda_2}} e^{\lambda_1 - 1}$$

Dividing, $\sigma^2 = \dfrac{1}{2\,(-\lambda_2)}$ or, $(-\lambda_2) = \frac{1}{2\sigma^2}$

Also, $e^{\lambda_1 - 1} = \dfrac{\sqrt{-\lambda_2}}{\sqrt{\pi}} = \sqrt{\dfrac{1}{2\sigma^2 \cdot \pi}} = \dfrac{1}{\sqrt{2\pi\sigma^2}}$

Hence, $p(x) = \dfrac{1}{\sqrt{2\pi\sigma^2}} e^{-x^2/2\sigma^2}$ (9.12h)

We thus conclude that for a given source power, the entropy attains a maximum when the source distribution $p(x)$ is Gaussian with zero mean and variance σ^2. The corresponding maximum entropy can be calculated as

$$h_{\max}(x) = \int_{-\infty}^{\infty} p(x) log_2 \frac{1}{p(x)} dx$$

Now

$$\log_2 \frac{1}{p(x)} = \log_2 \left[\sqrt{2\pi\sigma^2} \cdot e^{x^2/2\sigma^2}\right] = \log_2 \left(\sqrt{2\pi\sigma^2}\right) + \frac{x^2}{2\sigma^2} \log_2^e$$

so,

$$h_{\max}(x) = \int_{-\infty}^{\infty} p(x) \left[\log_2 \left(\sqrt{2\pi\sigma^2}\right) + \frac{x^2}{2\sigma^2} \log_2^e\right] dx$$

$$= \log_2 \sqrt{2\pi\sigma^2} \int_{-\infty}^{\infty} p(x)\, dx + \frac{\log_2^e}{2\sigma^2} \int_{-\infty}^{\infty} x^2 p(x)\, dx$$

$$= \log_2 \sqrt{2\pi\sigma^2} + \frac{1}{2} \log_2^e = \frac{1}{2} \log_2 \left(2\pi\sigma^2\right) + \frac{1}{2} \log_2^e$$

$$= \frac{1}{2} \log_2 \left(2\pi\sigma^2 e\right) \qquad (9.12i)$$

This relative entropy may have negative values when $2\pi e\sigma^2 < 1$

We may conclude finally that

$$h(x) \leq \frac{1}{2} \log_2 \left(2\pi e\sigma^2\right). \qquad (9.21j)$$

9.3 Channel Capacity: Shannon's Theorem

The effectiveness of a communications system in carrying information from a source to the destination is measured by a large number of parameters. These include the channel bandwidth, signal-to-noise ratio, probability of bit error, etc. Shannon's theorem deals with the rate of signal transmission through a communication channel. In principle, one can transmit the information without any error through a communication channel of capacity C if the source information rate R is less than C. Although the channel noise corrupts the signal (even the noise overrides the signal in some cases), error-free signal transmission is possible by assigning appropriate codes to the messages. A receiver circuit can detect the errors in the coded data and also can correct them.

Suppose a source sends M messages, all are equally likely, at the rate R. The messages are to be transmitted through a channel having capacity C. If $R \leq C$ is satisfied, there exists a coding technique such that the transmitted messages can be received in a receiver with an error probability as small as possible. The probability of error will increase towards unity for large M if the source information rate R exceeds the specified value of C. R. Hartley developed a relationship between the channel bandwidth B (in hertz), the transmission time T (in seconds), and the information capacity C (in bits/second). Stated mathematically,

$$C \propto B \times T. \tag{9.13}$$

Shannon showed that for the case of a white, bandlimited Gaussian channel, the channel capacity C could be calculated such that if the information rate R is less than C, the error probability would approach zero. The basic formula for the capacity of the channel is given by

$$C = B\log_2\left(1+\frac{S}{N}\right) \text{ bits/second,} \tag{9.14}$$

where S/N is the signal-to-noise power ratio at the input to the receiver. Since S is the average transmitted power and C is the bit rate in bit per second, we may write $S = CE_b$ (since $C = \frac{1}{T_b}$). In analog systems, an optimum system is one that achieves the largest S/N ratio. If a S/N ratio of 30 dB (=1000) is required for a channel having bandwidth B, the Shannon limit for the information capacity is obtained as $C = B\log_2(1+1000) \approx 10B$ bits/second. Transmission at such a large rate is possible if the symbol carries more than 1 bit of information.

Although Shannon's channel capacity theorem is restricted to only Gaussian channel, the performance of a non-Gaussian channel can be inferred from the law. In fact, this theorem gives a lower bound on the system performance with a non-Gaussian channel. The derivation of Eq. (9.14) is provided in Section (9.4.2).

Equation (9.14) states that a noiseless Gaussian channel ($S/N \to \infty$) has infinite channel capacity. This certainly is not the best noise performance of the system. An increase in bandwidth increases the noise power and hence decreases the S/N ratio thus limiting the channel capacity C. The channel capacity thus approaches an upper limit with increasing bandwidth. Shannon's limits can be proved as follows.

Assuming a two-sided noise power spectral density $\eta/2$, the total noise power N within the bandwidth B is $N = \eta B$. Thus,

$$C = B\log_2\left(1+\frac{S}{\eta B}\right) = \frac{S}{\eta}\log_2\left(1+\frac{S}{\eta B}\right)^{\eta B/S}. \tag{9.15}$$

In the limit $B \to \infty$,

$$C_\infty = C(B\to\infty) = \frac{S}{\eta}\lim_{B\to\infty}\log_2\left(1+\frac{S}{\eta B}\right)^{\eta B/S}$$

$$= \frac{S}{\eta}\log_2 e = 1.44\frac{S}{\eta}, \tag{9.15a}$$

as $log_e(1+x)^{1/x} \to 1$ as $x \to 0$.

This means that when bandwidth approaches infinity, the capacity of the channel approaches the asymptotic value.

If the bit duration is T_b and energy per bit is E_b, then for optimum encoded signal, $S = E_b/T_b$.

Equation (9.15) may be expressed as

$$\frac{C}{B} = \log_2\left(1+\frac{E_bC}{\eta B}\right)$$

$$\text{or,} \quad \frac{E_b}{\eta} = \frac{2^{C/B} - 1}{C/B} \tag{9.16}$$

For the normalised channel capacity $\frac{C}{B} = 1$, we have $E_b/\eta = 1(= 0dB)$. As $C/B \to \infty$,

$$\frac{E_b}{\eta} = \frac{2^{C/B}}{C/B} = \exp\left(\frac{C}{B}\ln 2 - \ln C/B\right) \tag{9.16a}$$

Thus E_b/η increases exponentially as $C/B \to \infty$. On the other hand, for $C/B \to 0$,

$$\frac{E_b}{\eta} = \lim_{C/B \to 0} \frac{2^{C/B} - 1}{C/B} = \ln 2 \tag{9.16b}$$

or, $E_b/\eta(dB) = 1.6dB$ when $C/B \to 0$.

This allows maximum information rate for minimum probability of error. Thus $\frac{1}{T_b} = C$ and so $E_b/\eta = \ln 2$.

Example 9.6

Find the channel capacity of a teleprinter channel that has a bandwidth of 300 Hz and a S/N ratio of 30 dB.

Solution

$C = B\log_2\left(1 + \frac{S}{N}\right)$.

Hence, $B = 300$ Hz, $S/N = 30$ dB

$\Rightarrow 10\log_{10}(S/N) = 30$

$\Rightarrow S/N = 1000.$

Thus, $C = 300\log_2(1001) = 2990$ bits/second.

Example 9.7

A computer keyboard has 110 characters and each character is sent using binary words. Find the number of bits needed to represent a character. The characters are to be sent over a voice-grade telephone line of bandwidth 3.3 kHz. How fast should they be sent if the desired S/N ratio is 20 dB? Find also the information content of each character if they have equal probability of being sent.

Solution

Here $2^n > 110$; the minimum value of n satisfying $2^n > 110$ is $n = 7$. Thus the character is 7-bit long.

$S/N = 20$ dB$= 100$ (numerical).

So, $C = B\log_2(1 + S/N)$ gives $C = 3300\log_2 101 = 21.972$ kbps.

For M equally likely messages,

$$H = \log_2 M = \log_2 110 = 6.78 \text{ bits/message.}$$

Example 9.8

A Gaussian channel has a bandwidth of 1 MHz. If the signal-to-noise power spectral density $S/\eta = 10^5$ Hz, find the channel capacity and also the maximum information transfer rate R.

Solution

Noise power $N = \eta B$.

$$\frac{S}{N} = \frac{S}{\eta}\frac{1}{B} = \frac{10^5}{10^6} = \frac{1}{10}.$$

Since the bandwidth is large, we take the expression for C_∞ which gives

$$C_\infty = 1.44\frac{S}{\eta} = 1.44 \times 10^5 \text{ bps.}$$

Shannon's ideal system

An ideal system is defined as one that does not lose channel capacity in the detection process. Consider the bandpass system shown in Fig. 9.2. A baseband signal $v_m(t)$ bandlimited to f_M is modulated and transmitted through the communication channel of bandwidth B. The noise $n(t)$ adds to the signal, the total noise power being $N_i = \eta B$ since we have taken a white Gaussian noise of two-sided power-spectral density $\eta/2$. The signal-to-noise power at the receiver input is $(S/N)_i = S_i/N_i$, where S_i is the signal power of $v_m(t)$. The bandpass channel capacity C_i is

$$C_i = B\log_2\left(1 + \frac{S_i}{N_i}\right), \tag{9.17}$$

which gives the maximum rate at which signal may be transmitted. The rate C_0 at which message is received at the receiver is

$$C_0 = f_M \log_2\left(1 + \frac{S_0}{N_0}\right). \tag{9.18}$$

where f_M denotes the bandwidth of the receiver baseband filter.

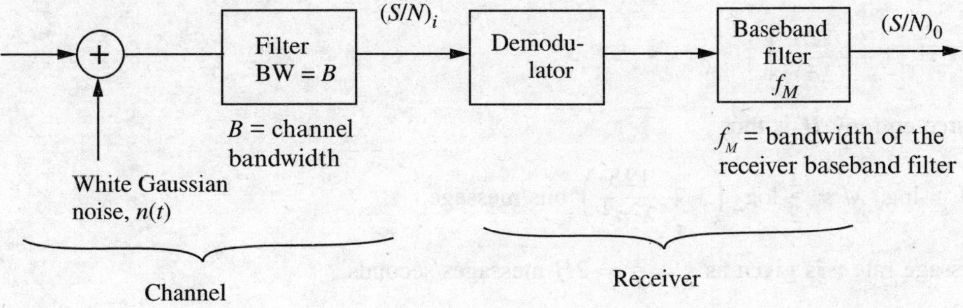

Fig. 9.2 *A bandpass receiver. The bandwidth of the channel is B.*

For the ideal system, $C_i = C_0$ and hence

$$B\log_2\left(1+\frac{S_i}{N_i}\right) = f_M \log_2\left(1+\frac{S_0}{N_0}\right).$$

Solving for S_0/N_0, we have

$$\frac{S_0}{N_0} = \left[1+\frac{S_i}{N_i}\right]^{B/f_M} - 1$$

$$= \left[1+\frac{f_M}{B}\frac{S_i}{\eta f_M}\right]^{B/f_M} - 1. \tag{9.19}$$

Equation (9.19) describes the ideal system performance and is plotted in Fig. 9.3 for different values of B/f_M. For $B/f_M = 1$, (9.18) gives $S_0/N_0 = S_i/\eta f_M$ and the relation is a straight line. The ratio S_0/N_0 increases rapidly for higher values of B/f_M, which means that the performance can be improved by sacrificing bandwidth. For ideal systems, $B/f_M = \infty$ and the curve increase exponentially.

Example 9.9
In an M-ary PCM transmitter, find out the source entropy.

An M-ary PCM system consists of M possible discrete amplitude levels separated by $\lambda\sigma$. Here λ is a constant and σ^2 is the noise variance such that $\lambda\sigma$ represents the noise margin sufficiently large to produce negligible error due to channel noise. Assuming M messages to be equally likely, the average signal power

$$S = \frac{2}{M}\left[\left(\frac{1}{2}\right)^2 + \left(\frac{3}{2}\right)^2 + \cdots + \left(\frac{M-1}{2}\right)^2\right](\lambda\sigma)^2.$$

Since, $1^2 + 3^2 + \cdots + (M-1)^2 = \frac{M}{6}(M^2-1)$, we have

$$S = \frac{2}{M}\left(\frac{\lambda\sigma}{2}\right)^2 \frac{M}{6}(M^2-1) = \frac{(M^2-1)(\lambda\sigma)^2}{12}.$$

Thus, $M = \left[1+\dfrac{12S}{(\lambda\sigma)^2}\right]^{1/2}$.

The source entropy H is thus

$$H = \log_2 M = \frac{1}{2}\log_2\left(1+\frac{12S}{\lambda^2\sigma^2}\right) \text{ bits/message.}$$

The message rate r is given as $r = \frac{1}{T} = 2B$ messages/seconds,

where $T = \frac{1}{2B}$ denotes the message interval and B is the highest frequency component in the signal. Thus channel capacity

$$C = rH = B\log_2\left(1 + \frac{12}{\lambda^2}\frac{S}{\sigma^2}\right).$$

Note that $\lambda = \sqrt{12}$ refers to Shannon's ideal system.

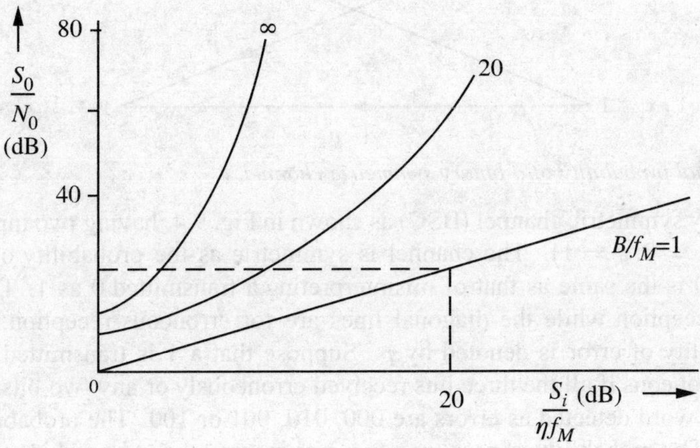

Fig. 9.3 *Variation of $(S/N)_0$ with $S_i/\eta f_M$ of an optimum system*

9.4 Coding and Redundancy

The channel noise causes mismatch between the input and output data sequence in a digital communications system. The bit error rate for a particular modulation scheme depends on the ratio of the signal-to-noise power. Practical considerations place an upper limit on the value of this S/N ratio. One way of achieving highly reliable data transmission at a high bit rate is by channel coding. A channel encoder in the transmitter accepts the input data sequence and converts it into codes in such a way that the overall detrimental effect of the channel is minimised; the encoder adds 'redundant' bits to the message bits according to some algorithm. The channel decoder in the receiver reconstructs the original data as closely as possible. The redundant bits carry no information but helps the receiver to detect and correct possible errors.

Consider, for example, the repetition code where a single message bit is encoded into a block of n identical bits. Suppose that we take $n = 3$, that is, a triplet of 0's is transmitted for a single 0 and a triplet of 1's for a single 1. The codes for 0 and 1 are 000 and 111, respectively. Assume that the channel noise is not high enough to produce more than one error in the triplet. The received message will be one of 001, 010, or 100 for a 0 bit and 110, 101 or 011 for a 1 bit. The decoder in the receiver then interprets the first group as a 0 bit and the second group as a 1 bit. With a group of three bits, there are eight possible binary combinations namely, 000, 001, 010, 011, 100, 101, 110, and 111; only two words, 000 and 111 are however valid words and the remaining six are invalid in the sense that they do not occur in our dictionary. Reception of any of these symbols is identified as error and the decoder makes the necessary correction based on the resemblance to 000 or 111. It is to be stressed here that not all the errors will be corrected. If two bits in the triplet are in error, we would be inclined to read a 0 as a 1 and vice versa.

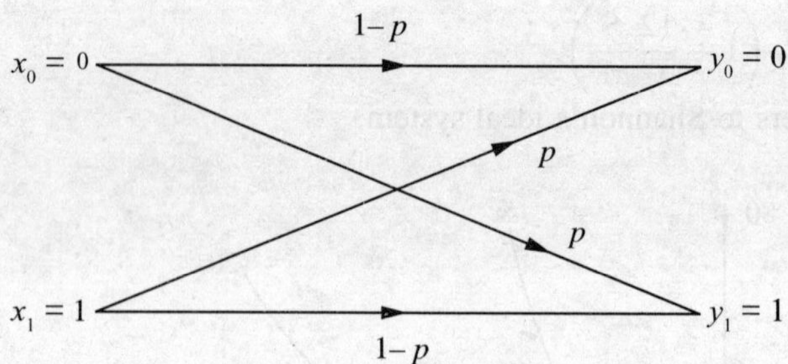

Fig. 9.4 *Transitional probability of a binary symmetric channel.*

Consider a binary symmetric channel (BSC) as shown in Fig. 9.4, having two inputs ($x = 0, x = 1$) and two outputs ($y = 0, y = 1$). The channel is symmetric as the probability of misinterpreting a transmitted 1 as a 0 is the same as that of misinterpreting a transmitted 0 as 1. The horizontal lines represent correct reception while the diagonal lines are for erroneous reception. In the figure, the conditional probability of error is denoted by p. Suppose that a 1 is transmitted by the triplet 111. The reception is erroneous if all the three bits received erroneously or any two bits in the triplet are in error. The received word detected as errors are 000, 010, 001 or 100. The probability of all the three bits in error is p^3. The probability of receiving two bits incorrectly and one bit correctly is $p^2(1-p)$ and there are three such errors. Thus the overall probability of incorrect reception is $p^3 + 3p^2(1-p)$. Intuitively, it is logical to say that if in a block of received bits in a word the number of 0's exceeds the number of 1's, the decoder decides in favour of a 0 and vice versa. If a block of n bits represents one message, the probability P_e of error is

$$P_e = \sum_{i=\frac{n}{2}\text{or}\frac{n+1}{2}}^{n} \binom{n}{i} p^i(1-p)^{n-i} \tag{9.20}$$

If $p = 10^{-12}$, then $P_e(n = 3) = 3p^2 = 3 \times 10^{-4}$; $P_e(n = 5) \approx 10^{-6}$. The probability of error thus decreases with increasing n.

Now suppose that the transition probability p is unacceptably high. The situation can be tackled by installing repeaters in between the transmitter and the repeater stations. The repeaters process the data it receives, takes decision about the bit received, and retransmits a fresh bit. A repeater improves the overall bit error rate as discussed below.

Consider now a cascade of two BSCs in between the transmitter and receiver as shown in Fig. 9.5. In this two-hop system, an error results if the bit is received incorrectly in either of the BSC. The probability of bit error is the sum of the probabilities of an error in the first hop followed by no error in the second and no error in the first followed by an error in the second. Errors in both the hops result in correct reception. Thus,

$$\begin{aligned} P_e &= p'(1-p') + (1-p')p' \\ &= 2p'(1-p') \approx 2p', \end{aligned} \tag{9.21}$$

where p' is the transition probability of each BSC. In equation (9.21), p' is usually less than $p/2$ since a new repeater station situated midway divides the spacing by half and reduces the bit error rate by more than a factor of 2.

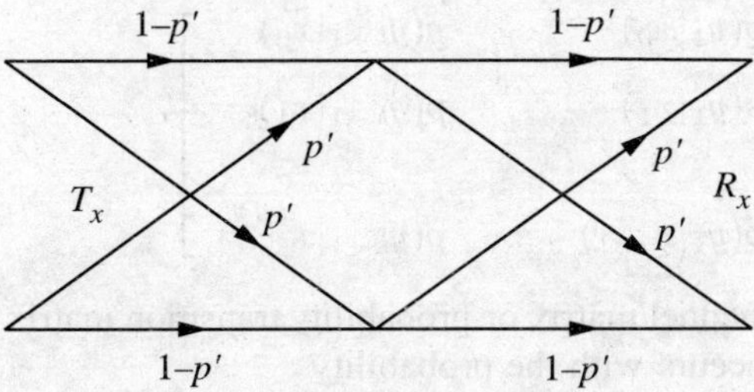

Fig. 9.5 *Cascade of two BSCs. The transition probability of each BSC is p ′.*

9.4.1 Discrete channel model

We have already come across the binary symmetric channel in the previous section. The BSC accepts two input symbols ($x = 0$ and $x = 1$) and gives out two output symbols ($y = 0$ and $y = 1$). That is, a binary bit stream enters at the transmitting end and another bit stream leaves out of the receiving end. We now extend this concept further by considering that the channel can accept an input symbol X and delivers an output symbol Y. Assume that X belongs to the at j-ary alphabet, i.e., $X = \{x_0, x_1, \ldots, x_{j-1}\}$ and Y belongs to the k-ary alphabet, i.e., $Y = \{y_0, y_1, \ldots, x_{k-1}\}$. The input and output can be related by a set of conditional probabilities

$$p(y_k|x_j) = P(Y = y_k|X = x_j) \text{ for all } j \text{ and } k. \tag{9.22}$$

This conditional probability $p(y_k|x_j)$ is defined as the channel transitional probability. This channel is called discrete memoryless channel (DMC) and is depicted in Fig. 9.6. The channel is discrete when both alphabet sizes are finite and memoryless in the sense that the current output solely depends on the current input and not on the inputs at previous stage(s). For a binary channel, $j = 2 = k$.

x_0, x_1, …, x_{j-1} — X → $p(y_k \mid x_j)$ → Y — y_0, y_1, …, y_{k-1}

Fig. 9.6 *A discrete memoryless channel*

The transitional probabilities are

$$p(1/0) = P(y = 1|x = 0) \text{ and } p(0|1) = P(y = 0|x = 1).$$

When these two probabilities are identical, $p(1/0) = p(0/1) = p$ (say), the channel is a symmetric channel.

Returning to the general discussions on DMC, suppose that the channel accepts an alphabet x_0. The possible outputs are $y_0, y_1, \ldots, y_{k-1}$, with the transition probabilities $p(y_0|x_0)$, $p(y_1|x_0)$, $\ldots, p(y_{k-1}|x_0)$. In the similar fashion, one can find the transition probabilities for all possible inputs. The conditional probabilities characterising a DMC can be conveniently arranged in the matrix form $P = [p_{kj}]$,

$$P = \begin{bmatrix} p(y_0|x_0) & p(y_1|x_0) & \dots & p(y_{k-1}|x_0) \\ p(y_0|x_1) & p(y_1|x_1) & \dots & p(y_{k-1}|x_1) \\ \dots & \dots & \dots & \dots \\ p(y_0|x_{j-1}) & p(y_1|x_{j-1}) & \dots & p(y_{k-1}|x_{j-1}) \end{bmatrix}. \tag{9.23}$$

The matrix P is called the channel matrix or probability transition matrix or simply transition matrix. The channel input $X = x_i$ occurs with the probability

$$p(x_i) = p(X = x_i), \quad i = 0, 1, \dots, j-1 \tag{9.24}$$

and these are the input *a priori* probabilities. Knowing $P(x_i)$ and the transition matrix P, we can calculate the probabilities of the output symbols $p(y_k)$ by the relation

$$p(y_k) = P(Y = y_k) = \sum_{j=0}^{j-1} P(y_k|x_j)p(x_j). \tag{9.25}$$

Mutual information

We now study the mutual information associated with the pair of events x_j and y_k. The self-information of a random variable X is defined as

$$H(X) = \sum_j p(x_j)I(x_j) = \sum_j p(x_j)\log_2 \frac{1}{p(x_j)}, \tag{9.26}$$

which gives a measure of the prior uncertainty about X. Imposing a condition on X, we may measure the conditional self-information of X, having observed Y, and this is defined as

$$H(X|Y) = \sum_j \sum_k p(x_j, y_k)\log_2 \frac{1}{p(x_j|y_k)}. \tag{9.27}$$

This is also called the average conditional self information or conditional entropy.

Thus, $H(X)$ defines the uncertainly about the channel input before observing the output and $H(X|Y)$ gives the same but having knowledge of the output. $H(X|Y)$ is also referred to as noise entropy. The difference $H(X) - H(X|Y)$ is that amount of uncertainty about the input that is resolved by knowing the channel output and is defined as the mutual information $I(X;Y)$ between the random variables X and Y. Thus,

$$I(X;Y) = H(X) - H(X|Y) = H(Y) - H(Y|X). \tag{9.28}$$

We may note the following important observations about $I(X;Y)$.

(i) Since the mutual information is non-negative, $I(X;Y) \geq 0$ and so $H(X) \geq H(X|Y)$. This implies that observance of Y cannot reduce the information about X.

(ii) Mutual information of a channel is symmetric, that is,
$I(X;Y) = I(Y;X)$.

(iii) Mutual information of a channel is related to the joint entropy $H(X,Y)$ by

$$I(X;Y) = H(X) + H(Y) - H(X,Y). \tag{9.29}$$

Channel capacity

For a discrete memoryless channel, the average information produced by the output Y about the input X is given by the mutual information:

$$I(X;Y) = \sum_k \sum_j p(x_j, y_k) \log_2 \frac{p(y_k|x_j)}{p(y_k)} = \sum_k \sum_j p(x_j, y_k) \log_2 \frac{p(x_j|y_k)}{p(x_j)}. \tag{9.30}$$

It is clearly seen that mutual information depends on the channel characteristics (e.g. noise, determined by the transition probability) and the input symbol probabilities $p(x_j)$. For a fixed transitional probability, a change in $p(x_j)$ results in different output symbols. We can maximise the mutual information over the set of input symbol probabilities and then $I(X;Y)$ depends only on the channel characteristics. This is called the channel capacity denoted by C. Thus

$$C = \max_{\{p(x_j)\}} I(X;Y) \tag{9.31}$$

and is measured in bits per channel use because each time the transmitter sends a symbol, it is said to *use* the channel. The right-hand side of (9.31) tells that the maximisation is obtained with respect to $p(x_j)$.

The mutual information as defined by Eq. (9.30) can be extended to continuous random variables. Let X and Y be two random variables with $p(x;y)$ the joint PDF and $p(x), p(y)$ being the marginal PDFs. The average mutual information between X and Y is defined as

$$I(X;Y) = \int_{-\infty}^{\infty}\int_{-\infty}^{\infty} p(x)p(y|x) \log_2 \frac{p(y|x).p(x)}{p(y)\,p(x)} dx\, dy \tag{9.31a}$$

There are certain instances where the random variable X is discrete but the random variable Y is continuous. Then, the variable X has some discrete outcomes x_i with the probability $p(x = x_i, i = 1, 2, \ldots, n)$ and Y is described by its marginal PDF $p(y)$. One can then write

$$p(y) = \sum_{i=1}^{n} p(y|x_i)\, p(X = x_i) \tag{9.31b}$$

The mutual information provided by the event $X = x_i$ for the occurrence of the event $Y = y$ is

$$I(x_i; y) = \log_2 \frac{p(y|x_i)\, p(X = x_i)}{p(y)\, p(X = x_i)} = \log_2 \frac{p(y|x_i)}{p(y)} \tag{9.31c}$$

The mutual information between X and Y is then

$$I(X;Y) = \sum_{v=1}^{n} \int_{-\infty}^{\infty} p(y|x_i) p(X = x_i) \log_2 \frac{p(y|x_i)}{p(y)} dy \tag{9.31d}$$

[Note the summation over discrete variable and integration over continuous variable].

Example 9.10

A binary symmetric channel is characterised by the transition probabilities $p(0|1) = p(1|0) = p$. Find the capacity of the channel.

Solution

For a binary channel, equation (9.30) reads as

$$I(X;Y) = \sum_{k=0,1}\sum_{j=0,1} p(x_j)p(y_k|x_j)\log_2 \frac{p(y_k|x_j)}{p(y_k)}$$

$$= \sum_{k=0,1}\left[p(x_0)p(y_k|x_0)\log_2 \frac{p(y_k|x_0)}{p(y_k)} + p(x_1)p(y_k|x_1)\log_2 \frac{p(y_k|x_1)}{p(y_k)}\right]$$

$$= p(x_0)p(y_0|x_0)\log_2 \frac{p(y_0|x_0)}{p(y_0)} + p(x_0)p(y_1|x_0)\log_2 \frac{p(y_1|x_0)}{p(y_1)}$$

$$+ p(x_1)p(y_0|x_1)\log_2 \frac{p(y_0|x_1)}{p(y_0)} + p(x_1)p(y_1|x_1)\log_2 \frac{p(y_1|x_1)}{p(y_1)}.$$

Now, assume input symbols are equiprobable, that is, $p(x_0) = p(x_1) = \frac{1}{2}$, where x_0 and x_1 are each 0 or 1. Also, the transitional probabilities are

$$p(y_0|x_0) = p(y_1|x_1) = 1 - p$$

$$p(y_1|x_0) = p(y_0|x_1) = p.$$

Thus, $C = \max I(X;Y) = \left[\frac{1}{2}(1-p)\log_2 2(1-p) + \frac{1}{2}p\log_2 2p\right] \times 2$

$$= (1-p)\log_2 2(1-p) + p\log_2 2p$$

$$= 1 - \left[p\log_2 \frac{1}{p} + (1-p)\log_2 \frac{1}{1-p}\right] = 1 - H(s).$$

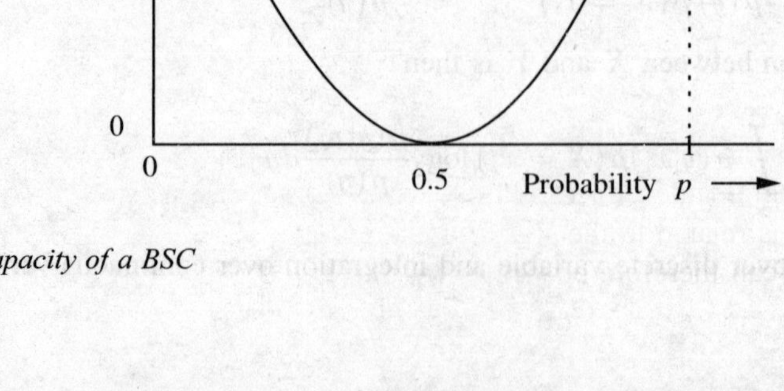

1.0

Capacity

0

0

0.5

Probability p ⟶

1

Fig. 9.7 *Channel capacity of a BSC*

Figure 9.7 depicts the plot of the channel capacity with the transition probability p. The following can be observed from the figure.

(i) For noiseless channel, $p = 0$ and the entropy $H(p)$ is zero. The channel capacity is maximum, equal to $C = 1$ bit/use. The information in each input symbol is also 1 bit. The channel thus delivers the full information at the output.

(ii) For $p = 1/2$, $H(p)$ attains its maximum value and hence channel capacity is zero. So the output provides nothing about the input. The channel is then useless. In fact, channel capacity decreases with increasing p in the range $0 < p < 1/2$.

(iii) When $p = 1$, the capacity is again 1 bit/use, a maximum. This is so because it is certain that every bit gets inverted. One just reverses whatever bit one gets to have the original input symbol.

9.4.2 Continuous channel capacity: Derivation of Shannon's law

The average mutual information for the discrete case has been defined in Eq. (9.30), in which we considered the source to transmit a symbol x_j that appears as another symbol y_k at the destination. In line with this, one can also define the average mutual information for continuous channels. Here, we assume that the source transmits a signal $x(t)$, which after contamination by noise, appears as another signal $y(t)$ at the destination. In this case,

$$I(X;Y) = \int_{-\infty}^{\infty}\int_{-\infty}^{\infty} f_{XY}(x,y)\log_2\frac{f_X(x|y)}{f_X(x)}dx\,dy \tag{9.32}$$

where the probabilities and conditional probabilities have been replaced by the density function and conditional density functions. Here $f_X(x)$ is source PDF, $f_{XY}(x,y)$ is the joint PDF. By knowing the destination entropy $H(Y)$ and the noise entropy $H(Y|X)$, one can calculate mutual information by Eq. (9.28). That is, $I(X;Y) = H(Y) - H(Y|X)$.

In the continuous channel, we may write conditional entropy $H(Y|X)$ as (vide Eq.9.27 for a comparison)

$$H(Y|X) = \int_{-\infty}^{\infty}\int_{-\infty}^{\infty} f_{XY}(x,y)\log_2\frac{1}{f_Y(y|x)}dx\,dy$$

$$= \int_{-\infty}^{\infty}\int_{-\infty}^{\infty} f_X(x)\,f_Y(y|x)\log_2\frac{1}{f_Y(y|x)}dx\,dy \tag{9.32a}$$

In order to have the expression for $H(Y|X)$, we assume that the channel is band-limited to B Hz and the noise $n(t)$ in the channel is independent additive white Gaussian-type of PSD $\eta/2$. So, the signal $y(t)$ at the channel output is related to the signal input $x(t)$ by $y(t) = x(t) + n(t)$. The transitional probability density of $y(t)$ can therefore be written as

$$f_Y(y|x) = f_Y(x+n) = f_N(y-x) \tag{9.32b}$$

where $f_N(y-x) = f_N(n)$ is the PDF of noise. Consequently,

$$H(Y|X) = \int_{-\infty}^{\infty}\int_{-\infty}^{\infty} f_X(x)\, f_Y(y/x) \log_2 \frac{1}{f_Y(y|x)}\, dx\, dy$$

$$= \int_{-\infty}^{\infty} f_X(x)\, dx \int_{-\infty}^{\infty} f_Y(y|x) \log_2 \frac{1}{f_Y(y|x)}\, dy$$

$$= \int_{-\infty}^{\infty} f_X(x) dx \int_{-\infty}^{\infty} f_N(y-x) \log_2 \frac{1}{f_N(y-x)}\, dy \tag{9.32c}$$

We let $u = y - x$ so that $du = dy$ and

$$H(Y/X) = \int_{-\infty}^{\infty} f_X(x) dx. \int_{-\infty}^{\infty} f_N(u) \log_2 \frac{1}{f_N(u)}\, du$$

$$= \int_{-\infty}^{\infty} f_X(x)\, dx. H(n) = H(n)$$

where $H(n)$ is the entropy of the noise sample $n(t)$ and $\int_{-\infty}^{\infty} f_X(x) dx = 1$ is used. Thus, mutual information becomes

$$I(X;Y) = H(Y) - H(n) \text{ bits/sample.} \tag{9.32d}$$

The channel capacity is thus

$$C_s = \max_{f_X(x)} I(X;Y) \text{bits/sample.} \tag{9.32e}$$

As the channel is band-limited to B Hz, the signal $y(t)$ is also band-limited to B Hz. The Nyquist rate of sampling should be $f_s = 2B$ sample/sec. In this case maximum rate of information transfer should be

$$C = C_s.2B \frac{\text{bits}}{\text{sample}} . \frac{\text{sample}}{\text{sec}}$$

$$= 2BC_s \text{ bits/sec} \tag{9.32f}$$

The implication of the above relation is that maximum information transfer without any error is possible if the information transfer rate $R \le C$.

One step forward, we further assume the following:

- the signal $x(t)$ from the source is with fixed power $S = \overline{x^2}$
- the channel noise $n(t)$ is with average noise power $N = \overline{n^2} = \frac{\eta}{2}.(2B) = \eta B$
- the signal and noise are independent so that they add in power i.e. for $y(t) = x(t) + n(t)$

$$\overline{y^2} = \overline{x^2} + \overline{n^2} = S + N \tag{9.32g}$$

In Section 9.2.2, we derived that the maximum entropy given by

$$H(X) \leq \frac{1}{2}\log_2(2\pi e\sigma^2)$$

where σ^2 is the signal power of $x(t)$. In the present case, the noise is with power N and so the noise entropy is given as

$$H(Y|X) \equiv H(n) = \frac{1}{2}\log_2(2\pi eN). \tag{9.32h}$$

Since the destination signal $y(t)$ has the power $\overline{y^2} = \overline{x^2} + \overline{n^2}$ or $\overline{y^2} = S + N$, we have

$$H(Y) \leq \frac{1}{2}\log_2(2\pi e(S+N)). \tag{9.32i}$$

The channel capacity in bits per sample is thus

$$\begin{aligned} C_s &= \max_{f_x(x)}[I(X;Y)] = \max_{f_x(x)}[H(Y) - H(Y/X)] \\ &= \max_{f_x(x)} H(y) - H(n) \\ &= \frac{1}{2}\log_2 2\pi e(S+N) - \frac{1}{2}\log_2 2\pi eN \\ &= \frac{1}{2}\log_2\left(\frac{S+N}{N}\right) = \frac{1}{2}\log_2\left(1 + \frac{S}{N}\right) \text{ bits/sample.} \end{aligned}$$

Thus, channel capacity per second is

$$C = 2BC_s = B\log_2\left(1 + \frac{S}{N}\right) \text{bits/sec.} \tag{9.32j}$$

The above is known as Hartley–Shannon law. This establishes an upper limit on the rate of information transmission over an AWGN-channel. We recognise here that S/N gives the signal- to-noise power ratio at the destination end, although S refer to the power of the transmitted signals $x(t)$. That means, it is implied that others than noise effect, there had been no detrimental causes to change the signal power.

9.5 Source Coding

Source coding is the process whereby efficient representation of the source symbols is accomplished by assigning binary codes. The most common examples of this assignment are binary, binary-coded decimal (BCD), binary-coded octal, two-out-of-five decimal and so on. The device that performs such representation is called source encoder. In selecting any such assignment, the source encoder considers the error detecting capability, compactness of codes (i.e., less redundant bits), ease of computation, etc. To generate efficient codes, the statistics of the 'source alphabet' must be known. Short codewords are assigned to frequent source symbols, whereas rare source symbols are represented by long codewords. This leads to the 'variable-length code'. As an illustrative example, consider that a source generates four symbols with probabilities 1/8, 1/8, 1/4, and 1/2. The source entropy is $H = (1/8\log_2 8) \times 2 + \frac{1}{4}\log_2 4 + \frac{1}{2}\log_2 2 = 1.75$ bits. One may select the binary codes 00, 01,10, and 11 to represent the four symbols, the average length of the codewords becoming 2 bits. This is one example of a fixed-length code if one selects 111, 110, 10, and 0 as the codewords, the average length is $\overline{N} = \sum_i p_i N_i = \frac{1}{8} \times 3 + \frac{1}{8} \times 3 + \frac{1}{4} \times 2 + \frac{1}{2} \times 1 = 1.75$ bits. We note that the most probable symbol is represented by the shortest codeword length.

Morse code is an example of variable length code in which the dots(.) and dashes $(-)$ are used to denote marks, and spaces respectively. 'E' is the most frequently used letter in English and so the Morse code encodes 'E' into a single dot. The letter 'Q' is less frequently used and encoded as (- - · -).

'Fixed-length codes' assign fixed number of bits to the source symbols, irrespective of their statistics. An example is the ASCII code which assigns 7-bit codewords to the source symbols to represent lower- and upper-script English letters (A to Z, a to z), decimal numbers (0 to 9), punctuation mark, commas, etc.

9.5.1 Properties of codes

The coded source symbols when transmitted in some order, appear as a sequence of bits. The codes must satisfy some properties so that the receiver can detect the possible symbols from the bit streams. We now demonstrate the properties by considering 4 source symbols A, B, C, and D, with the binary code assignments as shown in Table 9.1. Code 1 is a fixed-length code, whereas code 2 is a variable-length code. The message 'A BAD CD' in code 1 format would be encoded as 00 010011 1011, requiring a total 12 bits. But in code 2 format it appears as 0 1001 00 01, requiring 9 bits. Although a variable length code needs fewer number of bits, it does not qualify code 2 as the valid code – there is a decoding problem with it. Suppose that the sequence '0 1001 0001' is going to be decoded at the receiver. We could regroup the bit stream in different ways, namely, 01 00 1 00 01 which translates to DCBCD, or 0 1 0 01 0 00 1 which translates to ABADACB. A unique decoding of the codewords is not possible because there is no mark that indicates the 'end' or 'begin' of the codes. We do not face this problem with code 1 because it is a fixed length code and each group must include 2 bits together. The failure of unique decoding can be understood if one carefully looks at the codewords of the code 2 format. Letter 'A' is transmitted with the code '0' . But the receiver cannot decide if this is the whole codeword for 'A' or the partial codeword for 'C' or 'D'. Hence 'no codeword' will be the 'prefix' of any other codeword. This is the "prefix-free property". If this condition be satisfied, the possible codewords can be identified at the end of the codeword. Such codes are called 'uniquely decodable' or 'instantaneous codes'. Code 3 in Table 9.1 is such a code.

Table 9.1

Symbol	Code 1	Code 2	Code 3
A	00	0	0
B	01	1	10
C	10	00	110
D	11	01	111

9.5.2 Code length, entropy and coding efficiency

In the variable-length code assignments, sometimes we are interested in the 'average codeword length', $\overline{N}$, which is computed by the formula

$$\overline{N} = \sum_i p_i N_i. \tag{9.33}$$

Here N_i is the binary code length of the ith symbol whose probability is p_i. Shannon's theorem (also called noiseless coding) sets the minimum value of $\overline{N}$ as $N_{\min} = H$. We now define the 'coding efficiency' η as

$$\eta = \frac{N_{\min}}{\overline{N}} = \frac{H}{\overline{N}} \leq 1 \tag{9.34}$$

since $\bar{N} \geq H$. The 'redundancy' of a code is defined as

$$\text{Redundancy} = 1 - \eta = \frac{\overline{N} - H}{N}. \tag{9.35}$$

For a fixed-length code, the encoding scheme assigns a unique set of N bits to each symbol. If there are M possible symbols, the number of bits per symbol is

$$N = \log_2 M \tag{9.35a}$$

where M is some power of 2. For example, a set of 16 symbols ($16 = 2^4$) needs 4 bits assignment per symbol. When M is not a power of 2, then

$$N = [\text{largest integer in} \log_2^M] + 1 \tag{9.35b}$$

When source symbols are equally probable (when M is a power of 2), $N = H$ and so efficiency is 100 percent. However, when M is not a power of 2 but the source symbols are equally probable, N differs from entropy H by at most 1 bit per symbol.

When source symbols are not equally probable, the efficient encoding scheme is the variable-length code words. Normally, the symbols that occur more frequently are assigned short code words and those that are less frequent are assigned long code words. So, the assignment of code word lengths to the symbols depend on the probability of occurrence of the symbols and hence on the entropy of the symbols. Such encoding scheme is thus also called entropy coding.

Example 9.11

Consider a source that emits two symbols, s_1 and s_2, with the probabilities $p(s_1) = 3/4$ and $p(s_2) = 1/4$, respectively. Assign suitable binary codes to them and find the coding efficiency.

Solution

A compact code for the source is shown below.

s_i	$p(s_i)$	Code
s_1	3/4	0
s_2	1/4	1

The average code length, $\bar{N} = 1 \times \frac{3}{4} + 1 \times \frac{1}{4} = 1$ bit.

$$H = \frac{1}{4}\log_2 4 + \frac{3}{4}\log_2 (4/3) = 0.811 \text{ bits.}$$

So, coding efficiency, $\eta = \frac{H}{\bar{N}} = 0.811$.

Example 9.12

Find the coding efficiency for the two codes shown below.

Symbol	Prob.	Code 1	Code 2
A	9/16	0	00
B	3/16	10	01
C	3/16	110	10
D	1/16	111	11

Solution

The source entropy H is calculated as

$$H = \frac{9}{16}\log_2(16/9) + \frac{3}{16}\log_2(16/3) + \frac{3}{16}\log_2(16/3)\frac{1}{16} + \log_2(16) = 1.622 \text{ bits.}$$

Code 1: $\overline{N} = \frac{9}{16} \times 1 + \frac{3}{16} \times 2 + \frac{3}{16} \times 3 + \frac{1}{16} \times 3 = \frac{27}{16}$ bits.

$$\text{Hence } \eta = \frac{1.622}{27/16} = 0.961.$$

Code 2: This is a fixed-length code with $\overline{N} = 2$ bits.

$$\text{Hence } \eta = \frac{1.622}{2} = 0.811.$$

9.5.3 Prefix coding and decoding

The signal source produces a significant amount of information that is redundant. To obtain minimum sized codes, the redundant information should be removed from the signal prior to coding. This process is known as 'data compaction'. The code will then be 'exact'. One such coding technique is the 'prefix coding'. The general structure of the codeword of a symbol in this coding takes the form:

$$\text{Codeword} = [\underbrace{m_1\, m_2 \cdots m_i}\ m_{i+1} \cdots m_N]$$

where m_ks are 0's or 1's. The initial bit pattern $m_1 m_2 \cdots m_i$ forms the prefix of the codeword.

The decoding of a prefix code is as follows. The source decoder starts at the beginning of the input sequence, moves along a 'decision tree' and decodes one codeword at a time. A decision tree comprises an initial state, terminal state, and decision points. We now illustrate a decision tree decoding a prefix code shown in Table 9.2.

Table 9.2

Symbols	Code I	Code II
A	0	0
B	10	01
C	110	011
D	111	111

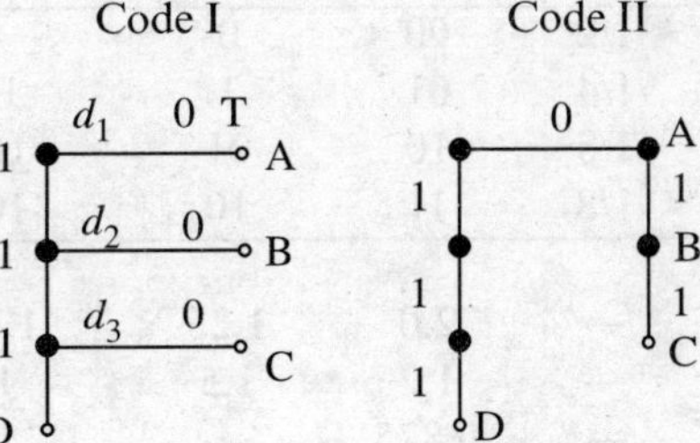

Fig. 9.8 *Decision tree of prefix coding*

Figure 9.8 shows the decision tree for the given codes I and II. The solid circles d_1, d_2, d_3 are the decision points (d_1 is the initial state as well) and open circles A, B, C, and D are the terminal states. The decoder always starts at the initial state d_1. If the received bit is a 0, move to the terminal state A, otherwise move to the decision point d_2. Depending on the received bit pattern, the decoder moves to the adjacent terminal point or moves one step further down. This process continues until each terminal state emits its symbol; the decoder then resets to its initial state.

The prefix condition requires that no codeword in the code is the prefix of any other codeword. However, Code II does not satisfy this condition. Symbol B is encoded as 01 which is the prefix of the code 011 for Symbol C.

The prefix codes guarantee that the codes are uniquely decodable (or decipherable) and no information is lost. A necessary and sufficient condition for the existence of a binary code is that the word lengths N_i must satisfy certain inequality known as 'Kraft–Mcmillan Inequality', which reach as

$$K = \sum_{i=1}^{M} e^{-N_i} \le 1, \tag{9.36}$$

where M is the number of source symbols. For a fixed length code, $N_i = \overline{N}$ and so $K = M.2^{-\overline{N}} \le 1$. Decipherability requires that $\overline{N} \ge \log_2 M$. Thus the coding efficiency

$$\eta = \frac{H}{\overline{N}} \le \frac{H}{\log_2 M}. \tag{9.37}$$

Example 9.13

Consider a source with possible symbols A, B, C, and D and the corresponding probabilities $p_{\mathrm{A}} = 1/2, p_{\mathrm{B}} = 1/4, p_{\mathrm{C}} = p_{\mathrm{D}} = 1/8$. Three different codewords are specified for the symbols. Justify the validity of the codewords.

Solution

Here, the source entropy is calculated as

$$H = \frac{1}{2}\log_2 2 + \frac{1}{4}\log_2 4 + \left(\frac{1}{8}\log_2 8\right) \times 2$$

$$= \frac{1}{2} + \frac{1}{4} \times 2 + \frac{1}{8} \times 3 \times 2 = \frac{1}{2} + \frac{1}{2} + \frac{3}{4} = 1.75 \text{ bits/symbol.}$$

Symbols	Prob.	Code 1	Code 2	Code 3
A	1/2	00	0	0
B	1/4	01	1	1 1
C	1/8	10	01	101
D	1/8	11	10	100
$\bar{N}$	$\rightarrow$	2.0	1.25	1.75
K	$\rightarrow$	1.2	1.5	1.0
η	$\rightarrow$	0.875	-	1.0

Code 1: Fixed-length code with $\bar{N} = 2$ bits.

$$K = \sum_i 2^{-Ni} = 2^{-2} \times 4 = 1.$$

$$\eta = \frac{H}{\bar{N}} = \frac{1.75}{2} = 0.875.$$

This is a valid code no doubt, but the coding efficiency η can be improved by a variable-length code.

Code 2: Variable-length code with $\overline{N} = \frac{1}{2} \times 1 + \frac{1}{4} \times 1 + \frac{1}{8} \times 2 + \frac{1}{8} \times 2$

$$= \frac{1}{2} + \frac{1}{4} + \frac{1}{4} + \frac{1}{4} = 1.25 \text{ bits.}$$

$$K = \sum_i 2^{-Ni} = 2^{-1} + 2^{-1} + 2^{-2} + 2^{-2} = 1.5 > 1.$$

Since Kraft inequality is not satisfied, the code is not uniquely decipherable and some information will be lost. For example, 'A BAD CD' is encoded as '0 1010 01 10' which may be decoded as 01 0 10 01 10, i.e., 'CAD CD'.

Code 3: Variable-length code with $\overline{N} = \frac{1}{2} \times 1 + \frac{1}{4} \times 2 + \frac{1}{8} \times 3 + \frac{1}{8} \times 3 = \frac{1}{2} + \frac{1}{2} + \frac{3}{4} = 1.75$

$$K = 2^{-1} + 2^{-2} + 2^{-3} + 2^{-3} = \frac{1}{2} + \frac{1}{4} + \frac{1}{8} + \frac{1}{8} = 1.$$

$$\eta = \frac{1.75}{1.75} = 1. \text{ This is the optimum code.}$$

For a given set of messages, the variable-length code achieves the shortest average code length $\overline{N}$. The best possible codes with variable lengths can be obtained following the two techniques namely 'Huffman codes' and 'Shannon-Fano algorithm'. Here the basic idea is to assign to each symbol a codeword of length roughly equal to the amount of information conveyed by the symbol. We discuss here the two source coding techniques.

9.5.4 Huffman coding

This code is optimum in the sense that the average number of bits in the codewords is a minimum. The steps of the coding algorithm are given below.

Step 1 Arrange the source symbols in the order of decreasing probabilities. The messages with equal probabilities can be selected in either order.

Step 2 Assign a 1 and a 0 to the bottom two symbols.

Step 3 Add the probabilities of the last two symbols. Rearrange the messages as in step 1, considering the last two symbols as a new symbol with added probability.

Step 4 Repeat steps 2 and 3 until only one probability is left.

Step 5 Obtain the prefix codeword by reading the levels starting from the right and moving back to the symbol.

We now illustrate the Huffman coding with an example.

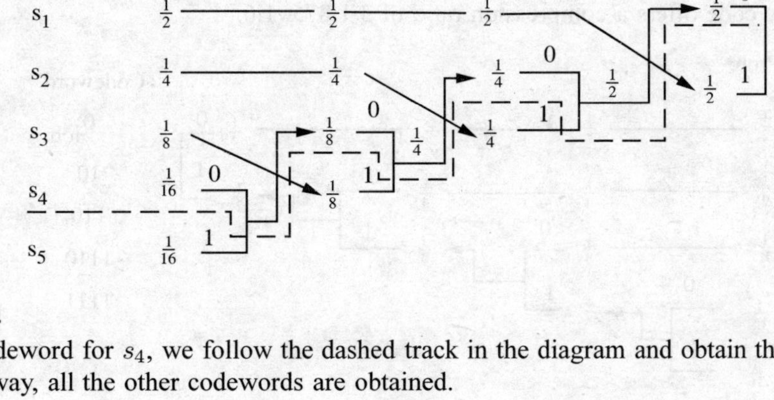

Symbol	Prob.
s_1	$\frac{1}{2}$
s_2	$\frac{1}{4}$
s_3	$\frac{1}{8}$
s_4	$\frac{1}{16}$
s_5	$\frac{1}{16}$

Example 9.14

To find the codeword for s_4, we follow the dashed track in the diagram and obtain the codeword as 0000. In this way, all the other codewords are obtained.

Symbol	Prob.	Codeword	Code length	Self information
s_1	1/2	1	1	1
s_2	1/4	01	2	2
s_3	1/8	001	3	3
s_4	1/16	0000	4	4
s_5	1/16	0001	4	4

For the symbols, the source entropy is

$$H = \frac{1}{2}\log_2 2 + \frac{1}{4}\log_2 4 + \frac{1}{8}\log_2 8 + 2 \times \frac{1}{16}\log_2 16 = \frac{15}{8} \text{ bits/symbol.}$$

The average codeword length

$$\overline{N} = \frac{1}{2} \times 1 + \frac{1}{4} \times 2 + \frac{1}{8} \times 3 + \frac{1}{16} \times 4 + \frac{1}{16} \times 4 = \frac{15}{8} \text{ bits.}$$

$$K = \sum_i 2^{-Ni} = 2^{-1} + 2^{-2} + 2^{-3} + 2^{-4} + 2^{-4} = \frac{1}{2} + \frac{1}{4} + \frac{1}{8} + \frac{1}{16} + \frac{1}{16} = 1.$$

Coding efficiency, $\eta = \dfrac{H}{\overline{N}} = 1$ (optimum code).

In the above, example, we have placed the new probability of the combined symbols at a position as high as possible in the list. Let us now repeat Example 9.14 with the combined symbols being placed as low as possible. The codewords obtained are 0, 10, 110, 1110, 1111. Again, the average code length and coding efficiency are the same as earlier.

Thus it is clear that the Huffman coding is not unique. There is arbitrariness in assigning 0's and 1's. Either of the tie-up probabilities may be assigned 0's or 1's; but this results in different codewords. Again, when the probability of the new symbol equals another probability in the list, we have a choice to put the combined probability higher or lower than the symbol.

We next discuss effective coding in terms of *compression ratio* defined as the ratio of the average number of bits per source symbol before coding to that after coding. In the above example, the average number of bits per symbol $\overline{N} = \frac{15}{8} = 1.875$ bits/symbol. This does not mean that we have to transmit such non-integer number of bits per symbol. Rather, it means that on the average we will have to transmit 1875 bits for 1000 source symbols. A fixed-length coding requires 3 bits per symbol. So the code offers a compression ratio of 3/1.875=1.6.

Symbol	Prob.				Codeword	Code length
s_1	$\frac{1}{2}$	$\frac{1}{2}$	$\frac{1}{2}$	$\frac{1}{2}$ 0	0	1
s_2	$\frac{1}{4}$	$\frac{1}{4}$	$\frac{1}{4}$ 0	$\frac{1}{2}$ 1	10	2
s_3	$\frac{1}{8}$	$\frac{1}{8}$ 0	$\frac{1}{4}$ 1		110	3
s_4	$\frac{1}{16}$ 0	$\frac{1}{8}$ 1			1110	4
s_5	$\frac{1}{16}$ 1				1111	4

9.5.5 Shannon–Fano code

In contrast to the Huffman code, Shannon–Fano code is performed in the forward direction. The Shannon–Fano algorithm for generating source code is presented below.

Step 1 Arrange the source symbols in the order of decreasing probabilities. The messages with equal probabilities can be selected in any arbitrary order.

Step 2 Divide the symbols in two partitions such that the sum of the probabilities in each partition is the same (or nearly the same).

Step 3 Assign bit 0 to all members in one partition and bit 1 to all members in the other partition.

Step 4 Repeat step 2 until each message is alone in each subdivision and then apply step 3.

We explain the algorithm by an example.

Example 9.15
Suppose that a source generates five messages with probabilities 1/2, 1/4, 1/8, 1/16, and 1/16. The encoding scheme is depicted in tabular form.

Symbols	Probabilities	Column 1	Column 2	Column 3	Column 4	Code
s_1	1/2	0				0
s_2	1/4	1	0			10
s_3	1/8	1	1	0		110
s_4	1/16	1	1	1	0	1110
s_5	1/16	1	1	1	1	1111

The entropy of the source is

$$H = \frac{1}{2}\log_2 2 + \frac{1}{4}\log_2 4 + \frac{1}{8}\log_2 8 + \frac{1}{16}\log_2 16 + \frac{1}{16}\log_2 16$$
$$= \frac{1}{2} + \frac{1}{2} + \frac{3}{8} + \frac{1}{4} + \frac{1}{4} = \frac{15}{8} = 1.875 \text{ bits.}$$

The average length of this code is

$$\overline{N} = 1 \times \frac{1}{2} + 2 \times \frac{1}{4} + 3 \times \frac{1}{8} + 4 \times \frac{1}{16} + 4 \times \frac{1}{16}$$
$$= \frac{15}{8} = 1.875 \text{ bits.}$$

Thus each bit on the average carries 1 bit of information and coding efficiency $\eta = 100\%$.

Example 9.16
Consider a discrete memoryless source with seven possible symbols $s_i, i = 1, 2, \ldots, 7$ with the corresponding probabilities $p_1 = p_2 = \frac{1}{4}$, $p_3 = p_4 = p_5 = \frac{1}{8}$, $p_6 = p_7 = 1/16$. Find the codewords for the seven symbols using Shannon–Fano algorithm.

Answer:

s_k	p	Column 1	Column 2	Column 3	Column 4	Code	No. of bits in the code
s_1	1/4	0	0			00	2
s_2	1/4	0	1			01	2
s_3	1/8	1	0	0		100	3
s_4	1/8	1	0	1		101	3
s_5	1/8	1	1	0		110	3
s_6	1/16	1	1	1	0	1110	4
s_7	1/16	1	1	1	1	1111	4

The average length of the codeword is

$$\overline{N} = \frac{1}{4}\times 2 + \frac{1}{4}\times 2 + \frac{1}{8}\times 3 + \frac{1}{8}\times 3 + \frac{1}{8}\times 3 + \frac{1}{16}\times 4 + \frac{1}{16}\times 4 = \frac{21}{8} = 2.625 \text{ bits}$$

$$\text{Source entropy, } H = \left(\frac{1}{4}\log_2 4\right)\times 2 + \left(\frac{1}{8}\log_2 8\right)\times 3 + \left(\frac{1}{16}\log_2 16\right)\times 2$$

$$= 1 + \frac{9}{8} + \frac{1}{2} = \frac{21}{8} \text{ bits/source symbol.}$$

So, coding efficiency, $\eta = \frac{H}{\overline{N}} = 1$.

In the fixed-length code of seven symbols, we require 3 bits per symbol and hence

$$\eta = \frac{21}{8\times 3} = 0.875.$$

9.5.6 Source coding theorem

So far we solved the coding problems with symbol probabilities occurring in power of 1/2. Let us now discuss the source symbols with arbitrary probabilities. In that case $\log_2 p_i$ is not an integer. Still, it is possible to construct a compact code that satisfies the prefix condition. The codeword length N_i of the ith symbol should satisfy the inequality

$$\log_2(1/p_i) \le N_i < \log_2(1/p_i) + 1, \tag{9.38}$$

that is, N_i is the first integer greater than $\log_2(1/p_i)$.

Equivalently, $\quad p_i \log_2(1/p_i) \le p_i N_i < p_i \log_2(1/p_i) + p_i,$

or, summing over the possible symbols,

$$H \le \overline{N} < H + 1. \tag{9.39}$$

Thus we can make the average number of code symbols per source symbol as small as but no smaller than the entropy of the source. This sets the lower bound of $\overline{N}$.

To prove the theorem, we proceed as follows.

$$
\begin{aligned}
H - \overline{N} &= \sum_i p_i \log_2(1/p_i) - \sum_i p_i N_i \\
&= \sum_i p_i \log_2(1/p_i) + \sum_i p_i \log_2 2^{-N_i} = \sum_i p_i \log_2\left(\frac{2^{-N_i}}{p_i}\right) \\
&= \log_2 e \sum_i p_i \ln\left(\frac{2^{-N_i}}{p_i}\right).
\end{aligned}
$$

By Stirling's inequality, $\ln x \le x - 1$, we have

$$
\begin{aligned}
H - \overline{N} &\le (\log_2 e) \sum_i p_i \left(\frac{2^{-N_i}}{p_i} - 1\right) = (\log_2 e) \sum_i (2^{-N_i} - p_i) \\
&= (\log_2 e) \left[\sum_i 2^{-N_i} - 1\right].
\end{aligned}
$$

According to Kraft inequality, $\sum_i 2^{-N_i} \le 1$ and so $H - \overline{N} \le (\log_2 e)\left(\sum_i 2^{-N_i} - 1\right) \le 0$, that is, $H \le \overline{N}$. To prove the upper bound, we select the codeword length N_i such that $2^{-N_i} \le p_i < 2^{-N_i+1}$. Summing over the possible source symbols,

$$
\sum_i 2^{-Ni} \le \sum_i p_i = 1,
$$

which is the Kraft inequality. Thus $2^{-N_i} \le p_i$ dictates the prefix condition. Now, take logarithm on both sides of $p_i < 2^{-N_i+1}$ to get

$$
\log_2 p_i < -N_i + 1 \quad \text{or,} \quad N_i < 1 - \log_2 p_i.
$$

Multiplying both sides by p_i and summing over i,

$$
\sum_i p_i N_i < \sum_i (p_i - p_i \log_2 p_i)
$$

$$
\text{or} \quad \sum_i p_i N_i < 1 + H
$$

that is, $\overline{N} < H + 1$.

Example 9.17

Consider a source which generates three symbols with probabilities 2/3, 2/9 and 1/9. Verify the source coding theorem. Repeat the same if the source symbols have probabilities in the form $(1/2)^\alpha$, i.e., 1/2 , 1/4 and 1/4.

Solution

Source entropy

$$H = \frac{2}{3}\log_2(3/2) + \frac{2}{9}\log_2(9/2) + \frac{1}{9}\log_2 9$$

$$= 0.3899 + 0.4822 + 0.3522$$

$$= 1.224 \text{ bits/source symbol.}$$

Symbols	p_i	Code
s_1	2/3	0
s_2	2/9	10
s_3	1/9	11

Average code-length, $\overline{N} = 1 \times \frac{2}{3} + 2 \times \frac{2}{9} + 2 \times \frac{1}{9} = 1.33$ bits. Thus we have $H < \overline{N} < H + 1$.

In the second case, $\overline{N} = 1\times\frac{1}{2}+2\times\frac{1}{4}+2\times\frac{1}{4} = 1.5$ bits and $H = \frac{1}{2}\log_2 2 + \frac{1}{4}\log_2 4 + \frac{1}{4}\log_2 4 = 1.5$ bits. Here we have $H \leq \overline{N} < H + 1$.

9.6 Channel Coding

All communication channels are affected by noise that causes discrepancies between the input and output data streams. For a typically noisy channel, the probability of error may reach a value as high as 10^{-2}, meaning that on the average, 1 bit out of 100 transmitted bits gets corrupted. For many applications, this level of reliability is far from acceptable. Typical acceptable bit error rate for various applications may range from 10^{-4} (as in speech telephony) to as low as 10^{-7} (as in video telephony or high-speed data computing).

Channel coding is the process of formatting the incoming data sequence into a channel input sequence and inverse formatting the channel output sequence into an output data sequence in a way that achieves a high level of reliability. The first formatting operation is performed in the transmitter by a channel encoder. The inverse formatting (or decoding) is done by the channel decoder. A number of redundant bits are added to the input sequence in a prescribed algorithm to attain an acceptable value of average probability of bit errors, thereby improving reliability. For a reliable transmission we may face the question: how may bits per second can one send over a given channel with bit error rate as low as possible? The answer is embodied in Shannon's second theorem known as channel coding theorem.

Let a discrete, memoryless source with an alphabet X and entropy $H(X)$ generates a symbols every T_s seconds. The average information rate of the source is $\frac{H(x)}{T_s}$ bits per second. Let us assume that a discrete, memoryless channel has a capacity of C bits per channel use and the channel is used once every T_c second. Then the channel capacity per unit time is $\frac{C}{T_c}$. There exists a coding scheme for which the source output can be transmitted over the channel and can be reconstructed with arbitrarily low probability of error if

$$\frac{H(X)}{T_s} \leq \frac{C}{T_c} \tag{9.40}$$

is satisfied. Otherwise it is not possible to transmit source information and reconstruct it with an arbitrarily low probability of error. The rate deciding parameter C/T_c is called the critical rate. The quantity T_c/T_s is called the code rate of the channel encoder.

Example 9.18

A discrete, memoryless binary source emits equally likely 1s and 0s once every T_s second. The channel encoder uses the channel of capacity C once every T_c second. Find the condition for reliable communication via a code.

Solution

The source entropy is given by

$$H(X) = p\log_2\frac{1}{p} + (1-p)\log_2\frac{1}{1-p} = 1 \text{ bit.}$$

Source information rate,

$$\frac{H(X)}{T_s} = \frac{1}{T_s} \text{ bits/second.}$$

Since channel is used once in every T_c second, the encoded symbol transmission rate is $\frac{1}{T_c}$ symbols per second.

By the channel coding theorem,

$$\frac{1}{T_s} \le \frac{C}{T_c} \quad \text{or,} \quad \frac{T_c}{T_s} \le C \quad \text{or} \quad r \le C.$$

This is the condition.

Error detection and correction relies on the use of redundant bits in the transmitted codewords. The receiver should know the encoding protocol so that the code structure provides the complete error information. When the structured code is sufficient to locate the position of the errors, the code is an error-correcting code. Normally, the number of bits per symbol in the error correcting codes is large because of the added redundant bits.

A number of codes have been developed to detect errors and these are used in Automatic Request for Repeat (ARQ) systems. One such code is the 'constant-ratio code' in which the ratio of 1's and 0's in the codeword remains the same. The '2 out-of-5' code follows a pattern of two 1's and three 0's in the symbols. The receiver declares an error if the ratio of 1's and 0's is changed. Once an error is detected in a group, the receiver sends a NAK (do not acknowledge) response to the transmitter and the word is retransmitted. For the correctly received data, the receiver sends a positive acknowledge (ACK) response. The transmitter then transmits the next word. This testing procedure continues for every word. This coding allows only odd numbers of errors to be detected.

The 'redundant code' is another important error detection code where the message symbol is repeated twice. These two sets of words are compared in the receiver. Data is erroneous when these two sets disagree bit by bit. The data is supposed to be retransmitted again. The efficiency of this code is obviously 50% as the second set of the transmitted word is cent percent redundant. The addition of redundant bits in the coded symbols needs increased transmission bandwidth. Also, the receiver section becomes complex for implementing the decoding operation.

Different error-correcting codes have been classified into either block codes or convolutional codes. The classification is based on the presence or absence of memory elements in the encoder. The following sections discuss a few important codes.

9.6.1 Block codes

This is a fixed-length' code characterised by (n, k) notation. The desired number of redundant bits is added to each message word. If r redundant bits are added to each message of k-bit length, the coded word is of length $n = k + r$. Such block code is designated as an (n, k) block code. For a (7,4) code, there are four information bits and three redundant bits. The total number of possible n-bit codeword is 2^n, out of which 2^k represents the possible messages and $(2^n - 2^k)$ possible n-bit words do not represent the message. It is essential that the k-bit message word and the n-bit coded word should take the same time for transmission. Denoting by T_b and T_c the bit durations of uncoded and coded words respectively, $nT_c = kT_b$. The dimensionless ratio $r = \frac{k}{n} = T_c/T_b$ is called the 'code rate'. When R_s is the bit rate of the information source, the channel encoder produces bits at the rate $R_0 = r.R_s$.

A block code is said to be a linear block code if any two codewords when added in modulo-2 addition produces a third codeword belonging to the code. When the message part in the coded word remains unaltered, the block code is referred to as 'systematic code'.

Distance between codewords

The distance between two equal-length codewords is defined as the number of locations in which the respective bits differ. Consider two 7-bit codewords $C_i = 1001101$ and $C_j = 1011011$ in which the third, fifth, and sixth positions differ in bits. Thus the distance between these two words, is $d_{ij} = 3$. The minimum of d_{ij} for all possible pairs is called the 'Hamming distance', denoted by $d_{\min}$. The Hamming distance indicates the potential of the code to detect and correct symbols in error. Suppose that there are d errors in a received codeword. The (n, k) block code can detect all the errors provided that $d \leq d_{\min} - 1$. The number of errors t that can be corrected is $(d_{\min} - 1)/2$ or $(d_{\min} - 2)/2$ for odd or even $d_{\min}$, respectively. If $d_{\min} = 3$ in a certain code, up to two errors can be detected and a single error can be corrected.

Parity-check codes

Parity-bits are the redundant bits added to the original (uncoded) data. In a 'single-parity check code', a single parity bit is added to a block of data bits; the system may be designed for 'even parity' or 'odd parity'. Figure 9.9 illustrates the formation of even parity in the 'serial structure' of the data bits. Here, a k-bit data block transforms into a block of $(k + 1)$ code bits. The rate of the code is $r = k/n = k/(k + 1)$. The decoding procedure at the receiving end consists of performing 'modulo-2 sum' of the code bits and testing whether the sum yields an even or odd result. For even parity there are even number of 1s in the codeword and the sum will be zero. But for odd parity, the result sums to 1. Suppose that the system is designed for even parity but the modulo-2 addition yields a 1. The decoder then detects error. It detects only the presence of an odd number of errors. If even number of bits is inverted, the parity test fails to give any indication of errors; this is a case of undetected errors.

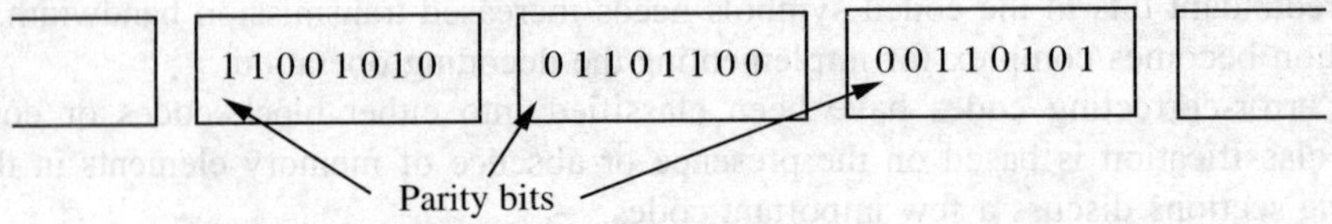

Fig. 9.9 *Parity checks for serial structure of words*

In a block of n bits, the probability of errors is given by the binomial distribution as

$$P(i,n) = {}^{n}C_i \, p^i (1-p)^{n-i}, \tag{9.41}$$

where 'p' is the probability that a bit is received in error due to channel impairment. Thus, in a single-parity test, the probability of undetected error (i.e., even number of bits being erroneous) is given by

$$P_u = {}^{n}C_2 \, p^2 (1-p)^{n-2} + {}^{n}C_4 \, p^4 (1-p)^{n-4} + \cdots \sum_{i=1}^{n/2 \text{ or } (n-1)/2} {}^{n}C_{2i} \, p^{2i} (1-p)^{n-2i}. \tag{9.42}$$

Example 9.19

In a (4,3) block code, error detection is based on a single parity test. The encoder is designed for even parity. (a) Compute all the codewords. (b) Which error patterns can be detected? (c) Calculate the probability of undetected errors.

Solution

Here, each message is $k = 3$ bit long and coded word is $n = 4$ bit long. So, parity bit $r = n - k = 1$.

(a)

Message	Parity	Codeword
0 0 0	0	0 0 0 0
0 0 1	1	1 0 0 1
0 1 0	1	1 0 1 0
0 1 1	0	0 0 1 1
1 0 0	1	1 1 0 0
1 0 1	0	0 1 0 1
1 1 0	0	0 1 1 0
1 1 1	1	1 1 1 1

(b) The code can detect any odd number of bits in error, that is, single- or triple-error patters can be detected. These errors may be anywhere within the codeword.

(c) $P_u = {}^4C_2 p^2 (1-p)^2 + {}^4C_4 p^4 (1-p)^0 = \dfrac{4!}{(4-2)!2!} p^2 (1-p)^2 + \dfrac{4!}{0!4!} p^4 = 6p^2(1-p)^2 + p^4.$

Rectangular code

In the rectangular code the message bits are arranged in an array comprising of M rows and N columns. The horizontal parity check bit is appended to each row and the vertical parity check bit is appended to each column. The number of message bits is $k = M \times N$ and the number of bits after coding is $n = (M+1) \times (N+1)$. The code rate is then

$$r = \frac{k}{n} = \frac{MN}{(M+1)(N+1)}. \tag{9.43}$$

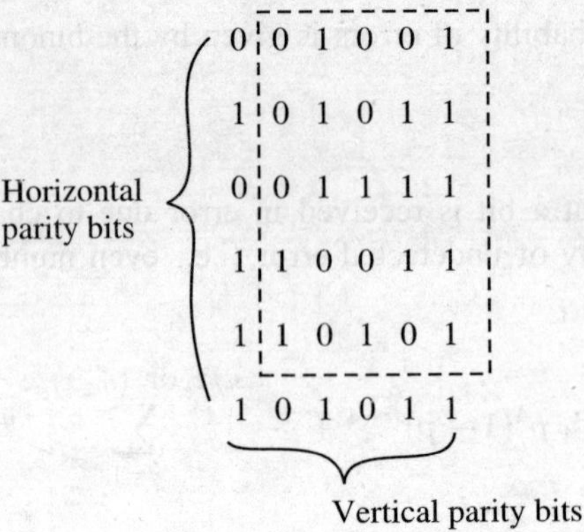

Fig. 9.10 *Rectangular code structure*

Figure 9.10 shows a (36, 25) rectangular code where $M = N = 5$. The 25 bits within the dotted square are the message bits. If there be a single error anywhere in the message array, both horizontal and vertical parity tests will disagree. The error is located at the intersection of the row and column that contains the disagreement. To get the correct message bit, simply invert the bit at that position.

Linear block codes

We have already introduced the (n, k) notation for the block codes. A message is a block of k-bits and a codeword is a block n-bits wide. With k-bits, there may be 2^k distinct message sequence. The encoder uniquely maps 2^k messages into a new set of 2^k-codewords of n-bit width. The remaining $(2^n - 2^k)$ words are invalid because these do not correspond to any of the messages. This mapping transformation is 'linear' for linear block codes. To begin with, we illustrate the mapping or code assignment in a (7,3) linear block code.

Here $n = 7$, and $k = 3$ so that $r = 4$. There are $2^k = 8$ messages. Table 9.3 shows the assignment of eight codewords to messages. This mapping can be accomplished via a 'look-up' table. But look-up table implementation is possible for a small k. When k is large, huge memory size is required to store the large number of coded words. The feasible solution is to generate the codewords when needed, instead of storing them. We will define a 'generator matrix' that can generate the codewords and represent the linear block code in matrix form.

Table 9.3 Linear mapping of message into codewords

Message (M)	Codeword (C)	Code vector
000	0000 000	V_1
001	1011 001	$V_2 \leftarrow$
010	1101 010	$V_3 \leftarrow$
011	0110 011	V_4
100	1110 100	$V_5 \leftarrow$
101	0101 101	V_6
110	0011 110	V_7
111	1000 111	V_8

Any n-bit codeword can be represented as a 'vector' in an n-dimensional vector space V_n. There are 2^n code vectors out of which 2^k are valid. In the vector space V_n, there are 'k basis vectors' that are linearly independent and all the other 2^k vectors can be generated with a linear combination of these basis vectors. The codewords (C) can be expressed as

$$C = m_1V_1 + m_2V_2 + \cdots + m_kV_k, \tag{9.44}$$

where $(m_1, m_2 \cdots m_k)$ is the k-bit message vector (M). We define the generator matrix (G) as a $k \times n$ array.

$$[G] = \begin{bmatrix} V_1 \\ V_2 \\ \vdots \\ V_k \end{bmatrix} = \begin{bmatrix} v_{11} & v_{12} & v_{13} & \cdots & v_{1n} \\ v_{21} & v_{22} & v_{23} & \cdots & v_{2n} \\ . & . & \cdots & \cdots & \cdots \\ v_{k1} & v_{k2} & v_{k3} & \cdots & v_{kn} \end{bmatrix}_{k\times n}. \tag{9.45}$$

The codeword $[C]$ is obtained by the matrix multiplication

$$[C]_{1\times n} = [M]_{1\times k}[G]_{k\times n}. \tag{9.46}$$

For the (7,3) code we find the basis vectors V_5, V_3 and V_2, and so the generator matrix is

$$[G] = \begin{bmatrix} \leftarrow & V_5 & \rightarrow \\ \leftarrow & V_3 & \rightarrow \\ \leftarrow & V_2 & \rightarrow \end{bmatrix} = \begin{bmatrix} 1\,1\,1\,0\,1\,0\,0 \\ 1\,1\,0\,1\,0\,1\,0 \\ 1\,0\,1\,1\,0\,0\,1 \end{bmatrix}_{3\times 7}. \tag{9.47}$$

Suppose we want to generate the codeword for the message, $M = [m_1\ m_2\ m_3] = [1\ 0\ 1]$.

$$C = [101] \begin{bmatrix} 1\,1\,1\,0\,1\,0\,0 \\ 1\,1\,0\,1\,0\,1\,0 \\ 1\,0\,1\,1\,0\,0\,1 \end{bmatrix} = [0101101]$$

Alternatively,

$$C = [101] \begin{bmatrix} V_5 \\ V_3 \\ V_2 \end{bmatrix} = 1 \times V_5 + 0 \times V_3 + 1 \times V_2 = V_5 + V_2$$

that is, the code vector is the linear combination of the basis vectors.

We have observed that the lookup table works fine for small number of bits k in the message word. However, the selection criterion of basis vectors in 2^k number of n-dimensional vectors has not been discussed.

Let the message vector be denoted by

$$[M] = [\ m_1 \quad m_2 \quad \ldots \quad m_k\ ] \tag{9.47a}$$

and the code vector C be given by

$$[C] = \begin{bmatrix} c_1 & c_2 & \dots & c_n \end{bmatrix} \tag{9.47b}$$

The encoding operation can be represented by a set of n equations of the form:

$$\left.\begin{array}{lcl} c_1 &=& m_1g_{11} + m_2g_{21} + \cdots + m_kg_{k1} \\ c_2 &=& m_1g_{12} + m_2g_{22} + \cdots + m_kg_{k2} \\ \cdot & & \cdot \quad \cdot \quad \cdot \\ c_n &=& m_1g_{1n} + m_2g_{2n} + \cdots + m_kg_{kn} \end{array}\right\} \tag{9.47c}$$

The above set of equations signifies that each bit of the coded word can be linearly dependent on the message words. The constant g_{ij} may be 0 or 1. In a matrix form, the above equations can be expressed as

$$\begin{bmatrix} c_1 \\ c_2 \\ \vdots \\ c_n \end{bmatrix}_{n\times 1} = \begin{bmatrix} g_{11} & g_{21} & \cdots & g_{k1} \\ g_{12} & g_{22} & \cdots & g_{k2} \\ \cdot & \cdot & \cdot & \cdot \\ g_{1n} & g_{2n} & \cdots & g_{kn} \end{bmatrix}_{n\times k} \begin{bmatrix} m_1 \\ m_2 \\ \vdots \\ m_k \end{bmatrix}_{k\times 1}$$

$$\text{or } \begin{bmatrix} c_1 & c_2 & \cdots & c_n \end{bmatrix}_{1\times n} = \begin{bmatrix} m_1 & m_2 & \cdots & m_k \end{bmatrix}_{1\times k} \begin{bmatrix} g_{11} & g_{12} & \cdots & g_{1n} \\ g_{21} & g_{22} & \cdots & g_{k2} \\ \vdots & \vdots & \vdots & \vdots \\ g_{k1} & g_{k2} & \cdots & g_{kn} \end{bmatrix}_{k\times n}$$

$$= \begin{bmatrix} m_1 & m_2 & \cdots & m_k \end{bmatrix} \begin{bmatrix} \rightarrow V_1 \leftarrow \\ \rightarrow V_2 \leftarrow \\ \vdots \\ \rightarrow V_k \leftarrow \end{bmatrix}$$

$$\text{or } [C] = [M][G] \tag{9.47d}$$

The matrix $[G]_{k\times n}$ above is called the generator matrix of the code. The row vectors V_i of the generator matrix must be linearly independent and represent the basis for the (n, k) code. Since the set of basis vectors is not unique, so also the generator matrix. This means there might be many code words for a given message word.

Systematic linear block codes

A linear block code is said to be in systematic form if the message bits are transmitted in unaltered from. In other words, part of the bit sequence in the codewords are the k-message bits intact and the remaining $(n - k)$ bits are the redundant bits. The codeword has the form:

$$[C] = (\underbrace{m_1m_2m_3\cdots m_k;}_{k-\text{message bits}}\ \underbrace{b_1b_2\cdots b_r)}_{r-\text{parity bits}} = [c_1c_2\cdots c_k;\ c_{k+1}c_{k+2}\cdots c_n] \tag{9.47e}$$

$[C]$ is a $1 \times n$ row matrix and the uncoded word $[M] = (m_1 m_2 \cdots m_k)$ is a $1 \times k$ row matrix.

The positions of the message bits and the redundant bits can be interchanged. The coded words $[c_1 c_2 \cdots c_n]$ can be represented by $c_i = m_i, i = 1, 2, \ldots, k$ and $c_i = b_{i-k}, i = k+1, k+2, \ldots, n$. The parity bits $[b_1 b_2 \cdots b_r]$ are obtained from the relation

$$[b_1 b_2 \cdots b_r] = [m_1 m_2 \cdots m_k][P]_{k\times r}, \tag{9.48}$$

where $[P]_{k\times r}$ is the parity array called the 'coefficient matrix' and has the form

$$[P]_{k\times r} = \begin{bmatrix} p_{11} & p_{12} & \cdots & p_{1r} \\ p_{21} & p_{22} & \cdots & p_{2r} \\ \cdots & \cdots & \cdots & \cdots \\ p_{k1} & p_{k2} & \cdots & p_{kr} \end{bmatrix}, \tag{9.49}$$

with $p_{ij} = 0$ or 1. We define the generator matrix as

$$[G]_{k\times n} = \begin{bmatrix} I_k & \vdots & P \end{bmatrix}$$

$$= \begin{bmatrix} \underbrace{\begin{matrix} 1 & 0 & \cdots & 0 \\ 0 & 1 & \cdots & 0 \\ . & . & \cdots & . \\ 0 & 0 & \cdots & 1 \end{matrix}}_{k\times k \text{ identity matrix}, I_k} & \underbrace{\begin{matrix} h_{11} & h_{12} & \cdots & h_{1r} \\ h_{21} & h_{22} & \cdots & h_{2r} \\ \cdots & \cdots & \cdots \\ h_{k1} & h_{k2} & \cdots & h_{kr} \end{matrix}}_{k\times r P^T \text{ matrix}} \end{bmatrix}_{k\times n}. \tag{9.50}$$

The merit of the block code depends on the ingenuity with which the h_{ij} have been selected. With the generator matrix defined in (9.50), the codewords can be formed as in (9.46). It is the part of the decoder to decode the received words and check for errors. It is the 'parity-check matrix' $[H]$ that governs such checking. For each $k \times n$ generator matrix $[G]$, there is always an $r \times n$ parity-check matrix such that rows of [G] are orthogonal to the rows of $[H]$. Mathematically,

$$[G]_{k\times n}[H]^T_{n\times r} = [0]_{k\times r} \tag{9.51}$$

where [0] is a $k \times r$ null matrix. Following (9.50) and (9.51), we have

$$[H]_{r\times n} = \begin{bmatrix} -P^T & \vdots & I_{n-k} \end{bmatrix} = \begin{bmatrix} h_{11} & h_{21} & \cdots & h_{k1} & \vdots & 1 & 0 & \cdots & 0 \\ h_{12} & h_{22} & \cdots & h_{k2} & \vdots & 0 & 1 & \cdots & 0 \\ \cdots & \cdots & \cdots & \cdots & \vdots & \cdots & \cdots & \cdots & \cdots \\ h_{1r} & h_{2r} & \cdots & h_{kr} & \vdots & 0 & 0 & \cdots & 1 \end{bmatrix} \tag{9.52}$$

$$\text{and } [H]^T = \begin{bmatrix} P^T \\ \cdots \\ I_{n-k} \end{bmatrix} = \begin{bmatrix} h_{11} & h_{12} & \cdots & h_{1r} \\ h_{21} & h_{22} & \cdots & h_{2r} \\ \cdots & \cdots & \cdots & \cdots \\ h_{k1} & h_{k2} & \cdots & h_{kr} \\ 1 & 0 & \cdots & 0 \\ 0 & 1 & \cdots & 0 \\ \cdots & \cdots & \cdots & \cdots \\ 0 & 0 & \cdots & 1 \end{bmatrix} \tag{9.53}$$

The coded words formed from $[C] = [M][G]$ has the property that $[C][H]^T = [M][G][H]^T = [M]_{1\times k}[0]_{k\times r} = [0]_{1\times r}$

or, equivalently, $[H][C]^T = [0]$. (9.54)

Example 9.20

Consider a (7.4) code with the generator matrix

$$G = \left[\begin{array}{cccc|ccc} 1 & 0 & 0 & 0 & 0 & 1 & 1 \\ 0 & 1 & 0 & 0 & 1 & 0 & 1 \\ 0 & 0 & 1 & 0 & 1 & 1 & 0 \\ 0 & 0 & 0 & 1 & 1 & 1 & 1 \end{array}\right] = \begin{bmatrix} I_4 & P \end{bmatrix}$$

The typical codeword can be expensed as

$$C = \begin{bmatrix} m_1 & m_2 & m_3 & m_4 & b_1 & b_2 & b_3 \end{bmatrix}$$

The parity-check bits are given by

$$\begin{bmatrix} b_1 & b_2 & b_3 \end{bmatrix} = \begin{bmatrix} m_1 & m_2 & m_3 & m_4 \end{bmatrix} \begin{bmatrix} 0 & 1 & 1 \\ 1 & 0 & 1 \\ 1 & 1 & 0 \\ 1 & 1 & 1 \end{bmatrix}$$

that is,

$$\begin{aligned} b_1 &= m_2 + m_3 + m_4 \\ b_2 &= m_1 + m_3 + m_4 \, . \\ b_3 &= m_1 + m_2 + m_4 \end{aligned}$$

The parity-check matrix for the code is then

$$[H]_{r\times n} = \begin{bmatrix} P^T & I_{n-k} \end{bmatrix} = \left[\begin{array}{cccc|ccc} 0 & 1 & 1 & 1 & 1 & 0 & 0 \\ 1 & 0 & 1 & 1 & 0 & 1 & 0 \\ 1 & 1 & 0 & 1 & 0 & 0 & 1 \end{array}\right]$$

It is now easy to verity that

$$G_{k\times n}H^T_{n\times r}=\left[\begin{array}{cccc|ccc}1&0&0&0&0&1&1\\0&1&0&0&1&0&1\\0&0&1&0&1&1&0\\0&0&0&1&1&1&1\end{array}\right]\left[\begin{array}{ccc}0&1&1\\1&0&1\\1&1&0\\1&1&1\\\hline 1&0&0\\0&1&0\\0&0&1\end{array}\right]=\begin{bmatrix}0&0&0\\0&0&0\\0&0&0\\0&0&0\end{bmatrix}_{4\times 3}$$

The (n, k) block encoder can be implemented using a k-bit shift register and $(n-k)$ adders as shown in Fig. 9.10 (a).

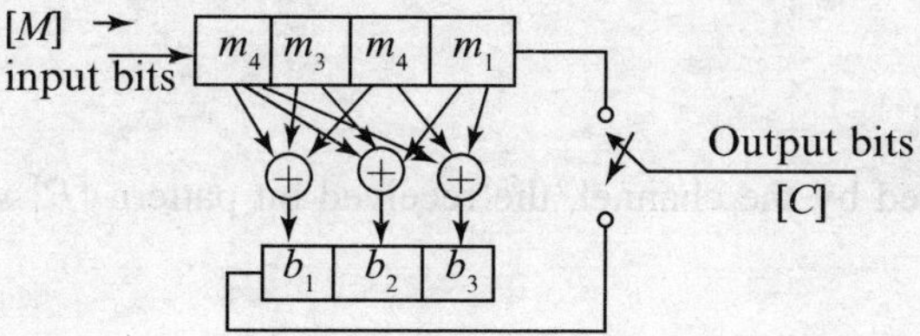

Fig. 9.10 (a) *Encoder for a (7.4) linear block code using linear shift register*

The logic expressions for the parity bits $[b_1, b_2, b_3]$ are implemented in modulo-2 addition. The k-input bits are shifted into the register stages, parity bits are then computed. The coded bits at the output line are the k-information bits $[m_1, m_2, m_3, m_4]$ followed by $n-k$ parity-check bits $[b_1, b_2, b_3]$.

Error Detection by Syndrome Testing

The coded signalling waveforms are transmitted on the AWGN channel. The n-bits of a received codeword are passed to the decoder and hard decision is made as to whether each transmitted bit in a codeword is a 0 or a1. The decoding rule should be so designed that there will be minimum probability of a codeword error. One such decoding algorithm is the minimum-distance coding algorithm, in which the decoder compares the received code word with the $2^k = M$ possible transmitted code words and calculates the Hamming distances. The decision goes in favour of the codeword that gives lowest Hamming distance.

Another efficient decoding algorithm is based on the parity-check matrix $[H]$. For a transmitted codeword $[C]$, the received codeword $[R]$ at the decoder output will be

$$[R] = [C] + [E] \tag{9.55}$$

Where $[E]_{1\times n}$is an n-dimensional error vector or error pattern. There are a total of $(2^n - 1)$ non-zero error patterns. We now compute the product

$$[R][H]^T = [C+E]_{1\times n}[H^T]_{n\times r} = CH^T + EH^T = EH^T = [S]_{1\times n-k} \tag{9.55a}$$

(since $CH^T = 0$).

The $(n-k)$-dimensional vector S is called the syndrome of the error pattern. If there are no errors, $S = [0]$. The vector S has non-zero components if the received word differs from the transmitted word. We note that S is a characteristic of the error pattern and it has no bearing on the transmitted coded words. The error vector is n-dimensional and has 2^n possible error patterns but there are only $2^{n-k} = 2^r$ syndromes. This means that many different error patterns may yield the same syndrome.

The error patterns may be single, double, triple, and so on. Most frequent error is the single error pattern, as follows:

$$\begin{array}{l}\text{Error pattern}\\ [\,1\ \ 0\ \ 0\ \ 0\ \ \cdots\ \ 0\,]\\ [\,0\ \ 1\ \ 0\ \ 0\ \ \cdots\ \ 0\,]\\ [\,0\ \ 0\ \ 1\ \ 0\ \ \cdots\ \ 0\,]\\ \ \ \cdot\ \ \ \cdot\ \ \ \cdot\ \ \ \cdot\ \ \ \cdot\\ [\,0\ \ 0\ \ 0\ \ 0\ \ \cdots\ \ 1\,]\end{array}$$

When an error is introduced by the channel, the received bit pattern $[R]$ will differ from the coded word $[C]$, that is,

$$[R] = [C] \oplus [E],$$

where $[E] = (e_1 e_2 \cdots e_n)$ is an error vector or error pattern. There are a total of $(2^n - 1)$ nonzero error patterns. The alternate expression for syndrome is

$$[S] = [H][R]^T. \tag{9.56}$$

$$\text{Now, } [S] = [H]([C] \oplus [E])^T = [H][C]^T \oplus [H][E]^T = [H][E]^T \tag{9.57}$$

since $[C][H]^T = [0]$ by (9.54).

Example 9.21

For a (7,4) linear block code, the parity-check matrix H is given as

$$[H] = \begin{bmatrix} 1 & 1 & 1 & 0 & 1 & 0 & 0 \\ 1 & 1 & 0 & 1 & 0 & 1 & 0 \\ 1 & 0 & 1 & 1 & 0 & 0 & 1 \end{bmatrix}_{3\times 7}.$$

Obtain the generator matrix $[G]$ and list all possible codewords.

Solution

Here $n = 7, k = 4$ so that $r = n - k = 3$. The generator matrix is of order $k \times n$, that is 4 by 7. The coefficient matrix is

$$[P]^T = \begin{bmatrix} 1 & 1 & 1 & 0 \\ 1 & 1 & 0 & 1 \\ 1 & 0 & 1 & 1 \end{bmatrix}.$$

Referring to (9.50), we can write

$$[G] = [I|P] = \begin{bmatrix} 1 & 0 & 0 & 0 & 1 & 1 & 1 \\ 0 & 1 & 0 & 0 & 1 & 1 & 0 \\ 0 & 0 & 1 & 0 & 1 & 0 & 1 \\ 0 & 0 & 0 & 1 & 0 & 1 & 1 \end{bmatrix}_{4\times 7}.$$

For an uncoded word $[M] = (m_1m_2m_3m_4)$, we have the coded words

$$[C] = [M][G] = (m_1m_2m_3m_4)\begin{bmatrix} 1 & 0 & 0 & 0 & 1 & 1 & 1 \\ 0 & 1 & 0 & 0 & 1 & 1 & 0 \\ 0 & 0 & 1 & 0 & 1 & 0 & 1 \\ 0 & 0 & 0 & 1 & 0 & 1 & 1 \end{bmatrix}$$

so that the parity bits are $b_1 = m_1 \oplus m_2 \oplus m_3$; $b_2 = m_1 \oplus m_2 \oplus m_4$; $b_3 = m_1 \oplus m_3 \oplus m_4$. Table 9.4 lists all the coded words.

Table 9.4

Message words	Coded words	Message words	Coded words
0000	0000 000	1000	1000 111
0001	0001 011	1001	1001 100
0010	0010 101	1010	1010 010
0011	0011 110	1011	1011 001
0100	0100 110	1100	1100 001
0101	0101 101	1101	1101 010
0110	0110 011	1110	1110 100
0111	0111 000	1111	1111 111

Example 9.22

In Example 9.21, let the transmitted bit stream $[C] = (1011001)$ be received as $[R] = (1100001)$. Find the syndrome $[S]$.

Solution

A bit by bit comparison of $[R]$ with $[C]$ reveals that there are 3-bits in error at the 4th, 5th, and 6th positions from the LSB.

Now,

$$[S] = [H][R]^T = \begin{bmatrix} 1110 & 100 \\ 1101 & 010 \\ 1011 & 001 \end{bmatrix}\begin{bmatrix} 1 \\ 1 \\ 0 \\ 0 \\ 0 \\ 0 \\ 1 \end{bmatrix} = \begin{bmatrix} 0 \\ 0 \\ 0 \end{bmatrix} = [0].$$

We note that the syndrome has a value $[0]$. It does not mean that syndrome testing fails. The syndrome test tells us whether the received word is a valid codeword or not. In the present example the errors are so large (3 bits in error out of 4-message bits) that the transmitted codeword has been changed to another valid codeword resulting in $[S] = [0]$. The error goes undetected.

Let us now discuss the methods of finding the error pattern. In the above example we have taken bit errors at the three positions. Can a syndrome-test locate the error pattern? We will answer the question in the following.

Suppose that the transmitted bit pattern is $[C] = [1010010]$ and the received word is $[R] = [1110\ 010]$ with a single error at the 6th position. The syndrome of this codeword is

$$S = HR^T = \begin{bmatrix} 1 & 1 & 1 & 0 & 1 & 0 & 0 \\ 1 & 1 & 0 & 1 & 0 & 1 & 0 \\ 1 & 0 & 1 & 1 & 0 & 0 & 1 \end{bmatrix} \begin{bmatrix} 1 \\ 1 \\ 1 \\ 0 \\ 0 \\ 1 \\ 0 \end{bmatrix} = \begin{bmatrix} 1 \\ 1 \\ 0 \end{bmatrix}.$$

Since $S = HE^T$, we have

$$\begin{bmatrix} 1 \\ 1 \\ 0 \end{bmatrix} = \begin{bmatrix} 1 & 1 & 1 & 0 & 1 & 0 & 0 \\ 1 & 1 & 0 & 1 & 0 & 1 & 0 \\ 1 & 0 & 1 & 1 & 0 & 0 & 1 \end{bmatrix} \begin{bmatrix} e_1 \\ e_2 \\ e_3 \\ e_4 \\ e_5 \\ e_6 \\ e_7 \end{bmatrix}, \text{ where } E = (e_1 e_2 \cdots e_7) \text{ is the error pattern.}$$

This leads to three simultaneous equations of seven unknowns.

$$1 = e_1 \oplus e_2 \oplus e_3 \oplus e_5$$

$$1 = e_1 \oplus e_2 \oplus e_4 \oplus e_6$$

$$0 = e_1 \oplus e_3 \oplus e_4 \oplus e_7.$$

Naturally, a unique solution is not possible. However, all the $(2^n - 1)$ possible error patterns are not of importance. A single bit error is most likely. The best possible error pattern involving single error is the following:

$$\begin{matrix} 1 & 0 & 0 & 0 & 0 & 0 & 0 \\ 0 & 1 & 0 & 0 & 0 & 0 & 0 \\ 0 & 0 & 1 & 0 & 0 & 0 & 0 \\ 0 & 0 & 0 & 1 & 0 & 0 & 0 \\ 0 & 0 & 0 & 0 & 1 & 0 & 0 \\ 0 & 0 & 0 & 0 & 0 & 1 & 0 \\ 0 & 0 & 0 & 0 & 0 & 0 & 1 \end{matrix}$$

That is, if the first bit is erroneous, the error pattern is $E = (1000000)$. With an error in the second bit position, the error pattern is $E = (0100000)$ and so on. Suppose that the first bit of the received word is erroneous. Then,

$$S = HE^T = \begin{bmatrix} 1 & 1 & 1 & 0 & 1 & 0 & 0 \\ 1 & 1 & 0 & 1 & 0 & 1 & 0 \\ 1 & 0 & 1 & 1 & 0 & 0 & 1 \end{bmatrix} \begin{bmatrix} 1 \\ 0 \\ 0 \\ 0 \\ 0 \\ 0 \\ 0 \end{bmatrix} = \begin{pmatrix} 1 \\ 1 \\ 1 \end{pmatrix}.$$

In this way we determine the syndrome to each single error sequence and form a look-up table (Table 9.5). We can use this look-up table to find the corresponding error pattern. The correct codeword is then estimated as

$$\widehat{T} = R \oplus E. \tag{9.58}$$

Table 9.5 *Look-up table*

Error pattern	Syndrome
1000000	111
0100000	110
0010000	101
0001000	011
0000100	100
0000010	010
0000001	001

Example 9.23

Assume that a codeword $T = 1001100$ is transmitted and the received codeword is $R = 1011100$. Using the syndrome look-up Table 9.5, locate the error pattern and find the correct word.

Solution

The syndrome of the received word is

$$S = HR^T = \begin{pmatrix} 1 & 1 & 1 & 0 & 1 & 0 & 0 \\ 1 & 1 & 0 & 1 & 0 & 1 & 0 \\ 1 & 0 & 1 & 1 & 0 & 0 & 1 \end{pmatrix} \begin{pmatrix} 1 \\ 0 \\ 1 \\ 1 \\ 1 \\ 0 \\ 0 \end{pmatrix} = \begin{pmatrix} 1 \\ 0 \\ 1 \end{pmatrix}.$$

The corresponding error pattern tabulated in Table 9.5 is $E = (0010000)$, that is, error at the third bit position. The correct word is

$$\widehat{T} = R \oplus E = 1011100 \oplus 0010000$$

$$= 1001\ 100.$$

Hamming codes

In an (n, k) block code, the number, $r = n - k$ of redundant bits is related to the block length k as

$$n = 2^r - 1, k = 2^r - 1 - r. \tag{9.59}$$

For $r \geq 3$, the codes are referred to as 'Hamming codes'. When $r = 3$, we have $n = 7$ and $k = 4$; a (7,4) code is formed. A value of $r = 4$ leads to a (15,11) code. For a (15,11) linear block code, the parity check matrix $[H]$ is given by

$$[H] = \begin{matrix} 15 & 14 & 13 & 12 & 11 & 10 & 9 & 7 & 6 & 5 & 3 & 8 & 4 & 2 & 1 \end{matrix}$$

$$\underbrace{\begin{bmatrix} 1 & 1 & 1 & 1 & 1 & 1 & 1 & 0 & 0 & 0 & 0 & 1 & 0 & 0 & 0 \\ 1 & 1 & 1 & 1 & 0 & 0 & 0 & 1 & 1 & 1 & 0 & 0 & 1 & 0 & 0 \\ 1 & 1 & 0 & 0 & 1 & 1 & 0 & 1 & 1 & 0 & 1 & 0 & 0 & 1 & 0 \\ 1 & 0 & 1 & 0 & 1 & 0 & 1 & 1 & 0 & 1 & 1 & 0 & 0 & 0 & 1 \end{bmatrix}}_{P \quad\quad I_4} \tag{9.60}$$

Note that the columns of the H-matrix are the binary representation of the numbers from 1 to 15. The identity matrix formed of 8-4-2-1 is separated from the rest. This is the systematic form of H-matrix. It may be mentioned here that the H-matrix is not 'unique'. All permutation of the columns of the P-matrix leads to a different H-matrix; all are valid matrices.

9.6.2 Cyclic codes

Cyclic codes are important class of linear block codes and have the property that any cyclic shift (end-around) of the codeword results in another valid codeword. Consider a codeword $[T] = 0110101$ that is subjected to cyclic shift.

–		$[T] =$	0110101
1st	Cyclic shift	$\longrightarrow$	1101010
2nd	Cyclic shift	$\longrightarrow$	1010101
3rd	Cyclic shift	$\longrightarrow$	0101011
4th	Cyclic shift	$\longrightarrow$	1010110
5th	Cyclic shift	$\longrightarrow$	0101101
6th	Cyclic shift	$\longrightarrow$	1011010
7th	Cyclic shift	$\longrightarrow$	0110101

At the seventh cyclic shift, the original codeword is obtained. Let us first develop the necessary mathematical background that yields the correct codeword after few cyclic shifts. We first express the codewords in polynomial form: the components of the codeword $[T] = (T_0 T_1 \cdots T_{n-1})$ can be treated as the coefficients of a polynomial $T(x)$.

$$T(x) = T_0 + T_1 x + T_2 x^2 + \cdots + T_{n-1} x^{n-1}. \tag{9.61}$$

For the codeword 1011010, the polynomial is $T = 1 + 0.x + 1.x^2 + 1.x^3 + 0.x^4 + 1.x^5 + 0.x^6 = (1 + x^2 + x^3 + x^5)$. Shifting all the bits by one position to left yields the code vector

$$T' = [T_1\ T_2\ T_3 \cdots T_{n-1}\ T_0].$$

The corresponding polynomial can be written as

$$T'(x) = T_1 + T_2x + T_3x^2 + \cdots + T_{n-1}x^{n-2} + T_0x^{n-1}. \tag{9.62}$$

For a comparison of (9.61) and (9.62), we multiply $T'(x)$ by x and sum to $T(x)$. Thus,

$$T(x) = T_0 + T_1x + T_2x^2 + \cdots + T_{n-1}x^{n-1},$$

$$xT'(x) = \quad T_1x + T_2x^2 + \cdots + T_{n-1}x^{n-1} + T_0x^n.$$

Since $T_1 \oplus T_1 = 0$ and so on, we get

$$T(x) + xT'(x) = T_0 + T_0x^n = T_0(1 + x^n)$$

or $\quad T(x) = xT'(x) + T_0(1 + x^n)$ (9.63)

since $T + C = 0$ implies $T = C$.

Similarly, for a second shift

$$T'' = [T_2\ T_3 \cdots T_{n-1}\ T_0\ T_1]$$

the associated polynomial is

$$T''(x) = T_2 + T_3x^1 + \cdots + T_{n-1}x^{n-3} + T_0x^{n-2} + T_1x^{n-1}. \tag{9.64}$$

Following the previous steps, we have

$$T(x) = T_0 + T_1x + T_2x^2 + \cdots + T_{n-1}x^{n-1}.$$

$$x^2T''(x) = T_2x^2 + T_3x^3 + \cdots + T_{n-1}x^{n-1} + T_0x^n + T_1x^{n+1}.$$

Summing, $T(x) + x^2T''(x) = T_0 + T_1x + T_0x^n + T_1x^{n+1} = T_0(1 + x^n) + T_1x(1 + x^n)$

or $\quad T(x) = x^2T''(x) + (T_0 + T_1x)(1 + x^n).$ (9.65)

It is informative to note from (9.63) and (9.65) that the polynomial $(x^n + 1)$ and its coefficient play important roles in cyclic codes. By iteration, the coefficient polynomial of $(x^n + 1)$ for the ith shifts can be written as $T_0 + T_1x + T_2x^2 + \cdots T_{i-1}x^{i-1}$. We next define an (n, k) cyclic code by a generator polynomial $g(x)$ of order $r = n - k$, of the form

$$g(x) = 1 + g_1x + g_2x^2 + \cdots + g_{r-1}x^{r-1} + x^r. \tag{9.66}$$

The values of coefficients $g_1, g_2, \cdots, g_{r-1}$ are such that $g(x)$ is a factor of $x^n + 1$. The codeword $T(x)$ is then found from the polynomial product

$$T(x) = g(x)A(x), \tag{9.67}$$

where $A(x)$ is the message polynomial representing k-bits:

$$A(x) = a_0 + a_1x + a_2x^2 + \cdots + a_{k-1}x^{k-1}. \tag{9.68}$$

The following example explains the procedure outlined above.

Example 9.24
Generate a (7,4) cyclic code.

Solution
Since $n = 7$, we take the factor $f(x) = x^7 + 1$. The uncode word $[A]$ is of $k = 4$ bits so that the associated polynomial is of degree $k - 1 = 3$. Take

$$A(x) = a_0 + a_1x + a_2x^2 + a_3x^3.$$

The factors of $f(x)$ are

$$f(x) = (1 + x)(1 + x + x^3)(1 + x^2 + x^3).$$

Since the polynomial $T(x)$ will be of $n - 1 = 6$, we can select either of the factor $(1 + x + x^3)$ and $(1 + x^2 + x^3)$. Selecting $g(x) = (1 + x^2 + x^3)$, we form the polynomial

$$T(x) = A(x)g(x) = (a_0 + a_1x + a_2x^2 + a_3x^3)(1 + x^2 + x^3)$$

$$= a_0 + a_1x + (a_0 + a_2)x^2 + (a_0 + a_1 + a_3)x^3 + (a_1 + a_2)x^4 + (a_2 + a_3)x^5 + a_3x^6.$$

Suppose the uncoded word is $[A] = 1101$, then

$$A(x) = 1 + x + x^3 \qquad \text{so that}$$

$$T(x) = (1 + x + x^3)(1 + x^2 + x^3) = 1 + x + x^2 + x^3 + x^4 + x^5 + x^6$$

and $[T] = [1\ 1\ 1\ 1\ 1\ 1\ 1]$.

The sixteen possible codewords are tabulated in Table 9.6 for the 4-bit messages.

Table 9.6 *Codewords of Example 9.19.*

Message word				Codeword	Message word				Codeword
a_0	a_1	a_2	a_3		a_1	a_2	a_3		
0	0	0	0	0000000	1	0	0	0	1011000
0	0	0	1	0001011	1	0	0	1	1010011
0	0	1	0	0010110	1	0	1	0	1001110
0	0	1	1	0011101	1	0	1	1	1000101
0	1	0	0	0101100	1	1	0	0	1110100
0	1	0	1	0100111	1	1	0	1	1111111
0	1	1	0	0111010	1	1	1	0	1100010
0	1	1	1	0110001	1	1	1	1	1101001

The codewords thus formed are non-systematic.

By the cyclic property, one can find the codeword for any number of cyclic shifts. A procedure for forming the codeword for the ith shift is as follows:

(i) Multiply the given codeword $T(x)$ by x^i.

(ii) Divide $x^iT(x)$ by $(x^n + 1)$.

(iii) Form the code by the remainder in the ascending order.

As an example, let the codeword be $[T] = [1101001]$.

The fifth cyclic shifts results in the codeword [0111010]. To verify this result, we first find $x^iT(x)$ with $i = 5$.

$$x^5T(x) = x^5(1 + x + x^3 + x^6) = x^5 + x^6 + x^8 + x^{11}.$$

Dividing by $x^7 + 1$, we get $x^4 + x$ and a remainder $= x^6 + x^5 + x^4 + x$.

The required code is $(x + x^4 + x^5 + x^6) = [0100111]$. Five shifts in the left yields the same result as two shifts in the right.

Matrix description of cyclic codes

For an (n, k) code, the generator matrix $[G]$ will have k rows and n columns. The matrix elements are given by the polynomial $g(x)$.

Let T be a cyclic code with

$$g(x) = g_0 + g_1x + g_2x^2 + \cdots + g_{r-1}x^{r-1}.$$

The generator matrix of $T(x)$ is given as

$$[G] = \begin{bmatrix} g_0 & g_1 & g_2 & \cdots & g_{r-1} & 0 & 0 & 0 \\ 0 & g_0 & g_1 & \cdots & . & g_{r-1} & 0 & 0 \\ 0 & & g_0 & \cdots & . & . & g_{r-1} & 0 \\ . & . & . & \cdots g_0 & & g_1 \cdots & \cdots & g_{r-1} \end{bmatrix}. \tag{9.69}$$

Example 9.25

Repeat Example 9.24 using generator matrix.

Solution

Here $[A] = [1101]$. We again take $g(x) = 1 + x^2 + x^3 = g_0 + g_1x + g_2x^2 + g_3x^3$, with $g_0 = 1, g_1 = 0.g_2 = 1, g_3 = 1$.

The generator matrix is thus

$$[G] = \begin{bmatrix} 1 & 0 & 1 & 1 & 0 & 0 & 0 \\ 0 & 1 & 0 & 1 & 1 & 0 & 0 \\ 0 & 0 & 1 & 0 & 1 & 1 & 0 \\ 0 & 0 & 0 & 1 & 0 & 1 & 1 \end{bmatrix}.$$

Hence, the coded word

$$[T]=[A][G]=(1101)\begin{bmatrix}1&0&1&1&0&0&0\\0&1&0&1&1&0&0\\0&0&1&0&1&1&0\\0&0&0&1&0&1&1\end{bmatrix}=(1111111).$$

Example 9.26

Find the generator matrix of all the codes of block length $n = 4$.

Solution

We first factorise $f(x) = x^4 + 1$.

$$x^4+1=(x+1)(x^3+x^2+x+1)$$
$$=(x+1)(x+1)(x^2+1).$$

If we take $g(x) = 1 + x$ as the generator polynomial, then $g_0 = 1, g_1 = 1$ so that $r = 1$. Then $k = n - r = 4 - 1 = 3$.

$$[G]=\begin{bmatrix}1&1&0&0\\0&1&1&0\\0&0&1&1\end{bmatrix}.$$

We can thus obtain $(n, k) = (4, 3)$ cyclic codes. For $g(x) = 1 + x^2, g_0 = 1, g_1 = 0, g_2 = 1$ and $r = 2$, so that $k = n - r = 4 - 2 = 2$. Then

$$[G]=\begin{bmatrix}1&0&1&0\\0&1&0&1\end{bmatrix}.$$ The code is a (4,2) cyclic code.

The parity-check matrix $[H]$ can also be generated by the relation

$$x^n+1=h(x)g(x)$$

and expressing $h(x)$ as $h_kx^k + h_{k-1}x^{k-1} + \cdots + h_1x + h_0$ and forming the matrix as above.

Systematic cyclic codes

In its systematic form, the message bits are kept unaltered. We shift the message bits into the right-most k-stages and append the parity bits in the left-most $(n - k)$-stages. That is,

$$\begin{array}{lll}T_0T_1\cdots T_{n-1} & p_0p_1\cdots p_{r-1} & a_0\cdots a_{k-1}\\ [n\text{-bit coded word}]= & [r\text{ parity bits;} & k\text{-message bits}].\end{array}$$

We define the message and parity check bit polynomials as:

$$A(x)=a_0+a_1x+a_2x^2+\cdots+a_{k-1}x^{k-1}$$

$$p(x)=p_0+p_1x+p_2x^2+\cdots+p_{r-1}x^{r-1}$$

$r = (n-k))$ stages right-shifted message polynomial is

$$x^{n-k}A(x) = a_0x^{n-k} + a_1x^{n-k+1} + \cdots + a_{k-1}x^{n-1}.$$

We require the codeword polynomial to be

$$T(x) = p(x) + x^{n-k}A(x). \tag{9.70}$$

Thus, $A(x)g(x) = p(x) + x^{n-k}A(x)$

or, dividing by $g(x)$

$$\frac{x^{n-k}A(x)}{g(x)} = A(x) + \frac{p(x)}{g(x)}. \tag{9.71}$$

Thus, $p(x)$ is nothing but the remainder obtained by dividing $x^{n-k}A(x)$ with $g(x)$.

Example 9.27
Find the (7,4) cyclic code in the systematic form using the generator polynomial $g(x) = 1 + x + x^3$ and for the message $[A] = [1011]$.

Solution
Here the message-bit polynomial is

$$A(x) = 1 + 0.x + 1.x^2 + 1.x^3 = (1 + x^2 + x^3). \text{ So,}$$

$$x^{n-k}A(x) = x^{7-4}(1 + x^2 + x^3) = x^3 + x^5 + x^6.$$

We divide $x^{n-k}A(x)$ by $g(x)$ to get x^3+x^2+x+1 with a remainder of 1. Thus $p(x) = 1+0.x+0.x^2$.

Hence the codeword is $[T] = [1001011]$.

9.6.3 Convolutional coding

A convolutional code is generated by convolving the message-bit sequence in some manner. The encoder generates redundant bits by modulo-2 convolution in accordance with the digital convolution equation

$$b_n = \sum_k s_k s_{n-k}, \tag{9.72}$$

where $\{s_k\}$ denotes the message sequence. The hardware for convolutional encoding consists of shift registers, logic circuits to implement modulo-2 operation, and a multiplexer to serialise the coded data. The functional unit is shown in Fig. 9.11. In contrast to block codes, the input bit stream is not grouped into words, rather each message bit influences a number of transmitted bit streams. The message bits shift through the tapped shift registers and undergo encoding process in the logic circuits. If the message stream is L bits long and the encoder employs M-stage shift registers, the coded output sequence will be of length $v(L+M)$ bits, where v denotes a number that represents a single message bit's influences on output bits. The code rate R_c is thus

$$R_C = \frac{k}{n} = \frac{L}{v(L+M)} \text{ bits/symbol.} \tag{9.73}$$

For the case $L >> M$, $R_c \approx 1/v$.

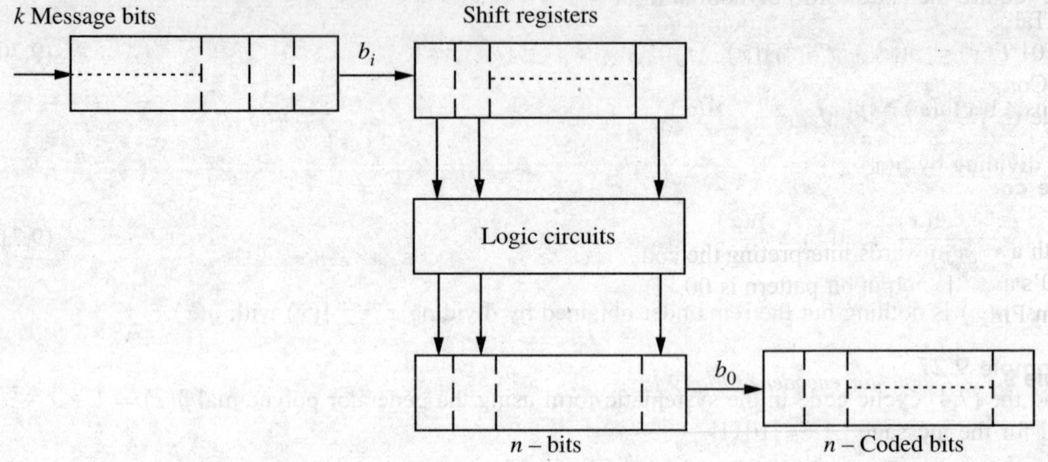

Fig. 9.11 *Convolutional encoding*

Figure 9.12 illustrates a convolutional encoder that employs three shift registers (SRI, SR2, and SR3) and two modulo-2 summers. The output bits of the summers are bit-interleaved via a commutator switch, producing $n = 2$ encoded bits:

$$v_1 = s_1 \oplus s_2 \oplus s_3, \tag{9.74}$$

$$v_2 = s_1 \oplus s_3.$$

Thus a single message bit ($k = 1$) spans over two ($n = 2$) output bits. So the code rate is $R_c = k/n = 1/2$. The encoder is shown in Fig. 9.12.

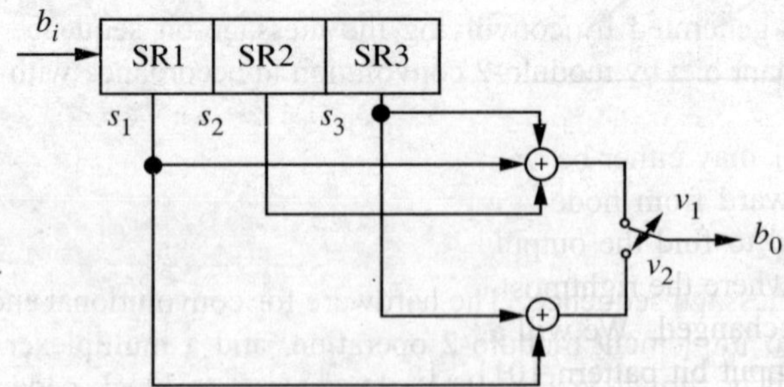

Fig. 9.12 *A rate-1/2 coder*

Initially all the shift registers are clear (reset to zeroes). The first message bit b_1 enters into SRI and the switch samples output data $v_1 v_2$. The second message bit b_2 enters into SRI while b_1 shifts to SR2. and the commutator switch takes the output bits v_1v_2. The process continues till the last data bit b_k enters in SR1. The encoding does not stop here. The last bit b_k now shifts down the shift registers and finally appears in the last shift register SR3. At each stage of such shifting process, enough 0's are inserted serially to ensure that all shift registers are again reset to zeroes.

Table 9.7 gives the coded output for a 4-bit message word 1011. The output bit stream is 11 10 00 01 01 11 00, a total of $v(L+M) = 2(4+3) = 14$ bits.

Convolutional encoding is normally studied in graphical representation. These are the code tree, code trellis, and state diagrams.

The code tree

With a view towards interpreting the code tree, we presume that all the shift registers have been reset to 0's and the output bit pattern is 00. This is the starting point or the initial state of the tree, labelled A in Fig. 9.13.

Table 9.7 *Output for encoder of Fig. 9.12*

	Input b_i	s_1	s_2	s_3	v_1	v_2	
		0	0	0	0	0	Initial state
Message bits, b_i	1	1	0	0	1	1	
	0	0	1	0	1	0	
	1	1	0	1	0	0	
	1	1	1	0	0	1	
	0	0	1	1	0	1	
	0	0	0	1	1	1	
	0	0	0	0	0	0	

The first message bit b_1 may either be 0 or 1. We will follow the convention that when a 0 input bit appears, we move upward from node A to find the output 00. And when the input bit is a 1, we will diverge downward to find the output 11. We will define the state in terms of the input bit pattern $\cdots b_{-3}b_{-2}b_{-1}b_0$, where the rightmost bit b_0 enters first in the shift register SR1. As b_0 shifts right, the status of SR1 is changed. We will define the state by the status of SR1 and SR2. Observe in Table 9.7 that for the input bit pattern 1011, the state changes from 00 to 10 to 01 to 10 to 11. We label the states 00, 01, 10, and 11 as A, B, C, and D respectively. Thus, for the first input bit being a_1, the coder output is 11 and its state is 10, labelled C. The code tree progressively diverge in this fashion for each new message bit.

There will be 2^k possible branches for the kth message bit. Since each message bit remains M-times the bit interval (M being the number of shift registers in the coder), each message bit influences a span of M groups of v-bits. The code can be formed by following the nodes encountered from the start to the end of the tree. When the input bit stream to the encoder is a 4-bit sequence 1011, the output code is

$$v_1 v_2 = 11\ 10\ 00\ 01\ 01\ 11\ 00.$$

Observe that the path through the tree encounters as many nodes as there are message bits.

State diagrams

Given a sequence of message bits, we can represent the output of the encoder through a state diagram. Following the encoder of Fig. 9.13, there are four states 00, 01,10, and 11, labelled as A, B, C, and D, respectively. Since the encoder starts with the registers containing all 0's and the first message bit entering the shift register SRI (which contains the most recent message bit), the state of the coder can be defined by the status of SRI and SR2. In Fig. 9.14, the four nodes A, B, C, and D denote the four states. The state-to-state transitions are represented by solid lines with arrows. The arrows are marked with the input bit and the coder output appears within brackets.

For example, suppose that the encoder is at state C (i.e., $s_1 s_2 = 10$). If a message bit 1 enters, the shift registers contain 110. The state of the coder is then 11, that is, state D, and the coder output is 01 (vide (9.74)). So, the state-to-state transition from C to D is labelled 1(01). Similarly, for a message bit 0, the register contents is 010. The state changes to 01 (that is, state B), resulting in coded output 10. So the label on this transition is 0(10). Note that each state has two incoming branches and two outgoing branches.

Code trellis

An equivalent construction is the code trellis shown in Fig. 9.15. The nodes on the left denote the possible current states (there will be 2^k states if the encoder has k-number of shift registers). The resulting next states are on the right side. The solid lines with arrows represent state-to-state transitions for each message bit, and the label on each line bears similar meanings as in the state diagram.

At the current state the number of paths diverging from each state is two and it increases as 2^k as the trellis grows with clock interval k. For , example, suppose that the coder is at a state C (say). The next message bit will change the state to either B or D for a 0 or 1 input bit, respectively. There are two possible state changes, A and C for B, and two state changes: B and D for D. Thus the total number of paths is four.

Example 9.28

Referring to Fig. 9.15, suppose that the coder is at the state B. Message bit stream 101100 appears in the coder. Find the trellis path.

Solution

Since there are message bits, we will have to consider clock cycles.

Clock cycle k ⟶ k $\quad k+1 \quad k+2 \quad k+3 \quad k+4 \quad k+5 \quad k+6$

A $\xrightarrow{1}$ C $\xrightarrow{0}$ B $\xrightarrow{1}$ C $\xrightarrow{1}$ D $\xrightarrow{0}$ B $\xrightarrow{0}$ A

The nodes that the coder encompasses are A → C → B → C → D → B → A.

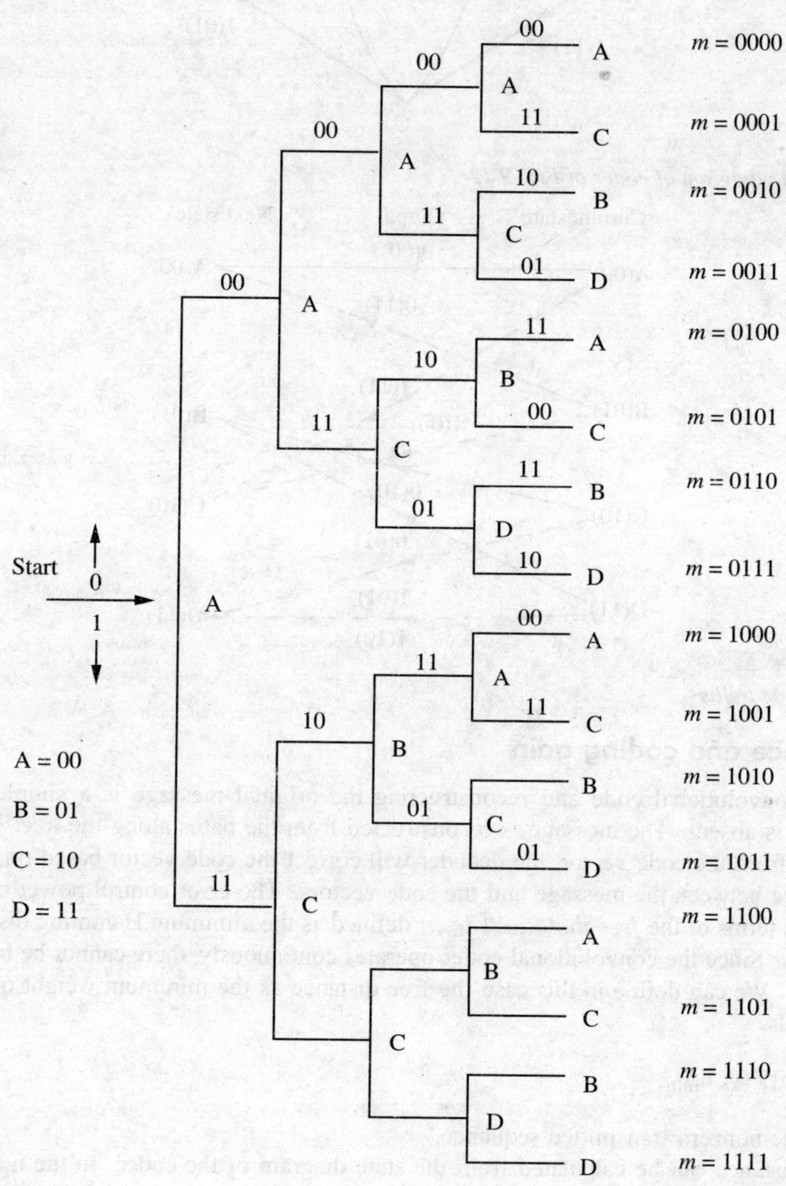

Fig. 9.13 *Code tree*

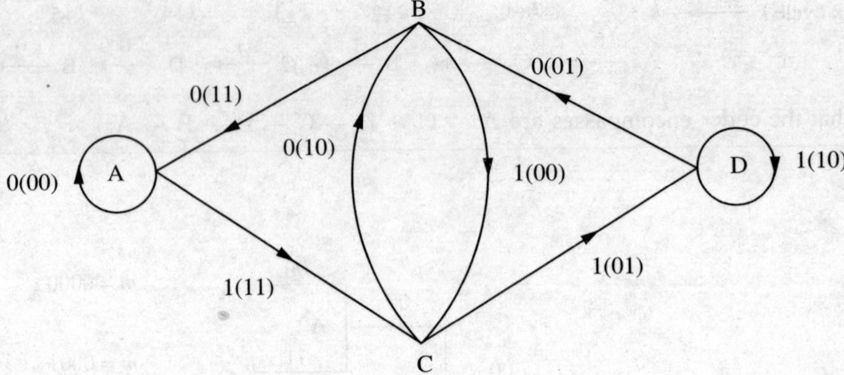

Fig. 9.14 *State diagram of coder of Fig. 9.12*

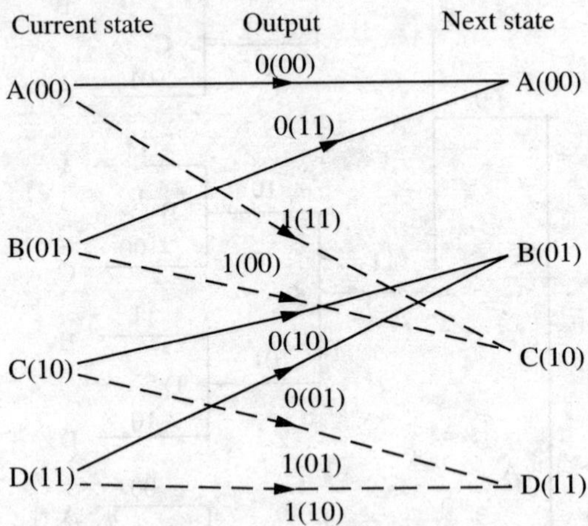

Fig. 9.15 *Code trellis*

Free distance and coding gain

Decoding a convolutional code and reconstructing the original message is a simple matter if the channel noise is absent. The message is reconstructed from the paths along the tree. If the received vector differs from the code vector, the decoder will correct the code vector based on the one-to-one correspondence between the message and the code vectors. The error control power of a block code is measured in terms of the *free distance* (d_{free}) defined as the minimum Hamming distance between the codewords. Since the convolutional coder operates continuously, there cannot be block formation of codewords. We can define in this case the free distance as the minimum weight of the entire bit sequence. Thus

$$d_{\text{free}} = [W(X)]_{\text{min}}, \tag{9.75}$$

where X is the nonzero transmitted sequence.

The free distance can be calculated from the state diagram of the coder. In the normal operating procedure, for any nonzero transmitted sequence, the coder starts from its initial state (state A) and finally returns to its initial state A . This may require sufficient number of 0's to be appended at the end of the message bits sequence to clear the shift registers and ensuring the return of the encoder to its initial state. In code trellis, each branch is then labelled with the weight of the corresponding codeword. In this way we will follow the path that accounts for the minimum sum of the branch weight.

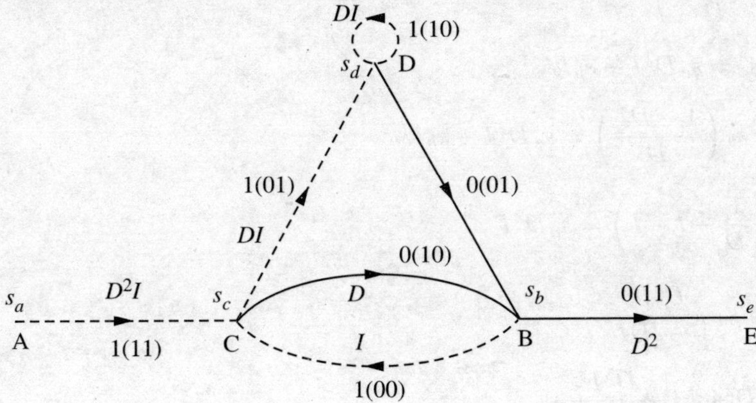

Fig. 9.16 *Modified state diagram for the coder of Fig. 9.12*

Another approach to calculate free distance is with the state diagram where the loop A–A has been removed, as shown in the modified state diagram (vide Fig. 9.16). Splitting of this loop is possible in the sense that it contributes nothing to the weight of the bit sequence. The B–A branch is renamed as the final B–E transition with state E denoting the end state. The figure looks like that of a signal-flow graph with a single input and a single output, connected through nodes and directed branches. Our task then boils down to the finding the *generating function* that relates the final and initial states by multiplication. For that, we assign *state variables* or node signals s_a, s_b, s_c, s_d, and s_e at the nodes A, B, C, D, and E respectively. Note that in the modified state diagram, the branch levels have been dictated by two variables D and I. The exponent of D signifies the branch weight and the exponent of I equals the number of nonzero message bits in that branch. For example, the label 1(11) on branch AC will be read as D^2I. The branch BC represents data =00 and message bit =1; it is labelled with $D^0I = I$ and so on.

The state equations are obtained as

$$s_b = s_c D + s_d D \tag{9.76a}$$

$$s_c = s_a D^2 I + s_b I \tag{9.76b}$$

$$s_d = s_c DI + s_d DI \tag{9.76c}$$

$$s_e = D^2 s_b. \tag{9.76d}$$

We have, $s_c = s_d \left(\dfrac{1 - DI}{DI}\right)$, i.e., $s_d = \dfrac{DI}{1 - DI} s_c$. Also, $s_d = \dfrac{s_b - s_c D}{D}$.

Eliminating s_d, $\dfrac{DI}{1 - DI} s_c = \dfrac{s_b - s_c D}{D}$

$$\text{or, } D^2 I s_c = s_b(1 - DI) - s_c D(1 - DI)$$
$$\text{or, } s_c(D^2 I + D - D^2 I) = s_b(1 - DI)$$
$$s_c = s_b\left(\frac{1 - DI}{D}\right).$$

From (9.76 b), $s_c = s_a D^2 I + s_b I$.

$$\text{Eliminating } s_c, \quad s_b\left(\frac{1 - DI}{D}\right) = s_a D^2 I + s_b I$$
$$\text{or,} \quad s_b\left(\frac{1}{D} - I - I\right) = s_a D^2 I$$
$$\text{or,} \quad s_b = s_a \frac{D^3 I}{1 - 2DI}.$$
$$\therefore \quad s_e = D^2 s_b = \frac{D^5 I}{1 - 2DI} s_a.$$

So, the transfer function $H(D, I)$ is found as

$$H(D, I) = \frac{s_e}{s_a} = \frac{D^5 I}{1 - 2DI}. \tag{9.77}$$

Using the binomial expansion of $(1 - 2DI)^{-1}$, one can write (9.77) in the form

$$\begin{aligned} H(D, I) &= D^5 I(1 - 2D)^{-1} \\ &= D^5 I(1 + 2DI + 4D^2 I^2 + 8D^3 I^3 + \cdots) \\ &= D^5 I + 2.D^6 I^2 + 4D^7 I^3 + 8D^8 I^4 + \cdots \\ &= \sum_{d=5}^{\infty} 2^{d-5} D^d I^{d-4}. \end{aligned} \tag{9.78}$$

The transfer function $H(D, I)$ gives the number of transmitted codewords that finally end on the B–E transition. There are 2^{d-5} valid paths in the code trellis with weight 'd'. The factor I^{d-4} indicates that there are $(d - 4)$ nonzero bits. The minimum path weight $d_{\min}$ is obtained for $d - 4 = 1$, that is $d_{\text{free}} = 5$.

It should be mentioned that for the computation of transfer function $H(D, I)$, the power series $(1 - 2DI)^{-1}$ must be convergent. But for some convolutional codes, $H(D, I)$ is not. In that case an infinite number of decoding errors result though there are finite number of channel errors (or transmission errors). The encoder is then subjected to 'catastrophic error propagation'.

Decoding methods

Two important methods of decoding convolutional codes are the *Viterbi algorithm* and the *sequential decoding*. Consider a convolutional coder with M shift registers and v outputs. We have described that a message bit has an effect only on Mv bits in the output sequence. Obviously, to reconstruct the first output bit we should be confined to examining only up to these Mv bits. For a coder with $M = 3$ and $v = 2$, the first message bit has influence on 6 output bit sequences. But the code tree reveals that there are $2^M = 8$ possible paths; each path is characterised by a specific codeword. For each path, we will take a count of the number of dissimilar bit positions between the received word and the codeword of the path. We next select that path which gives the least count. The decision is made at the node A. If the path under consideration diverges from A upwardly, the first message bit is a 0; otherwise, it is a 1. Similarly, the second and other message bits can be decided. This decoding procedure becomes a tiresome task for large M since each bit requires an examination of 2^M branches, each comprising of Mv sequences.

Sequential decoding

The decoding based on sequential algorithm is manageable even for large M. Sequential decoding executes *maximum likelihood decoding.* Consider a received sequence $R = T + E$, where E is the transmission error. The transmitted bit sequence T has a specific path in the code tree structure. But for nonzero E, the path of R diverges from the path of T and may not be a valid path for the code. The maximum likelihood decoding finds the best path that has the smallest Hamming distance from R. The sequential algorithm is explained below.

At the arrival of the first v code bits, the sequential decoder directs a path by taking the branch with the smallest Hamming distance (the first comparison is made with two branches that diverge from the starting node). In the way, if two branches from one node happen to have equal Hamming distance, the decoder selects one at random and progresses. After having selected a wrong path at some node, the Hamming distance rapidly increases and the decoder tackles such situations by retracing the path to the previous node and trying another path. Passing through the nodes, the decoder keeps on recording the number of bit mismatches between the received code and the bits encountered along the path. In this way the decoder finally searches out the correct path. Having found the path, the first message bit is decided on the basis of the divergence of this path from the starting node through M-nodes. The second message bit is determined similar to the above, but in this case the path is followed from the second node.

We now explain the decision strategy for a situation where the decoder has taken a wrong turn at a node and needs to backtrack the path to the previous node. This judgement is based on the selection of *surviving paths* that have not have exceeded some specified *threshold* or expected value of the Hamming distance. To explain this concept, consider the possible paths through the trellis of Fig. 9.17 that includes $k = 4$ nodes. The figure is obtained by repeating thrice the code trellis of Fig. 9.15.

Suppose we want to explore all the possible paths through the entire trellis from state $A(00)$ at node 1 back to state A at node 4. There are two available paths: $a_0a_1a_2a_3$ and $a_0c_1b_2a_3$. Suppose we consider that our received sequence starts with $\tau = 1\ 1\ 0\ 1\ 0\ 0\ 1$. Starting at node a_0, a 0 will cause an output 00 and will carry the decoder to node a_1 (again the same state A). But the received sequence in the branch a_0a_1 is 11 and hence two bit positions disagree, as indicated by the circled number. Similarly, in the branch a_0c_1, all the bit positions agree and hence the Hamming distance is 0. So the Hamming distance of the path $a_0a_1a_2a_3$ is $2 + 1 + 1 = 4$ and that of the path $a_0c_1b_2a_3$ is $0 + 1 + 1 = 2$. The latter has a minimum Hamming distance. So we *discard* path $a_0a_1a_2a_3$ and declare that the path $a_0c_1b_2a_3$ is the *survivor* at the node a_3. The path that we discard is indicated by cross ($\times$) mark. Likewise, we discard other paths arriving at other nodes b_3, c_3, and d_3. In fact, the number of surviving paths will be equal to the number of states.

The sequential decoder is designed to abandon a path when its Hamming distance exceeds some specified *threshold* or *discard level*. Let p be the transmission probability per bit. Since each branch is associated with v bits, the total number of bit difference at the mth node will be pmv on the average even when the correct path is followed. Figure 9.18 plots the bit errors versus number of nodes traversed. When the accumulated error exceeds the threshold at some node, the decoder decides to backtrack the branch and try again along some unexplored branch. Repeating this trial and error method, the decoder scans the entire nodes and finds the path with the lowest bit discrepancy. Note that the threshold line does not start at the origin. This is so because the first bit of the received bit sequence may be erroneous and the discard level may be crossed at the second node.

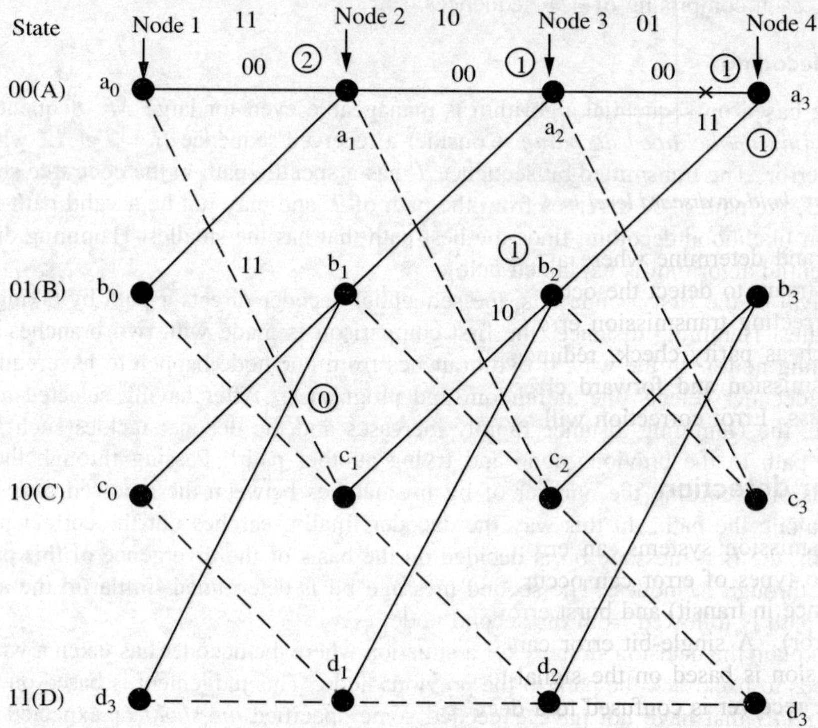

Fig. 9.17 *Illustrating sequential decoding*

9.7 Error Control

A data communication network frames a link between a receiver and a transmitter via a transmission medium. The distance between the source and the receiver can be as short as a few metres or as long as several thousand kilometres. The transmission medium can be as simple as a piece of copper wire or as complex as a satellite or fibre optic system. The great task before the designers of any communication system is to provide quality and reliable data at a rate acceptable to the receiver. But due to the non-ideal transmission characteristics of the communications system, data errors occur inevitably. There is then an urgent need to develop and implement procedures for error controls which include error detection and error correction. The purpose of error detection is to examine the

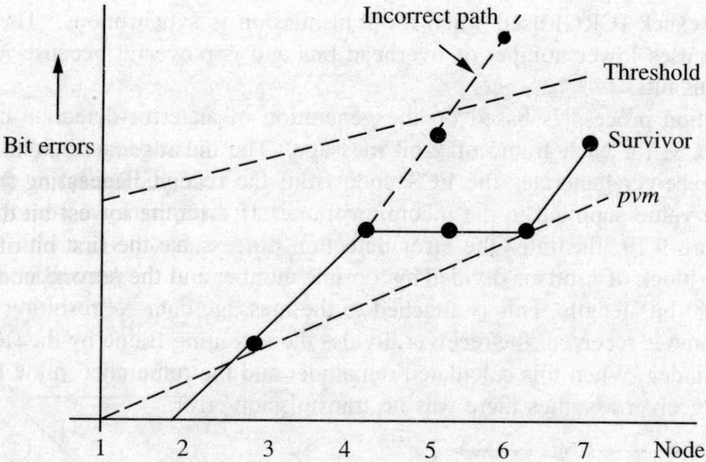

Fig. 9.18 *Threshold or discard level in sequential decoding*

received data and determine where a transmission error has occurred. Redundancy is introduced into the data stream to detect the occurrence of an error. Error correction on the other hand is the process of correcting transmission errors. We have explained earlier a number of error detection techniques such as parity check, redundancy, etc. There are basically two methods of correcting errors: retransmission and forward error correction. In this section, we first examine the error detection process. Error correction will be discussed in the next section.

9.7.1 Error detection

In digital transmission systems, an error occurs when a bit is flipped between transmission and reception. Two types of error can occur; single-bit errors (that takes place at random locations in the data sequence in transit) and burst errors (in which case a contiguous sequence of some bits are received in error). A single-bit error can occur in the presence of white noise in which case the receiver's decision is based on the signal-to-noise ratio. When noise effect degrades the signal-to-noise ratio, the receiver is confused to a decision of a single bit. Burst error are more common and difficult to handle. Impulse noise stands as a candidate for such error bursts. An impulse noise of only 1 μs duration on a data sequence rate of 10 Mbps results in an error burst of 1×10^{-6} s $\times 10 \times 10^6$ bits/s = 10 bits. However, it is not true that all the 10 bits in the cluster suffer errors.

Parity check is possibly the simplest method of error detection, As discussed in Section 9.6.1, a parity bit is attached to the block of data so that the coded character has an even number of 1s (even parity) or an odd number of 1s (odd parity). The receiver checks the received character for parity matching. When there is a mismatch between the parity, one bit is assumed to have erroneously flipped during transmission. However, if even number of bits are flipped due to error, parity check scheme fails and an undetected error occurs.

Cyclic redundancy check (CRC)

When noise impulses are long and causes errors in more than one bit (particularly at high data rate), a more efficient and powerful scheme need to be employed than the simple parity check. This is the cyclic redundancy check (CRC) used when the transmission is synchronous. The CRC scheme is efficient because it uses lower number of overhead bits and is powerful because it can detect more number of erroneous bits.

The error detection process is based on the generation of an error-detection code called frame check sequence (FCS) for each frame of k-bit message. The message data along with this FCS is transmitted. The receiver generates the FCS code from the received message data. This FCS is compared with the value supplied in the incoming frame. If discrepancy noted, the received frame is erroneous. Figure 9.19 illustrates the error detection process based on check sequence. In this method, a message block of k-bits is divided by a prime number and the remainder is the frame check sequence of $(n - k)$ bits length. This is attached to the message data X, resulting in a frame n bits long. When the frame is received, the receiver divides the incoming frame by the same prime number and finds the remainder. When this calculated remainder and the remainder in the transmitted frame are the same, the receiver assumes there was no transmission error.

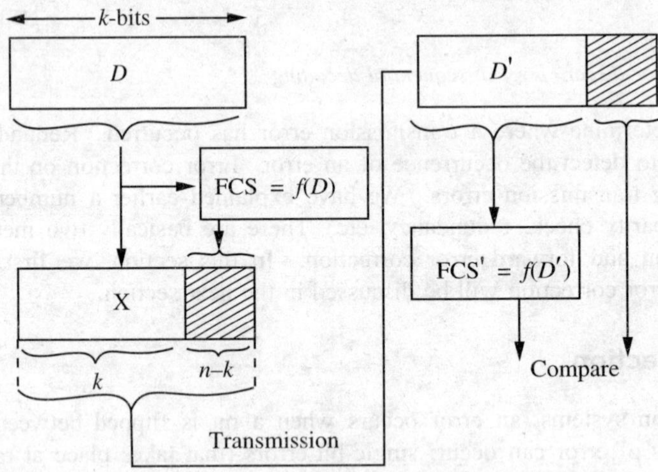

Fig. 9.19 *Error detection process on CRC scheme using FCS.*

We now explain procedure using polynomials, in which the data, frame check sequence and the prime divisor are all expressed as polynomials with binary coefficients. The coefficients corresponds to the binary numbers representing the data, FCS or the divisor. For the data $D = 1011010010$, we may write the appropriate polynomial $D(x) = x^9 + x^7 + x^6 + x^4 + x$ and data width $= k = 10$ bits. Suppose the binary pattern that is the divisor is $P = 101101$ and hence $P(x) = x^5 + x^3 + x^2 + 1$. The n-bit frame to be transmitted will be of the form

T	=	D	F
← n-bits →		← k-bits →	← $n - k$ bits →

The $(n - k)$-bit FCS is placed at the last $(n - k)$ bits of T. The pattern P is of $n - k + 1$ bits. The polynomial $T(x)$ when divided by $P(x)$ should yield no remainder. Placing F at the last $(n - k)$ bits of $T(x)$ implies that $D(x)$ in effect is shifted to the left by $n - k$ bits, resulting in the polynomial $x^{n-k}D(x)$ with the last $n - k$ positions padded with zeros. The polynomial $T(x)$ will be exactly divisible by $P(x)$, but $x^{n-k}D(x)$ when divided by $P(x)$ will result in a remainder polynomial $R(x)$. The remainder R is always at least one bit shorter than the divisor. Then,

$$\frac{x^{n-k}D(x)}{P(x)} = Q(x) + \frac{R(x)}{P(x)} = Q(x) + \frac{R(x)}{P(x)}, \tag{9.79}$$

where $Q(x)$ is a quotient polynomial. We now take the remainder $R(x)$ as our FCS. Then,

$$T(x) = x^{n-k}D(x) + R(x). \tag{9.80}$$

On reception, the receiver will divide $T(x)$ by $P(x)$:

$$\frac{T(x)}{P(x)} = \frac{x^{n-k}D(x) + R(x)}{P(x)} = \frac{x^{n-k}D(x)}{P(x)} + \frac{R(x)}{P(x)}$$

$$= Q(x) + \frac{R(x)}{p(x)} + \frac{R(x)}{P(x)} \quad \text{[by (9.79)]}$$

$$= Q(x) + \frac{R(x) + R(x)}{P(x)} = Q(x) \quad [\text{as division is modulo 2, } R(x) + R(x) = 0].$$

So, there is no remainder and there is no error. If transmission error takes place in the n-bit frame, the error can be represented by an error pattern E. The received frame can be represented as

$T' = T + E.$

The division T'/P will have a remainder.

Example 9.29
Using the message data $D = 1011010010$ and $P = 101101$, find the FCS and hence the transmitted code.

Solution
Here $k = 10$, $n = 15$ (say) and hence $n - k = 5$.

$D(x) = x^9 + x^7 + x^6 + x^4 + x$ and $P(x) = x^5 + x^3 + x^2 + 1$.

So, $x^5D(x) = x^{14} + x^{12} + x^{11} + x^9 + x^6$. The division $x^5D(x)/P(x)$ is shown below.

```
                       x^9+x                   <---- Q(x)
x^5+x^3+x^2+1 | x^14 + x^12+ x^11 + x^9 + x^6  <---- x^5D(x)
                x^14 + x^12+ x^11 + x^9
                ---------------------------
                                      x^6
                                      x^6 + x^4+ x^3 + x
                                      ------------------
                                            x^4+ x^3 + x   <---- R(x)
                                                       R = 11010
```

The transmitted polynomial is then

$$T(x) = x^5D(x) + R(x) = x^{14} + x^{12} + x^{11} + x^9 + x^6 + x^4 + x^3 + x,$$

that corresponds to the 15 bit frame

$$T = \underbrace{1011010010}_{\text{Data}}\underbrace{11010}_{\text{FCS}}.$$

If there are no errors, $T(x)$ will be received intact. When divided by $P(x)$,

$$\begin{array}{r|l} & x^9+x \\ \hline x^5+x^3+x^2+1 & x^{14}+x^{12}+x^{11}+x^9+x^6+x^4+x^3+x \\ & x^{14}+x^{12}+x^{11}+x^9 \\ \hline & x^6+x^4+x^3+x \\ & x^6+x^4+x^3+x \\ \hline & 0 \leftarrow \text{Remainder} \end{array}$$

9.7.2 Forward error correction (FEC) systems

There are many different error correcting codes, namely, convolutional codes and block codes. The power of a code depends on the overhead or redundant bits added to the codeword for the purpose of error-control. A popular error correcting code is the Hamming code. With FEC, Hamming bits are added to the message prior to transmission. The number of added bits is related through

$$2^r \geq k + r + 1 \tag{9.81}$$

where r = number of Hamming bits and k = number of bits in the message block.

Including the overhead bits, the total number of bits is the coded word is $n = k + r$.

Example 9.30

This example illustrates the error correction at a single bit position.

Consider a 12-bit message sequence M = 1100 1001 0110. Here $k = 12$. The number of Hamming bits r is given by $2^r \geq k + r + 1$.

For $r = 1$: $2^1 = 2 \geq 12 + 1 + 1 = 14$ not satisfied

So a single bit is insufficient.

For $r = 2$: $2^2 = 4 \geq 12 + 2 + 1 = 15$ not satisfied

Again, two bits are insufficient.

For $r = 5$: $2^5 = 32 \geq 12 + 5 + 1 = 18$ and so 5 bits are sufficient to correct the error. There will be $n = k + r = 12 + 5 = 17$ bits in the coded sequence. Now , arbitrarily place 5 Hamming bits in the message sequence:

17	16	15	14	13	12	11	10	9	8	7	6	5	4	3	2	1
1	H	1	0	H	0	1	H	H	0	0	1	0	H	1	1	0

To find the Hamming bits, we express all the bit positions that contain 1 as 5-bit binary numbers and XOR them:

Bit position	Binary number
2	0 0 0 1 0
3	0 0 0 1 1
XOR	0 0 0 0 1
6	0 0 1 1 0
XOR	0 0 1 1 1
11	0 1 0 1 1
XOR	0 1 1 0 0
15	0 1 1 1 1
XOR	0 0 0 1 1
17	1 0 0 0 1
XOR	1 0 0 1 0 = Hamming code

Thus, $b_{16} = 1$, $b_{13} = 0$, $b_{10} = 0$, $b_9 = 1$, $b_4 = 0$.

The 17-bit coded sequence becomes

↓ ↓ ↓↓ ↓
1 1 1 0 0 0 1 0 1 0 0 1 0 0 1 1 0

Suppose that a transmission error occurs in bit position 14. The received sequence is then

H H H H H
[R] = 1 1 1 1 0 0 1 0 1 0 0 1 0 0 1 1 0

We then extract the Hamming bits and XOR it with the binary words of each bit positions containing a1. The received data bits are 1s in bit positions 2, 3, 6, 11, 14, 15, and 17.

Bit position	Binary number
Hamming code	1 0 0 1 0
2	0 0 0 1 0
XOR	1 0 0 0 0
3	0 0 0 1 1
XOR	1 0 0 1 1
6	0 0 1 1 0
XOR	1 0 1 0 1
11	0 1 0 1 1
XOR	1 1 1 1 0
14	0 1 1 1 0
XOR	1 0 0 0 0
15	0 1 1 1 1
XOR	1 1 1 1 1
17	1 0 0 0 1
XOR	1 1 1 1 1 = binary 14

So, the bit position (14) was received erroneously. Once we know the erroneous bit position, we simply complement the bit to find the actual transmitted codeword.

The FEC system explained above works nicely for single-bit errors, but the drawback is that it cannot detect (and correct) multiple errors or errors in the Hamming bits. The system is somewhat slower as a number of logical steps are involved.

Performance of FEC systems

This section describes the performance of an FEC system depicted in Fig. 9.20. Let the message bits be at the rate r_b. The encoder converts the uncoded k-bit message words into n-bit codewords. Since the rate of information transmission is the same whether coded or uncoded, the bit rate r on the channel must be greater than r_b:

$$r = (n/k)r_b = \frac{r_b}{R_c}, \tag{9.82}$$

where $R_c = k/n$ is the code rate. Consequently, the bit duration T_c in the coded word is less than the bit duration T_b in the uncoded word. We will examine here the error performance of the system with FEC codes applied. Our discussion will be limited to the case with additive white noise of spectral density η. Assuming that the signal power P_s is the same in both the cases, the average energy per message bit E_b is then $E_b = P_s.T_\omega/k$, where T_ω is the duration of a word. The average energy per code bit is then $P_sT_\omega/n = R_cE_b$. The error probabilities p and p_c of the uncoded and coded bits are given as

$$p = \frac{1}{2}\,\text{erfc}\sqrt{\frac{E_b}{\eta}} = \frac{1}{2}\,\text{erfc}\sqrt{\frac{p_sT_\omega}{k\eta}}, \tag{9.83}$$

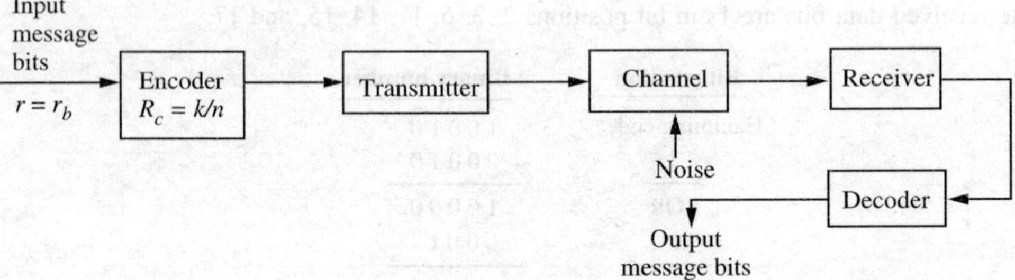

Fig. 9.20 *FEC system*

and $$p_c = \frac{1}{2}\,\text{erfc}\sqrt{\frac{R_cE_b}{\eta}} = \frac{1}{2}\,\text{erfc}\sqrt{\frac{p_sT_\omega}{n\eta}}, \tag{9.84}$$

where η = noise spectral density. The performance of the FEC system will be decided by the probability of output message-bit error P_c. In the uncoded case, a message word will be in error if one or more bits are erroneous. The probability that a message bit is not in error is $(1 - p)$ and the probability that all the k message bits are not in error is $(1 - p)^k$. So the probability of at least one bit in error is

$$P_\omega = 1 - (1 - p)^k \approx kp, \tag{9.85}$$

since $p << 1$. The quantity P_ω is the probability of a word error. Now the FEC code is applied which can correct up to t-errors per word of n bits. Obviously, decoding word error has the upper bound

$$P_d \leq \sum_{i=t+1}^{n} {}^{n}C_i p^i (1-p)^{n-i},$$

and since $p << 1$, one can have the approximation

$$P_d \approx {}^{n}C_{t+1} p^{t+1}. \tag{9.86}$$

This means that an uncorrected word typically has $(t+1)$ bit errors. Out of these $(t+1)$ errors, $\frac{k}{n}(t+1)$ errors are in the message bits and the rest $(1-\frac{k}{n})(t+1)$ errors are in the check bits. For N words of messages there are Nk bits, and so total number of message bits in error at the output is given by

$$\begin{aligned} P_c &= \frac{k}{n}(t+1)NP_d/Nk = \frac{1}{n}(t+1)P_d \approx \frac{t+1}{n}\, {}^{n}C_{t+1}\, p^{t+1} \\ &= \frac{(t+1)}{n}\frac{n!}{(t+1)!(n-t-1)!}p^{t+1} \\ &= \frac{(n-1)!}{t!(n-t-1)!}p^{t+1} = {}^{n-1}C_t\, p^{t+1}. \end{aligned} \tag{9.87}$$

So, the output error probability is given by

$$P_c = {}^{n-1}C_t \left(\frac{1}{2}\operatorname{erfc}\sqrt{\frac{R_c E_b}{\eta}}\right)^{t+1}, \tag{9.88}$$

in which we have used (9.84).

The uncoded transmission would have the error probability given in equation (9.83); it is reproduced here.

$$p = \frac{1}{2}\operatorname{erfc}\sqrt{\frac{E_b}{\eta}} \tag{9.89}$$

For large x, erfc $(x) = e^{-x^2} \div x\sqrt{\pi}$, so that

$$\begin{aligned} P_c &\simeq {}^{n-1}C_t \left[\frac{1}{2}\frac{e^{-(R_c E_b/\eta)}}{(R_c E_b/\eta)\sqrt{\pi}}\right]^{t+1} \\ &= {}^{n-1}C_t \left[\frac{1}{2\sqrt{\pi}}.\frac{e^{-(R_c E_b/\eta)}}{R_c E_b/\eta}\right]^{t+1}. \end{aligned} \tag{9.90}$$

Suppose we have a (7,4) block code which can correct a single error. Here $t = 1$ and $d_{\min} = 3$, code efficiency $R_c = k/n = 4/7$. A word will be erroneous if two or more digits in the word are in error. By (9.87),

$$P_c = {}^{7-1}C_2\, p_c^2 = \frac{6 \times 5}{2} p_c^2 = 15 p_c^2.$$

9.7.3 Automatic-repeat-request system

The error control in this system is based on resending a message when it is received in error and the receive terminal automatically calling for retransmission of the message. The 'automatic request for retransmission' (ARQ) proves most reliable method of error correction and requires fewer check bits resulting in a higher value of $R_c = k/n$. The receiver of the ARQ system is designed to detect errors but not to correct. If an error is detected in a word, the decoder issues a 'negative acknowledgment' (NAK) via a feedback path to the transmitter and requests retransmission of the word. If no errors are detected, the decoder simply issues a 'positive acknowledgment' (ACK) to successfully received word.

There are three basic ARQ schemes, namely, stop-and-wait, go-back N, and selective-repeat ARQ. In stop-and-wait ARQ, the source station transmits a single word and then waits for ACK signal. No other data can be sent until the receiver's response arrives at the source station. But if an error is detected at the receiving end, the receiver returns a NAK signal which means that the transmitter is requested to retransmit the same message. The transmitter again waits for receiver acknowledgment. The scheme is depicted in Fig. 9.21 (a). Note that the propagation delay t_d in the forward and return directions and the processing time at the receiver, results in an idle time $D \geq 2t_d$ between the transmitted words. The data is transmitted in this scheme at a slower rate and also, the transmitter has to maintain a copy of the transmitted word until an acknowledgment is received for that word, In Fig.9.21(a), on getting the ACK response for the word 1, the source transmits word 2. But this is received in error as indicated by the NAK pulse for word 2. The source then retransmits word 2 again. The ARQ system can be made faster in the go-back-N scheme. Here the transmitter sends words continuously and does not wait for the ACK/NAK response from the receiver end. When the receiver sends a NAK signal for some message the transmitter returns to that codeword and retransmits starting from that word. The receiver discards all words transmitted in the interim, correct or not, to preserve the word sequence. The go-back-N scheme is a significant improvement over the stop-and-wait scheme. Figure 9.21(b) illustrates this scheme. Note in Fig. 9.21(b) that a NAK response for word 3 is received by source at the time when it has already transmitted word 6. The source then goes back to the third word to retransmit the words sequentially starting from word 3. In the interim period, the words 4, 5 and 6 already received are discarded. The selective-repeat ARQ system, is again an improvement over the go-back-N scheme. Being the receipt of the NAK signal, that transmitter repeats only the particular word and then returns to its sequential order.

In all the three schemes, if a particular word is received without error, the word is not repeated. But a particular word received with errors needs be transmitted twice, thrice, etc., depending on the number of times the word is incorrectly received. A quantitative measure of the performance of ARQ systems is based on the average number of transmission of the word for every message word and is measured by the 'throughput efficiency' defined as

$$\text{Throughput efficiency, } \eta = \frac{\text{(Average number of message bits actually accepted by the receiver over a certain time)}}{\text{(Total number of bits transmitted including retransmission over the same time.)}}$$

The probability that a message word is transmitted only once is $(1 - p)$. The probability that a message word is transmitted twice is $p(1 - p)$. This is because the event consists of the message word rejected in first try (with probability p) followed by its correction (with the probability $1 - p$).

Similarly, that a message transmitted thrice has the probability $p^2(1-p)$ and so on. So, the average number of transmitted words per message word is

$$\overline{m} = 1 \times (1-p) + 2 \times p(1-p) + 3 \times p^2(1-p) + 4p^3(1-p) + \cdots$$

$$= (1-p)(1 + 2p + 3p^2 + 4p^3 + \cdots) = (1-p)^{-2} = \frac{1}{1-p}. \tag{9.91}$$

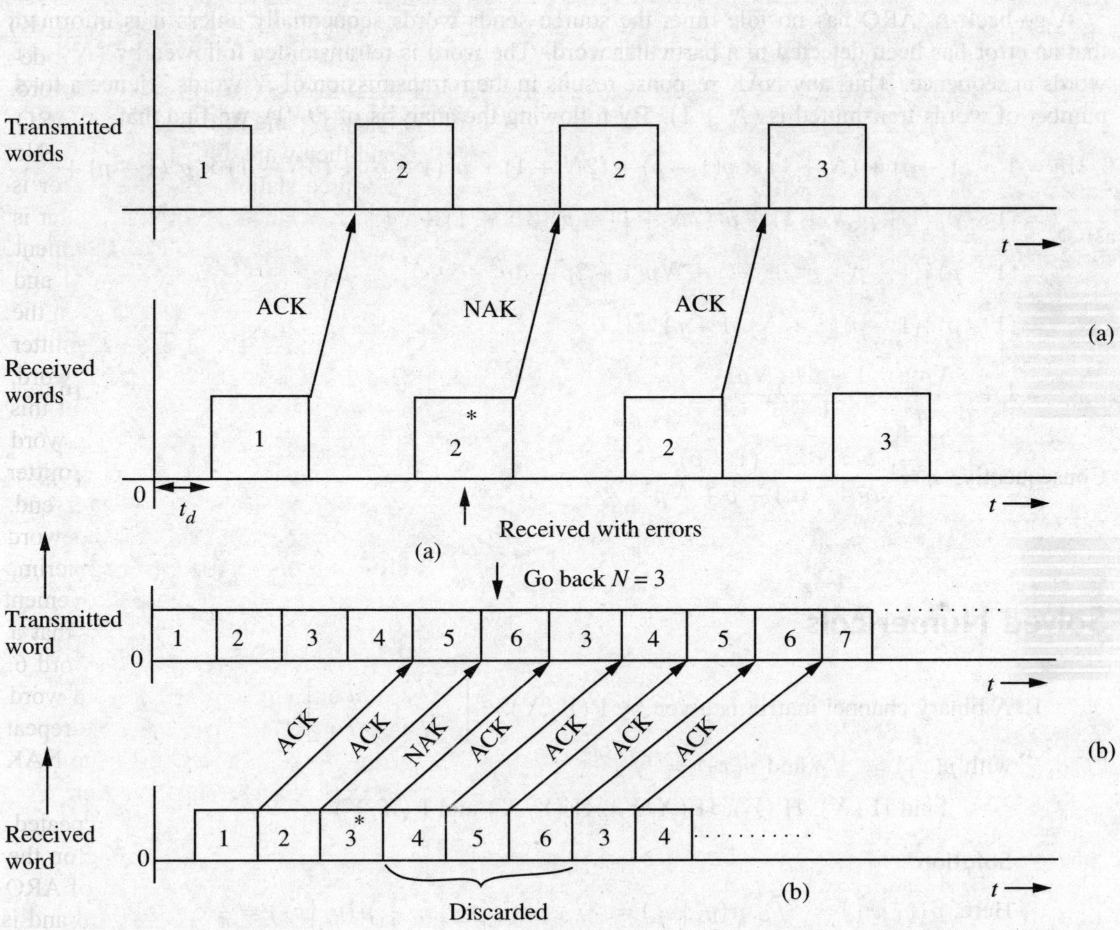

Fig. 9.21 *(a) Stop-and-wait ARQ (b) go-back-N ARQ*

Since a k-bit message is coded as an n-bit word, for every message word the system must transmit $n.\overline{m}$ bits on the average. So the throughput efficiency is given by (this is truly a selective-repeat case)

$$\eta = \frac{k}{n\overline{m}} = \frac{k}{n}(1-p). \tag{9.92}$$

For a stop-and-wait ARQ system, the total time needed to accept a word in a single attempt is

$(T_\omega + D)$ where $D \geq 2t_d$ is the round-trip delay. This delay reduces the efficiency by a factor $T_\omega/(T_\omega + D)$. Again the word duration is given by $T_\omega = \dfrac{n}{r}$, where r is the forward-transmission bit rate. Thus, the throughput efficiency is

$$\eta = \frac{k}{n}\frac{(1-p).T_\omega}{(T_\omega + D)} = \frac{k}{n}\frac{1-p}{(1+D/T_\omega)}. \tag{9.93}$$

A go-back-N ARQ has no idle time; the source sends words sequentially unless it is informed that an error has been detected in a particular word. The word is retransmitted followed by $(N-1)$ words in sequence. Thus any NAK response results in the retransmission of N words. Hence a total number of words transmitted is $(N+1)$. By following the analysis of (9.91), we find that

$$\begin{aligned}\overline{m} &= 1 \times (1-p) + (N+1) \times p(1-p) + (2N+1) \times p^2(1-p) + (3N+1) \times p^3(1-p) + \cdots \\ &= (1-p)\left[1 + p(N+1) + p^2(2N+1) + p^3(3N+1) + \cdots\right] \\ &= (1-p)\left[(1 + p + p^2 + \cdots) + Np(1 + 2p + 3p^2 + \cdots)\right] \\ &= (1-p)\left[(1-p)^{-1} + Np(1-p)^{-2}\right] \\ &= 1 + \frac{Np}{1-p} = \frac{1-p+Np}{1-p}. \end{aligned} \tag{9.94}$$

$$\text{Consequently, } \eta = \frac{k}{n\overline{m}} = \frac{k}{n}\frac{(1-p)}{1-p+Np}. \tag{9.95}$$

Solved Numericals

1. A binary channel matrix is given by $P(Y/X) = \begin{bmatrix} 2/3 & 1/3 \\ 1/4 & 3/4 \end{bmatrix}$

   with $p(x_1) = 4/9$ and $p(x_2) = 5/9$

   Find H (X), H (Y), H (X/Y), H (Y/X) and I $(X;Y)$

   Solution

   Here, $p(y_1|x_1) = 2/3,\ p(y_2|x_1) = 1/3,\ p(y_1|x_2) = \frac{1}{4},\ p(y_2|x_2) = \frac{3}{4}$

   So, marginal probabilities

   $$\begin{aligned} p(y_1) &= p(y_1|x_1)p(x_1) + p(y_1|x_2)p(x_2) = \frac{2}{3} \times \frac{4}{9} + \frac{1}{4} \times \frac{5}{9} = \frac{8}{27} + \frac{5}{36} \\ &= \frac{32+15}{108} = \frac{47}{108} \end{aligned}$$

   $$\begin{aligned} p(y_2) &= p(y_2|x_1)p(x_1) + p(y_2|x_2)p(x_2) = \frac{1}{3} \times \frac{4}{9} + \frac{3}{4} \times \frac{5}{9} = \frac{4}{27} + \frac{15}{36} \\ &= \frac{16+45}{108} = \frac{61}{108} \end{aligned}$$

$$H(X) = p(x_1)\log_2\frac{1}{p(x_1)} + p(x_2)\log_2\frac{1}{p(x_2)}$$

$$= \left(\frac{4}{9}\log_{10}\frac{9}{4} + \frac{5}{9}\log_{10}\frac{9}{5}\right)\Big/\log_{10}^{2}$$

$$= (0.1565 + 0.1418)/0.3010 = 0.9909 \text{ bits/msg.}$$

$$\mathrm{H}(Y) = p(y_1)\log_2\frac{1}{p(y_1)} + p(y_2)\log_2\frac{1}{p(y_2)}$$

$$= \left(\frac{47}{108}\log_{10}\frac{108}{47} + \frac{61}{108}\log_{10}\frac{108}{61}\right)\Big/\log_{10}^{2}$$

$$= (0.1572 + 0.1401)/0.3010 = 0.9876 \text{ bits/msg.}$$

$$\mathrm{H}(X\,|Y) = \sum_x\sum_y p(x,y)\log_2\frac{p(y)}{p(x,y)};\ \text{Now, } p(x,y) = p(y\,|x)\,p(x) \text{ by Bayes' theorem.}$$

Therefore,

$$\mathrm{H}(X\,|Y) = \sum_y\left(p(x_1,y)\log_2\frac{p(y)}{p(x_1,y)} + p(x_2,y)\log_2\frac{p(y)}{p(x_2,y)}\right)$$

$$= p(x_1,y_1)\log_2\frac{p(y_1)}{p(x_1,y_1)} + p(x_1,y_2)\log_2\frac{p(y_2)}{p(x_1,y_2)}$$

$$+ p(x_2,y_1)\log_2\frac{p(y_1)}{p(x_2,y_1)} + p(x_2,y_2)\log_2\frac{p(y_2)}{p(x_2,y_2)}$$

$$p(x_1,y_1) = p(y_1\,|x_1)p(x_1) = \frac{2}{3}\times\frac{4}{9} = \frac{8}{27}$$

$$p(x_1,y_2) = p(y_2\,|x_1)p(x_1) = \frac{1}{3}\times\frac{4}{9} = \frac{4}{27}$$

$$p(x_2,y_1) = p(y_1\,|x_2)p(x_2) = \frac{1}{4}\times\frac{5}{9} = \frac{5}{36}$$

$$p(x_2,y_2) = p(y_2\,|x_2)p(x_2) = \frac{3}{4}\times\frac{5}{9} = \frac{15}{36}$$

So,

$$\mathrm{H}(X\,|Y) = \frac{8}{27}\log_2\left(\frac{47/108}{8/27}\right) + \frac{4}{27}\log_2\left(\frac{61/108}{4/27}\right)$$

$$+ \frac{5}{36}\log_2\left(\frac{47/108}{5/36}\right) + \frac{15}{36}\log_2\left(\frac{61/108}{15/36}\right)$$

$$= \left[\frac{8}{27}\log_{10}\frac{47}{32} + \frac{4}{27}\log_{10}\frac{61}{16} + \frac{5}{36}\log_{10}\frac{47}{15} + \frac{15}{36}\log_{10}\frac{61}{45}\right]\Big/\log_{10}^{2}$$

$$= (0.04946 + 0.08610 + 0.06888 + 0.05504)/0.30103 = 0.8619$$

$$\text{H}(Y|X) = \sum_x \sum_y p(x_1 y)\log_2 \frac{p(x)}{p(x,y)}$$

$$= p(x_1, y_1)log_2 \frac{p(x_1)}{p(x_1, y_1)} + p(x_1, y_2)log_2 \frac{p(x_1)}{p(x_1, y_2)}$$

$$+ p(x_2, y_1)log_2 \frac{p(x_2)}{p(x_2, y_1)} + p(x_2, y_2)log_2 \frac{p(x_2)}{p(x_2, y_2)}$$

$$= \frac{8}{27}\log_2\left(\frac{4/9}{8/27}\right) + \frac{4}{27}\log_2\left(\frac{4/9}{4/27}\right)$$

$$+ \frac{5}{36}\log_2\left(\frac{5/9}{5/36}\right) + \frac{15}{36}\log_2\left(\frac{5/9}{15/36}\right)$$

$$= (0.05217 + 0.07068 + 0.08361 + 0.05205)/0.30103 = 0.8587$$

$$\text{I}(X, Y) = \text{H}(Y) - \text{H}(Y|X) = 0.9876 - 0.8587 = 0.1289$$

Also, $\text{I}(X; Y) = \text{H}(X) - \text{H}(X|Y) = 0.9909 - 0.8619 = 0.1290$

$\text{I}(X; Y) = \sum_x \sum_y p(x, y)\log_2\left(\frac{p(x, y)}{p(x)p(y)}\right)$ will also produce the same result.

(The student can verify this.)

2. Construct the table-look up to decode a (7.4) linear block code with single error correction. The 3×7 parity-check matrix for this code is given as

$$[H] = [\ P^T \ \vdots\ I\ ] = \begin{bmatrix} 1 & 0 & 1 & 1 & \vdots & 1 & 0 & 0 \\ 1 & 1 & 0 & 1 & \vdots & 0 & 1 & 0 \\ 1 & 1 & 1 & 0 & \vdots & 0 & 0 & 1 \end{bmatrix}.$$

Comment on the output of the decoder if the received code word is (1011001). Find all codewords and weight of each word.

Solution

Here $n = 7$ and $k = 4$, so there are $M = 2^k = 16$ different message words of 4-bit long. These will be 16-valid code words of 7- bit long.

The syndrome can be calculated by the formula $[S] = [E]\left[H^T\right]$ where $[E]$ is the vector representing error pattern. Now

$$[H^T] = \begin{bmatrix} 1 & 1 & 1 \\ 0 & 1 & 1 \\ 1 & 0 & 1 \\ 1 & 1 & 0 \\ \hdashline 1 & 0 & 0 \\ 0 & 1 & 0 \\ 0 & 0 & 1 \end{bmatrix}$$

The syndrome $[S]$ for the code is given in the table to follow.

E-error pattern							S-syndrome			
0	0	0	0	0	0	1	0	0	1	
0	0	0	0	0	1	0	0	1	0	
0	0	0	0	1	0	0	1	0	0	
0	0	0	1	0	0	0	1	1	0	
0	0	1	0	0	0	0	1	0	1	
0	1	0	0	0	0	0	0	1	1	
1	0	0	0	0	0	0	1	1	1	
0	0	0	0	0	0	0	0	0	0	← No error

The generator mating $[G]$ is given by

$$G = (I/P) = \begin{pmatrix} 1 & 0 & 0 & 0 & \vdots & 1 & 1 & 1 \\ 0 & 1 & 0 & 0 & \vdots & 0 & 1 & 1 \\ 0 & 0 & 1 & 0 & \vdots & 1 & 0 & 1 \\ 0 & 0 & 0 & 1 & \vdots & 1 & 1 & 0 \end{pmatrix}$$

The code-words for the 16-message words are tabulated in the following table.

Message word $M = (m_1\ m_2\ m_3\ m_4)$	Code-word $C = (m_1\ m_2\ m_3\ m_4\ c_5\ c_6\ c_7)$	Weight of Code $w(c)$
0 0 0 0	0 0 0 0 0 0 0	0
0 0 0 1	0 0 0 1 1 1 0	3
0 0 1 0	0 0 1 0 1 0 1	3
0 0 1 1	0 0 1 1 0 1 1	4
0 1 0 0	0 1 0 0 0 1 1	3
0 1 0 1	0 1 0 1 1 0 1	4
0 1 1 0	0 1 1 0 1 1 0	4
0 1 1 1	0 1 1 1 0 0 0	3
1 0 0 0	1 0 0 0 1 1 1	4
1 0 0 1	1 0 0 1 0 0 1	3
1 0 1 0	1 0 1 0 0 1 0	3
1 0 1 1	1 0 1 1 1 0 0	4
1 1 0 0	1 1 0 0 1 0 0	3
1 1 0 1	1 1 0 1 1 1 0	5
1 1 1 0	1 1 1 0 0 0 1	4
1 1 1 1	1 1 1 1 1 1 1	7

Syndrome decoding: Let the message word (1001) be transmitted, for which the transmitted code word is (1001 001); the received code word is $X = 1011001$(say) with an error pattern (0010000). The syndrome of X is

$$S = XH^T = (1011\ 001)\begin{bmatrix} 1 & 1 & 1 \\ 0 & 1 & 1 \\ 1 & 0 & 1 \\ 1 & 1 & 0 \\ \hline 1 & 0 & 0 \\ 0 & 1 & 0 \\ 0 & 0 & 1 \end{bmatrix} = (101)$$

The look-up table computed above predicts that the correct error pattern and the decoded pattern will be

$$\hat{X} = X + E = 1011001 + 0010000 = 1001001.$$

Let us now suppose that exactly two errors happen in the received word with a double-error pattern $E = (1000\,100)$ and let the received word be $X = 0001\,101$ (for the same transmitted code word $c = 1001\ 001$). In this case, the syndrome of X is

$$S = XH^T = (0001\ 101)\begin{bmatrix} 1 & 1 & 1 \\ 0 & 1 & 1 \\ 1 & 0 & 1 \\ 1 & 1 & 0 \\ 1 & 0 & 0 \\ 0 & 1 & 0 \\ 0 & 0 & 1 \end{bmatrix} = (011)$$

which corresponds to an error pattern of $E = 0100000$ (from the look-up table).

So, the decoded output word would be

$$X = X + E = 0001101 + 0100000 = 0101101.$$

Comparing this with the original transmitted code we find three errors (two transmission errors and one erroneous detection by the decoder).

$$C = \boxed{1}\boxed{0}0\,1\boxed{0}0\,1$$
$$\hat{X} = \boxed{0}\boxed{1}0\,1\boxed{1}0\,1$$

3. We want to transmit data using a (7, 4) linear block code over a binary symmetric channel with a bit error probability of $\alpha = 0.01$. The code is to detect two or fewer errors. Find the probability that no error is detected in the receiver.

Solution

The probability of detecting t errors in a block of n bits is given by the binomial $n_{c_t}\alpha^t(1-\alpha)^{n-t}$.

In the present case, the decoder is unable to detect if the errors are in three or more than three positions. Hence, the probability of such errors is given by the sum of probabilities of zero, one, or two errors. Thus,

$$P_e = \sum_{i=0}^{2} 7_{c_i}\alpha^i(1-\alpha)^{7-i} = 7_{c_0}(0.01)^0(0.99)^7 + 7_{c_1}(0.01)^1(0.99)^6 + 7_{c_2}(0.01)^2(0.99)^5$$
$$= 0.93206 + 0.06590 + 0.00199 = 0.99995$$

So, probability that no errors are detected is

$$P_{\text{no error}} = 1 - P_e = 1 - 0.99995 = 0.00005 = 5 \times 10^{-5} \text{ (excellent channel)}$$

4. For a (7, 4) cyclic code, find the code vectors for the following message vectors: 0100, 1010, 1101. Repeat the same by using generator matrix. Assume that the generator polynomial is $g(x) = 1 + x + x^3$.

Solution

For the message vector $A = 0100$, we have the message polynomial $A(x) = 0.x^0 + 1.x^1 + 0.x^2 + 0.x^3 = x$ so that the polynomial of the coded word is

$$T(x) = A(x)g(x) = x(1 + x + x^3) = x + x^2 + x^4$$
$$= 0.x^0 + 1.x^1 + 1.x^2 + 0.x^3 + 1.x^4 + 0.x^5 + 0.x^6$$

Hence, the code word is $T = 0110100$

For the message vector $A = 1010$, $A(x) = 1 + x^2$ and so

$$T(x) = (1 + x^2)(1 + x + x^3) = 1 + x + x^3 + x^2 + x^3 + x^5 = 1 + x + x^2 + x^5$$
$$= 1.x^0 + 1.x^1 + 1.x^2 + 0.x^3 + 0.x^4 + 1.x^5 + 0.x^6$$

The code word is $T = 1110010$.

Similarly for

$$A = 1101, A(x) = 1 + x + x^3$$
$$T(x) = (1 + x + x^3)(1 + x + x^3) = 1 + x + x^3 + x + x^2 + x^4 + x^3 + x^4 + x^6$$
$$= 1 + x^2 + x^6 = 1.x^0 + 0.x^1 + 1.x^2 + 0.x^3 + 0.x^4 + 0.x^5 + 1.x^6$$

So, $T = 1010001$.

Message vector				Code vector						
0	1	0	0	0	1	1	0	1	0	0
1	0	1	0	1	1	1	0	0	1	0
1	1	0	1	1	0	1	0	0	0	1

The generator matrix of the cyclic code is

$$[G] = \begin{bmatrix} g_0 & g_1 & g_2 & g_3 & 0 & 0 & 0 \\ 0 & g_0 & g_1 & g_2 & g_3 & 0 & 0 \\ 0 & 0 & g_0 & g_1 & g_2 & g_3 & 0 \\ 0 & 0 & 0 & g_0 & g_1 & g_2 & g_3 \end{bmatrix}$$

$$= \begin{bmatrix} 1 & 1 & 0 & 1 & 0 & 0 & 0 \\ 0 & 1 & 1 & 0 & 1 & 0 & 0 \\ 0 & 0 & 1 & 1 & 0 & 1 & 0 \\ 0 & 0 & 0 & 1 & 1 & 0 & 1 \end{bmatrix}$$

Now, [T] = [A] [G] gives the identical results as tabulated above.

5. Consider the $(n, k) = (k+1, k)$ systematic linear block code with the parity check digit $C_n = m_1 + m_2 + \cdots + m_k$. (a) Find the generator matrix for this code. (b) Construct all possible code words for message word length $k = 4$. (c) Find the error syndrome for error pattern $E = (e_1 e_2 e_3 e_4 e_5)$ for $k = 4$.

Solution

(a) Since $n = k + 1$, this code is a parity-check code where a single parity bit is added to the message word.

If the message word be $M = (m_1 m_2 \ldots, m_k)$, the codeword would be $C = (c_1 c_2 \ldots, c_k c_{k+1}) = (m_1 m_2 \ldots, m_k, c_{k+1})$ where $c_i = m_i$ for $i = 1$ to k. Thus, the generator matrix is given by

$$G = \left[\begin{array}{ccccc|c} 1 & 0 & 0 & \cdots & 0 & 1 \\ 0 & 1 & 0 & \cdots & 0 & 1 \\ \hline 0 & 0 & 0 & \cdots & 1 & 1 \end{array}\right]$$

with the left block labelled $I_{k \times k}$.

(b) For $k = 4$, [G] takes the form

$$G = \left[\begin{array}{cccc|c} 1 & 0 & 0 & 0 & 1 \\ 0 & 1 & 0 & 0 & 1 \\ 0 & 0 & 1 & 0 & 1 \\ 0 & 0 & 0 & 1 & 1 \end{array}\right]$$

and code word $[C]_{1 \times n} = [M]_{1 \times k} [G]_{k \times n}$

$k = 4$ implies that the code is (5, 4) parity code and is tabulated as follows. The parity bit is $c_5 = m_1 + m_2 + m_3 + m_4$ (even parity)

$M = (m_1\ m_2\ m_3\ m_4)$				$C = (m_1\ m_2\ m_3 m_4\ c_5)$				
0	0	0	0	0	0	0	0	0
0	0	0	1	0	0	0	1	1
0	0	1	0	0	0	1	0	1
0	0	1	1	0	0	1	1	0
0	1	0	0	0	1	0	0	1
0	1	0	1	0	1	0	1	0
0	1	1	0	0	1	1	0	0
0	1	1	1	0	1	1	1	1
1	0	0	0	1	0	0	0	1
1	0	0	1	1	0	0	1	0
1	0	1	0	1	0	1	0	0
1	0	1	1	1	0	1	1	1
1	1	0	0	1	1	0	0	0
1	1	0	1	1	1	0	1	1
1	1	1	0	1	1	1	0	1
1	1	1	1	1	1	1	1	0

(c) To find error syndrome $[S] = [E]\left[H^T\right]$, we first find $\left[H^T\right]$ as follows.

$$[H] = [\ P^T \quad I\ ] = [\ 1 \quad 1 \quad 1 \quad 1 \ \vdots\ 1\ ]$$

So, $[\ H^T] = \begin{bmatrix} 1 \\ 1 \\ 1 \\ 1 \\ 1 \end{bmatrix}$.

For an error pattern $[E] = (e_1 e_2 \ldots e_5)$

we have $[S] = (e_1 e_2 e_3 e_4 e_5) \begin{bmatrix} 1 \\ 1 \\ 1 \\ 1 \\ 1 \end{bmatrix} = (e_1 + e_2 + e_3 + e_4 + e_5)$

If $S = 0$ (single bit), there is no error, if $S = 1$, there are errors.

Exercises

Activity

9.1 A source has infinitely large number of outputs with the probability of occurrence given by $p(x_{\rm i}) = p(1-p)^{{\rm i}-1}$ where $i = 1, 2, 3, \ldots$ Find the average self information or entropy $H(x)$ of the source.

(Hint: $\sum\limits_{i=1}^{\infty} p(x_i) = 1; H(x) = \sum\limits_{i=1}^{\infty} p(x_i) \log_2 \frac{1}{p(x_i)}$, etc)

9.2 A continuous random variable x has uniform distribution with the PDF given by

$$f_X(x) = \begin{cases} {}^{1}\!/_{k}, & 0 \le x \le k \\ 0, & \text{otherwise} \end{cases}$$

Find the differential entropy $H(x)$ in terms of the parameter K. (Hint: Apply Eq. (9.12) as in Example 9.5).

9.3 Show that a random variable will have the largest possible differential entropy if the distribution is Gaussian. Also show that $H(x)\,|_{\max} = \log_2 \sqrt{2\pi e \sigma^2}$, where σ^2 is the variance of the random variable. (Hint: Consult Section 9.2.2)

9.4 Consider the following generator matrix $[G]$ of a (6, 3) linear block code.

$$G = \begin{pmatrix} 1 & 1 & 1 & 0 & 0 & 0 \\ 0 & 0 & 1 & 1 & 1 & 0 \\ 1 & 0 & 0 & 0 & 1 & 1 \end{pmatrix}$$

a) Generate all possible code words

b) Find the generator matrix of an equivalent systematic code.

9.5 Consider the polynomial $g(x) = (x^5 + x^3 + x + 1)$ be the generator polynomial of a cyclic code of block length 15.

(a) Find the generator matrix $[G]$

(b) Find the parity-check matrix $[H]$

(c) Write down the generator matrix in systematic form.

(Hint: $g(x)$ is the factor of $(x^{15} + 1)$. Find the other factor $h(x)$ such that $g(x)h(x) = (x^{15} + 1)$. So $h(x)$ is the polynomial representation of $[H]$-matrix).

9.6 The generator functions of a rate $-\frac{1}{3}$ convolutional encoder is given by $g_1 = (1, 1, 0, 1), g_2 = (1, 0, 1, 0), g_3 = (1, 0, 0, 1)$.Design the encoder. Find the impulse response function of this coder.

9.7 An information source emits M-alphabets. What is the range of source entropies?

(Hint: Since $0 < p_i \leq 1$ and each term of Eq. (9.26) is positive. So $H(x) > 0$. The upper bound of $H(x)$ is when $P_i = \frac{1}{M}$.)

9.8 The input alphabet set of a discrete memory-less channel is $x = \{x_0, x_1\}$ and the output alphabet set is $y = \{y_0, y_1, y_2\}$, with the channel as shown in the adjoining diagram. The transitional probabilities are also indicated.

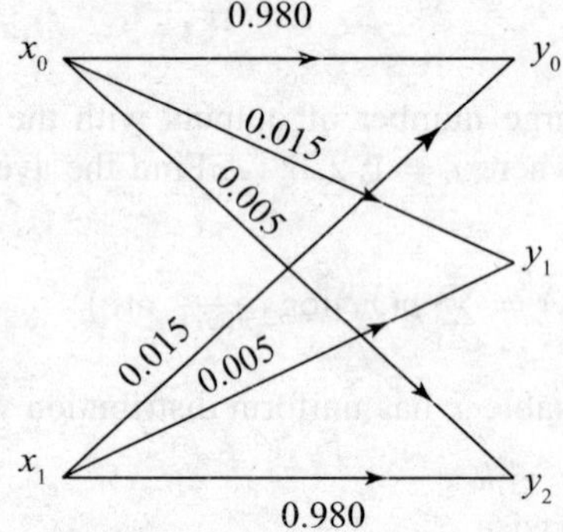

Find the channel matrix $[P(Y/X)]$. If the source alphabets are equally probable, what is the probability of the occurrence of $x_0 \to y_0$ and $x_1 \to y_2$ transitions.

9.9 The parity-check matrix for the (7, 4) systematic Hamming Code is given as

$$[H] = \left[\begin{array}{ccc|cccc} 1 & 0 & 0 & 1 & 1 & 0 & 1 \\ 0 & 1 & 0 & 1 & 0 & 1 & 1 \\ 0 & 0 & 1 & 0 & 1 & 1 & 1 \end{array}\right]$$

Find the code words of the (7, 3) dual code. Also decode the following vectors of the systematic (7, 3) dual code, on minimum distance decoding.

$[R] = [1010001], [R] = [1101011]$.

(Hint: Parity-check matrix of (n, k) code is the generator matrix of $(n, n - k)$ dual code)

9.10 The generator polynomial of a cyclic code in octal representation is $(23)_8$. Find the code word for the message polynomial $m(x) = x^3 + 1$.

Review questions

1. (a) Write the generation matrix for (7,4) systematic block code with single error correction capability. Use this to find out the code for a typical 4-bit message 1101. Given that the probability of a bit being in error is α, what is the probability of undetected error in the above case.

   (b) Consider a systematic (7,3) cyclic code generated by $G(P) = p^4 + p^3 + p^2 + p + 1$. Find $H(p), C(P)$ and X when $M = 101$ (the symbols have their usual significance).

   (c) Considering triple repetition coding, find out d_{min}. Hence calculate the number of errors that can be corrected and detected per word. (K.U. 2001)

2. (a) Explain the meanings of n and k in an (n, k) block code.
   A (7,4) Hamming code has the following $\overline{H}$ matrix.

$$\overline{H} = \begin{bmatrix} 1 & 1 & 1 & 0 & 1 & 0 & 0 \\ 1 & 1 & 0 & 1 & 0 & 1 & 0 \\ 1 & 0 & 1 & 1 & 0 & 0 & 1 \end{bmatrix}$$

   Find the generating matrix $\overline{G}$. Hence list all the codewords for the uncoded words $[a_1a_2a_3a_4]$.

   (b) What is the Hamming distance? How many errors can be detected and corrected if the Hamming distance is 3? Define and explain the significance of 'syndrome'.

3. (a) What do you mean by information content of a message and the entropy of the source of messages?

   (b) Show that for two messages the maximum value of average information per message occurs when the two messages are equally likely. Find the corresponding maximum value of entropy.

   (c) An analog signal is bandlimited to 4 kHz and sampled at Nyquist rate. The samples are quantised into 4 levels having probabilities $p_1 = 1/8, p_2 = 3/8, p_3 = 3/8$ and $p_u = 1/8$. Assuming that the quantisation levels are independent, find the information rate of the source. (WBUT 2004)

4. (a) What is entropy?

   (b) Show that the maximum entropy of a binary system occurs when $p = 1/2$.

   (c) State and explain five entropies in a communication system.

   (d) A transmitter has an alphabet of four letters (x_1, x_2, x_3, x_4) and the receiver has an alphabet of three letters (y_1, y_2, y_3). The joint probability matrix is given by

$$P(x,y) = \begin{bmatrix} 0.3 & 0.5 & 0 \\ 0 & 0.25 & 0 \\ 0 & 0.15 & 0.05 \\ 0 & 0.05 & 0.15 \end{bmatrix}.$$

Calculate all the entropies of the system. (K.U. 2003)

5. (a) Define Kraft–Mcmillan inequality.
   (b) State the algorithm for Shannon–Fano coding
   (c) What is coding efficiency?
   (d) Given ten message with probabilities

$p(m_1) = 0.49 \quad p(m_6) = 0.05 \quad p(m_2) = 0.14 \quad p(m_7) = 0.02 \quad p(m_3) = 0.14$

$p(m_8) = 0.02 \quad p(m_4) = 0.07 \quad p(m_9) = 0.005 \quad p(m_5) = 0.07 \quad p(m_{10}) = 0.005$

Find the Shannon–Fano code and calculate the coding efficiency. (K.U. 2003).

6. (a) Define block code.
   (b) What do you mean by a (7,4) block code?
   (c) What is the Hamming bound in an error correcting code?
   (d) In (n, k) linear block codes, write the general form of the G and H matrices.
   (e) The generator matrix of a(3,2) block code is $G = \begin{bmatrix} 1 & 0 & 1 \\ 0 & 1 & 1 \end{bmatrix}$. Find all the code vectors. (KU 2003)

7. A discrete memoryless source has the symbols with probabilities 0.5, 0.4 and 0.1.

   Determine the Huffman code and the coding efficiency. What is the average length of the codewords?

8. Consider a polynomial $g(x) = x^{10} + x^8 + x^5 + x^4 + x^2 + x + 1$ to represent the generator polynomial of a cyclic code with block length 15. Find the generator matrix G and the parity-check matrix H. Write down the generator matrix in the systematic form. How may errors this code can detect and correct?

9. (a) Derive an expression for the maximum channel capacity of a communications channel in the presence of Gaussian noise.
   (b) A signal is bandlimited to 10 kHz which is sampled and quantised into 8 levels with probabilities 0.25, 0.20, 0.15, 0.15, 0.10, 0.05, 0.05, 0.05. Calculate the entropy and rate of information in the involved PCM system. [RPE, CU 2002]

10. Define 'Information'. How can information be measured? A code is composed of dots and dashes. Assume that the dash is 4-times as long as the dots and has one-third the probability of occurrence.

    (i) Calculate the information in a dot and that in a dash.
    (ii) Calculate the average information in the dot-dash code.

(iii) Assume that a dot lasts for 10 ms and that the same time is allowed between symbols. Calculate the average rate of information transmission. [B.U.2002]

11. (a) A source produces six symbols with the following probabilities: 1/2, 1/4, 1/8, 1/16, 1/32, 1/32. Using Fano's procedure, generate the codes of the symbols. Calculate the channel capacity and the coding efficiency.
(b) Define 'mutual information'. Obtain its value form the Figure P9.1.
Given $p(x_1) = 0.35, \quad p(x_2) = 0.65$ [B.U. 2000]

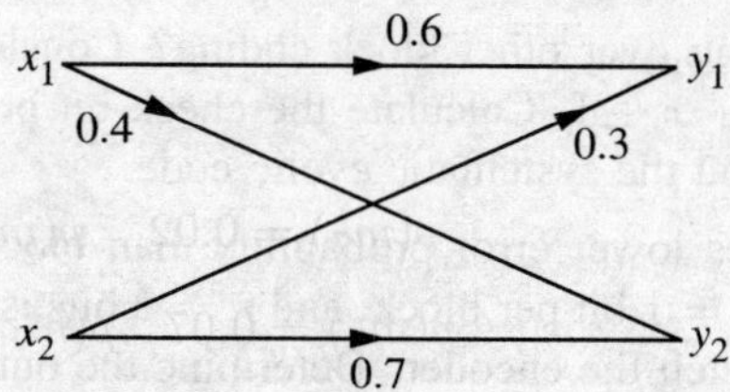

Fig. P9.1

12. (a) Explain Shannon–Fano algorithm to generate more efficient code with an example.
(b) Define entropy and obtain an expression to find the average information.

13. Find the capacity of an additive white Gaussian noise channel with a bandwidth of $1MHz$, average signal power 10 W, and noise power spectral density $\eta/2 = 10^{-9}$W/Hz.

14. Figure P.9.2 shows two channel models. Write down the transition probability matrix for them.

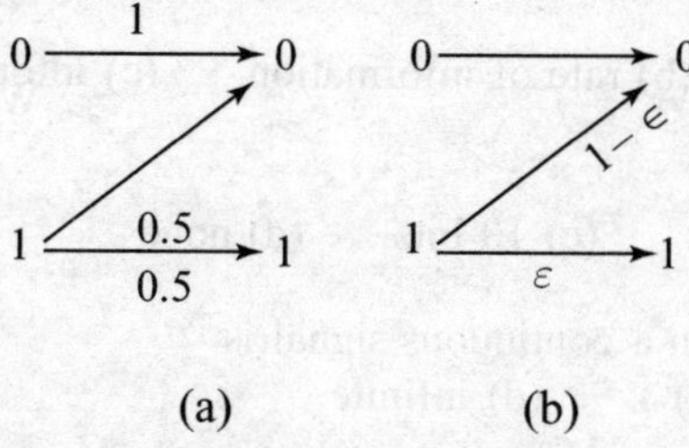

Fig. P9.2 *Two channels of review question 14.*

15. Two channels of Fig. P9.2 (b) are cascaded. Show that the resulting channel is equivalent to the channel with error probability $\varepsilon_1 = \varepsilon^2$.

16. (a) Define entropy of a binary memory-less source and lists its properties.
(b) What is the function of a source encoder? What do you mean by a prefix coding? State the algorithm for Huffman Coding.
(c) Five symbols emitted from a source have the probabilities $1/2, 1/4, 1/8, 1/16$, and $1/16$. Find the optimum code using Huffman algorithm. Calculate the coding efficiency and average length of the code. (MUST, Autumn 2009)

17. (a) A (7, 4) systematic linear block code is given by the following parity-check equations. $p_1 = m_1 \oplus m_2 \oplus m_3, p_2 = m_1 \oplus m_2 \oplus m_4$; and $p_3 = m_1 \oplus m_3 \oplus m_4$, where m_i's are the message bits and p_1, p_2, p_3 are the parity-check bits. Find the generator matrix $[G]$ and parity-check matrix $[H]$. Hence show that $[G][H^T]$ is a null matrix.
    (b) Write down the steps for constructing an (n, k) systematic cyclic code. Construct a systematic (7, 4) cyclic code with $g(x) = x^3 + x + 1$ as the generating polynomial. Assume that $[M] = [1001]$ is the message word. (MUST, Spring 2012)

18. (a) What is the advantage of cyclic coding over other block coding? Consider the cyclic (7, 4) code generated by $g(x) = x^3 + x + 1$. Calculate the check-bit polynomial for the message $[M] = [1101]$. Hence find the systematic cyclic code.
    (b) Why convolutional encoding produces lower error probability than block coding? A convolutional encoder K = 1 stage, $k = 1$ bit per block, and $n = 2$ bits is described by $g_1 = (1, 1, 1)$, and $g_2 = (1, 0, 1)$. Sketch the encoder. Determine the output sequence if the input is $[M] = [11001101]$. (MUST, Spring 2014)

Multiple-choice questions

1. Capacity of a channel refers to
   (a) bandwidth required for information
   (b) maximum rate of information transmission
   (c) number of digits in the coding
   (d) volume of information it can transmit.

2. Dit is a unit of
   (a) channel capacity (b) rate of information (c) information (e) entropy

3. One dit equals
   (a) 1 bit (b) 3.32 bits (c) 10 bits (d) none

4. Amount of information in a continuous signal is
   (a)zero (b) 1/2 (c) 1 (d) infinite

5. The maximum possible value of entropy is
   (a) 1 (b) 4 (c) 8 (d) infinite.

6. A source generates 4 messages. The entropy of the source will be maximum if
   (a) all message are equiprobable.
   (b) the probabilities are 1/2, 1/4, 1/8, and 1/8.
   (c) three message are equiprobable.
   (d) none of the above.

7. A message source generates 4 message with probabilities 1/2, 1/4,1/8, and 1/8. The entropy of the source is
   (a) 1.0 bit/message (b) 1.75 bits/message
   (c) 2.0 bits/message (d) 1.544 bits/message.

8. The probability of a message is $1/16$. The information in bits is
(a) 1 bit (b) 2 bits (c) 3 bits (d) 4 bits

9. A three-character set of messages have the probability 0.75, 0.15, and 0.10. The possible Huffmann code is
(a)1, 01,00 (b) 00, 01, 10 (c) 0, 01, 001 (d) 1,10,100

10. For an (n, k) code, the redundancy of the code is
(a) n/k (b) k/n (c) $(n-k)/k$ (d) $k/(n-k)$

11. For a code, H is the parity-check matrix. The syndrome of a received word is defined as
(a)$S = R[H]^{-1}$ (b) $S = RH^T$ (c) $S^T = RH$ (d) $S^T = RH^{-1}$

12. An (n, k) block code with minimum distance $d_{\min}$ has the error detection capability
(a)$e = 2^k - 1$ (b) $e = 2^n - 2^k$ (c) $e = 2^n - 1$ (d) $e = d_{\min}^{-1}$.

13. A (15,11) BCH code can correct
(a) single error (b) double error
(c) depends on generator polynomial (d) nothing can be said.

14. In a Hamming code of code length $n = 63$, the number of redundant bits is
(a) 4 (b) 6 (c) 3 (d) none

15. Defining error syndrome as $[S] = [R][H^T]$, where $[R]$ is the received vector, the dimensions of the syndrome matrix [S] is
(a) $n-k$ by 1 (b) $n-k$ by k (c) 1 by $n-k$ (d) none

10 Digital Modulation Techniques and Data Transmission

10.1 Introduction

Digital modulation is the process by which digital data is modulated onto a sinusoidal carrier waveform. A sinusoid of finite duration T can be thought of as a digital data. The most common digital modulation techniques are amplitude-shift keying (ASK), frequency-shift keying (FSK) and phase-shift keying (PSK). All the advantages of digital communications, namely noise immunity, ease of data processing and multiplexing and error detection and correction facility are available in digital carrier systems. A digital signal can modulate either the amplitude, phase, or frequency of a sinusoid carrier, and the modulated parameter is switched or keyed from one discrete value to another.

This chapter discusses the various digital modulation and demodulation techniques. At the receiving end, the signal is corrupted by noise and there is finite probability that the receiver misinterprets the data. We will attempt to calculate such error probabilities at the receiving end.

10.2 Amplitude-Shift Keying

Recall that in pulse amplitude modulation the amplitude of the pulse is varied in accordance with the baseband signal. A binary signal (that is, digital data) can modulate the carrier amplitude. If $d(t)$ is the binary data, the general analytic expression for the amplitude-shift keying (ASK) signal can be written as

$$s_{\text{ASK}}(t) = \tfrac{A_c}{2}[1 + md(t)] \cos \omega_c t, \tag{10.1}$$

where m plays the role of modulation index. For a bipolar data, $d(t)$ is either $+1$ (binary one) or -1 (binary zero). The binary ASK waveform is illustrated in Fig. 10.1. The commonly used value of m is 1 for which the carrier is turned on (for binary 1) and off (for binary zero), a process described as *on-off- keying* (OOK). Figure 10.1(b) depicts the OOK signal for the data 1011. Note that there is no phase change of the carrier at the bit transitions. The ASK signalling was used earlier in the telegraphy systems for transmission of discrete information. The OOK signalling finds applications in pulsed radar transmission systems.

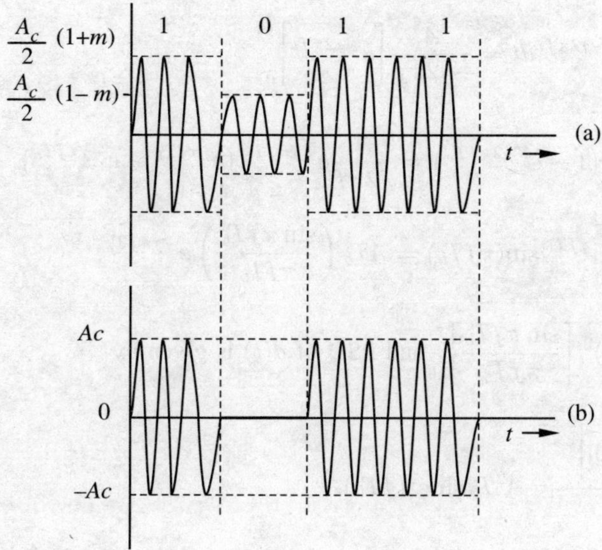

Fig. 10.1 *(a) ASK signal. (b) Binary ASK or OOK signal for the data stream 1011*

Spectrum of OOK

Let the digital data be represented by a unipolar NRZ signal, as shown in Fig. 10.2. The keyed operation involves NRZ rectangular pulses. The power spectral density (PSD) of a signal $s(t)$ can be obtained by the ratio of the energy spectral density to the symbol duration. Since the energy spectral density is the squared magnitude of the Fourier transform of the signal, the signal's PSD is given by

$$G(f) = |S(f)|^2/T_b, \tag{10.2}$$

where T_b is the mean time between bits and $S(f)$ is the Fourier transform of $s(t)$. The Fourier transform of the data stream $d(t)$ is obtained as follows:

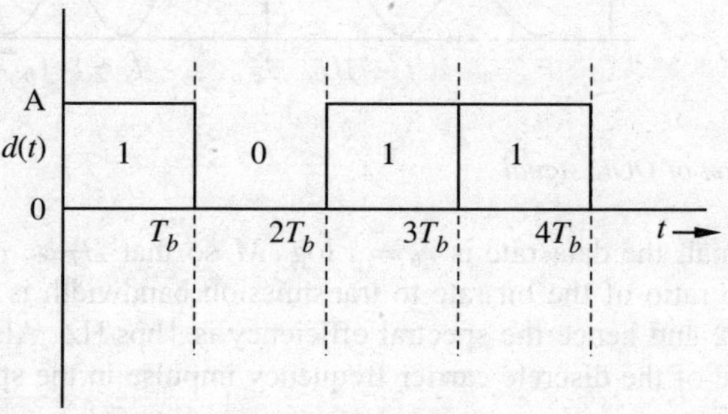

Fig. 10.2 *Unipolar NRZ data stream 1011*

$$\mathcal{F}[d(t)] = \int_0^{T_b} Ae^{-j2\pi ft}dt = \frac{A}{-j2\pi f}\left[e^{-j2\pi ft}\right]_0^{T_b}$$

$$= \frac{A}{j2\pi f}\left(1 - e^{-j2\pi fT_b}\right) = \frac{A}{j2\pi f}e^{-j\pi fT_b}\left(e^{j\pi fT_b} - e^{-j\pi fT_b}\right)$$

$$= \frac{A}{\pi f}e^{-j\pi fT_b}\sin(\pi fT_b) = AT_b\left(\frac{\sin \pi fT_b}{\pi fT_b}\right)e^{-j\pi fT_b}.$$

So, $\left|\mathcal{F}[d(t)]\right|^2 = (AT_b)^2\left[\frac{\sin \pi fT_b}{\pi fT_b}\right]^2$ and PSD of $d(t)$ is given by

$$G_d(f) = \frac{\left|\mathcal{F}[d(t)]\right|^2}{T_b} = A^2T_b\,\text{sinc}^2(fT_b). \tag{10.3}$$

The OOK signal given in (10.1) will have the power spectral density for $f > 0$ as

$$G(f) = A^2T_b\left(\frac{mA_c}{2}\right)^2 \text{sinc}^2(f - f_c)T_b + \frac{A_c^2}{4}\delta(f - f_c) \tag{10.4}$$

and is sketched in Fig. 10.3. The first zero-crossing occurs when $\pi(f - f_c)T_b = \pm\pi$, that is, $f = f_c \pm \frac{1}{T_b} = f_c \pm r$, where r is the data rate. Most of the signal power is contained within the first lobe in the range $f_c \pm r$. Energy considerations suggest that the transmission bandwidth B_T be taken as $B_T \approx 1/T_b = r$.

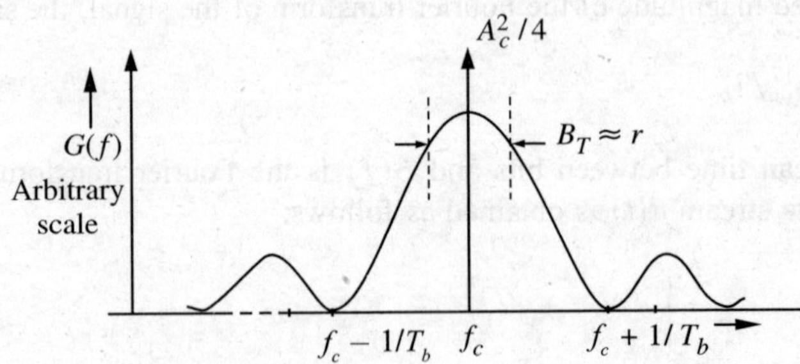

Fig. 10.3 *Power spectrum of OOK signal*

For the M-ary ASK signal, the data rate is $r_b = r\log_2 M$ so that $B_T \approx r_b/\log_2 M$. The spectral efficiency defined as the ratio of the bit rate to transmission bandwidth is $r_b/B_T \approx \log_2^M$ bps/Hz. For binary OOK, $M = 2$ and hence the spectral efficiency is 1bps/Hz. Also, OOK signal has poor power efficiency because of the discrete carrier frequency impulse in the spectrum.

Example 10.1

Write down the generalised expression for the M-ary ASK signal and sketch the signal-space diagram.

Solution

For the carrier $s(t) = A_c \cos\omega_c t \;\; 0 \le t \le T_b$, over the bit interval, the symbol energy per bit interval is $\frac{E_b}{T_b}$ = signal power $= \frac{1}{2}A_c^2$ so that $A_c = \sqrt{2E_b/T_b}$ and so the ASK waveform takes the form

$$s_{\text{ASK}}(t) = \sqrt{2E_b/T_b}\cos\omega_c t, \quad 0 \le t \le T_b.$$

For the M-ary case, the amplitude takes M-possible amplitude levels corresponding to M-possible data symbols. Thus, the M-ary ASK signal expression is

$$s_{\text{ASK}}(t) = \sqrt{\frac{2E_i(t)}{T_b}}\cos\omega_c t, \qquad \begin{array}{l} 0 \le t \le T_b \\ i = 1, 2, \ldots, M. \end{array}$$

The ASK signal being one-dimensional,

$$s_{\text{ASK}}(t) = \sqrt{E_i(t)}\sqrt{\frac{2}{T_b}}\cos\omega_c t = \sqrt{E_1(t)}\psi(t)$$

where $\psi(t)$ is the orthonormal basis function. The ASK amplitude $\sqrt{2E_i(t)/T_b}$ is not sent as M-ary levels, rather each symbol is mapped to $k = \log_2 M$ binary digits — this can be done in a number of ways. The Gray code is an acceptable scheme since adjacent codewords differ only in one bit position. The signal-space diagram for $M = 2, 4$, and 8 are shown in Fig. 10.4.

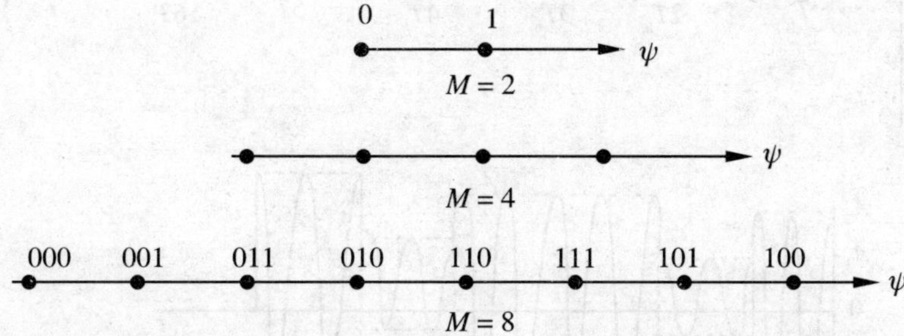

Fig. 10.4 *Signal-space diagram for M-ary ASK signals.*

10.2.1 M-ary digital PAM (ASK) signal

In Example (10.1), the binary ASK wave form has been shown to be $s_{ASK}(t) = A_c\cos 2\pi f_c t$, $0 \le t \le T_b$, where $A_c = \sqrt{2E_b/T_b}$, E_b is the energy per bit and T_b is the bit duration equivalent to the bit rate of $R = 1/T_b$ bits per second. As shown in Fig. 10.1, binary ASK waveform is represented by two constant levels; for logic 1, the level amplitude is $\frac{A_c}{2}(1+m)$ and for logic 0, it is $\frac{A_c}{2}(1-m)$. For M-ary ASK, the signal will be represented by M different amplitude levels. We may represent M-ary ASK signal waveform by

$$s_m(t) = A_m p(t) cos 2\pi f_c t, m = 1, 2, ..., M; \;\; 0 \le t \le T \tag{10.5}$$

where $A_m, m = 1, 2, \ldots, m$ represents the M-different amplitudes levels, given by

$$A_m = (2m - 1 - M)\, d_o, \quad m = 1, 2, ..., M \tag{10.5a}$$

where the adjacent signal amplitude levels differ by

$$|A_m - A_{m-1}| = |(2m - 1 - M)\, d_o - (2(m-1) - 1 - M)\, d_o| = 2d_o.$$

We consider here $M = 2^k$ where the k-bit block of symbols constitute one M-ary ASK symbol and duration of digital PAM is $T = kT_b$ or the symbol rate $R = 1 \big/ T_b = \frac{1}{kT_b} = R/k$. The pulse shape $p(t)$ depends on the type of modulation. For digital modulation, a rectangular pulse shape with non-return-to zero (NRZ) is generally adopted. That is why in Fig. 10.1, the binary ASK is plotted with rectangular-shaped NRZ pulse.

Let us first explain the difference between baseband ASK and bandpass ASK signals. For the simplest case of binary ASK, we consider two amplitude levels A_2 and A_1 for logic1 and logic 0 respectively. If the bit duration is T_b, then the binary date stream 101101 can be represented as shown in Fig. 10.4.1(a). The bandpass ASK is also shown, as in Fig. 10.4.1(b), in which the carrier signal is assumed to complete two oscillations in T_b period.

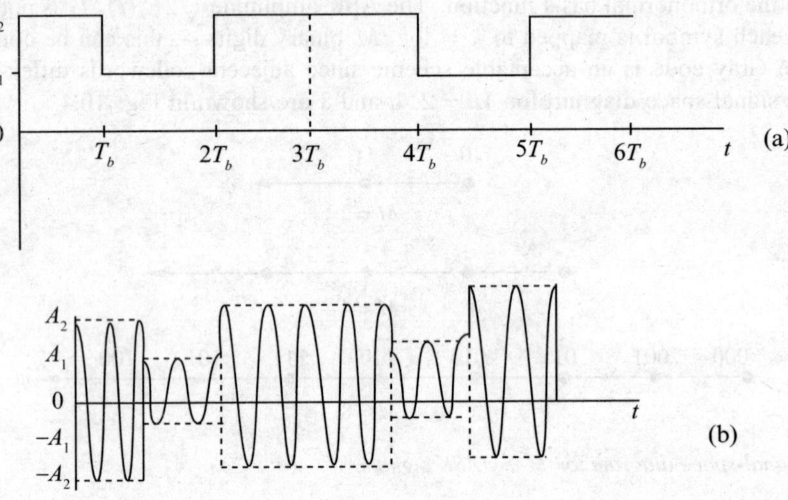

Fig. 10.4.1 *(a) Baseband binary ASK;* $s_m(t) = A_m p(t)$. $A_1 = \frac{A_c}{2}(1+m)$, $A_2 = \frac{A_c}{2}(1+m)$
(b) Bandpass binary ASK; $s_m(t) = A_m p(t) \cos 2\pi f_c t$, $A_1 = \frac{A_c}{2}(1+m)$, $A_2 = \frac{A_c}{2}(1+m)$

It is customary to represent the amplitude levels as being symmetrical about zero volts. In this representation, the baseband ASK may be represented as shown in Fig. 10.4.2(a). The corresponding bandpass ASK is sketched in Fig. 10.4.2(b).

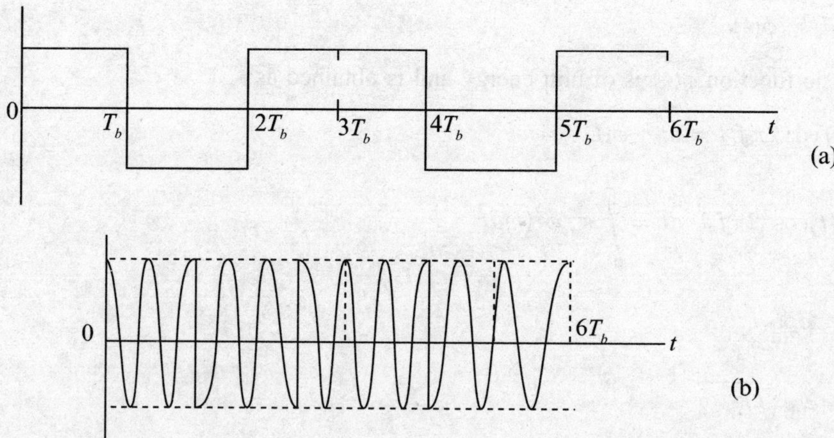

Fig. 10.4.2 *(a) Baseband binary ASK,* $s_m(t) = A_m p(t)$, $m = 1, 2$ *(b) Bandpass binary ASK;* $s_m(t) = A_m p(t) \cos 2\pi f_c t$ $m = 1, 2$.

For M-ary ASK waveform, symbol duration should be $T = kT_b$.

Since the amplitude level A_m is symmetrical about the origin, the mean of A_m is zero. To calculate the energy E_m of the $M-$PAM signals, we note that

$$E_m = \int_0^T s_m^2(t)dt = \int_0^T A_m^2 p^2(t) cos^2 2\pi f_c t \; dt$$

$$= \frac{1}{2} A_m^2 \int_0^T p^2(t)dt = \frac{1}{2} A_m^2 E_p \tag{10.5b}$$

where E_p is the pulse energy. So, each level representing symbols have different energies. The average energy E_{av} of the M-ary ASK may be an important measure. This is given as

$$E_{av} = \frac{1}{M}\sum_{m=1}^{M} E_m = \frac{1}{M}\sum_{m=1}^{M} \frac{E_p \cdot A_m^2}{2}$$

$$= \frac{E_p}{2M}\sum_{m=1}^{M} (2m - 1 - m)^2 d_o^2 = \frac{E_p d_o^2}{6}\left(m^2 - 1\right) \tag{10.5c}$$

The average power P_{av} is then

$$P_{av} = \frac{E_{av}}{T} = \frac{\left(m^2 - 1\right) d_o^2}{3} \cdot \left(\frac{E_p/2}{T}\right) \tag{10.5d}$$

Since ASK signals are one-dimensional, one can express them in terms of basic function $\phi(t)$ as

$$s_m(t) = s_m \phi(t) \tag{10.5e}$$

where the basic function $\phi(t)$ is of unit energy and is obtained as:

$$A_m p(t) \cos 2\pi f_c t = s_m \cdot \phi(t)$$

or, $\displaystyle\int_0^T A_m^2 p^2(t) \cos^2 2\pi f_c t \; dt = \int_0^T s_m^2 \phi^2(t) dt$

or, $\dfrac{A_m^2}{2} \cdot E_p = s_m^2$

that is, $s_m = A_m \sqrt{E_p/2}$

and hence

$$\phi(t) = \frac{s_m(t)}{s_m} = \sqrt{\frac{2}{E_p}} p(t) \cos 2\pi f_c t, \quad 0 \le t \le T \tag{10.5f}$$

The formula $s_m = A_m \sqrt{E_p/2}$ provides the signal-space diagram. For $M = 2$ (binary ASK), $m = 1$ and 2. So, $A_1 = -d$, $A_2 = +d$.

For $M = 4$ (4-ary ASK), $m = 1, 2, 3, 4$ and $A_1 = -3d$, $A_2 = -d$, $A_3 = d$, and $A_4 = 3d$.

For $M = 8$ (8-ary ASK), $m = 1, 2, 3, \ldots, 8$ and $A_1 = -7\,d$, $A_2 = -5\,d$, $A_3 = -3\,d$, $A_4 = -d$, $A_5 = d$, $A_6 = 3\,d$, $A_7 = 5d$ and $A_8 = 7d$.

The signal space is as sketched in Fig. 10.4 and is one-dimensional. The distance between adjacent signal points, known as Euclidean distance can be calculated as follows:

$$d_{\min} = |s_m - s_{m+1}| = \sqrt{\frac{E_p}{2}} |A_m - A_{m+1}| = 2d_o \cdot \sqrt{\frac{E_p}{2}}. \tag{10.5g}$$

In the present notation, the binary ASK signal has the waveform

$$s_m(t) = A_m p(t) \cos 2\pi f_c t, \; m = 1, 2 \tag{10.5h}$$

Using NRZ-rectangular pulse of unit amplitude, $p(t) = 1$ for $0 \le t \le T_b$,

$$s_m(t) = A_m \cos 2\pi f_c t, \; 0 \le t \le T_b$$

So,

$$\left.\begin{aligned} s_1(t) &= A_1 \cos 2\pi f_c t = -d_o \cos 2\pi f_c t \\ s_2(t) &= A_2 \cos 2\pi f_c t = d_o \cos 2\pi f_c t = -s_1(t). \end{aligned}\right\} \tag{10.5i}$$

Both the signals have same energy and are called antipodal signals.

10.3 Phase-Shift Keying (PSK)

The binary phase-shift keying (BPSK) signal is a sinusoidal, of fixed amplitude but contains phase-shifts of π radian when binary data changes from one level to another. If the sinusoid is of amplitude A, the analytic expression for BPSK is

$$s_{\text{BPSK}}(t) = A\cos\omega_0 t \tag{10.5}$$

or
$$\begin{aligned} s_{\text{BPSK}}(t) &= A\cos(\omega_0 t + \pi) \\ &= -A\cos\omega_0 t. \end{aligned} \tag{10.6}$$

Figure 10.5 shows a typical BPSK waveform with its abrupt phase changes of π radian at the symbol transition. If one adopts bipolar NRZ line code to represent the binary data $d(t)$, the general expression for the BPSK waveform is

$$s_{\text{BPSK}}(t) = Ad(t)\cos\omega_0 t, \tag{10.7}$$

in which the data stream $d(t)$ takes on $+1$ V for logic 1 and -1 V for logic 0.

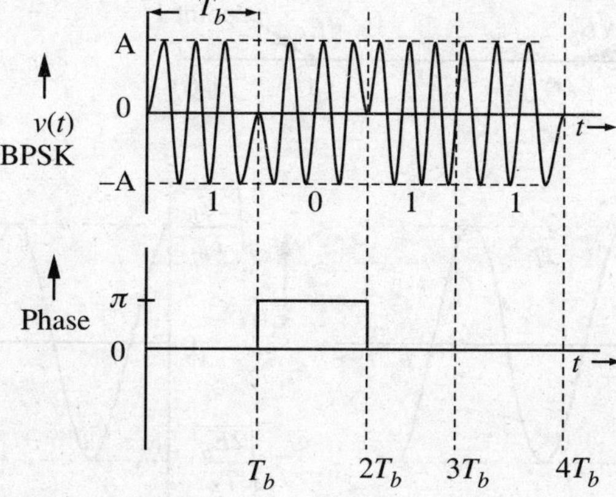

Fig. 10.5 *Illustration of BPSK waveform*

It is more often convenient to express the signal amplitude A in terms of symbol energy E_b. Since the symbol duration is T_b, we have $E_b/T_b = \frac{1}{2}A^2$ or $A = \sqrt{2E_b/T_b}$. Then the BPSK waveform takes the form

$$s_{\text{BPSK}}(t) = \sqrt{2E_b/T_b}\,d(t)\cos\omega_0 t. \tag{10.8}$$

The above representation of signal amplitude facilitates direct calculations of error probability as function of signal energy.

The signal is either

$$s_1(t) = \sqrt{\frac{2E_b}{T_b}}\cos\omega_0 t$$

$$\text{or} \quad s_2(t) = \sqrt{\frac{2E_b}{T_b}} \cos(\omega_0 t + \pi) = -\sqrt{\frac{2E_b}{T_b}} \cos \omega_0 t, \tag{10.9}$$

where $0 \leq t \leq T_b$.

The signal waveform can be represented as vectors or phasors on a polar plot. It is obvious from the above equation that there is one basis function $\phi(t)$ of unit energy given as

$$\phi(t) = \sqrt{\frac{2}{T_b}} \cos \omega_0 t. \tag{10.10}$$

In terms of $\phi(t)$, one can express the BPSK signal as

$$s_1(t) = \sqrt{E_b}\phi(t) \text{ and } s_2(t) = -\sqrt{E_b}\phi(t), \quad 0 \leq t \leq T_b. \tag{10.11}$$

Figure 10.6 shows the vector plot of (10.11), where the vector length gives the signal amplitude and the direction represents the signal phase. The minimum distance between the message points is $d = 2\sqrt{E_b}$.

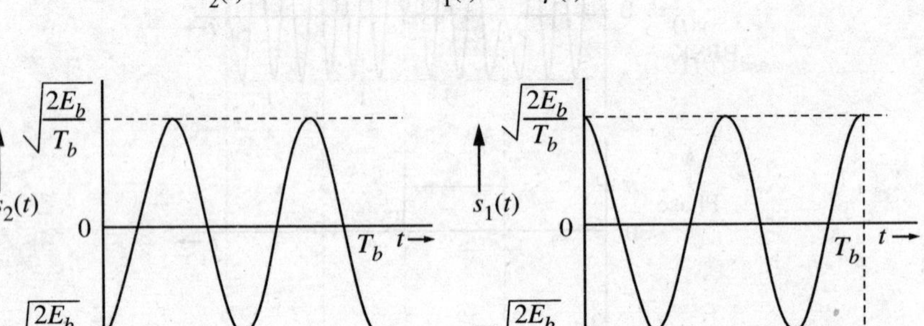

Fig. 10.6 *Signal-space diagram for BPSK waveform. The BPSK waveforms are also shown corresponding to the message points.*

Power spectrum of binary PSK

In the expression (10.8) of the BPSK signal, the binary input waveform $d(t)$ is an NRZ signal varying randomly with symbols 1 and 0. The power spectrum of the NRZ baseband signal is given by (vide Section 8.9)

$$G_b(f) = \left[\frac{\sin(\pi f T_b)}{\pi f T_b}\right]^2, \tag{10.12}$$

where T_b is the symbol duration. Figure 10.7 sketches the NRZ data stream 10110, which varies between $+1$ and -1 and has a duration of T_b. The BPSK signal (10.8) is the NRZ waveform multiplied by $\cos\omega_0 t$ and varies between $+\sqrt{2E_b/T_b}$ and $-\sqrt{2E_b/T_b}$. The power spectral density of such bandpass signal is obtained as

$$G_{\text{BPSK}}(t) = 2E_b\left[\left(\frac{\sin\pi(f-f_0)T_b}{\pi(f-f_0)T_b}\right)^2 + \left(\frac{\sin\pi(f+f_0)T_b}{\pi(f+f_0)T_b}\right)^2\right]. \tag{10.13}$$

The corresponding spectrum along with the baseband spectrum is plotted in Fig. 10.8. Note that the spectrum extends over all frequencies but falls off rapidly as inverse square of the frequency. A comparison with the spectrum of ASK signal (Fig. 10.3) reveals that the spectrum of BPSK has the same shape as the ASK spectrum but in the present case, the carrier impulse is absent. Thus, BPSK has improved power efficiency over ASK signal.

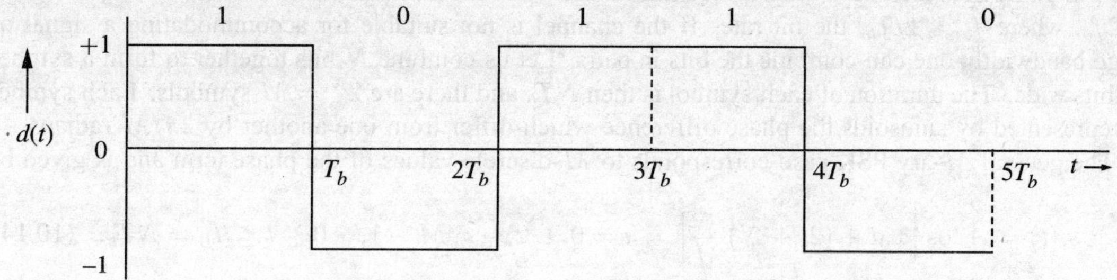

Fig. 10.7 *NRZ data stream 10110*

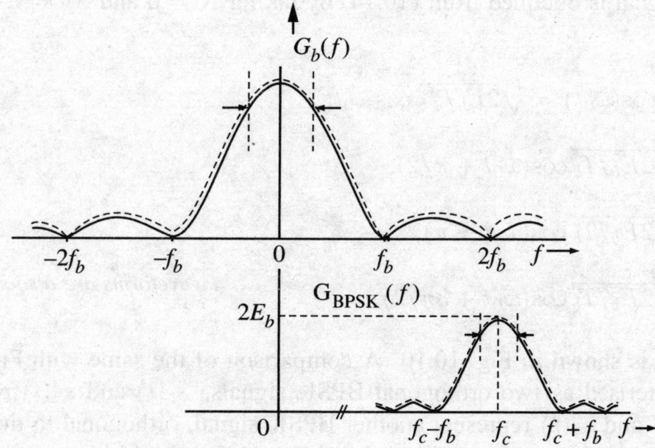

Fig. 10.8 *BPSK Power spectral density*

The generation of PSK waveform is quite straightforward. The output of the channel encoder is a digital data stream which is transformed to an NRZ line code. The BPSK signal is generated by applying $\cos\omega_0 t$ as carrier and baseband data $d(t)$ as modulating signal to a balanced modulator, as shown in Fig. 10.9.

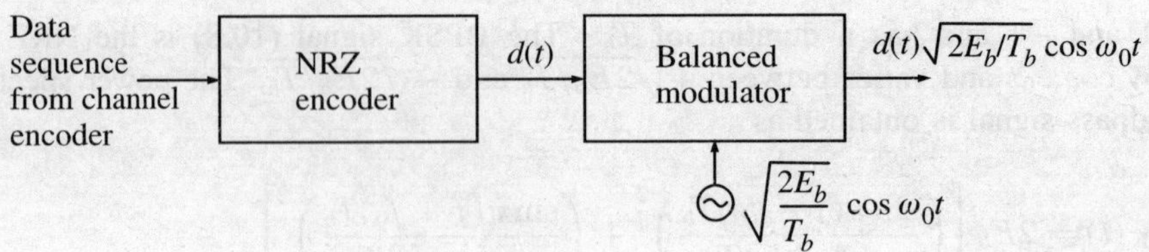

Fig. 10.9 *Digital phase modulation—BPSK generation*

10.3.1 Quadrature phase-shift keying (QPSK)

A more efficient bandwidth utilisation can be achieved if the signalling element contains more than one bit. In BPSK, each bit is transmitted individually and the nominal bandwidth of the BPSK signal is $2f_b$, where $f_b = 1/T_b$, the bit rate. If the channel is not suitable for accommodating a signal of large bandwidth, one can combine the bits in pairs. Let us combine N bits together to form a symbol N-bits wide. The duration of each symbol is then NT_b and there are $2^N = M$ symbols. Each symbol is represented by sinusoids the phase difference which differ from one another by $2\pi/M$ radians.

The general M-ary PSK case corresponds to M-discrete values of the phase term and is given by

$$s_i(t) = A\cos\left[\omega_0 t + (2i+N)\frac{\pi}{M}\right] \qquad i = 0, 1, 2, \cdots, M-1, \quad 0 \le t \le T_0 = NT_b. \qquad (10.14)$$

Here the integer N can be either -1, 0, or $+1$. The PSK signal for $N = 0$ and $M = 2$ is referred to as the BPSK, as given in (10.9). We will now consider here the 4-phase PSK also called quadri-phase PSK (QPSK). The signal is obtained from (10.14) by taking $N = 0$ and $M = 4$. The representative expressions are

$$\begin{aligned} s_1(t) &= A\cos(\omega_0 t) = \sqrt{2E_b/T_b}\cos\omega_0 t, \\ s_2(t) &= \sqrt{2E_b/T_b}\cos(\omega_0 t + \pi/2), \\ s_3(t) &= \sqrt{2E_b/T_b}\cos(\omega_0 t + \pi), \\ s_4(t) &= \sqrt{2E_b/T_b}\cos(\omega_0 t + 3\pi/2). \end{aligned} \qquad (10.15)$$

The QPSK signal set is shown in Fig. 10.10. A comparison of the same with Fig. 10.6 reveals that QPSK can be characterised as two orthogonal BPSK signals. $s_1(t)$ and $s_3(t)$ represent one set of BSK signal and $s_2(t)$ and $s_4(t)$ represent another BPSK signal, orthogonal to the previous one. So, the QPSK bit stream $d(t)$ is usually partitioned into an even and odd bit stream; They are referred to as in-phase (I) and quadrature-phase (Q) stream. The even bit stream $d_e(t)$ modulates a carrier $\cos\omega_0 t$ and the odd bit stream $d_o(t)$ modulates the quadrature carrier $\sin\omega_0 t$. If the original QPSK vector is of length A, the two components have the length $A/\sqrt{2}$ each, as shown in Fig. 10.11. Each BPSK component has a bit rate half that of the QPSK waveform. It is interesting to note that the QPSK signal can be transmitted using half the bandwidth of a BPSK signal.

The mechanism by which the partitioning of bit stream $d(t)$ generates a QPSK signal will now be illustrated. Consider that the original bit stream $d(t) = d_0, d_1, d_2, \cdots$ consists of bipolar pulses which can take $+1$ or -1 representing binary 1 and binary 0, respectively. Figure 10.12 shows the original data stream which is numbered $d_0, d_1, d_2, \cdots$, and the in-phase stream $d_e(t)$ and quadrature stream $d_o(t)$ as

$$d_e(t) = d_0, d_2, d_4, \cdots$$
$$d_o(t) = d_1, d_3, d_5, \cdots.$$

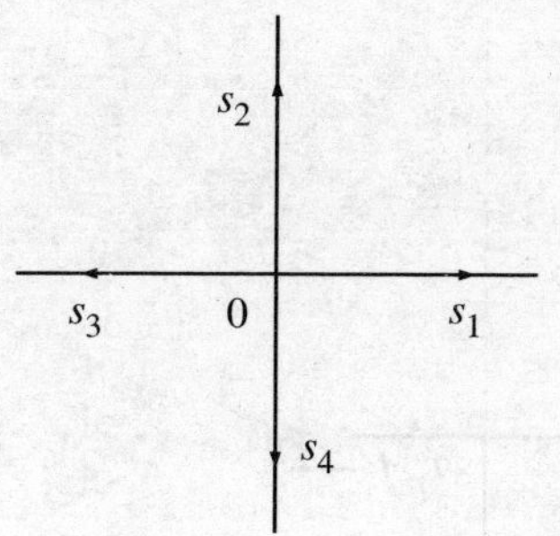

Fig. 10.10 *QPSK signal set*

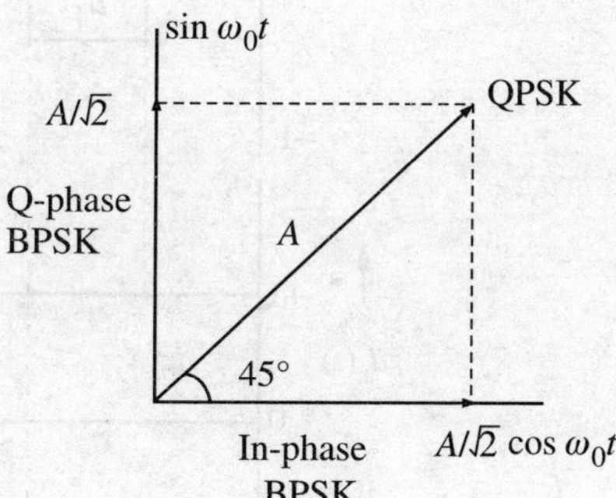

Fig. 10.11 *I and Q components of QPSK*

Note that in each bit stream every bit is of width $2T_b$, where T_b is the basic bit duration.

The QPSK signal is obtained by amplitude modulating these data streams into two orthogonal carriers, $\sqrt{E_b/T_b}\cos\omega_0 t$, and $\sqrt{E_b/T_b}\sin\omega_0 t$ and adding the results. Thus,

$$s_{\text{QPSK}}(t) = \sqrt{E_b/T_b}d_e(t)\cos\omega_0 t + \sqrt{E_b/T_b}d_o(t)\sin\omega_0 t. \tag{10.16}$$

The first term on the right-hand side of (10.16) produces a BPSK waveform, while the second term produces another BPSK waveform. Both the terms occupy the same spectral range but they can be detected separately because of the orthogonality of the carriers.

By substituting $d_e(t) = x\cos\theta(t)$ and $d_o(t) = x\sin\theta(t)$, we have

$$\begin{aligned} s_{\text{QPSK}}(t) &= \sqrt{E_b/T_b}\sqrt{d_e^2 + d_o^2}\cos(\omega_0 t - \theta(t)) \\ &= \sqrt{2E_b/T_b}\cos(\omega_0 t - \theta(t)). \end{aligned} \tag{10.17}$$

The value of $\theta(t) = \tan^{-1}\frac{d_o}{d_e}$ corresponds to one of the four possible combinations of $d_e(t)$ and $d_o(t)$ and hence the four possible output signals. Note that each signal has equal amplitude $\sqrt{2E_b/T_b}$. Figure 10.13 shows the signal space for QPSK signal.

Following (10.16), the QPSK modulator can be modelled as shown in Fig. 10.14. Since in QPSK, two binary digits are grouped together to form a 'symbol', one signal representing 'one symbol' of two bits with symbol time $T_s = 2T_b$. The odd and even pulse streams are both transmitted at the data rate $1/2T_b$ bits per second. After each time T_b, if the two pulse streams change, their transitions coincide. The incoming data sequence is first transformed into polar NRZ and then partitioned into two bit sequences – even and odd. These two even and odd data sequences modulate the appropriate carriers to produce two orthogonal BPSK signals. There are summed to generate the QPSK signal.

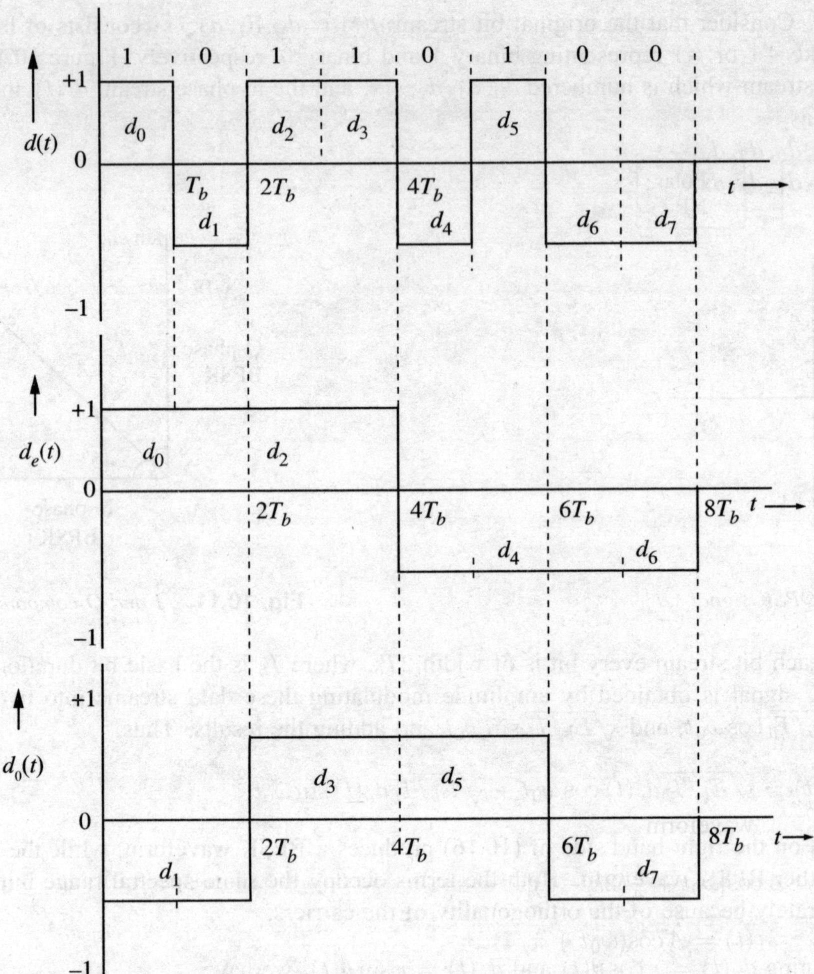

Fig. 10.12 *Partitioning data stream into even and odd stream*

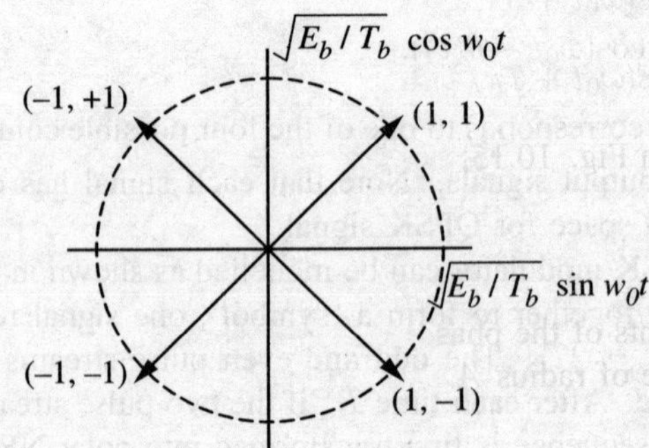

Fig. 10.13 *Signal space for QPSK signal*

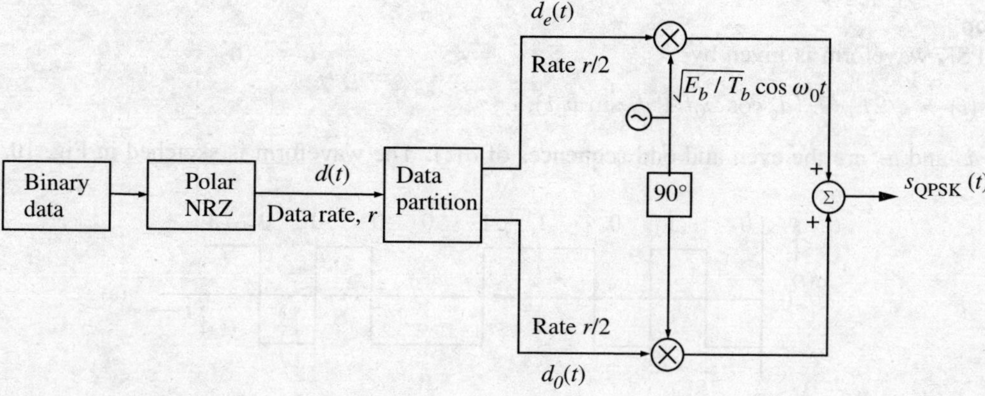

Fig. 10.14 *QPSK modulator*

A special form of QPSK called 'staggered' or 'offset QPSK' (OQPSK) has been developed. This is also based on bit partitioning and orthogonal transmission. The difference between the QPSK and OQPSK lies in the alignment of two baseband data streams. In OQPSK the timing of the bit streams in odd and even channels is shifted so that at the end of each bit interval T_b, either $d_e(t)$ or $d_o(t)$ changes but not simultaneously.
An OQPSK signal can be generated by introducing a delay T_b in the quadrature channel.

Example 10.2

Draw the signal-space diagram for QPSK with $\phi_k = \frac{\pi}{4}(2k-1)$

Solution

Given the QPSK waveform

$$S_k(t) = A\cos[\omega_0 t + \tfrac{\pi}{4}(2k-1)], \quad k = 1, 2, 3, 4,$$

$$k = 1: \quad s_1(t) = A\cos(\omega_0 t + \pi/4)$$

$$k = 2: \quad s_2(t) = A\cos(\omega_0 t + 3\pi/4)$$

$$k = 3: \quad s_3(t) = A\cos(\omega_0 t + 5\pi/4)$$

$$k = 4: \quad s_4(t) = A\cos(\omega_0 t + 7\pi/4).$$

The signal space is shown in Fig. 10.15.

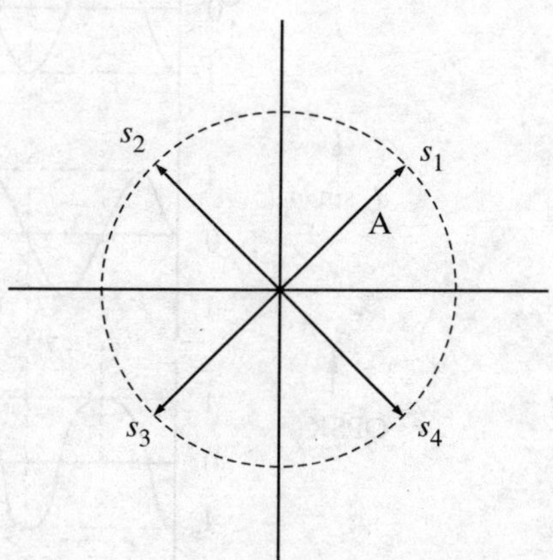

Fig. 10.15 *Signal-space diagram for QPSK signal*

Note that the vector end points of the phasor diagram (Fig. 10.15) fall on the circle of radius A.

Example 10.3

The bit stream $d(t)$ is to be transmitted using QPSK. If $d(t)$= 0101100101, sketch the QPSK waveform.

Solution

The QPSK waveform is given by

$$s(t) = \sqrt{2E_b/T_b}(d_e \cos \omega_0 t + d_o \sin \omega_0 t),$$

where d_e and d_o are the even and odd sequences of $d(t)$. The waveform is sketched in Fig. 10.16.

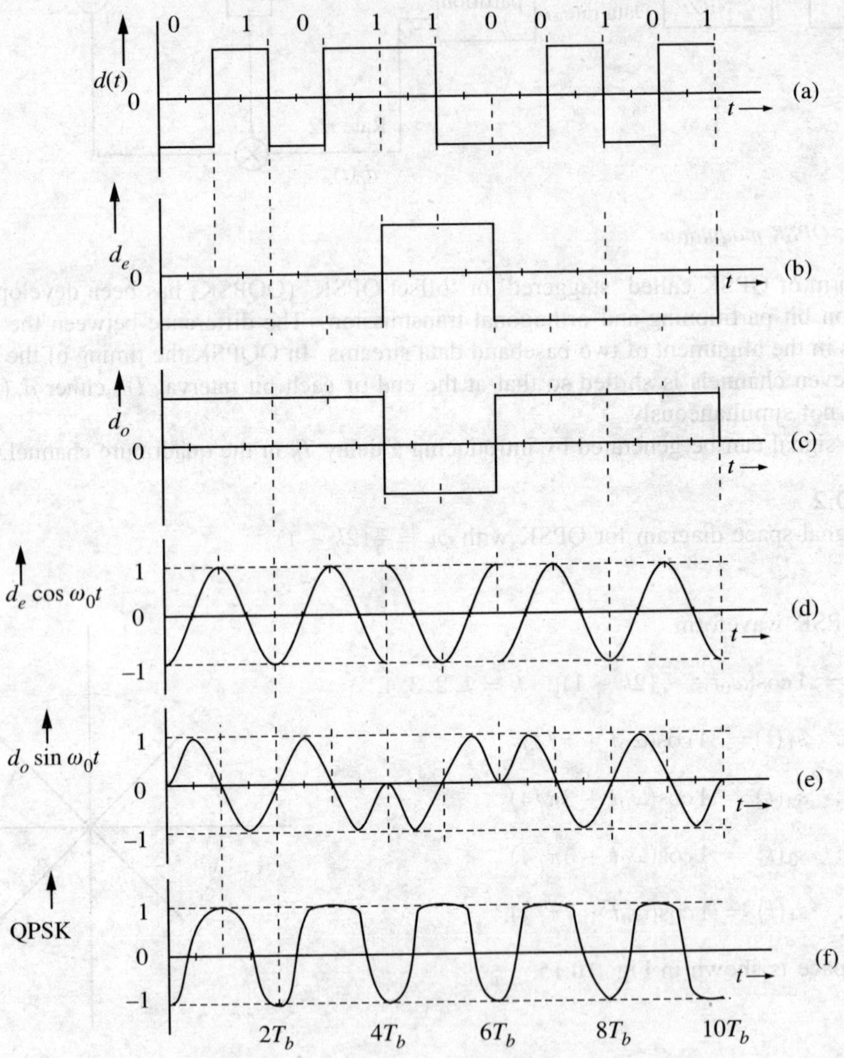

Fig. 10.16 *QPSK waveform: (a) polar data stream; (b)&(c) even and odd sequences; (d) even sequence modulating a carrier* $\cos \omega_0 t$; *(e) odd sequence modulating a carrier* $\sin \omega_0 t$; *(f) QPSK signal.*

10.3.2 Differential PSK (DPSK)

In digital phase modulation, the message information resides in the phase. The carrier signal needs be recovered from the received BPSK signal. In any attempt to determine the actual value of the phase of the incoming signal, there is always the synchronisation problem associated with the process. The differential phase-shift keying (DPSK) avoids the need to provide the synchronising carrier at the demodulator. Instead, the carrier phase of the previous signalling interval can be used as a phase reference. The message bits are regenerated from the changes in phase from symbol to symbol. Such systems include 'differential encoding' of the message sequence at the transmitter. Figure 10.17

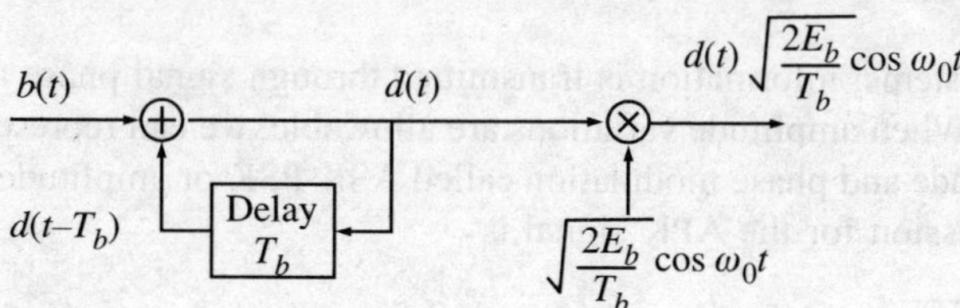

Fig. 10.17 *DPSK transmitter.* $[d(t) = b(t) \oplus d(t - T_b)]$

shows a logic circuit for differential encoding of a binary message data stream $b(t)$. The encoded bit sequence $d(t)$ is given by

$$d(t) = b(t) \oplus d(t - T_b), \tag{10.18}$$

where $\oplus$ represents modulo-2 addition. Data bits $d(t)$ are determined according to the following rule: the present code bit $d(t)$ is a one if the message bit $b(t)$ and prior coded bit $d(t - T_b)$ disagree, otherwise $d(t)$ is a zero. In the following discussion we shall use the symbolic notations $d(k)$ and $b(k)$ to represent $d(t)$ and $b(t)$, respectively, where k is the sample time index. The encoding starts with an arbitrarily chosen initial bit, say $d(0) = 1$. Table 10.1 illustrates the differential encoding of a binary message data $b(k) = 101101001010$.

Table 10.1 Differential encoding

Sample index k	0	1	2	3	4	5	6	7	8	9	10	11	12
Message, $b(k)$		1	0	1	1	0	1	0	0	1	0	1	0
Encoded message delayed, $d(k-1)$		1	0	0	1	0	0	1	1	1	0	0	1
Encoded message $d(k)$	1	0	0	1	0	0	1	1	1	0	0	1	1
Phase-shift $\theta(k)$	π	0	0	π	0	0	π	π	π	0	0	π	π
Regenerated msg.		1	0	1	1	0	1	0	0	1	0	1	0

The coded bit sequence $d(k)$ is converted into a phase-shift sequence $\theta(k)$, in which a 1 is characterised by a 180° phase-shift and a 0 by a 0° phase-shift.

The regeneration is based on phase comparison method. If phase $\theta(k)$ matches will $\theta(k-1)$, the detected bit $\widehat{d}(k)$ will be 1, otherwise, $\widehat{d}(k) = 0$. We have arbitrarily chosen the initial bit $d(0) = 1$. It can be verified that data will be correctly decoded if the encoding starts with $d(0) = 0$.

We have described differential encoding and phase-comparison detection in the absence of noise. Since phase comparison needs two successive bit intervals, noise in one-bit interval may cause errors in the two-bit estimation. So, in DPSK, errors tend to occur in groups of two bits. However, it is not always the fact that errors occur always in pairs.

10.3.3 Amplitude phase keying

In phase-shift keying systems, information is transmitted through signal phase and all the signals are of the same amplitude. When amplitude variations are allowable, we can represent the source symbols by the combined amplitude and phase modulation called ASK/PSK or amplitude phase keying (APK). The mathematical expression for the APK signal is

$$s_{\text{APK}}(t) = \sqrt{\frac{2E_i(t)}{T_b}} \cos[\omega_0 t + \phi_i(t)] \qquad \begin{matrix} 0 \le t \le T \\ i = 0, 1. \cdots, M-1 \end{matrix} \tag{10.19}$$

Here the signal amplitude and the phase change simultaneously at the transition points. A time-domain plot of such a modulation is sketched in Fig. 10.18(a). For simplicity, we have considered two amplitude levels but with 8 distinct phases, differing by 45°. Since the amplitudes are different, the signal vectors in the signal space do not fall on the circumference of a circle.

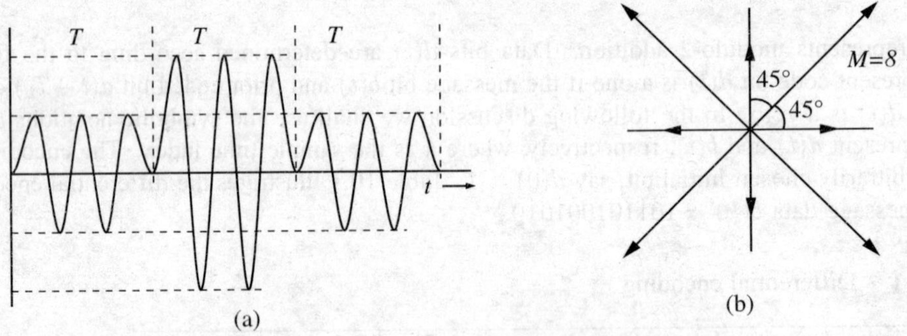

Fig. 10.18 *ASK/PSK signal with two distinct amplitudes. (b) Phasor diagram for $M = 8$.*

The APK signal involves direct modulation of carriers in quadrature. Let the carriers be $\cos \omega_0 t$ and $\sin \omega_0 t$. The in-phase and quadrature components are allowed to vary independently to form a scheme called quadrature amplitude modulation (QAM). It is useful to employ M-ary systems with keyed modulation in a bandwidth-limited channel. In binary systems, one bit of information is transmitted per symbol period. But in the M-ary case, $k = \log_2 M$ bits are transmitted during each symbol period thereby allowing a k-fold increase in the data rate within the same bandwidth. Since QAM consists of two independently amplitude-modulated carriers in quadrature, it can be viewed as ASK in two-dimensions and hence might be abbreviated as quadrature amplitude-shift keying (QASK).

M-ary QAM signal involves two orthogonal basis vectors:

$$\phi_1(t) = \sqrt{\frac{2}{T}} \cos \omega_0 t, \qquad 0 \le t \le T \tag{10.20a}$$

and $$\phi_2(t) = \sqrt{\frac{2}{T}} \sin \omega_0 t, \qquad 0 \le t \le T. \tag{10.20b}$$

For the kth symbol, M-ary QAM can be represented by

$$s_k(t) = \sqrt{\frac{2E_0}{T}} a_k \cos \omega_0 t + \sqrt{\frac{2E_0}{T}} b_k \sin \omega_0 t, \qquad 0 \le t \le T, \tag{10.21}$$

where E_0 is the energy with the lowest amplitude. The integers a_k and b_k have values $\pm 1, \pm 3$, $\cdots$, $(\sqrt{M} - 1)$. The average energy per M-ary symbol is

$$E = \frac{1}{2} \left(\sqrt{\frac{2E_0}{T}} \right)^2 \left(\overline{a_k^2} + \overline{b_k^2} \right) T = \frac{2}{3} E_0 (M - 1) \tag{10.22}$$

since $\overline{a_k^2} = \overline{b_k^2} = (M - 1)/3$.

For M even, we have a square constellation with M number of message points in the (ϕ_1, ϕ_2) plane.

Suppose that we form a symbol that consists of 4 bits (4-ary QAM). There are $2^4 = 16$ different symbols and hence 16 signals are needed to represent them. Each signal point corresponds to two quadrature carriers. So there will be 16 signal points arranged in a square signal constellation as shown in Fig. 10.19. The ordered pairs of the coordinates (a_k, b_k) of (10.21) for 16-ary QAM takes the form of a square matrix given by

$$(a_i \; b_i) = \left[\begin{array}{cc|cc} (-3, 3) & (-1, 3) & (1, 3) & (3, 3) \\ (-3, 1) & (-1, 1) & (1, 1) & (1, 3) \\ (-3, -1) & (-1, -1) & (1, -1) & (3, -1) \\ (-3, -3) & (-1, -3) & (1, -3) & (3, -3) \end{array} \right]. \tag{10.23}$$

The energy associated with the signal of lowest amplitude (i.e., for the points $(\pm 1, \pm 1)$) is given by $E_0 = \frac{1}{2}(A_c^2 + A_c^2) = A^2$. So the average energy of a signal is

$$E_s = \frac{2}{3} E_0 (M - 1) = \frac{2}{3} A_c^2 (16 - 1) = 10 A_c^2 \tag{10.24}$$

and so $A_c = \sqrt{0.1 E_s}$.

The minimum distance between two message points is

$$d_{\min} = 2A_c = 2\sqrt{0.1 E_s}. \tag{10.25}$$

Since each message is represented by 4 bits, $E_s = 4E_b$.

Since QAM consists of two independent amplitude modulations of carriers in quadrature, it represents a method of reducing the bandwidth required for data transmission. Consider that a data stream with data rate $r_b = 144$ Mbps is to be transmitted on a QAM scheme. The data stream is first partitioned into two channels, 72 Mbps in phase and 72 Mbps in quadrature. For an allowable DSB bandwidth of 36 MHz for both the channels, the spectral efficiency is $\frac{r_b}{B_T} = \frac{144\text{Mbps}}{36\text{MHz}} = 4$ bits per second per Hz.

Equation (10.19) reveals the combined amplitude and phase modulations. One can select any combination of M_1-ary ASK and M_2-ary of PSK to form an $M = M_1 M_2$-ary of QAM signal constellation. Let the ASK symbols be mapped to $n_1 = \log_2 M_1$ bits and the PSK symbols to $n_2 = \log_2 M_2$ bits. Then, the resulting QAM constellation uses simultaneous transmission of $n_1 + n_2 = \log_2 M$ bits per symbol. Thus, for a bit transmission rate of r_b bits per second, the symbol transmission rate can be reduced to $r_s = r_b/(n_1 + n_2)$ symbols per second.

There will also be proportionate reduction in the transmission bandwidth. The added advantage is that the QAM employs two carriers in quadrature and hence do not interfere with each other in the transmission path. They can be detected independently.

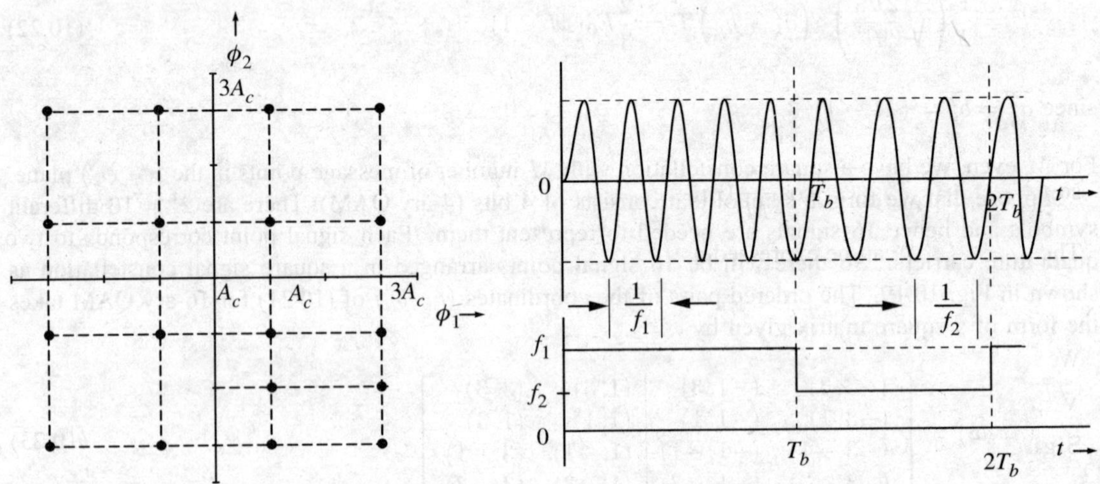

Fig. 10.19 *Signal space for 16-ary QAM*

Fig. 10.20 *(a) FSK modulated wave. (b) Frequency shift at the data transition times.*

10.3.4 *M*-ary PSK and signal constellation

The general expression for the M-ary PSK signal has been given in Eq. (10.14). We write the equation here again.

$$s_m(t) = p(t)\cos\left[2\pi f_c t + \frac{2\pi}{M}(m-1)\right],\ m = 1, 2, \ldots, M,\ 0 \le t \le T = kT$$

where $p(t)$ is the NRZ-rectangular pulse shape and k-bits are blocked to form a PSK symbol so that symbol duration is $T = kT_b$. The amplitude of the pulse is absorbed in $p(t)$. Expanding in phase and quadrature components,

$$s_m(t) = p(t)\cos\frac{2\pi}{M}(m-1)\cos 2\pi f_c t - p(t)\sin\frac{2\pi}{M}(m-1)\sin 2\pi f_c t \tag{10.26}$$

Since the signal amplitude is constant, independent of m, the M-different signals have equal energy given by

$$E = \int_0^T s_m^2(t)\,dt = \int_0^T p^2(t)\cos^2(2\pi f_c t + \theta_m)\,dt$$

$$= \frac{1}{2}\int_0^T p^2(t)\,dt = \frac{E_p}{2} \text{ for all } m. \tag{10.26a}$$

The signal $s_m(t)$ is considered two-dimensional because of the quadrature carriers $\cos 2\pi f_c t$ and $\sin 2\pi f_c t$. We can express the signal in terms of two ortho-normal basic functions $\phi_1(t)$ and $\phi_2(t)$ as

$$s_m(t) = s_{m1}\,\phi_1(t) + s_{m2}\,\phi_2(t) \tag{10.26b}$$

$$\left.\begin{aligned} \text{where}\quad \phi_1(t) &= \sqrt{\tfrac{2}{E_p}}\,p(t)\cos 2\pi f_c t \\ \text{and}\quad \phi_2(t) &= -\sqrt{2/E_p}\sin 2\pi f_c t \end{aligned}\right\} \tag{10.26c}$$

so that $s_{m1} = \sqrt{E_p/2}\ \cos\frac{2\pi}{M}(m-1)$ and $s_{m2} = \sqrt{\frac{E_p}{2}}\sin\frac{2\pi}{M}(m-1)$.

The signal $s_m(t)$ can be expressed as a two-dimensional vector with components s_{m1} and s_{m2}. Thus,

$$s_m = [\ s_{m1} \quad s_{m2}\ ] \tag{10.26d}$$

We note that $s_{m1}^2 + s_{m2}^2 = E_p/2$, a constant so that the signal points must stay on the circle of radius $\sqrt{E_p/2}$.

Signal-space diagram

$$s_m = \left[\sqrt{\tfrac{E_p}{2}}\cos\tfrac{2\pi}{M}(m-1) \qquad \sqrt{\tfrac{E_p}{2}}\sin\tfrac{2\pi}{M}(m-1)\right], \quad m = 1, 2, \ldots, M \tag{10.26e}$$

For BPSK signal, $M = 2$ and hence, $m = 1, 2$. That is,

$$s_1 = [\ s_{11} \quad s_{12}\ ]$$

$$s_2 = [\ s_{21} \quad s_{22}\ ]$$

where $s_{11} = \sqrt{\frac{E_p}{2}}$, $s_{12} = 0$; $s_{21} = -\sqrt{\frac{E_p}{2}}$, $s_{22} = 0$. The coordinates of the signal points are $\left(\sqrt{\frac{E_p}{2}}, 0\right)$ are $\left(-\sqrt{\frac{E_p}{2}}, 0\right)$ with respect to the coordinate axes $\phi_1(t)$ and $\phi_2(t)$. For $M = 2$, the signal space diagram has already been shown in Fig. 10.6 and is reproduced in Fig. 10.21 including $\phi_1(t)$ and $\phi_2(t)$. The distance between the signal points is $d_{\min} = \sqrt{2E_p}$.

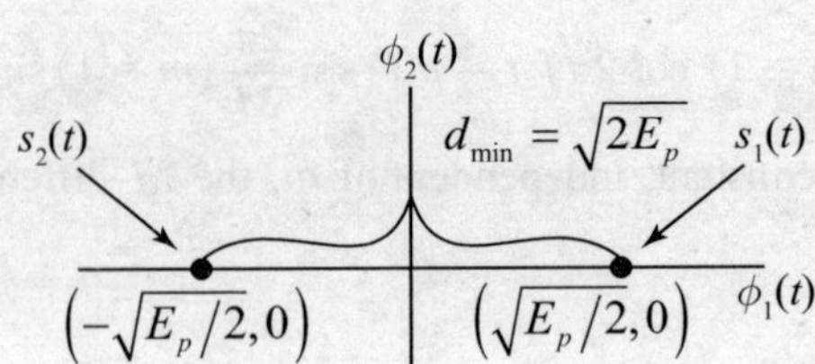

Fig. 10.21 *Signal-space diagram for BPSK. (vide Fig. 10.6 for comparison)*

For the QPSK case, $M = 4$ and $m = 1, 2, 3, 4$. So,

$$s_1 = [\ s_{11} \quad s_{12}\ ]; \qquad s_2 = [\ s_{21} \quad s_{22}\ ];$$
$$s_3 = [\ s_{31} \quad s_{32}\ ]; \qquad s_4 = [\ s_{41} \quad s_{42}\ ]$$

where, $s_{11} = \sqrt{\frac{E_p}{2}} \cos \frac{2\pi}{4} (1 - 1) = \sqrt{\frac{E_p}{2}}$, $s_{12} = 0$,

$s_{21} = 0$, $s_{22} = \sqrt{E_p/2}$,

$s_{31} = -\sqrt{\frac{E_p}{2}}$, $s_{32} = 0$

$s_{41} = 0$, $s_{42} = -\sqrt{E_p/2}$.

Taking $\phi_1(t)$ and $\phi_2(t)$ as coordinate axes normalised by $\sqrt{E_p/2}$, the coordinate of the four signal points are $s_1 = (1, 0)$, $s_1 = (0, 1)$, $s_1 = (-1, 0)$ and $s_1 = (0, -1)$ and are shown in Fig 10.22.

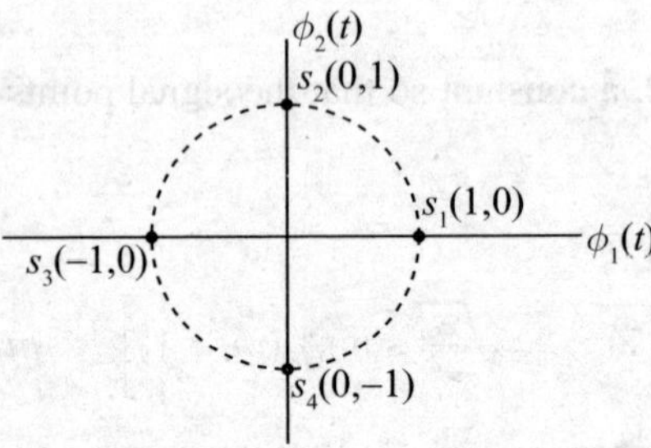

Fig. 10.22 *Signal-space diagram for a QPSK signal.*

Since one PSK signal consists of block of 2-bits, the symbols are encoded as $s_1 = 00$, $s_2 = 01$, $s_3 = 11$, and $s_4 = 10$ following the Gray encoding scheme. Note here that the four signal points are on a circle of radius $\sqrt{E_p/2}$.

For $M = 8$, the signal-space diagram consists of 8-points on the circle as shown in Fig.10.23. By using equation (10.26e), readers can find out the coordinates of all the eight signal points.

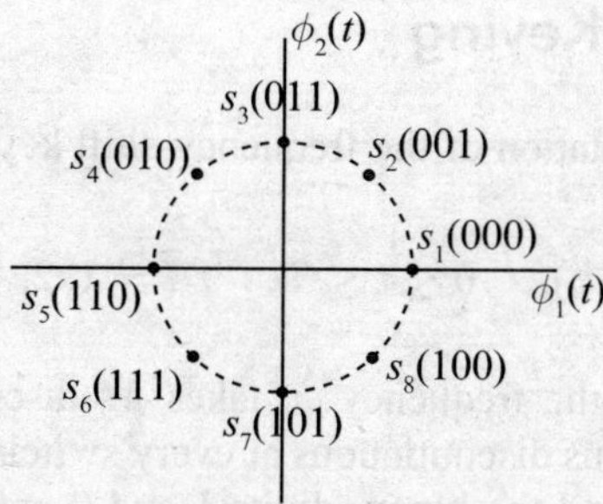

Fig. 10.23 *Signal-space diagram for 8-ary PSK signals.*

The Euclidean distance between any two consecutive signal points can be derived as follows:

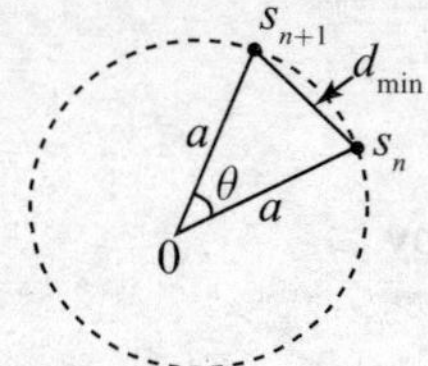

With reference to the adjoining figure, each of the vectors s_n or s_{n+1} is of length $a = \sqrt{E_p/2}$.

Since the phase angle of s_n is $\frac{2\pi}{M}(n-1)$;

$$\theta = \frac{2\pi}{M}[(n+1)-1] - \frac{2\pi}{M}(n-1) = \frac{2\pi}{M} \tag{10.26f}$$

Hence by law of triangle,

$$\begin{aligned} d_{\min}^2 &= a^2 + a^2 - 2a \cdot a \cdot \cos\theta \\ &= \frac{E_p}{2} + \frac{E_p}{2} - 2\sqrt{\frac{E_p}{2}}\sqrt{\frac{E_p}{2}} \cdot \cos\frac{2\pi}{M} \\ &= E_p\left(1 - \cos\frac{2\pi}{M}\right) \end{aligned}$$

That is, $d_{\min} = \sqrt{E_p\left(1 - \cos\frac{2\pi}{M}\right)}$ (10.26g)

For binary PSK, $M = 2$ and $d_{\min} = \sqrt{E_p\left(1 - \cos\frac{2\pi}{2}\right)} = \sqrt{2E_p}$.

For 4-ary PSK, $M = 4$ and $d_{\min} = \sqrt{E_p\left(1 - \cos\frac{2\pi}{4}\right)} = \sqrt{E_p}$.

For octal-PSK, $M = 8$ and $d_{\min} = \sqrt{E_p\left(1 - \cos\frac{2\pi}{8}\right)} = \sqrt{E_p\left(\frac{\sqrt{2}-1}{\sqrt{2}}\right)}$ and so on.

10.4 Frequency-Shift Keying

The general mathematical representation of the frequency-shift keying (FSK) modulation is given by

$$s_{\text{FSK}}(t) = \sqrt{\frac{2E_b}{T_b}}\cos(\omega_i t + \phi), \quad 0 \le t \le T_b \qquad i = 1, 2, \cdots, M. \tag{10.27}$$

Here ϕ is the phase constant and the frequency ω_i takes M discrete values selected by the digital signal. The modulating signal is thus discontinuous at every switching time. For binary FSK (BFSK) known as Sunde's FSK, $M = 2$ and the binary digits 1 and 0 are designated by two sinusoids that differ in frequency by a fixed constant. Symbol 1 may be represented by the sinusoidal waveform $s_1(t)$ given by

$$s_1(t) = \sqrt{\frac{2E_b}{T_b}}\cos 2\pi f_1 t, \quad 0 \le t \le T_b \tag{10.28}$$

$$= 0, \qquad \text{elsewhere.}$$

Similarly, symbol 0 may be represented by

$$s_2(t) = \sqrt{\frac{2E_b}{T_b}}\cos 2\pi f_2 t, \quad 0 \le t \le T_b \tag{10.29}$$

$$= 0 \qquad \text{elsewhere.}$$

Thus we observe that at the symbol transitions the change in frequency is quite abrupt (from f_1 to f_2), as shown in Fig. 10.20. At the switching times the frequency changes stepwise from f_1 to $f_2 < f_1$ but the amplitude of oscillation remains constant throughout. The phase has discontinuities at transition times. For BFSK, let the two frequencies f_1 and f_2 be related to the oscillation frequency f_0 by

$$f(t) = f_0 + f_d d(t), \tag{10.30}$$

where $d(t)$ is the bipolar data ($d(t) = \pm 1$) and f_d is the frequency shift. Thus, $f_1 = f_0 + f_d$ and $f_2 = f_0 - f_d$. Correspondingly, the BFSK oscillation can be represented as

$$s_{\text{BFSK}}(t) = A_c \cos[(\omega_0 + \omega_d d(t))t]. \tag{10.31}$$

When the data is logic 1, $d(t) = +1$ and the signal voltage is $A_c \cos(\omega_0 + \omega_d)t$, and when the data is logic 0, $d(t) = -1$ and the signal is $A_c \cos(\omega_0 - \omega_d)t$.

There is a special class of BFSK, called Sunde's FSK, for which the phase change is gradual at the bit transition times. This is called continuous phase FSK (CPFSK). Phase continuity will be maintained when the two frequencies f_1 and f_2 differ by the bit rate of $1/T_b$. One can express this BFSK as

$$s(t) = \sqrt{\frac{2E_b}{T_b}}\cos\left(\omega_0 t \pm \frac{\pi}{T_b}t\right), \quad 0 \le t \le T_b \tag{10.32}$$

since $f_1 - f_2 = 2f_d = \frac{1}{T_b}$ and $\omega_d = 2\pi f_d$.

After trigonometric manipulations,

$$s(t) = \sqrt{\frac{2E_b}{T_b}}\left[\cos\omega_0 t \cos\left(\pm\frac{\pi t}{T_b}\right) - \sin\omega_0 t \sin\left(\pm\frac{\pi t}{T_b}\right)\right]$$

$$= \sqrt{\frac{2E_b}{T_b}}\cos\left(\frac{\pi t}{T_b}\right)\cos\omega_0 t \mp \sqrt{\frac{2E_b}{T_b}}\sin\omega_0 t \sin\left(\frac{\pi t}{T_b}\right). \tag{10.33}$$

The plus sign of the second term is for transmitting symbol 0 and the minus sign for transmitting symbol 1.

Example 10.4

Find the minimum tone spacing for orthogonal FSK.

Solution

For any signal set to be orthogonal, they must be uncorrelated over a signal interval. For the two FSK signals of (10.28) and (10.29) to be orthogonal,

$$\int_0^T \cos(2\pi f_1 t)\cos(2\pi f_2 t)dt = 0$$

or $$\int_0^T [\cos 2\pi(f_1+f_2)t + \cos 2\pi(f_1-f_2)t]dt = 0$$

or $$\left[\frac{\sin 2\pi(f_1+f_2)t}{2\pi(f_1+f_2)} + \frac{\sin 2\pi(f_1-f_2)t}{2\pi(f_1-f_2)}\right]_0^T = 0$$

or $$\frac{\sin 2\pi(f_1+f_2)T}{2\pi(f_1+f_2)} + \frac{\sin 2\pi(f_1-f_2)T}{2\pi(f_1-f_2)} = 0$$

or $$\sin 2\pi(f_1-f_2)T = 0 \text{ (since } f_1+f_2 >> 1 \text{ and } \frac{\sin 2\pi(t_1+t_2)T}{2\pi(f_1+f_2)} \to 0)$$

or $$2\pi(f_1-f_2)T = n\pi, n \text{ being integer.}$$

Minimum spacing for the FSK signalling is thus for $n = 1$ and $f_1 - f_2 = \dfrac{1}{2T} = \dfrac{r_b}{2}$.

M-ary FSK system is characterised by a signal space that is M-dimensional with the axes representing sinusoids of different frequencies. In the PSK case we have defined two basis vectors $\phi_1(t)$ and $\phi_2(t)$ that are orthonormal in the signal space. But in the FSK, the orthogonality results from the special selection of the frequencies of the basis vectors. (Note that the signals $s_1(t)$ and $s_2(t)$ of (10.28) and (10.29), representing BFSK signal set, are both co-sinusoidal). We take here two basis functions, $\phi_1(t)$ and $\phi_2(t)$, each of unit normalised energy, given as

$$\phi_1(t) = \sqrt{\frac{2}{T_b}}\cos 2\pi m f_b t, \tag{10.34}$$

and $\phi_2(t) = \sqrt{\frac{2}{T_b}} \cos 2\pi n f_b t,$ (10.35)

where $f_b = 1/T_b$ and m, n are integers. The selection of these two harmonics, mf_b and nf_b, is significant as different harmonics $(m \pm n)f_b$ are orthogonal over the bit interval $T_b (= 1/f_b)$. If these two frequencies are representatives of the BFSK system, then $f_1 = mf_b$ and $f_2 = nf_b$; the corresponding BFSK waveforms are

$$s_1(t) = \sqrt{E_b}\phi_1(t), \tag{10.36}$$

$$s_2(t) = \sqrt{E_b}\phi_2(t). \tag{10.37}$$

The signal-space diagram for this BFSK signals is shown in Fig. 10.24. The distance between two message points is $d = \sqrt{2E_b}$.

Power spectrum of BFSK

We see in (10.32) and (10.33) that the binary FSK has independent in-phase and quadrature components. The in-phase component is a pure sinusoid ($\sqrt{2E_b/T_b} \cos\left(\frac{\pi t}{T_b}\right)$) and is independent of the input binary data. This component contributes baseband spectral impulses at $\pm\frac{1}{2T_b}$ of strength $\frac{1}{4}\left(\sqrt{\frac{2E_b}{T_b}}\right)^2 = E_b/2T_b$. The quadrature component $\sqrt{2E_b/T_b}\sin(\pi t/T_b)$ has power spectral density $\frac{8E_b\cos^2(\pi f T_b)}{\pi^2(4f^2T_b^2-1)^2}$. Thus the power spectrum of the binary FSK in the baseband is given by

$$G_{\text{BFSK}}(f) = \frac{E_b}{2T_b}\left[\delta(f - \frac{1}{2T_b}) + \delta(f + \frac{1}{2T_b})\right] + \frac{8E_b\cos^2(\pi f T_b)}{\pi^2(4f^2T_b^2 - 1)^2}. \tag{10.38}$$

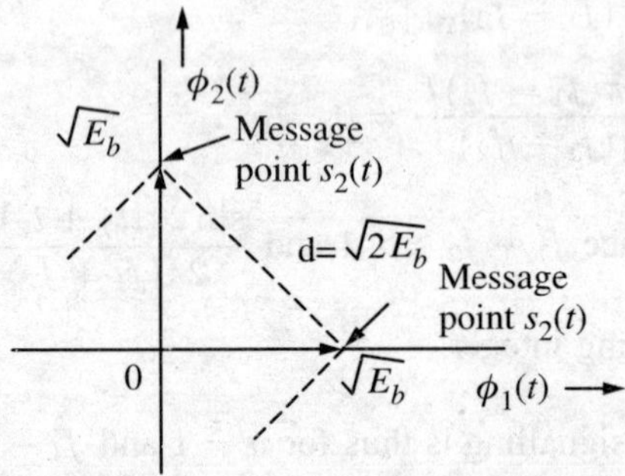

Fig. 10.24 *Signal-space diagram for orthogonal BFSK*

The corresponding bandpass spectrum is plotted in Fig. 10.25. The two impulses on the BFSK spectrum provides clock information at the receiver. The second term of (10.38) gives the power spectrum of a BFSK with continuous phase; the spectrum has a fourth-order roll-off.

The transmission bandwidth B_T of BFSK can be taken to be $B_T \approx 1/T_b$.

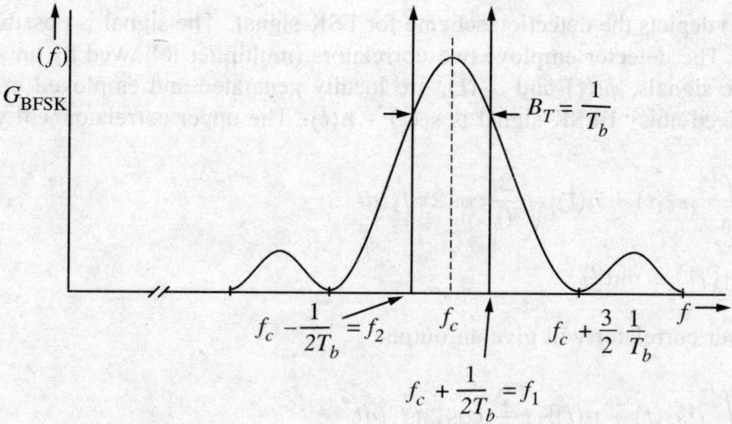

Fig. 10.25 *Power spectrum for binary FSK*

Example 10.5

Set up block diagrams to represent the generation and reception of BFSK signal.

Solution

We may use the scheme of Fig.10.26(a) to generate BFSK signal. By (10.28) and (10.29), for logic 1 a signal of frequency f_1 will be transmitted and for logic 0 a signal of frequency $f_2 < f_1$ will be transmitted. The incoming data sequence is first applied to a level shifter. It will give two dc levels corresponding to the binary data 0 or 1. The voltage controlled oscillator is designed to yield frequencies f_1 and f_2 in accordance with the input binary data.

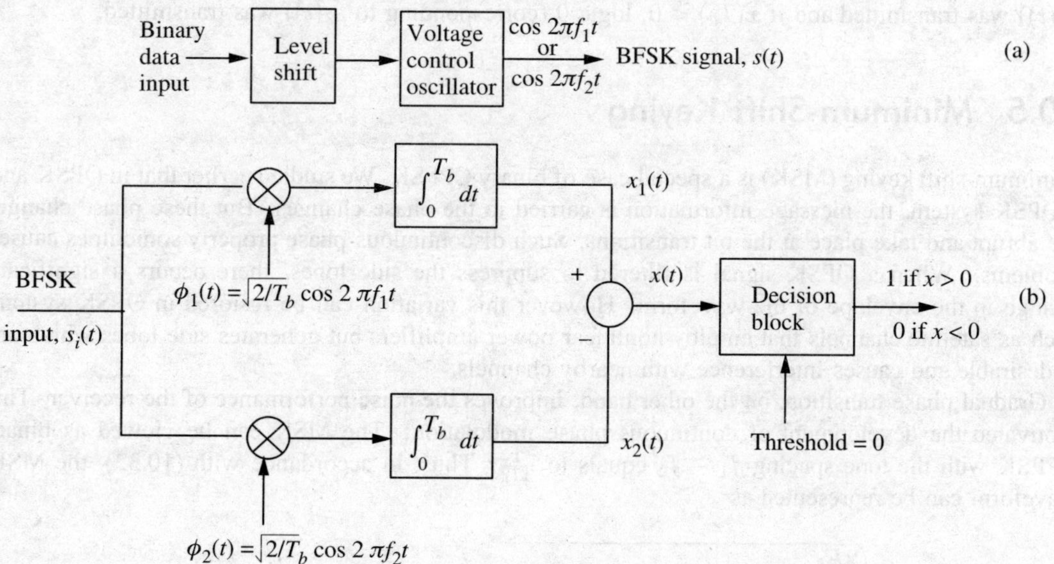

Fig. 10.26 *Block diagrams for BFSK: (a) transmitter, (b) receiver.*

Figure 10.26(b) depicts the detection scheme for FSK signal. The signal is possibly contaminated by noise voltage. The detector employs two correlators (multiplier followed by an integrator). Two coherent reference signals, $\phi_1(t)$ and $\phi_2(t)$, are locally generated and employed in each correlator. Suppose the received noisy BFSK signal is $s_1(t) + n(t)$. The upper correlator will yield an output

$$x_1(T_b) = \int_0^{T_b} [s_1(t) + n(t)]\sqrt{\frac{2}{T_b}} \cos 2\pi f_1 t dt$$

$$= a_1(T_b) + n_0(T_b).$$

Similarly, the lower correlator will give an output

$$x_2(T_b) = \int_0^{T_b} [s_2(t) + n(t)]\sqrt{\frac{2}{T_b}} \cos 2\pi f_2 t dt$$

$$= a_2(T_b) + n_0(T).$$

In the absence of noise, the input waveform $s_i(t)$ yields the output $a_i(T_b), i = 1, 2$, a signal component. Assuming a Gaussian noise, the noise component $n_0(T_b)$ is zero-mean Gaussian and so the output $x_i(T_b)$ will have a mean of either $a_1(T_b)$ or $a_2(T_b)$, in accordance with whether $s_1(t)$ or $s_2(t)$ was received. The decision block can be configured to follow the rule of threshold detection:

$$x(T_b) = x_1(T_b) - x_2(T_b) = a_1(T_b) - a_2(T_b) = a_1(T_b) \text{ or } -a_2(T_b).$$

Because of the orthogonal property of the basis function, either of $a_1(T_b)$ or $a_2(T_b)$ will give a nonzero value. Thus, by choosing a threshold level = 0 volt, if $x(T_b) > 0$, logic 1 (corresponding to $s_1(t)$) was transmitted and if $x(T_b) < 0$, logic 0 (corresponding to $s_2(t)$) was transmitted.

10.5 Minimum-Shift Keying

Minimum-shift keying (MSK) is a special case of binary CPFSK. We studied earlier that in QPSK and OQPSK system, the message information is carried in the phase changes. But these phase changes are abrupt and take place at the bit transitions. Such discontinuous-phase property sometimes causes problems. When a QPSK signal is filtered to suppress the side lobes, there occurs a significant change in the envelope of the waveform. However this variation can be restored in QPSK systems such as satellite channels that employ nonlinear power amplifiers but generates side lobes which are undesirable and causes interference with nearby channels.

Gradual phase transition, on the other hand, improves the noise performance of the receiver. This motivated the development of continuous phase modulation. The MSK can be viewed as binary CPFSK with the tone spacing $f_1 - f_2$ equals to $\frac{1}{4T_b}$. Thus, in accordance with (10.32), the MSK waveform can be represented as

$$s_{\text{MSK}}(t) = \sqrt{\frac{2E_b}{T_b}} \cos[2\pi(f_0 + d_k f_d)t + \theta_k],$$

$$= \sqrt{\frac{2E_b}{T_b}} \cos[\omega_0 t + \frac{\pi t d_k}{2T_b} + \theta_k]. \qquad kT_b \leq t \leq (k_b + 1)T_b$$

$$= \sqrt{2E_b/T_b} \cos[\omega_0 t + \theta(t)], \tag{10.39}$$

where $\theta(t) = \theta_k \pm \frac{\pi}{2T_b} t = \theta_k + \frac{\pi t}{2T_b} d_k.$ (10.40)

The quantity θ_k is a phase constant over the data interval and is 0 or π that varies the phase gently at the symbol transitions. Here, d_k is the bipolar data which assumes $+1$ for sending symbol 1 and -1 for sending symbol 0.

We will now describe the behaviour of the phase pattern. Rewriting (10.40) as

$$\theta(t) - \theta_k = \frac{\pi t d_k}{2T_b}, \tag{10.41}$$

it becomes clear that the phase changes by $\frac{\pi}{2}$ at the transition points. Figure 10.27 reveals the trellis pattern of the phase. Note that for even multiples of T_b the phase changes are $0, \pm\pi, \pm 2\pi, \cdots$ and for odd multiples of T_b, the phase changes are $\pm\pi/2, \pm 3\pi/2, \cdots$. The positive sign holds for sending a 1 and the negative sign holds for sending a 0. It is evident from Fig. 10.27 that the phase change over one bit interval is $\pm\pi/2$ radians.

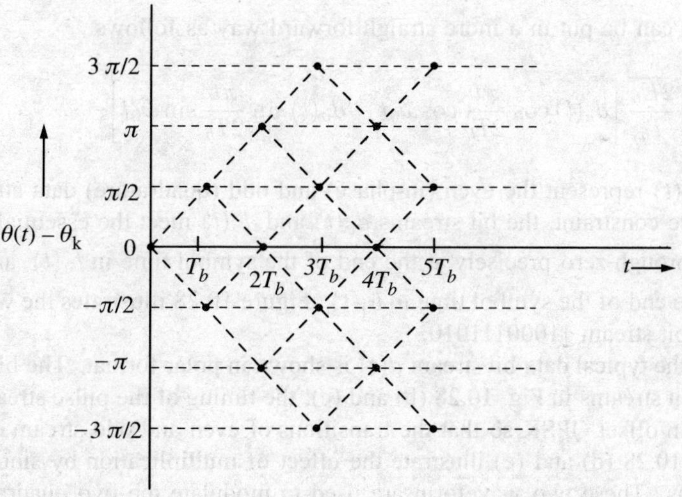

Fig. 10.27 *Phase trellis of MSK signal*

We may express the MSK signal in terms of the in-phase and quadrature-phase components as follows:

$$s_{\text{MSK}}(t) = \sqrt{\frac{2E_b}{T_b}} [\cos\omega_0 t \cos\{\theta(t)\} - \sin\omega_0 t \sin\{\theta(t)\}]. \tag{10.42}$$

The in-phase component is

$$s_I(t) = \sqrt{\frac{2E_b}{T_b}} \cos\left[\theta_k + \frac{\pi t}{2T_b} d_k\right] \cos \omega_0 t$$

$$= a_k \sqrt{\frac{2E_b}{T_b}} \cos\left(\frac{\pi t}{2T_b}\right) \cos \omega_0 t, \quad -T_b \le t \le T_b. \tag{10.43}$$

In the above equation, $a_k = \cos\theta_k = \pm 1$ and it is data-dependent. The value of $a_k = 1$ when $\theta_k = 0$ and -1 when $\theta_k = \pi$ radius. The quantity a_k is data-dependent because θ_k may take either 0 or π to maintain the phase-continuity property, which again depends on the past data bits.

The quadrature component is

$$S_Q(t) = \sqrt{\frac{2E_b}{T_b}} \sin\left[\theta_k + \frac{\pi t}{2T_b} d_k\right] \sin \omega_0 t$$

$$= b_k \sqrt{\frac{2E_b}{2T_b}} \sin\left(\frac{\pi t}{2T_b}\right) \sin \omega_0 t \quad 0 \le t \le 2T_b, \tag{10.44}$$

where b_k is again a data-dependent term. Thus in MSK the transmitted signal is

$$s_{\text{MSK}}(t) = \sqrt{\frac{2E_b}{T_b}} \left[a_k \cos\frac{\pi t}{2T_b} \cos\omega_0 t - b_k \sin\left(\frac{\pi t}{2T_b}\right) \sin\omega_0 t\right]. \tag{10.45}$$

The above equation can be put in a more straightforward way as follows.

$$s_{\text{MSK}}(t) = \sqrt{\frac{2E_b}{T_b}} \left[d_e(t) \cos\frac{\pi t}{2T_b} \cos\omega_0 t - d_o(t) \sin\frac{\pi t}{2T_b} \sin\omega_0 t\right], \tag{10.46}$$

where $d_e(t)$ and $d_o(t)$ represent the even (in-phase) and odd (quadrature) data streams. Because of the continuous phase constraint, the bit streams $d_e(t)$ and $d_o(t)$ meet the essential requirements that $\cos\left(\frac{\pi t}{2T_b}\right)$ passes through zero precisely at the end of the symbol time in $b_e(t)$, and $\sin\left(\frac{\pi t}{2T_b}\right)$ have zero-crossings at the end of the symbol time in $b_0(t)$. Figure 10.28 illustrates the waveforms of MSK signal for the data bit stream 11000111010.

In Fig. 10.28(a) the typical data bit stream $d(t)$ is shown in polar format. The bit stream is divided into even and odd bit streams in Fig. 10.28 (b) and (c); the timing of the pulse stream $d_e(t)$ and $d_o(t)$ is shifted by T_b as in offset QPSK so that the transitions of even and odd stream do not occur at the same time. Figure 10.28 (d) and (e) illustrate the effect of multiplication by sinusoids, resulting in smoother waveforms. These two waveform are used to modulate the two quadrature component of the carrier, $\cos\omega_0 t$ and $\sin\omega_0 t$.

Example 10.6

Show that the MSK signal of (10.45)is basically an FSK signal.

Solution

Rewrite (10.45) as

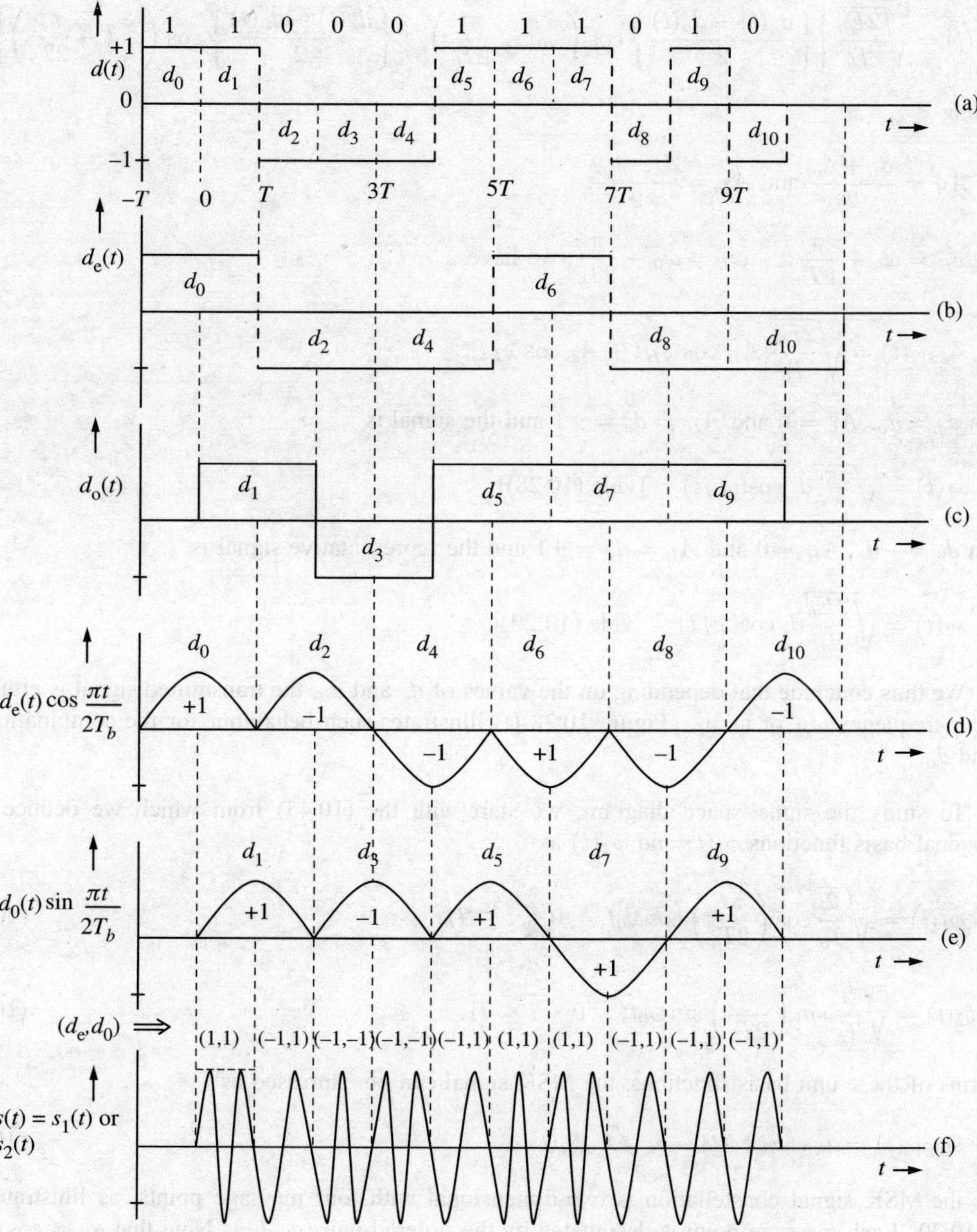

Fig. 10.28 *Minimum-shift keying: (a) Message data stream, (b) & (c) even stream and odd stream, (d) & (e) modified even and odd bit streams, (f) MSK waveform.*

$$s_{\text{MSK}}(t) = \sqrt{\frac{2E_b}{T_b}}\left[\left\{\frac{d_e(t) + d_o(t)}{2}\right\}\cos\left(\omega_0 t + \frac{\pi t}{2T_b}\right) + \left\{\frac{d_e(t) - d_o(t)}{2}\right\}\cos\left(\omega_0 t - \frac{\pi t}{2T_b}\right)\right].$$

Defining

$$A_H = \frac{d_e + d_o}{2} \text{ and } A_L = \frac{d_e - d_o}{2},$$

$$\omega_H = \omega_0 + \frac{\pi}{2T_b}, \qquad \omega_L = \omega_0 - \frac{\pi}{2T_b}, \text{ we have}$$

$$s_{\text{MSK}}(t) = \sqrt{\frac{2E_b}{T_b}}[A_H \cos\omega_H t + A_L \cos\omega_L t].$$

When $d_e = d_o$, $A_L = 0$ and $A_H = d_e = \pm 1$ and the signal is

$$s_1(t) = \sqrt{\tfrac{2E_b}{T_b}}d_e \cos(\omega_H t) \quad \text{[vide (10.28)]}.$$

When $d_0 = -d_e$, $A_H = 0$ and $A_L = d_e = \pm 1$ and the representative signal is

$$s_2(t) = \sqrt{\frac{2E_b}{T_b}}d_e \cos(\omega_L t) \quad \text{[vide (10.29)]}.$$

We thus conclude that depending on the values of d_e and d_o, the transmitted signal is either at angular frequency ω_H or at ω_L. Figure 10.28 (f) illustrates such behaviour for the combination of d_e and d_o.

To study the signal-space diagram, we start with the (10.45) from which we deduce two orthogonal basis functions $\phi_1(t)$ and $\phi_2(t)$ as

$$\phi_1(t) = \sqrt{\frac{2}{T_b}}\cos\left(\frac{\pi t}{2T_b}\right)\cos\omega_0 t, \quad 0 \le t \le T_b \tag{10.47}$$

$$\text{and } \phi_2(t) = \sqrt{\frac{2}{T_b}}\sin\left(\frac{\pi t}{2T_b}\right)\sin\omega_0 t, \quad 0 \le t \le T_b. \tag{10.48}$$

In terms of these unit basis functions, the MSK signal can be expressed as

$$s_{\text{MSK}}(t) = a_k\sqrt{E_b}\phi_1(t) - b_k\sqrt{E_b}\phi_2(t). \tag{10.49}$$

Thus the MSK signal constellation is two-dimensional with four message points, as illustrated in Fig. 10.29. Each message point is designated by the ordered-pair (a_k, b_k). Note that $a_k = \cos\theta_k = \pm 1$ and $\theta_k = 0$ corresponds to the right half plane of the ϕ_2-axis; $\theta_k = \pi$ corresponds to the left half plane. Again, it is evident from (10.43) and (10.44) that for $\theta_k = 0, b_k = 1$ for symbol 1; $b_k = -1$ for symbol 0. Similarly, for $\theta_k = \pi$, $b_k = 1$ for symbol 0, -1 for symbol 1. Table 10.2 lists the values of a_k and b_k for the input data stream.

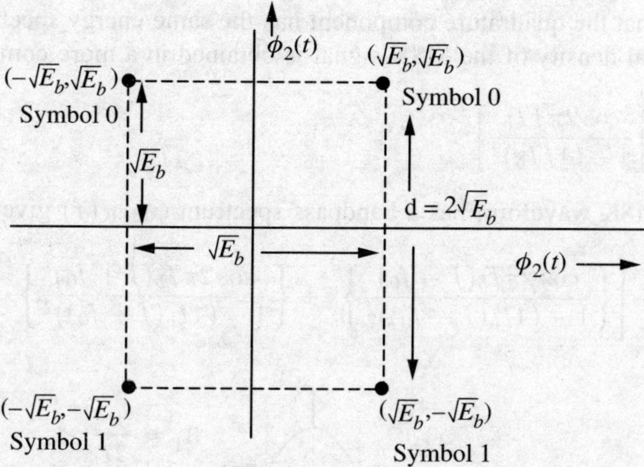

Fig. 10.29 *Signal-space representation of MSK signal*

Table 10.2 MSK constellation data

Transmitted symbol	θ_k	a_k	b_k	Coordinates of message point
0	0	1	-1	$(+\sqrt{E_b}+\sqrt{E_b})$
1	0	1	1	$(\sqrt{E_b},-\sqrt{E_b})$
0	π	-1	-1	$(-\sqrt{E_b},+\sqrt{E_b})$
1	π	-1	1	$(-\sqrt{E_b},-\sqrt{E_b})$

The smallest distance between signal points is

$$d = 2\sqrt{E_b} = \sqrt{4E_b} = \sqrt{2E_s} \tag{10.50}$$

since two bits are used to designate a symbol.

Power spectrum of MSK signal

In (10.43), we see that the in-phase component is obtained by multiplying the carrier $\cos\omega_0 t$ with the baseband signal

$$p(t) = a_k\sqrt{\frac{2E_b}{T_b}}\cos\left(\frac{\pi t}{2T_b}\right), \quad -T_b \le t \le T_b, \tag{10.51}$$

where $a_k = \pm 1$, depending on the value of phase θ_k. The energy spectral density of this waveform is given by (vide Example 10.7)

$$|P(f)|^2 = 2E_bT_b\left[\text{sinc}\ 2T_b\left(t-\frac{1}{4T_b}\right)+\text{sinc}\ 2T_b\left(t+\frac{1}{4T_b}\right)\right]^2. \tag{10.52}$$

It can also be proved that the quadrature component has the same energy spectral density. Hence the baseband power spectral density of the MSK signal is obtained in a more compact form as

$$G_B(f) = \frac{32E_b}{\pi^2}\left[\frac{\cos 2\pi f T_b}{1-(4fT_b)^2}\right]^2. \tag{10.53}$$

Correspondingly, the MSK waveform has a bandpass spectrum $G_{\text{MSK}}(f)$ given by

$$G_{\text{MSK}}(f) = \frac{8E_b}{\pi^2}\left[\left\{\frac{\cos 2\pi T_b(f-f_0)}{1-(4T_b(f-f_0))^2}\right\}^2 + \left\{\frac{\cos 2\pi T_b(f+f_0)}{1-(4T_b(f+f_0))^2}\right\}^2\right]. \tag{10.54}$$

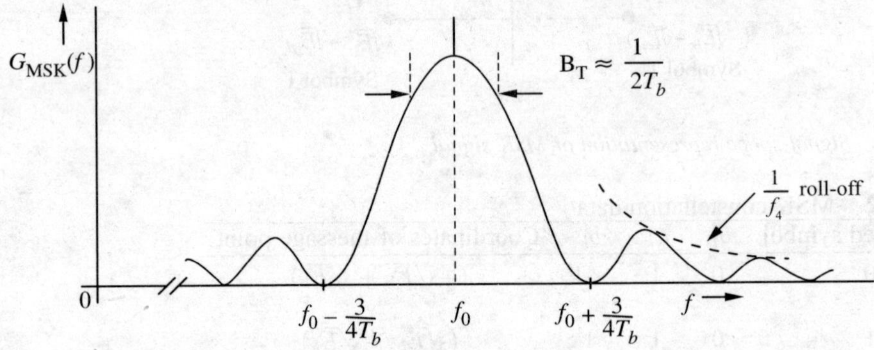

Fig. 10.30 *Spectrum of MSK signals*

The MSK bandpass spectral density is shown in Fig. 10.30. Observe that the spectrum is free of impulses and has a central lobe of width $\frac{3}{2T_b}$. For $(f - f_0) >> 1/T_b$, the MSK spectral density falls off rapidly as $1/f^4$ in contrast to a roll-off of $1/f^2$ in the QPSK case. It turns out that in MSK the interference outside the signal band is greatly suppressed in comparison with QPSK. The transmission bandwidth of the MSK can be taken as $B_T \approx \frac{1}{2T_b}$ and the spectrum efficiency is $(1/T_b)/B_T \approx 2\text{bps/Hz}$.

Example 10.7

Find the power spectral density of $p(t) = \cos\frac{\pi t}{2T_b}, \quad -T_b \le t \le T_b$.

Solution

Since the PSD is obtained by dividing the energy spectral density by the bit duration, the energy spectrum density is given by

$$|P(f)|^2 = |\int_{-\infty}^{\infty} p(t)e^{-j2\pi ft}dt|^2 = |\int_{-T_b}^{T_b} \cos\frac{\pi t}{2T_b}e^{-j2\pi ft}dt|^2.$$

Using the trigonometric identity,

$\cos x = (e^{jx} + e^{-jx})/2$, we have

$$P(f) = \int_{-T_b}^{T_b} \frac{1}{2}\left(e^{\frac{j\pi t}{2T_b}} + e^{\frac{-j\pi t}{2T_b}}\right) e^{-j2\pi f t} dt$$

$$= \frac{1}{2}\int_{-T_b}^{T_b} e^{j\pi(\frac{1}{2T_b}-2f)t} dt + \frac{1}{2}\int_{-T_b}^{T_b} e^{-j\pi(\frac{1}{2T_b}+2f)t} dt = I_1 + I_2,$$

where

$$I_1 = \frac{1}{2}\frac{e^{j\pi\left(\frac{1}{2T_b}-2f\right)t}}{j\pi\left(\frac{1}{2T_b}-2f\right)}\Bigg|_{-T_b}^{T_b} = \frac{e^{j\pi(\frac{1}{2}-2fT_b)} - e^{-j\pi(\frac{1}{2}-2fT_b)}}{2j\pi\left(\frac{1}{2T_b}-2f\right)}$$

$$= \frac{\sin\pi\left(\frac{1}{2}-2fT_b\right)}{\pi\left(\frac{1}{2T_b}-2f\right)} = \frac{\sin 2\pi T_b\left(f-\frac{1}{4T_b}\right)}{2\pi\left(f-\frac{1}{4T_b}\right)}$$

$$= T_b \operatorname{sinc} 2T_b\left(f-\frac{1}{4T_b}\right),$$

and $I_2 = \frac{1}{2}\frac{e^{-j\pi(\frac{1}{2T_b}+2f)}}{-j\pi(\frac{1}{2T_b}+2f)}\Bigg|_{-T_b}^{T_b}$

$$= \frac{e^{j\pi(\frac{1}{2}+2fT_b)} - e^{-i\pi(\frac{1}{2}+2fT_b)}}{2j\pi\left(\frac{1}{2T_b}+2f\right)} = \frac{\sin\pi\left(\frac{1}{2}+2fT_b\right)}{\pi\left(\frac{1}{2T_b}+2f\right)}$$

$$= \frac{\sin 2\pi T_b(f+\frac{1}{4T_b})}{2\pi(f+\frac{1}{4T_b})} = T_b \operatorname{sinc} 2T_b\left(f+\frac{1}{4T_b}\right).$$

Thus, $|P(f)|^2 = T_b^2\,[\operatorname{sinc} 2T_b(f-1/4T_b) + \operatorname{sinc} 2T_b(f+1/4T_b)]^2$

and hence the power spectral density

$$\frac{|P(f)|^2}{T_b} = T_b\,[\operatorname{sinc} 2T_b(f-1/4T_b) + \operatorname{sinc} 2T_b(f+1/4T_b)]^2$$

10.6 Gaussian Noise and Errors in Binary Systems

A baseband digital transmission system using binary encoding transmits a time sequence of binary digits (1's and 0's), usually in the form of shaped pulses. But in bandpass transmission the resulting shaped pulses modulate a sinusoidal carrier. The received signal being corrupted by noise, there is some possibility that the receiver makes an error concerning the transmitted bit. In this section, we will first study the detection of a binary coded signal in the presence of noise and investigate the error probabilities in baseband transmission. The idea will then be extended to the bandpass detection process to assess the performance of many other keyed modulations.

10.6.1 Probability of error and optimum filtering

Consider the linear signal processing of baseband signal shown in Fig. 10.31. In our discussion, we assume that the noise is Gaussian and has zero mean. The task of the receiver is to decide whether a 0 or a 1 has been transmitted. The received noisy signal is applied to a lowpass filter to remove excess noise. A sample-and-hold (S/H) network triggered at optimum times within each bit interval yields the sample value

$$v_0(kT) = s(kT) + n(kT). \tag{10.55}$$

The decision block is basically a comparator that compares successive samples with reference to a fixed threshold level V_t. The decision rule for bit regeneration is as follows: whenever the sampled values exceeds V_t, the comparator goes HIGH to indicate a 1; when it is below V_t, the comparator goes LOW to indicate a 0.

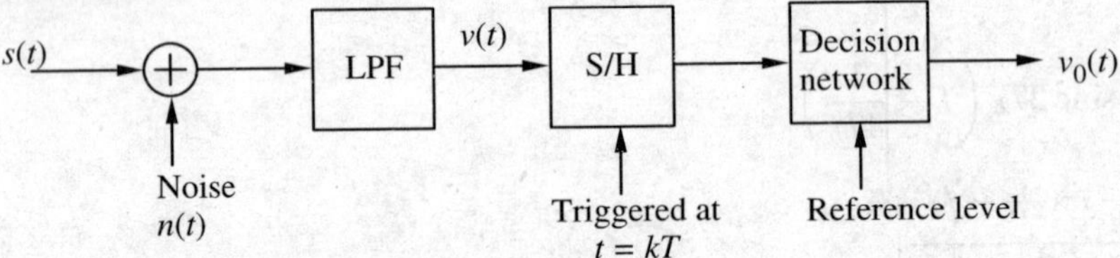

Fig. 10.31 *A baseband signal processor*

Consider that $s(t)$ is a polar signal in which a voltage level $+V/2$ represents the message bit 1 and $-V/2$ represents 0. The decision level should therefore be set at some intermediate level $-\frac{V}{2} < V_t < \frac{V}{2}$. We purposely choose the decision level to be at $V_t = 0$ so that if the sample value is positive the transmitted level was $+V/2$, and if the sample value is negative the transmitted level was $-V/2$.

The noise may have positive and negative excursions. Suppose a $+V/2$ level was transmitted and a negative noise excursion results in $v_0(kT) < 0$. Based on the decision rule, the receiver decides in favour of $-V/2$ level. The occurrence of such errors is illustrated in Fig. 10.32. Again suppose that the sample time is set at the middle of each bit interval. In the bit interval $T_b < t < 2T_b$ the noise is so severe that signal plus noise is positive at the sample instant although the transmitted message was a 0. As a result, an error occurs. Similarly, the regenerated bit is also erroneous over the interval $3T_b < t < 4T_b$ in which case the noise voltage is so negative that the signal plus the noise falls below the threshold level although the transmitted bit was a 1.

Suppose that the noise is Gaussian with zero mean and variance σ_n^2 so that the probability density function is

$$f_N(n) = \frac{1}{\sqrt{2\pi\sigma_n^2}} e^{-n^2/2\sigma_n^2}, \tag{10.56}$$

where the random variable N represent $n(kT)$ at the sample time. The density function of the signal sample $v(kT)$ obviously involves the noise PDF. So, in the polar case the specified level for message 0 is $-V/2$ characterised by the PDF

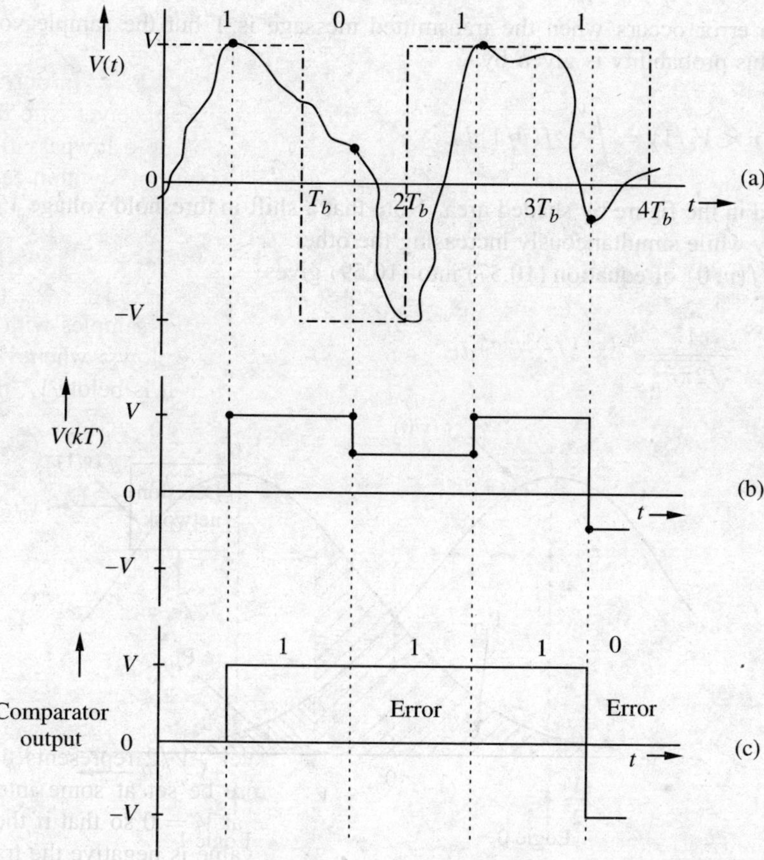

Fig. 10.32 *Baseband detection process: (a) signal added with noise; (b) S/H output; (c) regenerated data.*

$$f(v/0) = \frac{1}{\sqrt{2\pi\sigma_n^2}} e^{-(v+V/2)^2/2\sigma_n^2}. \tag{10.57}$$

Similarly, the PDF of v for message 1 is

$$f(v/1) = \frac{1}{\sqrt{2\pi\sigma_n^2}} e^{-(v-V/2)^2/2\sigma_n^2}. \tag{10.58}$$

These two PDFs are plotted in Fig. 10.33 which helps to find the error probabilities. Suppose that during some bit interval input signal voltage is held at $-V/2$ (message 0). If the noise sample $n(kT)$ is large in magnitude such that the total sample voltage $v(kT) = s(kT) + n(kT)$ exceeds V_t, then an error results whose probability is given by

$$P_{e0} = P(v > V_t/0) = \int_{V_t}^{\infty} f(v/0)dv. \tag{10.59}$$

This is indicated in the figure by shaded area.

Similarly, an error occurs when the transmitted message is 1 but the sample voltage $s(kT) + n(kT) < V_t$. This probability is given by

$$P_{e1} = P(v < V_t/1) = \int_{-\infty}^{V_t} f(v/1)dv. \tag{10.60}$$

This is indicated in the figure by shaded area. Note that a shift in threshold voltage V_t decreases one error probability while simultaneously increasing the other.

Substituting $f(v/0)$ of equation (10.57) into (10.59) gives

$$P_{e0} = \int_{V_t}^{\infty} \frac{1}{\sqrt{2\pi\sigma_n^2}} e^{-(v+V/2)^2/2\sigma_n^2} dv.$$

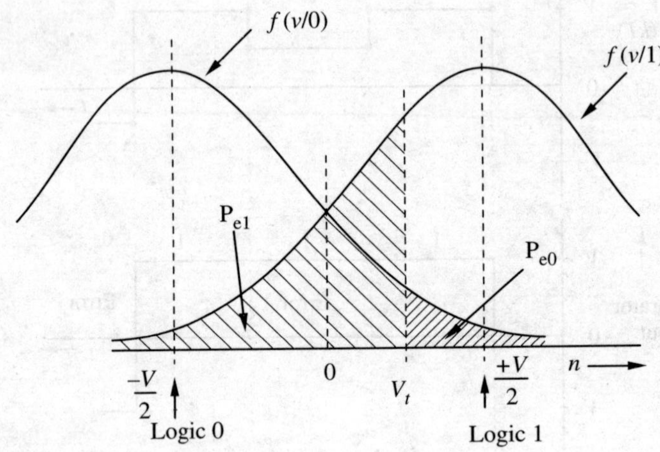

Fig. 10.33 *PDFs and error probabilities*

Let $\dfrac{v + V/2}{\sqrt{2}\sigma_n} = z$, then $dv = \sqrt{2}\sigma_n dz$.

When $v = V_t$, $z = \dfrac{V_t + V/2}{\sqrt{2}\sigma_n}$ so that

$$P_{e0} = \int_{\frac{V_t+V/2}{\sqrt{2}\sigma_n}}^{\infty} \frac{1}{\sqrt{2\pi\sigma_n^2}} e^{-z^2} \sqrt{2}\sigma_n dz = \frac{1}{\sqrt{\pi}} \int_{\frac{V_t+V/2}{\sqrt{2}\sigma_n}}^{\infty} e^{-z^2} dz$$

$$= \frac{1}{2} \operatorname{erfc}\left(\frac{V_t + V/2}{\sqrt{2}\sigma_n}\right). \tag{10.61}$$

Similarly, substituting $f(v/1)$ of (10.58) into (10.60), we get

$$P_{e1} = \int_{-\infty}^{V_t} \frac{1}{\sqrt{2\pi\sigma_n^2}} e^{-(v-V/2)^2/2\sigma_n^2} dv.$$

Let $\frac{v - V/2}{\sqrt{2}\sigma_n} = z$, then $dv = \sqrt{2}\sigma_n dz$.

When $v = V_t$, $z = \frac{V_t - V/2}{\sqrt{2}\sigma_n}$ so that

$$P_{e1} = \int_{-\infty}^{\frac{V_t - V/2}{\sqrt{2}\sigma_n}} \frac{1}{\sqrt{2\pi}\sigma_n} e^{-z^2} \sqrt{2}\sigma_n dz = \frac{1}{2} \operatorname{erf}\left(\frac{V_t - V/2}{\sqrt{2}\sigma_n}\right). \tag{10.62}$$

Since $f_N(n)$ has even symmetry, the two PDFs intersect at the midpoint so that $V_t = 0$. This is the optimum threshold for minimising the probability of error. In fact, the probability of error is the sum of the probabilities of all the errors. For the binary case, the probability of bit error is

$$P_e = P_{e0}P(0) + P_{e1}P(1), \tag{10.63}$$

where $P(0)$ and $P(1)$ refer to the source digit probabilities in sending logic 0 and logic 1. That is, given that symbol 0 was transmitted, an error occurs with the probability P_{e0}; or given that symbol 1 was transmitted, an error occurs with the probability P_{e1}. When the source digit probabilities $P(0)$ and $P(1)$ are equal (that is, $P(0) = P(1) = \frac{1}{2}$),

$$P_e = \tfrac{1}{2}(P_{e0} + P_{e1})$$

and by the symmetry of the PDFs, $P_{e0} = P_{e1}$ so that $P_e = P_{e0} = P_{e1}$ and

$$P_e = \frac{1}{2} \operatorname{erfc}\left(\frac{V}{2\sqrt{2}\sigma_n}\right). \tag{10.64}$$

The complementary error function erfc cannot be evaluated in closed form, rather it is presented in tabular form. A plot of P_e versus $\frac{V}{2\sqrt{2}\sigma_n}$ reveals that P_e decreases rapidly as $\frac{V}{2\sqrt{2}\sigma_n}$ increases, and its maximum value is $\frac{1}{2}$.

We have optimised the threshold level for minimising P_e but have not optimised the receiving filter for detecting pulses. When we know the transmitted pulse shape, we can optimise the filter by maximising the argument of $\operatorname{erfc}(x)$.

10.6.2 The matched filter

A matched filter is a linear filter designed for achieving minimum error probability. This can be reached if the signal-to-noise power ratio at its output is maximum for a given input signal.

Consider that a pulse $p(t)$ is the input signal to the optimum filter. Let $p_O(t)$ is the output signal of the filter. We wish to find the filter transfer function $H(f)$ that maximises the signal-to-noise power ratio. Now,

$$P_0(f) = H(f)P(f), \tag{10.65}$$

where $P(f)$ and $P_0(f)$ are the Fourier transforms of $p(t)$ and $p_0(t)$, respectively. The received pulse energy is

$$E = \int_{-\infty}^{\infty} |P(f)|^2 df, \tag{10.66}$$

using Parseval's energy theorem. The signal amplitude at the output of the filter is given by

$$p_0(T) = \mathcal{F}^{-1}[P_0(f)] = \int_{\infty}^{\infty} H(f)P(f)e^{j2\pi fT} df, \tag{10.67}$$

since at time $t = T$, the sampler output consists of the signal component $p_0(T)$ and noise component $n_0(T)$. The output noise power spectral density $G_{n_o}(f)$ is related to the input noise PSD through

$$G_{n_o}(f) = |H(f)|^2 Gn_i(f). \tag{10.68}$$

We can express the rms output noise power σ_n^2 as

$$\sigma_n^2 = \int_{-\infty}^{\infty} G_{n_o}(f)df = \int_{-\infty}^{\infty} Gn_i(f)|H(f)|^2 df. \tag{10.69}$$

The ratio of the instantaneous signal power to the average noise power (S/N) is given by

$$(S/N) = \frac{p_0^2(T)}{\sigma_n^2} = \frac{|\int_{-\infty}^{\infty} H(f)P(f)e^{j2\pi fT} df|^2}{\int_{-\infty}^{\infty} Gn_i(f)|H(f)|^2 df}. \tag{10.70}$$

Our goal is to maximise this ratio. This optimisation problem can be tackled through the use of Schwarz's inequality which states that

$$\left|\int_{-\infty}^{\infty} X(f)Y(f)df\right|^2 \leq \int_{-\infty}^{\infty} |X(f)|^2 df \int_{-\infty}^{\infty} |Y(f)|^2 df, \tag{10.71}$$

where $X(f)$ and $Y(f)$ are two arbitrary function of f. The equality holds if $X(f) = kY^*(f)$, where k is an arbitrary constant and $Y^*(f)$ the complex conjugate of $Y(f)$.

Identifying $X(f)$ as $H(f)\sqrt{Gn_i(f)}$ and $Y(f)$ as $P(f)e^{j2\pi fT}/\sqrt{Gn_i(f)}$ in (10.70), we may write

$$\frac{p_0^2(T)}{\sigma_n^2} = \frac{\int_{-\infty}^{\infty} |H(f)\sqrt{Gn_i(f)}|^2 df \int_{-\infty}^{\infty} |\frac{P(f)e^{j2\pi fT}}{\sqrt{Gn_i(f)}}|^2 df}{\int_{-\infty}^{\infty} Gn_i(f)|H(f)|^2 df}$$

$$\leq \int_{-\infty}^{\infty} \left|\frac{P(f)e^{j2\pi fT}}{\sqrt{Gn_i(f)}}\right|^2 = \int_{-\infty}^{\infty} \frac{|P(f)|^2}{Gn_i(f)} df. \tag{10.72}$$

So,

$$\frac{p_0^2(T)}{\sigma_n^2}|_{\max} = \int_{\infty}^{\infty} \frac{|P(f)|^2}{Gn_i(f)} df. \tag{10.73}$$

When $X(f) = kY^*(f)$, we have

$$H(f)\sqrt{Gn_i(f)} = k\left|\frac{P(f)e^{j2\pi fT}}{\sqrt{Gn_i(f)}}\right|^*$$

so that the optimum filter transfer function is given by

$$H_{\text{opt}}(f) = k\frac{P^*(f)e^{-j2\pi fT}}{Gn_i(f)}. \tag{10.74}$$

Thus $|H_{\text{opt}}(f)|$ is directly proportional to $|P(f)|$ and inversely proportional to $G_{ni}(f)$.

An optimum filter is called a 'matched filter' when the noise is 'white' with $G_{ni}(f) = \eta/2$. Equation (10.73) yields

$$\frac{p_0^2(T)}{\sigma_n^2}|_{\max} = \frac{2}{\eta}\int_{-\infty}^{\infty}|P(f)|^2 df = \frac{2}{\eta}E. \tag{10.75}$$

Thus, the maximum output S/N ratio depends on the pulse energy and PSD of the noise, not on the shape of the pulse. The impulse response of the matched filter is obtained by taking the inverse Fourier transform of (10.74). This leads to

$$\begin{aligned}h_{\text{opt}}(t) &= \mathcal{F}^{-1}\left[\frac{2k}{\eta}P^*(f)e^{-j2\pi fT}\right]\\ &= \frac{2k}{\eta}\int_{-\infty}^{\infty}P^*(f)e^{-j2\pi fT}e^{j2\pi ft}df\\ &= \frac{2k}{\eta}\int_{-\infty}^{\infty}P^*(f)e^{j2\pi f(t-T)}df = \frac{2k}{\eta}p(T-t).\end{aligned} \tag{10.76}$$

Since $p(t)$ is a real-valued signal, so $P^*(f) = P(-f)$. Thus the impulse response is found to have the same shape as the message pulse $p(t)$ reversed in time, delayed by the symbol duration T. Without delay, the filter response $p(-t)$ turns out to be unrealisable. This delay should be chosen sufficiently large enough to make the system causal. Let us study a specific example to reveal the significance of the statements.

Example 10.8

Consider that the received pulse is triangular in shape, that is, $p(t) = \Lambda(t/\tau)$. Find the characteristic of the matched filter when $k = \eta/2$. Then find the condition on the delay T to make the system a causal system.

Solution

From (10.76), $h_{\text{opt}}(t) = \dfrac{2k}{\eta}p(T-t) = p(T-t)$ since $k = \eta/2$.

Step 1 Rotate $p(t)$ around $t = 0$ axis to obtain $p(-t)$ (the + and – signs within the figure is to simply aid our eyes).

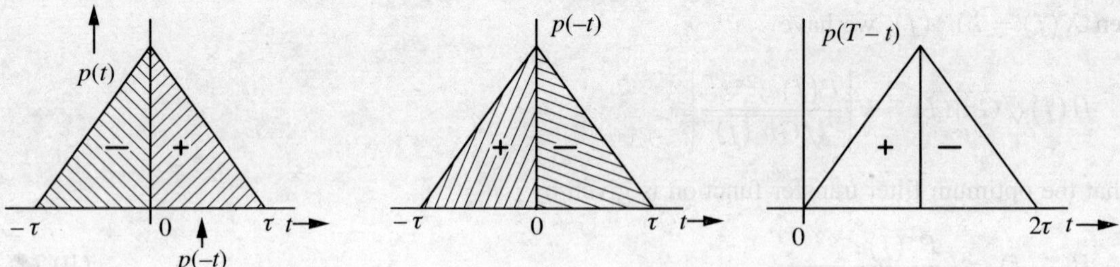

Step 2 Apply delay T. This delay should be $T = \tau$ so as to make $h_{\text{opt}}(t)$ a function of positive time.

We now study the implementation of a matched filter for a polar NRZ pulse shape. Such a receiver is shown in Fig. (10.34), called an 'integrate and dump' filter. Initially the capacitor C is made uncharged by pressing the dump switch S_1 momentarily. The active circuit integrates each incoming pulse noise so that

$$v_0(T) = \frac{1}{\tau}\int_0^T [s(t) + n(t)]dt, \tag{10.77}$$

where $\tau = RC$. The voltage at the end of the pulse due to the signal is

$$s_0(T) = \frac{1}{\tau}\int_0^T s(t)dt = \pm\frac{VT}{\tau}, \tag{10.78}$$

where V is the amplitude of the pulse. The voltage due to noise is

$$n_0(T) = \frac{1}{\tau}\int_0 n(t)dt. \tag{10.79}$$

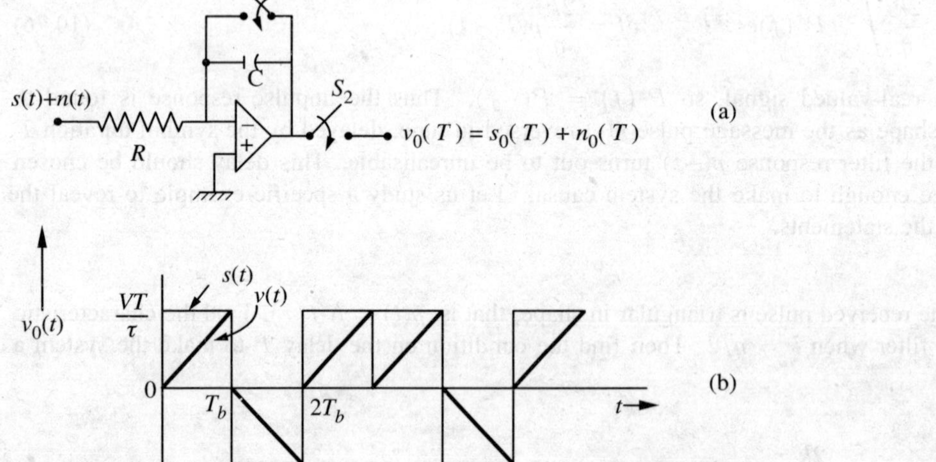

Fig. 10.34 *(a) Integrate-and-dump filter. (b) Polar binary output.*

That the filter enhances the signal with respect to the noise, one should be interested in calculating the ratio $s_0^2(T)/\overline{n_0^2(T)} = s_0^2(T)/\sigma_n^2$, where σ_n^2 is the rms noise output voltage. The integrator under discussion has the transfer function

$$H(f) = (1 - e^{-j2\pi fT})/j\omega\tau.$$
$$= (1 - \cos 2\pi fT + j\sin 2\pi fT)/j\omega\tau.$$

Hence

$$|H(f)|^2 = \frac{(1-\cos 2\pi fT)^2 + \sin^2 2\pi fT}{\omega^2\tau^2}$$
$$= \frac{2(1-\cos 2\pi fT)}{\omega^2\tau^2}$$
$$= \frac{4\sin^2 \pi fT}{4\pi^2 f^2\tau^2} = \left(\frac{T}{\tau}\right)^2 \left(\frac{\sin \pi fT}{\pi fT}\right)^2.$$

Thus, for an input noise PSD of $\eta/2$, the output noise PSD is $\left(\frac{T}{\tau}\right)^2 \left(\frac{\sin \pi ft}{\pi fT}\right)^2 \eta/2$ and the mean square noise is

$$\overline{n_0^2(T)} = \int_{-\infty}^{\infty} \frac{\eta}{2}\left(\frac{T}{\tau}\right)^2 \left(\frac{\sin \pi fT}{\pi fT}\right)^2 df$$
$$= \frac{\eta}{2}\left(\frac{T}{\tau}\right)^2 \frac{1}{\pi T}\cdot\pi \quad \left(\text{since } \int_{-\infty}^{\infty}\left(\frac{\sin x}{x}\right)^2 dx = \pi\right)$$
$$= \frac{\eta T}{2\tau^2}. \tag{10.80}$$

Thus, $$\frac{s_0^2(T)}{\overline{n_0^2(T)}} = \frac{V^2T^2}{\tau^2}\frac{2\tau^2}{\eta T} = \frac{2}{\eta}V^2T = \frac{2}{\eta}E, \tag{10.81}$$

where V^2T is the energy E of the signal. Equation (10.81) has the same form as (10.75). Interestingly, the signal voltage increases linearly with T, whereas the rms value of noise increases as $\sqrt{T}$, thereby emphasising the signal in relation to the noise.

10.6.3 Matched filter for binary modulation

Binary modulation involves two different signalling waveforms, $s_1(t)$ and $s_2(t)$, rather than one pulse shape $p(t)$. For example, in PSK signalling, $s_1(t) = A\cos\omega_0 t$ and $s_2(t) = -A\cos\omega_0 t$; in FSK signalling, $s_1(t) = A\cos\omega_L t$ and $s_2(t) = A\cos\omega_H t$. A bandpass binary receiver is depicted in Fig. 10.35. Here, both the signal and the noise are sampled at the end of the bit interval to yield $v_0(T_b) = s_0(T_b) + n_0(T_b)$.

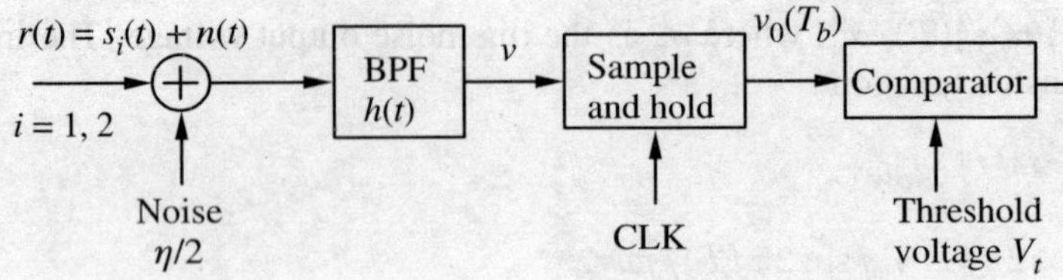

Fig. 10.35 *A receiver for bandpass signal*

Here, the signal-only-component at the output of the bandpass filter (BPF) is

$$s_{oi}(T) = s_i(t) * h(t)$$

$$= \int_0^{T_b} s_i(\lambda)h(T_b - \lambda)d\lambda, \quad i = 1, 2 \tag{10.82}$$

where $h(t)$ is the BPF impulse response which we need to optimise to produce minimum error probability. Our decision rule is as follows: $s_1(t)$ has been transmitted if $v_0(T)$ is closer to $s_{o1}(T)$; $s_2(t)$ has been transmitted if $v_0(T)$ is closer to $s_{o2}(T)$. The optimum threshold voltage should be $V_{\text{opt}} = \frac{1}{2}[s_{o1}(T) + s_{o2}(T)]$, which may be obtained from the plots of PDFs of $v_0(T)$ (shown in Fig. 10.36). Suppose that $s_1(t)$ was transmitted and so we expect an output voltage close to $s_{o1}(T)$. But if noise voltage is considerably large so that $s_{o1}(T) + n_o(T) \geq V_{\text{opt}}$, the decision would be in favour of $s_{o2}(T)$. This causes an error in regenerating the message bit. The noise voltage in this case would be

$$n_o(T) \geq V_{\text{opt}} - s_{o1}(T)$$

$$= \frac{1}{2}[s_{o1}(T) + s_{o2}(T)] - s_{o1}(T)$$

$$= \frac{1}{2}[s_{o2}(T) - s_{o1}(T)] \tag{10.83}$$

and the probability of error is

$$P_e = \int_{\left(\frac{1}{2}[s_{o2}(T) - s_{o1}(T)]\right)}^{\infty} \frac{1}{\sqrt{2\pi\sigma_n^2}} e^{-n_o^2(T)/2\sigma_n^2} d_{n_o}(T) = \frac{1}{2}\,\text{erfc}\left[\frac{s_{o2}(T) - s_{o1}(T)}{2\sqrt{2}\sigma_n}\right]. \tag{10.84}$$

So, to reduce P_e the BPF should maximise the ratio $\dfrac{s_{o2}(T) - s_{o1}(T)}{2\sqrt{2}\sigma_n}$, or equivalently the ratio $[s_{O2}(T) - s_{O1}(T)]^2/8\sigma_n^2$.

Now, by (10.82),

$$[s_{o2}(T) - s_{o1}(T)]^2 = \left|\int_{-\infty}^{\infty} [s_2(\lambda) - s_1(\lambda)]h(T_b - \lambda)d\lambda\right|^2. \tag{10.85}$$

Also, $\sigma_\eta^2 = \dfrac{\eta}{2}\displaystyle\int_{-\infty}^{\infty} |h(t)|^2 dt = \dfrac{\eta}{2}\int_{-\infty}^{\infty} |h(T_b - \lambda)|^2 d\lambda.$ (10.86)

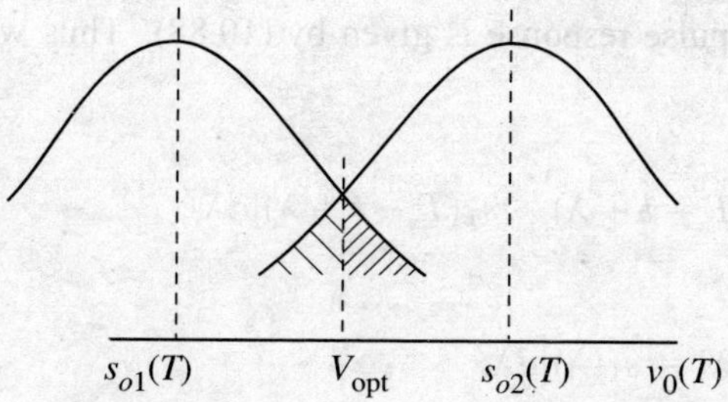

Fig. 10.36 *Conditional PDFs of $v_0(T)$*

Thus,
$$\frac{\left|s_{o2}(T) - s_{o1}(T)\right|^2}{8\sigma_n^2} = \frac{1}{4\eta}\frac{\left|\int_{-\infty}^{\infty}[s_2(\lambda) - s_1(\lambda)]h(T_b - \lambda)\right|^2 d\lambda}{\int_{-\infty}^{\infty}\left|h(T_b - \lambda)\right|^2 d\lambda}$$

$$\leq \frac{1}{4\eta}\int_{-\infty}^{\infty}[s_2(t) - s_1(t)]^2 dt, \tag{10.87}$$

where we have used Schwartz's inequality. The equality holds (that is, the left-hand side is maximum) if $h(T_b - t) = k[s_2(t) - s_1(t)]$, where k is a constant. Thus the BPF has the impulse response

$$h(t) = k[s_2(T_b - t) - s_1(T_b - t)]. \tag{10.88}$$

Figure 10.37 shows the matched filter representing the BPF.

$s_i(t)$, $i = 1, 2$ → $h(T_b - t)$ → $v_0(t)$, $i = 1,2$ → $t = T$ → $v_0(T) = s_{o2}(T) - s_{o1}(T)$

Matched to $s_2(t)$ $s_1(t)$

Fig. 10.37 *Matched filter for optimum binary detection*

10.6.4 Correlation detector

An equivalence of matched filter receiver is the correlation detector. Suppose that input to the matched filter is $r(t) = s_i(t) + n(t)$, $i = 1, 2$. The output of the causal filter $v_0(t)$ is obtained from the convolution integral

$$v_0(t) = r(t) * h(t)$$

$$= \int_0^t r(\lambda)h(t - \lambda)d\lambda. \tag{10.89}$$

For optimum filtering the impulse response is given by (10.88). Thus with $t = T$ (over one symbol interval),

$$v_0(t) = \int_0^T r(\lambda)k[s_2(T - t + \lambda) - s_1(T - t + \lambda)]d\lambda$$

$$= \int_0^T r(\lambda)k[s_2(\lambda) - s_1(\lambda)]d\lambda. \quad (10.90)$$

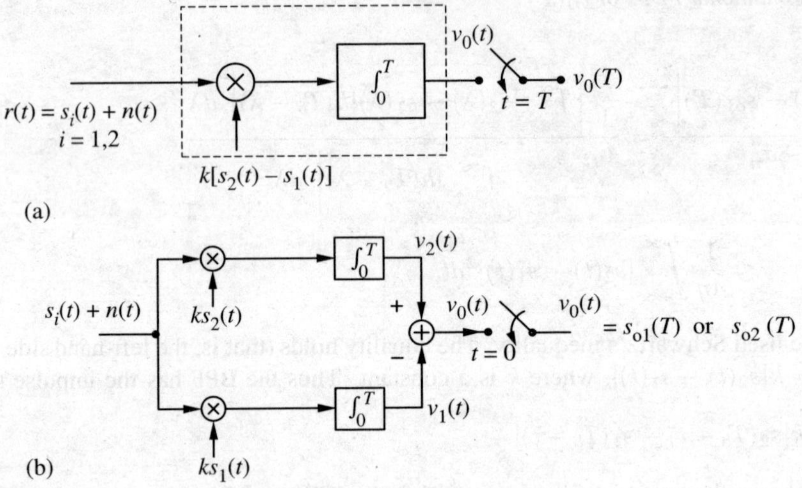

(a)

(b)

Fig. 10.38 *(a) Correlator, (b) bank of correlators.*

The above equation defines the correlation of the received signal and the noise with a replica of the transmitted waveform $s_2(t) - s_1(t)$. The correlation realisation of the matched filter based on (10.90) is shown in Fig. 10.38(a). The multiplier and integrator perform the operation of correlation. Note that at the beginning of the next bit interval, the stored energy in the filter must be discharged. Since we are dealing with two signals, $s_1(t)$ and $s_2(t)$, the system employs two correlators as shown in Fig. 10.38(b). The received signal, $s_i(t) + n(t)$ is correlated with each copy of $s_1(t)$ and $s_2(t)$, using two correlators in parallel. The signal $s_i(t)$ whose product integration with $r(t)$ gives the maximum output is the transmitted signal.

At this point we stress that optimum binary detection relies on the maximum ratio in (10.87). Consider the integration of (10.87):

$$\int_{-\infty}^{\infty} [s_2(t) - s_1(t)]^2 dt = \int_0^T s_2^2(t)dt + \int_0^T s_1^2(t)dt - 2\int_0^T s_2(t)s_1(t)dt$$

$$= E_2 + E_1 - 2E_{21},$$

where $E_2 = \int_0^T s_2^2(t)dt$ and $E_1 = \int_0^T s_1^2(t)dt$ are the energies associated with $s_2(t)$ and $s_1(t)$ over the bit interval. The average energy associated with a bit is $E_b = \frac{1}{2}(E_2 + E_1) = E_1 = E_2$, since the symbols are equi-probable. The term E_{21} is proportional to the cross-correlation coefficient ρ, which measures the similarity between the two signals $s_1(t)$ and $s_2(t)$ and is defined as

$$\rho = \frac{1}{E_b}\int_0^{T_b} s_1(t)s_2(t)dt, \quad -1 \le p \le 1. \tag{10.91}$$

In terms of E_b and ρ, (10.87) reads as

$$\left(\frac{s_{o2}(T) - s_{o1}(T)}{2\sqrt{2\sigma_n^2}}\right)^2\bigg|_{\max} = \frac{1}{4\eta}(2E_b - 2\rho E_b) = \frac{E_b(1-\rho)}{2\eta}. \tag{10.92}$$

The minimum error probability is then

$$\begin{aligned} P_e &= \frac{1}{2}\text{erfc}\left[\left(\frac{s_{o2}(T) - s_{o1}(T)}{2\sqrt{2}\sigma_n}\right)^2_{\max}\right]^{1/2} \\ &= \frac{1}{2}\text{erfc}\left[\frac{E_b(1-\rho)}{2\eta}\right]^{1/2}. \end{aligned} \tag{10.93}$$

The case $\rho = 1$ corresponds to signals $s_1(t)$ and $s_2(t)$ being perfectly similar over a bit interval and not distinguishable from one another. Such waveforms are discarded for digital signalling. The case $\rho = -1$ corresponds to antipodal signals where $s_1(t) = -s_2(t)$. Finally, $\rho = 0$ corresponds to orthogonal signals.

Example 10.9

A binary communications system receives equally likely signals $s_1(t)$ or $s_2(t) = \pm 2t/T, 0 \le t \le T$. The signal is corrupted by AWGN of PSD 10^{-6} volt^2/Hz. If the probability of error $P_e = 1.2 \times 10^{-4}$, find the minimum allowable interval T.

Solution

The received energy per bit

$$E_b = \int_0^T v^2(t)dt = \int_0^T \frac{4t^2}{T^2}dt = \frac{4}{T^2} \times \frac{1}{3}T^3 = \frac{4}{3}T \text{ joules.}$$

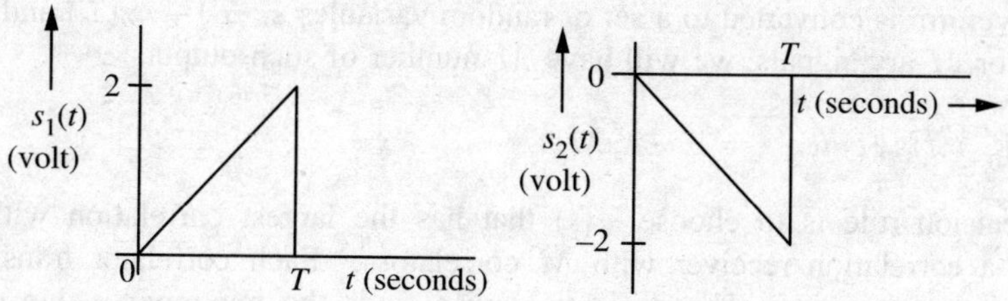

Since the two signals are antipodal, $\rho = -1$ and

$$P_e = \frac{1}{2}\text{erfc}\sqrt{\frac{E_b}{\eta}} = 1.2 \times 10^{-4} \quad \text{or} \quad \text{erfc}\sqrt{\frac{E_b}{\eta}} = 2.4 \times 10^{-4} = \text{erfc}(2.6).$$

Thus, $\sqrt{\dfrac{E_b}{\eta}} = 2.6$ or, $E_b = 6.76\eta = 6.76 \times 10^{-6}$ volt^2/Hz.

Therefore, $\dfrac{4}{3}T = 6.76 \times 10^{-6}$ or $T = 50.7\ \mu$s.

Example 10.10

Express the optimum threshold V_{opt} in terms of the signal energies E_1 and E_2.

Solution

Substituting (10.88) into (10.82),

$$s_{oi}(T) = \int_0^{T_b} s_i(t)k[s_2(t) - s_1(t)]dt, \quad i = 1, 2.$$

So, $s_{o1}(T) = \displaystyle\int_0^{T_b} k[s_1(t)s_2(t) - s_1^2(t)]dt = k(E_{12} - E_1)$

and $s_{o2}(T) = \displaystyle\int_0^{T_b} k[s_2^2(t) - s_2(t)s_1(t)]dt = k(E_2 - E_{12}).$

Thus, $V_{\text{opt}} = \frac{1}{2}[s_{o1}(T) + s_{o2}(T)]$

$$= \frac{1}{2}k(E_{12} - E_1 + E_2 - E_{12}) = \frac{1}{2}k(E_2 - E_1).$$

Example 10.11

We consider the more general M- ary detection. The received signal is the sum of the transmitted signal and the random noise (assume here AWGN).

$$r(t) = s_i(t) + n(t) \quad i = 1, 2, \cdots, M \quad (0 \le t \le T). \tag{10.94}$$

The received waveform is converted to a set of random variables $s_{oi}(T) + n_o(T)$ and then compared to a threshold. For M-ary signals, we will have M number of such outputs:

$$s_{oi}(T) = \textstyle\int_0^T r(t)s_i(t)dt \quad i = 1, \cdots, M. \tag{10.95}$$

A reasonable decision rule is to choose $s_i(t)$ that has the largest correlation with $r(t)$. Figure 10.39 illustrates a correlation receiver with M correlators. Each correlator transforms the message input $r(t)$ into a number. The decision device finds the maximum value of the numbers $s_{01}(T), s_{02}(T), \ldots, s_{0M}(T)$ and declares that $s_i(t)$ with the maximum of $s_{0i}(T)$ was transmitted.

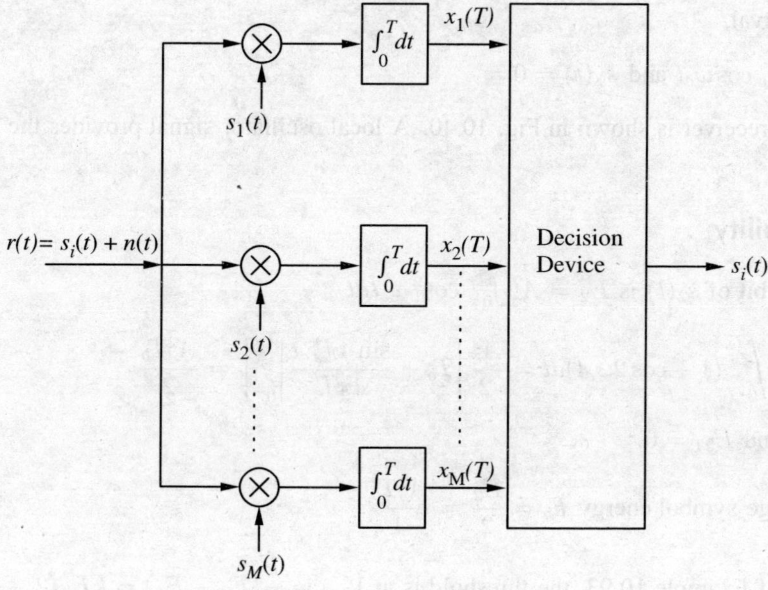

Fig. 10.39 *Correlation receiver for M-ary case*

10.7 Coherent Detection and Error Performance

Coherent detection employs the knowledge of the frequency and the phase of the carrier to detect a message. In this kind of digital systems, there is available at the receiver stored copies of each possible transmitted signal. These stored copies are used to regenerate the transmitted signals.

10.7.1 Coherent OOK systems

In on–off keying, the carrier is turned on and off by the digital data. Referring to (10.1) for the ASK signalling, the OOK waveforms are derived as

$$s_1(t) = A_c d(t) \cos \omega_c t, \quad s_2(t) = 0.$$

Received
signal + $n(t)$
$\int_0^T dt$
S/H
Decision
rule
$A_c \cos\omega_c t$
LO
Threshold

Fig. 10.40 *Correlation receiver for OOK signalling*

For any bit interval,

$$s_2(t) = A_c \cos\omega_c t \text{ and } s_1(t) = 0$$

The correlation receiver is shown in Fig. 10.40. A local oscillator signal provides the stored copy of $s_2(t)$.

Error probability

The energy per bit of $s_2(t)$ is $E_2 = A_c^2 \int_0^{T_b} \cos^2 \omega_c t dt$

$$= \frac{A_c^2}{2} \int_0^{T_b} (1 + \cos 2\omega_c t) dt = \frac{A_c^2}{2} \left[T_b + \frac{\sin 4\pi f_c t}{4\pi f_c} \bigg|_0^{T_b} \right] = \frac{A_c^2 T_b}{2}$$

while $E_1 = 0$ and $E_{21} = 0$.

Thus, the average symbol energy $E_b = \dfrac{E_2}{2} = \dfrac{A_c^2 T_b}{4}$.

By the results of Example 10.93, the threshold is at $V_{\text{opt}} = \dfrac{k}{2}(E_2 - E_1) = kE_b/2$.

The minimum error probability is given by (10.93). That is,

$$P_e = \frac{1}{2}\text{erfc}\left[\frac{E_b(1-\rho)}{2\eta}\right]^{1/2} = \frac{1}{2}\text{erfc}\left[\frac{E_b}{2\eta}\right]^{1/2}. \tag{10.96}$$

10.7.2 Coherent BPSK systems

The transmitted BPSK signals are given in (10.5) and (10.6). We re-write these as

$$s_2(t) = A_c \cos\omega_c t \text{ and } s_1(t) = -s_2(t).$$

For these antipodal signals,

$$E_2 = E_1 = \frac{A_c^2}{2} T_b \text{ and } E_{21} = \int_0^{T_b} (A_c \cos\omega_c t)(-A_c \cos\omega_c t) dt = -\frac{A_c^2}{2} T_b.$$

So the average symbol energy $E_b = \dfrac{A_c^2 T_b}{2}$, $E_{21} = -E_b$ and $\rho = -1$.

So, $P_e = \dfrac{1}{2}\text{erfc}[E_b/\eta]^{1/2}$. (10.97)

A coherent BPSK receiver needs only one matched filter since we are using antipodal signals.

Example 10.12

Find the error probability for the BPSK system that receives $s_1(t) = 10\cos\omega_0 t$ mV; $s_a(t) = -s_2(t)$ at a bit rate of 1 Mbps. Assume noise PSD of $\eta = 10^{-11}$ W/Hz.

Solution

Bit rate $r = 1/T$ so $T = 1/r = 10^{-6}$ seconds.

Energy per bit, $E_b = \frac{1}{2}A^2T = \frac{1}{2}(10 \times 10^{-3})^2 \times 10^{-6} = 50 \times 10^{-12}$ joules.

So, $P_e = \frac{1}{2}\text{erfc}\sqrt{\frac{50 \times 10^{-12}}{10^{-11}}} = \frac{1}{2}\text{erfc}(5) - 0.77 \times 10^{-12}.$

We will now discuss some of the detrimental effects caused by the correlator receiver itself or by the noise. For optimum coherent detection, it is essential that at the local oscillator of the correlator the signal must be synchronised accurately with the carrier. The local oscillator may, however, suffer from phase offset. Let there be a small phase error ϕ at the local oscillator. This reduces the effective signal level by factor $\cos\phi$, or equivalently, the signal energy becomes $E_b \cos^2\phi$. The error probability then becomes

$$P_e = \frac{1}{2}\text{erfc}\sqrt{\frac{E_b \cos^2\phi}{\eta}}. \tag{10.98}$$

Again, a timing error may affect the proper bit synchronisation. A correlation receiver performs integration followed by sampling. Recall that the integration starts at $t = 0$ and ends at $t = T_b$. If the receiver is not properly synchronised, this time interval way be from τ to $T + \tau$, the integration time being unchanged. Naturally, the sampling takes place at times $T_b(1 \pm \tau)$. This timing error is equivalent to a phase ϕ_τ, where $\cos\phi_\tau = \omega_0 T_b \tau$. Again, the signal amplitude reduces by a factor $\cos\phi_\tau$ and the error probability becomes

$$P_e = \frac{1}{2}\text{erfc}\sqrt{\left(\frac{E_b}{\eta}\cos^2\phi_\tau\right)}. \tag{10.99}$$

Example 10.13

Suppose that a BPSK system operates at a bit rate of 2 kbps and $\omega_0 = 100$ kHz. Compute the error probability for $E_b/\eta = 4$. If $\tau = 0.003T_b$, find the error.

Solution

$$T_b = \frac{1}{r_b} = \frac{1}{2} \times 10^{-3} \text{ seconds.}$$

$$P_e = \frac{1}{2}\text{erfc}\sqrt{\frac{E_b}{\eta}} = \frac{1}{2}\text{erfc}\sqrt{4} = \frac{1}{2} \times \text{erfc}(2) = \frac{1}{2} \times 4.38 \times 10^{-3} = 2.19 \times 10^{-3}.$$

Now, for a timing error of only 0.3% of the bit interval,

$$\theta_\tau = \omega_0 T_b \tau = 2\pi \times (100 \times 10^3) \times \frac{0.3}{100} \times \frac{1}{2} \times 10^{-3} \text{ radians} = 54^\circ.$$

So, $P_e = \frac{1}{2}\text{erfc}\sqrt{\frac{E_b}{\eta}\cos^2 54^\circ} = \frac{1}{2}\text{erf}(4 \times 0.345) = \frac{1}{2}\text{erfc}(1.38) = 4.8 \times 10^{-2}.$

10.7.3 Coherent BFSK systems

In binary FSK, the received signal is either

$$s_1(t) = A_c \cos 2\pi(f_0 - f_d)t$$

$$\text{or } s_2(t) = A_c \cos 2\pi(f_0 + f_d)t. \quad \text{[vide (10.31)]}$$

To calculate the error probability, we first find the signal energy per bit.

$$E_2 = \int_0^{T_b} A_c^2 \cos^2 2\pi(f_0 + f_d)t dt$$

$$= \frac{A_c^2 T_b}{2} + \int_0^{T_b} \frac{A_c^2}{2} \cos 4\pi(f_0 + f_d)t dt.$$

Since the difference frequency f_d is not necessarily a multiple of f_0, the second term gives a nonzero value, although small. Thus, $E_2 \approx \frac{A_c^2 T_b}{2}$, and similarly, $E_1 \approx \frac{A_c^2 T_b}{2}$. The average energy per bit $E_b \simeq \frac{A_c^2 T_b}{2}$. The correlation coefficient in this case is

$$\rho = \frac{1}{\sqrt{E_2 E_1}} \int_0^{T_b} A_c^2 \cos 2\pi(f_0 + f_d) \times \cos 2\pi(f_0 - f_d) dt$$

$$= \frac{1}{E_b} \frac{A_c^2}{2} \int_0^{T_b} [\cos 4\pi f_0 t + \cos 4\pi f_d t] dt$$

$$= \frac{A_c^2}{2E_b} \left[\frac{\sin 4\pi f_0 T_b}{4\pi f_0} + \frac{\sin 4\pi f_d T_b}{4\pi f_d} \right].$$

We may ignore the first term if we assume $\omega_0 >> \omega_d$.

$$\text{Thus, } \rho \simeq \frac{A_c^2}{2E_b} \frac{\sin 4\pi f_d T_b}{4\pi f_d} = \frac{\sin 4\pi f_d T_b}{4\pi f_d T_b} = \text{sinc}(4 f_d T_b). \tag{10.100}$$

For the case of Sunde's FSK, $f_d = \frac{1}{2T_b}$, and $\rho = \frac{\sin 4\pi f_d T_b}{4\pi f_d T_b} = \frac{\sin 2\pi}{2\pi} = 0.$

So, the error probability

$$P_e = \frac{1}{2} \text{erfc} \sqrt{\frac{E_b(1-\rho)}{2\eta}} = \frac{1}{2} \text{erfc} \sqrt{\frac{E_b}{2\eta}}, \tag{10.101}$$

the same as OOK. This is true for the case of orthogonal FSK.

For the other choice of f_d, a non-orthogonal case, the error probability is calculated as follows: Returning to (10.87), we see that

$$\left[\frac{|s_{o2}(T) - s_{o1}(T)|^2}{8\sigma_n^2}\right]_{\max} = \frac{1}{4\eta}\int_{-\infty}^{\infty}[s_2(t) - s_1(t)]^2 dt$$

$$\approx \frac{1}{4\eta}[E_2 + E_1 - 2E_b\rho] = \frac{1}{4\eta}2E_b(1-\rho)$$

$$= \frac{E_b}{2\eta}\left[1 - \frac{\sin 4\pi f_d T_b}{4\pi f_d T_b}\right]. \tag{10.102}$$

The maximum value of the left-hand side is attained when $\sin(4\pi f_d T_b) = -1$, that is for $4\pi f_d T_b = 3\pi/2$.

Thus,

$$\left[\frac{|s_{o2}(T) - s_{o1}(T)|^2}{8\sigma_n^2}\right]_{\max} = \frac{E_b}{2\eta}\left(1 + \frac{2}{3\pi}\right) = \frac{4.84}{8}\frac{E_b}{\eta}. \tag{10.103}$$

The probability of error is found to be

$$P_e = \frac{1}{2}\text{erfc}\left(0.6\frac{E_b}{\eta}\right)^{1/2}. \tag{10.104}$$

A comparison of (10.104) with (10.97) reveals that for the same E_b and η, error is larger in FSK. For an equal error probability, the BPSK has an energy advantage of $10\log(1/0.6) \approx 2$ dB over BFSK.

10.8 Non-coherent Detection and Error Performance

There are certain cases where recovery of carrier phase at the receiver is not possible. Non-coherent detection is the only way out in such situations. Multipath fading, rapidly varying delays in the transmission medium, etc., causes the phase of the received signal to shift in a random way so that the receiver is unable to track them. Synchronisation with the phase of the carrier may be too costly. Thus coherent detection process involves highly sophisticated receivers which may be avoided if the signal strength is high. Since incoherent detection is basically envelope detection, the study of the envelope of a sinusoid in the presence of bandpass noise seems appropriate when the signal strength is high.

10.8.1 Bandpass noise and envelope of sinusoids

Suppose we want to find the envelope of a sinusoidal carrier $A_c \cos\omega_0 t$ in the presence of bandpass noise $n(t)$. The quadrature carrier expansion of $n(t)$ is helpful here and can be written as

$$n(t) = n_i(t)\cos\omega_0 t - n_q(t)\sin\omega_0 t, \tag{10.105}$$

where $n_i(t)$ and $n_q(t)$ are, respectively, the in-phase and quadrature components of $n(t)$. Assume that $n(t)$ is a Gaussian random variable, of zero mean, variance σ^2 and so also are n_i and n_q. The envelop-and-phase representation of $n(t)$ is

$$n(t) = A_n \cos[\omega_0 t + \phi_n(t)], \tag{10.106}$$

where $A_n = \sqrt{n_i^2 + n_q^2}$; $\phi_n(t) = \tan \frac{n_q}{n_i}$.

The random variable A_n follows a Rayleigh distribution with

$$f_{A_n}(A_n) = \frac{A_n}{\sigma_n^2} e^{-A_n^2/2\sigma_n^2}, \tag{10.107}$$

and the phase ϕ_n has a uniform distribution over 2π radian, that is,

$$f_{\phi_n}(\phi_n) = \frac{1}{2\pi}, \quad 0 \le \phi \le 2\pi. \tag{10.108}$$

It is observed that $f_{A_n\phi_n}(A_n, \phi_n) = f_{A_n}(A_n) f_{\phi_n}(\phi)$ and A_n and ϕ_n are statistically independent. We consider the envelope of the sinusoidal carrier plus the bandpass noise. The sum of the carrier and the bandpass noise is

$$\begin{aligned} v(t) &= A_c \cos \omega_0 t + n_i(t) \cos \omega_0 t - n_q(t) \sin \omega_0 t \\ &= (A_c + n_i) \cos \omega_0 t - n_q(t) \sin \omega_0 t. \end{aligned} \tag{10.109}$$

Expressing it in the envelope and phase representation, we may write

$$v(t) = A \cos(\omega_0 t + \phi), \tag{10.110}$$

where the envelope $A = \sqrt{(A_c + n_i)^2 + n_q^2}$, and the phase $\phi = \tan^{-1} \dfrac{n_q}{A_c + n_i}$. The joint PDF of A and ϕ follows the distribution

$$f_{A\phi}(A, \phi) = \frac{A}{2\pi\sigma^2} \exp\left(-\frac{A^2 - 2A_c A \cos\phi + A_c^2}{2\sigma_n^2}\right), \quad 0 \le \phi \le 2\pi. \tag{10.111}$$

The envelope term A cannot be isolated from the phase term ϕ because of the term $A \cos \phi$ in the exponent on right-hand side. The PDF for the envelope is obtained as

$$\begin{aligned} f_A(A) &= \int_0^{2\pi} f_{A\phi}(A, \phi) d\phi \\ &= \frac{A}{2\pi\sigma_n^2} \exp\left(-\frac{A^2 + A_c^2}{2\sigma_n^2}\right) \int_0^{2\pi} \exp\left(\frac{A_c A \cos\phi}{\sigma_n^2}\right) d\phi \\ &= \frac{A}{\sigma_n^2} e^{-(A^2 + A_c^2)/2\sigma_n^2} I_0\left(\frac{A_c A}{\sigma_n^2}\right), \end{aligned} \tag{10.112}$$

where $I_0(x)$ is the modified zero-order Bessel function of the first kind, defined by

$$I_0(x) = \frac{1}{2\pi} \int_0^{2\pi} \exp(x \cos\theta) d\theta \tag{10.113}$$

and has the properties

$$I_0(x) \approx \begin{cases} e^{x^2/4} & x << 1 \\ \frac{e^x}{\sqrt{2\pi x}} & x >> 1. \end{cases} \tag{10.114}$$

Equation (10.112) is the Rician distribution followed by the envelope of sinusoids in presence of bandpass Gaussian noise. Under a large signal condition, this simplifies to

$$\begin{aligned} f_A(A) &= \frac{A}{\sigma_n^2} e^{-(A^2+A_c^2)/2\sigma_n^2} \frac{e^{A_c A/\sigma_n^2}}{\sqrt{2\pi A_c A/\sigma_n^2}} \\ &= \sqrt{\frac{A}{2\pi A_c \sigma_n^2}}\ e^{-(A-A_c)^2/2\sigma_n^2} \end{aligned} \tag{10.115}$$

which is essentially a Gaussian distribution with variance σ_n^2, centred at A_c. One important conclusion can be made here that the envelope of the noise follows the Rayleigh distribution; the envelop of the signal plus the noise follows the Rician distribution which, under large signal condition, moves to Gaussian distribution. With this background we will now attempt to study the non-coherent detection of various binary signalling schemes.

10.8.2 Non-coherent OOK

Over the bit interval, we write the expression for OOK signal as

$$s(t) = \frac{A_c}{2}[1 + d(t)]\cos(\omega_c t), \quad d(t) = -1, 1. \tag{10.116}$$

The signal energies are $E_1 = 0$ and

$$\begin{aligned} E_2 &= \frac{A_c^2}{2}\left[T_b + \frac{\sin \omega_c t}{2\omega_c}\bigg|_0^{T_b}\right] = \frac{A_c^2 T_b}{2}\left[1 + \frac{\sin 2\omega_c T_b}{2\omega_c T_b}\right] \\ &\approx \frac{A_c^2 T_b}{2}. \end{aligned}$$

The average signal energy per bit is

$$E_b = \frac{(E_2 + E_1)}{2} \approx \frac{A_c^2 T_b}{4}.$$

Figure 10.41 shows a non-coherent OOK receiving system. The regeneration is again based on the optimum threshold. The envelope detector removes the phase ambiguity and detects the peak amplitude of the envelope. However, the input to the envelope detector is the sinusoid plus the bandpass noise. When $d(t) = -1$, the sample value fed to the envelope detector is only the noise component $n(t)$. The envelope of $v(t)$ follows the Rayleigh distribution. When $d(t) = +1$, we have the envelope of both the signal and noise that follows the Rician distribution. These two conditional PDFs are plotted in Fig. 10.42.

Fig. 10.41 *The non-coherent OOK receiver*

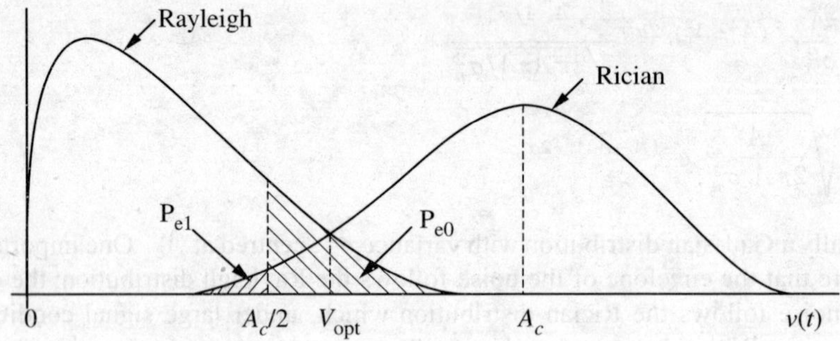

Fig. 10.42 *Plots of Rayleigh and Rician distributions*

The optimum threshold is at the intersection of these two curves. We observe that because of the lack of symmetry about the threshold, P_{e1} is not the same as P_{e0} when P_e is minimum.

For reasonable approximation, V_{opt} is taken to be nearly $A_c/2$ and the resulting error probabilities are calculated below.

The error probability P_{e0} (bit error when a transmitted 1 is misinterpreted as 0) is obtained by the area under the Rayleigh curve beyond the ordinate $A_c/2$. Thus,

$$P_{e0} = \int_{A_c/2}^{\infty} \frac{v}{\sigma^2} e^{-v^2/2\sigma^2}\, dv.$$

Let $v^2/2\sigma^2 = x$ so that $\frac{v}{\sigma^2} dv = dx$ and

$$P_{e0} = \int_{A_c^2/8\sigma^2}^{\infty} e^{-x}\, dx = e^{-A_c^2/8\sigma^2}. \tag{10.117}$$

The error probability P_{e1} is obtained as

$$P_{e1} = \int_0^{A_c/2} \sqrt{\frac{A}{2\pi A_c \sigma^2}}\, e^{-(A-A_c)^2/2\sigma^2}\, dA.$$

Since the tail of the Rician distribution below $A_c/2$ becomes less and is insignificant near the origin, we set $-\infty$ as the lower limit of the above equation. This manipulation helps P_{e1} to be expressed in a compact form. Thus,

$$P_{e1} = \int_{-\infty}^{A_c/2} \frac{A}{\sqrt{2\pi A_c \sigma_n^2}}\, e^{-(A-A_c)^2/2\sigma^2}\, dA$$

$$= \int_{-\infty}^{A_c/2} \frac{1}{\sqrt{2\pi\sigma_n^2}} e^{-(A-A_c)^2/2\sigma^2} dA \quad [A_c \approx A \text{ is taken as the amplitude}]$$

$$= \frac{1}{2}\text{erfc}\left(\frac{A_c}{2\sqrt{2}\sigma}\right)$$

$$= \frac{1}{2}\frac{e^{-(A_c/2\sqrt{2}\sigma)^2}}{\frac{A_c}{2\sqrt{2}\sigma}\cdot\sqrt{\pi}} = \frac{1}{\sqrt{\pi A_c^2/2\sigma^2}} e^{-A_c^2/8\sigma^2}. \tag{10.118}$$

In the last line above, we have used the approximation

$$\text{erfc}(x) \approx \frac{e^{-x^2}}{x\sqrt{\pi}}, \text{ for } x \gg 1.$$

On comparing (10.118) with (10.117), we find that $P_{e1} \gg P_{e0}$ when the signal power $\left(\frac{A_c^2}{2}\right)$ is large relative to the noise power.

The average error probability is

$$P_e = \frac{1}{2}(P_{e0} + P_{e1}) = \frac{1}{2}[e^{-A_c^2/8\sigma^2} + \frac{1}{2}\text{erfc}\left(\frac{A_c}{2\sqrt{2}\sigma}\right)]$$

$$\simeq \frac{1}{2}e^{-A_c^2/8\sigma^2} \tag{10.119}$$

10.8.3 Non-coherent FSK

Binary FSK forms a set of two signals with the same amplitude A_c but with different carrier frequencies, $f_1 = f_0 + f_d$ and $f_2 = f_0 - f_d$. The mathematical expression for the same can be found in (10.27). So the detector consists of two channels of bandpass filters (with centre frequencies f_1 and f_2) and envelope detectors as shown in Fig. 10.43. Each channel corresponds to a non-coherent detector of OOK signal. We assume here that the signals in the two channels are well separated in frequency and that they have negligible overlap. A frequency spacing of $f_2 - f_1 = 2f_d$ which is an integral multiple of $1/T_b$ ensures this condition. The bandwidth of the filters is adjusted to give a maximum output when an appropriate signal is received. When a signal $s_1(t)$ is received, the output of the upper channel $v_1(t)$ has a Rician distribution while the lower channel $v_2(t)$ has a Rayleigh distribution. The opposite condition prevails if signal $s_2(t)$ is received.

The outputs from the envelope detectors are denoted by $v_1(T)$ and $v_2(T)$. When there is no noise, the difference $v(T)$ becomes the signal amplitude $\sqrt{2E_b/T_b}$ when $s_1(t)$ is sent; $v(T) = -\sqrt{2E_b/T_b}$ when $s_2(t)$ is sent. In both the cases the distribution is a Rayleigh function. If there is noise, the $v(T)$ equals $\sqrt{2E_b/T_b}$ plus the noise when $s_1(t)$ is transmitted; it equals $-\sqrt{2E_b/T_b}$ plus the noise when $s_2(t)$ is transmitted. The distribution function is a Rician in both the cases. Thus, we conclude from this symmetry that the optimum threshold is $V_{\text{opt}} = 0$ regardless of the signal amplitude, and $P_{e1} = P_{e0} = P_e$.

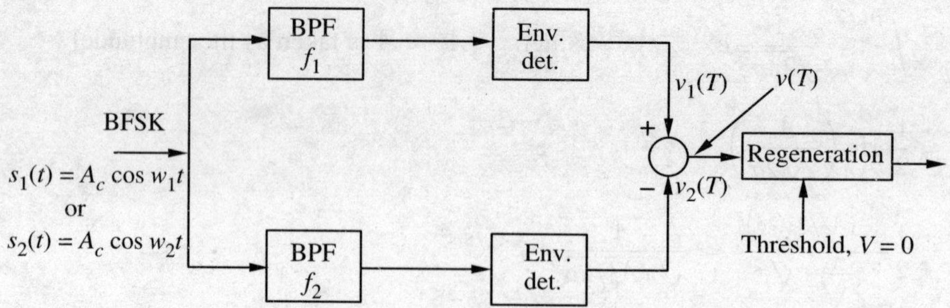

Fig. 10.43 *Non-coherent detector for BFSK. The BPF in each channel is nothing but the filter matched to the respective carrier frequencies*

Suppose that $s_1(t)$ is transmitted. The receiver makes an error if the envelope sample $v_2(T)$ (due to noise alone) exceeds the envelope sample $v_1(T)$ (due to signal + noise). The probability of this error is given by

$$\begin{aligned} P_e &= P(v_2 > v_1|s_1(t) \text{ is transmitted}) \\ &= \text{integration of } P(v_1|s_1) \text{ with respect to } v_1 \text{ from } v_2 \text{ to } \infty \text{ and averaging over all possible values of } v_2. \\ &= \int_0^{\infty} P(v_2|s_1) \left[\int_{v_2}^{\infty} P(v_1|s_1) dv_1 \right] dv_2. \end{aligned}$$

Note that $P(v_1|s_1)$ is the Rayleigh distribution and $P(v_2|s_1)$ is the Rician distribution. Thus,

$$P_e = \int_0^{\infty} \frac{v_2}{\sigma^2} e^{-(v_2^2+A_c^2)} I_0\left(\frac{v_2 A_c}{\sigma^2}\right) \left[\int_{v_2}^{\infty} \frac{v_1}{\sigma^2} e^{-v_1^2/2\sigma^2} dv_1 \right] dv_2,$$

where $A_c = \sqrt{2E_b/T_b}$ is the signal amplitude. The integral can be evaluated to yield finally

$$P_e = \frac{1}{2} e^{-A_c^2/4\sigma^2} \tag{10.120}$$

Although not explicit in (10.120), the error probability depends on the bandwidth of the BPF, as the filter output noise power $\sigma_0^2 = 2(\eta/2) \times BW$. Equation (10.120) reveals that error probability reduces as filter bandwidth decreases.

10.8.4 Differential PSK receiver

In Section 10.3.2, we studied the DPSK modulation which included differential encoding. The DPSK signal is given as

$$v_{\text{DPSK}}(t) = d(t) \sqrt{\frac{2E_b}{T_b}} \cos \omega_0 t, \tag{10.121}$$

where $d(t)$ is related to the message sequence $b(t)$ by $d(t) = b(t) \oplus d(t - T_b)$. In this system, the message information lies in the phase, and synchronous demodulation is used to recover the signal $b(t)$. The received signal is multiplied by the signal delayed by T_b, as shown in Fig. 10.44. In the absence of noise, the multiplier output is

$$\begin{aligned}
&[d(t)\sqrt{\frac{2E_b}{T_b}}\cos(\omega_0 t+\theta)] \times [d(t-T_b)\sqrt{\frac{2E_b}{T_b}}\cos\{\omega_0(t-T_b)+\theta\}] \\
&= \frac{E_b}{T_b}d(t)d(t-T_b)[\cos 2\omega_0[(t-T_b/2)+2\theta]+\cos\omega_0 T_b]. \qquad (10.122)
\end{aligned}$$

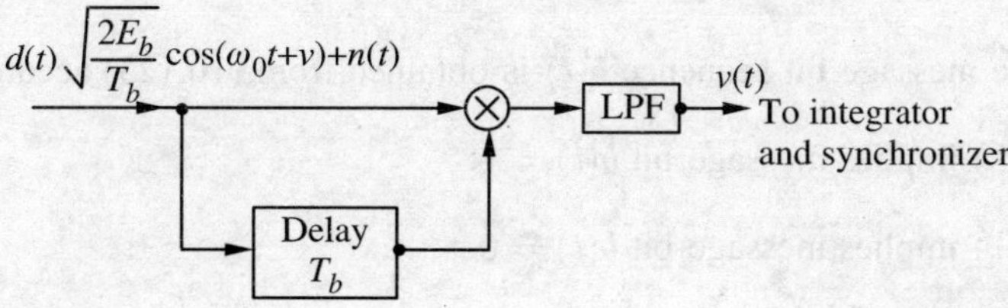

Fig. 10.44 *Receiver for DPSK signal*

We should select $\omega_0 T_b = 2\pi n$, n being an integer so that $\cos\omega_0 T_b = 1$. Lowpass filtering gives

$$\begin{aligned}
v(t) &= E_b/T_b, \quad d(t)=d(t-T_b)\text{: no phase change,} \\
&= -E_b/T_b, \quad d(t)\neq d(t-T_b)\text{: phase change.} \qquad (10.123)
\end{aligned}$$

The threshold can be set at $V=0$. The decision is based on the phase difference between successive received signals during two successive bit intervals. We define the BPSK signal set as

$$\begin{aligned}
x_1(t) &= \sqrt{\frac{2E}{T}}\cos\omega_0 t \quad 0\le t\le T, \\
x_2(t) &= \sqrt{\frac{2E}{T}}\cos(\omega_0 t+\pi), \quad 0\le t\le T. \qquad (10.124)
\end{aligned}$$

For the transmission of DPSK signal, we transmit each bit with the signal pair. The transmission of symbol 1 leaves the carrier phase unchanged and hence the transmitted DPSK signal $s_1(t)$ for $0\le t\le 2T_b$ can be written as

$$s_1(t)=\begin{cases}\sqrt{\frac{2E}{T}}\cos\omega_0 t, & 0\le t\le T_b \\ \sqrt{\frac{2E}{T}}\cos\omega_0 t, & T_b\le t\le 2T_b\end{cases}. \qquad (10.125)$$

The transmission of symbol 0 changes the carrier phase by π radians and the transmitted DPSK signal $s_2(t)$ for $0\le t\le 2T_b$ can be written as

$$s_2(t) = \begin{cases} \sqrt{\frac{2E}{T}} \cos \omega_0 t, & 0 \le t \le T_b \\ \sqrt{\frac{2E}{T}} (\cos \omega_0 t + \pi) & T_b \le t \le 2T_b. \end{cases} \tag{10.126}$$

In terms of the binary signal pair,

$$s_1(t) = (x_1, x_1) \text{ or } (x_2, x_2), \quad 0 \le t \le 2T_b$$

and $s_2(t) = (x_1, x_2)$ or (x_2, x_1), $\quad 0 \le t \le 2T_b$.

The generation of the message bit sequence $b(t)$ is obtained from (10.123) according to the rule:

$d(t) = d(t - T_b)$ implies message bit $b(t) = 1$.

$d(t) \neq d(t - T_b)$ implies message bit $b(t) = 0$.

This means if $v(t) > 0$, the receiver favours symbol 1 being transmitted and if $v(t) < 0$, the receiver favours symbol 0 being transmitted. The DPSK receiver can be designed using two-channel coherent detector as shown in Fig. 10.45.

The upper channel is the in-phase channel that takes care of $s_1(t) = (\sqrt{\frac{2E}{T}} \sin \omega_0 t, \sqrt{\frac{2E}{T}} \cos \omega_0 t)$ and $s_2(t) = (\sqrt{\frac{2E}{T}} \cos \omega_0 t, \sqrt{\frac{2E}{T}} \sin \omega_0 t)$. The lower channel, called the quadrature channel, processes the signals

$$s_1(t) = \left(\sqrt{\frac{2E}{T}} \sin \omega_0 t, \sqrt{\frac{2E}{T}} \sin \omega_0 t \right) \text{ and}$$

Fig. 10.45 *DPSK receiver*

$$s_2(t) = \left(\sqrt{\frac{2E}{T}} \sin \omega_0 t, -\sqrt{\frac{2E}{T}} \sin \omega_0 t \right).$$

As we shall prove in Example 10.11 that $s_1(t)$ and $s_2(t)$ are orthogonal over $0 \le t \le 2T_b$, DPSK can be thought of as a special case of non-coherent orthogonal modulation with $T = 2T_b$ and $E = 2E_b$. So, the probability of error is given by

$$P_e = \frac{1}{2}\exp\left(-\frac{E_b}{\eta}\right). \tag{10.127}$$

Example 10.14

Show that the DPSK signals given in (10.125) and (10.126) are orthogonal over $0 \le t \le 2T_b$.

Solution

We find the correlation between $s_1(t)$ and $s_2(t)$ for all the combinations of x_1 and x_2.

$$\text{Correlation} = \int_0^{2T_b} s_1(t)s_2(t)dt = \int_0^{T_b} s_1(t)s_2(t)dt + \int_{T_b}^{2T_b} s_1(t)s_2(t)dt$$

$$= \int_0^{T_b} x_1^2(t)dt + \int_0^{T_b} x_1x_2dt = \int_0^{T_b} x_1^2dt - \int_0^{T_b} x_1^2dt = 0$$

[since x_1 and x_2 are antipodal signals]

$$\text{Correlation} = \int_0^{T_b} x_1x_2dt + \int_0^{T_b} x_1^2dt = 0$$

$$\text{Correlation} = \int_0^{T_b} x_2x_1dt + \int_0^{T_b} x_2^2dt = 0$$

$$\text{Correlation} = \int_0^{T_b} x_2^2dt + \int_0^{T_b} x_2x_1dt = 0$$

10.9 Error Performance of M-ary Systems

In binary signalling, we transmit each bit individually. For example, in OOK, logic 1 and logic 0 are transmitted by turning ON and OFF the carrier; in BPSK symbols 1 and 0 are transmitted by sinusoids of fixed amplitude but differing in phase by 180°; in BFSK the symbols are transmitted with sinusoids of two different frequencies. In M-ary signalling, k bits are lumped together to form a k-bits long symbol. There will be $M = 2^k$ possible waveforms, each extending over the time kT_b. Binary signalling is a special case where $k = 1$ and there are only two possible symbols.

10.9.1 M-ary PSK

In M-ary PSK, the phase of carriers takes on M possible values. The general expression for M-ary PSK was stated in (10.14). We now re-write the same here in the form

$$s_m(t) = \sqrt{2E/T}\cos(\omega_0 t + \phi_m), m = 0, 1, \cdots, M-1, \tag{10.128}$$

where $\phi_m = (2m+1)\frac{\pi}{M}$. (10.129)

The energy per symbol is $E = kE_b$ since k-bits are lumped together to form a symbol and $2^k = M$. In terms of in-phase and quadrature components, the modulated wave can be written as

$$s_m(t) = (\sqrt{2E/T}\cos\phi_m\cos\omega_0 t - (\sqrt{2E/T}\sin\phi_m)\sin\omega_0 t. \tag{10.130}$$

Figure 10.46 illustrates the PSK signal-space diagram for $M = 2$, and 4. In the binary case $M = 2$, $m = 0, 1$ and so $\phi_0 = \pi/2$ and $\phi_1 = 3\pi/2$, giving a set of two antipodal signals: $s_0(t) = \sqrt{2E/T}\cos(\omega_0 t + \pi/2) = -\sqrt{E}\sqrt{\frac{2}{T}}\sin\omega_0 T$ and $s_1(t) = \sqrt{E}\frac{\sqrt{2}}{T}\sin\omega_0 t$.

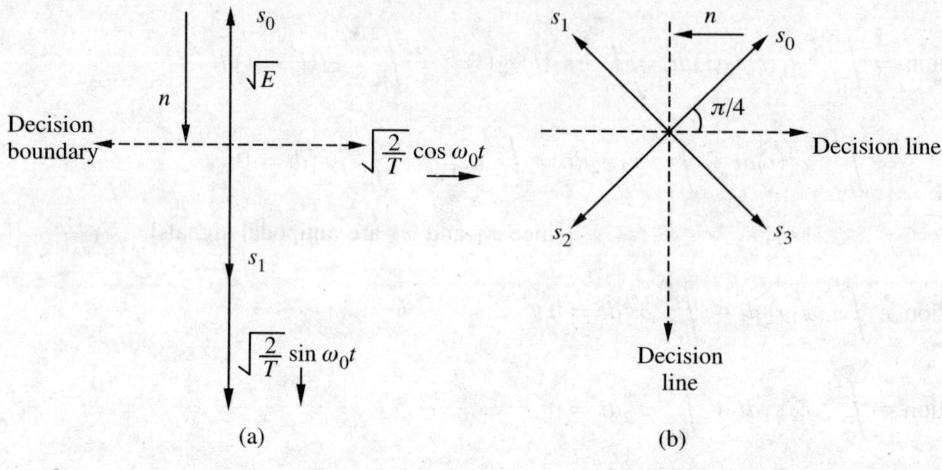

Fig. 10.46 *Signal-space diagram for (a) binary PSK (b) 4-ary PSK*

The 4-ary case corresponds to $M = 4, m = 0, 1, 2, 3$ and $\phi_0 = \frac{\pi}{4}, \phi_1 = \frac{3\pi}{4}, \phi_2 = \frac{5\pi}{4}, \phi_3 = 7\frac{\pi}{4}$. This gives a set of four signals, 90° spaced, as shown in Fig. 10.46(b). The decision line divides the signal space into two (for $M = 2$) or four ($M = 4$) regions. The noise vector n is drawn from the head of a signal vector (s_0, say) to the nearest decision line. Note that the minimum noise vector that crosses the decision boundary is smaller in Fig. 10.46(b) than in Fig. 10.46(a). The optimum receiver for M-ary PSK produces signal samples $s_I(kT) = \sqrt{2E/T}\cos\phi_m$ and $s_Q(kT) = \sqrt{2E/T}\sin\phi_m$; the phase generation is based on $\phi_m = \tan^{-1}(s_Q/s_I)$ as shown in Fig. 10.47. But in the presence of noise, the symbol is reconstructed from the noisy samples $(s_I + n_i)$ and $(s_Q + n_q)$ and is based on $\phi_m = \tan^{-1}\left(\frac{s_Q + n_q}{s_I + n_i}\right)$. Since we generalise our treatment for M-ary case, let us consider the M-ary signal space shown in Fig. 10.48. The decision boundaries are angularly equi-spaced at $2\pi/M$ radians. The noise vector n is also sketched to indicate the minimum noise that would force the detector to make a symbol error. Note that as M increases, signal points come closer even and small amount of noise may make a symbol error. This error can be removed by increasing the signal power. We will consider here the case of a large signal-to-noise power ratio for finding the error probability. Suppose that a signal $s_1(t)$ is transmitted. The nearest two message points, s_0 and s_2, are potential candidates for being mistaken for s_1 due to noise n. Each signal point is at a distance $\sqrt{E}$ from the origin, where $E = E_b(\log_2 M)$ is the symbol energy. Referring to Fig. 10.49, the distance between the signal points is given by

$$d_{12} = \frac{d}{\sin\phi} = \frac{d}{\cos 2\pi/M} = \frac{\sqrt{E}\sin 2\pi/M}{\cos\pi/M} = 2\sqrt{E}\sin\frac{\pi}{M}. \tag{10.131}$$

So, the symbol error probability

$$P_e \leq 2 \times \frac{1}{2}\text{erfc}\left(\frac{d_{12}}{2\sqrt{\eta}}\right) = \text{erfc}\left(\frac{E}{\eta}\sin^2\pi/M\right)^{1/2}. \tag{10.132}$$

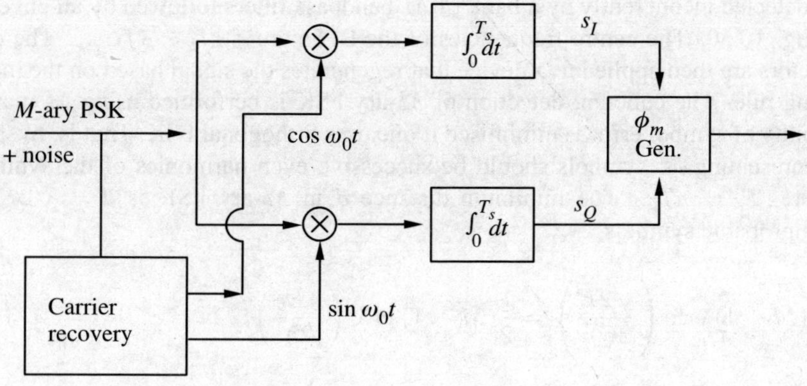

Fig. 10.47 *Coherent M-ary PSK receiver*

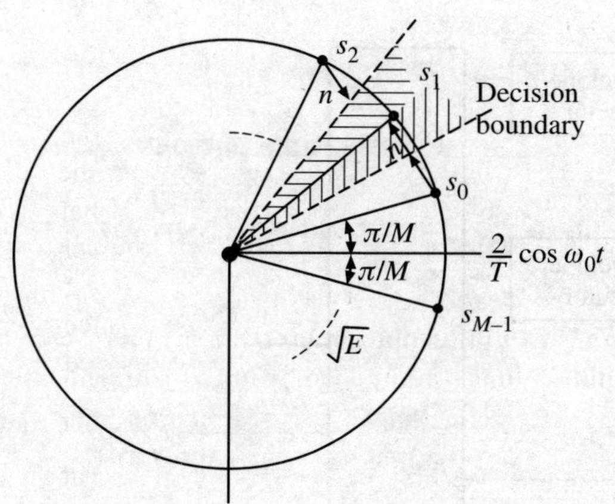

Fig. 10.48 *Signal-space diagram for M-ary PSK*

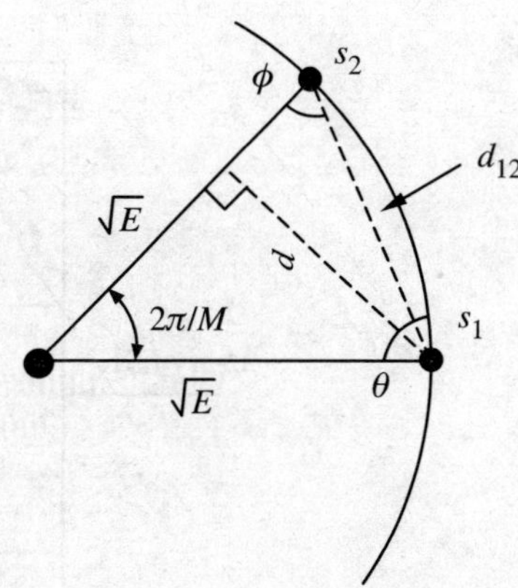

Fig. 10.49 *Minimum distance d_{12} between two signal points s_1 and s_2*

For large M, the equation reduces to

$$P_e = \text{erfc}\left(\frac{E\pi^2}{\eta M^2}\right)^{1/2} = \text{erfc}\left(\frac{nE_b\pi^2}{\eta(2^n)^2}\right). \tag{10.133}$$

10.9.2 Orthogonal M-ary FSK

For M-ary FSK, the transmitted signals are defined by (vide (10.27))

$$s_i(t) = \sqrt{\frac{2E}{T}} \cos(\omega_i t + \phi), \quad 0 \le t \le T \qquad i = 0, 1, 2, \cdots, M-1. \tag{10.134}$$

Over the symbol interval the transmitted signal is of frequency f_0, or f_1, or $\cdots$, f_{M-1}. The received signal can be detected incoherently by a bank of M bandpass filters followed by an envelope detector as shown in Fig. 10.50. The centre frequencies of the BPFs are $f_0, f_1, \cdots f_{M-1}$. The output of the envelope detectors are then applied to a device that regenerates the signal based on the maximum likelihood decoding rule. The coherent detection of M-ary FSK is performed using M matched filters.

The probability of symbol error is minimised if one uses orthogonal FSK. That is, to say, the carrier frequencies representing M symbols should be successive even harmonics of the symbol frequency $f_s = \frac{1}{T_s}$ where, $T_s = nT_b$. The minimum distance d in M-ary FSK is $d = \sqrt{2E} = \sqrt{2nE_b}$. Assuming equiprobable symbols,

$$P_e \le \frac{1}{2}(M-1) \text{ erfc}\left(\frac{\sqrt{2E}}{2\sqrt{n}}\right) = \frac{1}{2}(M-1) \text{ erfc}\left(\frac{nE_b}{2\eta}\right)^{1/2}. \tag{10.135}$$

It is interesting to observe that with increasing M (and hence n), P_e increases due to a factor $\frac{1}{2}(M-1)$ and decreases due to the argument of erfc. The goal is to realise as small a P_e as possible. The value of (E_b/η) plays an important role in reducing P_e for increasing M.

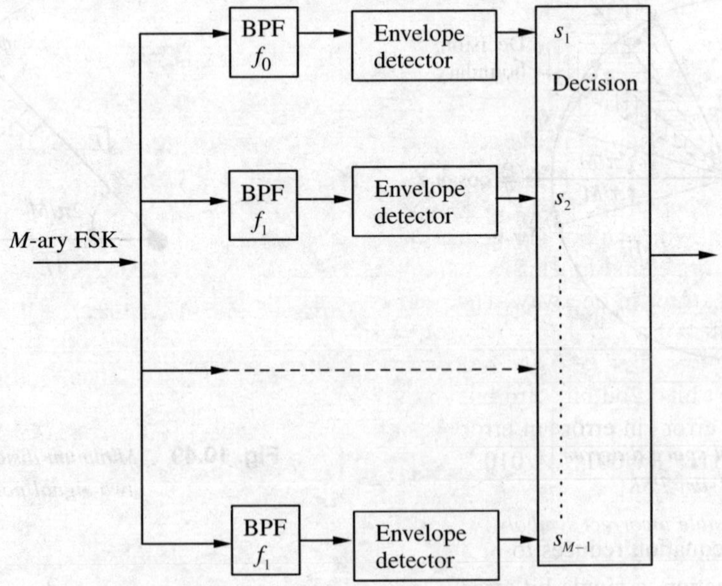

Fig. 10.50 *Incoherent decoder of M-ary FSK*

Example 10.15

Find the constraint on E_b/η that reduces P_e with increasing M.

Solution

We expect from (10.135) that the erfc decreases at a faster rate than the rate of increase of $\frac{1}{2}(M-1)$.

For large x, erfc $(x) = \dfrac{e^{-x^2}}{\sqrt{\pi x}}$ and hence

$$P_e \le \frac{2^n - 1}{2} \frac{e^{-\left(\frac{nE_b}{2\eta}\right)}}{\sqrt{\pi}\left(\frac{nE_b}{2\eta}\right)^{1/2}} = \frac{e^{n \ln 2} e^{(-nE_b/2\eta)}}{2\sqrt{\pi}\left(\frac{nE_b}{2\eta}\right)^{1/2}} = \frac{e^{-n\left(\frac{E_b}{2\eta} - \ln 2\right)}}{(2\pi n E_b/\eta)^{1/2}}.$$

Thus, P_e decreases when $E_b/2\eta > \ln 2$ or $E_b/\eta > 2\ln 2 = 1.4$.

10.9.3 Bit error probability and symbol error probability

In M-ary signalling, each symbol accommodates N bits and we need to transmit $M = 2^N$ distinct symbols over the period $T_s = nT_b$. This method is called symbol-by-symbol encoding while the binary case is referred to as bit-by-bit encoding since the two waveforms represent two possible values of a single bit. In this section, we will investigate weather M-ary signalling improves or degrades error performance. An N-bit symbol correctly received implies that all the N-bits are received correctly. When an N-bit symbol is received with error, this error may be in 1 bit, or 2 bits, or N-bits. An error in any of the bit positions of a symbol will be misinterpreted as any one of the $(M-1)$ incorrect symbols with equal probability.

As an example, let us consider that message symbols are transmitted on orthogonal 8-ary FSK waveforms. The message symbols comprise the bit patterns 000, 001, 010, 011,100,101,110, and 111, corresponding to the symbols $s_0, s_1, \cdots, s_7$ respectively.

Suppose that symbol $s_3 = 011$ was transmitted. An error in the left bit position leads to an incorrect decision, with symbol 111 being decoded. There are a total of 7 possible signals which are erroneous, as illustrated in Fig. 10.51. A single-bit error occurs in three ways, a two-bits error in three ways and a 3-bits error in one way. The probability of symbol error P_e is then a weighted average.

Transmitted symbol	Received symbols with						
	1st bit in error	2nd bit in error	3rd bit in error	1st 2bits in error	2nd, 3rd bits in error	1st, 3rd bits in error	3-bits in error
011	111	001	010	101	000	110	100

Fig. 10.51 *Possible incorrect symbols*

In a realistic situation, a single-bit error is most likely to occur in a symbol than more than one bit being erroneous. We assume that the M-ary symbols are coded in such a manner that the binary sequence corresponding to the adjacent symbols differ only in one bit position. This feature ensures that if a symbol is misread, it would be almost inevitable that either one of its two nearest neighbours was transmitted. Figure 10.52(b) illustrates the binary-to-M-ary code assignment using Grey code for an 8-PSK system. For comparison, uncoded binary assignment is also shown in Fig. 10.52(a). For a given symbol error, the most probable bit error is one, and therefore s_2 may be mistaken as either of s_1 or s_3. The average probability of symbol error P_e is thus

$$P_e = P(\text{ 1st bit in error or 2nd bit in error or } \cdots \text{ or } N\text{th bit in error})$$

$$\leq P\text{ (1st bit in error)} + P\text{ (2nd bit in error)} + \cdots + P(\ N\text{th bit in error}).$$

Each term on the right side of the above inequation is the probability of a bit error P_b (say). Thus

$$P_e \leq NP_b \quad \text{or} \quad \frac{P_e}{N} \leq P_b. \tag{10.136}$$

This gives a lower bound on P_b.

Further, in M-ary signalling, for a given transmitted symbol there are $(M-1)$ erroneous symbols that are equally likely and occur with probability $\frac{P_e}{(M-1)}$. If we study any bit position of all the symbols, we find that there are $M/2$ number of 1's and $M/2$ number of 0's. So, the bit error rate is $P_b = \frac{M/2}{M-1}P_e$, which is an overestimate. The closed form expression for P_b is then obtained as

$$\frac{P_e}{N} \leq P_b \leq \frac{M/2}{M-1}P_e. \tag{10.137}$$

For the BPSK case, $N = 1$ and $M = 2$ and so $P_e = P_b$. For QPSK, $N = 2$ and $M = 4$ and so $P_e = \frac{2}{3}P_e$. For large M, the bit error rate approaches the limit of $P_b \approx \frac{1}{2}P_e$.

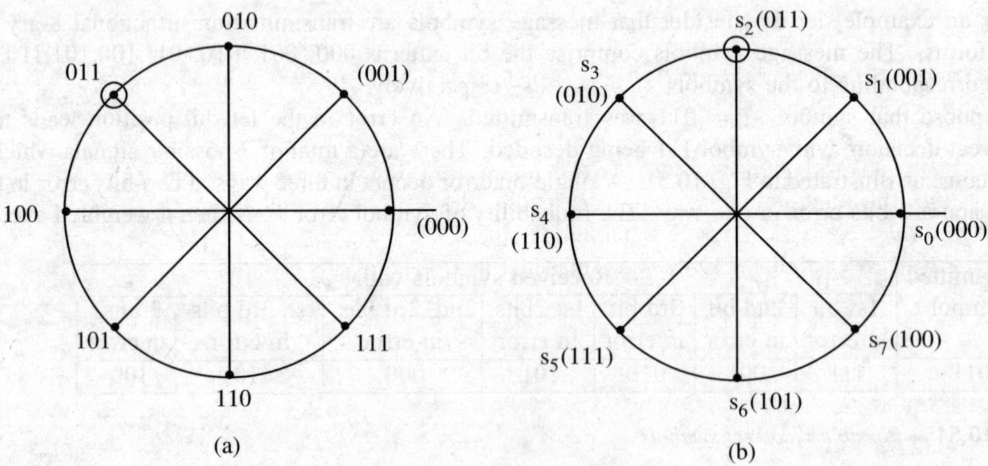

Fig. 10.52 *(a) Binary-coded M-ary, (b) Grey-coded M-ary PSK.*

Solved Numericals

1. Digital data is transmitted at the rate of 100 kbps (kilo bits per second) using binary PAM (on-off keying) signal. Assume that probability of occurrence of a 0 in the data stream is 0.4 and the average signal energy is 10 nanowatts. Estimate the amplitude of the rectangular pulse. Repeat the same if the data stream is transmitted using antipodal pulses.

Solution

A rectangular pulse of amplitude A and duration T has the energy $E = A^2T$.

We consider a 0-bit represents signal $s_1(t) = 0$ Volt and a 1-bit represents signal $s_2(t) = A$ volts. Since the data rate is 100 kbps, so duration of each bit $T = \frac{1}{r_b} = \frac{1}{100 \times 10^3} = 10\,\mu s$.

Given that the probability of occurrence of 0-bit is $p(0) = 0.4$ and that for 1-bit is $p(1) = 1 - p(0) = 0.6$.

Average signal energy

$$E_{av} = p(0) \times 0 + p(1) \times A^2T = 0.6A^2T = 10 \times 10^{-9} \text{ (given)}$$

or, $$0.6A^2 \times 10 \times 10^{-6} = 10 \times 10^{-9}$$

or, $$A^2 = \frac{10}{6} \times \frac{10 \times 10^{-9}}{10 \times 10^{-6}} = \frac{1}{600}\text{volt}^2$$

So, amplitude of the pulse $A = \frac{1}{\sqrt{600}} = 40.82\,\text{mV}$

For antipodal pulses, we assume that

0-bit $\equiv s_1(t)$ is transmitted with pulse of amplitude $A/2$

1-bit $\equiv s_2(t)$ is transmitted with pulse of amplitude $-A/2$

In either case, the energy of a pulse would be $E_P = (A/2)^2 \cdot T = \frac{A^2T}{4}$

Now, the average signal energy

$$E_{av} = p(0) \cdot \frac{A^2T}{4} + p(1) \cdot \frac{A^2T}{4} = \frac{A^2T}{4}$$

Thus,

$$\frac{A^24}{4} = 10 \times 10^{-9}$$

or, $$\frac{A^2 \times 10 \times 10^{-6}}{4} = 10 \times 10^{-9}$$

or, $$A^2 = \frac{4 \times 10 \times 10^{-9}}{10 \times 10^{-6}} = \frac{4}{1000}\text{volt}^2$$

So, pulse amplitude, $\frac{A}{2} = \frac{1}{10\sqrt{10}} = 31.62\,\text{mV}$

2. In a binary data transmission scheme, the base band signal generates data at the rate of 1Mbps. This data stream modulates a 10 MHz carrier to produce an orthogonal and minimum bandwidth BFSK signal. Find the frequencies of the BFSK signal.

Solution

For the signals to be orthogonal, the frequency separation must be $n/(2T)$ and for minimum bandwidth, we choose integer $n = 1$.

Let the carrier frequency be f_c. Then,

$$f_1 = f_c - \Delta f \text{ and } f_2 = f_c + \Delta f$$

Thus $f_2 - f_1 = 2 \cdot \Delta f = \frac{1}{2T} = \frac{r_b}{2}$.

$$\Delta f = \frac{r_b}{4} = \frac{10^6}{4} = 250\,\text{kHz}$$

So, $f_2 = f_c + \Delta f = 10\,\text{MHz} + 250\,\text{kH} = 10.25\,\text{MHz}$

and $f_1 = f_c - \Delta f = 10\,\text{MHz} - 250\,\text{kH} = 9.75\,\text{MHz}$.

3. A binary digital data at 10 Mbps is transmitted on binary FSK modulation. Assume that a 0 is transmitted using $s_1(t) = ACos(2\pi f_1 t)$ and a 1 is transmitted using $s_2(t) = A\text{Cos}(2\pi f_2 t)$ where $f_1 = 45$ MHz and $f_2 = 55$ MHz. Show that $s_1(t)$ and $s_2(t)$ are orthogonal.

Solution

For orthogonal FSK, the frequency separation must be $\dfrac{n}{2T} = \dfrac{nr_b}{2}$, where n is an integer.

$$f_2 - f_1 = \frac{nr_b}{2}$$

Since $r_b = 10$ Mbps $= 10 \times 10^6$ bits/sec, we have

$$f_2 - f_1 = (55 - 45)\text{MHz} = 10\,\text{MHz} = 10 \times 10^6.$$

Thus, $f_2 - f_1 = n \cdot \dfrac{r_b}{2}$ is satisfied for the integer $n = 2$.

Hence the signals $s_1(t)$ and $s_2(t)$ are orthogonal. This can be proved by direct computation also, as follows:

Integrating the product $s_1(t)\ s_2(t)$ over one symbol duration that is, over $T = \dfrac{1}{r_b} = 0.1\,\mu\,\text{sec}$, we have

$$I = \int_0^T Cos(2\pi f_1 t)Cos(2\pi f_2 t)dt = \frac{1}{2}\int_0^T [\text{Cos}2\pi(f_1 + f_2)t + \text{Cos}2\pi(f_2 - f_1)t]dt$$

$$= \frac{1}{2}\frac{\sin[2\pi(f_1 + f_2)T]}{2\pi(f_1 + f_2)} = \frac{1}{2}\frac{\sin[2\pi(f_2 - f_1)T]}{2\pi(f_2 - f_1)}$$

We have $2\pi(f_1 + f_2)T = 2\pi \times 100 \times 10^6 \times 0.1 \times 10^{-6} = 20\pi$ and

$$2\pi(f_2 - f_1)T = 2\pi \times 10 \times 10^6 \times 0.1 \times 10^{-6} = 2\pi$$

Thus, $\sin 2\pi(f_2 + f_1)T = 0$ and $\sin 2\pi(f_2 - f_1)T = 0$

So, orthogonality condition is satisfied. The carrier frequency is $f_c = \frac{f_2 + f_1}{2} = 50\,\text{MHz}$.

4. An M-ary PSK signal is represented by $s(t) = p(t)Cos\left[2\pi f_c t + \frac{2\pi}{M}(M-1)\right]$ where $p(t)$ is the pulse shape given by

$$p(t) = \begin{cases} A, & 0 \le t \le T \\ 0, & \text{otherwise.} \end{cases}$$

Find the minimum distance between the signal points on signal constellation diagram for $M = 4, 8$ and 16. Comment for very large M. Assume that the pulse amplitude $A = 25\,\text{mV}$ and the data rate is $r_b = 1\,\text{Mbps}$.

Solution

By Eq. (10.26 g), the minimum distance is given by

$$d_{\min} = \sqrt{E_p\left(1 - \text{Cos}\frac{2\pi}{M}\right)} = \sqrt{2E_p\text{Sin}^2\frac{\pi}{M}}.$$

Now, $E_p = \int\limits_0^T A^2\,dt = A^2 T$, where $T = \dfrac{1}{r_b} = 10^{-6}$ sec.

$$= (25 \times 10^{-3})^2 \times 10^{-6} = 625 \times 10^{-12} \text{ Joule}$$

For $M = 4$ $\quad d_{\min} = \sqrt{2 \times 625 \times 10^{-12} \times \sin^2 \times \dfrac{\pi}{4}}$

$$= \sqrt{2 \times 625 \times 10^{-12} \times \frac{1}{2}} = 25 \times 10^{-6} \text{ m.}$$

For $M = 8$ $\quad d_{\min} = \sqrt{2 \times 625 \times 10^{-12} \times \sin^2 \times \dfrac{\pi}{8}}$

$$= \sqrt{2 \times 625 \times 10^{-12} \times 0.14644} = 13.52 \times 10^{-6} \text{ m.}$$

For $M = 16$ $\quad d_{\min} = \sqrt{2 \times 625 \times 10^{-12} \times \sin^2 \times \dfrac{\pi}{16}}$

$$= \sqrt{2 \times 625 \times 10^{-12} \times 0.03806} = 6.89 \times 10^{-6} \text{ m.}$$

For large values of M, we may write $\sin\theta \simeq \theta$ and so

$$d_{\min} \approx \sqrt{2E_p \cdot \frac{\pi^2}{M^2}} = \frac{\pi}{M}\sqrt{2E_p}.$$

5. The error probability is expressed either by the Q-function or the error function (or its complementary function). Establish the relation between the Q-function and the complementary error function $erfc(x)$ defined as

$$Q(x) = \frac{1}{\sqrt{2\pi}}\int\limits_x^\infty e^{-u^2/2}du \text{ and } erfc(x) = \tfrac{2}{\sqrt{\pi}}\int\limits_x^\infty e^{-u^2}du.$$

Solution

Taking $\dfrac{u}{\sqrt{2}} = v$, we may write

$$Q(x) = \frac{1}{\sqrt{2\pi}}\int\limits_{x\sqrt{2}}^\infty e^{-v^2}.\sqrt{2}\,dv = \frac{1}{\sqrt{\pi}}\int\limits_{x/\sqrt{2}}^\infty e^{-v^2}dv = \frac{1}{2}erfc\left(\frac{x}{\sqrt{2}}\right).$$

Alternatively, $erfc(x) = \frac{2}{\sqrt{\pi}} \int_{\sqrt{2}x}^{\infty} \overline{e}^{u^2} du$, take $u = v/\sqrt{2}$ so that

$$erfc(x) = \frac{2}{\sqrt{\pi}} \int_{\sqrt{2}x}^{\infty} e^{-v^2/2} \frac{1}{\sqrt{2}} dv = 2 \cdot \frac{1}{\sqrt{2\pi}} \int_{x\sqrt{2}}^{\infty} e^{-v^2/2} dv = 2Q(x\sqrt{2})$$

The above two relations are very important in converting one form of error probability into another form. The following table may be useful for ready conversion.

Parameters	In terms of complementary error function $erfc(x)$	In terms of Q-function $Q(x)$
Bit error probability for binary case (Eq. 10.64)	$P_e = \frac{1}{2} erfc\left(\frac{v}{2\sqrt{2}\sigma_n}\right)$	$P_e = Q\left(\frac{v}{2\sigma_n}\right)$
Error probability in matched filter reception (Eq. 10.84)	$P_e = \frac{1}{2} erfc\left(\frac{s_{o2}(T) - s_{o1}(T)}{2\sqrt{2}\sigma_n}\right)$	$P_e = Q\left(\frac{s_{o2}(T) - s_{o1}(T)}{2\sigma_n}\right)$
Error probability in correlation detector (Eq. 10.93)	$P_e = \frac{1}{2} erfc\left(\sqrt{\frac{E_b(1-\rho)}{2\eta}}\right)$	$P_e = Q\left(\sqrt{\frac{E_b(1-\rho)}{\eta}}\right)$
Minimum error probability in coherent 00K system (Eq. 10.96)	$P_e = \frac{1}{2} erfc\left(\sqrt{\frac{E_b}{2\eta}}\right)$	$P_e = Q\left(\sqrt{\frac{E_b}{\eta}}\right)$
Error probability in coherent reception of BPSK signal (Eq. 10.97)	$P_e = \frac{1}{2} erfc\left(\sqrt{\frac{E_b}{\eta}}\right)$	$P_e = Q\left(\sqrt{\frac{2E_b}{\eta}}\right)$
Error probability in coherent BFSK system (Eq. 10.101)	$P_e = \frac{1}{2} erfc\left(\sqrt{\frac{E_b}{2\eta}}\right)$	$P_e = Q\left(\sqrt{\frac{E_b}{\eta}}\right)$
Symbol error probability for M-ary PSK (Eq. 10.132). The same for large M (Eq. 10.133)	$P_e = erfc\left(\sqrt{\frac{E}{\eta}\text{Sin}^2\frac{\pi}{M}}\right)$ $P_e = erfc\left(\sqrt{\frac{E\pi^2}{\eta M^2}}\right)$	$P_e = 2Q\left(\sqrt{\frac{2E}{\eta}\text{Sin}^2\frac{\pi}{M}}\right)$ $P_e = 2Q\left(\sqrt{\frac{2E}{\eta} \cdot \frac{\pi^2}{M^2}}\right)$
Error probability for orthogonal FSK (Eq. 10.135)	$P_e = \frac{1}{2}(M-1) efrc\left(\sqrt{\frac{nE_b}{2\eta}}\right)$	$P_e = (M-1)Q\left(\sqrt{\frac{E}{\eta}n}\right)$

Exercises

Activity

10.1 The complex envelope of a particular ASK modulated signal is of the form

$$v(t) = \sum_{n=-\infty}^{\infty} a_n f(t - nT_b),$$

where, $f(t)$ is an NRZ rectangular-shaped pulse $f(t) = A$ for $0 \le t \le T_b$ and a_n is unipolar binary data $a_n = \{1, 0\}$. Find the power spectral density of the complex envelope.

10.2 The data $d(t)d(t)$ consists of the bit sequence 1000 1010 1100. Assume that the bit rate r_b equals the carrier frequency f_0. Sketch the binary PSK waveform.

10.3 Given the input binary sequence 101011001110, transmitted serially at 10 kbps, sketch the in phase and quadrature components (d_e and d_o) of the modulated wave for QPSK modulation. What is the symbol rate for d_e and d_o?

10.4 The bit stream $b(t) =$ 1001 1010 1011 1100 is to be transmitted using DPSK modulation based on the scheme

$$d(k) = b(k) \oplus d(k-1).$$

Find $d(t)$. If the data $b(t)$ is in error in the fifth position from the left, find which bit positions of $d(t)$ are erroneous.

10.5 The bit stream 0110100010010111 is to be transmitted using BFSK. Sketch the waveform assuming $f_L = f_b$ and $f_H = 2f_b$, where $f_b = 1/T_b$, T_b being the bit duration.

10.6 The bit sequence 101110011010 is to be transmitted using minimum-shift keying (MSK). Sketch the in-phase and quadrature components of the MSK signals. Also, sketch the MSK waveform.

10.7 (a) What is meant by a matched filter. Derive its relation to impulse response.
(b) Find the transfer function of the matched filter for a signal that is either $s_1(t) = \cos 2\pi f_0 t$ or $s_2(t) = 0$ for an interval n/f_0, where n is any integer. The system has white noise with $G_n(f) = \eta/2$. B.P.T.U 2004.

10.8 A system transmits data at the rate of 100 kbps maintaining a probability of bit error of $P_{be} = 10^{-3}$, using M-ary PSK modulation over an AWGN channel. The system bandwidth is 50 kHz. Find the value of M. What is the value of E_b/η for the given P_{be}?

10.9 A digital PAM at the symbol rate of $r_b = 10^8$ bps, employs $M = 8$-level antipodal signalling. The peak-to-peak amplitude of the PAM signal is $A_c = 5$ Volts. Find the average symbol energy if all the symbols are equiprobable.

10.10 A data sequence $b(k) = 1010100010111011$ is transmitted using differential PSK modulator (Fig. 10.17). Find the encoded message sequence if the symbols assigned for the modulator are 0-radian for a bit 0 and π-radian for a bit 1. Assuming that there is a phase-shift of π-radian of the recovery carrier, show that the differential decoder can recover the data sequence correctly.

Review questions

1. (a) Write down an expression for the binary phase-shift keying signal. Derive its power spectrum and plot it. Sketch a scheme for the generation of BPSK signal.

   (b) What is a matched filter? Derive the relation of its transfer function to the impulse response. (MUST, Autumn 2009)

2. (a) What is a phase-shift keying signal? The bit stream $b(t) = 10110010$ is transmitted using binary PSK signalling. Sketch the waveform.

   (b) Write down the expression for $QPSK$ signalling. Sketching the signal-space points, show that $QPSK$ is a combination of two orthogonal $BPSK$ signals. (MUST, Spring 2011)

3. (a) For the M-ary Amplitude-Shift Keying (ASK) signal, draw the signal-space diagram for $M = 4$.

   (b) Write short notes on bit error probability (BEP) and symbol error probability (SEP).

4. (a) What is the advantage of CPFSK over ordinary FSK? Compare the frequency spacing of Sunde's FSK and Minimum-Shift Keying.

   (b) Under what condition will an ASK signal be an on-off keying (OOK) signal?

5. (a) What is a matched filter?

   (b) Show that the output signal-to-noise ratio of the matched filter is $2E_b/N_0$, where E_b is the signal energy per bit and N_0 is the PSD of white noise.

6. Write down the expression for the quadrature-carrier amplitude modulation (QAM) signal. Express the QAM signal as a combination of $PAM - PSK$ signal and comment on its signal constellation.

Multiple-choice questions

1. Codec (coder–decoder) uses

   (a) PCM only (b) PCM and Delta modulation
   (c) Delta modulation only (d) None of these.

2. A signal is band-limited to B Hz. In PCM, samples are taken at least every

   (a) B sec (b) 2 B sec (c) $\frac{1}{B}$ second (d) $\frac{1}{2B}$ second.

3. In QAM both phase and .... of the carrier signal is varied.

   (a) Bit rate (b) Amplitude (c) Frequency (d) None

4. If the baud rate of a 4-ary PSK is 400, the bit rate is

   (a) 400 bps (c) 100 bps (c) 800 bps (d) None

5. In a BPSK modulation on a carrier frequency of 80 MHz and input bit rate of 10 Mbps, the minimum Nyquist bandwidth is

   (a) 10 MHz (b) 20 MHz (c) 40 MHz (d) None

6. ASK signal has the bandwidth of

   (a) Half of the baseband signal (b) Same as the baseband signal
   (c) Double the baseband signal (d) Depends on carrier frequency

7. Let W_{BPSK} and W_{BFSK} be the bandwidth of BPSK and BFSK signals, respectively. Then,

   (a) $W_{BPSK} > W_{BFSK}$ (b) $W_{BPSK} < W_{BFSK}$
   (c) $W_{BPSK} = W_{BFSK}$ (d) Nothing can be said

8. Maximum bandwidth is occupied by which modulation?

   (a) ASK (b) BPSK (c) BFSK (d) None

9. In QPSK modulation, each symbol consists of

   (a) 1 bit (b) 2 bits (c) 3 bits (d) 4 bits

10. In minimum-shift keying (MSK), the difference between the higher and lower frequency is

    (a) ½ of the bit rate (b) same as bit rate (c) two times the bit rate (d) None

11. Differential PSK (DPSK) uses the line code

    (a) Differential code (b) Manchester code (c) Polar NRZ (d) Bipolar NRZ

12. How many carrier frequencies are used in BPSK modulation?

    (a) 2 (b) 1 (c) No carrier (d) none

13. In coherent BFSK transmission at a bit-rate of 8 kbps, the best possible values of the transmitted frequencies would be

    (a) 24 kHz, 30 kHz (b) 32 kHz, 40 kHz (c) 24 kHz, 40 kHz (d) 32 kHz, 42 kHz

14. Coherent orthogonal BFSK modulation is used to transmit two equiprobable symbol waveform $s_1(t) = \alpha\, Cos 2\pi f_1(t)$ and $s_2(t) = \alpha\, Cos 2\pi f_2(t)$, where $\alpha = 4$ mV. Assume an AWGN channel with two-sided noise PSD $\dfrac{N_0}{2} = 0.5 \times 10^{-12} \mathrm{W}/Hz$. Using an optimal receiver and the relation $Q(v) = \dfrac{1}{\sqrt{2\pi}} \int_v^{\infty} e^{-u^2/2} du$, the bit error probability for a data rate of 5000 kbps is [GATE 2014]

    (a) $Q\,(2)$ (b) $Q\,(4)$ (c) $Q\left(2\sqrt{2}\right)$ (d) $Q\left(4\sqrt{2}\right)$

15. At a given probability of error, binary coherent FSK is inferior to binary coherent PSK by

    (a) 6 dB (b) 3 dB (c) 2 dB (d) 0 dB

11 Data Encryption and Decryption

11.1 Introduction

Security of information transmission has been an important area of concern for humankind since the earliest of times. It is important not only in personal matters but in other spheres like commerce, military, business. Cryptography, evolved to deal with secure communications, endeavours to hide or make messages unintelligible to all except the intended users. The term is derived from the Greek word Kryptos meaning 'hidden'. The subject cryptography deals with the techniques of ensuring secrecy, authenticity, and integrity of information in communications.

'Secrecy' refers to the denial of information extraction from the channel (eavesdropping) to unauthorised persons. Authentication ensures prevention of unauthorised persons injecting information into the channel (spoofing). In other words, authenticity verifies the validation of source signature. The term 'integrity' refers to the fact that the message was not modified by intruders. All the above three services are provided in a cryptosystem by means of a process called 'encryption coding'[1]. Encryption is the transformation of an original 'plaintext' message into coded from, called the 'ciphertext'. This transformation must be invertible so that the original message can be decrypted in an inverse transformation. [2] Two very popular encryption schemes are the data encryption standard (DES) and the Rivest–Shamir–Adleman (RSA) systems. Due to the availability of fast computing devices today, most modern cryptosystems involve mathematical ciphers based on modular arithmetic. In the next section, we highlight some of the important concepts of modular mathematics.

11.1.1 Modular arithmetic

Consider any positive integer N and any integer M. If we divide M by N, we get an integer quotient Q and an integer remainder R, obeying the relation

$$M = QN + R, \quad 0 \leq R < N.$$

[1] The first element of a cryptosystem is the original set of information called the 'plaintext', which can be made unintelligible to all but the authorised reader. This is the information that we want to keep private. For example, a person's bank account number, security number, etc.

[2] The transformation is the algorithm commonly known as 'cipher, which makes the information unintelligible to the unintended person. This is the 'ciphertext', which can be sent out over a public communications channel without the fear of anybody's interference.

For example, $M = 13, \quad N = 7; \quad 13 = 1 \times 7 + 6; \quad R = 6.$

$$M = -13, \quad N = 7; \quad -13 = (-2) \times 7 + 1; \quad R = 1.$$

We define M mod N to be the remainder when M is divided by N. That is, M mode $N = R$ means that the integer M when divided by a positive integer N returns a residue R. Thus,

$$13 \bmod 7 = 6, \quad -13 \bmod 7 = 1.$$

There are certain algebra that finds some short cuts to evaluate modulo results. We may use the symbolic notation N/M to mean that N divides M with no remainder. Here N is a divisor or M. Two integers A and B are said to be congruent modulo N if (A mod N)=(B mod N) and this is written as $A \equiv B \bmod N$. For example,

$$83 \bmod 11 = 5; \ 5 \bmod 11 = 5; \quad \text{So,} \quad 83 \equiv 5 \bmod 11.$$

$$17 \bmod 7 = 3; \ -11 \bmod 7 = 3; \quad \text{So,} \quad 17 \equiv -11 \bmod 7.$$

The modulo operator has the following properties:

1. $A \equiv B \bmod N$ if $N/(A - B)$.
2. $A \equiv B \bmod N$ implies that $B \equiv A \bmod N$.
3. $A \equiv B \bmod N$ and $B \equiv C \bmod N$ imply that $A \equiv C \bmod N$.

Addition rule

- $[(A \bmod N) \pm (B \bmod N)] \bmod N = (A \pm B) \bmod N.$

  (a) First add the two numbers, then reduce.

  (b) First reduce each term, then add.

  For example, we find (11+17) mode 6 as below.

  (a) $(11 + 17) \bmod 6 = 28 \bmod 6 = 4$

  (b) $(11 + 17) \bmod 6 = [(11 \bmod 6) + (17 \ \bmod 6)] \ \bmod 6 = (5 + 5) \bmod 6$
  $= 10 \bmod 6 = 4$

Multiplication rule

- $[(A \bmod N) \times (B \bmod N)] \bmod N = (A \times B) \bmod N.$

  (a) First multiply the two numbers, then reduce.

  (b) First reduce each term, then multiply.

  As an illustration, we find $13 \times 17 \bmod 6$.

  (a) $(13 \times 17) \bmod 6 = (221) \bmod 6 = 5$

(b) $(13 \times 17) \bmod 6 = [(13 \bmod 6) \times (17 \bmod 6)] \bmod 6 = (1 \times 5) \bmod 6 = 5.$

Exponentiation rule

- Apply repeated multiplication and follow multiplication rule.

Algorithm for exponential rule:

(a) $17^3 \bmod 7 = 17 \times 17 \times 17 \bmod 7 = [(17 \bmod 7) \times (17 \bmod 7) \times (17 \bmod 7)] \bmod 7$

$= (3 \times 3 \times 3) \bmod 7 = 27 \bmod 7 = 6.$

(b) $2^{10} \bmod 11 \equiv 1024 \bmod 11 = 1.$

Alternately,

$$2^{10} \bmod 11 = (2^5)^2 \bmod 13 = 32 \times 32 \bmod 11$$

$$= 10 \times 10 \bmod 11 = 100 \bmod 11 = 100 \bmod 11 = 1.$$

(c) $11^{13} \bmod 7 \equiv (11^2)^6 \times 11 \bmod 7 = 2^6 \times 4 \bmod 7 = 4 \times 4 \bmod 7 = 4.$

Alternately, $11 \bmod 7 = 4$

$$11^2 = 121 = 2 \bmod 7$$

$$11^4 = 2^2 = 4 \bmod 7$$

$$11^8 = 4^2 = 2 \bmod 7$$

$$11^{13} = (11^8 \times 11^4 \times 11) \bmod 7 = [(2 \bmod 7) \times (4 \bmod 7) \times (11 \bmod 7)] \bmod 7$$

$$= (2 \times 4 \times 4) \bmod 7 = 4.$$

The reduction of a number M by an integer N returns the remainder R. For example, the reduction of 23 by 19 is 4. We thus say that 23 is equivalent to 4 modulo 19 or $23 \equiv 4 \bmod 19$.

Another example: $15 \equiv 4 \bmod 11$.

Two important short cuts are

$(a)(b) \bmod N \equiv (a \bmod N)(b \bmod N)$

$a^m \bmod N \equiv (a \bmod N)^m.$

Examples:

$(12)(9) \bmod 7 = (12 \bmod 7)(9 \bmod 7) = (5)(2) \bmod 7 = 10 \bmod 7 = 3$

and $(11)^{13} \bmod 7 = (11 \bmod 7)^{13} = 4^{13} \bmod 7$

$$= (4^2)^6 \times (4^1) \bmod 7 = 2^6 \times 4 \bmod 7 = 1 \times 4 \bmod 7 = 4.$$

11.2 Model of a Cryptographic Channel

Figure 11.1 illustrates the model of a cryptographic channel which enciphers a plaintext message x into ciphertext y. The transformation can be modelled by the analytic expression

$$y = f(x, k), \tag{11.1}$$

where f denotes the encryption transformation. The parameter k is a set of symbols or characters called the key, which dictates the specific encryption transformation. The ciphertext is transmitted over public channel and the recipient deciphers the ciphertext by applying an inverse transformation as follows:

$$x = f^{-1}(x, k). \tag{11.2}$$

In the 'secret-key encryption', the key k must be the same for the sender and the recipient. The key is transmitted to the authorised users over a secure channel (another communication channel). Any intruder (cryptanalyst) can decrypt the message if the key information is known to him. Therefore, to ensure secure communication the key must be kept secret so that an intruder is unable to produce an estimate of the plaintext, $\widehat{x}$. Periodical change of the key during data transmission may avert intruder's interception.

Suppose we define a transformation $y = f(x, k)$ as an exclusive -OR operation. Let the plain text message x be an 8-bit word.

$$\begin{array}{rcl} x & : & 0\,1\,0\,0\,1\,0\,1\,0 \\ k & : & 1\,0\,1\,1\,1\,1\,0\,1 \\ \hline y = x \oplus k & : & 1\,1\,1\,1\,0\,1\,1\,1 = \text{encrypted message} \end{array}$$

On receiving this ciphertext, the recipient performs the decryption algorithm by using the same key.

$$\begin{array}{rl} y: & 1\,1\,1\,1\,0\,1\,1\,1 \\ k: & 1\,0\,1\,1\,1\,1\,0\,1 \\ \hline x: & 0\,1\,0\,0\,1\,0\,1\,0 = \text{original plain text.} \end{array}$$

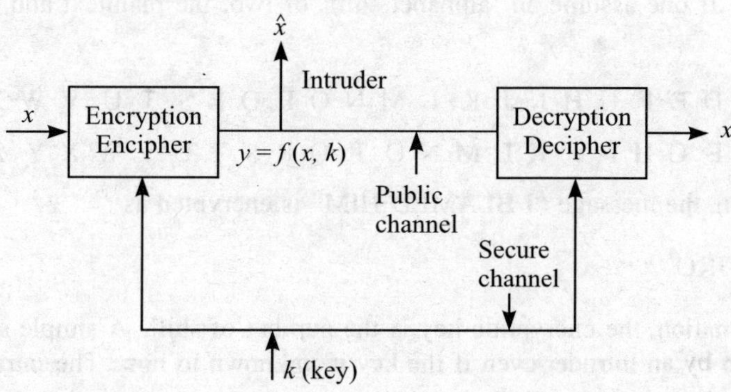

Fig. 11.1 *Illustrates the model of conventional cryptosystem.*

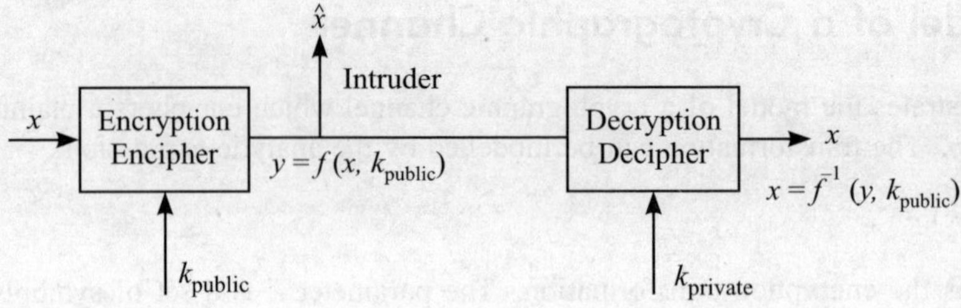

Fig. 11.2 *Public-key system*

In public-key cryptosystem, shown in Fig. 11.2, the sender and the recipient have two sets of private and public keys, k_{public} and k_{private}. The ciphertext is then

$$y = f(x, k_{\text{public}}). \tag{11.3}$$

The recipient decrypts the message using his own 'private key' k_{private} as

$$x = f^{-1}(y, k_{\text{private}}). \tag{11.4}$$

Equations (11.3) and (11.4) dictates that the inverse of each function can be computed if we know the other key.

11.3 Block and Stream Ciphers

Cryptographic ciphers fall into two broad categories: block ciphers and stream ciphers. A stream cipher is one that encrypts a data stream one bit or one byte at a time. On the other hand, in block encryption the plaintext (serial data) is divided into blocks of fixed number of data bits. Each block is enciphered using the same secret key and the cipher block is finally transmitted in a serial form. Typically, a block size of 64 bits or 128 bits is used. The cipher block is of equal length. To clarify the block cipher, let us look at two elementary operations, 'substitution' and 'permutation' which denote the key and the transformation. In substitution, each plaintext letter is converted into a new letter by some rule. If one assume an 'alphabet shift' of two, the plaintext and ciphertext symbols read as

Plaintext A B C D E F G H I J K L M N O P Q R S T U V W X Y Z

Ciphertext C D E F G H I J K L M N O P Q R S T U V W X Y Z A B

With this substitution, the message "I BLAMED HIM" is encrypted as

K–DNCOGF–JKO

In the shift transformation, the encryption key is the number of shift. A simple shift transformation can easily be broken by an intruder even if the key is unknown to him. The intruder simply checks at most 25 different shifts to read out the message.

Now we put the characters in fixed block lengths of four symbols and permute the text by some rule. If one follows a right-circular shift (end-around shift in each block) by two, the plaintext word ABCD becomes CDAB and the message 'I blamed him' encrypts as follows:

Original message:	I BLAMED HIM
After alphabet shift:	K- DNCOGF- JKO (hyphen indicates blank space)
Block formation	K-DN COGF - JKO
After permutation	DNK- GFCO KO-J

Thus the message 'I-BLAMED-HIM' is hidden in the block cipher 'DNK-GFCOKO-J'.

In the data stream encryption there is no fixed block size and each individual plaintext bit x_i is enciphered with a time-varying key element k_i of the key-stream. The most popular stream cipher is the one based on modulo-2 addition, where the secret key is used to generate data stream (key stream) that is added to the plaintext bit-wise to generate the ciphertext bit. If the message is represented by the sequence $\{x_1x_2x_3\ldots\}$ and the key is represented by the key stream $\{k_1k_2k_3\ldots\}$, the resultant cipher $\{y_1y_2y_3\ldots\}$ is as follows:

$$y_i = x_i \oplus k_i \quad i = 1, 2, 3, \cdots. \tag{11.5}$$

For decryption, the ciphertext y_i is subjected to modulo-2 addition with the key-stream bits to recover the plaintext bit x_i as

$$x_i = y_i \oplus k_i \quad i = 1, 2, 3\cdots. \tag{11.6}$$

The following example illustrates the encryption and decryption based on stream ciphers.

Message	:	0 1 1 1 0 1 0 0	
Key	:	0 1 0 0 1 0 0 1	
Cipher	:	0 0 1 1 1 1 0 1	← Encryption
Cipher	:	0 0 1 1 1 1 0 1	
Key	:	0 1 0 0 1 0 0 1	
Message	:	0 1 1 1 0 1 0 0	← Decryption

11.4 Secrecy of Cipher Systems

The secrecy of the cipher systems lies in the secrecy of the key parameter. Except the secret key, the mechanism of encryption in a cipher systems is known to an intruder (cryptanalyst). The intruder may take advantage of certain statistical structure of the text language. For example, the letter 'e' occurs more frequently than others; letters x and y occur rarely, and the letter 'q' is always followed by a u. The knowledge of some computer formats, namely, the words 'login' and some common formats of the message ('Dear sir', for example) also facilitates the cryptanalyst to break the cipher code. The following attacks may be attempted by the intruder in order to retrieve the message context.

(a) Ciphertext-only attack is the weakest cryptanalytic threat on a system. The intruder has some knowledge of the mechanism and the language used; only a part of the data is available to him.

(b) Known plaintext attack is more serious threat than above. The intruder has the knowledge of some plaintext and its corresponding ciphertext.

(c) Chosen plaintext attack is a kind of attack attempted by the intruder when he has gained enough knowledge of the plaintext matter and can select the important plaintext. In this attack the cryptanalyst submits a plaintext message in the hope to get the ciphertext in return.

(d) Chosen ciphertext attack is another threat in which the enemy chooses an arbitrary ciphertext and decrypts the correct plaintext.

The safety of secure communication fails if the enemy cryptanalyst can break the codes of encryption, thereby obtaining the unique solution to the cryptogram. The intruder, therefore, makes an attempt at finding the secret key with the knowledge of ciphertext and *a priori* probabilities of the plaintext matter. The amount of information in a message is related to the probability of occurrence of the message. Let $x = (x_1 x_2 ... x_n)$ and $y = (y_1 y_2 ... y_n)$ denote an n-bit plaintext and an n-bit ciphertext messages, respectively. The uncertainty about x is expressed by its entropy $H(x)$. The uncertainty that x was sent, having received y is expressed by the conditional entropy (also called equivocation) $H(x/y)$. We define the mutual information between x and y as

$$I(x; y) = H(x) - H(x|y). \tag{11.7}$$

This gives a measure of secrecy of the cryptosystem. Perfect security of the cryptosystem means that $I(x, y) = 0$, that is, $H(x) = H(x/y)$ implying that the plaintext x and its cryptogram y are statistically independent. The intruder here cannot find a way to link the ciphertext to the plaintext matter.

11.4.1 Uniticity distance and perfect security

We already stated that perfect security requires that the mutual information between the plaintext x and the ciphertext must be zero. To decrypt x, we need both y and k together. We recall that

$$H(x|y) \le H(x, k|y) = H(k|y) + H(x|y, k). \tag{11.8}$$

The term $H(x|y, k)$ is zero and so (11.8) reduces to

$$H(x|y) \le H(k|y) \le H(k). \tag{11.9}$$

The quantity $H(k/y)$ is the key equivocation which determines the size of the key required to decrypt the ciphertext.

For a messages of length n, the average number of bits in each character, r (also called true rate of a language), is given as

$$r_1 = \frac{H(x)}{n}, \tag{11.10}$$

where $H(x)$ is the plaintext entropy. For a ciphertext alphabet size of L, the maximum entropy of the language (also called absolute rate) is given by

$$r_2 = \log_2 L \tag{11.11}$$

provided that the characters are equally likely to occur. In this context, we may define the 'redundancy' r of the message information in an n-bit cipher as the difference between the absolute rate and the true rate. Thus,

$$r = r_2 - r_1 = \log_2 L - \frac{H(x)}{n} \tag{11.12}$$

Returning to the intruder's attempt in breaking the code, one can expect that it is possible to break the cipher if the enemy cryptanalyst studies a large number of ciphertext. In this connection we can define the 'uniticity distance' as the smallest number of ciphertext such that the key equivocation $H(k|y_1y_2 \dots y_N)$ is close to zero. In other words, the uniticity distance is the amount of ciphertext needed to uniquely determine the key and thereby break the ciphertext. In some cryptosystems it is possible to approximately calculate the uniticity distance as given by Shannon and Hellman. The derivation of the uniticity distance is based on a particular kind of 'random cipher', which states that for each key k and ciphertext y, the deciphering operation $f^{-1}(y, k)$ yields an independent variable distributed over the possible message space. The resulting uniticity distance n_0 is approximately given by

$$n_0 = \frac{H(k)}{r \log L}, \tag{11.13}$$

where $H(k)$ is the entropy of the key. If the ciphertext message is formed from a finite alphabet of L symbols and the key is constructed with N symbols chosen from the ciphertext symbol, then,

$$H(k) \leq \log_2 L^N = N \log_2 L. \tag{11.14}$$

The equality sign applies for the key being completely random. Then,

$$n_0 = \frac{N \log_2 L}{r \log_2 L} = \frac{N}{r}. \tag{11.15}$$

To illustrate the application of (11.15), consider the English language with $r_2 = \log_2 26 = 4.7$ bits/character and $r_1 = 1.5$ bits/character so that $r = 4.7 - 1.5 = 3.2$ bits/character. Thus percentage redundancy $= \frac{r}{r_2} = \frac{3.2}{4.7} = 0.68 \approx 0.7$. Following (11.15), $n_0 = 1.43k$, which means that the intruder can break the cipher by studying $1.43k$ number of ciphertext bits. Thus the ciphertext sequence can be breakable at best in principle. But it is computationally secure in the sense that for very large key size, it needs an exhaustive test requiring large computing times. For example, if the key is 56 bits-wide, there will be 2^{56} possible keys. One may reach the goal after 2^{56} trials on the average. If we assume that each test requires a computation time of 1 μs, the total search time exceeds 1000 years.

11.4.2 Permutation and substitution

In a permutation or transposition cipher, the plaintext is divided into groups of fixed size and permutation is applied to each group, that is, the positions of the plaintext alphabets in a group are simply rearranged following certain permutation rule dictated by the secret key. This rearrangement or permutation can be performed in the P-box structures shown in Fig. 11.3. The operation can be straight, expanded, or compressed. Each P-box contains some input lines ($I_0, I_1, \ldots$) and some output lines ($O_0, O_1 \cdots$).

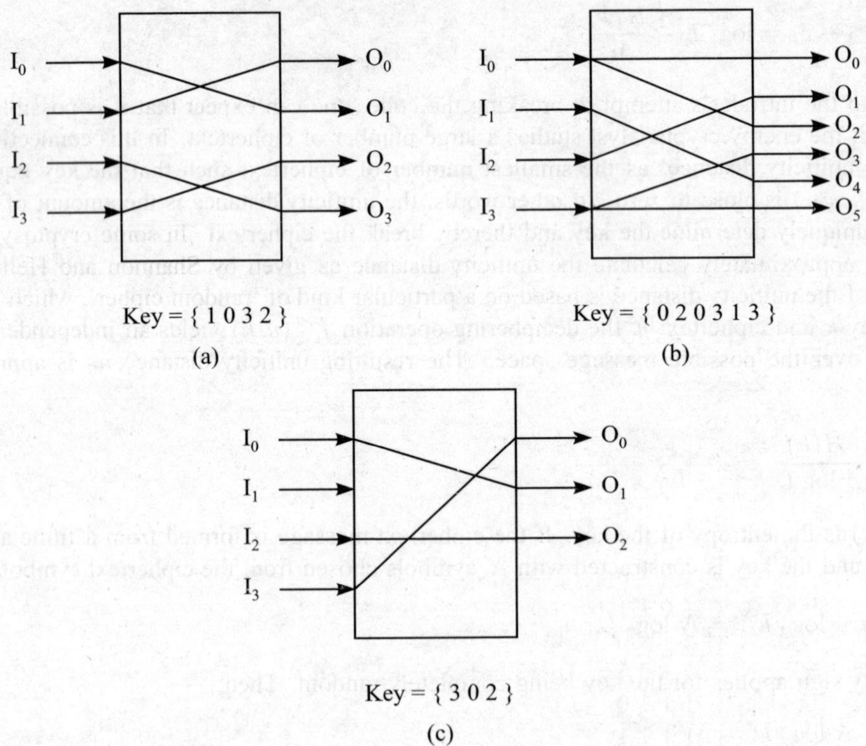

Fig. 11.3 *Illustrates three permutations operations: (a) straight, (b) expanded, and (c) compressed.*

For straight permutation, each output alphabet comes from any of the input alphabets. In the P-box of Fig. 11.3(a), the output O_0 comes from input I_1, output O_1 comes from input I_0, output O_2 comes from input I_3 and output 0_3 comes from input I_2. The key here is the sequence { 1 0 3 2 }. The word BULK might appear as UBKL. In an expanded permutation shown in Fig. 11.3(b), the P-box maps a single input to more than one output. For example, outputs O_0 and O_2 come from the single input I_0. For the key { 0 2 0 3 1 3 } the word BULK might appear as the ciphertext BLBKUK. In the compressed permutation some inputs are left out. Input I_1 of Fig. 11.3(c) is left out. The permutation operation has one major disadvantage; the enemy cryptanalyst having access to the plaintext attack, can find ways to reveal the internal connections of the P-box by the trick message, as illustrated in Fig. 11.4. If the trick message $I_0I_1I_2I_3 = 1000$ be applied at the input, the ciphertext $O_0O_1O_2O_3 = 0100$ revealing that the input line I_0 is connected to the output line O_1. In this way the intruder can find all the interconnections by transmitting a sequence of such messages, namely, 0100, 0010, 0001.

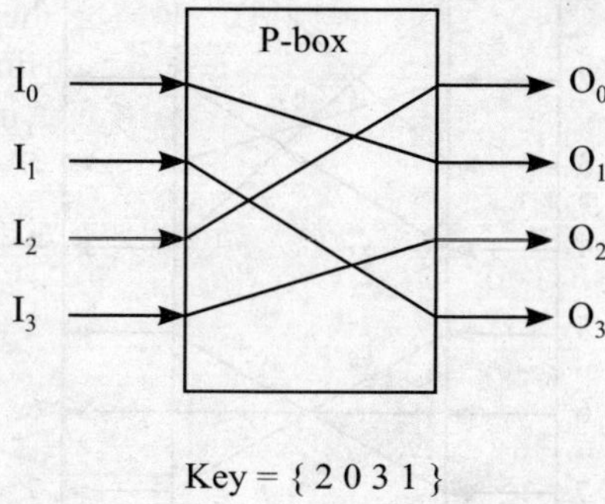

Fig. 11.4 *Illustrates the hardware connections to be revealed to an intruder by some trick message*

Example 11.1

Find the ciphertext that the P-box of Fig. 11.4 performs for all input plaintext messages. The P-box is characterised by the permutation rule $\{2031\}$.

Solution

Input	Output	Input	Output
0 0 0 0	0 0 0 0	1 0 0 0	0 1 0 0
0 0 0 1	0 0 1 0	1 0 0 1	0 1 1 0
0 0 1 0	1 0 0 0	1 0 1 0	1 1 0 0
0 0 1 1	1 0 1 0	1 0 1 1	1 1 1 0
0 1 0 0	0 0 0 1	1 1 0 0	0 1 0 1
0 1 0 1	0 0 1 1	1 1 0 1	0 1 1 1
0 1 1 0	1 0 0 1	1 1 1 0	1 1 0 1
0 1 1 1	1 0 1 1	1 1 1 1	1 1 1 1

In substitution encryption technique, each plaintext letter is replaced by a fixed letter in the alphabet set. A simple example is as follows.

Plaintext	A B C D	E F G H	I J K L	M N O P	Q R S T	U V W X	Y Z
Substitute text	E I M Q	U Y A F	J N R V	Z B G K	O S W C	H L P T	X D

The plaintext message 'CIPHERTEX' might be transformed into the substituted ciphertext message 'MJKFUSCUTC'. The substitution box (S-box) structure is shown in Fig. 11.5. It consists of an n to 2^n encoder followed by a P-box that permutes the 2^n characters into another set according to some key sequence. The characters are then converted back to an n-bit ciphertext in a 2^n-to-n decoder. For the given key sequence $\{3, 0, 4, 2, 1, 7, 6, 5\}$, the input and output symbols are as follows.

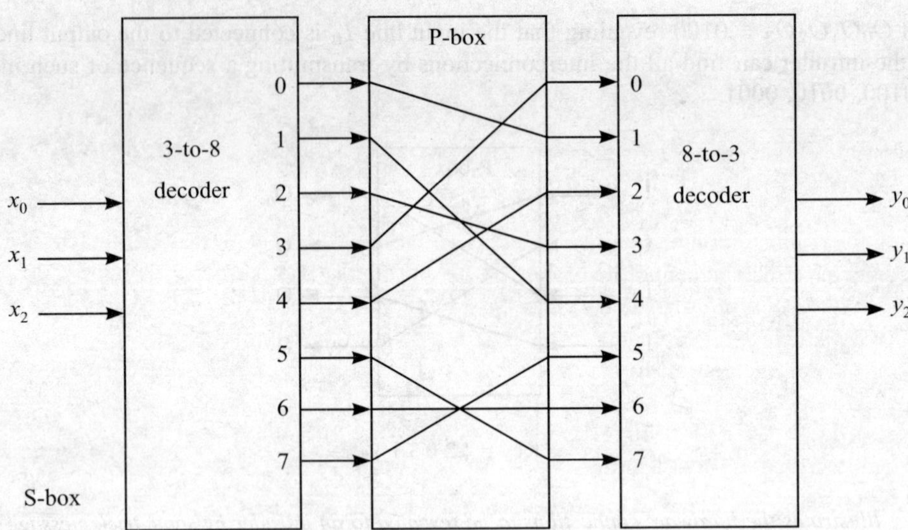

Key = {3, 0, 4, 2, 1, 7, 6, 5}

Fig. 11.5 *Substitution box*

Input $x_2x_1x_0$	0 0 0	0 0 1	0 1 0	0 1 1	1 0 0	1 0 1	1 1 0	1 1 1
Output $y_2y_1y_0$	0 0 1	1 0 0	0 1 1	0 0 0	1 0 1	1 1 1	1 1 0	1 0 1

For an input symbol $x_2x_1x_0 = 000$, the 3-to-8 decoder activates the output line marked '0' which permutes into line 1 in the P-box. The 8-to-3 decoder gives the output $y_2y_1y_0 = 001$. In this way all the other symbols can be generated. Unlike the P-box, the S-box performs a nonlinear transformation which can be verified as follows. Let the S-box perform the transformation defined by the function $f(X)$. We transform two symbols X_1 and X_2 and superpose the results giving

$$Y = f(X_1) \oplus f(X_2). \tag{11.16}$$

Next we superpose the symbols and transform giving

$$Y' = f(X_1 \oplus X_2). \tag{11.17}$$

If the two ciphertexts Y and Y' are identical for all the inputs, the transformation is indeed linear; otherwise it is nonlinear. In an attempt to verify the statement, let $X_1 = 001$ and $X_2 = 101$. Then,

$$Y = f(011) \oplus f(101) = 000 \oplus 111 = 111.$$

$$Y' = f(011 \oplus 101) = f(110) = 110.$$

So the substitution performs nonlinear transformation.

The P-box and S-box ciphers offer little encryption protection. To increase the security of the cipher, a more complex structure is required. This is the 'product cipher' system which is a combination of the S-box and P-box transformations. The Date Encryption System (DES) is based on product cipher.

11.4.3 Confusion and diffusion

A good ciphertext must counter security problems. Elaborate attempts have been made in designing practical cipher systems that cannot be broken by the intruders. The enemy cryptanalyst tries to reveal the plaintext. Two encryption concepts, namely, 'confusion' and 'diffusion' are adopted that aim at frustrating a statistical analyst. In the method of confusion encryption, the data transformations are so designed to make the relationship between the cipher and the plaintext so complex that most of the key is required to decrypt even for small sequence of the ciphertext. This will frustrate the intruder who may most likely give up trying to break the cipher. Diffusion encryption renders the ciphertext bits statistically similar, so that a statistical analysis fails in decrypting the message.

We have discussed earlier that a cryptanalyst can take advantage of the knowledge of the frequency distribution of the various letters in a given language. If this statistics is reflected in the ciphertext, the intruder may be able to find the encryption key used. If the statistical structure of the plaintext be removed in the ciphertext, the cryptanalyst would find no linkage between the two texts and thereby would not be able to break the key. This can be done by having each plaintext digit to affect a number of ciphertext digits. In other words, a given ciphertext digit is affected by many plaintext digits. In this way the letter frequencies in the ciphertext will be rendered more or less uniform compared to the plaintext.

11.5 Data Encryption Standard

The data encryption standard (DES) is a block encryption system based on a single key concept. It was originally conceived by Shannon in 1940, developed by 'Diffie and Hellman' to become one of the most widely used algorithms of today. The DES is based on the 'product cipher' which uses the P-box and S-box ciphers alternately.

The DES algorithm operates on 64-bit blocks of plaintext data and encrypts with a 56-bit key. If the plaintext block is 128-bit wide, each 64-bit block would be enciphered separately. The resulting ciphertext is 64-bit wide, as shown in Fig. 11.6. Since there are 2^{56} possible keys, it requires tremendous effort to break the cipher. Decryption is done by running the data in the reverse way using the inverse of the S-boxes. The DES algorithm though complex, is however very straightforward. Figure 11.7 illustrates the flowchart of the DES process. The steps of the DES algorithm are as follows.

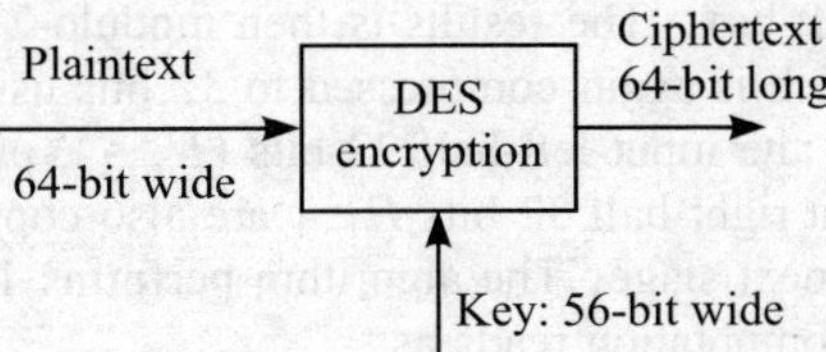

Fig. 11.6 *DES system*

(a) The algorithm starts with an 'initial permutation' (IP) stage of the 64-bit plaintext, using a P-box. The permuted plaintext bits are split into two 32-bit left-half bits (L_0) and 32-bit right-half bits (R_0).

(b) Each half is expanded into 48 bits and substituted with a 48-bit subkey. The halves are then shifted using a permutation through an exclusive-OR. This process is known as a round. To complete the algorithm, this process is repeated 16 times.

(c) After 16-rounds, the two 32-bit outputs undergo a permutation which is the inverse of the initial permutation. This results in a 64-bit ciphertext.

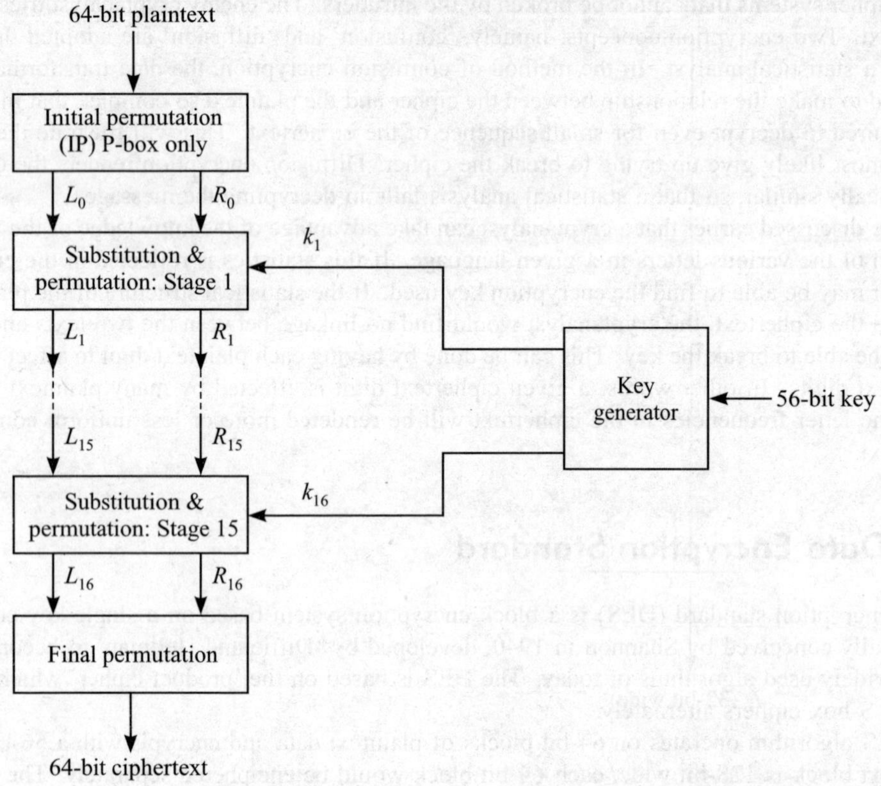

Fig. 11.7 *Data encryption standard flowchart*

The heart of the scheme is the 'substitute and permute' block, explained in Fig. 11.8.

At the ith block, the input right-half 32 bits (R_{i-1}) from the previous stage are expanded and transformed to 48-bits using a P-box. The results is then modulo-2 summed with a 48-bit subkey k_i. This forms 48-bit word which is again compressed to 32 bits using an S-box. The 32-bit word is then modulo-2 summed with the input left-half 32-bits (L_{i-1}), producing the right-half 32 bits, R_i for the next stage. The input right-half 32 bits R_{i-1} are also copied unchanged, which becomes the left-half 32 bits L_i for the next stage. The algorithm performs 16 such rounds of computation. Mathematically, the ith-round computation reads as

$$L_i = R_{i-1} \tag{11.18}$$

and $R_i = L_{i-1} \oplus f(R_{i-1}, k_i),$ (11.19)

where i runs from 1 through 16. After sixteen iterations, the result is input into a final permutation to produce a 64-bit ciphertext.

For decryption, the above-stated algorithm is used but the subkey sequence in this case is taken in the reverse order.

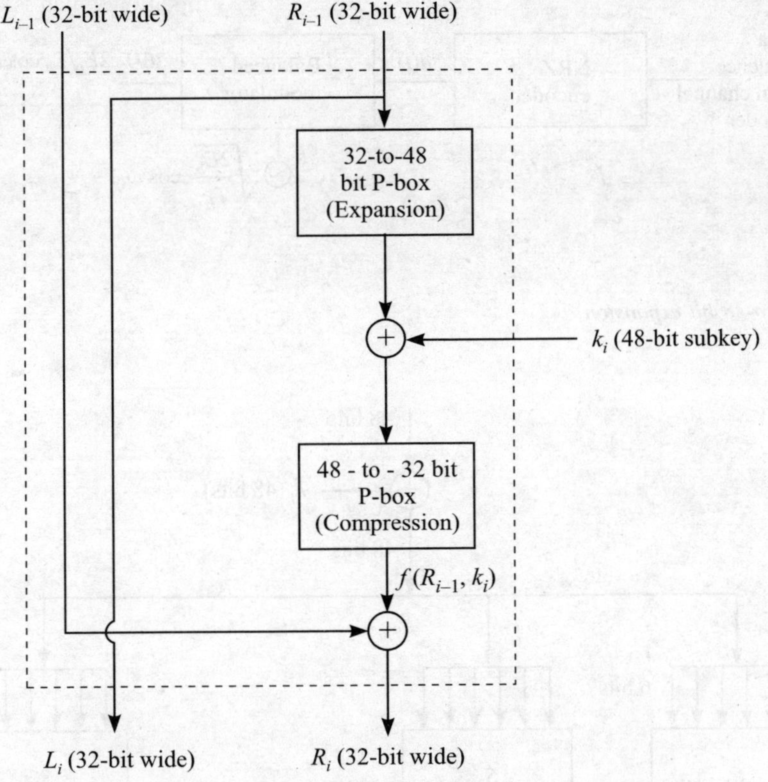

Fig. 11.8 *Standard building block: substitute and permute*

The 32-to-48- bit expansion block of Fig. 11.9 works as follows. The 32-bit block R is first grouped to eight 4-bit words. That is,

$$R = \underbrace{R_1 R_2 R_3 R_4}_{\text{1st 4bit word}} \quad \underbrace{R_5 R_6 R_7 R_8}_{\text{2nd 4bit word}} \quad \cdots\cdots \quad \underbrace{R_{29} R_{30} R_{31} R_{32}}_{\text{8th 4bit word}}$$

Each of the 4-bit words are expanded into 6-bit words by repeating the edge bits. The second 4-bit word, $R_5R_6R_7R_8$, after expansion takes the form $R_4R_5R_6R_7R_8R_9$. The complete 48-bit word reads as

$$R' = R_{32}R_1R_2R_3R_4R_5 \vdots R_4R_5R_6R_7R_8R_9 \vdots \cdots \vdots R_{28}R_{29}R_{30}R_{31}R_{32}R_1.$$

Another important block of Fig. 11.8 is the 48-to-32 compression block. The 48-bit expanded R_{i-1} and the subkey k_i are added modulo-2 and the resultant data is divided into eight 6-bit words as shown in Fig. 11.10.

Each 6-bit word is the input to the S-box which produces a 4-bit word according to the look-up table. Eight such 4-bit words are permuted to produce to desired 32-bit function $f(R_{i-1}, k_i)$.

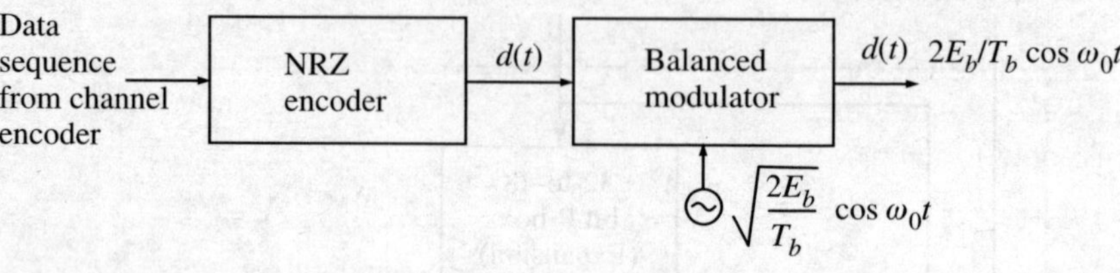

Fig. 11.9 *32-to-48 bit expansion*

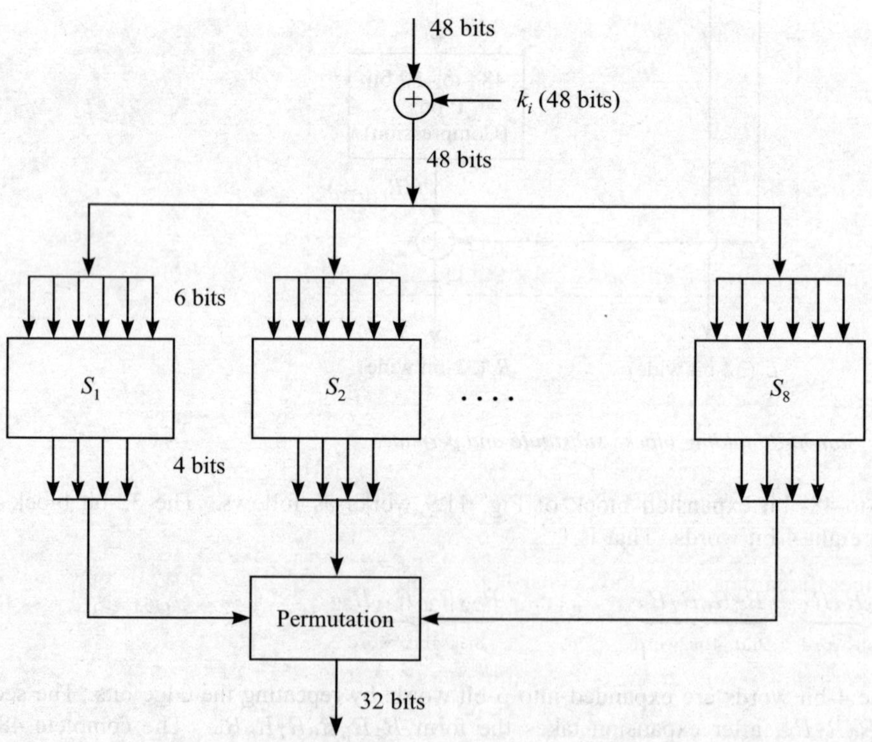

Fig. 11.10 *48-to-32 bits S-blocks*

11.5.1 Key generation

The key for the DES is a 56 bit number in a hexadecimal format. The 48-bit subkeys are produced for every round of the algorithm. The procedure to generate subkeys k_i is shown in Fig. 11.11. The key selection also uses 16 iterations. We start with the input key, a 64-bit block with 8 parity bits in every eighth positions, that is, in positions 8, 16, ..., 64. A 64-to-56-bit P-box (permuted choice 1) discards these parity bits and permutes the remaining 56-bit key. This key is then split into two 28-bit words, C and D. Each 28-bit word is left-circularly shifted by one or two bit positions and the resultant words are then input into a 56-to-48-bit P-box, which generates a subkey according to the permuted choice 2. The process is repeated 16 rounds for 16 different subkeys. The two 28-bit blocks serve as input to the next round.

Example 11.2

For an 8-bit block of plain text, let the initial permutation be defined as:

IP							
2	6	3	1	4	8	5	7

Find the inverse permutation IP^{-1} so that $IP^{-1}(IP(X)) = X$.

Solution

Let the 8-bit data be $d_1d_2d_3d_4d_5d_6d_7d_8$.

Then, $IP(d_1d_2d_3d_4d_5d_6d_7d_8) = d_2d_6d_3d_1d_4d_8d_5d_7$.
bit position : 1 2 3 4 5 6 7 8

Since the final permutation will bring back the bits in sequence, IP^{-1} would be as follows:

IP^{-1}							
4	1	3	5	7	2	8	6

Example 11.3

For an n-bit general substitution block cipher, prove that the size of the key is $n \times 2^n$.

Solution

We remember that a classical S-box consists of a n to 2^n decoder, a P-box (2^n input and 2^n output) and a 2^n to n encoder. For a plaintext block of n-bits, there are 2^n possible different blocks. So an n-bit input to the decoder produces one of 2^n possible input states which are permuted by the P-box into a unique, one of the 2^n possible output states. Each of these states is represented by n-bit ciphertext. Thus for a given word, the cipherword is determined by the specific mapping from among all possible mappings. This is the key that determines such transformations. The size of the key is then the number of possible positions at the output, which is $n \times 2^n$ bits.

Example 11.4

Let $y = E_k(x)$ where x is the plaintext and y is the encrypted version of x. The encryption is denoted by E_k (k is the specific key). Show that $y' = E_{k'}(x')$, where prime quantities are the bit-wise complement of the original ones.

Solution

If we assume that the encryption is a simple X-OR operation, then $y = x \oplus k$.

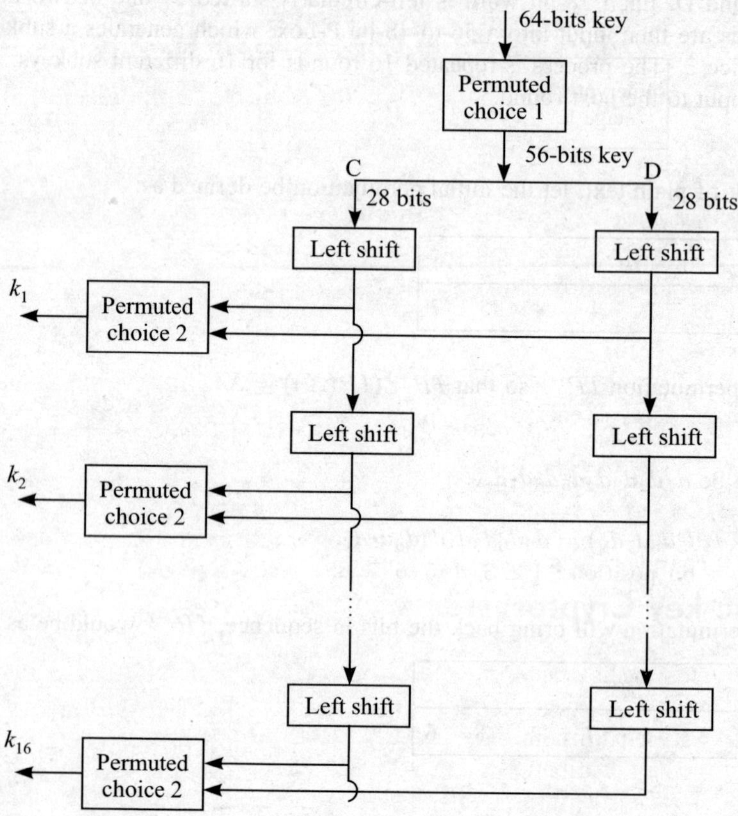

Fig. 11.11 *Key schedule calculation*

We will have to show that $y' = x \oplus k'$.

Now, $y = x \oplus k = x'k + xk'$

$$\text{or,}\quad y' = (x'k + xk')' = (x'k)'(xk')' = (x + k')(x' + k)$$

$$= xk + x'k' = [(xk)'(x'k')']'$$

$$= [(x' + k')(x + k)]' = (xk' + x'k)' = x' \oplus y'$$

Hence the result.

11.5.2 Multiple encryption

Multiple encryption with DES and multiple keys is an approach to improve cryptographic strength. Triple DES (3DES) using two keys is a popular standards. The encryption algorithm employs three encryption stages and two different keys as shown in Fig.11.12. Stages 1 and 3 truely perform encryption with key k_1. The middle block performs the decryption with the key k_2.

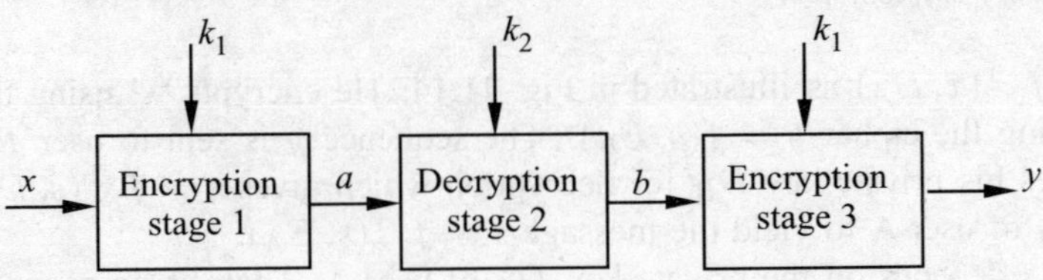

Fig. 11.12 *Triple encryption*

For a plaintext x, the ciphertext is generated as

$$y = E_{k_1}[D_{k_2}[E_{k_1}(x)]], \tag{11.20}$$

following an encrypt–decrypt–encrypt (EDE) sequence. The decryption will be the reverse of encryption. That is,

$$x = D_{k_1}[E_{k_2}[D_{k_1}(y)]. \tag{11.21}$$

11.6 Public-key Cryptosystems

In conventional cryptosystems, secure communications between subscribers can be established if the key is kept safeguarded. It is necessary to exchange the key information among all authorised users since the same key is used for both encryption and decryption. Naturally, the users can employ a secure channel for the key distribution.

The public-key cryptosystem uses two different keys, one for encryption and the other for decryption. The encryption key is publicly revealed, thereby simplifying the key management. But the decryption key is kept secret. Figure 11.13 illustrates the public-key cryptosystem. Suppose that user A wishes to send message x to another user B. All authorised users' encryption keys are public; user A first searches the public directory for the encryption key of user B. Let it be E_B. User A then encrypts the plaintext with the key E_B from ciphertext $y = f(x, E_B)$. This ciphertext is then passed on the user B over a public channel. Since no one can know the decryption key D_B of user B, it is user B who can decrypt y to obtain $x = f^{-1}(y, D_B)$. The two transforms $y = f(x, E_B)$ and $x = f^{-1}(y, D_B)$ are designed in such a way that the inverse of y can only be computed only if we know the other private key D_B.

The public-key cryptosystem can also be used for signature authentication. Suppose user A sends a signed message for user B. He encrypts his message using his private key D_A (decryption algorithm),

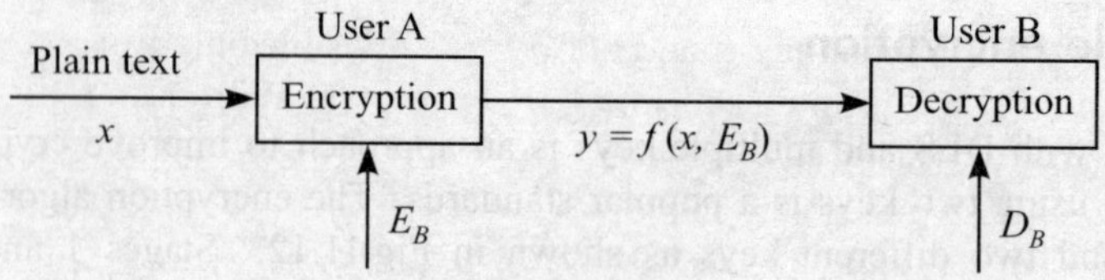

Fig. 11.13 *Public-key cryptosystem*

which yields $s = f^{-1}(x, D_A)$ as illustrated in Fig. 11.14. He encrypts 's' using the encryption key E_B of user B giving the cipher $y = f(s, E_B)$. The sequence y is sent to user B. After receiving y, user B then uses his private key D_B for decryption which gives $s = f^{-1}(s, D_B)$. He then uses encryption key E_A of user A to yield the message $x = f^{-1}(s, E_A)$.

It is noted that s depends on the private key D_A of user A. There is no scope of forgery in this case since the encryption and decryption require private keys of the parties; user B can be sure that the message came from user A. Also, user A can be sure that the message can only be read by B.

As we have seen, public-key cryptosystems rely on one key for encryption and a different key for decryption. Each user places his public key in a public register or other accessible file. The other key is kept private and this key is related to the public key of the user.

The triple DES which is a minor variation of the DES standard, is far more secure than DES and is currently in use today. This scheme consists of three DES encryption stages with the second DES stage operating in the decrypt mode. Each stage employs a different 56-bit key, thereby giving the effect of a $3 \times 56 = 168$-bit key length.

Two public-key algorithms with improved security are the Rivest–Shamir–Adleman (RSA) algorithm, and the Diffie–Hellman algorithm, which are described below.

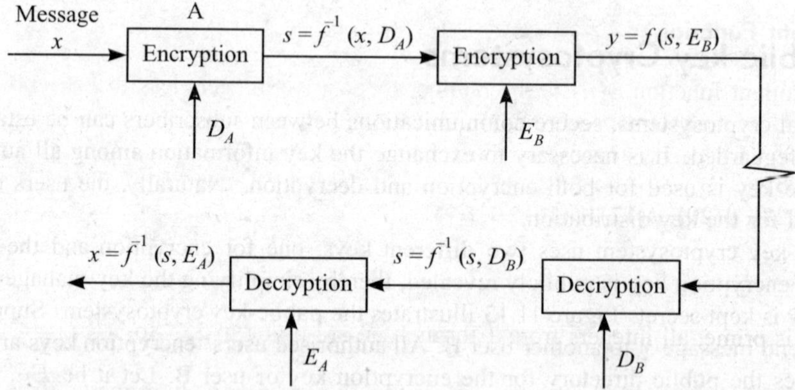

Fig. 11.14 *Message authentication and secrecy in public-key cryptosystems*

11.6.1 Important concepts from number theory

Several public-key algorithms are based on number theory, generation of prime numbers and discrete logarithms. Before discussing the public-key encryption applications, we give here a brief overview of prime numbers and discrete logarithms.

An integer $n > 1$ is a prime number if and only if n is divisible only by ± 1 and $\pm n$. The integers 2, 3, 5, 7, 11, 13, 17, ... are primes. All integers can be factored out in terms of prime numbers. For example,

$$70 = 2 \times 5 \times 7 \text{ (divisors are all prime numbers).}$$

$$373527 = 9 \times 343 \times 121 = 3^2 \times 7^3 \times 11^2.$$

Fermat's Theorem

If n is a prime and a is a positive integer not divisible by n, then

$$a^{n-1} \equiv 1 \bmod n. \tag{11.22}$$

As an illustration, let $n = 23, a = 7$.

$$7^2 = 49 \equiv 3 \bmod 23$$

$$7^4 \equiv 9 \bmod 23$$

$$7^8 \equiv 81 \equiv 12 \bmod 23$$

$$7^{16} \equiv 144 \equiv 6 \bmod 23$$

$$a^{n-1} = 7^{22} = 7^{16} \times 7^4 \times 7^2 \equiv (6 \times 9 \times 3) \bmod 23 = 1 \bmod 23$$

Euler's Totient Function

The Euler's totient function $\phi(n)$ is the number of positive integers less than n and also relatively prime to n.

Example 11.5
Determine $\phi(23), \phi(20), \phi(17), \phi(10)$.

Solution

(a) Since 23 is prime, all integers from 1 through 22 are prime to 23. So, $\phi(23) = 22$.

(b) To find $\phi(20)$, we first find all integers less than 20 and relatively prime to 20:

$$1, 3, 7, 9, 11, 13, 17, 19$$

There are 8 numbers and so $\phi(20) = 8$.

(c) Since c 17 itself is a prime so, $\phi(17) = 16$.

(d) To find $\phi(10)$, the integers less than 10 but relatively prime to it are 1, 3, 7, 9, i.e., $\phi(10) = 4$.

Euler's Theorem

The theorem can be stated as: For every a and n that are relatively prime,

$$a^{\phi(n)} \equiv 1 \bmod n, \tag{11.23}$$

where $\phi(n)$ is the Euler's Totient function.

Let $a = 3, n = 17$. Then, $\phi(n) = \phi(17) = 16$.

$a^{\phi(n)} = 3^{16}$

$3^1 \equiv 3 \bmod 17$

$3^2 = 9 \bmod 17$

$3^4 = 9^2 = 81 \equiv 13 \bmod 17$

$3^8 = 13^2 = 169 \equiv 16 \bmod 17$

$3^{16} = 16^2 = 256 \equiv 1 \bmod 17$

11.6.2 The Rivest–Shamir–Adleman System

The Rivest-Shamir–Adleman (RSA) public-key system is based on the selection of two prime numbers p and q which are kept secret. Depending on their product n, a pair of positive integers e and d are computed in the standard procedure. The number pair (n, e) forms the public key for encryption, while the number pair (n, d) is the decryption key. We explain here the steps for encryption and decryption in the RSA scheme.

1. Choose two prime numbers p and q which are kept hidden (private) and compute the product

   $$n = pq. \tag{11.24}$$

   The number n can be made public. The larger the value of p and q, the harder the code is to break.

2. Compute the Euler's totient function $\phi(n) = (p - 1)(q - 1)$.

3. Then find a composite integer k (k is an integer but not a prime) that satisfies

   $$k \equiv 1 \bmod \phi(n). \tag{11.25}$$

   k must have at least two factors other than k and 1.

4. Pick two of these factors such that their product is k, and call them e and d for 'encoding'

and 'decoding'. The number pair (n, e) constitutes the public key and the number pair (n, d) constitutes the 'private key'.

5. The plaintext message x is encrypted as

$$y = E(x) \equiv x^e \bmod n, \tag{11.26}$$

where E is the encoding function.

6. For a ciphertext y, the decoding function is

$$x = D(y) \equiv y^d \bmod n. \tag{11.27}$$

The security of the RSA algorithm lies in factoring large prime number. The calculations of p and q from their product n takes a long computation time for large n

Example 11.6

Suppose we choose the primes $p = 2$ and $q = 11$. Apply RSA scheme to encrypt the letter 'B' (each letter is transformed to its numerical equivalent of alphabet, starting with A=1). Choose appropriate decryption key to decipher the message.

Solution

$n = pq = (2)(11) = 22$, and $\phi(n) = (p-1)(q-1) = 10$.

$$k = 1 \bmod 10 \Rightarrow k = 11, 21, 31, 41, 51, \ldots$$

Since k has to be a composite integer, not prime, $k = 21$ is chosen.

$$k = 21 = (3)(7). \quad \text{Let} \quad e = 3 \text{ and } d = 7.$$

So, encryption key = $(n, e) = (22, 3)$, decryption key = $(n, d) = (22, 7)$

Now, letter B is represented by the number 2. So, it is encoded as follows:

$$y = E(2) = 2^3 \bmod 22 = 8.$$

To decode, we use

$$x = D(8) = 8^7 \bmod 22 = (8^3)^2 \times 8 \bmod 22$$

$$= (6)^2 \times 8 \bmod 22 = 14 \times 8 \bmod 22 = 2.$$

Note here that the next composite integer k in this problem is 51, which can be factored as 3×17 (both being prime numbers). If we select $e = 3$, then $d = 17$, but since $\phi(n) = 10$, we cannot accept this value of 'd' as 'e' and 'd' must be less than $\phi(n)$.

Example 11.7

Encrypt and decrypt the message number $x = 3$. Use the prime numbers $p = 5$ and $q = 11$. Choose $e = 7$ and calculate the value of decryption key, d.

Solution

$$n = pq = (5)(11) = 55,$$

$$\phi(n) = (p-1)(q-1) = 4 \times 10 = 40.$$

Now, $k \equiv 1 \bmod 40 \Rightarrow k = 41, 81, 121, 161, 201$, etc.

The lowest composite number is $k = 161$, which is not prime and is divisible by $e = 7$. So, the decryption key is $de = k \Rightarrow d = \frac{161}{7} = 23$.

The encrypted message is

$$y = x^e \bmod n = 3^7 \bmod 55 = 3^4 \times 3^3 \bmod 55.$$

$$= 26 \times 27 \bmod 55 = 42.$$

The decoded message is

$$\widehat{x} = y^d \bmod n = (42)^{23} \bmod 55 = (42 \bmod 55)^{23}$$

$$= 13^{23} \bmod 55 = (13^2)^{11} \times 13 \bmod 55 = (4)^{11} \times 13 \bmod 55$$

$$= (4^3)^3 \times 4^2 \times 13 \bmod 55 = (9)^3 \times 4^2 \times 13 \bmod 55$$

$$= 14 \times 43 \bmod 55 = 52 \bmod 55 = 3.$$

In the above two examples on RSA encryption–decryption, we observe the complexities in computation; raising an integer to an integer power and performing $\bmod n$ operation. Also, the key generation again involves computation. The product $n = pq$ is publicly known and the attacker may try to find p and q by exhaustive methods. So one has to select the primes p and q sufficiently large.

11.7 Diffie–Hellman System

The Diffie–Hellman public-key distribution system was developed by Whitfield Diffie, Martin E, Hellman and Ralph C. Markle in 1976. The protocol, although oldest in its class, is considered to be one of the most secure public-key exchange system. The security here relies on the complexity in computing discrete logarithm over finite fields.

The Diffie–Hellman system is based on two parameters, α and n; all users are presumed to know both α and n. The parameter n is a large prime number and α is an integer less than n which has the property

$$p = \alpha^k \bmod n, \tag{11.28}$$

where $1 \leq p \leq n-1$. Equation (11.28) defines the discrete exponential. Conversely, k is the discrete logarithm of p to the base α given by

$$k = (\log_\alpha p) \bmod n. \tag{11.29}$$

It is easy to compute p by knowing k but the reverse of finding k from p is difficult. The encryption scheme of sending a message M from user j to user i is described below.

- User i randomly selects a large integer number a_i, (which is his private key)
- User i's public key is computed as

  $$b_i = \alpha^{a_i} \bmod n \qquad \text{(vide (11.28))}.$$

  The users have their own public keys computed from their private keys and deposited in the public directory so that every other user can access it when needed. The sender must search the public key of the recipient from the public directory.

- For sending a message M to the user i, sender j first fetches b_i (the public key of i) and encrypts the key b_i with his private key as

  $$\begin{aligned} K_{ji} &= (b_i)^{a_j} \bmod n \\ &= (\alpha^{a_i})^{a_j} \bmod n \\ &= \alpha^{a_i a_j} \bmod n. \end{aligned}$$

  This is the secret key. In a similar fashion, user i computes K_{ij}. These two secret keys are identical and can be used for message encryption. An intruder attempting to find the secret key has only the information of b_i and b_j available in the public directory.

  The intruder may apply the formula

  $$K_{ji} = [(b_i)^{\log a_j}] \bmod n,$$

  which seems difficult to compute as it involves discrete logarithm.

- User j computes the ciphertext pairs by generating a random number k less than n:

  $$C_1 = \alpha^k \bmod n$$

  $$C_2 = M \times (b_i^k \bmod n),\ b_i \text{ is user } i\text{'s public key}$$

  and transmits the ciphertext (C_1, C_2) to user i.

- Upon receiving the ciphertext (C_1, C_2), user i computes the plaintext message as follows:

  $$M = \frac{C_2}{C_1^{a_i} \bmod n}.$$

Example 11.8

Suppose that in a Diffie–Hellman system the parameters are $\alpha = 5$ and $n = 11$. Show the encryption and decryption steps for a plaintext message $M = 17$. Assume that the private key of user A is 3. .

Solution

User j generates the public key of user i as

$$b_i = \alpha^{ai} \bmod n = (5)^3 \bmod 11 = 125 \bmod 11 = 4$$

Let user j randomly select a value of $k = 1$ (less than n).

Ciphertext $C_1 = \alpha^k \bmod n = 5^1 \bmod 11 = 5$.

Ciphertext $C_2 = M \times (b_i^k \bmod n) = 17 \times (4' \bmod 11) = 17 \times 4 = 68$.

So, user i receives the ciphertext pair $(C_1, C_2) = (4, 68)$ and computes the message as

$$\widehat{M} = \frac{C_2}{C^{a_i} \bmod n} = \frac{68}{5^3 \bmod 11} = \frac{68}{4} = 17.$$

Privacy using the combination of secret key and public key

One can obtain improved security by combining the advantages of secret key with public-key methods. Here the public key is used to encrypt the message. Figure 11.15 shows the encryption using this combination of methods.

The encryption/decryption process is as follows:

- Sender chooses a secret key and encrypts it with the public key of the receiver.
- The user decrypts that encrypted secret key with his private key.
- The sender encrypts the actual message with this shared secret key. The same key is used by the receiver for decryption.

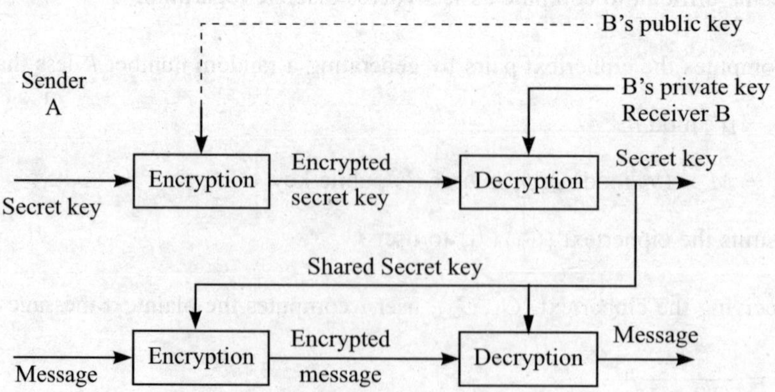

Fig. 11.15 *Privacy using combination of secret key and public key*

Digital signature

When we send a document electronically, we can sign the message just as a person uses his signature to sign a paper document. The privacy, authentication and integrity of a document can be achieved using this electronic signing, called digital signature. Digital signature can be achieved using secret key encryption, where private key is used for encryption and the public key is used for message decryption.

There are two ways of signing of a document. One can sign the entire document or one can sign a condensed form of the document called the message digest. Signing the digest is very efficient if the message is very long. The sender here generates a miniature version of the document and signs it. The digest is always of fixed size although the message is of variable length. The digest is formed from the original message by the operation of a Hash function. The sender encrypts the digest using his private key and appends it to the original message and finally sends to the receiver, as shown in Fig. 11.16.

The receiver B gets the original message and the encrypted digest. He then operates the same Hash function to the message part and generates a second digest. This is compared to the original digest. If the two digests are the same, the message has not been altered any way, thereby ensuring security. Fig. 11.17 illustrates the operation performed at the receiver's end.

Since Hash code is encrypted using symmetric encryption, the receiver decrypts the Hash code separately. A new Hash code is also generated from the received message. On comparison, if the two Hash codes agree, then the message is received intact.

Hash function can be used in a number of ways. One of the ways is shown in Fig.11.16 where the hash value h is encrypted and appended to the message. The message M is of variable length but the hash value $h = H(M)$ is of fixed length. The Hash function is used for creating message digest or finger print and is a useful tool for message authentication. The function should have the following properties:

- The function can act on a block of message of any size but the hash value will be of fixed length.
- $H(M)$ is relatively easy to compute for a given M, but for a given h, it is computationally infeasible to derive M (one-way property).
- Hashing should e a one-to-one mapping, that is, a given function executing on two identical messages, would result in identical digests.

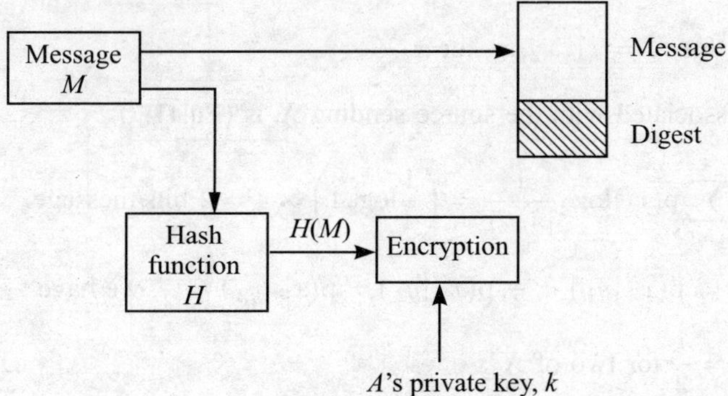

Fig. 11.16 *Use of Hash function to provide message authentication*

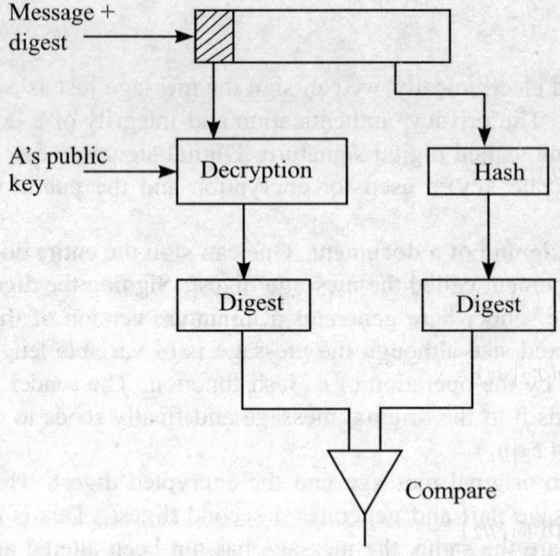

Fig. 11.17 *The receiver computes the Hash value and compares*

Solved Numericals

1. A message set consists of four equally likely massages $X = \{x_1, x_2, x_3, x_4\}$. Let us associate to this X, another equally likely message set $Y = \{y_1, y_2\}$ with the following relationship:

   y_1 appears only when x_1 and x_3 appear

   y_2 appears only when x_2 and x_4 appear

   Find the equivocation of message X conditioned on Y.

   Solution

   Since the symbols message set X are equally probable, we have

   $$\mathrm{p}(x) = \frac{1}{4} \text{ for } X = x_1, x_2, x_3, \text{ or } x_4$$

   Thus, entropy associated with the source sending X is (Eq.11.7)

   $$H(X) = \sum_{i=1}^{4} \mathrm{p}(x_i)\log_2 \frac{1}{\mathrm{p}(x_i)} = \left(\frac{1}{4}\log_2 4\right) \times 4 = 2 \text{ bits/message.}$$

   Since $\mathrm{p}(x_1\,|y_1) = \mathrm{p}(x_3\,|y_1) = \frac{1}{2}, \mathrm{p}(x_2\,|y_2) = \mathrm{p}(x_4\,|y_2) = \frac{1}{2}$, we have

   $$\mathrm{P}(X\,|Y) = \frac{1}{2} \text{ for two of } X\text{'s}$$

   $$= 0 \text{ for rest two of } X\text{'s,}$$

We may find (from Eq.9.27)

$$H(X|Y) = \sum_{x_1 y} \text{p}(x,y)\log_2 \frac{1}{\text{p}(x|y)}, \text{ where p } (x,y) \text{ is the joint probability of } X \text{ and } Y$$

$$\text{p}(x,y) = \text{p}(x|y)\text{p}(y) = \frac{1}{2}\text{p}(x|y) \text{ since } p(y_1) = p(y_2) = \frac{1}{2}$$

So,

$$\text{p}(x_1,x_2) = \frac{1}{2}\text{p}(x_1|y_1) = \frac{1}{2}\cdot\frac{1}{2} = \frac{1}{4}, \text{p}(x_1,y_2) = \frac{1}{2}\text{p}(x_1|y_2) = 0$$

$$\text{p}(x_2,y_1) = \frac{1}{2}\text{p}(x_2|y_1) = 0, \text{ p}(x_2,y_2) = \tfrac{1}{2}\text{p}(x_2|y_2) = \tfrac{1}{2}\cdot\tfrac{1}{2} = \tfrac{1}{4}$$

$$\text{p}(x_3,y_1) = \frac{1}{2}\text{p}(x_3|y_1) = \frac{1}{2}\cdot\frac{1}{2} = \frac{1}{4}, \text{p}(x_3,y_2) = 0$$

$$\text{p}(x_4,y_1) = 0, \text{ p}(x_4,y_2) = \frac{1}{2}\cdot\frac{1}{2} = \frac{1}{4}.$$

Thus,

$$H(X/Y) = \sum_y \left[\text{p}(x_1,y)\log_2\frac{1}{\text{p}(x_1,y)} + \text{p}(x_2,y) + \log_2\frac{1}{\text{p}(x_2,y)}\right.$$

$$\left. + \text{p}(x_3,y)\log_2\frac{1}{\text{p}(x_3,y)} + \text{p}(x_4,y)\log_2\frac{1}{\text{p}(x_4,y)}\right]$$

$$= \text{p}(x_1,y_1)\log_2\frac{1}{\text{p}(x_1,y_1)} + \text{p}(x_1,y_2)\log_2\frac{1}{\text{p}(x_1,y_2)}$$

$$+ \text{p}(x_2,y_1)\log_2\frac{1}{\text{p}(x_2,y_1)} + \text{p}(x_2,y_2)\log_2\frac{1}{\text{p}(x_2,y_2)}$$

$$+ \text{p}(x_3,y_1)\log_2\frac{1}{\text{p}(x_3,y_1)} + \text{p}(x_3,y_2)\log_2\frac{1}{\text{p}(x_3,y_2)}$$

$$+ \text{p}(x_4,y_1)\log_2\frac{1}{\text{p}(x_4,y_1)} + \text{p}(x_4,y_2)\log_2\frac{1}{\text{p}(x_4,y_2)}$$

$$= \frac{1}{4}\log_2 4 + \frac{1}{4}\log_2 4 + \log_2 4 + \log_2 4 = 2\log_2 2 = 2 \text{ bits/message.}$$

2. The alphabet set of a language consists of 20 letters, with the following probability of occurrence in the text:

   4 letters with the probability of $\frac{1}{16}$ each

   12 letters with the probability of $\frac{1}{32}$ each

   4 letters with the probability of $\frac{3}{32}$ each

   Find the unicity distance when a 20-letter alphabet is permuted to generate the key sequence. Assume average number of bits per character as $r_1 = 1.5$ bits.

Solution

Plain text entropy is

$$H(X) = \left(\frac{1}{16}\log_2 16\right) \times 4 + \left(\frac{1}{32}\log_2 32\right) \times 12 + \left(\frac{3}{32}\log_2 \frac{32}{3}\right) \times 4$$

$$= \frac{1}{4} \times 4 + \frac{5}{32} \times 12 + \frac{3}{8} \times 3.415 = 1 + 1.875 + 1.2806$$

$$= 4.1556 \text{ bits/chartacter}$$

For the alphabet size of $L = 20$, maximum entropy is given by (Eq. 11.11)

$$r_2 = \log_2 L = \log_2 20 = 4.3219 \text{ bits/character.}$$

Redundancy for an n-bit cipher is (Eq. 11.12)

$$r = r_2 - r_1 = 1.3219 - 1.5 = 2.8219 \text{ bits/character}$$

So, percentage redundancy

$$= \frac{r}{r_2} \times 100 = \frac{2.8219}{4.3219} \times 100 = 65.3\%$$

Let us find the key entropy $H(k)$ (by Eq.11.14). Since key is generated by permuting 20-letter alphabets, that is, with a key sequence of $k_1k_2k_3 \ldots k_{20}$, each k_i can be any of 20 alphabets. Hence, we have a total number of key sequence of 20^{20}, each with equal probability. Thus,

$$H(k) = \log_2 (20)^{20} = 86 \text{ bits}$$

So, $n_0 = \frac{H(k)}{r} = \frac{86}{2.82} \approx 31$ characters.

This means that the intruder can break the cipher by studying 31 number of cipher text bits.

3. In the RSA algorithm, let $p = 47$ and $q = 59$. If encryption key is chosen as $e = 17$, find the decryption key d.

Solution

$n = pq = 47 \times 59, = 2773$

$$\phi(n) = (p - 1)(q - 1) = 46 \times 58 = 2668$$

Find composite integer k (Eq.11.25)

$$k = 1 \bmod \phi(n) = 1 \bmod 2668$$

So, $k = 2669, 5337, 8005, 10673, 13341, 16009, 18677, 21345$

The lowest composite number divisible by $e = 17$ is 2669.

Since $d \cdot e = k$, we have,

$$d = \frac{k}{e} = \frac{2669}{17} = 157$$

4. RSA scheme is used to encrypt/decrypt the number '8' with $p = 13$, $q = 31$ and $e = 7$. Find the decryption key and the cipher of '8'. Also decode the plain text number from the cipher.

Solution

Here, $n = pq = 13 \times 31 = 403$,

$$\phi(n) = 12 \times 30 = 360$$

Now, $k = 1 \bmod 360 \Rightarrow k = 361, 721, 1081, \ldots$

The lowest number is $k = 721$(which is divisible by $e = 7$)

So, decryption key is $d = \frac{k}{e} = \frac{721}{7} = 103$

The encrypted message (for $M = 8$) is

$$C = M^e \bmod n = 8^7 \bmod 403 = 512 \times 4096 \bmod 403$$
$$= 343 \quad (= \text{cipher of } M = 8)$$

Decrypted message

$$\hat{M} = c^d \bmod n = (343)^{103} \bmod 403 = 8$$

5. Apply Diffie–Helman protocol to encrypt the plain-text message $M = 7$. The public parameters are $\alpha = 7$ and $n = 17$. Assume the private key of the recipient is $a_i = 3$. Repeat the same for the message $M = 9$.

Solution

The public key of the recipient is generated first and this is

$$b_i = \alpha^{ai} \bmod n = (7)^3 \bmod 17 = 3.$$

To encrypt the message, the sender randomly chooses a value of k (less than n). Let this value be chosen as $k = 5$.

Sender computes the cipher text pair as

$$C_1 = \alpha^k \bmod n = 7^5 \bmod 17 = 11$$
$$C_2 = M \times (b_i^k \bmod n) = 7 \times (3^5 \bmod 17) = 7 \times 5 = 35$$

The recipient receives the cipher text $(C_1, C_2) = (11, 35)$. The decryption is done as follows:

$$\hat{M} = \frac{C_2}{(C_1{}^{a_i} \bmod n)} = \frac{35}{11^3 \bmod 17} = \frac{35}{5} = 7 = M$$

For the message, $M = 9$, we have $C_2 = 9 \times 5 = 45$, so the cipher text pair is (11, 45) and decrypted message is

$$\hat{M} = \frac{45}{5} = 9$$

Exercises

Activity

11.1 An RSA scheme is used to encrypt the message $M = 3$ with the prime numbers $p = 5$ and $q = 7$. Find the encryption key if the decryption key d be 11.

11.2 For the prime numbers $p = 7$ and $q = 11$, list all the private/public keys for an RSA system.

11.3 An encryption technique of 'n alphabetic shift' produces the ciphertext. How many different attempts must be made to break the code?

11.4 The facility of the secret key cryptography is to be provided to a group of n-people. If each pair of people are to avail the secure communication, how many distinct keys are required?

11.5 Generate the private and public keys using the pair of primes 17 and 31 in the RSA algorithm.

11.6 (a) A block of 64- bits are encrypted using DES algorithm with a key that is 56-bit long. Find the unicity distance.

(b) What is the effect on unicity distance if the key-length is doubled?

11.7 RSA algorithm is based on the keys 'e' and 'd' such that ed modulo $-\phi(n) = 1$, where $n = pq$, $\phi(n) = (p-1)(q-1)$. Let the message M be an integer in the range $(0, n-1)$ such that M and n are prime to each other. Prove that

$(M^{\mathrm{e}}$ modulo-$n)^{\mathrm{d}}$modulo-$n = M$.

11.8 (a) For the RSA algorithm, p and q are prime integers. One can obtain the decryption key d. Is this key unique?

(b) For $p = 7$ and $q = 11$, list four possible decryption keys.

11.9 Caesar cipher is an example of monoalphabetic substitution following the rule

$C = E(p) = (p + 3)$ mod26, where $E(p)$ is encryption.

That is, the plain text p is shifted by 3 units right (on mod 26). Find the cipher text c for the plain text message

$p =$ good luck.

11.10 The formula for decryption from cipher text to plain text is given by

$\phi = D(c) = (c - 3)$ mod 26.

Decrypt the cipher text of Example 9 above.

Review questions

1. What are the requirements for a useful cryptosystem? Explain the terms 'confusion' and 'diffusion' in connection cryptosystems.

2. What is the difference between a block cipher and a stream cipher? What is a product cipher?

3. Describe the steps of the message encryption in data encryption standard (DES). What are multiple encryption standards? Explain the operation of a 3DES.

4. Explain the terms 'message digest', and Hash function. State the properties of a Hash function.

5. Explain briefly the Diffie–Hellman key exchange

6 (a) Define the terms:

   (i) Cipher (ii) Keys (iii) Block encryption (iv) Stream encryption

   (b) What are the purposes of encryption coding?

   (c) State and explain the data encryption standard (DES) method. Draw the flow-chart of a DES system.

   (d) Briefly discuss RSA system of encryption coding. [KU 2003]

7. (a) What are the requirements of a cryptosystem? Describe the major differences between symmetric-key cryptosystem and public-key cryptosystem.

   (b) Describe the steps for message encryption using the RSA scheme. [WBUT 2005]

8. "The larger the size of the key space, the more secure a cipher" – Justify or comment on the statement.

9. (a) The key elements of security are authentication, integrity and non-reputation. Explain the terms with examples.

   (b) Write short notes on asymmetric cryptography.

10. Explain different attacks on the crypto system. What is meant by brute-force attack on DES system?

Multiple-choice questions

1. Using a 4-bit long key 1101, the stream of binary data 01111011 encrypts as

   (a) 01101010 (b) 10100110 (c) 10101001 (d) 10011010

2. Using character + x encryption technique with $x = 3$, the word RSA stands as

   (a) WXF (b) TUC (c) ASR (d) RSA

3. Regarding the number N_m of message bits and the number of bits N_h in the hash value, which of the following is correct?

   (a) $N_m = N_h$ (b) N_m fixed block, N_h fixed block

   (c) N_m may be variable but N_h fixed (d) both N_m and N_h are variable.

4. Number of iterations in DES algorithm is

   (a) 64 (b) 56 (c) 16 (d) none

5. The Euler Totient function $\phi(253)$ has the value

   (a) 252 (b) 220 (c) 317 (d) none

6. The value of 7^5 mod 851 is

   (a) 5 (b) 7 (c) 790 (d) 638

7. In a certain encryption A encrypts as $A, \ldots, I$ encrypts a I, J encrypts as A, ..., S encrypts as J, T encrypts as $B, \ldots, Z$ encrypts as H. The message CODING encrypts as

   (a) CODING (b) DOCGNI (c) GNIDOC (d) CFDIEG

8. Working with an 8-bit data as in a simplified DES scheme, the initial permutation IP follows the rule $IP(d_1 d_2 d_3 d_5 d_6 d_7 d_8) = (d_3 d_6 d_1 d_4 d_7 d_2 d_5 d_8)$. The reverse permutation IP^{-1} should be

   (a) (6 4 2 8 3 1 7) (b) (1 4 7 3 6 2 5 8)

   (c) (7 2 5 8 3 6 1 4) (d) (3 6 1 4 7 2 5 8)

9. The major disadvantage of symmetric encryption is

   (a) It is a more complex and time-consuming computation.

   (b) The problem of secure transmission of secret key.

   (c) Encryption is less secure.

   (d) None of the above.

10. Cipher means

    (a) Encrypted message by sender

    (b) Decrypted message by recipient

    (c) Algorithm for encryption/decryption

    (d) None.

11. Data Encryption Standard (DES) is an example of

    (a) block cipher (b) stream cipher

    (c) bit by bit cipher (d) none

12. The .... provides a one-time session key for both the sender and recipient.

    (a) RAS (b) DES (c) Diffie–Hellman (d) none

13. In an asymmetric-key cipher, the receiver uses the .... key.

    (a) Private key (b) Public key

    (c) Either of (a) or (b) (d) Both (a) and (b)

14. Same key is used by both the parties in the following encryption.

    (a) Symmetric-key (b) Asymmetric key

    (c) Both (a) and (b) (d) None

15. The integer x such that $5x = 4 \bmod 3$ is

    (a) 1 (b) 2 (c) 3 (d) None

12 Spread-Spectrum Modulation

12.1 Introduction

Spread spectrum (SS) is a technique where the bandwidth of the message signal is spread compared to the minimum bandwidth required to transmit the message. The spectral spreading reduces the power spectral density significantly so that to a receiver turned to a specific AM or FM broadcast, the spread signal appears simply as noise and becomes unnoticed. A spread spectrum system should fulfil the following essential requirements:

- The message signal should undergo two modulations that produces wideband spectrum having bandwidth much in excess of the minimum bandwidth needed to send the message.
- The spreading, that is, the second modulation, is caused by means of a spreading signal.
- The receiver should recover the message by despreading the received signal with the use of the same spreading signal.

Spread spectrum systems are transparent to the interfering signals (caused by jammers) and hence offer reliable communications. We begin with an example to illustrate the effect of spectrum spreading. Consider an AM signal that has a bandwidth of 10 kHz centred about a carrier frequency f_c and of power P_s. We assume that the spread spectrum signal is with the same carrier that has a wide bandwidth of 1 MHz. The power spectrum reduces by a factor $\dfrac{P_s \times 10^4}{10^6} = \dfrac{P_s}{100}$, a 20 dB below the AM signal. Thus the AM receiver simply recognises the spread signal as noise. Another attribute of the spread-spectrum signal is the multiple access, a technique whereby a number of users can share the communication resources in a controlled manner. A popular multiple access technique is the code-division multiple access (CDMA), in which several users have different code signals but share a single RF communication link. In this chapter, we will present two important spread spectrum techniques, direct sequence (DS) and frequency hopping (FH).

We will first consider the generation of spreading signal in greater detail and study its properties.

12.2 Pseudonoise Sequence

Pseudonoise (PN) sequence, also called pseudorandom number, is a wideband waveform. It is a binary sequence with properties similar to those of white noise in the sense that it cannot be predicted but its future variations can only be described in a statistical manner. However, a PN sequence is not random at all, it is deterministic and periodic. The sequence has the interesting properties that it has high value of autocorrelation between identical codes and low values of cross-correlation between different codes. This correlation property is used extensively in the SS systems. A friendly receiver knows the code and finds a high correlation of the received signal. But an enemy receiver does not know the code and might use a different one and finds a low correlation.

The PN sequence is generated with a logic circuit called linear feedback shift register (LFSR). An example of a 4-bit LFSR is shown in Fig. 12.1 consisting of a 4-stage shift register for storage and shifting , a multi-input exclusive OR circuit and a feedback path. Assume that the logic circuit starts in a state not all 0's. The sequence is classified as maximal length if for an n-stage LFSR, the sequence repeats after pth clock pulses, where $p = (2^n - 1)$.

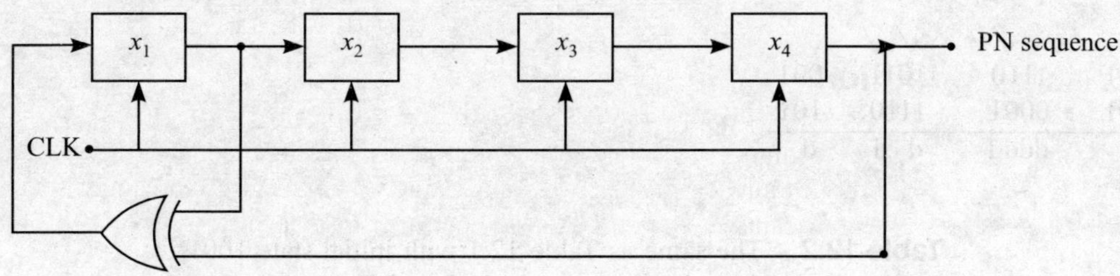

Fig. 12.1 *A 4-stage maximal-length LFSR that generates PN signal*

At each clock pulse, the contents of shift registers shift one stage to the right. Stages x_1 and x_4 are exclusive-ORed, the result is fed back to stage x_1. The output of any particular shift register is pseudonoise sequence. Table 12.1 list the codewords for a PN generator with initial state 1000. There are 15 states and the states are said to be of 'maximal length'.

Table 12.1 Fifteen nonzero codewords for the PN generator of Fig. 12.1, with initial state 1000.

CLK	x_1	x_2	x_3	x_4
1	1	0	0	0
2	1	1	0	0
3	1	1	1	0
4	1	1	1	1
5	0	1	1	1
6	1	0	1	1
7	0	1	0	1
8	1	0	1	0
9	1	1	0	1
10	0	1	1	0
11	0	0	1	1
12	1	0	0	1
13	0	1	0	0
14	0	0	1	0
15	0	0	0	1
16	1	0	0	0

The resulting codeword would be the output of any shift register (x_4, say) and appears as 00011110101 1001. In general, for an n-stage shift register, the sequence will go through $2^n - 1$ distinct states and it will have 2^{n-1} zeroes and $(2^{n-1}+1)$ ones, randomly distributed. (This is known as balance property.) The 4-stage register under consideration has 7 zeroes and 8 ones. Since there are $(2^4 - 1) = 15$ nonzero initial states of the LFSR, a total of 15 nonzero codewords are possible. Next consider the initial state 1001. The 15 possible states are given in Table 12.2 and the resulting codeword is 1001 0001 1110 101. These twocode words have the distance 8 between them. It can be proved that any two of the 15 different codewords would have a distance of 8 between them. It can be proved that any two of the 15 different codewords would have a distance of 8 between them.

√	√√√√	√√	√
0001	1110	1011	001
1001	0001	1110	101
d	dddd	d d	d

Table 12.2 The same as Table 12.1 with initial state 1001

CLK	x_1	x_2	x_3	x_4
1	1	0	0	1
2	0	1	0	0
3	0	0	1	0
4	0	0	0	1
5	1	0	0	0
6	1	1	0	0
7	1	1	1	0
8	1	1	1	1
9	0	1	1	1
10	1	0	1	1
11	0	1	0	1
12	1	0	1	0
13	1	1	0	1
14	0	1	1	0
15	0	0	1	1
16	1	0	0	1

Now, coming to the design of a maximal length LFSR, the required inputs to the exclusive-OR gates are found as follows.

For a linear structure of n-stage shift register, we can write

$$x_{n+1} = g_n x_n + g_{n-1} x_{n-1} + \cdots + g_2 x_2 + g_1 x_1, \tag{12.1}$$

where g_1 defines the feedback connection from the ith shift register output to the exclusive-OR gates. Wherever a connection exists, $g_i = 1$, otherwise it is 0. The state of the shift registers keeps on changing every clock period. At the next clock (12.1) modifies to

$$x_{n+2} = g_n x_{n+1} + g_{n-1} x_n + \cdots\cdots + g_2 x_2. \tag{12.2}$$

In this way we have n-such equations. In matrix form, (12.1) reads as

$$[x] = [X][g], \tag{12.3}$$

$$\text{where, } x = \begin{bmatrix} x_{n+1} \\ x_{n+2} \\ \vdots \\ x_{2n} \end{bmatrix}, \quad g = \begin{bmatrix} g_1 \\ g_2 \\ \vdots \\ g_n \end{bmatrix} \quad \text{and } X = \begin{bmatrix} x_1 & x_2 \cdots x_n \\ x_2 & x_3 \cdots x_{n+1} \\ \vdots & \\ x_n & x_{n+1} \cdots x_{2n-1} \end{bmatrix}.$$

$$\text{Hence, } [g] = [X]^{-1}[x]. \tag{12.4}$$

In our case $n = 4$, the contents of the shift registers at the four clock times are 1001, say. Thus,

$$\begin{aligned} g_1.1 + g_2.0 + g_3.0 + g_4.1 &= 0 \\ g_1.0 + g_2.1 + g_3.0 + g_4.0 &= 0 \\ g_1.0 + g_2.0 + g_3.1 + g_4.0 &= 0 \\ g_1.0 + g_2.0 + g_3.0 + g_4.1 &= 1. \end{aligned} \tag{12.5}$$

There are four unknowns and four equations. Solving, we get $g_4 = 1, g_3 = 0, g_2 = 0$, and $g_1 = 1$. Thus for maximal length, exclusive-OR inputs are from the first and 4th shift registers.

It can be proved that g_i's are the coefficients of a primitive polynomial over a finite field. We will not explore here any more of finite field theory; rather, we furnish here the list of coefficients of primitive polynomials of lower orders that generate maximal-length shift register connections. The list is tabulated in Table 12.3.

Table 12.3 Shift register connections for generating PN sequence

N	Shift register outputs
2	1, 2
3	1, 3
4	1, 4
5	2, 5
6	1, 6
7	1, 7
8	2, 3, 4, 8
9	4, 9
10	3, 10

The linear feedback shift register based PN sequence generator provides a periodic sequence based on the feedback connections. The n-bit shift register produces a maximum length (ML) sequence given by $P = 2^n - 1$. The ML-sequence should have the following properties:

1. Balance property, which dictates that the number of ones should be one or more than the number of zeros.

2. Run property, which determines the single or similar types of digits in the sequence. The sequence will have one half-of its runs of length 1, one-quarter of its runs of length 2 and so on.

3. All the 2^n-possible states (except the all-zero state) will be generated in the sequence.

4. Autocorrelation property of the PN sequence will exhibit only a single autocorrelation peak.

Eample 12.1

Design a 3-bit maximal length LFSR to be used for generating 7-bit PN sequence. List all 7 random codes.

Solution

From the look-up table 12.3, we find that the connections are from the first and third shift register stages; they are sketched in Fig. 12.2.

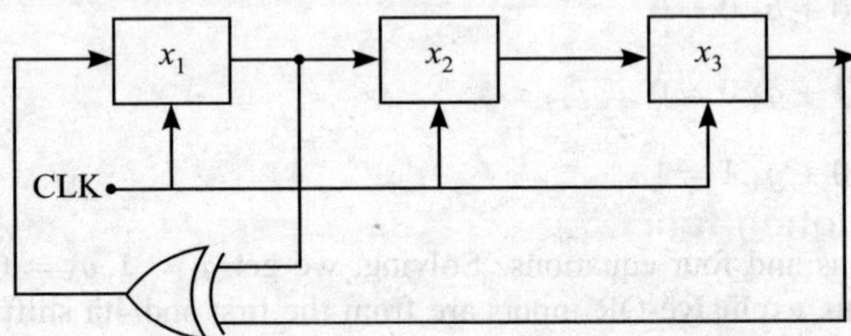

Fig. 12.2 *Linear feedback shift register for generating 7-bit PN sequence*

We list below the sequences of the contents of shift registers at different clock periods for 7 initial states.

Table 12.4 PN sequence for different initial stages of shift registers

Initial state $x_1x_2x_3 = 001$				Initial state $x_1x_2x_3 = 010$				Initial state $x_1x_2x_3 = 011$			
CLK	x_1	x_2	x_3	CLK	x_1	x_2	x_3	CLK	x_1	x_2	x_3
1	0	0	1	1	0	1	0	1	0	1	1
2	1	0	0	2	0	0	1	2	1	0	1
3	1	1	0	3	1	0	0	3	0	1	0
4	1	1	1	4	1	1	0	4	0	0	1
5	0	1	1	5	1	1	1	5	1	0	0
6	1	0	1	6	0	1	1	6	1	1	0
7	0	1	0	7	1	0	1	7	1	1	1
8	0	0	1	8	0	1	0	8	0	1	1

Codeword 1 00 1110 010 0111 1101 001

(Continued...)

Table 12.4 *(Continued...)*

Initial state $x_1x_2x_3 = 100$				Initial state $x_1x_2x_3 = 101$				Initial state $x_1x_2x_3 = 110$				Initial state $x_1x_2x_3 = 111$			
CLK	x_1	x_2	x_3	CLK	x_1	x_2	x_3	CLK	x_1	x_2	x_3	CLK	x_1	x_2	x_3
1	1	0	0	1	1	0	1	1	1	1	0	1	1	1	1
2	1	1	0	2	0	1	0	2	1	1	1	2	0	1	1
3	1	1	1	3	0	0	1	3	0	1	1	3	1	0	1
4	0	1	1	4	1	0	0	4	1	0	1	4	0	0	1
5	1	0	1	5	1	1	0	5	0	1	0	5	0	0	1
6	0	1	0	6	1	1	1	6	0	0	1	6	1	0	0
7	0	0	1	7	0	1	1	7	1	0	0	7	1	1	0
8	0	0	1	8	1	0	1	8	1	1	0	8	1	1	1
Codeword	0011 101				1010 011				0111 010				1110 100		

The codewords are

1001 110 0100 111 1101 001 0011 101 1010 011 0111 010 1110 100

The distance between any two words is 4. In general, the distance between any two code words is 2^{n-1}.

12.2.1 Autocorrelation function of PN sequence

The PN sequence generator produces a pseudorandom binary wave $g(t)$, as shown in Fig. 12.3(a).

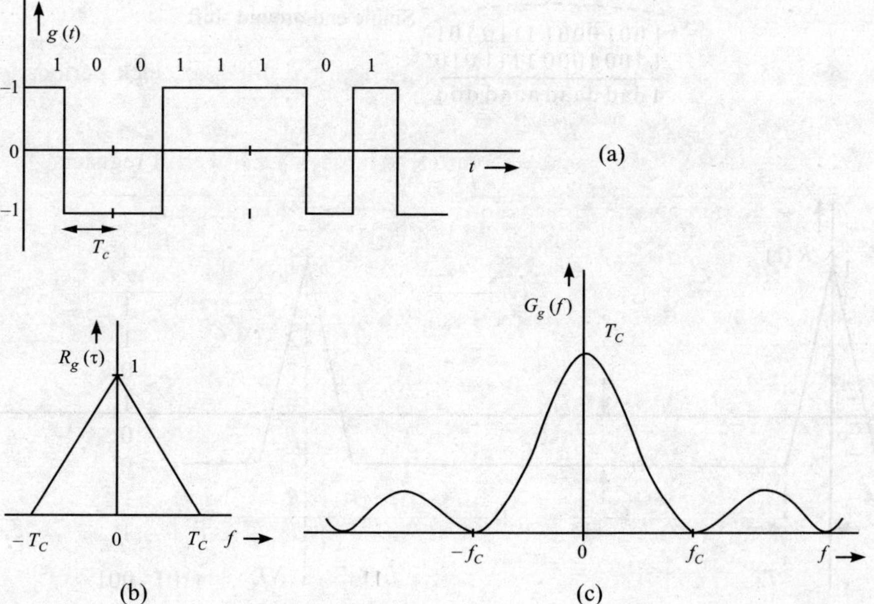

Fig. 12.3 *(a) Pseudorandom binary waveform. (b) its correlation function. (c) power spectral density.*

Each fundamental pulse in referred to as a 'chip'. The duration of each chip is T_c and chip rate is $f_c = 1/T_c$. The pulse amplitudes are $a_k = \pm 1$. A maximal length PN sequence repeats in clock pulses $N = 2^n - 1$.

The correlation function would be triangular in shape (vide Section 8.9) and is given by

$$R_g(\tau) = \left(1 - \frac{|\tau|}{T_c}\right), \quad |\tau| < T. \tag{12.6}$$

The corresponding power spectral density (PSD) is

$$G_g(f) = T_c\left(\frac{\sin \pi f/f_c}{\pi f/f_c}\right)^2. \tag{12.7}$$

Figure 12.3(b) and (c) plots the correlation function and the power spectral density.

Example 12.2
Sketch the PN autocorrelation function for the 15-bit sequence 1001 0001 1110 101.

Solution

For the digital waveform, the autocorrelation function is defined to be $R_g(\tau) = \frac{1}{N}\sum_{i=1}^{N} a_k a_{k+i}$. When the kth and $(k+1)$th amplitudes agree, $a_k a_{k+1} = 1$; otherwise $a_k a_{k+1} = -1$. The sequence has the repetition period $N = 15$.

Single end-around shift

1 001 0001 1110 101
1 100 1000 1111 010
a dad daad aaad ddd

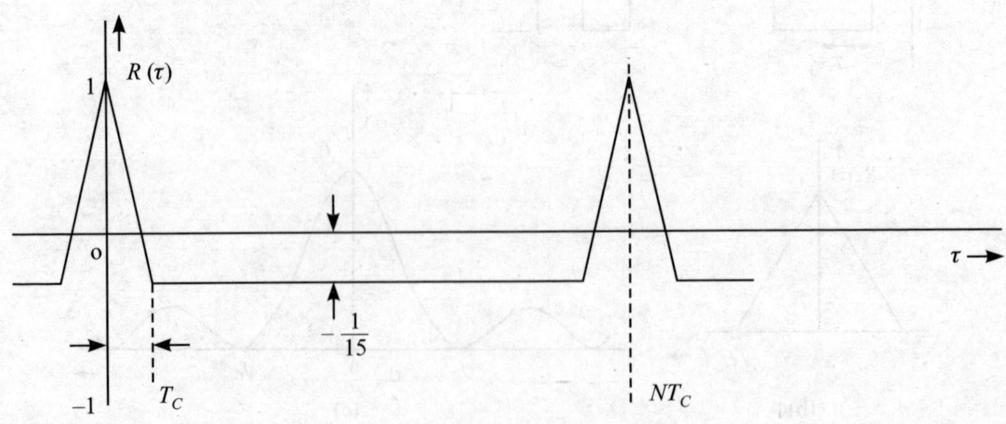

Fig. 12.4 *PN autocorrelation function*

When the digits agree, we level by 'a' and when they disagree, we level by 'd'. Thus, for a single shift ($\tau = 1$),

$$R(1) = \tfrac{1}{15}(7-8) = -\tfrac{1}{15} = -1/N.$$

```
 1 001 000 1 1110 101   double end-around shift
 0 110 010 0 0111 101
 --------------------
 d ddd ada d daad aaa
```

For a double end-around shift, $R(2) = \frac{1}{15}(7-8) = -1/N$, and this will be true for the other shifts. When there is no shift, $\tau = 0$ and $R(0) = 1$. Figure 12.4 shows the plot of $R(\tau)$. We see that for any cyclic shift, the number of agreements differs from the number of disagreements by one count. This is called the correlaltion property of the PN sequence.

12.3 Direct Sequence Spread Spectrum (DS/SS)

The basic concept of the spread spectrum (SS) technique is to spread the spectrum of the massage signal. In direct sequence (DS)/SS scheme, this spectral spreading is obtained in two stages: A carrier wave is first modulated with a message signal using digital modulation. The modulated signal is again modulated with a high speed spreading signal $g(t)$ that effectively masks the message and spread the spectrum of the modulated wave. Figure 12.5 depicts a direct sequence (DS) modulator. The transmitted wave can be written as

$$s(t) = \sqrt{2P_s}x(t)g(t)\cos\omega_0 t, \tag{12.8}$$

where $g(t)$ is an antipodal pulse stream with pulse amplitudes $+1$ and -1, $x(t)$ is the message input and P_s is the power of a modulated carrier.

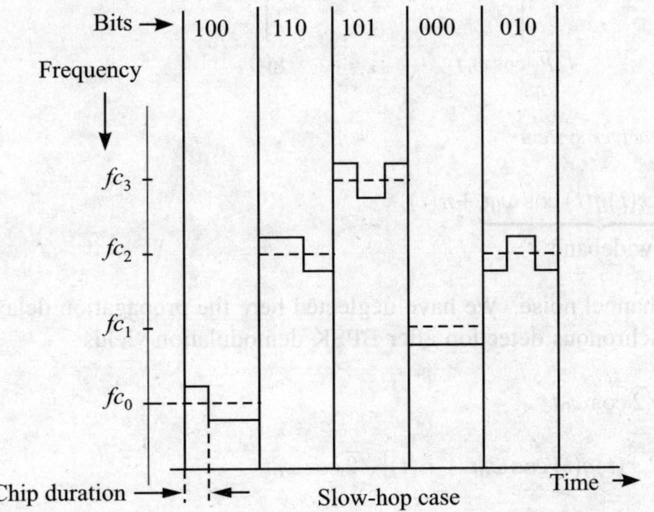

Fig. 12.5 *Direct sequence transmitter*

Since the function of the pseudorandom number $g(t)$ is to spread the spectrum, the positions of the BPSK modulator and the multiplier block in Fig. 12.5 can be interchanged. That is, the message $x(t)$ is first multiplied by a wideband PN sequence $g(t)$, and the product signal $x(t)g(t)$ then modulates the carrier as shown in Fig. 12.6. Here the bit rate of $g(t)$, called the chip rate, is $f_c >> f_b$, where f_b is the data bit rate.

The signal $g(t)$ chops the message $x(t)$ into 'chips' causing a bandwidth expansion by a ratio f_c/f_b. For practical systems, this ratio ranges from 10 to 10,000. Since the power contained in $x(t)$ and $x(t)g(t)$ is the same, the spectral density of $s_1(t)$ is reduced by the factor f_b/f_c and it becomes difficult to distinguish it from the background noise to an unauthorised receiver. However, the authorised receiver can recover the message from the DS/SS signal, as be seen in Fig. 12.7. At the receiver, it is necessary to regenerate the synchronised replica of the sinusoidal carrier $\cos \omega_0 t$ and the spreading pulse $g(t)$. The received signal is

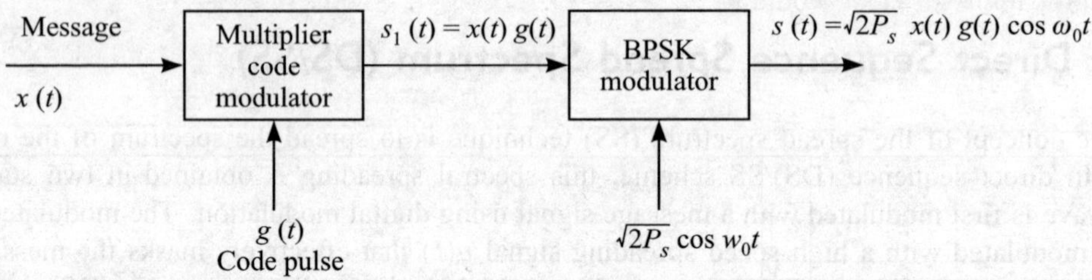

Fig. 12.6 *Simplified BPSK DS transmitter*

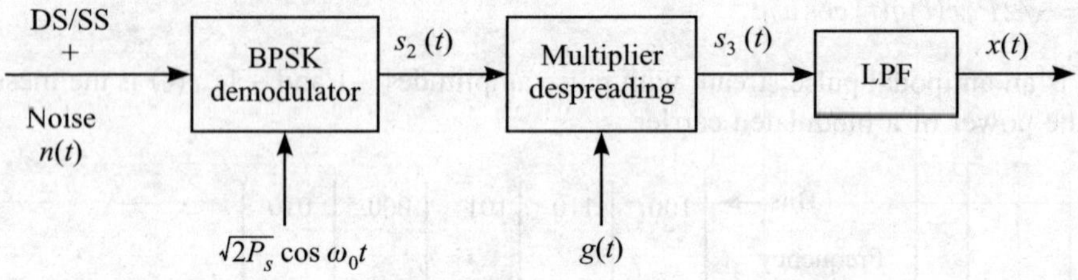

Fig. 12.7 *DS/SS receiver system*

$$r(t) = \underbrace{\sqrt{2P_s}x(t)g(t)\cos\omega_0 t}_{\text{wideband}} + n(t), \tag{12.9}$$

where $n(t)$ is the channel noise. We have neglected here the propagation delay from the transmitter to the receiver. Synchronous detection after BPSK demodulation yields

$$\begin{aligned} s_2(t) &= r(t)\sqrt{2}\cos\omega_0 t \\ &= [\sqrt{2P_s}x(t)g(t)\cos\omega_0 t + n(t)]\sqrt{2}\cos\omega_0 t \\ &= 2\sqrt{P_s}x(t)g(t)\cos^2\omega_0 t + n(t)\sqrt{2}\cos\omega_0 t \end{aligned}$$

$$= \sqrt{P_s}x(t)g(t)[1 + \cos 2\omega_0 t] + \sqrt{2}n(t)\cos\omega_0 t. \tag{12.10}$$

This signal is multiplied by a locally generated PN sequence $g(t)$ to yield

$$\begin{aligned} s_3(t) &= s_2(t)g(t) \\ &= \sqrt{P_s}x(t)g^2(t)[1 + \cos 2\omega_0 t] + \sqrt{2}n(t)g(t)\cos\omega_0 t \\ &= \sqrt{P_s}x(t)[1 + \cos 2\omega_0 t] + \sqrt{2}n(t)g(t)\cos\omega_0 t \end{aligned} \tag{12.11}$$

since $g^2(t) = 1$. Note that the noise is chipped at the receiver (that is, reversal of polarities at random times) producing spreaded spectrum. Finally, the lowpass filter removes the high frequency components, leaving the signal $x(t)$ with a scale factor.

Since the channel noise is white Gaussian, the chopping has no effect on the noise spectral density and the probability of error would be

$$P_e = \frac{1}{2}\,\text{erfc}\,\sqrt{E_b/\eta}. \tag{12.12}$$

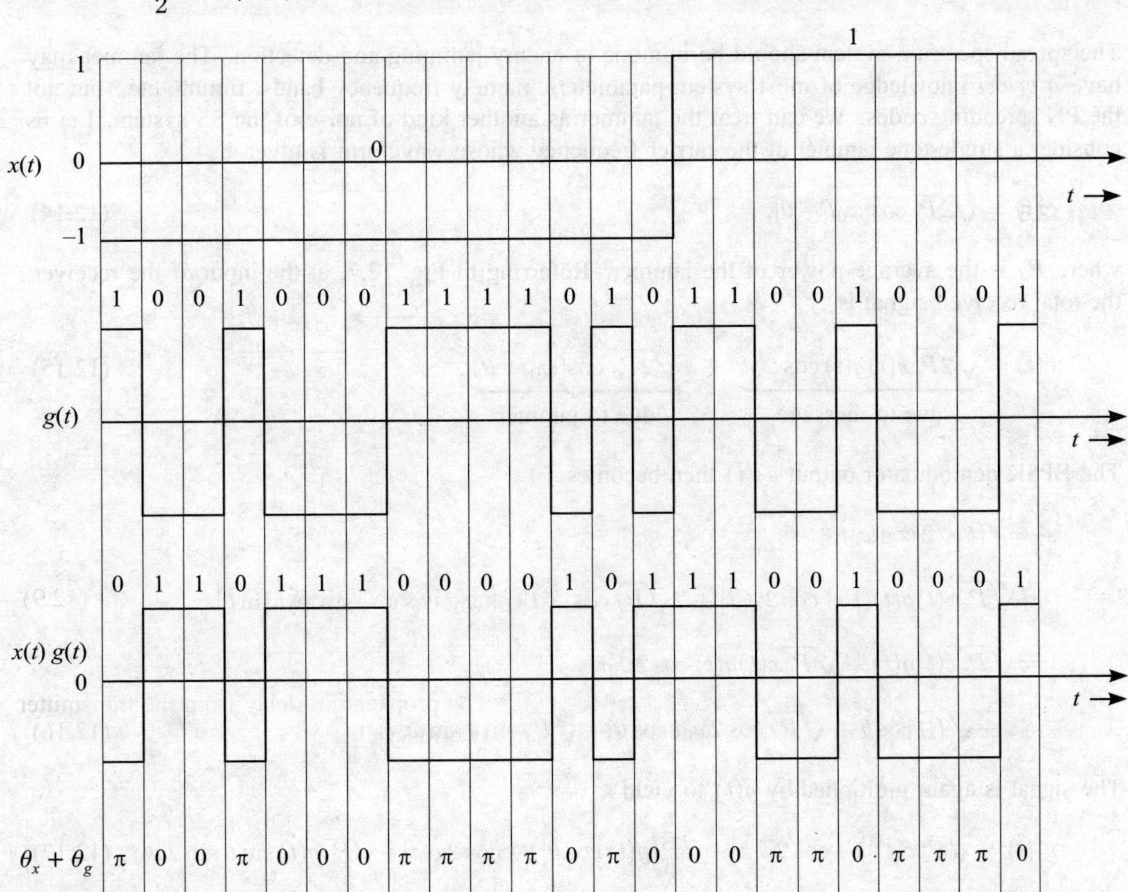

Fig. 12.8 *Spread spectrum waveforms and phase of the transmitted carrier*

It is important to study the waveform of DS/SS signal. If $x(t)$ is to represent that digital message, $x(t) = \pm 1$. Again the pseudorandom signal has $g(t) = \pm 1$. The transmitted DS/SS signal can be written as (from (12.8))

$$\begin{aligned} s(t) &= \sqrt{2P_s}x(t)g(t)\cos\omega_0 t \\ &= \sqrt{2P_s}\cos[\omega_0 t + \theta_x(t) + \theta_g(t)], \end{aligned} \tag{12.13}$$

with carrier phase $\theta_x(t) + \theta_g(t)$; $\theta_x(t)$ is due to the digital message $x(t)$ and $\theta_g(t)$ is due to the PN sequence $g(t)$. Obviously, the carrier phase is π radians when the product $x(t)g(t)$ equals -1 and zero when $x(t)g(t)$ equals $+1$. Figure 12.8 shows the waveforms of $x(t)$, $g(t)$, $x(t)g(t)$ and the related phase. One can observe that the signal $x(t)$ is hidden within $x(t)g(t)$. From the fast-varying $x(t)g(t)$ and the phase; the receiver will have to recover the slow-varying message $x(t)$. This is only possible if the exact phase change of 0 or π can be produced in the receiver by the code signal $g(t)$.

12.3.1 Single-tone interference in DS/SS systems

The spread spectrum system should be immune to enemy jamming and detection. The jammer may have *a priori* knowledge of most system parameters, namely frequency bands, timing, etc., but not the PN spreading codes. We can treat the jammer as another kind of noise of the SS system. Let us consider a single-tone jammer at the carrier frequency whose waveform is given by

$$z(t) = \sqrt{2P_j}\cos[\omega_0 t + \theta], \tag{12.14}$$

where P_J is the average power of the jammer. Referring to Fig. 12.7, at the input of the receiver, the total received signal is

$$r(t) = \underbrace{\sqrt{2P_s}x(t)g(t)\cos\omega_0 t}_{\text{due to message}} + \underbrace{\sqrt{2P_J}\cos(\omega_0 + \theta)}_{\text{due to jammer}}. \tag{12.15}$$

The BPSK demodulator output $s_2(t)$ then becomes

$$\begin{aligned} s_2 &= r(t)\sqrt{2}\cos\omega_0 t \\ &= \sqrt{P_s}x(t)g(t)[1 + \cos 2\omega_0 t] + 2\sqrt{P_J}\cos\omega_0 t[\cos\omega_0 t\cos\theta - \sin\omega_0 t\sin\theta] \\ &= \sqrt{P_s}x(t)g(t) + \sqrt{P_s}x(t)g(t)\cos 2\omega_0 t \\ &\quad + \sqrt{P_J}\cos\theta + \sqrt{P_J}\cos 2\omega_0 t\cos\theta - \sqrt{P_J}\sin\theta\sin\omega_0 t. \end{aligned} \tag{12.16}$$

The signal is again multiplied by $g(t)$ to yield

$$s_3(t) = \sqrt{P_s}x(t)[1+\cos 2\omega_0 t] + \sqrt{P_J}g(t)\cos\theta[1+\cos 2\omega_0 t] - \sqrt{P_J}g(t)\sin\theta\sin 2\omega_0 t. \tag{12.17}$$

Note that the interfering signal appears as a broadband signal. The lowpass filter suppresses the out-of-band spectrum giving,

$$v_0(t) = \sqrt{P_s}x(t) + \sqrt{P_J}g(t)\cos\theta. \tag{12.18}$$

A figure of merit of the SS systems that provides error performance analysis is the signal-to-jammer power ratio. The power spectral density $G_J(f)$ of the jammer can be obtained as

$$G_J(f) = \frac{P_J\overline{\cos^2\theta}}{f_c}\left(\frac{\sin\pi f/fc}{\pi f/f_c}\right)^2$$

$$\approx \frac{P_J}{2f_c}, \quad |f| \le f_b \tag{12.19}$$

(since in the signal frequency range $\pm f_b$, the function $\dfrac{\sin\pi f/f_c}{\pi(f/f_c}\approx 1$ and $\overline{\cos^2\theta} = \frac{1}{2}$, as θ is random variable).

Thus the jammer spectral density reduces by $2f_c$. In the case of digital message transmitted via BPSK, the probability of error at the output of the receiver would be

$$P_c = \frac{1}{2}\text{ erfc}\left(\frac{E_b}{\eta}\right)^{1/2},$$

where $\eta/2$ is taken as $G_J(f)$. Thus,

$$P_e = \frac{1}{2}\text{ erfc}\sqrt{\frac{P_sT_b}{2G_J}} = \frac{1}{2}\text{erfc}\sqrt{\frac{P_sT_b}{P_J/f_c}}$$

$$P_e = \frac{1}{2}\text{ erfc}\sqrt{\frac{P_s}{p_{J,\text{eff}}}}, \tag{12.20}$$

where $P_{J,\text{eff}} = \dfrac{P_J}{(f_c/f_b)} = \dfrac{P_J}{G_P}$, (12.21)

which signifies that the jamming power has been effectively reduced by the factor $G_P = f_c/f_b$, called the 'processing gain'. This defines the gain in the signal-to-jamming ratio by the use of the spread spectrum. The larger the gain, the better the jammer rejection capability.

The ratio P_J/P_s is termed the 'jamming margin' and usually expressed in decibels.

We can express the ratio of signal energy E_b to the jammer energy $N_J(N_J = P_J/f_c)$ as

$$\frac{E_b}{N_J} = \frac{G_P}{P_J/P_s}$$

or $\quad 10\log_{10}(P_J/P_s) = 10\log_{10}(G_p) - 10\log_{10}(E_b/N_J) \quad \text{dB}$ (12.22)

Example 12.3

A spread spectrum system has the following parameters. Message bit rate $f_b = 3$ kbps, PN sequence chip rate $f_c = 3.072 \times 10^6$. If the average probability of error is not to exceed 10^{-5}, find the jamming margin.

Solution

The processing gain is

$$G_P = \frac{f_c}{f_b} = \frac{3.072 \times 10^6}{3 \times 10^3} = 1024.$$

The PN sequence then repeats after 1024 bits. This means it requires 10 stages of shift register to design the PN sequence generator. From the table of erfc function, we find that to achieve $P_e \sim 10^{-5}$,

$$E_b/N_J = 10.$$

Hence the jamming margin $= 10\log_{10}(1024) - 10\log_{10}(10)$ dB

or $\quad 10\log(P_J/P_s) = 30.103 - 10 = 20.103$ dB

or $\quad P_J/P_s = 102.4.$

We get an important result that if the interfering jammer power is 102.4 times larger than the received signal power, even then the message bits can be detected reliably.

We have not yet considered the effect of thermal noise in calculating P_e. The error probability in the presence of jammer and wideband white noise would be

$$P_e = \frac{1}{2}\,\text{erfc}\sqrt{\frac{E_b}{N_J + N_0}}, \tag{12.23}$$

where N_J is the thermal noise power spectral density. This increases the error probability. If one wishes to maintain the same P_e, the processing gain needs be adjusted. The presence of the jammer increases the white noise spectral from N_0 to $N_0 + P_J/f_c$.

12.3.2 Code-division Multiple Access

Multiple access allows the sharing of a channel with a specified RF bandwidth among other spread spectrum users simultaneously without interfering with one another. We consider here code-division multiple access (CDMA) in direct sequence systems. In this scheme, each one of the M-users would have their own spreading code, $g_m(t), m = 1, 2, \cdots, M$, orthogonal to one another. This minimises interference between the user groups

Figure 12.9 illustrates a typical DS/CDMA modulator and demodulator. Consider that $x_m(t)$ is a data modulated signal of the mth user with its spreading code $g_m(t)$. The spread signal $x_m(t)g_m(t)$ is transmitted over the channel. At the receiver, the signal is the linear summation:

$$r(t) = \sum_{m=1}^{M} x_m(t) g_m(t), \tag{12.24}$$

where we have neglected the signal delay. Suppose that the receiver is for the first user, who has the despreading code $g_1(t)$. The output of the receiver would be

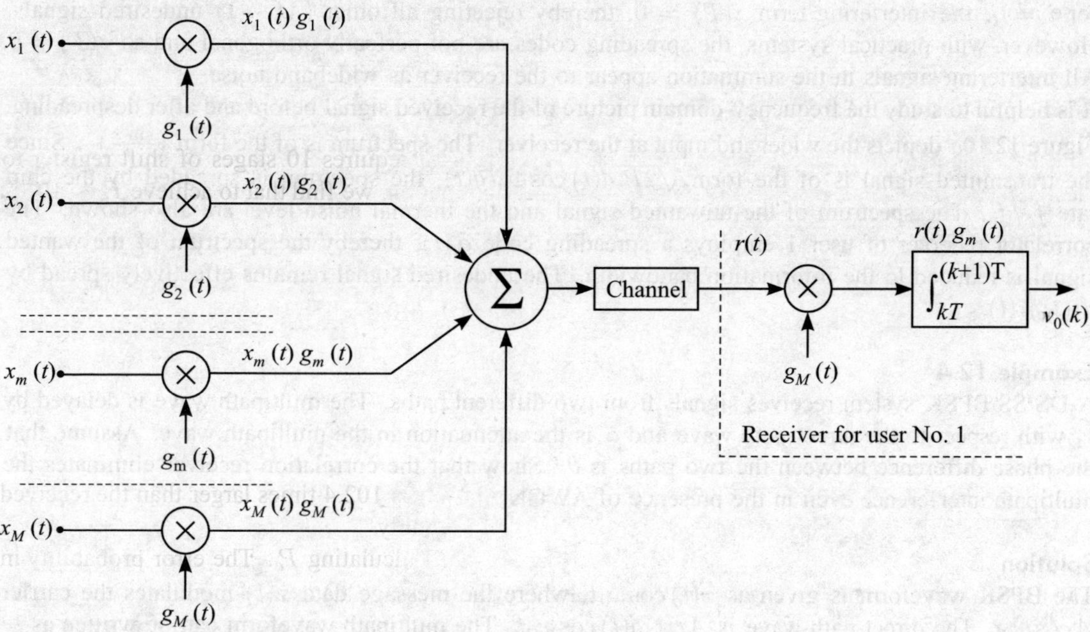

Fig. 12.9 *Code-division multiple access*

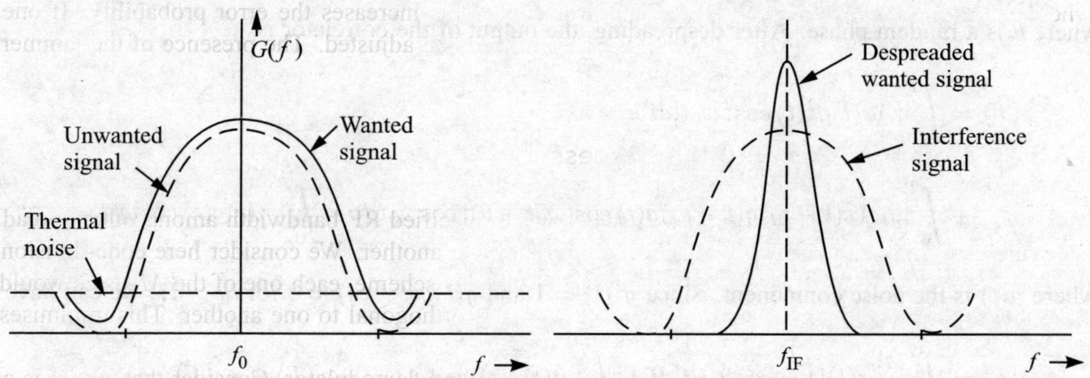

Fig. 12.10 *(a) Spectrum at the receiver input (b) spectrum after despreading by PN code.*

$$v_0(T) = \int_0^T r(t)g_1(t)dt$$

$$= \int_0^T x_1(t)g_1(t)g_1(t)dt + \sum_{m=2}^{M} \int_0^T x_m(t)g_m(t)g_1(t)dt = x_1(T) + z(T), \quad (12.25)$$

where $z(T)$ is the undesired interference due to the other $(M-1)$ users. Since $\int_0^T g_i(t)g_j dt = 0$ for $i \neq j$, the interfering term $z(T) = 0$, thereby rejecting all other $(M-1)$ undesired signals. However, with practical systems, the spreading codes are not perfectly orthogonal and so $z(T) \neq 0$. All interfering signals in the summation appear to the receiver as wideband noise.

It is helpful to study the frequency-domain picture of the received signal before and after despreading. Figure 12.10a depicts the wideband input at the receiver. The spectrum is of the form $\left(\frac{\sin x}{x}\right)^2$. Since the transmitted signal is of the form $\sqrt{2P_s}d(t)\cos\omega_0 t g(t)$, the spectrum is spreaded by the chip rate f_c/f_b. The spectrum of the unwanted signal and the thermal noise level are also shown. The correlator receiver of user 1 employs a spreading code $g_1(t)$, thereby the spectrum of the wanted signal is reduced to the information bandwidth. The undesired signal remains effectively spread by $g_1(t)g_1(t)$.

Example 12.4

A DS/SS BPSK system receives signals from two different paths. The multipath wave is delayed by t_d with respect to the direct-path wave and α is the attenuation in the multipath wave. Assume that the phase difference between the two paths is θ. Show that the correlation receiver eliminates the multipath interference even in the presence of AWGN.

Solution

The BPSK waveform is given as $x(t)\cos\omega_0 t$, where the message data $x(t)$ modulates the carrier $A\cos\omega_0 t$. The direct-path wave is $Ax(t)g(t)\cos\omega_0 t$. The multipath waveform can be written as

$$\alpha Ax(t-t_d)g(t-t_d)\cos(\omega_0(t-t_d)+\phi) = \alpha Ax(t-t_d)g(t-t_d)\cos(\omega_0 t+\theta),$$

where θ is a random phase. After despreading, the output of the correlator is

$$z(T) = \int_0^T [Ax(t)g^2(t)\cos^2\omega_0 t]dt$$

$$+ \int_0^T [\alpha Ax(t-t_d)g(t-t_d)g(t)\cos(\omega_0 t+\theta)]\cos\omega_0 t dt + \int_0^T n(t)g(t)\cos\omega_0 t,$$

where $n(t)$ is the noise component. Since $g^2(t) = 1$ and $g(t)g(t-t_0) \approx 0$ for $t_d > T_c$, we can have

$$z(T) = \int_0^T \frac{A}{2}x(t)[1+\cos 2\omega_0 t]dt + \int_0^T n(t)g(t)\cos\omega_0 t = \frac{A}{2}x(T) + n_0(T).$$

Thus the DS/SS system eliminates multipath interference.

12.3.4 Distance measurement using DS/SS

A popular application of DS spread spectrum is ranging, based on the time delay between the transmitted and the received signals. Suppose that a DS/SS signal $s(t) = \sqrt{2P_s}g(t)\cos\omega_0 t$ is transmitted to the target. The reflected signal from the target is received t_d seconds later where $t_d = 2D/c$, with D being the distance of the target from the transmitter and c the speed of light. Let the reflected signal received be $r(t) = \alpha s(t - t_d)$, where α accounts for the attenuation of the medium.

Thus,

$$r(t) = \alpha\sqrt{2P_s}g(t - t_d)\cos(\omega_0 t + \theta), \tag{12.26}$$

where $\theta = -\omega_0 t_d$. The PN sequence $g(t - t_d)$ is extracted from the received signal and correlated with the transmitted PN sequence $g(t - \tau_d)$, where τ_d is an adjustable delay. Figure 12.11 shows the direct sequence ranging system. The correlation function is

$$R_g(\tau_d) = \int_0^{LT_c} g(t - t_d)g(t - \tau_d)dt, \tag{12.27}$$

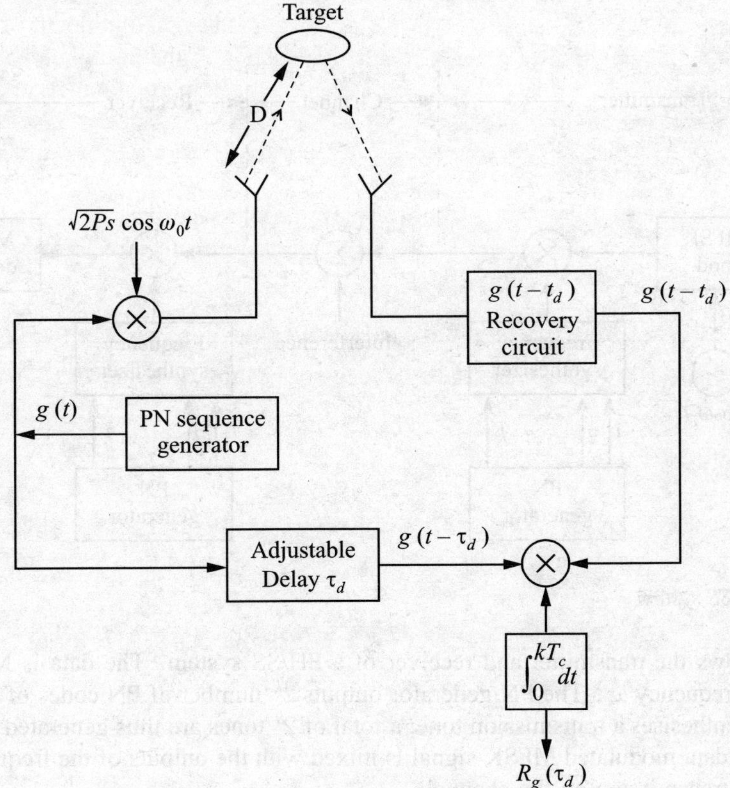

Fig. 12.11 *Ranging using DS/SS*

where L denotes the length of the PN sequence and T_c is the chip duration. The correlation function in (10.27) will attain maximum value if $t_d = \tau_d$. So the range D of the target would be

$$\frac{2D}{c} = \tau_d \quad \text{or} \quad D = \frac{1}{2}c\tau_d. \tag{12.28}$$

The resolution or precision of measurement is the chip duration $\pm T_c$ and hence

$$D = \frac{1}{2}c(\tau_d \pm T_c). \tag{12.29}$$

12.4 Frequency-Hop Spread Spectrum

Frequency-hop spread spectrum (FH/SS) is the process of randomly changing the frequency of the data-modulated carrier. A commonly used data modulation in this case is the M-ary FSK (MFSK), in which case the message data modulates a fixed frequency carrier. The carrier frequency is driven to hop from one frequency to another. Thus FH system includes a two-step modulation processes—data modulation (FSK, for example) followed by the frequency-hop modulation. Numerous carrier frequencies are generated in a frequency synthesiser driven by a PN generator. Since message bandwidth is spread out over wider frequency range, the jammer has reduced probability of interfering with the message.

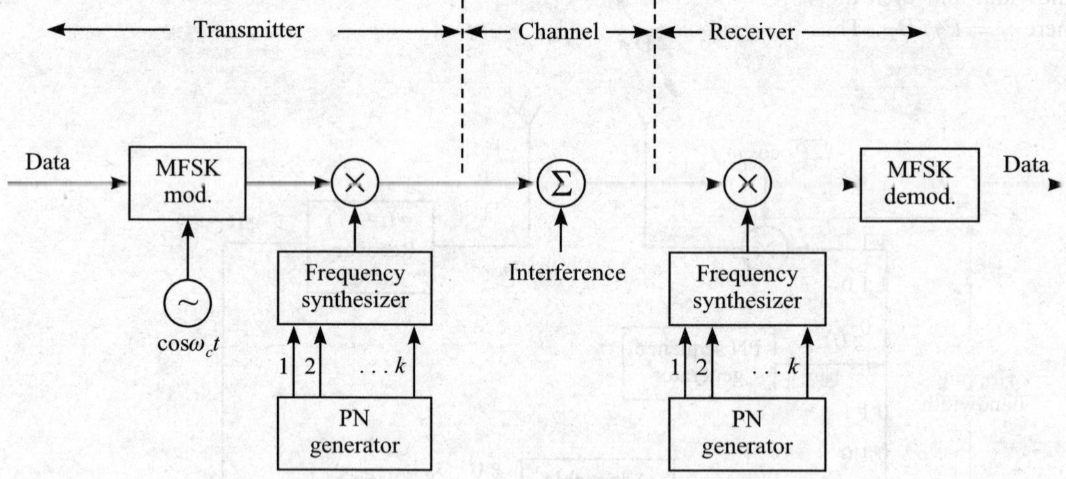

Fig. 12.12 *FH/SS system*

Figure 12.12 shows the transmitter and receiver of a FH/SS system. The data is MFSK modulated to some carrier frequency ω_c. The PN generator outputs 2^k number of PN codes of word size k-bits. Each PN code synthesises a transmission tone; a total of 2^k tones are thus generated by the frequency synthesiser. The data modulated MFSK signal is mixed with the outputs of the frequency synthesiser for hopping and transmission on the channel.

The received signal is first developed by using a mixer with the input coming from the same frequency synthesiser. The developed signal is then applied to the demodulator for data recovery.

For illustration, let us consider a FH/SS system with one symbol per frequency hop. The system uses a PN generator with $2^k = 2^3 = 8$ outputs. Consequently, there would be 8 different carrier frequencies, $f_{c0}, F_{c1}, \cdots, f_{c7}$, as illustrated in Fig. 12.13. If one uses binary FSK, and if carrier frequency f_{c0} is selected, bit 1 is transmitted using $f_{c0} + f_d$ and bit 0 is transmitted using $f_{c0} - f_d$, where f_d is the frequency drift. The maximum frequency is $f_{c7} + f_d$ and the minimum frequency is $f_{c0} - f_d$ so that the transmission bandwidth (or hopping bandwidth) is $B_T = f_{c7} - f_{c0} + 2f_d$. If the bit rate is r_b, then $f_d = r_b/2$ for Sunde's FSK. Thus $f_{c0} + f_d = f_{c1} - f_d$ and as on. This gives $B_T = 14f_d + 2f_d = 16f_d = 8r_b$. Further, the system transmits one symbol per frequency hop, giving the hopping rate $1/f_h = r_b$. At a particular hop if the PN code is 010, the carrier frequency is f_{c2}. Figure 12.13 also gives the plot of the output frequencies for the input messages 1 0 11 0 1 using the PN sequence: 010 101 000 011 111 101.

Frequency hopping systems are of two types: slow hop and fast hop. In slow-frequency hopping, there are one or more symbols per hop; in fast hop, there are several frequency hops per message symbol.

An individual FH/MFSK tone of shortest duration is also called a 'chip' (do not confuse it with the term 'chip' defined in DS/SS system). It is very essential to know the difference between the hop rate (R_h) and the symbol and the symbol rate (R_s). The chip rate R_c for this SS system is the maximum of R_h and R_c. The relation between the symbol rate and the bit rate is $R_b = KR_s$, where $K = \log_2 M$ number of bits together form a symbol.

Due to the spreading of the spectrum, the average jammer power P_J is reduced by the hopping bandwidth and then the jammer is effectively behaving as AWGN of power spectral density $\eta/2$, where $\eta = P_J/B_T$. Thus the symbol energy-to-noise spectral density

$$E_s/\eta = \frac{P_s/R_s}{P_J/B_T} = \frac{P_s/P_J}{R_s/B_T} = \frac{P_s}{P_J/(B_T/R_s)}.$$

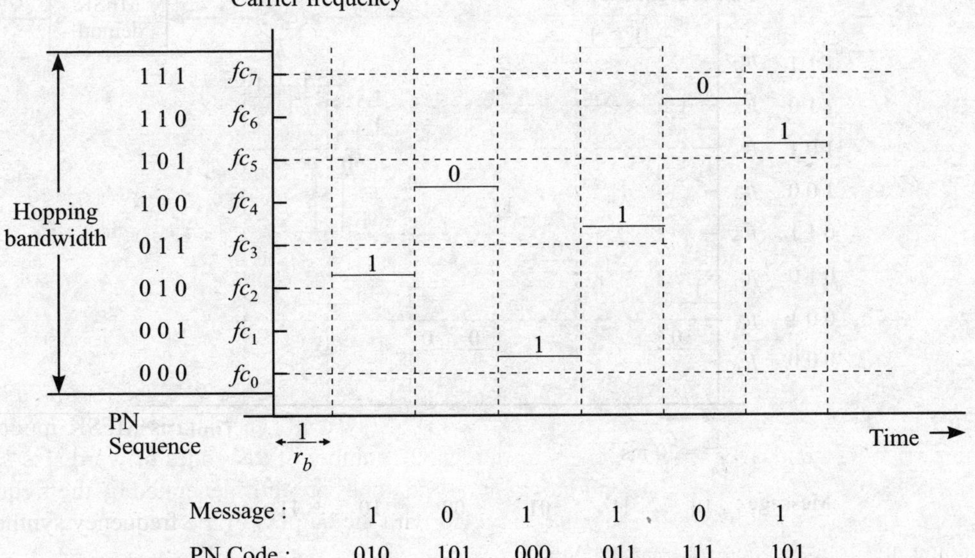

Fig. 12.13 *FH/SS spectral outputs*

Using a reasoning similar to the DS/SS, we define the processing gain of the slow hop frequency spectrum by

$$G_p = B_T/R_s = 2^k,$$

where k is the length of the PN sequence applied to the frequency synthesiser.

Example 12.5

A slow-hop frequency spectrum has the following parameters: Number of bits per symbol = 4; number of symbol per hop = 5. Find the processing gain.

Solution Here, $K = 4$, so the number of M-ary FSK tones is $M = 2^K = 16$.

Number of symbols per hop is $k = 5$.

So, processing gain $2^k = 2^5 = 32$.

Example 12.6

Consider a slow-hop SS with binary FSK that transmits two symbols per frequency hop and has a PN generator with $k = 3$ outputs. The bit rate is $r_b = 100$ bits/s. Binary message sequence 10 11 01 00 10 11 is transmitted using the PN sequence $\{001, 111, 011, 001, 101, 110\}$. Plot the output frequencies for the message.

Solution

The symbol rate is $r_s = r_b/\log_2 M = r_b = 100$ bits/s. Symbol duration, $T_s = 1/100 = 10$ ms. As there are two symbols per hop, so the hopping rate is $f_h = \frac{1}{2Ts} = \frac{1}{20}\ ms = 50$ Hz. Figure 12.14 shows the output frequencies for the message inputs.

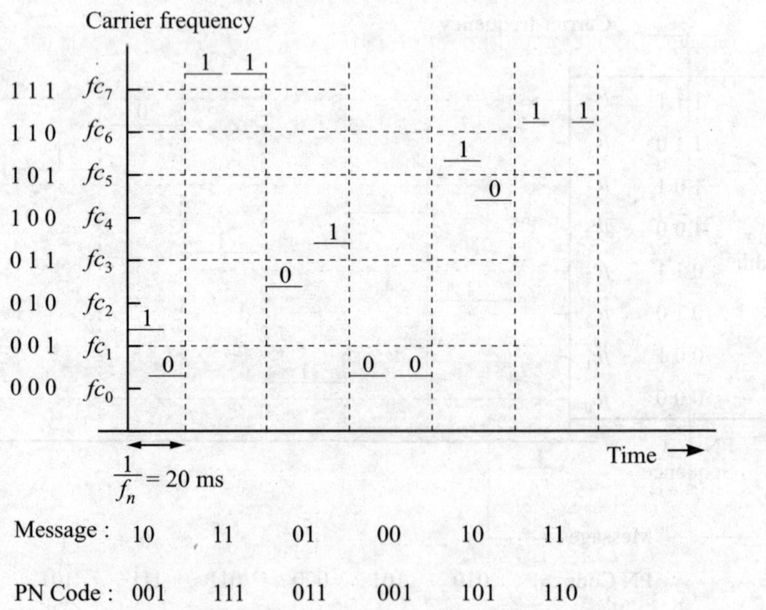

Fig. 12.14 *Frequency hop example, showing two symbols per hop.*

Example 12.7

Sketch the output frequency for the input message for fast-hop and slow-hop spread spectrum systems.

Solution Figures 12.15 and 12.16 depict the slow-hop and fast-hop cases respectively.

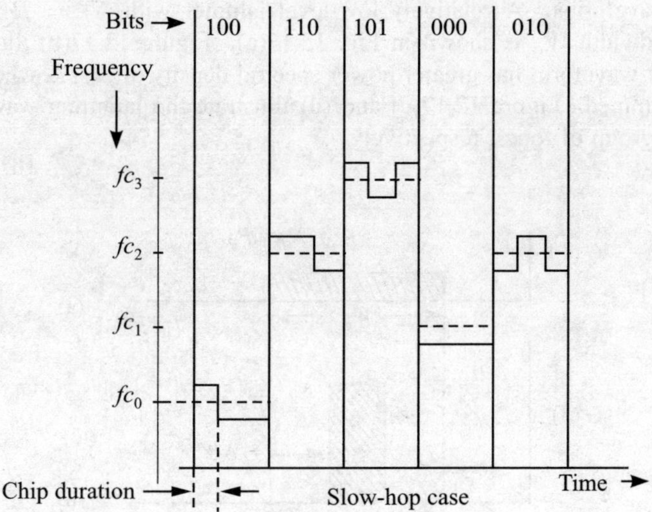

Fig. 12.15 *Slow-frequency hop, indicating two frequencies per symbol*

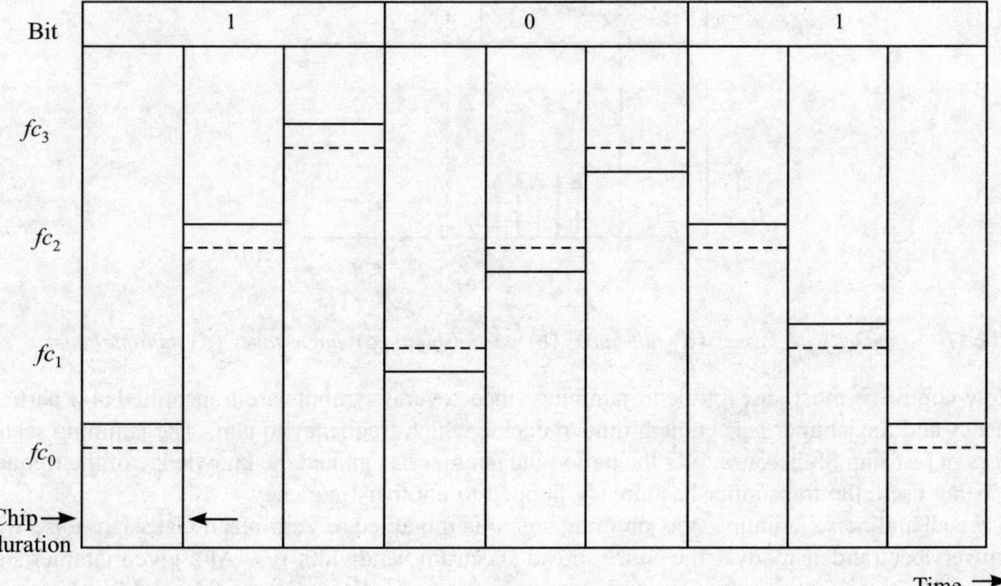

Fig. 12.16 *Fast frequency-hop example*

12.4.1 Jamming in FH/SS systems

There are many different interference or noise signals called jamming waveforms that are used for jamming communications processes. The jammer waveforms may be full-band or white noise, partial-band, single-tone or multiple tone. Figure 12.17 shows the plots of the power spectral densities of these jamming waveforms. A relatively low-level jammer with $N_J = P_J/W_c$ occupying the spread-spectrum bandwidth W_c is shown in Fig. 12.17(a). Figure 12.17(b) illustrates partial-band jamming; the jammer waveform has greater power spectral density of N_J/Δ where Δ is the fraction of the band being jammed. Figure 12.17(c) and (d) illustrate the jamming waveforms of jammers using single tone or group of tones, respectively.

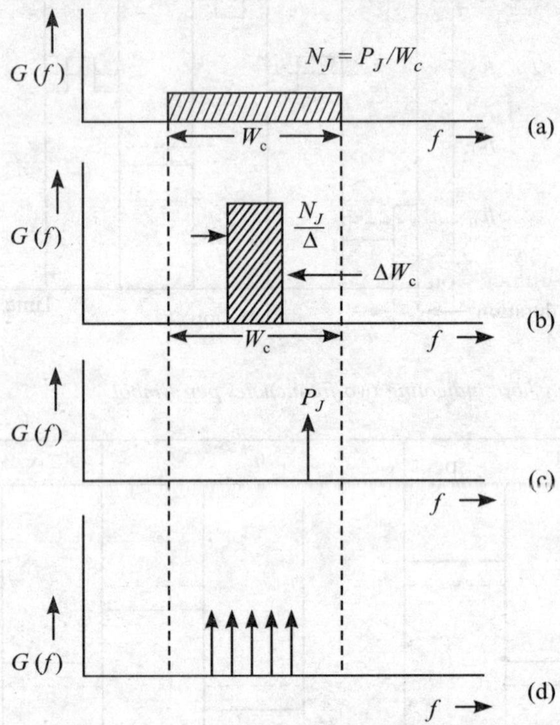

Fig. 12.17 *Jammer waveforms: (a) full-band; (b) partial-band (c) single-tone; (d) multiple-tone.*

Slow-hop SS is most susceptible to jamming since several symbols are transmitted at a particular frequency and the jammer gets enough time to decide which frequency to jam. The jamming scenario changes in fast-hop SS because over the period the jammer has gained the knowledge of the frequency band being used, the transmitter has already hopped to another frequency.

In broadband noise jamming, the jamming signal is modelled as zero-mean Gaussian noise with a flat power spectral density over the entire spread spectrum bandwidth W_c. At a given jammer signal power of P_J, the jammer power spectral density is $N_J = P_J/W_c$. For an M-ary FSK system, the bit error probability P_e in the presence of white noise is

$$P_e = \frac{1}{2} e^{-E_b/2N_0}, \tag{12.30}$$

where $N_0/2$ is the two-sided power spectral density of white noise. The jammer power increases this noise power spectral density from N_0 to $N_0 + N_j$ so that

$$P_e = \frac{1}{2} e^{-E_b/2(N_0+N_J)}. \tag{12.31}$$

In the partial-band noise jamming, the jammer noise power spectral density is concentrated within $\Delta \times W_c$, where $0 < \Delta < 1$. A transmitted symbol will be received unjammed with probability $1 - \Delta$ and the probability of being jammed is Δ. So the average bit error probability is given by

$$P_e = (1 - \Delta)\frac{1}{2} e^{-E_b/2N_0} + \Delta \frac{1}{2} e^{-E_b/2(N_0+N_J\Delta)}. \tag{12.32}$$

When the jamming power is large compared to the thermal noise that is $N_J \gg N_0$, we can simplify (12.32) to

$$P_e \approx \frac{\Delta}{2} e^{-\Delta E_b/2N_J}. \tag{12.33}$$

The choice of Δ by the jammer would be such that it maximises P_e.

Example 12.8
Find the fraction Δ for which P_e in (12.33) is maximum. Also find this maximum P_e.

Solution
By differentiating (12.33) with respect to Δ and setting $\frac{dP_e}{d\Delta} = 0$ that results in maximum P_e, one can have,

$$0 = \frac{dP_e}{d\Delta} = \frac{\Delta}{2}\left[-\frac{E_b}{2N_J}.e^{-\Delta E_b/2N_J}\right] + e^{-\Delta E_b/2NJ}.\frac{1}{2}$$

$$\frac{1}{2} e^{-\Delta E_b/2N_J}\left[1 - \frac{\Delta E_b}{2N_J}\right] = 0$$

which gives $\Delta = \dfrac{2}{E_b/N_J}$ and consequently,

$$(P_e)_{\max} = \frac{2}{E_b/N_J}\frac{1}{2}e^{-1} = \frac{e^{-1}}{E_b/N_J}.$$

Single-tone jamming does not introduce large errors provided $P_J \ll P_s$ and there are large enough number of frequency slots to hop. If, incidentally, for a particular frequency hop the jamming falls within that slot during the same hop, the detector would be confused while deciding the correct symbol.

In multiple-tone jamming, the jammer divides its total power P_J into various random-phase CW tones. The tones are distributed over the same frequency band W_c. A number of FH/SS users within the same band in the CDMA channel also causes jam-like interference.

12.5 Spread Spectrum Benefits

The spread spectrum communication techniques are aimed at several application areas namely, anti-jam communications, high resolution ranging, multiple access and the others. Spread spectrum is truely a bandpass communication. Over the signal bandwidth, the presence of any component other than the signal can interfere with the signal itself. At the spread spectrum modulator, the information signal bandwidth is spreaded by multiplying by spreading signal. The demodulator again multiplies the received signal by the same spreading signal and recovers the orignal signal by filtering. As the interference signal gets multiplied only once, its spectrum is spreaded at receiver, giving reduced power spectral density and hence goes unnoticed.

The ever present white Gaussian noise can also be treated as an interfering signal but in this case, we consider only the noise components that are present within the signal bandwidth. Signal spreading in the presence of white noise will not offer an improvement in noise performance which can be understood by the following example.

Example 12.9

Show that the received E_b/η remains the same after spreading.

Solution

Refer to Fig.12.18 where we have shown the power spectral densities of a communication signal and white noise before and after spreading. For the ease of calculation, we have considered ideal rectangular shapes. As shown in the figure, the signal bandwidth is W Hz and the processing gain is f_c/f_b. Assume a transmitted data rate of r bits per second.

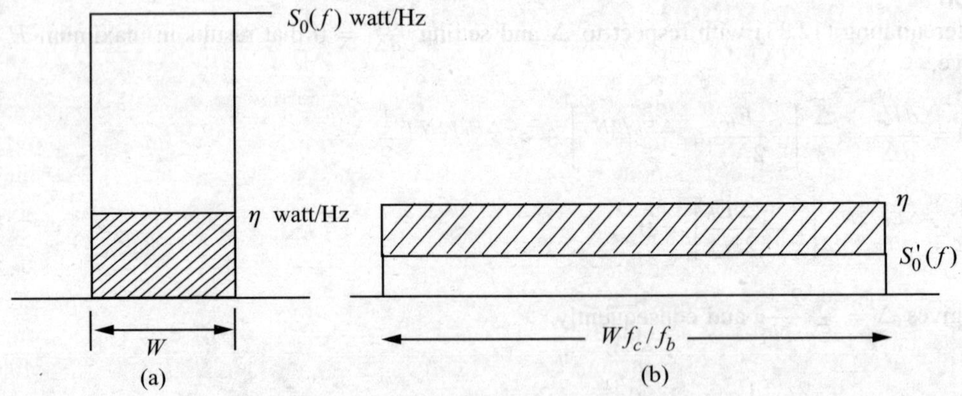

Fig. 12.18 *Signal and noise PSDs: (a) before spreading, (b) after spreading.*

The total average signal power is

$$S = S_0 \frac{\text{watt}}{\text{Hz}} \times W \text{ Hz} = S_0 W \text{ watts}$$

The total average noise power is

$$N = \eta \frac{\text{watt}}{\text{Hz}} \times W \text{ Hz} = \eta W \text{ Watts.}$$

So the received E_b/η is (note that E_b refers to signal energy per bit and η is noise PSD)

$$\frac{E_b}{\eta} = \frac{\frac{S}{r}\frac{\text{watts}}{\text{bits/sec}}}{\eta\frac{\text{watts}}{\text{Hz}}} = \frac{S}{r\eta}. \tag{12.34}$$

After spreading, the PSD of the signal is reduced by f_c/f_b, as the bandwidth is increased by the same factor, So, the total average power of the signal remains the same. With the noise assumed AWGN, the PSD does not change but the total average noise power is

$$N' = \eta \frac{\text{watts}}{\text{Hz}} \times \frac{f_c}{f_b} \text{ W Hz} = \eta \frac{W f_c}{f_b} \text{ watts.}$$

So, after spreading the $E_b/\eta r$ is

$$\frac{E_b}{\eta} = \frac{S/r}{\eta} = \frac{S}{r\eta}.$$

Since the bit error is determined by the ratio E_b/η, there is no improvement in error performance. The signal appears burried in noise (as shown in the figure), but it is the scheme in the DS/SS that despreads the signal, thereby returning the original message spectrum with high PSD and hence the signal is detected.

12.6 Synchronisation

Successful demodulation of the received spread spectrum signal is possible if we know the transmitted carrier frequency and phase together with a perfect alignment between the transmitter and received PN codes. Usually, the despreading process takes place before the carrier synchronisation and so the carrier phase is unknown at this point. The constraint of phase synchronisation can be removed if one utilises non-coherent detection. We then have to align the phase of the transmitted and locally generated spreading codes. The sources of non-synchronisation are broadly due to clock instabilities of the transmitter and receiver and the Doppler frequency offset of the incoming signal. The latter is very pertinent in mobile or satellite SS systems. The alignment of the receiver PN code to the incoming PN code is usually accomplished in two steps: The first step is called the acquisition, where the two PN codes are brought into coarse alignment. The second step is called the tracking, which continuously maintains a fine alignment of the code.

Acquisition

In this method of code synchronisation, the transmitted PN code and the locally generated replica are correlated. The result is then compared with a threshold. Figures 12.19 and 12.20 show the serial search acquisition system for DS/SS and FH/SS systems, respectively. In the DS/SS serial acquisition system, the locally generated PN code signal is multiplied with the incoming PN code.

The product is then integrated over an interval of lT_c seconds (assuming that both the codes contain the same number of PN chips). The output signal is then compared with a threshold V. When there is alignment, the integrator produces a value higher than V and the phase of PN code is reached; the tracking system then takes over. Otherwise, the comparison circuitry generates an acquisition signal that increments the phase of the PN generator (the increment is usually a half-chip). The FH/SS serial acquisition of Fig. 12.20 works in the similar way. The PN code generator controls the frequency synthesiser.

Assume that the locally generated PN code is initially offset by a phase difference of N_c number of chips. Here N_c is a random number which may very between 0 and $L - 1$, where L is the PN code period in the chips. In a single search, the phase increment is half the chip period. To examine a sequence of N_c chips, the number of search is $2N_c$ chip period.

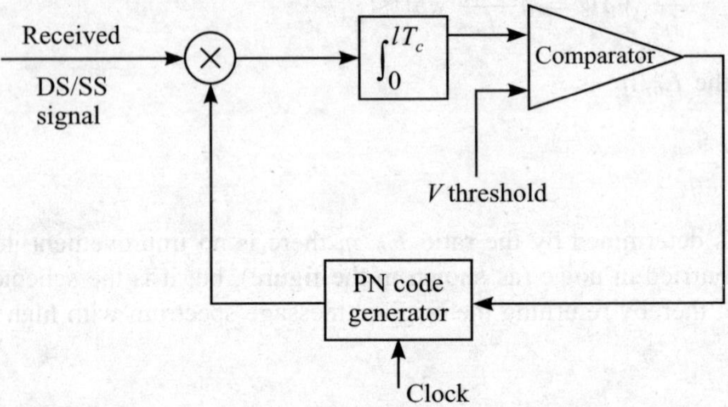

12.19 *Direct-sequence serial search*

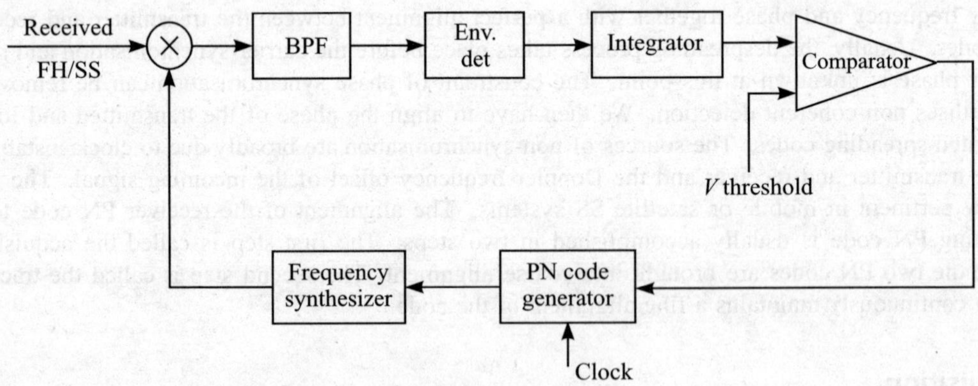

12.20 *Frequency hopping serial search acquisition*

Solved Numericals

1. A frequency hopping system is with a hopping bandwidth of $B_T = 500$ MHz and a frequency step size (frequency drift) f_d of 125 Hz. Find the minimum number of PN chips required for every frequency allocation.

Solution

Number of frequency tones in the range is $= \frac{B_T}{t_d}$

$$= \frac{500\,\text{MHz}}{125\,\text{Hz}} = 4 \times 10^6$$

If PN codes are of word size k-bits, a total of 2^k PN codes will synthesise 2^k different tones. Thus,

$$2^k = 4 \times 10^6$$

or, Number of chips, $k = log_2(4 \times 10^6)$

Now, $2^{22} = 4194304 \approx 4.19 \times 10^6$

Thus, $k = 22$ chips.

2. A DS/SS BPSK system has the following parameters.
Data rate $r_b = f_b = 3$ kbps, write noise power spectral density $\eta = 10^{-12}$ W
The data is received with an error probability of $P_e = 10^{-8}$. For a single-tone jammer, one requires an P_e of 10^{-7}; find the processing fain G_p. Given that the power received by the jammer is 10 times higher than the signal power.

Solution

The error probability for BPSK reception is given by (Eq. 12.12)

$$P_e = \frac{1}{2} erfc\left(\sqrt{\frac{E_b}{\eta}}\right) = Q\left(\sqrt{\frac{2E_b}{\eta}}\right)$$

For a received error $P_e = 10^{-8}$ for wideband white noise,

$$10^{-8} = Q\left(\sqrt{\frac{2E_b}{\eta}}\right) \quad \text{or} \quad \sqrt{\frac{2E_b}{\eta}} = 5.65 \text{ (from the table of } Q\text{-function)}$$

$$\text{Or, } E_b = (5.65)^2 \frac{\eta}{2} = \frac{10^{-12}}{2}(5.65)^2 = 15.96 \times 10^{-12} \text{ Joule}$$

Since $P_s T_b = E_b$, we have signal power $P_s = \frac{E_b}{T_b} = f_b E_b = r_b E_b = 3 \times 10^3 \times 15.96 \times 10^{-12} =$ 47.88 nW

Jamming power, $P_J = 10 \times P_s = 470.88$ nW

So, in presence of both the white noise and jamming signal, we may write Eq. (12.23) as,

$$Pe = \frac{1}{2} erfc\left(\sqrt{\frac{E_b}{\eta}}\right) = Q\left(\sqrt{\frac{2E_b}{\eta + \eta_J}}\right)$$

whose $\eta_J = PSD$ of jamming signal = $\frac{P_J}{f_c}$,(f_c being the chip rate)

Since $P_e = 10^{-7}$, we have

$$10^{-7} = Q\left(\sqrt{\frac{2P_s/f_b}{\eta+\eta_J}}\right) = 5.2 \text{ (from the table of Q-function)}$$

So,

$$\left(\frac{2P_s/f_b}{\eta+\eta_J}\right) = (5.2)^2$$

or,

$$\eta_J = \frac{2P_s/f_b}{(5.2)^2} - \eta = \frac{2 \times 15.96 \times 10^{-12}}{(5.2)^2} - 10^{-12} = 0.1804 \times 10^{-12}$$

So, $P_J/f_c = 0.1804 \times 10^{-12}$

or, $f_c = \frac{P_J}{\eta_J} = \frac{10 \times 47.88 \times 10^{-9}}{0.1804 \times 10^{-12}} = 2654 \times 10^{-12} = 2.654 \times 10^6$ Hz

Thus processing gains

$$G_P = \frac{f_c}{f_b} = \frac{2.654 \times 10^6}{3 \times 10^3} = 885$$

3. A FH/SS system receives message signals at the rate $r_b = 3$ kbps with a received signal power $P_s = 5.5 \times 10^{-8}$ W. The PN sequence is 10- bit long. Find the maximum error probability of the received symbols. Assume the system is in the partial-band jamming.

Solution

In the case of partial-band noise jamming, the error probability is given by Eq. (12.32).

That is, $P_e \approx \frac{\Delta}{2} \cdot e^{-\Delta.E_b/2N_J}$

For maximum error probability of the received symbols, we must have $\Delta = \frac{2}{E_b/N_J}$ (vide Example 12.8)

Jammer PSD is $N_J = {}^{P_J}/_{W_c}$ where W_c is the spread spectrum bandwidth and is given by

$G_p = \frac{W_c}{r_b}$ where the processing gain G_p is obtained as $G_p = 2^k = 2^{10} = 1024$. Thus,

$$W_c = G_p \cdot r_b = 3 \times 10^3 \times 1024 = 3074 \times 10^6 \text{ Hz.}$$

Jammer PSD, $N_J = \frac{P_J}{W_c} = \frac{P_J}{2^k \cdot r_b}$

Received signal power, $P_s = E_b \cdot r_b$ so that

$$E_b \Big/ N_J = \frac{P_s/r_b}{P_J/(2^k r_b)} = \frac{P_s}{P_J} \cdot 2^k = \frac{2^k}{10} = \frac{1024}{10} \text{ (since } P_J = 10P_s)$$

Thus, $\Delta = \frac{2 \times 10}{1024} = 0.0195 \simeq 2\%$

We may now compute maximum error probability

$$P_e(\max) = \frac{\Delta}{2}e^{-1} = \frac{\Delta}{2e} = \frac{0.0195}{2 \times 2.7182} = 3.6 \times 10^{-3}$$

4. In a code-division multiple access system M-users have their distinct spreading codes $g_m(t)$, $m = 1, 2, \ldots M$ on the same level. Codes being orthogonal to each other, minimises interference between users. One user may be affected with the interfering terms of all other $(M - 1)$ users. As applied to FH/SS, find the probability that $(M - 1)$ users will interfere with the M-th user.

Solution

If we assume that there are $F = 2^k$ frequencies in the CDMA channel, the probability that one interfering user will be in a given slot is $\frac{1}{N}$. So the probability of the interferer not in the slot is $\left(1 - \frac{1}{N}\right)$. So, the probability that the rest of the $(M - 1)$ users will not interfere with the M-th user is $\left(1 - \frac{1}{N}\right)^{M-1}$. Thus, the probability that $(M - 1)$ users will interfere with the M-th user is $P_i = \left(1 - \frac{1}{N}\right)^{M-1} \approx 1 - \left(1 - \frac{M-1}{N}\right) = \frac{M-1}{N}$ (The final step above is true if $\left(\frac{1}{N}\right) \ll M - 1$ *or* $N \gg M$, that is, number of frequencies in the CDMA channel is much higher than the number of users)

The above error due to interferers is conditional in the sense that:

1. A user interfering with any other user in the same frequency slot has the probability of 1/2 and is denoted as $p(e\,|i) = {}^1\!/_2$.

2. A user is not affected by any other users but error is due to white noise only. Such error probability is denoted as $p(e\,|i^c) = \frac{1}{2}e^{-E_b/2N_0}$, which is the same as error probability in M-ary FSK and is given by Eq. (12.30).

So, M-th user may incur error either because of interference of the rest of the users or because of the white noise and these two noise effects are disjoint. The total error probability is given by

$$\begin{aligned} P_e &= P(e\,|i)P(i) + P(e\,|i)P(i^c) \\ &= \frac{1}{2} \cdot P_i + \frac{1}{2}e^{-E_b/2N_0} \cdot (1 - P_i) \\ &= \frac{1}{2}\left(\frac{M-1}{N}\right) + \frac{1}{2}e^{-E_b/2N_0}\left(1 - \frac{M-1}{N}\right) \end{aligned}$$

The term $P(e\,|i)$ may be interpreted as the transmitted symbol received jammed by the interferer. The term $P(e\,|i^c)$ is interpreted as probability of the signal being unjammed by the interferer but affected with noise.

5. In a DS/SS distance measuring system, what chip rate is required to achieve a resolution (or precision) of measurement of 0.01 kilometre? Find the distance resolution if a 10 MHz clock is used.

Solution

By Eq. (12.28) or (12.29), the resolution of distance measurement is

$\Delta D = \frac{1}{2} c T_c$ where c is velocity of light and T_c is the chip duration. Since ΔD is 0.01 km, we have

$$T_c = \frac{2\Delta D}{c} = \frac{2 \times 0.01 \times 10^3}{3 \times 10^8} \text{ sec}.$$

So chip rate is

$$f_c = \frac{1}{T_c} = \frac{c}{2 \times \Delta D} = \frac{3 \times 10^8}{2 \times 0.01 \times 10^3} \simeq 1.5 \times 10^7 = 15 \text{ MHz}.$$

When 10 MHz clock is used, clock period

$$T_c = \frac{1}{10 \times 10^6} = 0.1 \mu \text{ sec}.$$

and so distance resolution

$$\Delta D = \frac{1}{2} c T_c = \frac{1}{2} \times 3 \times 10^8 \times \frac{1}{10 \times 10^6} = 15 \text{ metres}.$$

6. For the generation of maximum length (ML) PN sequence, the shift resistor connections are provided in Table 12.3. Find the 4-bit ML SR Connections for generating 15-bit long PN sequence. Also verify all the properties of PN sequence.

Solution

Using Table 12.3, for $n = 4$ the shift register outputs are taken from (1, 4). The resulting PN sequence is shown below.

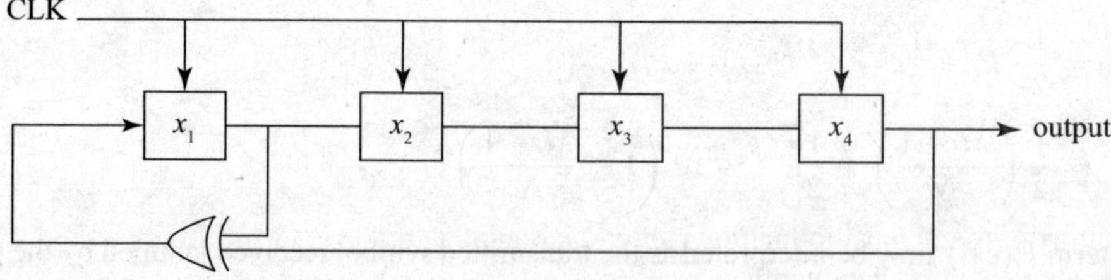

As in Table 12.4, we can find the PN sequence based on 15 different initial states of the SR contents x_1, x_2, x_3, x_4. The update of the SR contents is governed by the rule.

$$x_1' = x_1 + x_4, x_2' = x_1, x_3' = x_2, \text{ and } x_4' = x_3.$$

where $x_i'$is the content of the i th SR at the next clock for the initial state $x_1\ x_2\ x_3\ x_4 = 0001$, 0010, and so on.

We find the PN sequence as in the following table. (Only four sequences are derived with four initial states. Students are advised to find all 15 such sequences.)

Initial state $x_1\ x_2\ x_3\ x_4 = 0001$	Initial state $x_1\ x_2\ x_3\ x_4 = 0010$	Initial state $x_1\ x_2\ x_3\ x_4 = 0011$	Initial state $x_1\ x_2\ x_3\ x_4 = 0100$
0 0 0 1	0 0 1 0	0 0 1 1	0 1 0 0
1 0 0 0	0 0 0 1	1 0 0 1	0 0 1 0
1 1 0 0	1 0 0 0	0 1 0 0	0 0 0 1
1 1 1 0	1 1 0 0	0 0 1 0	1 0 0 0
1 1 1 1	1 1 1 0	0 0 0 1	1 1 0 0
0 1 1 1	1 1 1 1	1 0 0 0	1 1 1 0
1 0 1 1	0 1 1 1	1 1 0 0	1 1 1 1
0 1 0 1	1 0 1 1	1 1 1 0	0 1 1 1
1 0 1 0	0 1 0 1	1 1 1 1	1 0 1 1
1 1 0 1	1 0 1 0	0 1 1 1	0 1 0 1
0 1 1 0	1 1 0 1	1 0 1 1	1 0 1 0
0 0 1 1	0 1 1 0	0 1 0 1	1 1 0 1
1 0 0 1	0 0 1 1	1 0 1 0	0 1 1 0
0 1 0 0	1 0 0 1	1 1 0 1	0 0 1 1
0 0 1 0	0 1 0 0	0 1 1 0	1 0 0 1
0 0 0 1	0 0 1 0	0 0 1 1	0 1 0 0
Sequence is 1000 1111 0101 100	Sequence is 0100 0111 1010 110	Sequence is 1100 1000 1111 010	Sequence is 0010 0011 1101 011

In this way all other sequences can be obtained.

Observe the following.

Each sequence has eight 1's and seven 0's satisfying the balance property.

Exercises

Activity

12.1 Explain why frequency modulation cannot be regarded as a spread spectrum technique although it spreads the signal bandwidth.

12.2 Spread spectrum techniques can provide error performance benefits against interfering jaming signal. But this is not so for the AWGN. Explain why?

12.3 Define 'chip' and 'processing gain' parameter in connection with the spread spectrum modulation. What is the difference between fast-hopping and slow-hopping spread spectrum?

12.4 Distringush between the processing gains of DS and FH spread spectrum.

12.5 (a) Mention two basic criteria to be followed by the transmitted signal of a communications system so that it may be considered a spread spectrum system. Mention various types of spread spectrum systems.

(b) If a direct sequence spread spectrum signal is interfered with a sinusoidal signal of normalised power P_J at carrier frequency f_0, then prove the error probability

$$P_e = \frac{1}{2}\ \text{erfc}\ \sqrt{P_s/P_J/(f_c/f_b)}.$$

The symbols have their usual significance. [KU 2002]

12.6 Name the various classes of periodic sequences. Describe Euler totient function. Use this function to find the number of sequences formed for $m = 2$ in m sequences. Considering a basic chip function, show how a PN signal can be generated from PN sequences using proper correlation function. Plot the signal correlation.

12.7 A PN sequence generator consists of 4 linear feedback shift registers. The chip rate is 10^6 chip per second. Find (a) the sequence length (b) chip duration and (c) PN sequence period for maximal length.

12.8 (a) Let a DSB-SC signal be derived from a single tone. How many lines will be there in the spectrum of the spreaded signal? Also, how many lines are there in the de-spread signal?

(b) If the message bit rate is made a sub-multiple of the PN-sequence chip rate, what advantage will be there in the system?

12.9 In a spread spectrum communication system, the bit rate is $r_\text{b} = f_\text{b} = 1/T_\text{b}$ and chip rate $f_\text{c} = 1/T_\text{c}$. The system operates on BPSK modulation. The E_b/N_0 is taken as 10. Find the bandwidth of the system and maximum number of simultaneous users. Take $r_\text{b} = 3\,\text{kbps}$ and $f_\text{c} = 1\text{MHz}$.

(Hint: $p_\text{e} = \frac{1}{2}erfc\left(\sqrt{\frac{E_\text{b}}{N_0}}\right) \approx 10^{-5}$)

12.10 A sequence 101100011 is arriving at the bit rate of 75 bits/sec to a DS-SS BPSK transmitter. Let $g(t)$ be the PN sequence generated by the shift register stage as shown, with the initial state 1111 and at a clock rate of 300 Hz.

(a) Find the final transmitted sequence.

(b) What is the bandwidth of the transmitted signal?

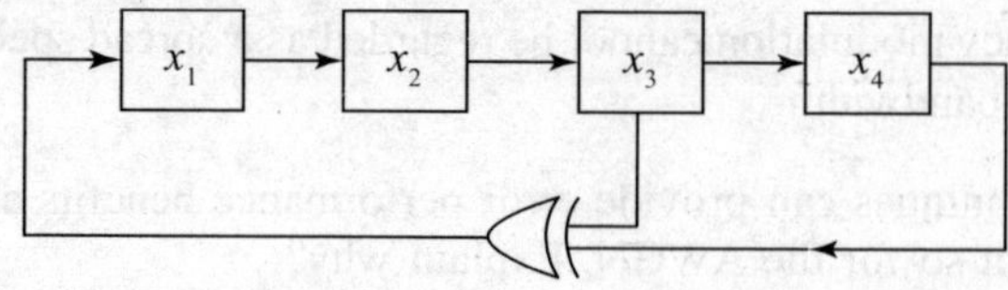

(c) Find the processing gain.
(Hint: With the initial state $x_1x_2x_3x_4 = 111\,1$, the sequence is obtained as $g(t) = 1111000100110\,1$ 0 with clock rate $= 300$ Hz.

1	1	1	1
0	1	1	1
0	0	1	1
0	0	0	1
1	0	0	0
0	1	0	0
0	0	1	0
1	0	0	1
1	1	0	0
0	1	1	0
1	0	1	1
0	1	0	1
1	0	1	0
1	1	0	1
1	1	1	0
1	1	1	1

$$\frac{T_c \text{ of } g(t)}{T_b \text{ of } x(t)} = \frac{1/300}{1/75} = \frac{1}{4}$$

So in 1 bit duration of $x(t)$, there are 4 sequence bits.

Processing gain $G_p = \frac{f_c}{f_b} = \frac{1/T_c}{1/T_b} = \frac{T_b}{T_c} = \frac{300}{75} = 4$, etc)

Review questions

1. In an FH/SS with 8-ary FSK modulation, the PN generator consists of 10 stages of LFSR with a maximal length sequence. Each state of the register corresponds to a new centre frequency within the hopping band. Assume the minimum step size between the centre frequencies as 200 Hz; clock frequency is 2 kHz, and data rate is 1.5 kbps. Find the following.

   (a) hopping bandwidth (b) chip rate

   (c) processing gain and (d) number of chips in each data symbol.

   (Hint: Hopping bandwidth B_T = Number of states $\times$ Minimum step between frequencies $= (2^{10} - 1) \times 200\,\text{Hz}$

   Symbol rate R_s and bit rate R_b are related as $R_b = KR_s$ where $K = \log_2 M = \log_2 8 = 3$. So, $R_s = \frac{R_b}{3} = \frac{1500}{3} = 500$ symbols/second.

   $$\text{Chip per symbol} = \frac{\text{Chip rate}}{\text{Symbol rate}} = \frac{2000}{500} = 4\frac{\text{chip}}{\text{symbol}}$$

   $$\text{Processing gain, } G_p = \frac{B_T}{R_s} = \frac{1023 \times 200}{500} \simeq 400, \text{etc)}$$

2. N-users share a frequency band through a CDMA system. Each user has a bit rate r_b and uses DS/SS with BPSK modulation. Find the minimum chip rate required to obtain a bit error probability of P_e.

   (Hint: DS/SS with BPSK modulation, $P_e = \frac{1}{2} erfc\left(\sqrt{\frac{E_b}{\eta}}\right) = Q\left(\sqrt{\frac{2E_b}{\eta}}\right)$ with P_e, one computes (E_b/η).

3. Discuss the advantages of CDMA spread spectrum in respect of privacy, fading environment, jamming and flexibility.

4. In spread spectrum communication, if receiver spreading sequence is not synchronised with the transmitter spreading sequence, what will happen to the received data? Discuss analytically.

5. In the spread-spectrum technique, the information-carrying signal is spread over a wide spectrum. Does it mean that (a) the same information is sent repeatedly over individual frequency components (within the spread spectrum band), OR (b) the data is divided and sent as multiple discrete signals over various frequencies OR (c) it is a continuous smearing of the input signal over the spreaded spectrum? Explain your answer.

6. The adjoining table explains the FH/SS scheme for one complete period of PN sequence. Find the following.

Time	0 1	2 3	4 5	6 7	8 9	10 11	12 13	14 15	16 17	18 19
Input data	0 1	1 1	1 0	1 1	0 1	1 0	0 1	1 0	1 1	1 0
Frequency	f_1	f_3	f_2	f_3	f_1	f_2	f_1	f_2	f_3	f_2
PN sequence	1 1 1		1 0 0		0 1 0		0 1 1		0 1 0	

   (a) Period of PN sequence.

   (b) Number of bits per symbol.

   (c) Number of FSK frequencies.

   (d) Length of PN sequence per hop.

   (e) Total number of carrier frequencies.

Multiple-choice questions

1. A PN sequence produced by a LFSR with n stages is called a maximal-length-sequence if the sequence has a period of exactly

   (a) $2^{n-1} + 1$ (b) $2^n - 1$ (c) $2^n + 1$ (d) none

2. For a DS/SS, the processing gain $Gp = f_c/f_b$ and the PN sequence length N are related as

   (a) $G_p = N$ (b) $G_p > N$ (c) $G_p < N$ (d) $G_p = 1/N$.

3. In a slow-hop MFSK signal, the length of PN segment per hop is $k = 3$. Two 2-bit symbols are transmitted per hop. How many bits are transmitted per PN-sequence period?

   (a) 32 (b) 16 (c) 64 (d) 8

4. A PN sequence generator consists of an n-stage shift register. The balance property of the PN sequence states that the number of zeroes and ones in one period are given by

   (a) $2^{n-1}+1, 2^{n-1}$ (b) $2^{n-1}, 2^{n-1}+1$ (c) $\frac{2^n-1}{2}, \frac{2^n-1}{2}$ (d) none

5. The processing gain of a DS/BPSK spread spectrum system is 1024. The number of shift register stages needed to generate the PN sequence is

   (a) 8 (b) 10 (c) 12 (d) none

6. In FH/SS, privacy of data can be maintained if the hopping frequency is

   (a) zero (b) low (c) high (d) infinity.

7. For avoiding jamming, which spread spectrum is better?

   (a) DS-SS (b) FH-SS (c) both (a) and (b) (d) None

8. Pseudorandom signal cannot be predicted but pseudorandom sequence can be predicted. Find the best answer.

   (a) True, False (b) False, True

   (c) True, True (d) False, False

9. For a maximal length PN sequence generator, if the initial state of the shift registers be 10101, after how many clock pulses does the sequence repeat?

   (a) 15 (b) 16 (c) 31 (d) 32

10. The PN sequence produced by an LFSR depends on

    a) Number of shift registers

    b) Initial conditions of shift registers

    c) Feedback tap connections

    d) both (b) and (c)

11. To reduce the error probability, the processing gain G_p should be

    (a) increased (b) decreased (c) independent of G_p (d) None

12. In an FH/SS system with a channel bandwidth $B = 4\,\text{kHz}$ and transmission bandwidth $B_\text{T} = 100\,\text{kHz}$, what is the minimum number of bits in the PN sequence?

    (a) 2 (b) 3 (c) 4 (d) 5

    (Hint $2^\text{k} = B_\text{T}/R_\text{S} = \frac{100}{4} = 25;\; k = \log_2^{25} = 4.64 \simeq 5\,\text{bits}$)

13. A 56 kbps data is transmitted using spread spectrum technique. If the SNR of received signal is required to be $\frac{1}{100}$, the channel bandwidth required would be

(a) 7.8 MHz (b) 1.3 MHz (c) 3.9 MHz (d) None

(Hint: Apply $C = B\log_2\left(1 + \frac{S}{N}\right)$.)

14. A digital medium can carry a data rate of 50 Mbps. How many 64-kbps voice channels can be sustained by the medium if one uses DS/SS technique with Barker sequence (11-bit code).

    (a) 14 (b) 7 (c) 28 (d) none

    (Hint: Barker sequence increases the bit rate by 11 times. So, number of different channels $= \frac{50 \times 10^6}{11 \times 64 \times 10^3}$ channels).

15. A pseudorandom number is generated by the iteration formula $N_{i+1} = (3 + 7N_i)$, mod 13, $i = 0, 1, 2, 3\ldots$ If you take $N_0 = 1$, how many random numbers will be generated?

    (a) 12 (b) 6 (c) 13 (d) none

13 Equalisation and Pulse Shaping

13.1 Introduction

Filtering properties of communication channels have been discussed in chapter 2. In practical communications systems, the frequency response of the channel is not known with sufficient precision. There are various reactive circuit elements present throughout the system whose frequency-response characteristics are time-variant. For example, in baseband telephone systems, the channel has distributed reactances all along the link which distorts the transmitted pulses. In wireless systems such as radio channels, the frequency characteristics are not known *a priori*. When the signal in the form of pulses are transmitted over the channel, the shape of the pulses is not preserved due to such channel nonlinearity. This results in spectral distortion that tends to spread the pulses beyond its normal interval. The tail of a pulse can smear into the adjacent symbol interval. This is known as intersymbol interference (ISI), which may lead to high errors in the detection of pulses. Both the transmitter and the channel distort the signal. When a receiver is designed to compensate for the ISI, it is often called an 'equaliser' or an 'equalising filter'. If not compensated, the ISI problem poses a great obstacle to the transmission of high-speed data through bandlimited channels.

Equalisation methods are partitioned into three broad categories. The first one is based on 'maximum likelihood sequence estimation' (MLSE), where the detector makes a relatively simple decision from the distorted pulse sequence. There is no need to reshape the distorted pulses. The receiver chooses that signal which will maximise the likelihood (or probability) of observing the received vector. The second method uses linear filters with adjustable coefficients to compensate the distorted pulses. The filter equaliser can further be classified into two categories; one that contains only feed-forward elements (transversal equaliser) and the other that contains both feedback and feedforward elements (decision feedback equaliser). The nature of operation of the equalisers may be automatic or adaptive.

13.2 Intersymbol Interference (ISI)

To study the ISI problem, we need to transmit a pulse at every T time interval, the nth pulse being $a_n p(t-nT)$. Here a_n represents the information message sequence and $p(t)$ is an arbitrary pulse. The channel is assumed to be ideal with frequency response $H_c(f)$ limited to $|f| \le W$, where W is the channel bandwidth. We start with the case of transmission of time-limited pulses. Since time-limited pulses cannot be bandlimited, part of the pulse spectra beyond W is suppressed by the channel. This causes pulse distortion. Next we consider the transmission of bandlimited pulses over bandlimited channel. Since bandlimited pulse can not be time-limited, various pulses will overlap with the adjacent pulses in the time domain, causing ISI. Thus the ISI problem cannot be avoided in either way.

As an example, a medium range telephone channel has non-uniform amplitude response. It increases upto 1.3 kHz and then falls gradually. The impulse response has significant contribution up to 10 ms. So a pulse entering at $t = 0$ would last for next 10 ms. A data rate of 1 kHz results in ISI extending over 10 symbols.

We now develop a mathematical treatment for the ISI. Actually, ISI is inherent in the system with finite transmission bandwidth. We consider a lowpass transmitted signal of the form

$$x(t) = \sum_{n=0}^{\infty} a_n p(t - nT). \tag{13.1}$$

The pulse is assumed to be bandlimited to W. The channel also has a bandlimited frequency response $H_c(f)$, i.e., $H_c(f) = 0$ for $|f| > W$. The channel output is obtained by the convolution theorem as

$$p_0(t) = \int_{-\infty}^{\infty} p(\tau) h_c(t - \tau) d\tau, \tag{13.2}$$

where $h_c(t)$ is the impulse response of the filter. The received signal $r(t)$ can be represented by

$$r(t) = \sum_{n=0}^{\infty} a_n p_0(t - nT). \tag{13.3}$$

Now, suppose that the receiver section consists of a filter followed by a sampler as shown in Fig. 13.1. The output of the receiving filter is

$$y(t) = \sum_{n=0}^{\infty} a_n v(t - nT) + z(t), \tag{13.4}$$

where

$$v(t) = \int_{\infty}^{\infty} p_0(t) h_r(t - \tau) d\tau \tag{15.5}$$

is the response of the receiving filter to the input received pulse $p_0(t)$. The quantity $z(t)$ in (13.4) is the cumulative additive white Gaussian noise (AWGN) at the filter output. After sampling, one gets

$$y(kT) = \sum_{n=0}^{\infty} a_n v(kT - nT) + z(kT), \tag{13.6}$$

or in simple notation,

$$y_k = \sum_{n=0}^{\infty} a_n v_{k-n} + z_k \tag{13.7}$$

.

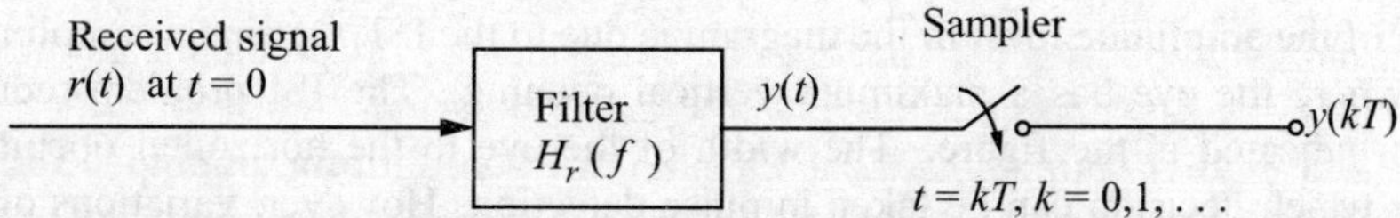

Fig. 13.1 *Receiver section. The filter has the frequency response $H_r(f)$ and the sampling rate is $1/T$ samples per second.*

where $k = 0, 1, 2, \cdots$. More conveniently, the sample at kT can be expressed as

$$y_k = a_k v_0 + \sum_{\substack{n=0 \\ n \neq k}}^{\infty} a_n v_{k-n} + z_k = v_0 \left[a_k + \frac{1}{v_0} \sum_{\substack{n=0 \\ n \neq k}}^{\infty} a_n v_{k-n} \right] + z_k. \tag{13.8}$$

The first term within brackets represents the desired information-bearing message sequence at the kth instant. The second term gives the ISI, that is, the kth sample is the superposition of all the other samples. The term z_k accounts for the effect of AWGN. Note that the signal degradation due to ISI has got an origin different from noise.

One might attempt to measure the performance of the receiver to the baseband signal, presuming signal degradation due to ISI and noise impairment as being present. One such method is the study of the eye diagram viewed on an oscilloscope; this is presented in the next section.

13.3 Eye Diagram

Eye diagram is the display of the signal of interest on an oscilloscope viewed within a particular signal interval. The distorted signal of the channel output is applied to the vertical input of the oscilloscope with the horizontal sweep rate fixed at $1/T$, where T is the bit interval of the random pulse sequence. The oscilloscope thus superimposes all the traces with T interval. The resulting pattern resembles the human eye and hence the name. Figure 13.2 illustrates the eye diagram for binary polar pulses. In the presence of ISI, the eye will tend to close the opening; this effectively reduces the noise tolerance. However, a sampling taken at the point of maximum eye opening provides the least number of errors in pulse detection. This is the point of optimum sampling. Figure 13.3 gives a clear eye diagram indicating all the pertinent parameters.

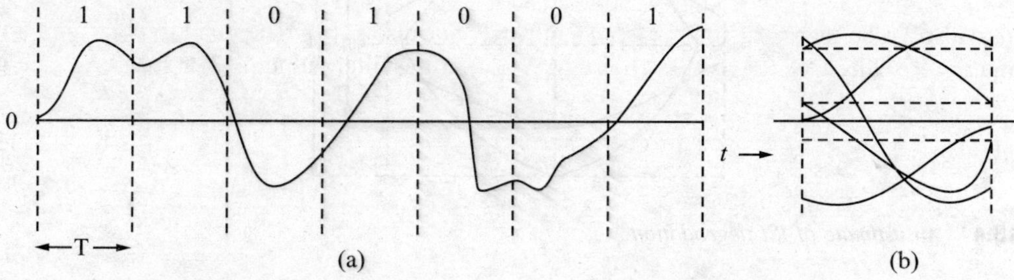

Fig. 13.2 *Eye diagram: (a) Representative polar sequence. (b) Eye diagram on the scope with time-base at the data rate.*

The depression of the amplitude level in the diagram is due to the ISI. Optimum sampling corresponds to the instant where the eye has a maximum vertical opening. The ISI problem reduces the noise margin M_N as indicated in the figure. The width of the eye in the horizontal opening determines the region over which decision can be taken in pulse detection. However, variations of the incoming pulse position cause timing jitter and this reduces the sensitivity to timing error. This has the effect on clock recovery circuitry.

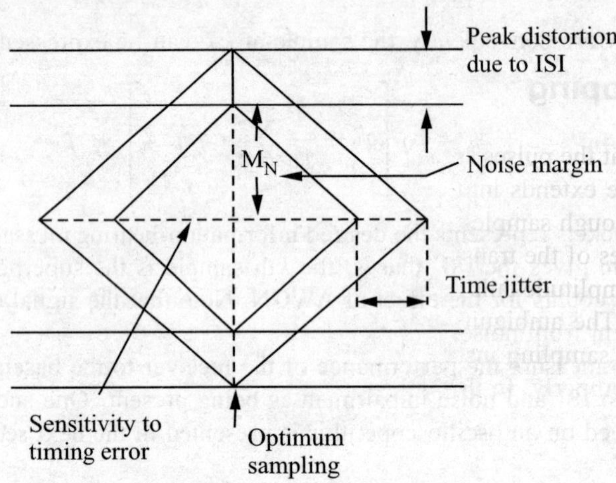

Fig. 13.3 *Eye diagram indicating parameters relating to ISI*

Example 13.1
What will be the eye pattern if the channel provides no filtering effect?

Solution The absence of filtering means that the channel behaves ideally with infinite bandwidth. The transmitted rectangular pulse shape will be received undistorted. When the signal is cut up every pulse interval and superposed, the eye diagram will be a rectangular box.

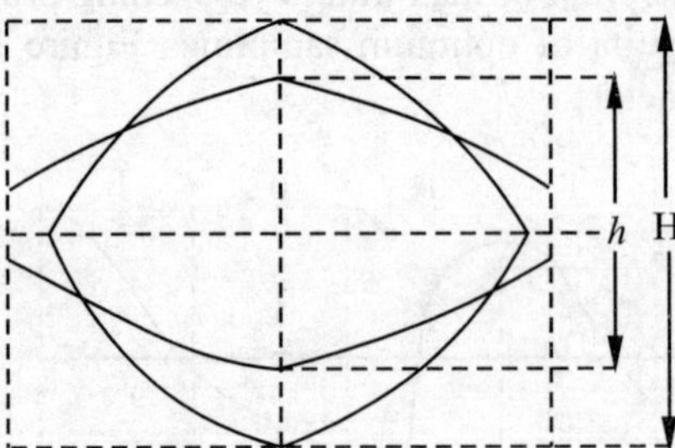

Fig. 13.4 *An estimate of ISI degradation*

ISI-degradation due to filtering imperfection can be measured as follows. In Example 13.1, we have explained that without any filtering imperfection, the eye pattern will look like a rectangular box. Pulse degradation due to ISI is a reduction in the size of the ideal eye, as shown in Fig. 13.4.

The ideal vertical eye opening is H and the degraded eye opening is h. The ISI degradation can be expressed as

$$\text{ISI} = 20\log_{10}\frac{h}{H}\ \text{dB}.$$

For a vertical opening of 80%, ISI degradation is $-20\log_{10}\frac{80}{100}\ \text{dB} = 1.94\ \text{dB}$.

13.4 Pulse Shaping

We discussed earlier that the pulses spread in time while propagating through some channels of finite bandwidth. Every pulse extends into every other pulse causing ISI. These undesirable tails add up to contribute to ISI although samples are being taken at correct locations. Figure 13.5 illustrates this effect. The output pulses of the transmitting filter have prolonged tails corresponding to the symbols x_1, x_2 and x_3. Large amplitude tails nearest to the main lobe of x_3 introduce ISI and may even result in incorrect detection. The ambiguity in pulse detection can be avoided if the pulse can be shaped such that exactly at all sampling instants, the tails pass through zero. Here is the need for shaping the transmitted pulses properly. In this context Nyquist proposed different criteria that are discussed below.

13.4.1 Nyquist criterion for zero ISI

With reference to Fig. 13.5, it can be said that the ISI can be avoided completely if the tails of all other pulses have zero amplitudes at the centre of a given pulse. The term $\sum_{\substack{n=0\\n\neq k}}^{\infty} a_n v_{k-n}$ of (13.8) is the ISI component. It is instructive to study the spectral properties of the pulse shape $v(t)$ and hence the transmitted pulse $p(t)$ that results in nil ISI. The condition of zero ISI is obtained by choosing a pulse shape $p(t)$ such that

$$v(t = kT) = \begin{cases} 1 & k = 0 \\ 0 & k \neq 0. \end{cases} \tag{13.9}$$

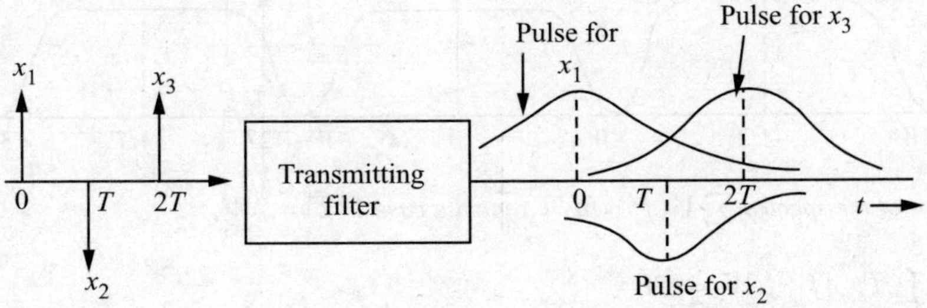

Fig. 13.5 *Pulse tails that contribute to ISI*

It can be proved that the spectrum of $v(t)$ should satisfy the condition that

$$\sum_{m=-\infty}^{\infty} V(f+m/T) = T. \tag{13.10}$$

Stated in words, the sum of the spectra formed by repeating $V(f)$ every $1/T$ is a constant equal to T. This principle is referred to as the Nyquist pulse shaping criterion for zero ISI.

We now interpret the condition under (13.10) for three cases. Suppose that the channel has a bandwidth W so that the data transmission rate can have a maximum of $2W$ bits/second.

Case 1 $T < 1/2W$ or $1/T > 2W$

The spectra $\sum_m V(f+\frac{m}{T})$ consists of non-overlapping replicas of $v(f)$ separated by $1/T$ (vide Fig. 13.6). Here no such pulse spectra $v(f)$ can be designed to satisfy (13.10).

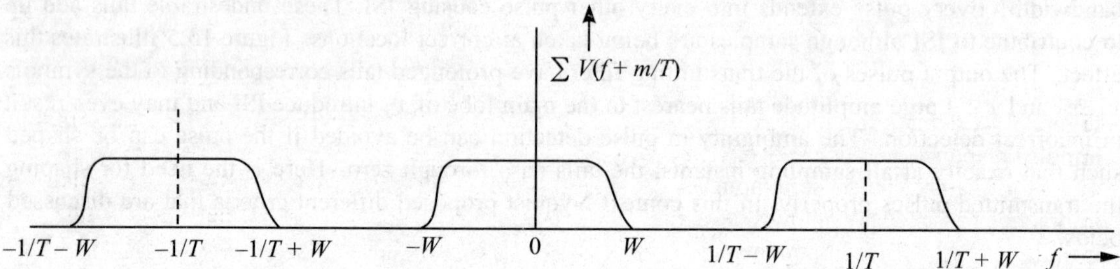

Fig. 13.6 *Plot of the spectra* $\sum V(f+m/T)$ *for the case* $1/T > 2W$

Case 2 $T = \frac{1}{2}W$ or $\frac{1}{T} = 2W$

This is the Nyquist rate. The spectra $\sum V(f+m/T)$ is sketched in Fig. 13.7 and shows that the replicas are separated from one another by $1/T$. Here the only choice of $V(f)$ that satisfies (13.10) is that it should be rectangular, that is,

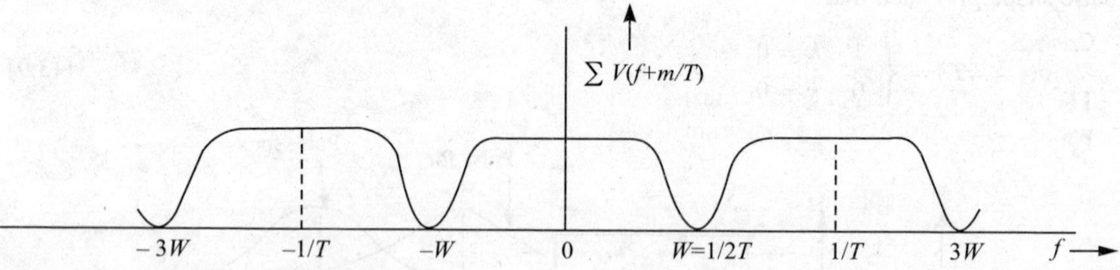

Fig. 13.7 *Plot of the spectra* $\sum V(f+m/T)$ *for the case* $1/T = 2W$.

$$V(f) = \begin{cases} T & |f| < W = \frac{1}{2}T \\ 0 & \text{otherwise.} \end{cases} \tag{13.11}$$

This filter is known as Nyquist filter. The corresponding pulse is the impulse response of the Nyquist filter and is given as

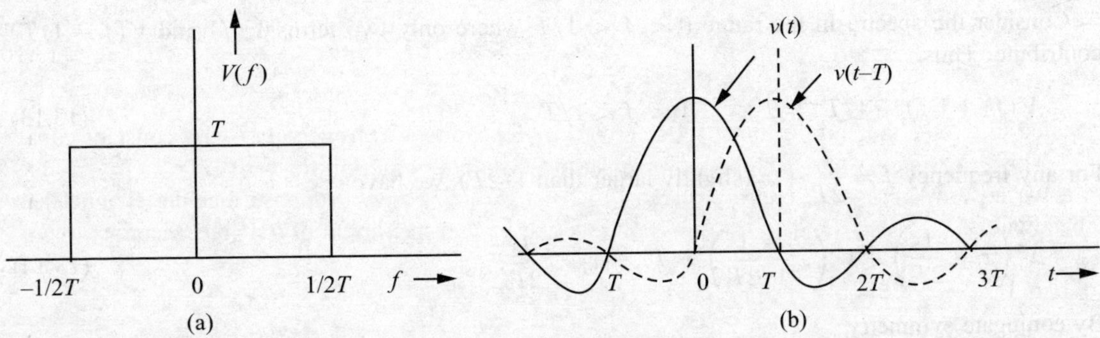

Fig. 13.8 *(a) Minimum bandwidth pulse spectrum. (b) Impulse response.*

$$v(t) = \frac{\sin(\pi t/T)}{\pi t/T} = \text{sinc }(\pi t/T). \tag{13.12}$$

Figure 13.8 depicts the plots of the transfer function of the Nyquist channel and the corresponding impulse response.

Unfortunately, the pulse is noncausal as its tails are infinitely long and starts at $-\infty$. Also, it is impractical as it decays slowly at a rate $1/t$; any small time error (called time jitter) results in an infinite series of ISI components.

Even though the Nyquist pulses $v(t)$ have long tails, the tails pass through zero amplitudes at the sampling points $T, 2T, 3T, \cdots$. For ideal Nyquist filtering, the maximum possible symbol rate per hertz of bandwidth is calculated as below.

Transmission of R symbols per second requires a theoretical bandwidth of $R/2$ Hz. So symbol transmission rate per hertz of bandwidth is 2 symbols/second/Hz without ISI. For an M-ary signalling set, the coding assigns k bits to each symbol such that $2^k = M$. So the data rate is k times faster than the symbol rate. For an 8-ary PAM signal set, $k = 3$ and so the bandwidth efficiency is 6 bits/s/Hz.

Case 3 $T > 1/2W$ +or $1/T < 2W$

The spectra of $\sum V(f + m/T)$ consists of overlapping replicas of $V(f)$ as shown in Fig. 13.9. Here we have the choices of the spectrum $V(f)$ that satisfies (13.10).

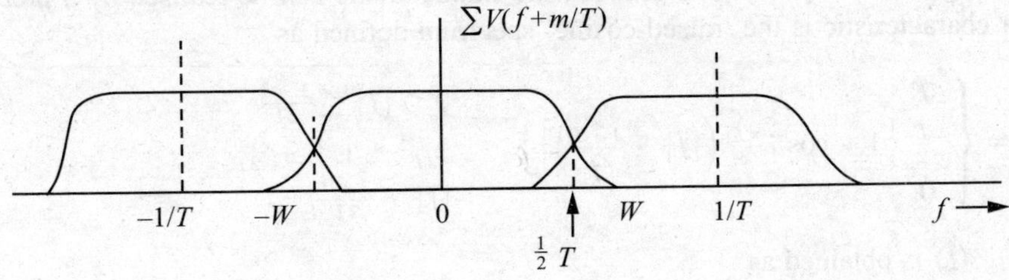

Fig. 13.9 *Spectra of $\sum V(f + m/T)$ for the case $T > 1/2W$*

Consider the spectra in the range $0 < f < 1/T$ where only two terms $V(f)$ and $V(f - 1/T)$ contribute. Thus,

$$V(f) + V(f - 1/T) = T, \qquad 0 < f < 1/T. \tag{13.13}$$

For any frequency $f = \dfrac{1}{2T} + x$ (slightly larger than $1/2T$), we have

$$V\left(x + \frac{1}{2T}\right) + V\left(x - \frac{1}{2T}\right) = T, \quad |x| < \frac{1}{2T}. \tag{13.14}$$

By conjugate symmetry,

$$V\left(\frac{1}{2T} + x\right) + V^*\left(\frac{1}{2T} - x\right) = T. \tag{13.15}$$

For the pulse $v(t)$, we may take $V(f)$ of the form $V(f) = |V(f)|e^{-j2\pi f t_d}$, where t_d is the time delay.

Thus,

$$\left|V(\frac{1}{2T} + x)\right| + \left|V(\frac{1}{2T} - x)\right| = T, \quad |x| < \frac{1}{2T}. \tag{13.16}$$

The desirable spectral property of $v(f)$ that satisfies the above equation has the shape as shown in Fig. 13.10. Here W is the actual bandwidth and W_0 is the minimum Nyquist bandwidth at the half-amplitude point, i.e., $|V(\frac{1}{2T})| = 0.5|V(0)|$. The difference $(W - W_0)$ is the bandwidth in excess of the theoretical minimum bandwidth. The rate at which the tail of $V(f)$ bends is expressed by the roll-off factor β, defined as $\beta = (W - W_0)/W_0$, and takes values in the range $0 \leq \beta \leq 1$. The total bandwidth of $V(f)$ is the sum of the minimum bandwidth due to Nyquist and the excess bandwidth. That is,

$$W = \frac{1}{2T} + \beta\frac{1}{2T} = \frac{(1+\beta)}{2T} = \frac{(1+\beta)R}{2}, \tag{13.17}$$

where R is the data rate.

For the case when $\beta = 1/2$, the excess bandwidth is 50% and total bandwidth is $3/(4T)$. For $\beta = 1$(100% excess bandwidth), the total bandwidth is $W = 1/T = 2W_0$.

As the rolling-off of $V(f)$ is quite gradual, the characteristic can be realised by a practical filter. One popular characteristic is the 'raised-cosine' spectrum defined as

$$V_{RC}(f) = \begin{cases} T & 0 \leq |f| \leq \frac{1-\beta}{2T} \\ \frac{T}{2}\left\{1 + \cos\left[\frac{\pi T}{\beta}\left(|f| - \frac{1-\beta}{2T}\right)\right]\right\} & \frac{1-\beta}{2T} \leq |f| \leq \frac{1+\beta}{2T} \\ 0 & |f| > \frac{1+\beta}{2T}. \end{cases} \tag{13.18}$$

The pulse $v_{RC}(t)$ is obtained as

$$V_{RC}(t) = \left[\frac{\sin(\pi t/T)}{\pi t/T}\right]\frac{\cos(\pi\beta t/T)}{(1 - 4\beta^2 t^2/T^2)}. \tag{13.19}$$

When $\beta = 1$, (13.18) reduces to $V_{RC}(f) = \frac{T}{2}\{1 + \cos(\pi f T)\}$ which is a cosine function raised by its peak amplitude.

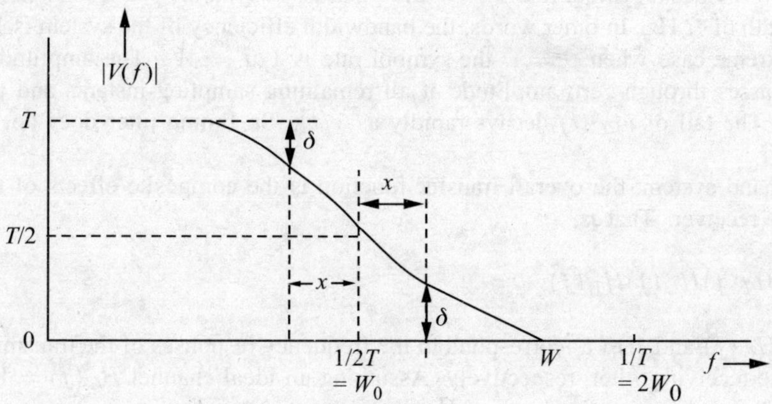

Fig. 13.10 *Spectrum of $V(f)$ that follows equation (13.16)*

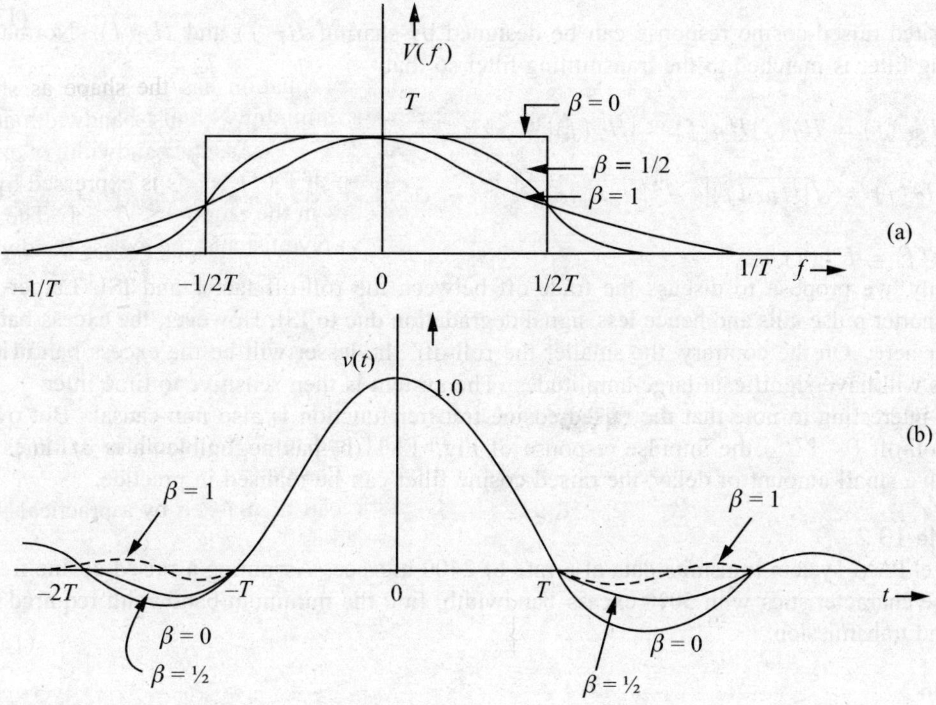

Fig. 13.11 *(a) Raised-cosine pulse and (b) its spectrum*

Figure 13.11 illustrates the raised-cosine characteristics for $\beta = 0, 1/2$ and 1. Note that the tails of $v_{RC}(t)$ pass through zero values at all sampling points regardless of β. The case of $\beta = 0$ corresponds to rectangular spectrum and the pulse shape is $v(t) = \frac{\sin(\pi t/T)}{\pi t/T}$, a sinc function. In this case the system provides a symbol rate $1/T = 2W$, that is, R symbols per second can be transmitted using a bandwidth of R Hz. In other words, the bandwidth efficiency of the system is 1 symbol/s/Hz. In the other extreme case when $\beta = 1$, the symbol rate is $1/T = W$. The amplitude is maximum at $t = 0$ and passes through zero amplitude at all remaining sampling instants and points midway between them. The tail of $v_{RC}(t)$ decays rapidly as $1/t^3$. So timing jitter does not create serious problems.

In any baseband system, the overall transfer function is the composite effects of the transmitter, channel and the receiver. That is,

$$H(f) = H_T(f)H_C(f)H_R(f), \tag{13.20}$$

where $H_T(f), H_C(f)$ and $H_R(f)$ correspond to the frequency responses of the transmitting filter, the channel, and the receiving filter, respectively. Assuming an ideal channel $H_C(f) = 1$ for $|f| \leq W$, we have the raised-cosine transfer function $H_{RC}(f)$

$$H_{RC}(f) = H_T(f)H_R(f). \tag{13.21}$$

The desired raised-cosine response can be designed by sharing $H_T(f)$ and $H_R(f)$. Normally, the receiving filter is matched to the transmitting filter so that

$$H_{RC}(f) = H_T(f)H_R(f) = |H_T(f)|^2.$$

Thus, $H_T(f) = \sqrt{|H_{RC}(f)|}e^{-j2\pi f t_0}$, (13.22)

and $H_R(f) = H_T^*(f)$.

Finally, we propose to discuss the trade-off between the roll-off factor and ISI. Larger roll-off yields shorter pulse tails and hence less signal degradation due to ISI. However, the excess bandwidth is larger here. On the contrary, the smaller the roll-off, the lesser will be the excess bandwidth but the tails will have significant large amplitudes. The system is then sensitive to time jitter.

It is interesting to note that the raised-cosine transfer function is also non-causal. But owing to faster roll-off ($\sim 1/t^3$) the impulse response of Fig. 13.11(b) has negligible value at times $t < 0$. So, with a small amount of delay, the raised-cosine filter can be realised in practice.

Example 13.2

A 4-level PAM system transmits data at a rate of 2400 bits/sec. Assuming a raised-cosine frequency response characteristics with 50% excess bandwidth, find the minimum bandwidth required for the baseband transmission.

Solution

Here $M = 4$ so $2^k = 4$ or $k = 2$. Each symbol consists of two bits. Hence symbol rate is $R_s = R/\log_2 M = 2400/2 = 1200$ symbols/second.

$$\text{Minimum bandwidth } W = \frac{1+\beta}{2T} = \frac{(1+\beta)R_s}{2} = 900 \text{ Hz}$$

Example 13.3
An 8-ary PAM system transmits digital data over a channel of bandwidth $W = 2400$ Hz. The bit rate is 14,400 bits/second. Use raised-cosine roll-off filter characteristics.
(a) What is the symbol rate? (b) Find the roll-off factor.

Solution
Bit rate $R = 14,400$ bits/second $= 1/T$. For 8-ary PAM system, $M = 8$ so $k = 3$. Each symbol consists of 3 bits. Thus,

(a) symbol rate $R_s = R/\log_2 M = 14,400/3 = 4,800$ symbols/second.

(b) $W = (1+\beta)\dfrac{R_s}{2}$

Roll-off factor, $\beta = \dfrac{2W}{R_s} - 1 = \dfrac{2 \times 2400}{4,800} - 1 = 0$ (rectangular pulse shape).

13.4.2 Controlled ISI and partial response signalling

In the previous section we attempted to eliminate ISI completely and this was possible provided the symbol rate $1/T$ was somewhat smaller than the Nyquist rate of $2W$ symbols/second. We will now try to design the bandlimited signal to achieve this highest symbol rate by allowing a known amount of ISI (possibly at one or two time instants).

The condition of zero ISI was expressed as $v(kT) = 0$ for $k \neq 0$ (vide (13.9)). We now consider a special form of pulse shape that causes zero ISI with all pulses except the succeeding pulse. That is,

$$v(kT) = \begin{cases} 1 & k = 01 \\ 0 & \text{otherwise.} \end{cases} \tag{13.23}$$

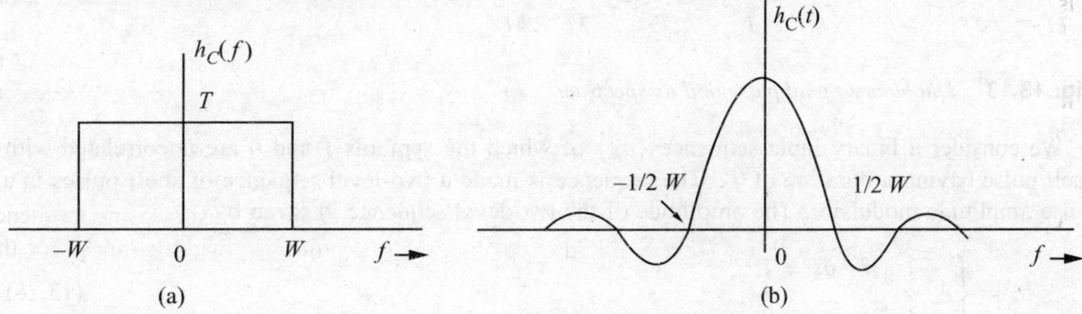

Fig. 13.12 *(a) Ideal channel transfer function. (b) Channel impulse response.*

The generation of $v(t)$ from its samples $v(kT)$ needs attention. This is done using the interpolation formula. Suppose that the channel is an ideal lowpass filter of bandwidth W Hz as shown in Fig. 13.12. At the Nyquist rate of $T = 1/2W$ the impulse response is

$$h_C(t) = \text{sinc } (2\pi W t). \tag{13.24}$$

The output of the channel is obtained in the linear interpolation scheme as

$$v(t) = \sum_k v(kT) h_C(t - kT) = \sum_k v(kT) \text{ sinc } [2\pi W(t - kT)]. \tag{13.25}$$

The terms for $k = 0$ and 1 in the summation give nonzero values. Thus,

$$v(t) = \text{sinc } 2\pi W t + \text{sinc } 2\pi W(t - T). \tag{13.26}$$

We see that the pulse consists of two sinc (Nyquist) pulses displaced by T units with respect to each other. The response to an input pulse is spread over few signalling intervals.

This pulse is referred to as a 'duo-binary' pulse. Figure 13.13 shows the pulse and its amplitude spectrum. Fourier transform of (13.26) gives

$$V(f) = \begin{cases} \frac{1}{2W} \cos \frac{\pi f}{2W} & |f| < W \\ 0 & \text{otherwise.} \end{cases} \tag{13.27}$$

The spectrum decays smoothly to zero at $f = W$. The duo-binary encoder is shown in Fig. 13.14.

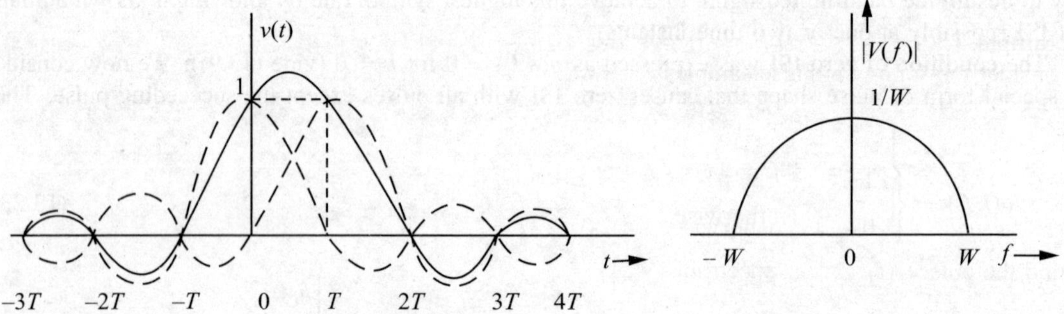

Fig. 13.13 *Duobinary signal pulse and its spectrum*

We consider a binary input sequence $\{a_k\}$ of which the symbols 1 and 0 are uncorrelated with each pulse having a duration of T. The sequence is made a two-level sequence of short pulses in a pulse amplitude modulator. The amplitude of the two-level sequence in given by

$$b_k = \begin{cases} +1 & \text{if} \quad a_k = 1 \\ -1 & \text{if} \quad a_k = 0. \end{cases} \tag{13.28}$$

When applied to a duo-binary encoder of Fig. 13.14, the sequence converts to a three-level sequence $\{c_k\}$, having outputs $-2, 0$, and $+2$. That is,

$$c_k = b_k + b_{k-1}. \tag{13.29}$$

The sequence $\{c_k\}$ is obviously correlated and this correlation introduces controlled amount of inter-symbol interference.

Note in Fig. 13.13 that the duo-binary pulses pass through zeros at two sampling points $t = -T$ and $t = +2T$. The main lobe has a width of $3T$ in contrast to $2T$ of the Nyquist pulse. This widening in the width of the pulse results in narrowing of the bandwidth. This scheme of permitted ISI is known as correlative or partial response signalling. In general, the number of nonzero samples taken should be extended to more than two. The nonzero power spectral density of the duo-binary scheme is undesirable in some communication channels that cannot transmit dc components. We can remove this deficiency by using the modified duo-binary method, as explained in Example 13.4.

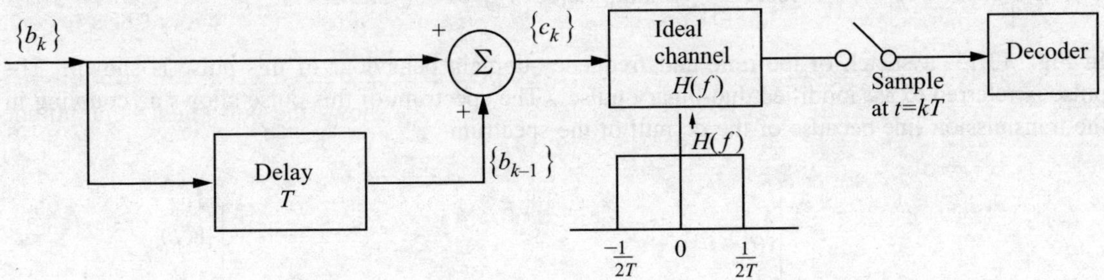

Fig. 13.14 *Duobinary scheme*

Example 13.4

Consider the following pulse samples:

$$v(kT) = \begin{cases} 1 & k = -1 \\ -1 & k = 1 \\ 0 & \text{otherwise.} \end{cases}$$

Find the pulse $v(t)$ and its spectrum $V(f)$.

Solution

We have, by interpolation formula

$$v(t) = \sum_k v(kT)\,\text{sinc}[2\pi W(t - kT)].$$

The two terms for $k = -1$ and 1 survive. $\quad [v(-T) = 1$ and $v(T) = -1]$.

So, $v(t) = v(-T)\sin\text{c}[2\pi W(t+T)] + v(T)\,\text{sinc}[2\pi W(t-T)]$

$$= \text{sinc}\frac{\pi}{T}(t+T) - \text{sinc}\,\frac{\pi}{T}(t-T), \quad \text{since} \quad T = \frac{1}{2W}. \tag{13.30}$$

Taking the Fourier transform,

$$\begin{aligned} V(f) &= \frac{1}{2W}(e^{j\pi f/W} - e^{-j\pi f/W}) \\ &= \frac{j}{2W}\sin(\pi f/W), \qquad |f| \leq W \\ &= 0, \qquad \text{otherwise.} \end{aligned} \tag{13.31}$$

The above relation is obtained by noting that

$$\mathcal{F}\Big[\operatorname{sinc}(Wt)\Big] = \frac{\pi}{W}\ \operatorname{rect}\left(\frac{w}{2W}\right) \text{ and } \mathcal{F}[x(t-t_0)] = X(f)e^{-j\omega t_0}.$$

In Fig. 13.15, a sketch of the time and frequency-domain behaviour of this pulse is shown. The pulse is referred to as 'modified duo-binary pulse'. The spectrum of this pulse allows ac coupling in the transmission line because of the dc null of the spectrum.

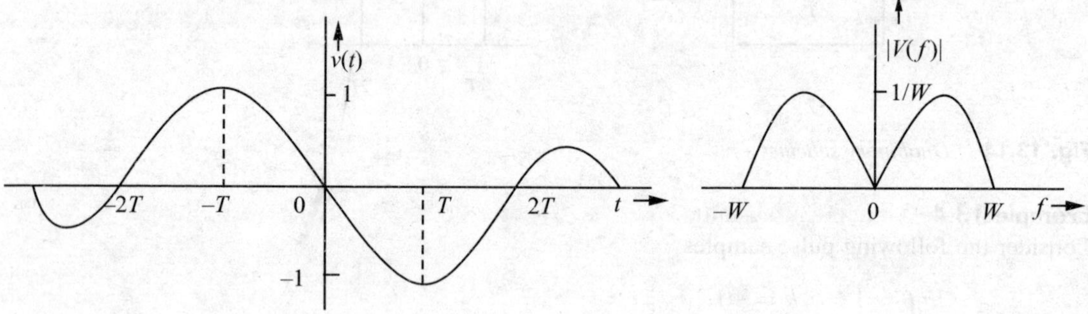

13.15 *Modified duo-binary pulse and its frequency response*

From (13.30) we find that the impulse response of the modified duo-binary consists of two sinc (Nyquist) pulses, separated by $2T$ seconds from each other. The response of Fig. 13.15 shows that it has three distinct levels, $+1, 0$ and -1 at the sampling instants.

Detection of duo-binary signals

The baseband transmitter sends a discrete message sequence $\{a_n\}$ via a line code with the pulse shape $p(t)$. This lowpass transmitted signal takes the form of (13.1). The received signal is first passed through the receiving filter and then an equalising filter to produce the output

$$y(t) = \sum_{n=0}^{\infty} a_n v(t - nT) + z(t) \quad \text{(vide equation (13.4)).}$$

Here the pulse response $v(t)$ is due to the combination of the receiving filter and the equalising filter (vide equation (13.5)). Ignoring the transmission delay through the channel and the noise effect, one can write

$$y(kT) = \sum_{n=0}^{\infty} a_n v(kT - nT),$$

or in simple notation, the signal part of the filter output is

$$y_k = \sum_{n=0}^{\infty} a_n v_{k-n}.$$

For the duo-binary pulse, $y_k = a_k v_0 + a_{k-1} v_1 = a_k + a_{k-1} = d_k$ since v_0 and v_1 are 1 (Fig. 13.13). As $a_k = \pm 1$ with equal probability, the signal part of the filter output y_k takes on three possible values, -2, 0, +2. Thus by knowing y_k and a_{k-1}, the value a_k can be found by subtraction $y_k - a_{k-1}$. If the initial value a_0 of the data, sequence is known, all other values of a_k can be predicted from y_k. The receiving scheme is fine if there be no additive noise. The noise adds to the data making erroneous detection. If, for example, a_{k-1} is received error, it influences d_k to be erroneous.

This has the effect of error propagation into subsequent symbols. This can be overcome by precoding the data at the transmitter. The precoding is governed by the rule

$$e_k = a_k + e_{k-1}, k = 1, 2, \ldots$$

where $\{a_k\}$ is the original data sequence. The transmitted data sequence $\{b_k\}$ can be obtained from the precoded sequence $\{e_k\}$ by the rule $b_k = 2e_k - 1$. That is,

$$b_k = \begin{cases} -1 & e_k = 0 \\ +1 & e_k = 1. \end{cases}$$

These symbols$\{b_k\}$ are then transmitted using line codes. On reception, the signal component of the filter output is

$$\begin{aligned} d_k &= b_k + b_{k-1} = (2e_k - 1) + (2e_{k-1} - 1) \\ &= 2(e_k + e_{k-1} - 1). \end{aligned}$$

That is, $e_k + e_{k-1} = \frac{1}{2} d_k + 1$

and this is the predicted value of the original data sequence. Thus, for $d_k = \pm 2$, the predicted value is $\widehat{b}_k = \frac{1}{2}(\pm 2) + 1 = 0$ and for $d_k = 0$, $\widehat{b}_k = 1$. So, one can formulate the rule

$$\widehat{b}_k = \begin{cases} 1 & y_k < 1 \\ 0 & |y_k| > 1. \end{cases}$$

The pre-coding and decoding is illustrated in Table 13.1

Table 13.1 *Precoding in duo-binary signalling*

Data sequence, a_k		0	1	0	1	1	0	1	0	1	1	1	0
Precoded sequence, e_k	0	0	1	1	0	1	1	0	0	1	0	1	1
Transmitted sequence, b_k	−1	−1	1	1	−1	1	1	−1	−1	1	−1	1	1
Received sequence, d_k		−2	0	2	0	0	2	0	−2	0	0	0	2
Decoded sequence, $\widehat{b}_k$		0	1	0	1	1	0	1	0	1	1	1	0

13.4.3 Pulse generation

The pulse that arrives at the input of the receiver is distorted due to the non-ideal nature of the channel. If not compensated, this will result in errors in the detection process. The equalising filter in the receiver cancels the ISI. The desired duo-binary partial response characteristics can be generated using a transversal filter as shown in Fig. 13.16, which involves tapped delay lines with weights $C_0, C_1, \ldots, C_{N-1}$.

The tap-weights $C_0, C_1, \ldots, C_{N-1}$ are proportional to the samples $v(kT) = v\left(\frac{k}{2W}\right)$, $k = 0, 1, \ldots, N-1$. The output of the transversal filter will be a staircase approximation of the input pulse $v(t)$. To smooth out the pulse waveform, the signal is applied to an ideal lowpass filter having impulse response $\text{sinc}(2\pi Wt)$. The resulting output (which is the detector input) signal has a shape identical to the Nyquist or duo-binary form.

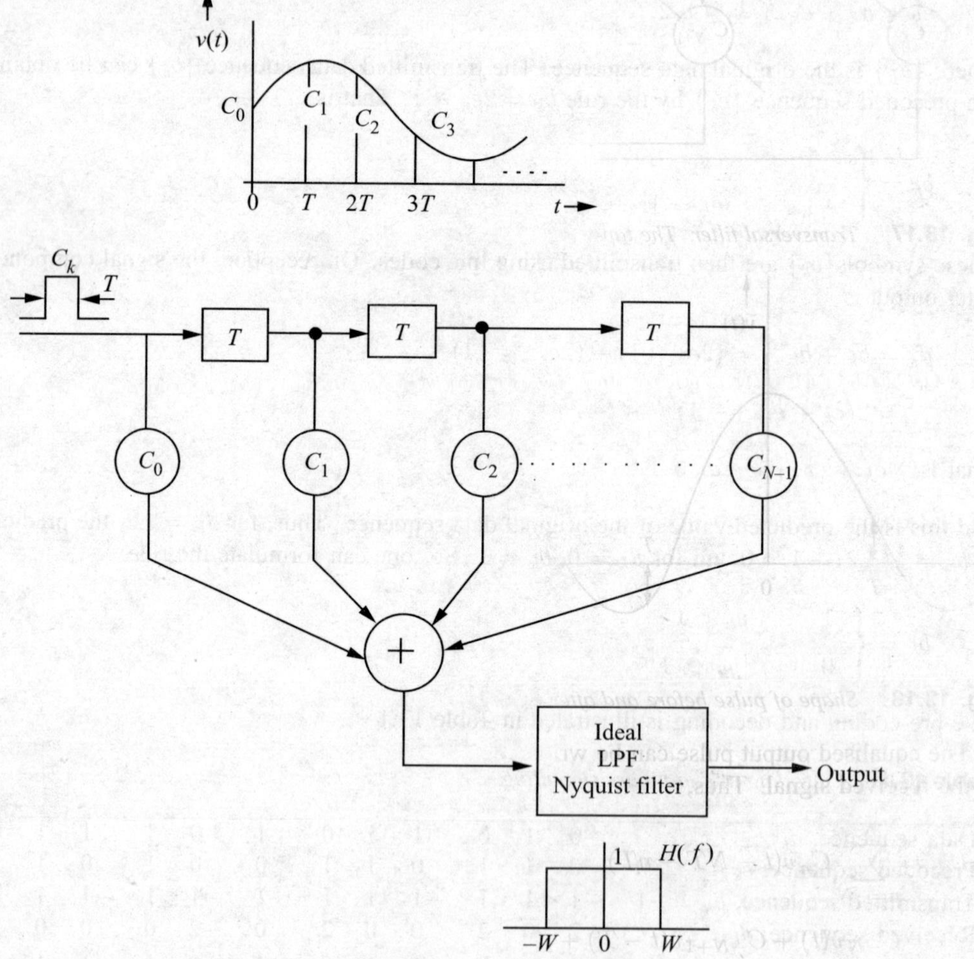

Fig. 13.16 *Pulse generation by transversal filter*

13.5 Transversal Equaliser: Zero-forcing Equaliser

The most popular equaliser filter is the transversal filter, shown in Fig. 13.17. It consists of $(2N+1)$ taps and a total delay of $2NT$. The transmitter forwards samples at the rate $1/T$ symbols/s and the receiver reproduces samples at the same rate. In the design, a maximum of N samples are considered to contribute to ISI. An unequalised pulse $\tilde{v}(t)$ is applied at the input of the filter. The pulse $\tilde{v}(t)$ is assumed to have peak at $t = 0$ and ISI on both sides as shown in Fig. 13.18. The name of this equaliser implies that the intersymbol interference is forced to zero at all sampling instants $t = kT$, except for $k = 0$ where the symbol of interest is assumed to occur.

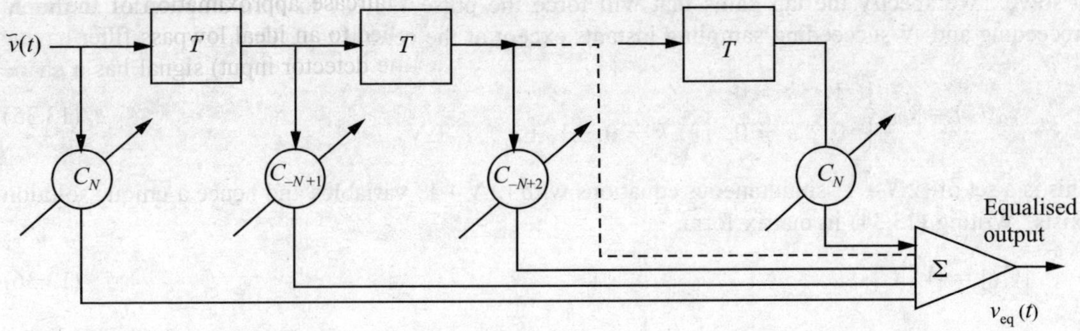

Fig. 13.17 *Transversal filter. The tap-weights $C_{-N}, \ldots, C_N$ can be adjusted by algorithm*

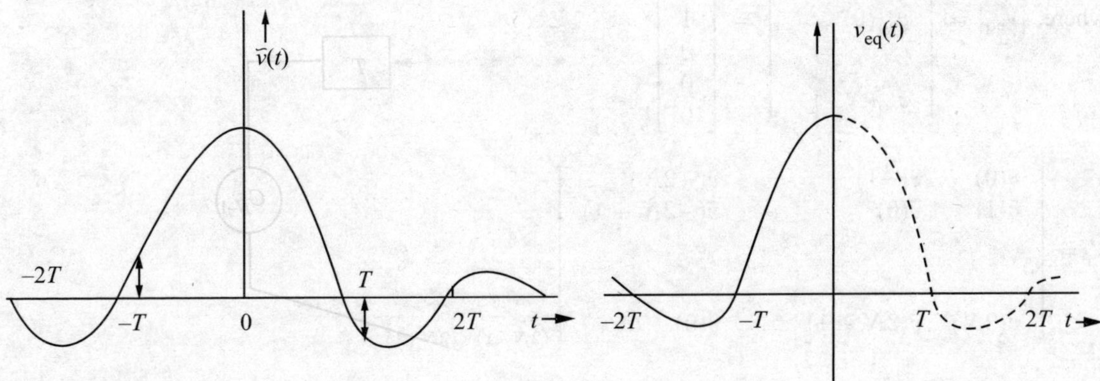

Fig. 13.18 *Shape of pulse before and after equalisation*

The equalised output pulse can be written as the linear weight of the current and previous values of the received signal. Thus,

$$
\begin{aligned}
v_{\text{eq}}(t) &= \sum_{n=-N}^{N} C_n \tilde{v}(t - NT - nT) \\
&= C_{-N}\tilde{v}(t) + C_{-N+1}\tilde{v}(t-T) + \ldots + C_{N-1}\tilde{v}(t - 2NT + T) + C_N\tilde{v}(t - 2NT). \qquad (13.32)
\end{aligned}
$$

At the sampling time $t_k = kT + NT$, the equalised output is the discrete convolution

$$v_{\text{eq}}(t_k) = \sum_{n=-N}^{N} C_n \tilde{v}(kT - nT). \tag{13.33}$$

Using a short hand notation $\tilde{v}(k)$ for $\tilde{v}(kT)$, the above equation reads as

$$v_{\text{eq}}(t_k) = \sum_{n=-N}^{N} C_n \tilde{v}(k-n). \quad \begin{array}{c} k = -2N, \ldots, 2N \\ n = -N, \ldots, N \end{array} \tag{13.34}$$

Note that (13.34) represents a set of infinite number of simultaneous equations, which is impossible to solve. We specify the tap gains that will force the pulse $V_{\text{eq}}(t_k)$ to zero amplitudes at the N preceeding and N succeeding sampling instants except at the central tap. Then,

$$v_{\text{eq}}(t_k) = C_n \begin{cases} 1, & k = 0 \\ 0, & k \neq 0, \text{ i.e., } k = 0, \pm 1, \pm 2, \ldots, \pm N. \end{cases} \tag{13.35}$$

This is a set of $(2N+1)$ simultaneous equations with $(2N+1)$ variables and hence a unique solution exists. Writing (13.34) in matrix form,

$$[V_{\text{eq}}] = [\tilde{V}][C] \tag{13.36}$$

$$\text{where, } [V_{\text{eq}}] = \begin{bmatrix} v_{\text{eq}}(-NT) \\ \vdots \\ v_{\text{eq}}(0) \\ \vdots \\ v_{\text{eq}}(NT) \end{bmatrix} = \begin{bmatrix} 0 \\ 0 \\ 0 \\ 1 \\ 0 \\ 0 \\ 0 \end{bmatrix}_{2N \times 1}$$

$$\tilde{V} = \begin{bmatrix} \tilde{v}(0) & \tilde{v}(-1) & \ldots & \tilde{v}(-2N) \\ \tilde{v}(1) & \tilde{v}(0) & \ldots & \tilde{v}(-2N+1) \\ \cdots & \cdots & \cdots & \cdots \\ \vdots & & & \\ \tilde{v}(2N) & \tilde{v}(2N-1) & \ldots & \tilde{v}(0) \end{bmatrix}_{(2N-1)\times(2N+1)}$$

$$\text{and } [C] = \begin{bmatrix} C_{-N} \\ \vdots \\ C_0 \\ \vdots \\ C_N \end{bmatrix}.$$

The equaliser design that forces N zeros on either side of the peak of $v_{\text{eq}}(t_k)$ is the zero-forcing equaliser. The desired tap gains to achieve this can be found by solving (13.36):

$$[C] = [\tilde{V}]^{-1}[V_{\text{eq}}]. \tag{13.37}$$

Example 13.5

For a single Nyquist criterion pulse, the received pulse $\tilde{v}(t)$ has the following samples:

$$\tilde{v}(0) = 1, \quad \tilde{v}(+T_b) = 0.1, \quad \tilde{v}(-T_b) = 0.3,$$

$$\tilde{v}(2T_b) = -0.2, \quad \tilde{v}(-2T_b) = 0.4.$$

Design a 3-tap equaliser that forces the ISI to be zero.

Solution

The received pulse shape is sketched in Fig. 13.19.

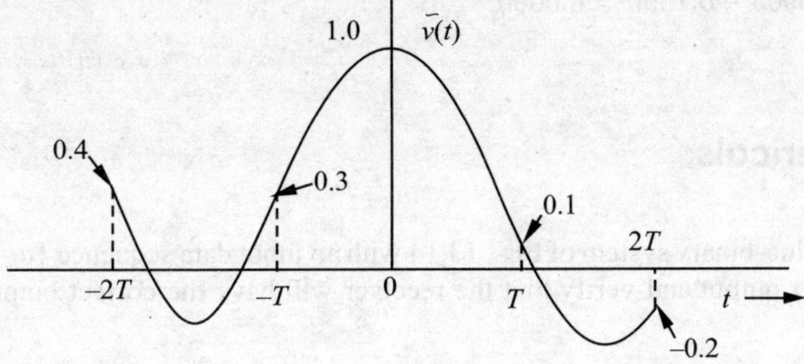

Fig. 13.19 *Received pulse shape of Example 13.5*

For the three-tap equaliser, (13.36) reads as

$$\begin{bmatrix} V_{\text{eq}}(-T) \\ V_{\text{eq}}(0) \\ V_{\text{eq}}(T) \end{bmatrix} = \begin{bmatrix} \tilde{v}(0) & \tilde{v}(-1) & \tilde{v}(-2) \\ \tilde{v}(1) & \tilde{v}(0) & \tilde{v}(-1) \\ \tilde{v}(2) & \tilde{v}(1) & \tilde{v}(0) \end{bmatrix} \begin{bmatrix} C_{-1} \\ C_0 \\ C_1 \end{bmatrix}.$$

To removed ISI, one needs $v_{\text{eq}}(-T) = 0, v_{\text{eq}}(+T) = 0$, and $v_{\text{eq}}(0) = 1$. Substituting these values and the given data,

$$\begin{pmatrix} 0 \\ 1 \\ 0 \end{pmatrix} = \begin{pmatrix} 1 & 0.3 & 0.4 \\ 0.1 & 1 & 0.3 \\ -0.2 & 0.1 & 1 \end{pmatrix} \begin{pmatrix} C_{-1} \\ C_0 \\ C_1 \end{pmatrix}.$$

Solving for C_{-1}, C_0 and C_1, we get the tap gains

$$C_{-1} = -0.2585, C_0 = 1.0735, C_1 = -0.1590.$$

Equation (13.36) can be used to compute the equalised pulse samples $\{V_{\text{eq}}\}$ corresponding to the sample times $k = 0, \pm 1, \pm 2$. Equation (13.34) reads as

$$v_{\text{eq}}(-1) = C_{-1}\tilde{v}(0) + C_0\tilde{v}(-1) + C_1\tilde{v}(-2)$$

$$= -0.2585 \times 1 + 1.0735 \times 0.3 - 0.1590 \times 0.4$$

$$= -0.32210 + 0.32205 = 0.0000.$$

$$v_{\text{eq}}(0) = C_{-1}\tilde{v}(-1) + C_0\tilde{v}(0) + C_1\tilde{v}(-1)$$

$$= -0.2585 \times 0.1 + 1.0735 \times 1 - 0.1590 \times 0.3$$

$$= -0.07355 + 1.0735 = 1.000.$$

$$v_{\text{eq}}(1) = C_{-1}\tilde{v}(2) + C_0\tilde{v}(1) + C_1\tilde{v}(0)$$

$$= -0.2585 \times (-0.2) + 1.0735 \times 0.1 - 0.1590 \times 1$$

$$= 0.15905 - 0.1590 = 0.0000.$$

Solved Numericals

1. Consider the duo-binary system of Fig. 13.14 with an input data sequence $\{a_k\} = \{110110101\}$. Find the coder output and verify that the receiver will have the correct output.

Solution

As discussed in Section 13.42, the input data sequence $\{a_k\}$ is with two symbols 1 and 0. The two-level sequence $\{b_k\}$ can be obtained as $b_k = \pm 1$ for ($a_k = 1$ or $a_k = 0$).

When applied to duo-binary scheme, the resulting sequence $\{c_k\}$ is a three-level sequence $c_k = b_k + b_{k-1}$, with $c_k = -2, 0, or +2$. The received sequence $\{d_k\}$ would be governed by $d_k = 2(e_k + e_{k-1}) = (b_k + b_{k-1})$ where e_k is the error propagation given as $e_k = a_k + e_{k-1}$.

The decoded sequence $\left\{\hat{b}_k\right\}$ thus can be found, as follows.

		$k=0$	1	2	3	4	5	6	7	8	
Input data, a_k		1	1	0	1	1	0	1	0	1	
Precoded sequence, e_k	0	1	0	0	1	0	0	1	1	0	
Transmitted sequence, b_k	-1	$+1$	-1	-1	$+1$	-1	-1	$+1$	$+1$	-1	
Received sequence, d_k		0	0	-2	0	0	-2	0	2	0	
Decoded sequence, $\hat{b}_k$		1	1	0	1	0	0	1	0	1	$= a_k$

2. A raised-cosine pulse at the rate of 150 kHz is transmitted through a low-pass filter of bandwidth 100 kHz. Find the roll-off factor.

Solution

The minimum bandwidth W is given by

$$W = \frac{(1+\beta)\,R}{2},$$

where R is the data rate and β is the roll-off factor.

Here, $R = 200$ kHz and $W = 75$ kHz so that

$$\beta = \frac{2W}{R} - 1$$

$$= \frac{2 \times 100}{150} - 1 = \frac{1}{3} = 33.3\%$$

3. A zero-forcing 3-tap equaliser is used to force the ISI to become zero. The received distorted pulse samples have the voltages $\tilde{v}(-2T_b), = -0.1\, \tilde{v}(-T_b) = 0.3$, $\tilde{v}(0) = 0.9$, $\tilde{v}(T_b) = 0.1\, \tilde{v}(2T_b) = -0.2$. Find the filter tap coefficients c_{-1}, c_0 and c_1. Verify that the pulses are equalised in the filter.

Solution

By Eq. (13.36) , we have

$$\begin{bmatrix} v_{eq}(-T) \\ v_{eq}(0) \\ v_{eq}(T) \end{bmatrix} = \begin{bmatrix} \tilde{v}(0) & \tilde{v}(-1) & \tilde{v}(-2) \\ \tilde{v}(1) & \tilde{v}(0) & \tilde{v}(-1) \\ \tilde{v}(2) & \tilde{v}(1) & \tilde{v}(0) \end{bmatrix} \begin{bmatrix} c_{-1} \\ c_0 \\ c_1 \end{bmatrix}$$

For equalised pulses $v_{eq}(-T) = v_{eq}(-1) = 0$, $v_{eq}(+T) = v_{eq}(1) = 0$ and $v_{eq}(0) = 1$. Thus

$$\begin{bmatrix} 0 \\ 1 \\ 0 \end{bmatrix} = \begin{bmatrix} 0.9 & 0.3 & -0.1 \\ 0.1 & 0.9 & 0.3 \\ -0.2 & 0.1 & 0.9 \end{bmatrix} \begin{bmatrix} c_{-1} \\ c_0 \\ c_1 \end{bmatrix}$$

That is,

$$\begin{aligned} 0.9c_{-1} + 0.3c_0 - 0.1c_1 &= 0 \\ 0.1c_{-1} + 0.9c_0 - 0.3c_1 &= 1 \\ -0.2c_{-1} + 0.1c_0 - 0.9c_1 &= 0 \end{aligned}$$

Solving, we have, $c_{-1} = -0.4388, c_0 = 1.2382, c_1 = -0.2345$

By Eq. (13.34), we have the equalised pulse

$$v_{eq}(k) = \sum_{n=-1}^{1} c_n \tilde{v}(k-n) = c_{-1}\tilde{v}(k+1) + c_0\tilde{v}(k) + c_1\tilde{v}(k-1)$$

Thus,

$$v_{eq}(-1) = c_{-1}\tilde{v}(0) + c_0\tilde{v}(-1) + c_1\tilde{v}(-2)$$

$$= -0.4388 \times 0.9 + 1.2382 \times 0.3 + (-0.2345) \times (-0.1)$$

$$= -0.39492 + 0.37146 + 0.02345 = -0.00001$$

$$v_{eq}(0) = c_{-1}\tilde{v}(1) + c_0\tilde{v}(0) + c_1\tilde{v}(-1)$$

$$= -0.4388 \times 0.1 + 1.2382 \times 0.9 + (-0.2345) \times 0.3$$

$$= -0.04388 + 1.11438 - 0.07035 = 1.00015$$

$$v_{eq}(1) = c_{-1}\tilde{v}(2) + c_0\tilde{v}(1) + c_1\tilde{v}(0)$$

$$= -0.4388 \times (-0.2) + 1.2382 \times 0.1 - 0.2345 \times 0.9$$

$$= 0.08776 + 0.12382 - 0.21105 = 0.00053$$

Thus the equalised pulses have the samples $\{-0.00001, 1.00015, 0.00053\} \approx \{0, 1, 0\}$.

4. In a low-pass channel, the received signal consists of a pair of echoes given by

$$r(t) = s(t) = \frac{\alpha}{2}s(t - t_0) + \frac{\alpha}{2}s(t + t_0).$$

Find the transmission channel characteristics $H_c(f)$. Design a tapped-delay line equaliser for $H_c(f)$ with $\alpha = 0.4$.

Solution

Taking the Fourier transform, we may write

$$R(f) = S(f) + \frac{\alpha}{2}S(f)\left[e^{-i2\pi f t_0} + e^{+i2\pi f t_0}\right]$$

$$= S(f) + \alpha S(f) \cdot \mathrm{Cos}2\pi f t_0$$

or, $$R(f) = S(f)\left[1 + \alpha\mathrm{Cos}2\pi f t_0\right]$$

Thus, channel transfer function is given by

$$H_c(f) = \frac{R(f)}{S(f)} = 1 + \alpha\ \cos 2\pi f t_0, \quad |\alpha| < 1| \text{ and } |f| < B|,$$

where B is the channel bandwidth.

The equaliser characteristic should be

$$H_{eq}(f) = \frac{1}{H_c(f)} = \frac{1}{1 + \alpha \mathrm{Cos}2\pi f t_0}$$

$$= (1 + \alpha \cos 2\pi f t_0)^{-1}$$

$$= 1 - \alpha \cos 2\pi f t_0 + \alpha^2 \cos 4\pi f t_0 - \alpha^3 \cos 6\pi f t_0 + \cdots$$

Assuming small echoes, $|\alpha| < 1|$ and

$$H_{eq}(f) = 1 - \alpha\cos 2\pi f t_0 = 1 - \frac{\alpha}{2}\left\{e^{i2\pi f t_0} + e^{-i2\pi f t_0}\right\}$$

$$= -\frac{\alpha}{2}e^{i2\pi f t_0} + 1 - \frac{\alpha}{2}e^{-i2\pi f t_0}$$

For a 3-tap delay line filter, we may write.

$$c_{-1} = -\frac{\alpha}{2},\ c_0 = 1,\ c_1 = -\frac{\alpha}{2} = -0.2.\ (\text{as } \alpha = 0.4)$$

Exercises

Activity

13.1 Discuss the causes of intersymbol interference. A channel has a frequency response $H(f) = 1 + 0.4\cos\pi f T$. Determine the equalising filter response that yields zero ISI and has a 50% excess bandwidth.

13.2 A radio channel suffering from multipath distortion has the response $r(t)$ to a signal $s(t)$ as

$$r(t) = a_1 s(t - t_1) + a_2 s(t - t_2),$$

where a_1 and a_2 are constant representing attenuation and t_1 and t_2 are the delays. Design a 3-tap equaliser filter to equalise the multipath distortion.

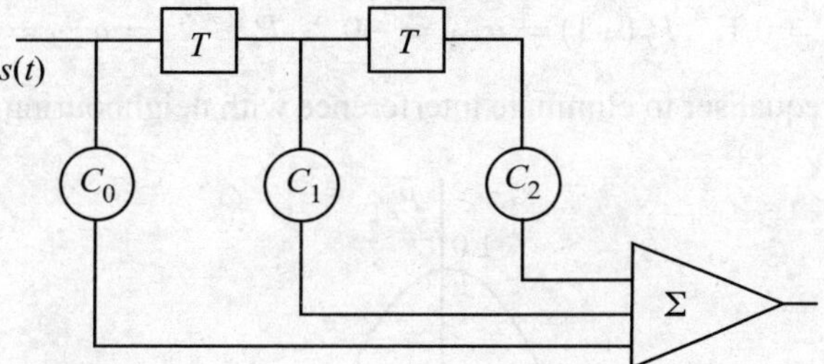

13.3 In a non-ideal bandlimited channel, a single impulse transmission results in a sequence of samples $\{0.1, 0.2, -0.1, 1.0, 0.3, -0.1, 0.1\}$. Design a filter that suppresses the ISI at the first sampling points on either side of the impulse.

13.4 In a duo-binary scheme, show that the additive white Gaussian noise in a data propagates into successive data sequence. Explain how can this be removed by the use of a precoder at the transmitter.

13.5 The binary data sequence $\{b_k\}$ = 1001 1010 1011 is subjected to precoding and then used to modulate a duo-binary transmitting filter. Assuming an initial bit 0 for the precoded sequence, tabulate the transmitted, received and decoded sequence.

13.6 What is meant by noise immunity? How can eye pattern of a received signal provide information about the error probability?

13.7 Why is channel equalisation necessary? What channel parameters are responsible for inter-symbol interference (ISI) to take place?

13.8 Explain the zero-forcing equaliser for reshaping the received pulses.

13.9 Explain how the channel and the receiver section in a digital communication can be expected to distort the transmitted waveform. Explain the purposes of pulse-shaping filter, matched filter and adaptive equaliser in the digital link.

13.10 A Nyquist filter response $V_N(f)$ has a bandwidth from $-\frac{1}{T}$Hz to $+\frac{1}{T}$Hz, and is symmetrical about $f = 0$ Hz such that $V_N(f) + V_N\left(\frac{1}{T} - f\right) = $ constant for $0 \leq f \leq {}^1/_T$. Verify that a pulse $s(t)$ satisfying

$$S(f) + S\left(\frac{1}{T} - f\right) = \text{Constant for } 0 \leq f \leq {}^1/_T, \text{ will have zero-crossings}$$

at $t = 0, \pm T, \pm 2T, \ldots$ to eliminate ISI.

Review questions

1. (a) What is an equaliser?

(b) Show how a zero-forcing equaliser eliminates or minimises interference with neighbouring pulses at their respective sampling instants at the time of receiving a digital signal.

(c) For a received pulse shown in figure below, let $P_r(0) = a_0 = 1$, $P_r(1) = a_1 = -0.3$, $P_r(2) = a_2 = 0.1$, $P_r(-1) = a_{-1} = -0.2$, $P_r(-2) = a_{-2} = 0.05$.

Design a three-tap equaliser to eliminate interference with neighbouring pulses at their sampling instants. [K.U. 2002]

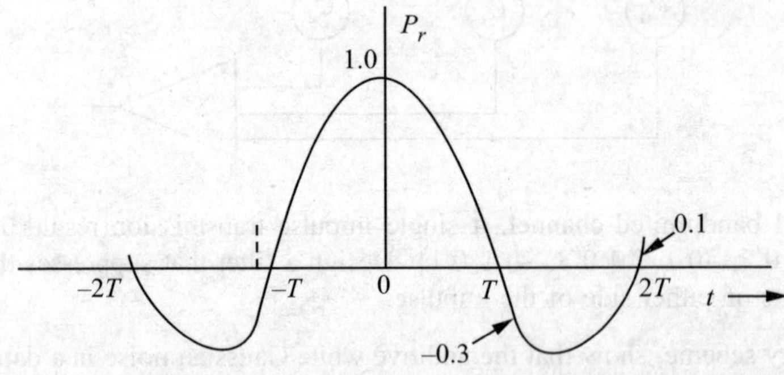

2. (a) What are the causes of intersymbol interference? How is it taken care of ? What is optimum pulse shape? How is the corresponding rate of transmission linked to the channel bandwidth?

(b) Obtain the expressions for the tap gains of a three-tap zero-forcing equaliser when

$\tilde{p}_1 = e, \tilde{p}_0 = 1, \tilde{p}_1 = \delta$, and $\tilde{p}_k = 0$ for $|k| > 1$.

(The symbols used have their usual meanings.) [R.P.E.C.U 2002]

3. What is meant by a binary channel? What does ISI stand for? How is it taken care of ? What is the maximum possible transmission rate through a bandlimited channel? What is the required pulse-shape in this case? Is it the pulse shape chosen in practice? Give reasons for your answers.

   Given that

$$
\begin{aligned}
p(f) &= \frac{1}{B}\cos\frac{\pi f}{2B}, \quad |f| \leq B \\
&= 0, \quad \text{otherwise},
\end{aligned}
$$

   determine $p(t)$. [RPE., C.U. 2003]

4. (a) What are the desirable features of a line code? Show that the envelope $p(t)$ goes down as $\frac{1}{|t|^3}$ with full-cosine roll-off pulse spectrum.

   (b) How is intersymbol interference taken care of in a digital link? Explain the operation of a zero-forcing type of an equaliser. [R.P.E.C.U. 2004]

5. (a) What is Nyquist criterion for zero intersymbol interference?

   (b) What are the limitations of the ideal solution and how can it be solved with the help of raised cosine function?

   (c) A communications channel of bandwidth 75 kHz is required to transmit binary data at a rate of 0.1 Mbps using raised cosine pulse. Determine the roll-off factor $\propto$.

6. What is the eye pattern of digital signal? Discuss the method of getting eye pattern of digital data on a CRO. What are the important information regarding the quality of transmission obtained from eye pattern? [WBUT 2004]

7. Intersymbol interference causes an increase in error rate for a given SNR. How would you explain loss of noise margin arising from ISI?

8. Inaccuracies in decision timing (due to jitter) may cause loss of noise margins. Explain how this happens.

9. Show that the transfer-function of a duo-binary filter can be obtained as

$$H_{\text{duo}}(f) = 2T\text{Cos}(\pi f T).$$

10. Explain how precoding technique helps to minimise error propagation.

Multiple-choice questions

1. A fruitful method for finding the effect of pulse degradation as they travel to the receiver is the study of

   (a) Voltage standing wave ratio

   (b) Eye pattern

(c) Transmission delay

(d) None

2. Signal distortion is calculated as the

(a) ratio of relative power of two signals at the same point.

(b) ratio of same signal power at two different points.

(c) Both (a) and (b).

(d) None.

3. The raised-cosine pulse $p(t)$ is used for zero ISI. The expression for $p(t)$ with unity roll-off factor is

$p(t) = \dfrac{\sin(4\pi Bt)}{4\pi Bt(1 - 16B^2t^2)}$. The value of $p(t)$ at $t = \dfrac{1}{4B}$ is

(a) 0.5 (b) –0.5 (c) 0 (d) ∞

4. The raised-cosine spectrum the bandwidth W is given by $W = \dfrac{1+\beta}{2T}$ (Eq. 13.17), where β is the excess bandwidth. For an excess bandwidth of 75% and $W = 3500$ Hz, the maximum possible symbol rate $\dfrac{1}{T}$ is

(a) 2000 symbols/sec (b) 4000 symbols per sec

(c) same as W (d) None

5. In digital data sequence, to avoid intersymbol interference the shape of the pulse would be

(a) Sine pulse (b) Sine-pulse

(c) cosine-shaped pulse (d) Rectangular pulse

6. Minimum Nyquist bandwidth for rectangular spectrum in raised cosine filter is

(a) $2/T$ (b) $T/2$

(c) $^1/_{2T}$ (d) $2T$

7. Roll-off factor signifies the relation between

(a) Absolute bandwidth and minimum Nyquist bandwidth

(b) Absolute bandwidth and excess bandwidth

(c) Minimum Nyquist bandwidth and excess bandwidth

(d) Both (a) and (b)

8. A pulse shaping filter should be

   (a) causal

   (b) with proper roll-off factor

   (c) Both causal and with proper roll-off factor

   (d) None

9. Which of the following method deliberately introduces small amount of ISI into the data stream?

   (a) Duo-binary (b) correlative coding

   (c) Partial response signalling (d) All the above

10. Duo-binary filter is primarily a

    (a) Sine filter (b) cosine filter

    (c) Raised cosine filter (d) None

Appendix: Error Functions

Error functions are used in the applications of engineering and science particularly when we deal with the Gaussian or Normal probability density functions.

The error function is represented by erf (x) and is given as

$$\text{erf}\,(x) = \frac{1}{\sqrt{2\pi}} \int_0^x e^{-u^2} du \quad \text{A.1}$$

The complementary error function erf $c\,(x)$ is defined as

$$\text{erf}\,c\,(x) = 1 - \text{erf}\,(x) = \frac{1}{\sqrt{2\pi}} \int_x^\infty e^{-u^2} du \quad \text{A.2}$$

The definition of Q-function is

$$Q\,(\text{x}) = \frac{1}{\sqrt{2\pi}} \int_x^\infty e^{-u^2/2} du \quad \text{A.3}$$

Error function is related to Q-function as

$$\text{erf}\,(x) = 2Q\left(\sqrt{2}x\right) \quad \text{A.4}$$

OR

$$Q\,(x) = \frac{1}{2} erfc\left(\frac{x}{\sqrt{2}}\right) \quad \text{A.5}$$

Table A.1 Values of $Q(x)$ versus x (for $x = 0.00$ to 2.99 in steps of 0.01)

	0.00	0.01	0.02	0.03	0.04	0.05	0.06	0.07	0.08	0.09
0.0	.5000	.4960	.4920	.4880	.4840	.4801	.4761	.4721	.4681	.4641
0.1	.4602	.4562	.4522	.4483	.4443	.4404	.4364	.4325	.4286	.4247
0.2	.4207	.4168	.4129	.4090	.4052	.4013	.3974	.3936	.3897	.3859
0.3	.3821	.3783	.3745	.3707	.3669	.3632	.3594	.3557	.3520	.3483
0.4	.3446	.3409	.3372	.3336	.3300	.3264	.3228	.3192	.3156	.3121
0.5	.3085	.3050	.3015	.2981	.2946	.2912	.2877	.2843	.2810	.2776
0.6	.2743	.2709	.2676	.2643	.2611	.2578	.2546	.2514	.2483	.2451
0.7	.2420	.2389	.2358	.2327	.2296	.2266	.2236	.2206	.2177	.2148
0.8	.2119	.2090	.2061	.2033	.2005	.1977	.1949	.1922	.1894	.1867
0.9	.1841	.1814	.1788	.1762	.1736	.1711	.1685	.1660	.1635	.1611
1.0	.1587	.1562	.1539	.1515	.1492	.1469	.1446	.1423	.1401	.1379
1.1	.1357	.1335	.1314	.1292	.1271	.1251	.1230	.1210	.1190	.1170
1.2	.1151	.1131	.1112	.1093	.1075	.1056	.1038	.1020	.1003	.0985
1.3	.0968	.0951	.0934	.0918	.0901	.0885	.0869	.0853	.0838	.0823
1.4	.0808	.0793	.0778	.0764	.0749	.0735	.0721	.0708	.0694	.0681
1.5	.0668	.0655	.0643	.0630	.0618	.0606	.0594	.0582	.0571	.0559
1.6	.0548	.0537	.0526	.0516	.0505	.0495	.0485	.0475	.0465	.0455
1.7	.0446	.0436	.0427	.0418	.0409	.0401	.0392	.0384	.0375	.0367
1.8	.0359	.0351	.0344	.0336	.0329	.0322	.0314	.0307	.0301	.0294
1.9	.0287	.0281	.0274	.0268	.0262	.0256	.0250	.0244	.0239	.0233
2.0	.0228	.0222	.0217	.0212	.0207	.0202	.0197	.0192	.0188	.0183
2.1	.0179	.0174	.0170	.0166	.0162	.0158	.0154	.0150	.0146	.0143
2.2	.0139	.0136	.0132	.0129	.0125	.0122	.0119	.0116	.0113	.0110
2.3	.0107	.0104	.0102	.00990	.00964	.00939	.00914	.00889	.00866	.00842
2.4	.00820	.00798	.00776	.00755	.00734	.00714	.00695	.00676	.00657	.00639
2.5	.00621	.00604	.00587	.00570	.00554	.00539	.00523	.00508	.00494	.00480
2.6	.00466	.00453	.00440	.00427	.00415	.00402	.00391	.00379	.00368	.00357
2.7	.00347	.00336	.00326	.00317	.00307	.00298	.00289	.00280	.00272	.00264
2.8	.00256	.00248	.00240	.00233	.00226	.00219	.00212	.00205	.00199	.00193
2.9	.00187	.00181	.00175	.00169	.00164	.00159	.00154	.00149	.00144	.00139

Table A.2 Values of $Q(x)$ versus x (for $x = 3.00$ to 5.95 in steps of 0.05)

x	$Q(x)$	X	$Q(x)$	X	$Q(x)$
3.00	1.35E-03	4.00	3.17E-05	5.00	2.87E-07
3.05	1.14E-03	4.05	2.56E-05	5.05	2.21E-07
3.10	9.68E-04	4.10	2.07E-05	5.10	1.70E-07
3.15	8.16E-04	4.15	1.66E-05	5.15	1.30E-07
3.20	6.87E-04	4.20	1.33E-05	5.20	9.96E-08
3.25	5.77E-04	4.25	1.07E-05	5.25	7.61E-08
3.30	4.83E-04	4.30	8.54E-06	5.30	5.79E-08
3.35	4.04E-04	4.35	6.81E-06	5.35	4.40E-08
3.40	3.37E-04	4.40	5.41E-06	5.40	3.33E-08
3.45	2.80E-04	4.45	4.29E-06	5.45	2.52E-08
3.50	2.33E-04	4.50	3.40E-06	5.50	1.90E-08
3.55	1.93E-04	4.55	2.68E-06	5.55	1.43E-08
3.60	1.59E-04	4.60	2.11E-06	5.60	1.07E-08
3.65	1.31E-04	4.65	1.66E-06	5.65	8.03E-09
3.70	1.08E-04	4.70	1.30E-06	5.70	6.00E-09
3.75	8.84E-05	4.75	1.02E-06	5.75	4.47E-09
3.80	7.23E-05	4.80	7.93E-07	5.80	3.32E-09
3.85	5.91E-05	4.85	6.17E-07	5.85	2.46E-09
3.90	4.81E-05	4.90	4.79E-07	5.90	1.82E-09
3.95	3.91E-05	4.95	3.71E-07	5.95	1.34E-09

Bibliography

Carlson AB and Crilly PB, *Communication Systems: An Introduction to Signals and Noise in Electrical Communication*, 5e, Tata McGraw-Hill Education Private Limited, 2007.

Couch LW II, *Digital and Analog Communication Systems*, 7e, Pearson Education Asia, 2006.

Ghosh PK, *Theory of Probability and Stochastic Processes*, Universities Press (India) Private Limited, 2012.

Glover IA and Grant PM, *Digital Communications*, 2e, Prentice Hall Europe, 2009.

Haykins S, *Digital Communications*, John Wiley and Sons, 1988.

Haykins S and Moher M, *An Introduction to Digital and Analog Communications*, 2e, John Wiley and Sons Inc., 2007.

Jagannatham AK, *Principles of Modern Wireless Communication Systems Theory and Practice*, McGraw-Hill Education (India) Private Limited, New Delhi, 2016

Kennedy GS, Davis B and Prasanna SRM, *Electronic Communication System*, 5e, Tata McGraw-Hill Education Private Limited, New Delhi, 2014.

Lathi BP and Ding Z, *Modern Digital and Analog Communications Systems*, 4e, Oxford University Press, 2010.

Proakis JG and Salehi M, *Digital Communications*, 5e, McGraw-Hill Higher Education, 2014.

Reddy MY, *Probability Theory and Stochastic Processes*, 4e, Universities Press (India) Private Limited, 2013.

Roddy D and Coden J, *Electronic Communications*, 4e, Prentice Hall of India Private Limited, New Delhi, 1995.

Schweber W, *Electronic Communication System: A Complete Course*, 3e, Prentice Hall International Inc., 2002.

Singal TL, *Wireless Communications*, McGraw-Hill Education (India) Private Limited, New Delhi, 2014

Sklar B and Ray PK, *Digital Communications: Fundamentals and Applications*, 2e, Indian Adaptation, Dorling Kindersley (India) Pvt. Ltd, 2009.

Stallings W, *Cryptography and Network Security: Principles and Practices*, 3e, Pearson Education, 2002.

Taub H and Schilling DL, *Principles of Communication Systems*, 2e, Tata McGraw-Hill Education Private Limited, New Delhi, 1990.

Tomasi W, *Electronic Communications System: Fundamentals Through Advanced*, 5e, Pearson Education India, 2011.

Index

D

T

U

V

W

Z